Physical and Chemical Separation in Water and Wastewater Treatment

Physical and Chemical Separation in Water and Wastewater Treatment

Norihito Tambo and Koichi Ogasawara

IWA PUBLISHING

Published by **IWA Publishing**
Republic – Export Building, 1st Floor
2 Clove Crescent
London E14 2BE, UK
Telephone: +44 (0)20 7654 5500
Fax: +44 (0)20 7654 5555
Email: publications@iwap.co.uk
Web: www.iwapublishing.com

First published 2020

British Library Cataloguing in Publication Data
A CIP catalogue record for this book is available from the British Library

ISBN: 9781789061291 (Paperback)
ISBN: 9781789061307 (eBook)
ISBN: 9781789061314 (ePub)

Contents

Introduction

The results of Norihito Tambo's 55 years of water research were first summarized in 'History and Technologies of Urban and Regional Water Systems' (Tambo, 2012).

After over 60 years during which both authors have worked to expand and disseminate modern water systems, more attention is now being paid to the operation and management of the resulting systems, and to the detailed measurements and informatics of system and process components rather than to the basics of technological development and operation of water systems.

In addition, if developing countries have mature, off-the-shelf technologies, the self-affirmative idea of water businesses being established spreads to other industries, without any sense of tension. Water systems are technical systems. Industrial societies that are founded on technology but cannot continue to develop innovatively will eventually be bankrupt. Even if technical manuals are available, a technological society in which new manuals cannot be created every day will quickly decline. The ability to create new developments can only be possessed by well-founded and experienced engineers and technicians.

In this book, we would like to review our studies of the original theories and experimental results of water treatment engineering. We hope that the book will help students, young researchers and people who come from other fields to learn about water treatment. We would like to encourage more detailed discussions of the day-to-day work of active researchers/engineers all over the world through this English version of our original publications, on which we worked during the whole of the latter half of the 20th Century.

Norihito Tambo is Former President (the 15th) and Professor Emeritus of Hokkaido University and was the 5th President of the Open University of Japan. He is the Chairman of the National University Research and Education Evaluation Committee, Tokyo, Japan. He was also the 2nd President (the first single president) of the International Water Association (IWA).

Koichi Ogasawara (co-author and a former graduate student and assistant of Professor Tambo) was awarded the Japan Society of Civil Engineers' Publishing Award with Professor Tambo for their 'Water Purification Technologies' published 35 years ago (Tambo & Ogasawara, 1985).

Author(s): Norihito Tambo and Koichi Ogasawara
doi: 10.2166/9781789061307_xi

REFERENCES

Tambo, N. (2012). 'History and Technologies of Urban and Regional Water Systems'. Kashima Press, Tokyo (in Japanese).

Tambo, N. and Ogasawara, K. (1985). 'Water Purification Technologies'. Gihodo Publishing, Tokyo (in Japanese).

Chapter 1

Earth: The water planet

1.1 THE PRESENCE OF WATER ON EARTH

The Earth is called the 'water planet'. It sits in the only planetary system (the solar system, centered on the sun), in which water can exist in large quantities in liquid form. Situated in the so-called 'habitable zone', the Earth is 149.6 million kilometers from the sun (this distance being known as one astronomical unit, or 1AU), and has a mass which is adequate enough to keep water and other gases around it by gravity. Figure 1.1 shows the relative distances from the sun and magnitudes of all the solar planets.

1.1.1 Planets of the solar system

Mercury is a rocky, terrestrial planet and is the closest planet to the sun with an average distance of 0.387AU: about 0.580 billion km. Its surface temperature ranges from −170 to +430°C. On Mercury, water occurs as a gaseous molecule, due to the largely high temperature environment which exists there. Because of the

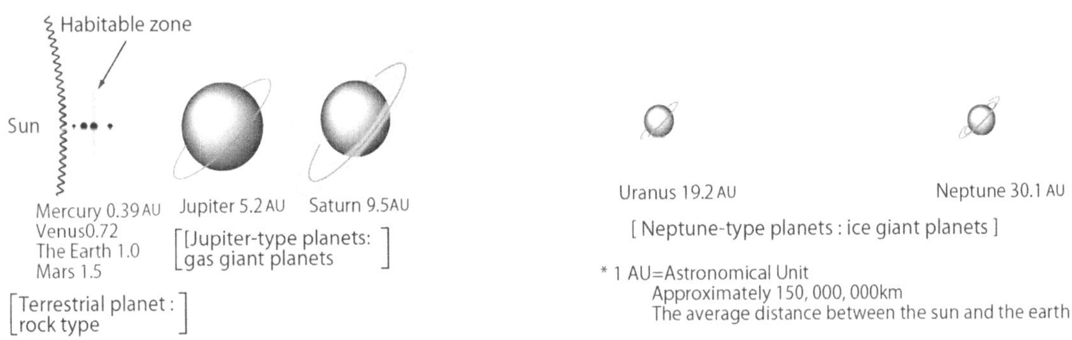

Figure 1.1 The Earth and the solar system.

Author(s): Norihito Tambo and Koichi Ogasawara
doi: 10.2166/9781789061307_0001

small mass of the planet (0.055 times that of the earth), its gravity is small, so water molecules are not able to be held in the atmosphere and they dissipate into outer space.

The Moon has an even smaller mass so that it cannot hold water and other gases around it, despite being in the same habitable zone as the Earth.

Mercury, Venus (average distance from the Sun: 0.7233AU; average surface temperature +480 C; mass ratio to the Earth 0.81), the Earth (1.00AU; average surface temperature +17°C) and Mars (1.5237AU; average surface temperature −45°C; mass ratio 0.11) are referred to as the inner planets inside the asteroid belt. There is a tiny probability of water on rock-type (Earth-type) planets. However, the temperature is too high on Venus and it is too cold on Mars for water to exist as a liquid. Mars seems to have water in the form of ice on the ground but water molecules in the atmosphere are thought to have been gradually lost because of the small gravity force of the planet.

Jupiter (5.2025AU; −120°C), Saturn (9.5549AU; −180°C), Uranus (19.2148AU; −210°C) and Neptune (30.1104AU; −220°C) are called the outer planets. Although the density of these planets is 0.13 to 0.30 times smaller than that of the Earth, their mass is larger, being 318 times larger than that of the Earth (in the case of Jupiter) and 17.1 times larger than that of the Earth (in the case of Neptune), and they are thought to exist as high-density low-temperature gaseous bodies. Jupiter and Saturn are classified as Jupiter-type planets (gas giant planets), and Uranus and Neptune are classified as Neptune-type planets (ice giant planets). These planets are thought to have a core in which the body of gas freezes but do not have a solid surface like the Earth, instead being covered with a body of gas.

1.1.2 Earth, the water planet

Among the planets of the solar system, only the Earth continues to hold a large amount of liquid water on its surface; because of the distance from the sun (1AU), the surface temperature remains at an average +17C and water can therefore exist as a liquid. At the same time, the gravitational acceleration (G = 1) due to the Earth's mass (6×10^{24} kilograms, which is about nine times larger than that of Mars) has enough force to restrain gas molecules, and water molecules that have become water vapor are also stopped from flying out into space.

Another factor that constitutes a condition for water to exist as a liquid at normal temperatures on the Earth is the so-called greenhouse gases including atmospheric water vapor, carbon dioxide, methane, and nitrogen dioxide. This greenhouse gas 'jacket' wrapped around the earth stores some of the heat radiated into outer space on the earth's surface. Without these gases, the atmosphere would become −8°C, resulting in a frozen Earth. The greenhouse effect of these gases has raised the surface temperature to a normal temperature of +17°C. If the concentration of greenhouse gases becomes too high, it could produce a very hot atmosphere with a surface temperature as high as +480°C, like Mercury. The Earth became a water planet under the three subtle conditions described above, and became a habitable planet, a planet for the well-being of living things, by distributing solar energy to various processes on Earth using water as a medium. Thus, while physics and chemistry are universal basic sciences which lead us to space, biology is limited as a science that only works on a habitable planet. In search of the traces and existence of living things in outer space, biological science is now being studied in an attempt to achieve the same universality as physics and chemistry (Figure 1.2).

1.1.3 Origin of the Earth's water

How does liquid water, which characterizes the Earth as a habitable planet, appear and exist on Earth? Although the Earth is said to be a water planet, the mass of the ocean (1.4×10^{21} kg) is only 0.023% of the total mass (weight) of the Earth (6.0×10^{24} kg). Table 1.1 shows the percentage of water on the

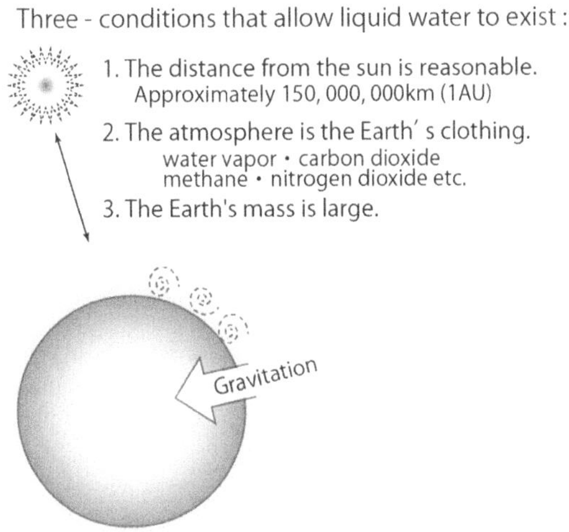

Figure 1.2 The conditions under which water can exist in liquid form.

planet. Seawater accounts for 97.3% of the total water mass. Fresh water accounts for only 2.7% of the total, most of which is confined in glaciers in the Polar Regions, whilst only 0.01 to 0.02% of total water resources used in ecosystems and human social systems are river water, lake water and shallow groundwater. A graph of Table 1.1 appears in Figure 1.3.

A variety of studies are underway on the origins of how the Earth's water was acquired. Broadly speaking, it is thought that water was generated and accumulated in the Earth's formation process by collision coalescence of planetesimals, in which water-containing minerals in the form of hydroxyl groups and ice became materials. The origin of the overwhelmingly large percentage of sea water are derived from chondrite meteorites from when the Earth was formed by planetesimal collisions 4.6 billion years ago, or from H_2O which is the main component of comets, and from primitive circumstellar disks of gases in the process of forming the dynamics of the solar system.

Table 1.1 The water mass of the Earth.

Existence State of the Water	Each Mass	Presence Ratio
Total mass	136×10^7 km^3	100%
Sea water	132.3×10^7 km^3	97.3%
Fresh water	3.67×10^7 km^3	2.7%
Vapor (in the atmosphere)	13×10^3 km^3	0.001%
River water	1.25×10^3 km^3	0.0001%
Lake water	125×10^3 km^3	0.009%
Groundwater	835×10^4 km^3	0.61%
Soil component	67×10^3 km^3	0.005%
Glaciers in Polar Regions	2.92×10^7 km^3	2.14%

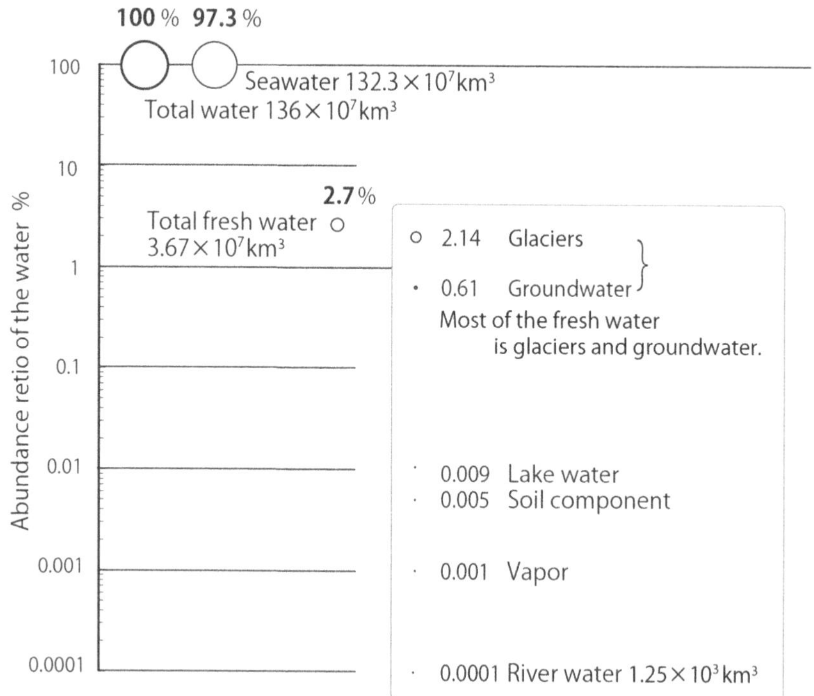

Figure 1.3 The presence ratio of the Earth's water.

1.1.4 Water circulation

The fresh water on the Earth is constantly recycled by solar energy to maintain the life and environment of the planet. Figure 1.4 schematically illustrates the hydrological cycle that forms the basis for the maintenance of humans and ecosystems. Water vapor returns to the ground as precipitation in a short period of 10 days of mean retention time in the sky, and it then becomes surface water/groundwater to be used to grow organisms.

Figure 1.5 shows the total global water fluxes produced by Spiedel and Agnew (1982). The ocean has an overwhelmingly large storage volume that forms the basis for water circulation. The ocean has a mean depth of 3500 m, and has a retention time of nearly 3200 years in relation to evaporative volume. When considering water systems, the size of the global circulation of water (flux, mm/year, etc.) is required as basic knowledge.

The basic sustainable water resources that we can rely on are river water and groundwater which are recharged in a relatively short period of time in the course of the hydrological cycle. The average residence time of deep groundwater is said to be as about 600 years. We should not rely on non-circulating groundwater, such as deep groundwater and fossil water, and glacial water with a residence time of 100 to 10,000 years or more. Quite a few parts of the modern global granary area have relied heavily over a short period of 30 to 50 years on the use of such stock (non-recycling) types of water resources taken from deep groundwater with a long residence time, and are now faced with phenomena such as reduced pumping capacity, depressed groundwater levels, and exhaustion.

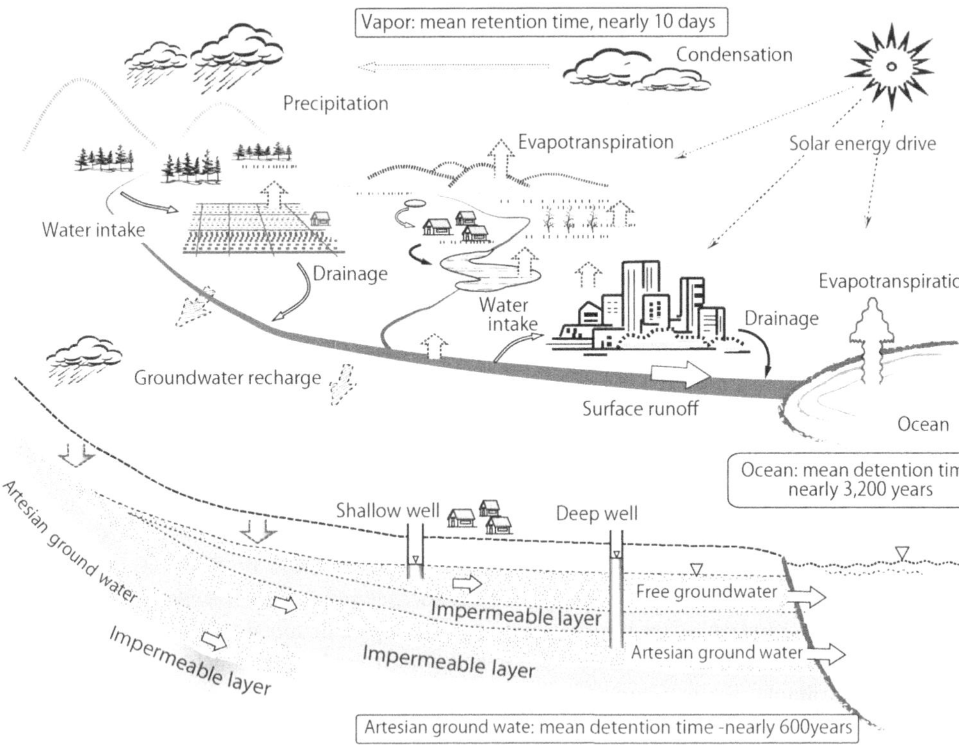

Figure 1.4 The hydrological cycle.

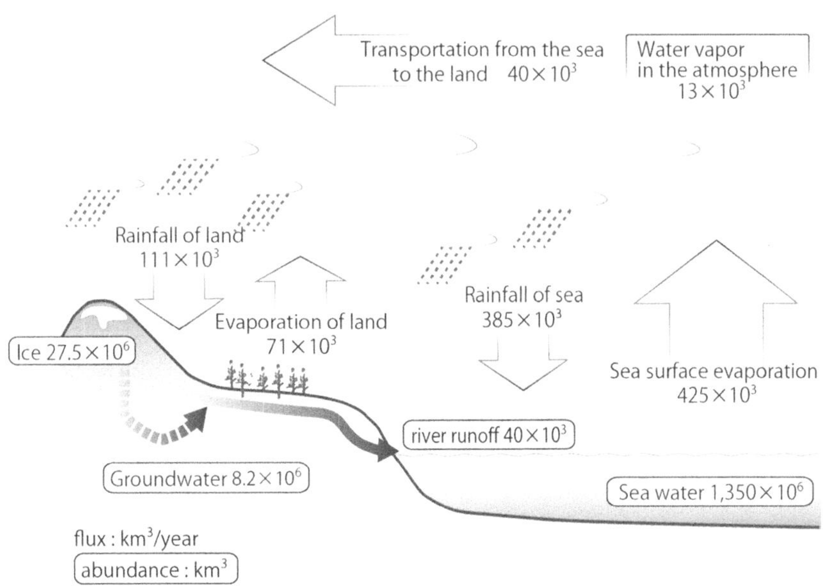

Figure 1.5 The global water circulation flux.

Water exists in flow (in daily circulation) and in stock (as deep groundwater). The short-term temporary use of deep groundwater that has been stored for a long period of time is the same as the non-recyclable use of stored fossil energy resources collected in a short period of time. Groundwater to be used for human activities should be limited to recyclable groundwater that flows in relatively shallow geological areas. Societies that seek sustainability should not consider the unreasonable use of water in agricultural and urban areas which consumes fossil water or deep storage groundwater that is close to being non-recyclable in a short period of time.

1.2 WATER CIRCULATION

1.2.1 The spectrum of water circulation

On Earth, as shown schematically in Figure 1.6, the various water uses consist of hierarchical water cycles, but modern water systems have not dealt explicitly with this hierarchy. The water cycle (circulation) on Earth has cycles of different qualities and quantities (gaps), with different cycle times (cycles) and cycle spaces (scales).

The first water cycle, which is the basis for water utilization/elimination (water metabolism), is the Hydrological Cycle as shown in Figure 1.6. The term 'hydrological circulation' is generally used without limitation in many cases to refer to the general hydrological circulation of the first cycle. The main source of our urban and regional water use systems is the second cycle of water circulation, which overrides the first cycle. Modern water supplies and sewerage systems are second cycle technologies. In modern aquatic engineering and technologies, third and fourth cycles are much poorer than first and second cycles. For example, we can consider why Minamata disease (caused by severe environmental pollution from mercury, in Japan) occurred. The chemical company involved did not take a strict material balance operation during the production process in the fourth cycle. In the fourth cycle on the production site, where quality and quantity compositions were limited and it was easy to track a few components involved in the process, it was possible to assess the material balance, which is the prime

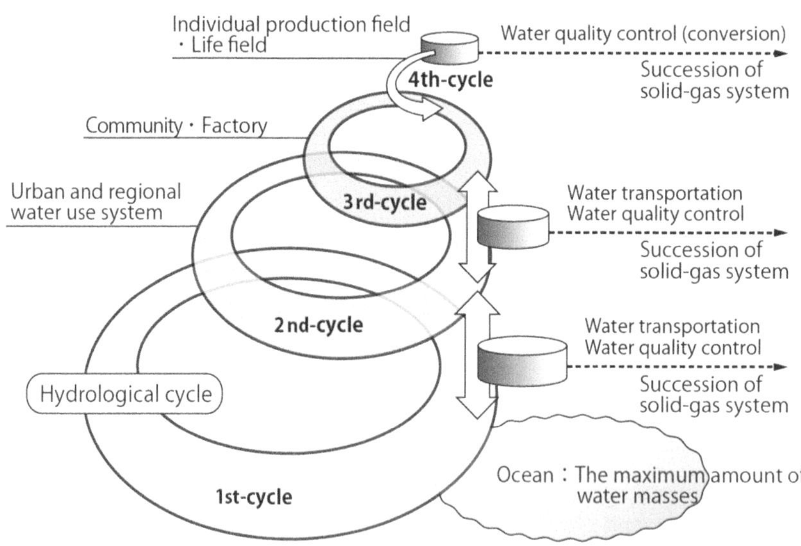

Figure 1.6 Hierarchical water cycles.

basis of environmental and chemical engineering. Mercury, which should originally have been stopped in the fourth cycle, was not correctly treated in the second and third cycles. The mercury was finally accumulated in coastal and inner bay ecological concentration systems, which are water circulation routes of the first cycle and which are difficult to control, resulting in such tremendous human health damage. This is a typical example in which the hierarchy of the cycle was not clearly specified, and clarification of the material balance of the system, which is the basis of chemical environmental engineering, was not guaranteed.

It is now half a century since Rachel L. Carson published her book *Silent Springs* (Carson, 1962) and warned that trace organics and inorganics threaten the existence of modern water systems. Modern urban water systems of the bulk transportation type continue to expose vulnerabilities with respect to the risks of organic and inorganic trace pollution, in which drinkable water is supplied collectively for all usage, from drinking to miscellaneous purposes. Various wastes are mixed with clean drinkable tap water to create artificial waste 'sewage' and removed collectively downstream to keep urban aquatic environments from harmful deterioration. Basically, it is extremely difficult to handle the trace pollutants that must be processed in the fourth cycle in the second cycle, i.e. the urban/regional water system, with the largest water quantity and the highest level of treatment. This is the historical water used in the second cycle, and the discharge port is open to the first cycle side. In addition, the third and fourth cycle treatments for pesticides and inorganic trace contaminants from natural origin are extremely difficult to control in situations where the conventional use of water in urban and agricultural aquatic systems is designed and operated only up to the second cycle. As the quality requirements of urban water and environmental water become more stringent, it is necessary to constantly carry out advanced treatment of the entire flow rate passing through a city in the second cycle by paying attention to all the trace components of the second cycle. A high degree of water quality improvement is a fatal difficulty inherent in modern integrated water transport systems, which must be constantly carried out for the greatest (treated) water volume.

Collecting rainwater and using it in homes and communities is the third cycle, but it is somewhat connected to the second cycle and also connected to the fourth cycle. Treatment and disposal of excreta was originally a matter for the fourth cycle. However, in modern sewage systems, treatment and disposal of excreta is directly connected to the second cycle and is linked to the first cycle. There is a need to discuss more precisely how to think about the responsibility sharing and crossover problems of these cycles in future water systems, and in which cycle we should handle the balance of income and expenditure of those components.

1.2.2 Water treatment processes

Modern water supply and sewerage systems have been developed to protect the health of people and maintain the comfort of the living environment as a second cycle system for managing hydrological circulation, directly in connection with the first (hydrological) cycle by attaching drinking outlets and opening expulsion (waste) outlets. As shown in Figure 1.7, in urban water systems, the second cycle mainly consists of a water supply and sewerage system which is connected to the hydrological cycle. A water treatment operation is almost always inevitable in order to satisfy the health requirements of citizens and the quality requirements of the water environment at the same time.

Water treatment can be described as an operation conducted to increase or maintain the relative value of water environments by transferring impure substances to a solid or gas state rather than by allowing a substance to exist in a water system. The basic principles of the operation of environmental engineering are the maintenance of the strict material balance of the components concerned and minimization of

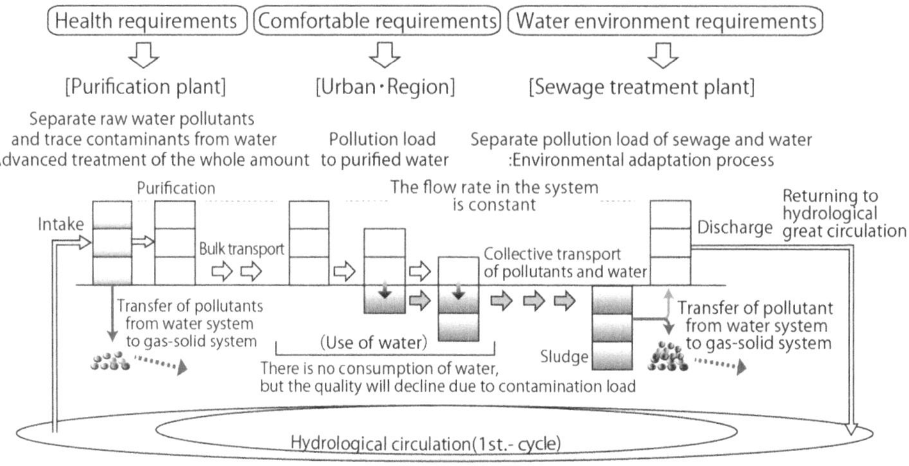

Figure 1.7 A modern urban water system (second cycle).

Table 1.2 Natural water quality recovery and natural water processing systems and modern process counterparts.

The Natural World	Ability of Nature	Artificial Idea	New Knowledge Research	Water Processing System Development
Gravel layer	Capture of suspended matter	Sand filtration	Biofilm of the sand layer surface ⇩	⇨ Slow sand filtration
			Trial of the artificial filtration film	⇨ Rapid sand filtration
Lake River	Sedimentation	Storage	Calm consecutive flows	Plain sedimantation basin
			Particle size growth	Chemical sedimantation basin
			Settling device	Plate settler etc.
	Organic matter decomposition		Aeration: microbial growth Highly-cocentrated microbe reaction	Biological membrane contact treatment Suspended biological treatment
	Chemical oxidation		Metal oxidation	Deferriatin Demanganizatin
Organism	Infiltration Dialysis	Seawater desalinatin Ultrapure water production	Impurities separation and membrane property	Membrane filtration

energy-consumption and space utilized. Therefore, the design of water-treatment systems should satisfy these principles.

Nevertheless, in the long history of urban water systems starting in Rome and leading to modern times, most of the foundations of urban water systems have been left to hydrological circulation (first cycle) and deployed mainly as a bulk water transport system, based on the assumption that high-quality natural water is secured. Therefore, the degree to which urban water systems rely on water treatment was much smaller in the past. There was almost no water treatment in Rome or in Edo (old Tokyo). As populations have grown and human populations have connected with each other more intensely in modern times, transport systems alone have become difficult and uneconomical to ensure the quality of water in cities and regions. While learning about the mechanisms of natural quality restoration capabilities, we have attempted to artificialize and streamline these mechanisms (making them more compact) and have begun to develop water quality conversion systems, as shown in Table 1.2. An extreme example of this is Singapore's energy-intensive urban/regional water metabolism system, which is built on a rainwater collection system with a water treatment system at the core, plus reclaimed sewage water and desalinated seawater. For more information, please refer to Tambo (2012).

1.3 BASINS AND WATER SYSTEMS

1.3.1 Basin composition

When considering water problems, the hydrological cycle (the first cycle) is the foundation for the establishment of all water systems in cities and regions. Water returning to the ground as rain or snow falls on mountains, forests and fields where it is used by the eco-system and agriculture, and flows to the sea after cities have repeatedly used it. River basins (or watersheds) consist mainly of rivers and groundwater that flow in clear directions from the most upstream mountains down toward the ocean. They are natural systems. Humans around the world have relied for long periods of time on watersheds, and have fostered civilizations in the context of the interdependence of watersheds. The Edo period in Japan (15th to 17th Century) was built based on a basin, and it seems that basin helped foster its own cultural and industrial development. Modern Japan, by comparison, has formed a centralized modern nation by piercing watersheds with modern methods of transportation such as railways and highways. Today, Japan acquires food and raw materials from across the ocean and also exports products.

A basin can simply be described as consisting of three basic characteristic spaces (regions) constituting the environment (as shown in Figure 1.8): urban areas; productive green (agricultural) areas; and natural conservation areas. In his keynote speech at the first Environmental Engineering Symposium of the Science Council of Japan (1986), Tambo proposed that the activities of mankind would be easier to understand by characterizing them in these three areas. The basin is a universal foundation of ecosystems, including human activities, from upstream to coastal/oceanic areas, and these characteristic spaces are integrated by rivers, with travel upstream and downstream.

Humans operate their economic activities and earn goods in urban/industrial areas and productive green spaces. Since urban/industrial areas do not produce food but do not grow vegetation and are primarily designed to acquire goods, most of the food that supports these activities comes from the productive green areas, which produce organic matter with high productivity. Solar energy is the main source of energy in the productive area. In the second half of the 20th Century, the introduction of large amounts of fossil energy into agricultural land has led to the advancement of the 'Green Revolution' and an increase in food production, thereby supporting the rapid increase in the global population. Forests in productive green areas are overwhelmingly dependent on solar energy, unlike agricultural land, which uses a large amount of fossil energy.

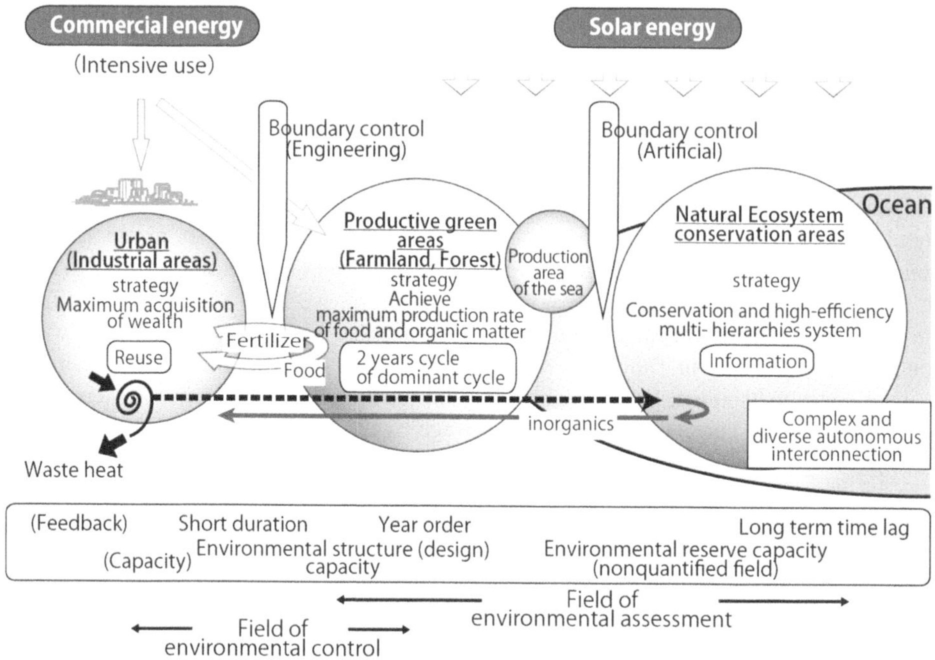

Figure 1.8 The three basic characteristic spaces constituting the environment.

Green areas of production are areas where we want to harvest as many products as possible from a single grain of seeds. In modern times, we try to promote the growth of plants as efficiently as possible by appropriately adding commercial energy onto a solar energy base. This tends to suggest that civilization is centered on two spaces: urban/industrial areas and organic matter production areas (green areas). However, it should be remembered that natural ecosystem conservation areas exist at the outer edges of these two areas as indispensable areas for the survival of mankind. This area is different from the urban/industrial areas and productive green areas established mainly for human activities. It is an area that maintains the delicate multilayered ecosystem of nature that is scarcely involved in human activities. It differs from the former two in the composition of the ecosystem. The best strategy is to avoid unnecessary contact with people because of the diverse areas of natural ecosystems that have evolved over long periods of time – billions of years – with multiple use of solar energy and water sources, and which are based on delicate equilibria.

Generally speaking, a natural conservation area is very often situated upstream of a river basin. If a water conservation forest and a good natural ecosystem does not exist on the periphery of the urban and productive green area, a sound human societies will not be established. The coastal areas and oceans on the most downstream side of a basin are also large ecosystem conservation areas. Coastal areas are production areas for marine organisms, being natural ecosystems that are fluidly connected to the environment of the oceans and deep seas. Human activities in land-based cities/industrial areas and productive green areas have a major impact on overall marine conditions. However, even when observing and monitoring oceans that flow around the world in a complex manner, it is extremely difficult to control land areas clearly by feeding back this information. A closed discussion of self-reliant urban water metabolism systems in a basin is inevitably necessary, as will be described later. 'Integrated basin management', including conservation areas, is needed, with urban/industrial areas and productive green area water systems at the core of activities.

1.3.2 Metabolism of urban areas

Metabolic functions in urban areas provide citizens with sufficient materials and energies to carry out household activities, vocational activities, recreation and other activities, and dispose of the waste and residue without hindering the lives of citizens. How to ensure this is the subject of environmental engineering. The organic matter metabolism in cities (urban and industrial areas) is covered by the mass balance (food and fertilizers) of a year's units between 'productive green areas' of organic matter production (as in Figure 1.8). The existence of 'urban/industrial areas' and 'productive green areas' is ensured by the stable existence of 'natural ecosystem conservation areas' that support them. 'Hydrological circulation' is the connection of three characteristic regions on the Earth (consisting of cities/industrial areas, green production spaces, and natural ecosystem conservation areas), geographically connected in a basin. The key to human survival in the future is how to control human metabolism in order to achieve coexistence between growing human (over-)population and natural ecosystems in a watershed. The difference between a world of 10 billion people (a global population which is nearly saturated) and the earlier world of 1 billion people (whose relationship between living things and people has scarcely been preserved since the beginning of modern times) has made a significant difference to the very foundation of each basin.

In modern times, cities were often formed in a series upstream and downstream of a great river, resulting in the emergence of river basin urban clusters. In response to pollution discharged into rivers in upstream cities, the treatment of sewerage discharge became indispensable, and it was started only after the middle of the 20th Century. In the majestic River Rhine which flows through the center of the continent of Europe, water pollution in industrial areas is particularly severe, and strong water quality management is now required in order to eliminate the harmful competition for water use, and the evolution of an advanced basin water management system was started in the 1960s.

As population density increases from $10^2/km^2$ to $10^3/km^2$, interventions by downstream populations on the water use of those upstream in a watershed becomes serious, and modern water supply and sewerage systems and conventional agricultural water systems begin to have controls to ensure the coexistence of local communities, necessitating the development of a variety of new basic technologies for the next era. We will think about this problem on the basis of the following three conditions:

- Minimum human drinkable water demand (50 l/person/day) and other water demand supplies;
- Ensuring agricultural water;
- Preservation of natural ecosystem diversity.

Water systems in the future should be committed to the design and operation of the system in a manner that maximizes independence within the hydrological cycle, assuming the idea that all responsibilities for every source and outcome of the metabolism imposed on it are self-imposed. Responsible water use and elimination (water metabolism) in cities and regions occurs in accordance with the polluter (user) pays principle.

1.3.3 Water environment district and environmental lakes

This section describes the basic concept of the water environment district as shown in Figure 1.9, which was first proposed by Tambo in the 1970s as the principle measure to realize the self-dependent water metabolism of cities.

The concept of the water environment district proposes that each city in a high population density basin should be independent of each other, consuming minimal energy and taking a minimal water load from the environment, and constructing a basic metabolic component to coexist with natural river channels (streams),

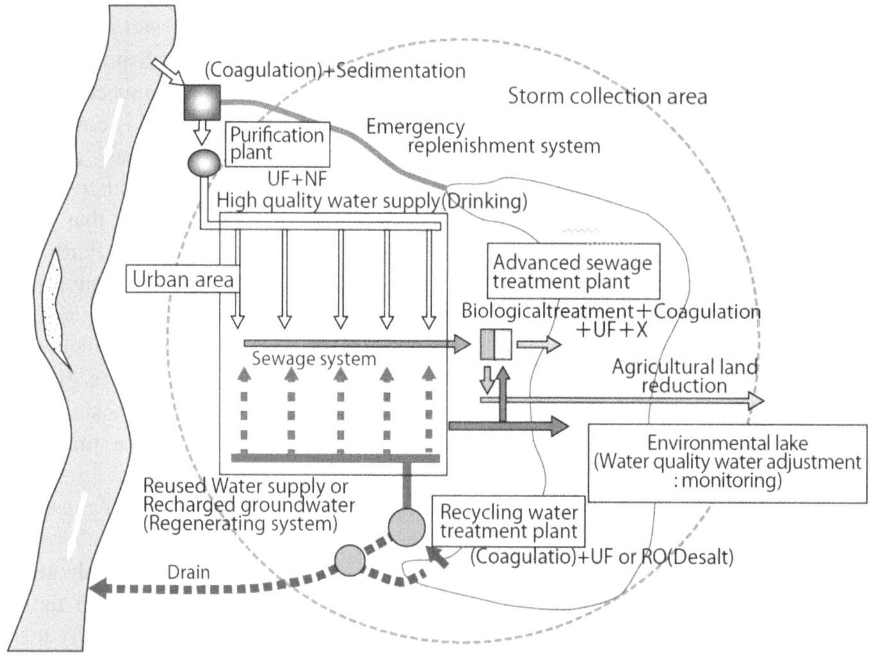

Figure 1.9 The water environment districts.

downstream cities and agricultural industries, just as living organisms do in their own bodies. Closed water metabolism (non-discharge policy) and drive energy minimization (using minimum energy), urban water metabolism (which allows the 'Polluters Pay' principle to be used) and achieving public-private cooperation (Public Private Partnership), are all needed as the base of new sustainable civilizations in an age in which we must move toward harmonious coexistence under global environmental constraints.

Modern water supply and sewerage systems are environmental sanitation facilities designed to 'supply clean water abundantly and inexpensively (water supply)', 'collect wastewater, prevent rainwater flooding, and discharge rainwater and sewage to public water areas (sewerage)', and 'treat water downstream to prevent pollution (roles added to sewerage in the 20th Century)'. Supplying drinkable water to every demand is the principle of water supply systems, which supply the best water quality to everyone (e.g. total water demand: 300 l/person/day), and therefore can only do this with enough quality and quantity of raw water. Sewerage is a system for promptly removing sewage (wastewater), which is produced by mixing various types of waste discharged by citizens into the drinkable water supply, together with rainwater, into the downstream area. In the 20th Century, the role of water pollution prevention was greatly increased. Nevertheless, both water supply and sewerage systems are based on the quantitative features for lump-sum transport and are operated on a flow rate basis, with water supply and sewerage tariffs charged per m^3. Unlike in today's water supply and sewerage systems, the composition of the water metabolism system in the next ra (the age of quality standards: the next generation of symbiosis) will be 'a municipal water metabolism system with a distribution and recirculation system

Table 1.3 Changes in the water metabolism system.

Modern Water Metabolism System	**System on the Basis of Collective Transportation and Quantity**
Water works	Supply (unrelated to use) of water which is clean, of good quality and cheap. When good quality aquatic resources are abundant, it works well.
Sewage works	Collects dirty water and rain water, and prevents contamination and stormwater flooding. Became a heavily burdened system in the 20th Century.
Next Generation Water Metabolism System	**System on the Basis of Quality and Symbiosis**
	Supply of water of appropriate quality and necessary quantity, depending on survival activity. Using lowest amount of transportation and processing energy, depending on use.

feeding water, supplying optimal quality and quantity of water required for each living activity, being operated with the smallest transport and treatment energies according to demand, and having a highly independent environmental boundary control (water treatment) and with a storage buffer of the required size: a reservoir for water quality monitoring'. In such a case, it is necessary to keep as much of the clean river water as possible in the river channels for downstream people and natural ecosystems, and to ensure that cities take up the minimum amount of clean river water necessary for human health and survival. As has been required since the time of the Dublin Declaration and the Rio de Janeiro Summit, one criterion is that cities can preferentially obtain only a fixed amount of drinkable water (high-quality water), such as 50 l/d per citizen, from natural clean water areas. Citizens collect and clarify rainwater and spent wastewater in the region and use it for all non-drinking demands. There will also be reuse of spent wastewater (drainage) stored in nearby environmental lakes (or groundwater basins) with appropriate 'biological treatment plus physical-chemical treatment' of recycled groundwater from wells for non-drinking water supplies. With the development of new advanced water treatment technologies in the 21st Century, high levels of reclaimed water can be stored in reservoirs and used repeatedly for use other than drinking (Table 1.3).

The water in environmental lakes can ensure a healthy habit for fish and other aquatic organisms, and integrated quality control can be carried out so that the bottom mud is not anaerobic; results of water treatment can be guaranteed to be healthy by constant daily observation. It would be better to return the water to groundwater, and then pump it up in wells, etc., and use it for other purposes than drinking. In addition, daily and hourly fluctuations in the flow rate used and flow rate discharged by cities can be adjusted by environmental lakes. It is also possible to recover all rain falling in urban areas in a city water system. In order to prevent salt accumulation in environmental lakes, it is possible to overflow pond water into flood rivers during floods or strong rainfall, or to demineralize water using nanofiltration/reverse osmosis (NF/RO) membranes. In environmental lakes, flow rate and water quality can be monitored continuously and accurately in urban areas, while time varying adjustments of reclaimed flow rate and demand flow rate can be made. Daily technical management of environmental lakes will enable clear, high-level management of urban wastewater masses for each basin water system, without the need to deploy a large number of monitoring systems that are not economically or technically easy down-stream of a wide natural system. Even if climate change causes fluctuations or shortages in water resources, it is not difficult to keep 50 liters of unused initial high-quality water for drinking. Therefore, the remaining 250 liters of non-potable water required by each person per day can be supplied by increasing the rate of water regeneration and if no drought occurs.

REFERENCES

Carson R. (1962). Silent Spring. https://archive.org/details/silentspring0000cars. Houghton Mifflin Company.

Speidel D. Hi. and Agnew A. F. (1982). The Natural Geochemistry of Our Environment, 1st edn, West view press, Boulder Colorado, USA.

Tambo N. (2012). History and Technologies of Urban Water Metabolism Systems. Kashima Press, Tokyo (in Japanese).

Chapter 2

Properties and impurities of water

2.1 THE WATER MOLECULE

2.1.1 Structure and nature of water

Water, this mysterious thing which accounts for 70% of the Earth's global surface, has been investigated by many outstanding, cutting-edge studies of its various structures and behaviours – and many are still ongoing. Here, we focus on the basic science required to develop quality conversion technologies for impurity control and removal, which we will be discussing later. Thus, we want to discuss the individual characteristics of water – knowledge of which is required as a basis for the various water treatment operations used at each stage of treatment.

Water, of course, is a compound with the molecular formula H_2O (with a molecular weight of 18), consisting of two atoms of hydrogen (H) and one atom of oxygen (O). Hydrogen and oxygen have various isotopes (elements with different mass numbers, made up of a proton number plus a neutron number). There are various types of water, which combine various types of hydrogen isotope, such as light hydrogen H: protium (mass number 1), deuterium D (mass number 2: proton 1 + neutron 1), tritium T (mass number 3: proton 1 + neutron 2) with various oxygen isotopes, such as light water heavy water (D_2O), and tritium water (T_2O).

Oxygen also has three stable isotopes: mass number 16, 17 and 18. Then, when they are combined with three kind hydrogen isotopes, total 9 kind of light water, heavy water and triple water are present.

The main nuclide of oxygen^{16}O [8 protons and 8 neutrons] combined with H[protium] consists normal water which accounts for 99.726% of the natural species.

2.1.1.1 Structure of water molecules

The structure of a water molecule is shown schematically in Figures 2.1(a, b). When approximated by a sphere, the diameter is about 3 Å. The molecular diameter of water at room temperature is particularly small compared to various other single elements as slightly larger than the monoatomic neon (2.8 Å) and

Author(s): Norihito Tambo and Koichi Ogasawara
doi: 10.2166/9781789061307_0015

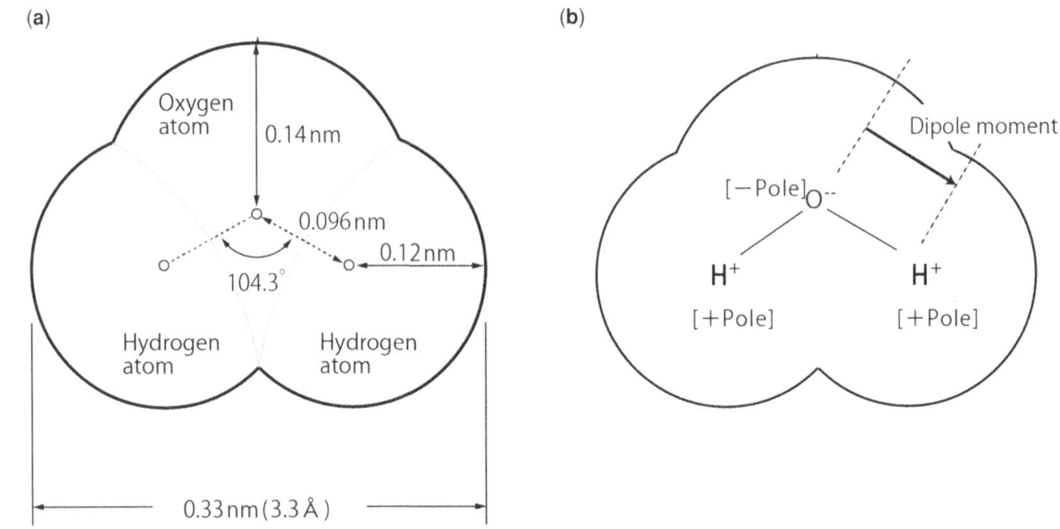

Figure 2.1 (**a**) The shape of a water molecule (H_2O); (**b**) The shape and the electric dipole moment of a water molecule (H_2O).

slightly smaller than argon (3.4 Å). This is one reason for the ability of water to act as a solvent to dissolve various small molecules in water as solvents.

Water molecules are charged negatively on the oxygen electron-side and positively on the hydrogen-atom side, and exhibit a dipole state in which one water molecule has both positive and negative poles. This dipole moment causes water to exhibit a high dielectric constant of 80.4 at room temperature (20° C). Alcohols exhibit 16 to 31, glass 5.4 to 9.9, wood 2.5 to 7.7, natural gum 2.0 to 3.5, and air 1.0.

The specific dielectric constant changes with temperature in dimensionless quantities that indicate the relative magnitude of the electric field response (polarization) of a material, assuming that the value of the atoms or molecules in a vacuum respond to an external electric field which can be compared with the air dipole moment which is defined as unity: 1. The hydrogen atoms of water molecules are not only tightly bonded to their own oxygen atoms but also loosely bonded to the oxygen atoms of neighbouring water molecules. The electrons of hydrogen atoms sometimes move toward the oxygen atoms of adjacent water molecules, thereby attracting adjacent water molecules. Such loose bonds are called hydrogen bonds. The combination of covalent O–H and hydrogen O••H bonds, which share electrons in the outermost orbitals, causes water molecules to exhibit an H–O••H component which governs the physical properties of liquid water. The strength of hydrogen bonding is much weaker (about 8 to 40 kJ/mol) than covalent bonding (about 150 to 50 kJ/mol) in which molecules share electrons. However, hydrogen bonding has a greater bonding force than van der Waals attraction between particles (which is described later in detail in Section 4.1.3 on coagulation treatment and is capable of loosely attracting molecules). This causes water to exhibit a certain degree of viscosity which changes greatly depending on temperature. For example, when water at a temperature of 0°C rises to 20°C and the movement of water molecules increases, the distance between molecules decreases by half. Table 2.1 shows the relationship between water temperature and viscosity. The viscosity coefficient for water in the tropics at 25°C is half that of water at 0°C in colder regions, and tropical water is less viscous (Figure 2.2).

Table 2.1 Density and viscosity of water at different temperatures.

Temp. (°C)	Density (g/cm^3)	Viscosity (Pa·s)	Kinematic Viscosity (cm^2/s)
(Ice) 0	0.9168	—	—
(Water) 0	0.9998	0.01792	0.01792
5	1.0000	0.01519	0.01519
10	0.9997	0.01308	0.01308
15	0.9991	0.01140	0.01141
20	0.9982	0.01005	0.01007
25	0.9971	0.00894	0.00897
30	0.9957	0.00801	0.00804
40	0.9922	0.00656	0.00661
50	0.9881	0.00549	0.00556
60	0.9832	0.00469	0.00477
70	0.9778	0.00406	0.00415
80	0.9718	0.00357	0.00367
90	0.9653	0.00317	0.00238
100	0.9584	0.00284	0.00296

Kinematic viscosity = Viscosity/Density.
Viscosity Pa·s = g/cm·s.

By comparison, the change in density of liquid water with temperature change is not as great as the change in that of viscosity. However, even with this small change, the fact that ice, as a solid form of water, is lighter than liquid water leads to the major characteristics of the Earth's hydrosphere.

Liquid water is sometimes said to consist of clusters of several water molecules. This is thought to be an agglomeration of water molecules by hydrogen bonding. In reality, there are no steady clusters and

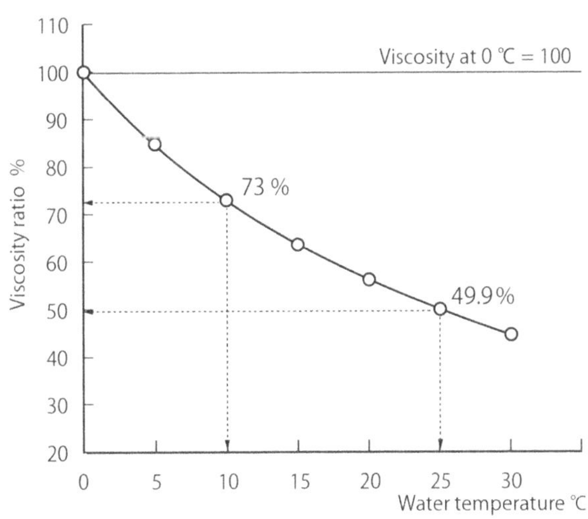

Figure 2.2 Water viscosity and water temperature.

Table 2.2 Boiling and melting points for various hydrocarbons.

	Molecular Weight	Boiling Point °C	Melting Point °C
Water H_2O	18	100	0
Methane CH_4	16	−161	−183
Ethane C_2H_6	30	−89	−183
Carbon dioxide CO_2	44	−56.6	−78.5

molecules appear to gather in fluctuation cycles of about 10^{-12} seconds. If we look at the phenomena in much shorter times, there may certainly be clusters of a certain size. Water clusters appear to be a dynamic phenomenon that repeatedly separate and concentrate at high speed.

Because of this hydrogen bond, water has a higher boiling point (100°C) and melting point (0°C) than other hydrocarbons such as methane, which has a similar molecular weight (−161°C and −183°C), ethane (−89°C and −183°C) and carbon dioxide (−78.5°C and −56.6°C) – all of which have boiling and melting points below zero (Table 2.2).

2.1.1.2 Water and ice density and structure

The peculiar nature of water has major impacts on global ecosystems. One peculiarity is that, whilst the density of most solids is normally higher than that of the liquid form, the density of solid ice is lower than that of liquid water, and maximum density appears at 4°C liquid water. This means that ice forms on the surface of a pond in winter and that fish can survive in a pond due to the protection of the surface ice cover. Figure 2.3 shows the change in density of ice and water by temperature, at atmospheric pressure.

As shown in Figure 2.4, when the water at 0°C becomes ice at 0°C, the volume increases by 10%. This is why pipes break out when water in a water supply line freezes, or why a bottle breaks easily when water freezes.

Let's briefly explains how this happens. As shown in Figure 2.5, water molecules form a tetrahedral structure by hydrogen-bonding and arrange themselves in an orderly manner (Arakawa, 1991). There are

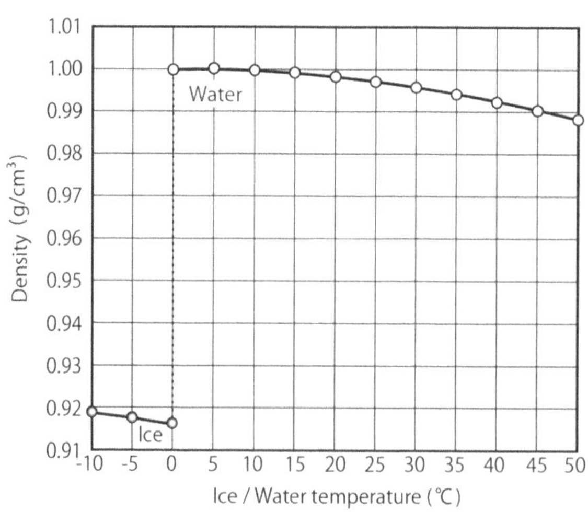

Figure 2.3 Change in density of ice and water, by temperature.

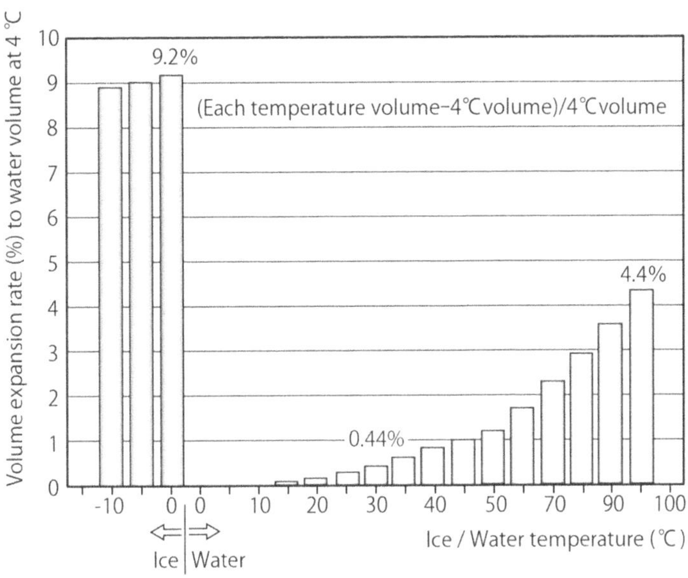

Figure 2.4 Expansion by volume of water (%) by temperature, relative to water volume at 4°C.

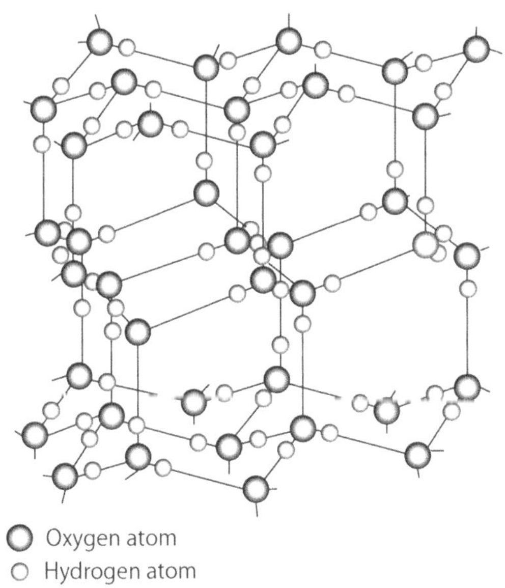

Figure 2.5 A schematic diagram of the structure of ice (Arakawa, 1991).

many gaps in ice crystals. When the temperature rises and thermal motion becomes intense and the water becomes liquid, the crystalline array breaks down, and water molecules become arranged in a nested structure and the density increases until the temperature rises to 4°C (Figure 2.6). From then on, further activation of the thermal motion of water molecules due to the increase in temperature causes the

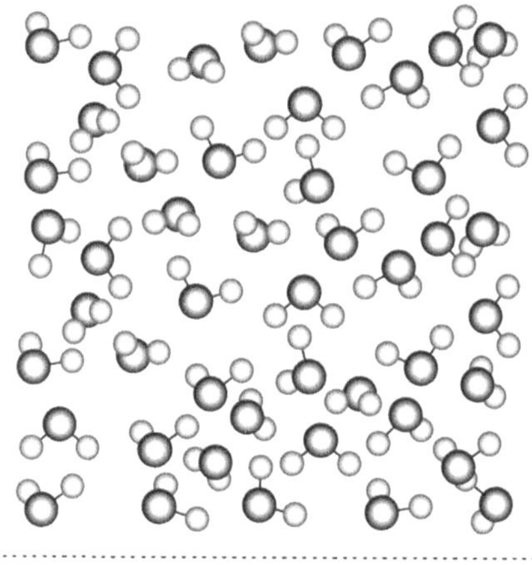

Figure 2.6 A schematic diagram of the structure of water.

intermolecular distance to open and the density to decrease as in normal liquids of other materials. The density of water at room temperature is about 1 g/cm^3, and 1 m^3 of water has a weight of about 1 ton. It is for this reason that water is used to represent flow rate, as it often has the same mass as volume. It is known that ice crystals of up to ten different structures can be produced under different conditions. The most common structure of ice is the tetrahedral coordination of the hydrogen-bonded O–H••O shown in Figure 2.5, with a wide range of voids between water molecules. Higher pressure can lead to dense ice with fewer voids or irregularly shaped ice with distorted hydrogen bonds.

Fish and aquatic organisms can continue to survive in cold winter and polar oceans because they have water in which they do not freeze, even if the water surface is covered with ice. In addition, there are many warm water masses of 4°C with the highest density at the bottom of the water, and the ecosystem of the hydrosphere is the basis for sustaining a stable state. In spring, when ice melts and the cold water on the surface warms to 4°C and reaches its maximum density, sinking begins, the density gradient of the body of water disappears, and there is an unstable situation in which easy mixing (turnover) of the water mass occurs due to the force of the wind blowing on the water surface. This mixing raises the nutrients that have settled in the bottom layer to near the surface of the water and promotes phytoplankton growth under sunlight. The number of animal plankton that feed on phytoplankton increases, and fish and underwater mammals that feed on these plankton begin to thrive. This is the natural peculiarity of temperature change on the density of water and ice (see Figure 2.7).

2.1.1.3 Three forms of water and latent heat/sensible heat

When temperature is further increased to 100°C under atmospheric pressure, the kinetic energy of the thermal motion of water molecules becomes larger than the energy of the attraction between molecules, hydrogen bonds are broken and water molecules are dispersed as a gas (water vapor). To break the hydrogen bonds that hold the liquid water structure in this final step requires the input of as much as 539 calories of heat per gram of water. This is called the latent heat of evaporation. Similarly, 80 calories of latent heat of melting per gram of water are required to activate the vibration of water molecules by

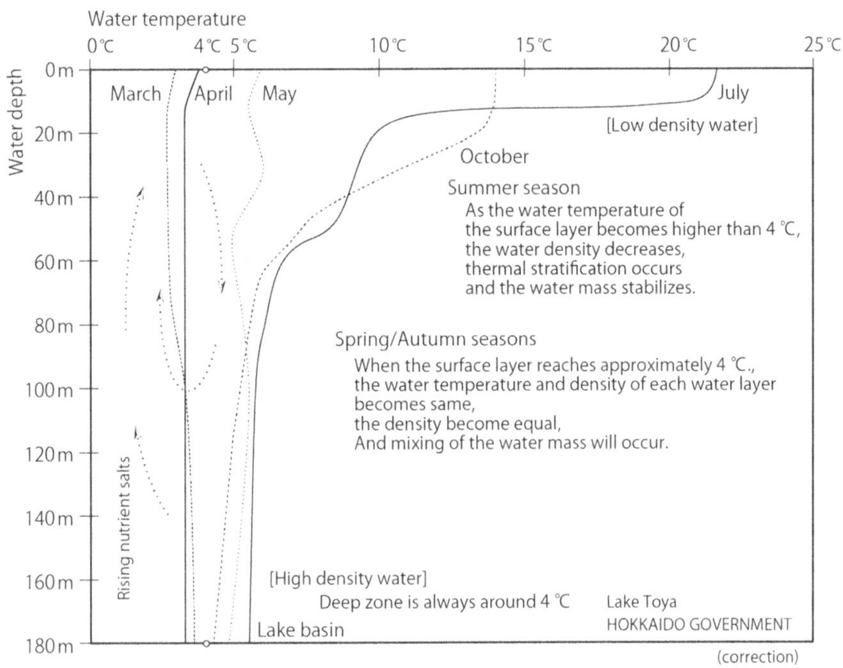

Figure 2.7 The turnover of a water mass.

adding heat to solid water molecules that are tightly bound as ice, allowing the molecules to move as liquid water. This is the amount of heat required to convert ice at 0°C to water at 0°C. The amount of heat required to change phases without changing temperatures is called latent heat. The amount of heat that serves to raise the temperature of water under the same conditions is called sensible heat, and the amount of sensible heat required to raise the temperature of 1 g of water by 1°C is a calorimetric unit (Figure 2.8). The calorimetric unit is 1 cal = 4.1819 J (20°C) in SI units (Degremont & Lyonnaise des Eaux, 1991).

Depending on temperature and pressure, water assumes three states: gas, liquid or solid, as shown in Figure 2.9. The latent heat of vaporization and melting at 1 ATM is as described above, but at temperatures as low as −10°C, ice does not break up into liquid but is converted directly into water vapor (gas). The direct change of solids into gas is called sublimation. The critical point up to which the two states of water exist is at a temperature of 374.1°C and a pressure 218 atmospheres; when the temperature and pressure exceed these points, the liquid/gas state becomes indistinguishable, resulting in a single-phase supercritical water. At room temperature, the dielectric constant drops to about 10, which is usually as high as 80. The dielectric constant approaches the value of the organic solvent and has the property of dissolving organic substances such as oil, which has a dielectric constant of about 2 to 3. Conversely, inorganic substances dissolved in ordinary water become insoluble. The closer the values are, the better the mixture is. Injection of oxygen into super critical water causes submerged combustion, which enables wet combustion of contaminants and greatly changes the solubility of various components. A variety of new technologies are being tried in the supercritical region and the nearby subcritical region. These conditions are also conditions for a variety of geochemical reactions occurring on the deep seabed. Supercritical boilers are widely used in thermal power stations to improve the thermal efficiency of power generation (Tambo, 1980).

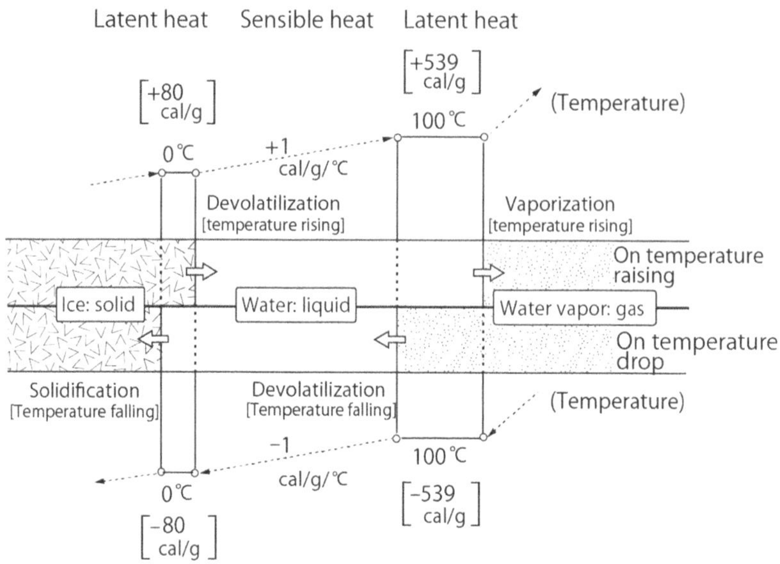

Figure 2.8 Latent heat and sensible heat of water.

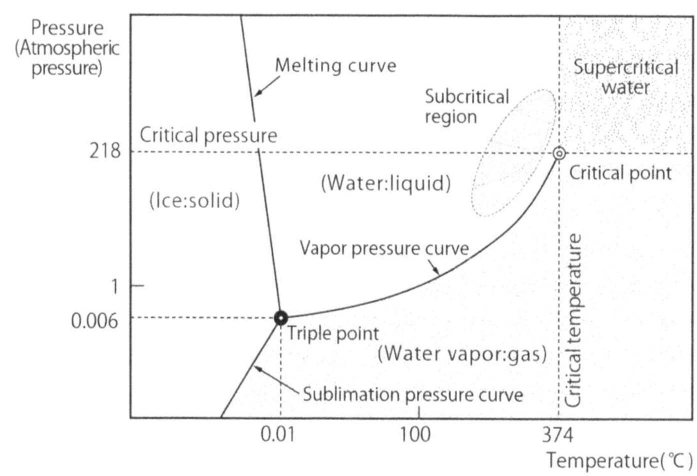

Figure 2.9 The three states of water.

2.1.2 Natural water conditions and impurities (Tambo & Ogasawara, 1985)

2.1.2.1 Hydrogen ion (concentration) index: pH

As a key factor in considering the nature of water, pH (the hydrogen index) is often used. As an indicator of the degree of acidity or alkalinity of water, pH is an important indicator underlying the geochemical/biochemical phenomena that govern the interrelationship between water and its impurities. Part of the water is separated into cations and anions of H^+ (hydrogen ions) and OH^- (hydroxyl ions).

The molarity of each of the slightly dissociated H^+ and OH^- in water are multiplied by 10^{-14} $(mol/\ell)^2$ at all times. In other words, $[H^+] \times [OH^-] = 10^{-14}$ $(mol/\ell)^2$, and pH is defined as the reciprocal of the exponent of the molarity of $[H^+]$ ions in this case, and pH is the value of x in $[H^+] = 10^{-x}$ mol/ℓ. For example, if $H^+ = 10^{-5}$ mol/ℓ, the pH is 5, and then OH^- will be only $10^{-14}/10^{-5} = 10^{-9}$ mol/ℓ, and H^+ will be 10^4 (10,000) times OH^-. H^+ is considered to be acidic, and OH^- is considered to be alkaline. In this case, H^+ is the predominant acidic water. When both H^+ and OH^- are equal to 10^{-7} mol/ℓ, the product is 10^{-14} $(mol/\ell)^2$, so it is neither acidic nor alkaline, and has a neutral pH = 7. The pH scale is logarithmic so that, as the value rises or falls by 1, the concentration changes by a factor of 10. Pure water, such as distilled water, tends to be neutral at pH 7. However, when distilled water is exposed to the atmosphere, carbon dioxide (CO_2) in the atmosphere dissolves in the water to form bicarbonate (HCO_3^-), resulting in weakly acidic water, such as pH 5.6. Therefore, distilled water is weakly acidic. If the pH of rainfall is lower than this pH value, it is considered that acid substances such as sulfuric acid and hydrochloric acid in the air are additionally dissolved, and acid rain has occurred.

The definition of the hydrogen ion index is given by Equation 2.1.

$$pH = \log_{10}\left(\frac{1}{[H^+]}\right) \quad \text{—(2·1)} \tag{2.1}$$

$[H^+]$: Hydrogen ion concentration mol/l

Equation 2.1. The definition of pH

In addition, as described above, there is a relationship of pH + pOH = 14, where pOH is the hydroxide ion index.

A similar term used is the alkalinity of water. Specifically, this is a representation of the pH buffering capacity of water. The amount of alkali required to reduce the pH to 4.8 is converted to mg/ℓ of calcium carbonate ($CaCO_3$). Carbon dioxide in the air is dissolved in water to form a weak acid. However, hard water in which hardness components such as lime are dissolved requires a large amount of strong acid to lower the water to a pH of 4.8. Alkalinity is a numerical value representing the quantitative magnitude of resistance to acid. Assuming that pH is the height of a mountain (altitude: strength index), the alkalinity corresponds to the steps leading up to the top or to the length of the mountain approach (distance: volume index). Water has a high alkalinity if it contains a weak acid such as organic acid, or a high hardness. Distilled water has a very low alkalinity. In the fermentation of wine or Japanese sake and other alcoholic beverages, organic acid generated during the course of the fermentation process often causes pH to fall below the appropriate pH range for alcoholic fermentation. However, the use of very hard (alkaline) water can prevent the pH from falling prior to the conversion to vinegar. As a result, water with high alkalinity can promote the alcohol reaction, which is a rare feature of famous breweries. Today, it is possible to automatically maintain fermentation conditions at the appropriate pH for any water for an extended period.

2.1.2.2 Redox (OR) potential

Another key determinant of the basic state of water is oxidation-reduction (OR) potential. A strength indicator (intensity factor) is used to determine whether water is in an oxidized or reduced state, and to what extent. As will be discussed in detail later (e.g., in Chapter 5), this plays a crucial role in assessing the presence of metals and other inorganic ions in quality conversion processing, in controlling corrosion, and in environmental assessment and setting of processing conditions for microbial reactions

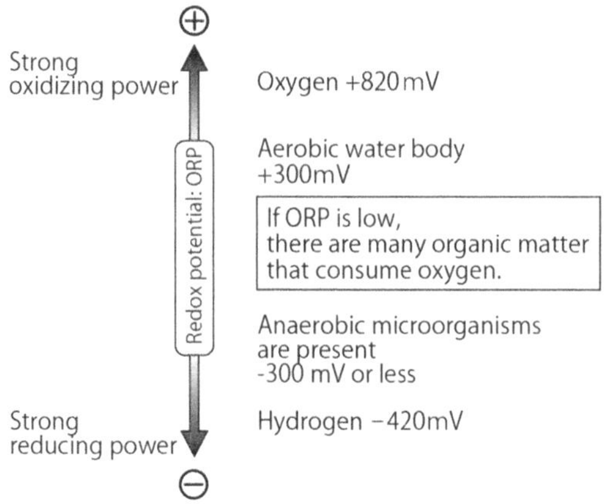

Figure 2.10 Redox potential.

related to organic components. Many commercially available hydrogen ion density meters (pH meters) are capable of measuring OR potential by replacing the electrode with a redox potential (ORP).

A low OR potential in environmental conditions means that many organic substances are consuming oxygen. In aerobic waters, redox potentials are as high as +300 mV. For the survival of anaerobic microorganisms, conditions of −300 mV or less are recommended, and values of −330 mV are recommended for culture media for anaerobic bacteria. Since the redox potential of oxygen is +820 and −420 mV for hydrogen, the redox potential of water ranges from +820 to −300 mV (Figure 2.10).

The definition of OR potential is somewhat complicated. The expression for a redox potential E under certain conditions of a specific substance (e.g., iron) is known as the Nernst Equation, and is expressed in volts from the ratio of the redox activity [ox] to the redox activity [red] of the specific substance and the standard redox potential E_0 inherent in the substance system (Equation 2.2):

$$E = E_0 + \frac{2.303RT}{nF}\log\left(\frac{[ox]}{[red]}\right) \fallingdotseq E_0 + \frac{0.059}{n}\log\left(\frac{[ox]}{[red]}\right)_{25^\circ C} \quad —(2\cdot 2) \tag{2.2}$$

E_0 : Standard oxidation-reduction potential V
R : Gas constant 8.314 $JK^{-1}mol^{-1}$
T : Absolute temperature K
n : Number of electrons transferred by redox reaction
F : Faraday constant 96,485coulomb/mol (Charge of electron per mol)
[ox]·[red] : Activities of oxides and reductants

Equation 2.2 The Nernst Equation

Concentrations and activities of oxides and reductants can be described in terms of concentrations of reactants, C mol/ℓ, for dilute systems. However, when concentrations of solutions are high, interactions occur between molecules and the reaction effects are reduced. The molar concentration C multiplied by the activity factor (Activity coefficient, $\gamma < 1$) is used to calculate the equation. Activity is the proportion of a specific substance that contributes to the reaction.

The standard redox potential E_0 indicates the potential difference V between target solids and solutions, using a Standard Hydrogen Electrode (SHE) (hydrogen ion activity of 1 at 25°C) as 0.0 mV, and is also referred to as the standard Electrode potential. Measurement is performed using an electrode such as standard calomel electrode (SCE) or silver–silver chloride electrode, which is easy to handle. For example, the difference between a calomel electrode potential and a hydrogen electrode potential is adjusted to 248 mV to obtain the E_0. When dealing with phenomena in water environments, we use a pH = 7 potential as the intermediate redox potential E_0' as a reference and calculate the intermediate redox potential E′ indicated by the target substances. This is sometimes simply referred to as the oxidation-reduction potential (ORP). Figure 2.11 shows the redox potential of several key substances.

For a reaction that exhibits a large standard redox potential (E_0), the more distant the metal is from $E_0 = 0.000$ V in the hydrogen electrode reaction, the more stable the metal exists in an aqueous solution. In other words, a substance with a larger positive value has a stronger oxidizing power (weaker reducing power). Substances with large negative values have a strong reducing power (weak oxidizing power). As Figure 2.11 shows, gold has a large $E_0 = +1.498$ V, is highly stable, and is referred to as a noble metal. Aluminium, on the other hand, exhibits a low normal redox potential of $E_0 = -1.662$ V, is soluble in water, and is referred to as a base metal. By comparing the standard redox potential E_0, we can classify the situations that correspond to redox reactions of different substances. Substance A, which exhibits a high E_0, oxidizes substance B, which exhibits a low E_0. A is an oxidizing agent and B is a reducing agent.

2.1.2.3 *Water quality and impurities*

As noted above, even when we refer to the basic nature of water, we have an index that considers the presence of impurities such as pH and alkalinity. The oxidation-reduction potential, ORP,

Electrode reaction	E_0(V)
$Cs^+ + e^- \leftrightharpoons Cs(s)$	−3.026
$Al^{3+} + 3e^- \leftrightharpoons Al(s)$	−1.662
$2H_2O + 2e^- \leftrightharpoons H_2(g) + 2OH^-$	−0.828
$Zn^{2+} + 2e^- \leftrightharpoons Zn(s)$	−0.762
$Cr^{3+} + 3e^- \leftrightharpoons Cr(s)$	−0.74
$Fe^{2+} + 2e^- \leftrightharpoons Fe(s)$	−0.440
$Cd^{2+} + 2e^- \leftrightharpoons Cd(s)$	−0.403
$2H^+ + 2e^- \leftrightharpoons H_2(g)$	0.000
$Cu^{2+} + 2e^- \leftrightharpoons Cu(s)$	+0.340
$O_2(g) + 2H_2O + 4e^- \leftrightharpoons 4O^-(aq)$	+0.70
$Fe^{3+} + e^- \leftrightharpoons Fe^{2+}$	+0.771
$O_2(g) + 4H^+ + 4e^- \leftrightharpoons 2H_2O$	+1.229
$Cl(g) + 2e^- \leftrightharpoons 2Cl^-$	+1.36
$Au^{3+} + 3e^- \leftrightharpoons Au(s)$	+1.498

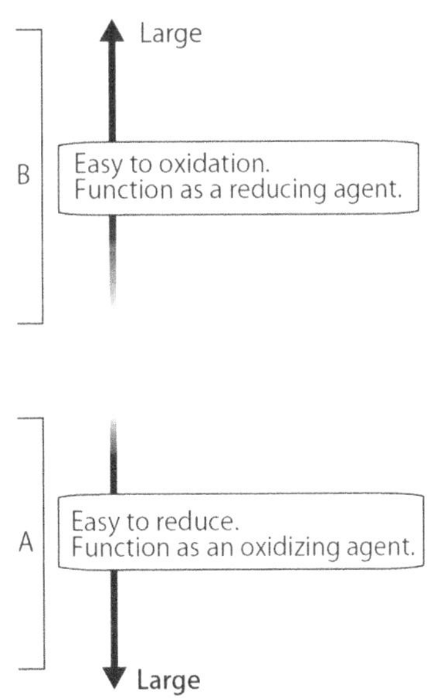

(s): solid (g): gas (aq): liquid

Figure 2.11 Standard Redox Potential, E_0(V).

Origins \ Classification		Suspended matter	Colloidal matter	Soluble matters			
				Gaseous	Nonionic	Cationic	Anionic
Inorganic origin	Inorganic soil Rock	Sand,clay other inorganic soil	Clay, SiO_2 Fe_2O_3, Al_2O_3 MnO_2	CO_2		Ca^{2+}, Mg^{2+} Na^+, K^+, Fe^{2+} Mn_2^{+}, Zn^{2+}	HCO_3^-, Cl^-, SO_4^{2-}, NO_3^- CO_3^{2-}, OH^-, $HSiO_3^-$, $H_2BO_3^-$ HPO_4^{2-}, $H_2PO_4^-$, F^-
	Air			N_2, O_2, CO_2 SO_2		H^+	HCO_3^-, SO_4^{2-}
Organic origin	Animal carcasses and plants	Organic topsoil Organic waste	Plant pigment Organic waste	N_2, O_2, CO_2 NH_3, H_2S CH_4, H_2 Odor component	Plant pigment Organic waste	Na^+, NH_4^+, H^+	HCO_3^-, Cl^-, NO_2^- NO_3^-, OH^-, HS^-
	Animal and plant organisms	Fish, algae microorganisms	Virus,Algae Bacteria				
	Chemical industrial products				Mainly petroleum compounds such as DDT, BHC, PCB etc.		
Dimensions of impurities		mm ~ μm	μm ~ nm	nm ~ Å			

Figure 2.12 Impurities in water and their origin.

which indicates the oxidation and reduction states, is also defined by any contaminating substances. Indeed, except in special cases, we do not handle pure water. The water actually used is pure water in which various impurities are dissolved. Classifying impurities in water by their origins and clues considering their processability are illustrated in Figure 2.12. In general, the term 'water quality' refers to the quality of water in a narrow sense, which is characterized by these types and concentrations of impurities.

The presence of impurities in water can be divided into three states: solids, liquids and gases. Depending on the size of the impurities, the impurities are roughly divided into suspensions, colloids and solutes. In environmental surveys and water quality tests of a water supply, the size of impurities is divided into two components: suspended and soluble matter. Ingredients that pass through 0.45 μm filter paper (filter) are soluble and any ingredients that remain on the filter paper will be suspended. Suspended ingredients correspond to groups of substances that are often expressed as turbidity in water. These suspended ingredients are mainly derived from the outflow of organic and inorganic soils due to rain and the discharge of organic and inorganic ingredients from municipal and industrial wastewater.

A substance (particle) with a size of about 1 μm to 1 nm is called a colloid. According to the definition of a general water quality test, particles that cannot pass through 0.45 μm filter paper are classified as suspensions. Following this classification, colloidal components (which are particles with a size of 1 μm to 1 nm), fall on the soluble component side. However, as described below, water quality control and water treatment operations have positioned the coagulation of colloidal particles as one of the central operations. For water quality control/water quality conversion engineering, colloids are considered particles and are treated as a suspension (a colloidal suspension) in order to control their behaviour. This is because coagulation operation can cope with a wide range of particulate impurity treatment and can be clearly distinguished from other treatment such as adsorption, which targets only impurities with much smaller sub-colloid size (soluble) particles.

Figure 2.13 shows how to display various units of length.

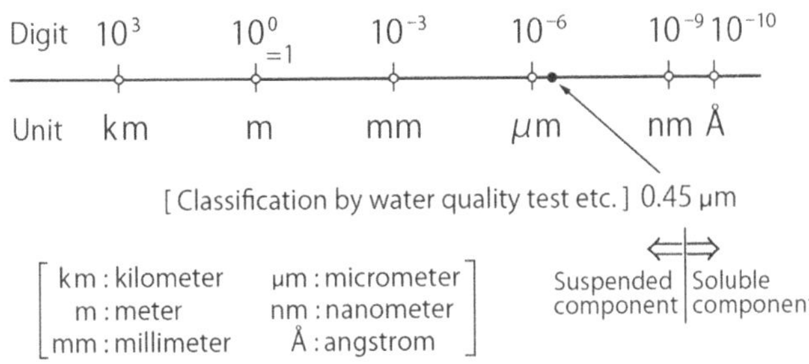

Figure 2.13 Units of length and how to express them.

2.1.2.4 Colloidal suspensions

There are many types of colloidal components in water. Crude colloids such as mineral turbidity (1 μm) including clay particles, organic coloured components such as humic components in peat land water (10 nm), and microorganisms such as bacteria (μm) and viruses (nm) belong to this group. These are all typical components for water and sewage treatment.

Taking such a colloidal suspension in a beaker and applying a thin beam of light causes a Tyndall phenomenon in which the light path scatters. Under a microscope, colloidal particles collide with water molecules that move thermally, causing a Brownian motion phenomenon that moves randomly. This phenomenon makes it difficult to see colloids with ordinary optical microscopes and requires ultra- and electron microscopes.

Such colloidal particles have the same type of charge (surface charge, surface potential) on their surface which causes the particles to repel each other and disperse and suspend stably in water. Many colloidal particles in natural water are negatively charged at neutral pH; these include clays, powdered glasses, silicas, bacteria, many organic matters and blood cells but not metal hydroxide particles with positive potential. The particles have a surface potential of −20 to +30 mv (zeta-electric potential, one of the typical potentials that define the charge status of the surface of the particles) and are stably suspended by mutual repulsion. Generally, when the zeta potential changes from +10 to −10 mv (which is close to 0 mv: the iso-electric point), the repulsion force between particles decreases and van der Wall's forces bring colloids to coagulate with each other, as will be discussed in more detail in Chapters 3 and 4 (Figure 2.14).

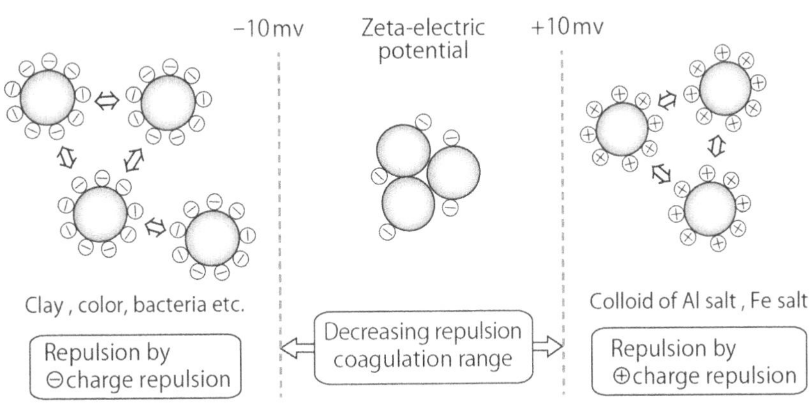

Figure 2.14 The dispersion and coagulation of colloids.

Hydrated colloids of aluminium and iron salts have different positive and negative charges and different sizes depending on pH, and exhibit positive charges in the insoluble region on the acidic side and negative charges on the alkaline side. Coagulation treatment, the main operation of the rapid filtration system is an operation in which aluminium or iron polyvalent positively charged complexes neutralize the charges of negative colloidal particles such as clay, colour colloids and bacteria, which are stably present in the natural system, to lose their repulsive force, and to agglomerate and remove the particles. Coagulation is described in detail below.

2.1.2.5 Aqueous solution: soluble component

It was stated above that it is appropriate to treat colloidal solutions as suspensions in water treatment operations. But what is a solution with soluble components in a real sense? It would be better to assume that water molecules (with diameters as small as 3 Å) and ions and colloidal macromolecules are uniformly mixed in a water body. As noted above, the small diameter of the water molecule, which is 3 Å in this case, is beneficial for making homogeneous solutions with various impurity molecules.

Among these components, most of the inorganic components are dispersed in water as positive or negative ions. Salt (NaCl) is said to be dissociated into Na^+ ions and Cl^- ions in water, but it is actually a hydrated ion bound to water molecules such as $Na(H_2O)_8{}^+$ and $Cl(H_2O)_4{}^-$. The Na^+ and Cl^-, which constitute the salt crystals, are connected by strong ionic bonds and it is not easy to separate them. However, when the salt is put into water, the ions hydrogen-bond with water molecules and are easily separated into positively charged hydrated Na ions and negatively charged hydrated chloride ions. Such a hydration reaction plays a major role in many water treatments and biological reactions.

Ions carry the charge they carry by thermal movement through water, so the more soluble inorganic ion components they contain, the greater the electrical conductivity of water. Therefore, by measuring the electrical conductivity of water, the concentration of dissolved inorganic matter can be estimated. Incidentally, pure water is assumed to have values of approximately 0.05–1.00 μs/cm, distilled water 1–10 μs/cm, Municipal Water (Japanese, soft water) 100–250 μs/cm and recycled sewage water 400–600 μs/cm (where 1 μs = 1 micro-siemens, and 1 μs/cm is a unit of electrical conductivity).

Not all of the components dissolved in water are ionic, and many non-ionic soluble components occur in natural water, both inorganic and organic. Typical water-insoluble substances include a wide range of organic substances that are biologically persistent (methane, petroleum, DDT, BHC, PCB, etc.) and organic substances that are not decomposed easily.

Methane (CH_4) and ethane (C_2H_6) are hydrophobic substances (groups) that are only slightly soluble in water. However, alcohols such as methanol (CH_3OH) and ethanol (C_2H_5OH), in which one hydrogen of methane or ethane is replaced with a hydroxyl:(OH), are dissolved easily in any proportion of water. This is because the OH is a hydrophilic group that is susceptible to hydrogen bonding with water molecules. However, OH-bearing alcohols with a much higher carbon number are not as easily dissolved e.g., n-butyl alcohol, of which only 7.7 g can be dissolved in 100 ml water. When hydrophobic CH_2 in the chain structure increases to $n = 15$, the hydrophobic CH_3 and hydrophilic OH groups are located at both ends of the long chain, one of which binds to water molecules (dissolves), the other with oil and fats (such as stains in clothing), and the compound serves as a detergent (surfactant) that removes oil and fats from clothing and dissolves in water (Figure 2.15). In addition, sugars are dissolved in large quantities by forming hydrogen bonds with water molecules because the glucose ($C_6H_{12}O_6$) of a monosaccharide contains as many as five OH groups, even if it contains many carbons (Figure 2.16). Substances with highly polar components, such as SH^- and $NH_2{}^-$, are also soluble in water.

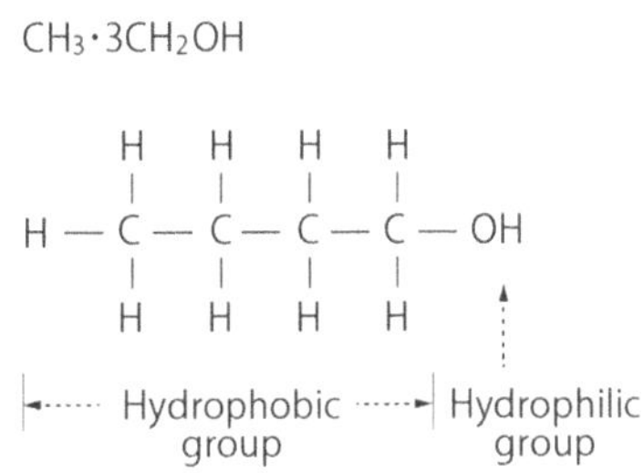

Figure 2.15 The structure of butyl alcohol.

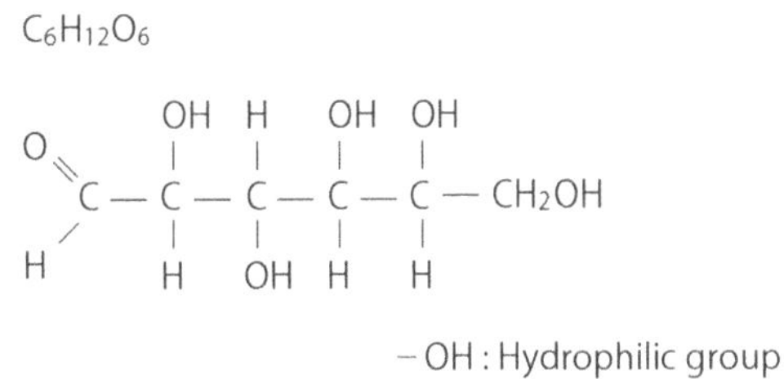

Figure 2.16 The structure of glucose.

Another dissolved component is dissolved gas in water. The major components of air are oxygen, nitrogen and carbonic acid, all of which are low in solubility, and 21% of oxygen in air is soluble in water only at 10 mg/ℓ, which is consumed easily by many microorganisms and animals that breathe aerobically in water, resulting in anaerobic and insufficiently oxygenated water, which is the primary phenomenon of water pollution, making it difficult for many waterborne organisms to survive. When oxygen in water is lost and the water becomes anaerobic, methane and sulfide hydrogens are produced, leading to the destruction of aerobes and of healthy natural water environments.

The degree to which gas molecules can dissolve in water depends on temperature and the pressure of the gas. It is known that the partial pressure p[atm] of the gas phase of the components to be dissolved and the molar fraction x of the solute gas in the liquid phase depend on Henry's law, $p = Hx$, and the magnitude of solubility is quantified by knowing the proportional constant of each gas (Henry's constant H [atm/molar fraction]). Table 2.3 show the Henry's constants for various gases, showing that the solubility of a gas decreases with increasing temperature, as opposed to that of a solid.

Table 2.3 Henry's constants for various gases dissolved in water.

$H \times 10^{-4}$ (atm/mol fraction)

	0°C	10°C	20°C	30°C
Air	4.32	5.49	6.64	7.71
O_2	2.55	3.27	4.01	4.75
CO_2	0.0728	0.104	0.142	0.186
N_2	5.29	6.68	8.04	9.24
H_2S	0.0268	0.0367	0.0483	0.0609
CH_4	2.24	2.97	3.76	4.49

As can be seen, oxygen (which is present at 20% in the atmosphere) can be dissolved in water nearly twice as much as nitrogen (which is present at 78% – almost four times more than oxygen) and, as a result, oxygen is 1.66 times more soluble in water when viewed in molar terms. As a result, gas solubility [mg/ℓ] is 23.2 for N_2, 54.3 for O_2, 2318 for CO_2, 5112 for H_2, and 32.5 for CH_4 in water at 10°C and at atmospheric pressure. The ratio of N_2 to O_2 in water differs greatly from the ratio of 4/1 in the atmosphere, and the ratio is largely reversed. The very high solubility of carbon dioxide and hydrogen sulphide sometimes poses serious problems to underwater ecosystems.

2.2 WATER AND HUMANS

2.2.1 Human water metabolism

2.2.1.1 What is metabolism?

All living things interact with the outside world (the environment) to support their survival. This process is called metabolism. Depending on the substances involved, there are water metabolisms carbonic metabolisms, nitrogen metabolisms, amino acid metabolisms, etc., and the substance balance of various metabolisms includes the process of substance change in a living body (Tambo, 2012).

The smallest element in a city and local water systems is the uptake of water required by an individual from the external environment as food or drink, and the discharge of unnecessary water as urine, sweat, etc. to the outside of the body. Therefore, the first focus is to discuss the balance in a metabolism between living organisms and the environment. The behaviour of water itself is important; water plays the most important fundamental roles in the maintenance of living organisms and biological groups, as a medium for connecting organs and spaces inside and outside living organisms, while also preserving and transporting chemical substances, thermal energies, and other elements, and whilst also intertwining the process of change (Tambo, 1985).

2.2.1.2 Human water balance

Humans contain 90% water as foetuses, 75% as new-borns, 60% as adult males, and 50% or less in adult males over the age of 60, accounting for 50–80% of body weight with a decreasing percentage with increasing age. Because women are les muscular than men, the fluid to body weight ratio is lower. In the human body, fluid is divided into intracellular fluid (about 40% of body weight) and extracellular fluid (about 20% of body weight). Extracellular fluid is divided into interstitial fluid in the cell–cell space (about 15% of body weight) and intravascular fluid in blood and lymphatic vessels (about 5% of body weight). In adult males weighing 60 kg, the total volume of fluid is about 36 litres, which means that the volume of intracellular fluid is about 24 litres and that of extracellular fluid is about 12 litres. Adults weighing 60 kg uptake about 2.6 litres of fluid per daily. The breakdown of ingestion is as large as 1.0 ℓ from food, 1.2 ℓ from drinking water and 0.4 ℓ from metabolic water (water produced during the body's metabolic process). The breakdown of discharge is 1.5 ℓ from urine, 0.1 ℓ from faeces, 1.0 ℓ from evaporation (0.4 ℓ from pulmonary respiration and 0.6 ℓ from the skin) and sweating for temperature control (Hishinuma & Kitamura, 2005).

Approximately 2.6 litres of fresh water consumed daily is supplemented with water that is regenerated by the kidneys – about 150 litres each day. About 99% of the body's water comes from regenerated (recirculated) water, and 1.5 litres are discharged as urine. In an adult male, for example, if the total volume of fluid is 36 litres, the total internal water will be recovered and circulated at a rate of $150/36 = 4.1$ times per day after filtration. If 2.6 litres of fresh water are supplied per day, 36 litres of body water will be replaced in about 14 days (i.e. the retention time in the body will be about 14 days)

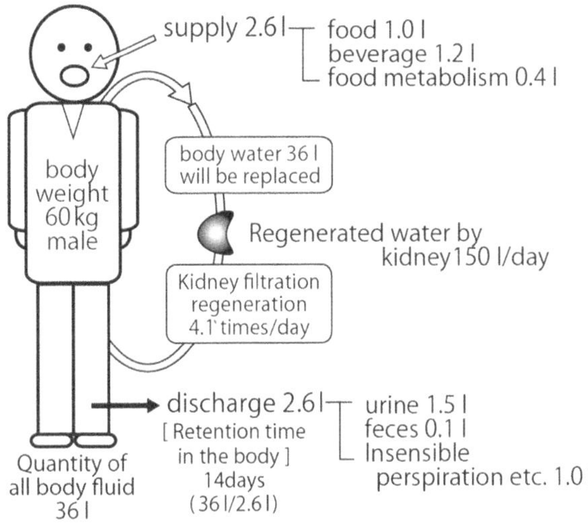

Figure 2.17 The human body water balance (Hishinuma & Kitamura, 2005).

and the total number of cycles will be less than 60. If the intracellular fluid volume changes by 10% or more, physiological disorders are said to occur. Usually the change in intracellular fluid volume is not great, hence, the body's main replacement occurs with the extracellular fluid of 12 litres with a retention time of less than 5 days. It seems that humans can live for days without supplemental water (Figure 2.17).

2.2.1.3 Socialization of water metabolism

As population density increases, multiple families gather from a dispersed isolated way of living to organize into communities of 10 or 20 people in order to improve their abilities in fishing, farming and defence. Water use and disposal mechanisms also start to have a social dimension in light of the relationship between individuals' wells and toilets and the relationships between humans in a population's residential area. In order to avoid harm to their peers, people seek a location for water intake upstream and discard waste outside the boundaries of the group's lives. Larger communities create an upstream–downstream relationship in a region. Social systemization of water metabolism occurs to enhance the benefits, order and safety of village communities. As shown in Figure 2.18, we distribute clean water drawn from upstream and distribute it to all the villagers, and discard waste into downstream channels (outside the control boundary) below the living area because it is difficult to disperse dirt and dust in front of neighbours. These initiatives are the beginning of water supply and sewerage systems (Montgomery J. M., Consulting Engineers, Inc. 1985).

The metabolic functions of cities provide citizens with sufficient materials and energies to carry out household, vocational and recreational activities, etc., and dispose of the waste and residue without hindering citizens' lives. The object of environmental engineering is how to preserve this.

2.2.2 Poisons and risks (water quality standards)

2.2.2.1 Classic water quality standards for human were epidemiological safety assurance

Until the 1970s, water supply (drinking water) quality standards set control targets for pollutants that are transmitted in natural water circulation, and which cause direct human health and epidemic disasters through municipal water facilities such as water supply and sewage works. Wastewater treatment

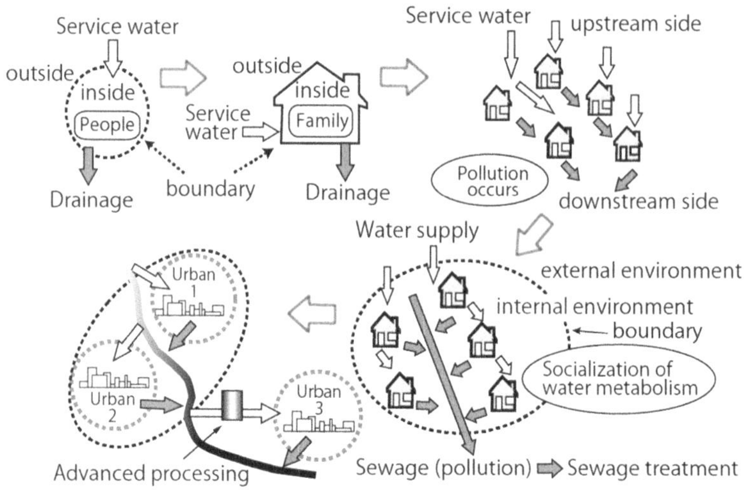

Figure 2.18 The social system of water metabolism.

(discharge) standards were decided by taking environmental requirements into account, as well as the conditions of use, amongst which human drinking standards are top-level targets. This issue relates to the control of biochemical oxygen demand (BOD) and dissolved oxygen deficiency (DOD), the circulation/control of nitrogen, phosphorus and sulphur, and the control of acute chronic poisons, which will be described below.

For epidemiological safety, the metabolic excretion of as many as 60 trillion cells (1 trillion per kilogram of body weight) by human organs are important, not forgetting the additional 100 trillion bacteria and enzymes which coexist in the human body. Many epidemics caused by pathogenic bacteria and viruses often hit humans in particular. The 14th Century black death and 18th Century cholera and typhus outbreaks are pandemics that triggered the process of consolidation of the Eurasian continental regions by the Mongolian Empire, and the establishment of the global hegemony of the British Empire.

By the 1960s, the most important factor in water quality safety was protecting people from widespread epidemics and protecting them from gastrointestinal infectious diseases. Until the end of the 1970s, the final stage of this phase, municipal water supply quality standards were much lower than are today's which take into account the environmental risks of trace pollutants. Briefly, the goal was to focus on the elimination of dangerous gastrointestinal infectious diseases caused by waterborne transmission of faecal bacteria and others. Another goal was to give a common awareness of heavy metals that could cause acute toxicity and cumulative toxicity damage. Organic chemicals such as surfactants and phenols were slightly visible and were intended to suppress the generation of foams and offensive odours. This is similar to the first appearance of trace contamination by chemicals in the next era.

After Rachel Carson's 'Silent Springs' (1962) came out, drinking water standards changed in response to the trihalomethane issue but the basic component of water quality standards has not yet been addressed. We still wait for a major change.

2.2.2.2 Water quality standards in the 1960s and 1970s

Let us look back at the Japanese drinking water quality standards (the Ministry of Health and Welfare of Japan No. 11, 1966) which were used until the late 1980s (Table 2.4).

Table 2.4 Former Japanese water quality standards for drinking water (taken from 1979).

[26 items, At 1979 years]	The units are under ppm			The Ministry of Health and Welfare of Japan	
Ammonium nitrogen and Nitrous nitrogen	Not to be detected together	Copper Iron	1.0 ppm 0.3 ppm	Phenolic compound Anion surfactant	0.005 ppm 0.5 ppm
Nitrate nitrogen	10 ppm	Manganese	0.3 ppm	Hydrogen-ion concentration	pH 5.8~8.6
Chlorine ion	200 ppm	Zinc	1.0 ppm	Odor	Not abnormal
Organic matter (Potassium permanganate consumption)	10 ppm	Lead	0.1 ppm	Taste	Not abnormal
General bacteria	100/ml	Sexivalent chrome	0.05 ppm	Color	5 degree
Coliforms	Not to be detected	Arsenic	0.05 ppm	Turbidity	2 degree
Cyanide	Not to be detected	Fluorien	0.8 ppm		
Mercury	Not to be detected	Calcium · Magnesium (Hardness)	300 ppm		
Organic phosphours	Not to be detected	Evaporation residue	500 ppm		

The first category in the water quality standards was that water should not include microbiologicals or substances that are suspected to have been contaminated by pathogenic organisms:

(1) The typical required level for enterobacteria was 'no coliforms detected'.
(2) In order to ensure the general environmental conditions of the water source area and that there is no possibility of faecal contamination, it was required that '100 or fewer bacteria per litre of water' be used.
(3) In addition, the concentrations of ammonia nitrogen, nitrite nitrogen, nitrate nitrogen, chloride ions, and potassium permanganate consumption (which are all indirect indices of organic matter, derived from excretions) should be below a certain level.

The second category related to 'acute/sub-acute poisons,' which required that cyanide ions, mercury and organophosphorus (pesticides) should not be detected. The reference values for these items should have been lower than the limit of detection under the official analytical method and (at that time) were referred to as undetected.

The third category was water-based metals, which generate colour or turbidity in water:

(1) Copper, iron, manganese, zinc and other substances that 'inhibit water use'.
(2) Lead, hexavalent chromium, arsenic and fluorine, which are known to cause 'disorders due to chronic toxicity,' were not to be more than the allowable limit values.
(3) Historically, water with high salt concentrations, high calcium and magnesium concentrations, and disturbances in mineral content, such as precipitation of sediments and poor soap dissolution, were all subject to control.

Table 2.5 Current categories of water quality standards.

	Category	Characteristic
Evaluation since the water quality standard establishment	① Pathogenic organisms	Index about pathogenic organisms pollution Expect environment without feces pollution
	② Acute/Sub-acute poisons	Influence human body after ingestion in a short time The continuous monitoring of the discharge source is necessary
	③ Inhibit water use Chronic toxicity	Coloration and precipitation/sedimentation by metals Chronic toxicity by heavy metals
	④ Normal water without abnormalities	Cleanness in the sense of human Fundamental properties
New evaluation	⑤ Toxicological evaluation by lifetime ingestion	Knowledge accumulation of micro-pollutants for health risk Advance of the measurement technology Introduced in 1992

The fourth and last category determined the conditions that humans think of as 'normal water without abnormalities', such as pH 5.8 to 8.6, that water is not strongly acidic or highly alkaline, that there is no unusual odour or taste, and that it is has a low degree of turbidity or colour.

Today, under present standards, as under the drinking water quality standards enacted in 1957 which had 27 listed items, waterborne disease control against faecal contamination has continued to be a very important factor. As will be discussed later, the present water quality standard has added toxicological evaluation by lifetime ingestion (exposure) risk of trace contaminants to this category (Table 2.5).

2.2.2.3 Expansion of standards for health risk control by trace contamination

With publication of Rachel Carson's 'Silent Spring' in 1962, the issues of accumulated toxicities of organic and inorganic trace pollutants (micro-pollutants) – which are carcinogenic and mutagenic – became known. In particular, in the 1970s, soil and complex contaminations caused by pesticides, plasticizers, medicines etc. (which had not been considered until then) became recognized as significant problems.

Modern water supply and sewage systems were the first to exert their power in the control of gastrointestinal infectious diseases. However, the issue of the risks of carcinogenic trace ingredients for the human body is different from that of classical hygiene. In order to avoid the danger of infectious diseases, the necessity of new health risk control over a long period of time was declared. This was a new problem that required the emergence of 'risk recognition and evaluation' problems and the creation of new technologies and systems to cope with them. The trihalomethane issue was a risk-emergent event in which municipal water itself produced hazardous substances across these two safety issues.

Dr J. J. Rook (1972, 1974) at the Rotterdam Water Works in the Netherlands, located far downstream on the international river Rhine, detected chloroform ($CHCl_3$) – a type of trihalomethane – and, in addition to this chlorination, the river Rhine water was also generating new chloroform at the purification plant. This information transported a strong shock to every water treatment plant worldwide in which chlorination had been traditionally used as a standard safety measure. In water treatment processes such as sand filtration, bacteria in water can be removed by one or two orders of magnitude, but in order to further enhance the safety of drinking water, it had been the standard measure of modern municipal water works to inject chlorine to filtrate water and control microbial contamination of bacteria in distribution systems. Suspensions (turbidity) and organic components (humic colour substances) are naturally derived and account for the majority of the components to be treated in water treatment. As will be described later, one reason why the rapid filtration system was developed in the US and subsequently widely used worldwide is that a slow sand filter system does not remove humic colloids derived from microorganisms, which are notable stable organic components of natural water systems (particularly the low-molecular-weight part of colour). Rapid sand filtration has been widely used as a standard method; however, the removal of microbial contamination such as bacteria is not adequate compared with the slow sand filter method, and consequently a chlorination process is added at the end of the water treatment processes. The two-basic operation processes of rapid sand filtration and chlorination, were considered to be the main operational methods of ensuring epidemiological safety measures. However, the removal of microbial metabolic wastes (humic substances) contained in raw water were narrowly separated from water. This was due to the issues of the artificial formation of chloroform and other trace chlorinated organic compounds. Moreover, it was thought to be a sort of suicide of modern water purification system trains, which should be safe, in the sense that they created risks by their own operation and had a major impact on the operation of urban water systems.

In 1986, the US Environmental Protection Agency (USEPA) raised the list shown in Table 2.6 and proposed 'The Safe Drinking Water Act Amendments of 1986' to create new safety standards for drinking water, to control drinking water quality by way of fully revised conventional water quality standards.

Microbial contamination, turbidity and minerals (inorganics) are indices of traditional sanitary engineering. However, volatile organic chemicals, organic compounds such as pesticides and radioactive materials have been suggested as objectives to change the history of water quality control technologies, such as how to cope with environmental risks (such as trace organic contamination) which were raised by 'Silent Springs'. This proposal triggered a probabilistic assessment of risk for maximum allowable contaminant levels (Maximum Contaminant Level; MCL). Since then, the worldwide review of issues has led to it now being in the mainstream of global drinking water quality assessment and control objectives, including in UNHC (WHO) standards, US Environmental Protection Agency (USEPA) standards, as well as in European (EU) and Japanese standards.

Many items are evaluated at levels ranging from $\mu g/\ell$ to ng/ℓ at much lower concentrations compared to classic historical safety standards for mg/ℓ levels. With regard to human exposure to cumulative toxic substances such as carcinogens, we have started to make a concerted effort to regulate water quality and produce water by moving to lifetime intake risk assessment. One person in 1 million (a 10^{-6} risk) or one person in 100,000 (10^{-5} risk) developing cancer after drinking a daily 2 l of water for 70 years are acceptable risk levels to be discussed. The list of pollutants of interest gradually increases as environmental surveys and observations progress, and even endocrine disruptors (popularly known as environmental hormones), which are difficult to evaluate, have been targeted.

2.2.2.4 Mechanism of risk assessment

It is extremely difficult to evaluate the chronic and cumulative toxicities of trace organic constituents, and the four principles for considering the risk of drinking water presented in 1977 by the US National Academy of Science are as follows:

(1) Using appropriate test methods, the results of animal experiments can be used for human safety assessment.

Table 2.6 U.S. Environmental Protection Agency list of contaminants in 1986.

Contaminants Required to be Regulated Under the Safe Drinking Water Act Amendments of 1986		
Volatile Organic Chemicals	Nitrate	Carbofuran
Trichloroethylene	Selenium	Alachlor
Tetrachloethyrene	Silver	Epichlorohydrin
Carbon tetrachloride	Fluoride	Toluene
1,1,1,-Trichloroethane	Aluminum	Adipates
1,2-Dichloroethane	Antimony	2,3,7,8,-TCDD (Dioxin)
Vinyl choloride	Molybdenum	1,2,3-Trichloroethane
Methylene chloride	Asbestos	Vydate
Benzen	Sulfate	Simazine
Monochlorobenzene	Copper	Polyaromatic hydrocarbons (PAHs)
Dichlorobenzene	Vanadium	Polychlorinated biphenyls (PCBs)
Trichlorobenzene	Sodium	Atrazine
1,1-Dichloroethylene	Nickel	Phthalates
trans-1,2-Dichloroethylene	Zinc	Acrylamide
cis-1,2-Dichloroethylene	Thallium	Dibromochloropropane (DBCP)
	Beryllium	1,2-Dichloropropane
Microbiology and Turbidity	Cyanide	Pentachlorophenol
Total coliforms		Pichloram
Turbidity	**Organics**	Dinoseb
Giardis lamblia	Endrin	Ethylene dibromide(EDB)
Viruses	Lindane	Dibromomethane
Standard plate count	Methoxychior	Xylene
Legionella	Toxapene	Hexachlorocyclopentadiene
	2,4-D	
Inorganics	2,4,5-TP	**Radionuclides**
Arsenic	Aldicard	Radium-226 and -228
Barium	Chlordane	Beta particle and photon radioactivity
Cadmium	Dalapon	Radon
Chromium	Diquat	Gross alpha particle activity
Lead	Endothall	Uranium
Mercury	Glyphosphate	

Note: MCL = Maximum Contaminant Level.

(2) There is no way to set tolerances for the long-term effects of toxic substances.
(3) Solutions involving administering a large amount of a toxic substance to a laboratory animal is a suitable method for detecting carcinogenicity to humans.
(4) Toxicities of substances should be discussed in terms of their risk to humans rather than whether they are safe or not.

These principles mean that it is difficult to establish a clear maximum permissible concentration when discussing toxic effects that accumulate and develop over time, such as carcinogenicity (Figure 2.19). It is not absolutely safe, but can only be discussed in the current scientific context as a risk-hedging issue of how much risk is allowed. The same can be said of the allowable intake of trace radioactive material.

Principle of chronic toxicity and
accumulative toxicity evaluation method of trace organic components

National Academy of Science,USA 1977

① Results of animal experiments
Safety evaluation based on animal experiment results is effective.

② Long-term exposure tolerance
There is no method to set the allowable value of the long-term influence of toxic substances.

③ Method of animal experiment
Animal experiments using toxic substances at high doses are reasonable.

④ Evaluation of the human body influence
Evaluation based on the presence or absence of danger rather than the safety of toxicity.

Figure 2.19 The concept of water quality risk assessment.

2.2.2.5 Development and dissemination of instrumentation

The development of measuring instruments and the improvement of accuracy are rapidly advancing. Although measurement itself includes issues that have been under consideration, it is becoming increasingly accepted by the public to evaluate the risks of trace pollutants. Although extremely expensive measuring instruments have been devised, advances in mass spectrometer systems that are more versatile have gradually matured as the means for measuring organic trace components, enabling precise analysis of inorganic materials. In light of the risks that began in the 1980s, water quality control has been taken up and used in a concrete manner by the rapid development and diffusion of various measuring methods. However, 75% of the drinkable water supplied by municipalities used for non-drinking purposes that are largely unrelated to the new quality standard lists containing such trace components. Although it is good to discuss the technical issues of assessment and water quality improvement in detail and innovatively, in modern water supply and sewerage systems, the water in the uppermost watersheds is mostly trapped by large cities downstream in order to ensure as a clean raw water as possible free from trace pollutants. As a result, the flow rate of rivers below the midstream decreases dramatically, and sewage discharge flows almost exclusively through river channels. Despite municipal water managers' historical common sense of continuing to provide drinking water to the whole coverage and the totally new water quality management system for humans, beginning with the 'Safe Drinking Water Act Amendments' proposed by USEPA in 1986, it is questionable as to what extent water can continue to be compromised in a limited water climate, and the future legacy of mankind must be considered (Figure 2.20).

2.2.2.6 Assessment of risk by animal testing

The first commitment to risk assessment is that, if appropriate, the results of animal studies can be used for human safety assessment. Safety is scientifically an opposition to danger, and the degree of risk is often estimated from clear statistics from studies of human diseases, etc., but, in many cases, it is also estimated from results from well-designed experiments with animals. In particular, the risks of a wide variety of chemicals on larger animals (ultimately humans) can be inferred from experiments with a large number of small animals. Experiments in which a large number of rats and mice, which (of course) are mammals like humans, are commonly used but fish such as 'medaka' (Japanese killifish) are also used in experiments in which a large number of populations are required. The Salmonella Ames test, proposed

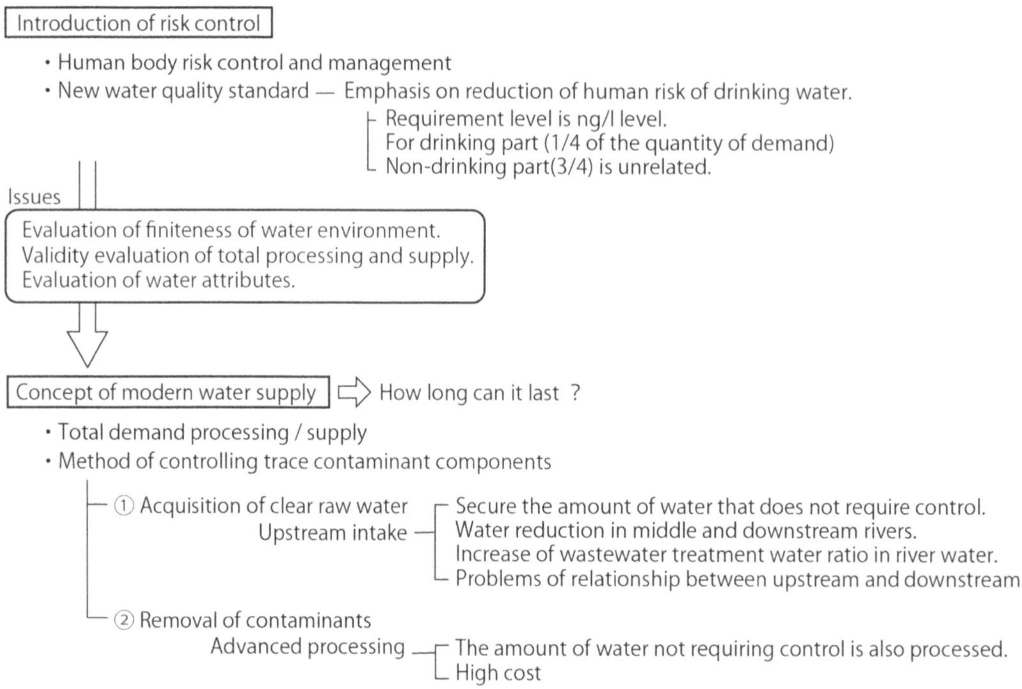

Figure 2.20 Modern water supply and risk control.

by Bruce Ames, is widely used as an environmental mutagenicity test (Ames *et al.*, 1987). It is also used as a screen test for carcinogenicity, but the mechanisms of carcinogenesis and mutagenicity differ, and both are not immediately addressed. In a famous example, furylfuramide (AF2), which was used as a preservative for tofu (bean curd), is highly mutagenic in the Ames test and was discontinued from use, but the discussion of carcinogenicity does not appear to have been concluded.

Mutagens are physical factors such as chemicals and radioactive radiation that possible cause carcinogenesis or adversely affect progeny through germ cells. These substances and factors are present in the environment and are collectively called environmental mutagens.

For acute-toxicity-like substances, a relatively small dose can be determined to cause death or disability in humans after ingestion. For example, for acute toxicants such as potassium cyanate, it is relatively easy to multiply the danger values (lethal doses) obtained in animal experiments by a certain safety factor to determine the maximum permissible contamination level (MCL) in humans. In many cases, small animals, such as mice and rats, are dosed with varying amounts of poisons to convert the amount of poison ingested when an accident occurs to 1 kg of body weight, and then extrapolated to the normal body weight of humans (50 kg for Japanese) to allow for a reasonable safety factor (the safety factor being approximately 10 to 100 times).

However, when discussing the chronic toxicity of carcinogenicity and accumulation, it is necessary to estimate the risk by conducting long-term administration experiments in a large number of animals at environmental concentrations that cause the phenomenon. But it is nearly impossible to conduct an enormous number of such experiments and observations over a long period of time and to directly establish the numerical value. Therefore, the proposed opinion of the American Academy of Sciences is that 'there is no method to directly determine the allowable value of the effects of long-term poison ingestion on the human body'. Therefore, we have to use the results of high-concentration toxicity studies in animals with large doses of toxicants to estimate (within a limited amount of time) the danger of carcinogens in trace amounts and over a long period of time' (Calabrese *et al.*, 1989).

Unlike acute toxicity, toxicity hazards caused by trace toxic components are of a time-consuming nature, and in order to determine the dose–response relationship between 'chronic and cumulative toxicity toxicants' and 'hazards' in animal experiments, the magnitude of toxicity donation (dose) must include not only the concentrations but also the cumulative intakes, taking into account the duration of donation. A cumulative dose (accumulated dose) of intake time is required, and the relationship between exposure (cumulative dose) = environmental concentration × exposure time (or = intake × intake time)] and the occurrence of the disorder is experimentally determined. However, in order to find a response under the constraints of the number of animals that can be used in the experiment and the duration of the experiment, it is necessary to conduct biological experiments at concentrations far higher than is actually problematic.

It is difficult to determine the probability of developing cancer in one of 100 mice (10^{-2} risk), and even if the probability of developing cancer in one of 100 mice is 10^{-3} or 10^{-4}, it is nearly impossible to perform dissections after exposure of hundreds or even thousands of mice to determine whether there is a single carcinogenic mouse. In practice, even at the expense of hundreds of millions of dollars or euros, and at the expense of a large number of rats, only a few high-exposure studies with probabilities of 10^{-1} to 10^{-2} are available, as shown in Figure 2.21.

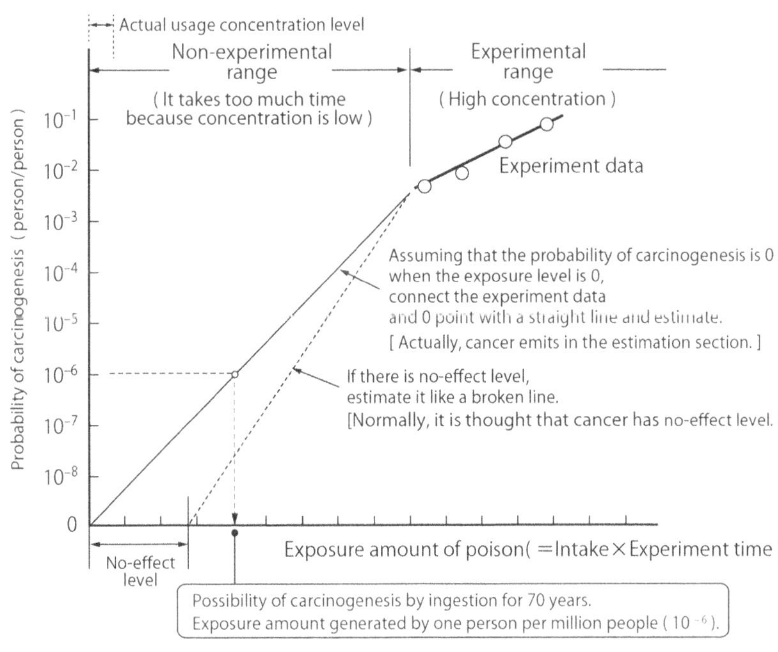

Figure 2.21 The concept of animal experiments to investigate carcinogenic risks.

For drinking water, the acceptable risk level for carcinogens is as follows: 'One person per million (10^{-6}) or less) is expected to be cancerous after drinking 2 litres of water per day for 70 years (MCL: Maximum contaminant level)'; therefore, in Figure 2.21 the exposure is estimated to correspond to the cancer probability of 10^{-6} on the exposure-carcinogenesis probability line (dose-response function), which is a function of the experimental values (the results of the high concentration range with effects at practicable times) and the origin of zero exposure and zero carcinogenesis, which is shown by connecting the origins of the high concentration range with a function. Furthermore, the permissible risk level for carcinogenicity is returned to the concentration at which the probability (risk) of obtaining this exposure is to be obtained, and the allowable concentration of drinking water is multiplied by the 'safety factor' to make the permissible concentration of drinking water a linear connection between experimental values and the origin. There are two interpolation methods, as shown by the straight line and the dashed line in Figure 2.21, between which there are various interpolation functions using risk models that consider the mechanisms of various genetic disorders. When drawing an interpolation line from an experimental risk, such as 10^{-2} for exposure to carcinogenesis (dose–response), values very close to the original (such as 10^{-6}) indicate a tolerable exposure that varies considerably depending on the method of interpolation. Either a no observed adverse effect level (NOAEL) or a lower observed adverse effect level (Minimum Effect Level) is derived from animal experiments to determine an acceptable daily intake, but the estimation from the interpolation line does not distinguish between the two.

Tolerances for the carcinogenicity of trace chemicals and radioactive materials are suggested by assumptions made using a variety of experimental and risk function tests to determine the extent of risk rather than whether it is safe (see Figure 2.22). Thus, tolerance is a difficult matter that can only be deduced and assumed using assumptions.

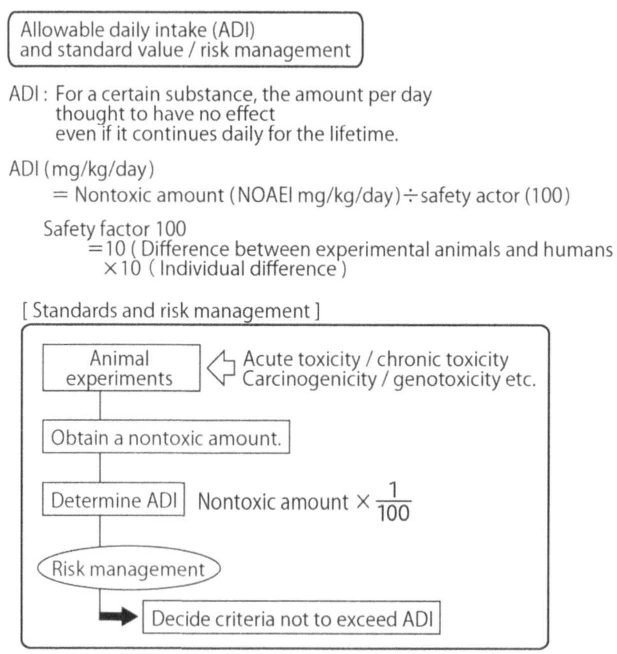

Figure 2.22 Allowable Daily Intake (ADI) and risk management.

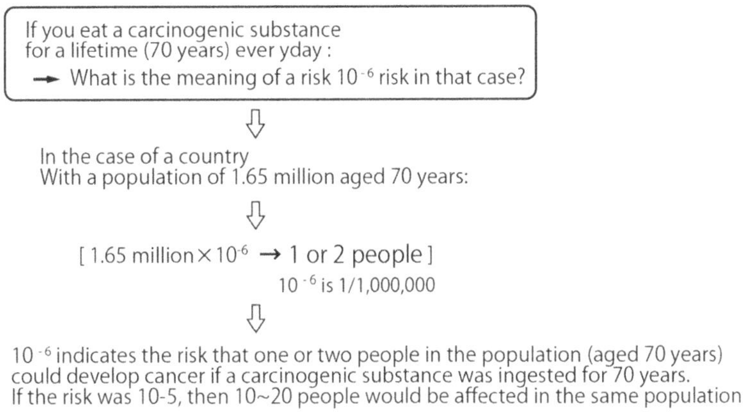

Figure 2.23 An example of a level of risk.

2.2.2.7 An approach to permissible risk level

When considering the allowable risk of carcinogens, WHO and USEPA base their analysis of the risk of carcinogenesis on 10^{-6} (1 in 1 million lifetime consumers). The authors believe that no one can explain why this is used explicitly, with an argument. Assuming the worst, everyone is anxious when they hear of a risk of 10^{-3} of their likely becoming cancerous (equivalent to 1 in 1000); a risk of 10^{-8} equates to a danger of one Japanese national out of the whole population but this level is also considered to be an unreasonable figure. But risks of 10^{-5} to 10^{-6} are less than the risk of an automotive accident, and will therefore probably be admitted as acceptable figures (Figure 2.23).

On the other hand, a 'quota rate' of the total exposure (total intake) to a substance in the environment that is consumed by humans from drinking water must be assumed. Animal studies usually evaluate total intake, so human activity requires knowledge of diet, respiration, other intakes and the rate of water intake. Simply obtaining a numerical value in such a case is a great experiment in physiology. Since it is not always possible to determine experimental values for all substances, in many cases, for convenience, it is assumed that 10% of the total intake is through drinking water.

We expect a margin for a 'safety factor' based on the concept of an uncertainty factor. For non-carcinogenic substances, we expect a 10-fold uncertainty (safety margin) between experimental animals and humans, and an additional 10-fold uncertainty in the aggregate of 100-fold uncertainty when using results from long-term exposure studies (chronic) with suspected carcinogenicity. A 1000-fold safety margin is obtained when there are no human studies or long-term exposure studies. For extremely potent carcinogens, the permissible value is assumed to be 0. However, the lower limit of measurement has been gradually decreasing due to the progress of the measurement method. Practically, the permissible value is referred to as 0 or less in the specification of the test method. When multiple carcinogens are present, it is difficult to accurately assess the additive/synergistic effects of the carcinogens, and the lack of a general method of assessing 'complex' situations has led to the implementation of single-substance regulations in practice.

2.2.2.8 Calculation of the maximum allowable value

The maximum permissible concentration (maximum contaminant level) of a chemical in drinking water is given by Equation 2.3, in which the maximum contaminant level for a 50 kg adult drinking 2 litres of water each day for 70 years with a certain probability of developing cancer (e.g., 1 in 1 million) is calculated.

$$\text{Maximum contaminant level (mg/l)} = \frac{\text{NOAEL(mg/kg/day)} \times 50\ \text{(kg-Adult weight)}}{\text{Uncertainty (safety margin)} \times 2\ \text{(l/day)}} \quad \text{—(2·3)} \tag{2.3}$$

NOAEL : No observed adverse effect level

Equation 2.3. The Maximum permissible concentration equation

This is a drinking water standard. In the United States, it is regarded as the target maximum permissible concentration (Maximum Contaminant Level Goal; MCLG). Depending on the ingredients, however, the maximum permissible contamination concentration (Maximum Contaminant Level; MCL) may be

Table 2.7 Drinking water quality standards in Japan (April 2015 –).

Standard Items (51) and Standard Value [mg/l]				The Ministry of Health, Labour and Welfare of Japan	
Common Bacteria	100/ml	Dichloromethane	0.02	Sodium	200
E. coli	Not to be detected	Tetrachloroethylene	0.01	Manganese	0.05
Cadmium	0.003	Trichloroethylene	0.01	Chloride ion	200
Mercury	0.0005	Benzene	0.01	Calcium, Magnesium (Hardness)	300
Selenium	0.01	Chlorate	0.6		
Lead	0.01	Chloroacetic acid	0.02	Anionic surface active agent	0.2
Arsenic	0.01	Chloroform	0.06	Total residue	500
Chromium (VI)	0.05	Dichloroacetic acid	0.03	Geosmin	0.00001
Nitrite nitrogen	0.04	Dibromochloromethane	0.1	2-Methylisobolneol	0.00001
Cyanide ion and Cyanogens chloride	0.01	Bromate	0.01	Nonionic surface active agent	0.02
Nitrate and Nitrite	10	Total trihalomethanes	0.1	Phenols	0.005
		Trichloroacetic acid	0.3		
Fluoride	0.8	Bromodichloromethane	0.03	Organic substances (Total Organic Carbon)	3
Boron	1.0	Bromoform	0.09	pH Value	5.8~8.6
Carbon tetrachloride	0.002	Formaldehyde	0.08	Taste	Not abnormal
1,4-dioxane	0.05	Zinc	1.0	Odor	Not abnormal
cis-1,2-Dichloroethylene and trans-1, 2-Dichloroethylene	0.04	Aluminium	0.2	Color	5 degree
		Iron	0.3	Turbidity	2 degree
		Copper	1.0		

Total trihalomethanes: Total of Chloroform, Dibromochloromethane, Bromodichloromethane and Bromoform.

considered separately from the MCLG to be a practical method of treatment. For carcinogens, both values are the same. The World Health Organization (WHO), the European Union (EU) and Japan treat water quality standards themselves.

2.2.2.9 *Japanese drinking water quality standards*

Japanese drinking water quality standards, which incorporate risk management considerations, were revised as a 50 item list with the addition of new items by the Ministry of Health and Labour under Ordinance No. 101 (2003). Subsequently, there were changes to the list, and the current criteria are shown in Table 2.7.

In addition to the water quality standards in 2003, the 'Water Quality Control Target Setting List' was set as an list that should be taken into consideration for water quality management, which may be detected in municipal water supply, and the 'Items to consider' was set as a list of actually detected substances for which no toxicological evaluation is available or their appearance in municipal water is unknown. Thus, the water quality standards are regularly monitored and examined in a way that complements the water quality standards.

Items such as antimony, uranium, nickel, nitrite nitrogen, 1.2 dichloroethane, toluene, chlorine dioxide, 60 pesticides, residual chlorine, hardness, manganese, turbidity, heterotrophic bacterium, aluminium, etc., are listed for setting water quality control targets. If a trend is likely to reach danger levels, it is added to and controlled by the Water Quality Standards. Forty-seven items (silver, barium, molybdenum, acrylamide, vinyl chloride, organotin, bromo-chloroacetic acid, acetaldehyde, aniline, etc.) that cannot be classified directly into the water quality standard list and the control target list are listed as an item to be examined, and information is collected. There is some uncertainty about the Japanese categorization of composition groupings when compared with the USEPA constituent groupings shown in Table 2.6 (Yasuhiro, 1995).

Figure 2.24 shows the evolution of Japanese water quality standards to date.

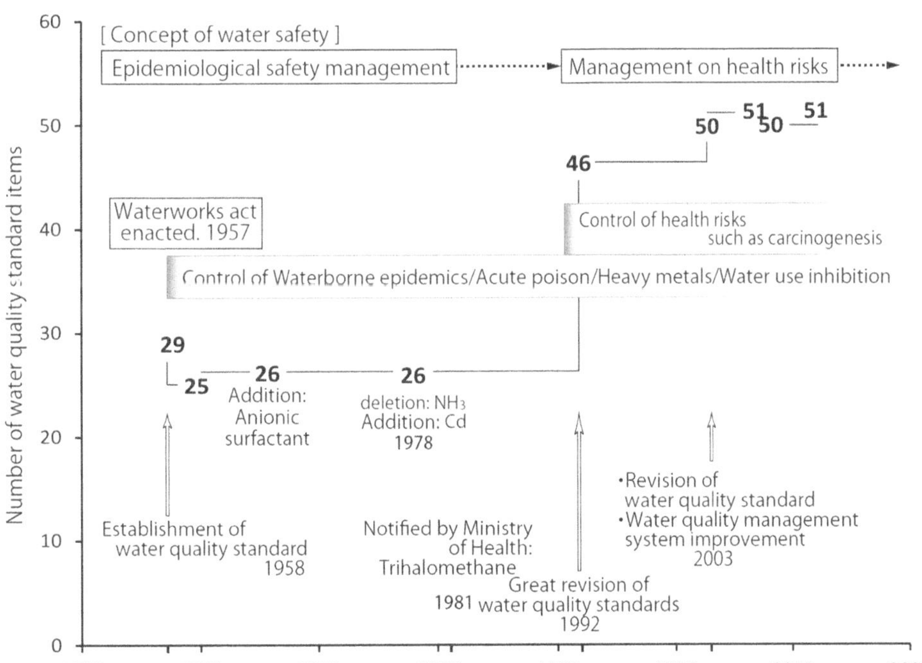

Figure 2.24 The evolution of Japanese drinking water quality standards.

2.3 WATER AND ECOSYSTEMS

2.3.1 Dissolved oxygen control

2.3.1.1 Insufficient dissolved oxygen in water

The most fundamental (universal) pollution phenomena in natural water systems is a deficit of dissolved oxygen (DO) in water. Oxygen can dissolve in water at only 10 parts per million (0.001%, 10 mg/ℓ) while in air the figure is more than 20%. This makes it easy to lack DO concentrations that underlie aerobic activities that are universally viable in water. In addition, the shortage of DO is caused by biodegradable organic matter (biochemical oxygen demand, or biological oxygen demand (BOD) component), which is the largest component in wastewater and causes typical pollution phenomena in water in rivers and coastal sea water areas. This local oxygen shortage became the first issue of water environmental control in water pollution. This is the beginning of the DO deficit concept in water.

Lack of DO in water adversely affects natural aquatic ecosystems that would otherwise be aerobic. At the extreme, DO is lost and the body of water becomes anaerobic and kills common aquatic organisms. Therefore, efforts to evaluate the degree of water quality degradation based on the notion of a shortage of DO began in the 1940s.

Table 2.8 The standards related to the conservation of the living environment (rivers) in Japan.

Item Class	Water Use	Hydrogen Ion Concentration pH	Biochemical Oxygen Demand BOD	Suspended Solids SS	Dissolved Oxygen DO	Total Coliform MPN/100 ml
AA	Water supply Class 1, conservation of natural environment and uses listed in A-E	6.5≦pH≦8.5	≦1 mg/l	≦25 mg/l	≧7.5 mg/l	≦50
A	Water supply Class 2, fishery Class 1, bathing and uses listed B-E	6.5≦pH≦8.5	≦2 mg/l	≦25 mg/l	≧7.5 mg/l	≦1000
B	Water supply Class 3, fishery Class 2, bathing and uses listed in C-E	6.5≦pH≦8.5	≦3 mg/l	≦25 mg/l	≧5 mg/l	≦5000
C	Fishery Class 3, industrial water Class 1, and uses listed in D-E	6.5≦pH≦8.5	≦5 mg/l	≦50 mg/l	≧5 mg/l	—
D	Industrial water Class 2, agriculture water and uses listed in E	6.0≦pH≦8.5	≦8 mg/l	≦100 mg/l	≧2 mg/l	—
E	Industrial water Class 3, Conservation of environment	6.0≦pH≦8.5	≦10 mg/l	Floating matter such as gabage should not be observed.	≧2 mg/l	—

Consider how large and small DO concentrations in water govern the availability of aquatic organisms. The effect of oxygen shortage in water is similar to that of human survival on 8000 m-high mountains such as Everest, where oxygen shortage (given that atmospheric pressure is about 30% above sea level and the partial pressure of oxygen decreases) exposes humans to danger. In addition, since oxygen (which makes up more than 20% of the air) can be dissolved in water at only about 10 mg/ℓ even in saturated conditions, a reduction in DO is a critical event that directly adversely affects the viability of aquatic organisms. Therefore, water quality control is carried out to keep DO concentrations in water above a certain level by considering the balance between the oxygen demand of various organic and inorganic components in water, the shortage of DO in water that has been consumed and reduced, and the amount of oxygen dissolved in water from the air (re-aeration). In Japan, the raw water in rivers used for water supply is controlled at environmental standards of type 'AA' or 'A' (see Table 2.8), with a DO level of 7.5 mg/ℓ or more.

The current environmental standards for water pollution in public waters in Japan, including the environmental standards for the conservation of the living environment of rivers with respect to various usages, excluding lakes and marshes, are set out in Table 2.8.

2.3.1.2 Streeter–Phelps equation

The concepts of DO assessment/control in aquatic environments were formulated by Dr H. W. Streeter, then Director of the Cincinnati Institute at the US Public Health Service (USPHS: the predecessor of the US Environmental Protection Agency: USEPA) and Prof. E. B. Phelps, founder of the Institute of Public Health and Engineering at the University of Florida, where Tambo (one of the authors of this book) was enrolled in the early 1960s. BOD is now used as an indicator of biological pollutants and is used worldwide as a basic method to express the oxygen balance of water systems (Phelps, 1944).

Equation 2.4 is the basic expression for the DO drop curve (oxygen sag curve) that describes how the DO content in water changes with time when a BOD load is applied to a point in the body of water, as shown in Figure 2.25. It is said to be the Streeter-Phelps sag curve.

$$\frac{dD}{dt} = K_1 L - K_2 D \quad \text{---}(2 \cdot 4) \qquad (2.4)$$

D : Saturation deficit from saturated dissolved oxygen concentration(D_{SATU}) (mg/l)
t : Time (days)
L : BOD of water (mg-O_2/l)
K_1 : Deoxygenation coefficient (t^{-1})
K_2 : Reaeration coefficient (t^{-1})

Equation 2.4. The Streeter-Phelps equation

The key to this equation is that it expresses water pollution levels in terms of oxygen deficit (OD) and uses BOD, which corresponds to the amount of organic matter that induces oxygen deficit, as an indicator of pollutants, to deal with the balance of oxygen.

Biodegradable organic matter (L: BOD mg/ℓ) in water is taken up by microorganisms, and oxygen in water is used when microorganisms grow, resulting in reactions that generate CO_2, H_2O, humic substances and other microbial metabolic wastes. The rate at which organic matter is oxidized is a first order reaction with biodegradable organic matter concentrations, K_1L, in the first term on the right side of Equation 2.4.

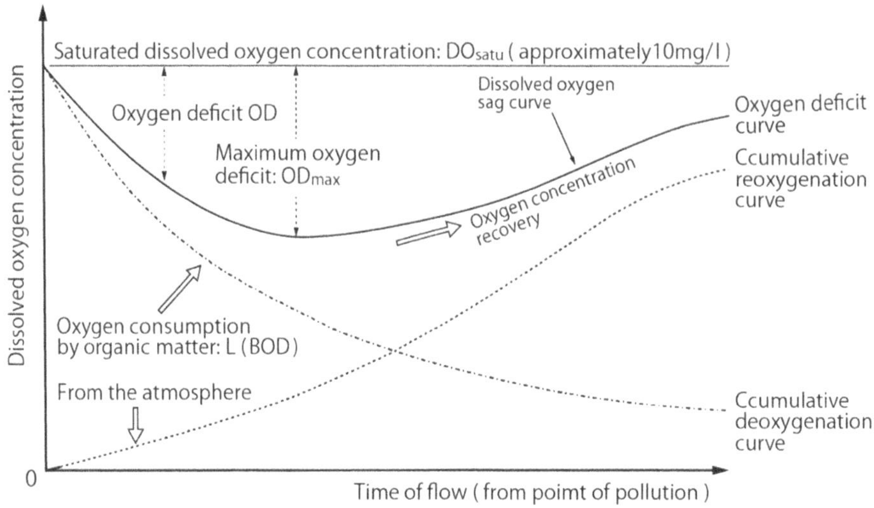

Figure 2.25 The Streeter-Phelps dissolved oxygen sag curve.

The rate of decomposition of organic matter is proportional to the rate of oxygen consumption (Equation 2.5), and the amount of organic matter oxidized and decomposed by aerobic microorganisms in water is considered to be proportional to the amount of oxygen used. A simple and sophisticated formulation was made by replacing a wide variety of biodegradable organic matter in water with simple inorganic matter and calling it 'oxygen consumption', as determined by a BOD test, and by directly linking this oxygen consumption with complex oxygen deficit (shortages) in water.

$$\text{Organic decomposition rate}\left[\frac{d\left(\text{Amount of organic matter}\right)}{dt}\right] = \text{Constant} \times \text{Oxygen consumption rate}\left[\frac{d\left(\text{Required oxygen amount}\right)}{dt}\right] \quad \text{—(2·5)} \qquad (2.5)$$

t : time

Equation 2.5. The Relationship between the decomposition rate of organic matter and the oxygen consumption rate

The basic idea of water pollution assessment using the Streeter-Phelps equation is based on the fact that local DO shortages in running water are the first significant event of water environmental degradation. In river environments such as in Japan, where rivers flow rapidly and the re-aeration factor K_2 is very large, the oxygen shortage of river water is a concrete issue only for the mouth of a river where flows are susceptible to tides, where large quantities of organic waste components flow into the river, and where rivers have slow flow and water stagnates. In Japan, where most flows are in rapid rivers and DO is nearly saturated, environmental standards have been established for BODs as a comprehensive indicator of biodegradable organic matter concentrations, regardless of the concepts of the oxygen balance payments. Types AA, A, and B of the environmental standard have BOD values of 1, 2, or 5 mg/ℓ or less, respectively, which only indicate that a decrease in biodegradable organic matter in the river, such as $dL/dt = -K'L$, is the primary response of BOD. This is a decay equation for organic components that is independent of Streeter-Phelps oxygen deficit concepts. Here, K' is the BOD damping factor (t^{-1}) and, generally, damping phenomena such as sedimentation, adsorption and biodegradation are treated as one phenomenon.

2.3.1.3 BOD test, COD, TOC

The BOD test originated in the United Kingdom at the end of the 19th Century. The BOD test is performed by taking samples into culture bottles of about 100 to 200 ml and incubating them at 20°C for 5 days to determine the consumption of oxygen in water by microorganisms. Both the test temperature and the test duration are simulated indicators for British river water temperature and retention time and are written as BOD_5 (5-day BODs). In order to estimate the total amount of biodegradable organic matter, extreme BOD (ultimate BOD, total oxygen demand or 100 day biological culture BOD is used, but the degradation takes a long time. As the test takes time, not only does the first-stage BOD (a carbon-based oxygen consumption) contribute to the reduction of inorganic substances but so does the second-stage BOD (in which oxygen is consumed by the nitrogen system in a nitration reaction) (see Figure 2.26).

As noted above, oxygen is only soluble in water at about 10 mg/ℓ, so samples with concentrated organic matter immediately consume any free oxygen. Therefore, the original sample is diluted in several steps so as to set the required oxygen demand at 5–6 mg/ℓ, so that the test can proceed at all times with a saturated oxygen content (10 mg/ℓ) in water. A single BOD_5 test requires several or more samples at different dilutions for oxygen determination at $T = 0$ and $T = 5$ days.

If the amount of organic matter per se is indicated, the measurement of total organic carbon content (TOC mg/l) (which can directly grasp the material balance) is considered the most clear and direct indicator. Total organic carbon (TOC) measurement systems, which are capable of continuous measurement, have been produced and marketed. The water quality evaluation method for organic components based on TOC is described in detail in the 'Water Quality Matrix' evaluation method developed by Tambo etc., covered later in Chapter 3, Section 3.5.

Chemical Oxygen Demand (COD) is widely used as an alternative or rapid method of BOD. There are two CODs, which use either dichromic acid or potassium permanganate as chemical oxidants. The former

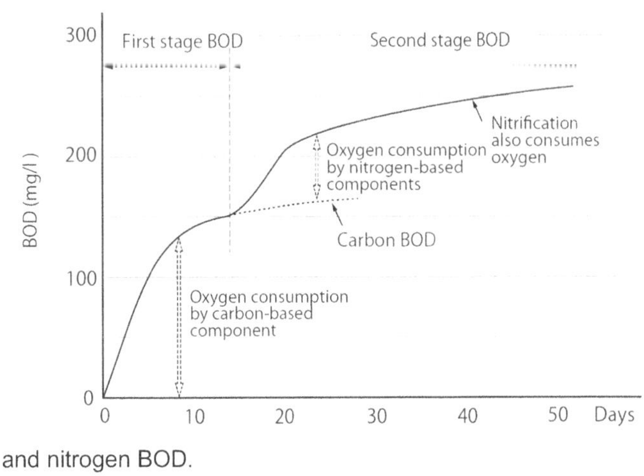

Figure 2.26 Carbon and nitrogen BOD.

oxidizes most of the organic matter because of its strong oxidizing power. However, the latter oxidizes only a part of the organic matter because of its weak oxidizing power, and the value is smaller than for the former. In both cases, the oxidizing agent consumed in the oxidative decomposition of organic matter in water is expressed in terms of the amount of oxygen, and the unit is mg/ℓ or ppm.

Of the COD studies used at present, dichromate oxygen demand ($COD_{cr)}$ approximately corresponds to TOCs. This method is used as a global standard. In Japan, potassium permanganate COD_{Mn} has traditionally been used to indicate the concentrations of organic matter in lakes and marshes, but only some of the oxidized matter in the total amount of organic matter can be expressed this way.

2.3.1.4 Eckenfelder's BOD-based aerobic biological treatment process design and operations

Environmental standards include suspended matter (suspended solids; SS), DO, BOD (COD in sea areas), *Escherichia coli* and pH. Suspended ingredients make water seem unsightly, indicate the growth of microorganisms, and are the main control in the discharge of sewage treatment. They are removed from water and allowed to separate into sludge form, the main standards of which are measured by SS and BOD. The treatment process of organic matter is explained with reference to an example of the standard activated sludge method, which is currently widely used as the main method of sewage treatment, as shown in Figure 2.27. Many suspended solids are removed by sedimentation. The suspended component also contains matter generated by biological coagulation, whilst the soluble component is enriched in microbial flocs by aerobic microbial reactions, some of which are used as nutrients in microbial cell production and are broken down into water and carbon dioxide to be utilized as energy. Solidified BOD content (bio-flocs) are separated by final sedimentation tanks and finally separated from the water system as excess suspended BOD components.

Because of the oxygen shortage formulated by Streeter-Phelps and the generation of pollution in natural waters, to prevent pollution, BOD components that cause oxygen deficit in receiving waters must be removed prior to discharge. For this reason, water quality control based on sedimentation (separation) for suspended organic matter (microorganism flocculation) in aerobic microbial reactors has become widely practiced. In the early 1960s, W. W. Eckenfelder and others compiled theories for the design of treatment with aerobic microorganisms using the inflow BOD in a sewage system as the main index. Treatment systems will be described in detail later in Chapter 6.

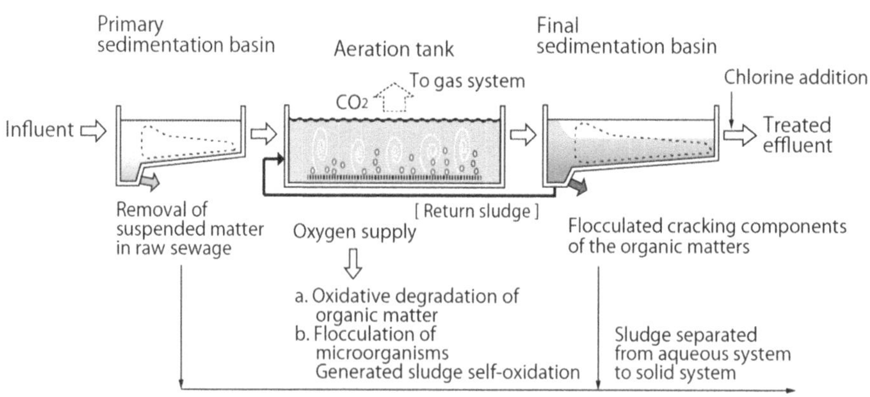

Figure 2.27 The conventional activated sludge method.

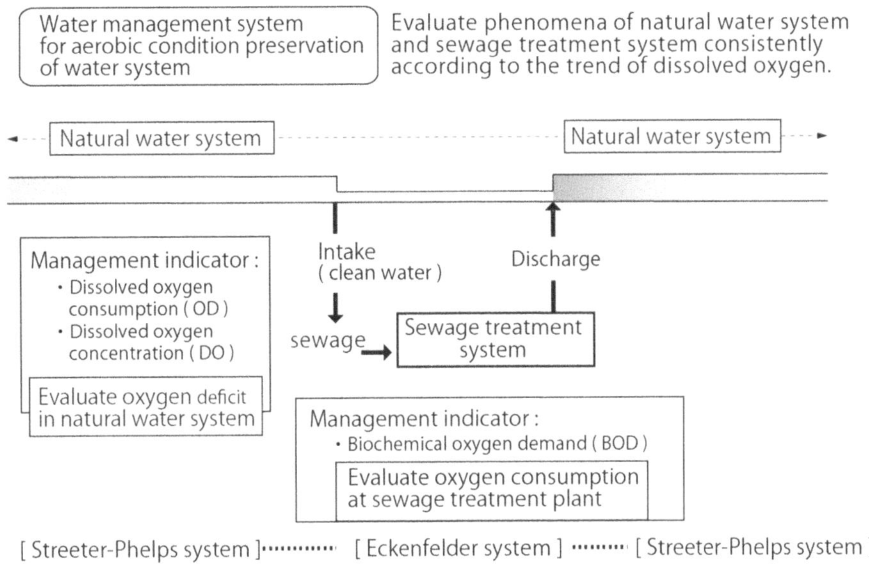

Figure 2.28 The Streeter-Phelps-Eckenfelder system using oxygen as an indicator.

We used reduction by oxidative decomposition of organic matter (by BOD reduction), sludging of soluble and suspended BOD components by the growth of microorganisms (microorganism flocculation), and auto-oxidation of the produced sludge and the oxygen required for oxidative decomposition of the BOD component (feed rate), so that we could evaluate the design of treatment plant functions and the material balance using inflow sewage BOD as the main indicator. In a similar way to oxygen deficit (OD) (oxygen consumption) and DO concentration management in natural water systems, BOD can be handled consistently in a sewage treatment plant in terms of both water pollution phenomena and sewage treatment in terms of oxygen equivalents. The system of design and operation of oxygen standards for water quality management and sewage treatment technologies in natural watersheds is the basic system for water environmental management to maintain aerobic conditions throughout a water system at present. This could be called the 'Streeter-Phelps-Eckenfelder system' (Figure 2.28).

2.3.2 Material circulation and pollution

2.3.2.1 Natural circulation of carbons, nitrogen, etc., and pollution phenomena

Oxygen can be described as the key to anthropogenic pollution, measured in terms of BOD and OD, because oxygen is closely involved in the global circulation of water, carbon, nitrogen and sulfur. If an inconvenient phenomenon occurs during the circulation which could affect human life, we recognize it as pollution and try to solve the problem in terms of environmental technology.

Among organic nitrogen pollutant loads, ammonia nitrogen (NH_3-N) accounts for the majority. Oxidation proceeds in natural systems where oxygen is present at all times and eventually oxidizes to nitrate nitrogen (NO_3-N) and stabilizes it. Phosphorus and sulphur, which are released during the decomposition of organic matter, also cause various problems in the natural circulation process.

Humans are masses of organic matter, and it is said that the number of cells forming the basis of human body alone is about 60 trillion. Each organ of the body is made up of cells. The basis for various metabolic events involving the human body and the environment is the metabolism of various cells. When we look at metabolic activity at the cellular level, anabolism and catabolism are performed continuously and cyclically throughout living bodies and environments.

The anabolic action is an endothermic reaction in which cells acquire light, chemical energy, etc. to synthesize organic matter, and the catabolic action is an exothermic reaction in which organic matter and other substrates are decomposed to extract energy. These two reactions combine to form a circulation loop. Typical anabolic effects are photosynthesis in plants and organic synthesis (mineral oxidation reduction) by bacterial and other inorganic redox reactions. The former is the most fundamental reaction of ecosystem composition for plants on Earth and for the animals that eat them. The latter is a reaction in which various bacteria operate in the natural circulation, such as the oxidation of ammonia to nitrite and nitrate, the oxidation of divalent iron and manganese to trivalent hydroxides, the oxidation of sulfide hydrogen to sulfur, the reduction of sulfates to sulfide hydrogen, and the oxidation to sulfuric acid.

2.3.2.2 Synthesis (utilization) and decomposition of organic materials

Endothermic reactions include those such as that shown in Equation 2.6, which provides a schematic overview of photosynthetic reactions, sunlight and carbon dioxide:

$$6CO_2 + 6H_2O = C_6H_{12}O_6 \text{ (Organic matter)} + 6O_2 - 650 \text{ (cal/mol)} \quad \text{— (2·6)} \qquad (2.6)$$

Equation 2.6. The Examples of the photosynthetic reaction

Carbon and water produce large quantities of organic matter and oxygen on the planet. In catabolism, cells are used as nutrients to decompose organic matter by oxidation or reduction of organic matter using the energy possessed by the cells. There are two types of decomposition: aerobic degradation (aerobic decomposition) and anaerobic degradation (anaerobic decomposition). The former is the dehydrogenation of cells by oxygen in the air or DO in water. The latter involves a compound other than oxygen (hydrogen acceptor), in a process which is called fermentation or decay (Figure 2.29).

In aerobic environments, environmental control is carried out using the decomposition process of carbon as an index of oxygen consumption (Table 2.9).

Figure 2.30 shows how the concentrations of the components in the organic matter decomposition process change over time under aerobic conditions. Biodegradable components (BOD) present in water are available to microorganisms in water, resulting in cell formation and an increase in the number of

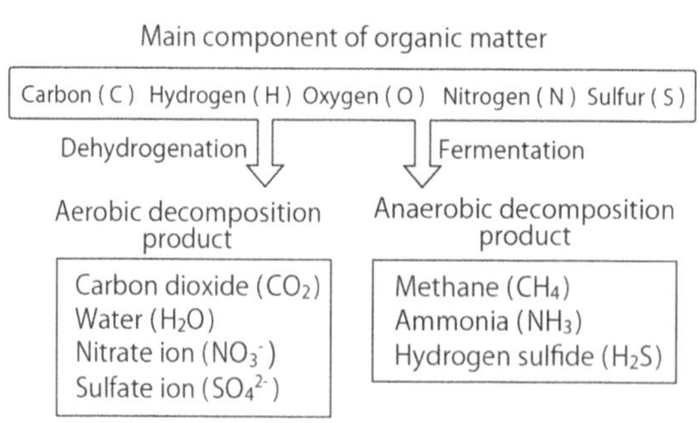

Figure 2.29 Aerobic and anaerobic decomposition.

Table 2.9 The decomposition process of carbon.

	Reaction	Energy
Assimilation	Synthesis reaction	Accumulation endothermal reaction
Catabolism	Decomposition reaction [Aerobic / Anaerobic]	Release, exothermic reaction

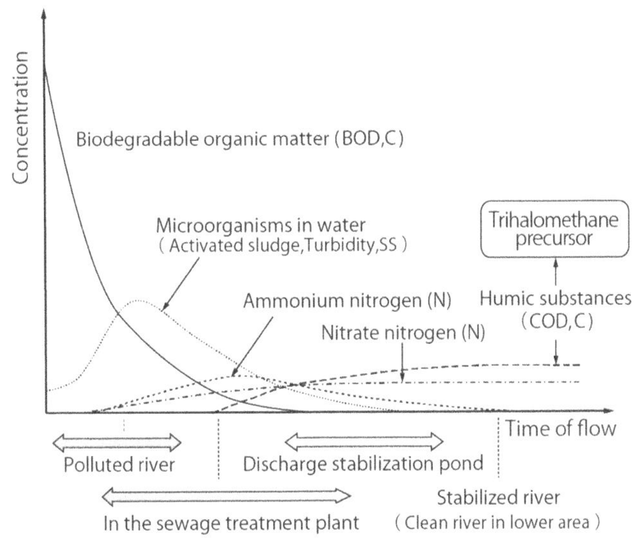

Figure 2.30 The behavior of biodegradable organic matter.

microorganisms (activated sludge, turbidity, SS). During the process, some carbon dioxide is released into the air. Increased microbial production gradually degrades over time and begins to decline. Ammonia is produced during the metabolic process of the microorganisms, but ammonia production decreases with decreasing amounts of the underlying microorganisms. The resulting ammonia eventually oxidizes to a nitrate nitrogen, which stabilizes the ammonia. At the same time, as microorganisms decompose, stable organic wastes such as humic substances (as indicated by COD and carbon concentrations, etc.) increase to equilibrium values. Our experimental results show that this reaction produces humic substances with 3–5% of the carbon content (mg carbon/ℓ) of the initial substrate (BOD component) (Tambo *et al.*, 1981). Chlorination of water containing organic coloured waste compositions produces trihalomethanes such as chloroform. A summary of the biochemical states and stabilities of water is shown below in the horizontal axis of Figure 2.30. Figure 2.31 shows this transition in a flow chart.

The next most common mineral elements in cells (excluding those from water) are nitrogen (7–10%) and phosphorus (2–3%), followed by sulfur and chlorine. The behaviour of these elements has a major impact on the water environment. Next, let's look at how environmental nitrogen and sulfur behave in the first and second cycles of water circulation.

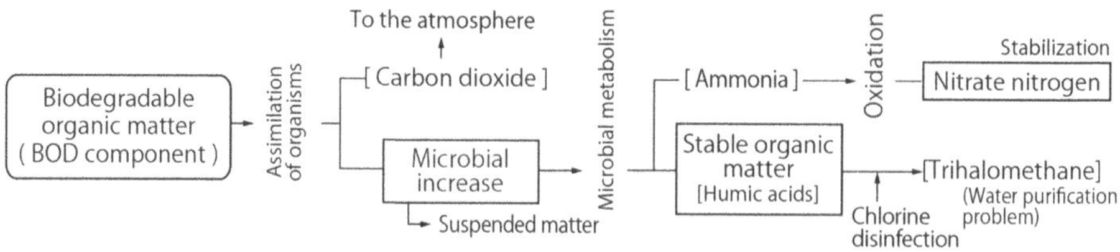

Figure 2.31 Flowchart depicting the transition of biodegradable organic matter.

2.3.2.3 Circulation of nitrogen

An overview of the circulation of nitrogen in the environmental system is shown in Figure 2.32.

Under aerobic conditions, such as in the air or in water, where there is plenty of oxygen, organic nitrogen and ammonia react with oxygen to oxidize to nitrite and then to nitrate. This reaction is caused by the aerobic inorganic bacteria of nitrite bacteria and nitrate bacteria (Nitrobacter). Such reactions in water consume dissolved oxygen. This nitrogen oxidation is slower than the oxidation of organic carbon. As shown in Figure 2.26, the long-term BOD curve is a two-stage oxygen consumption curve in which oxygen consumption from the nitrogen system overlaps the first-stage oxygen consumption curve from the carbon system. Nitrogen fertilizers work effectively on crops mainly in the form of nitrate nitrogen (NO_3-N).

When nitrate nitrogen is placed under anaerobic conditions without oxygen, it undergoes nitrogen removal (denitrification) and is reduced to N_2(gas) and released into the atmosphere. Nitrogen removal occurs in anaerobic soil at the bottom of paddy fields, relieving the accumulation of nitric acid in the soil. River water and groundwater in watersheds where water flows through cities and agricultural land in continental countries and which is repeatedly used tends to have high nitrate nitrogen concentrations. The danger of methemoglobinemia, which causes cyanosis, is known not only when young children drink water with higher nitrate nitrogen concentrations but also when they drink the milk of mothers

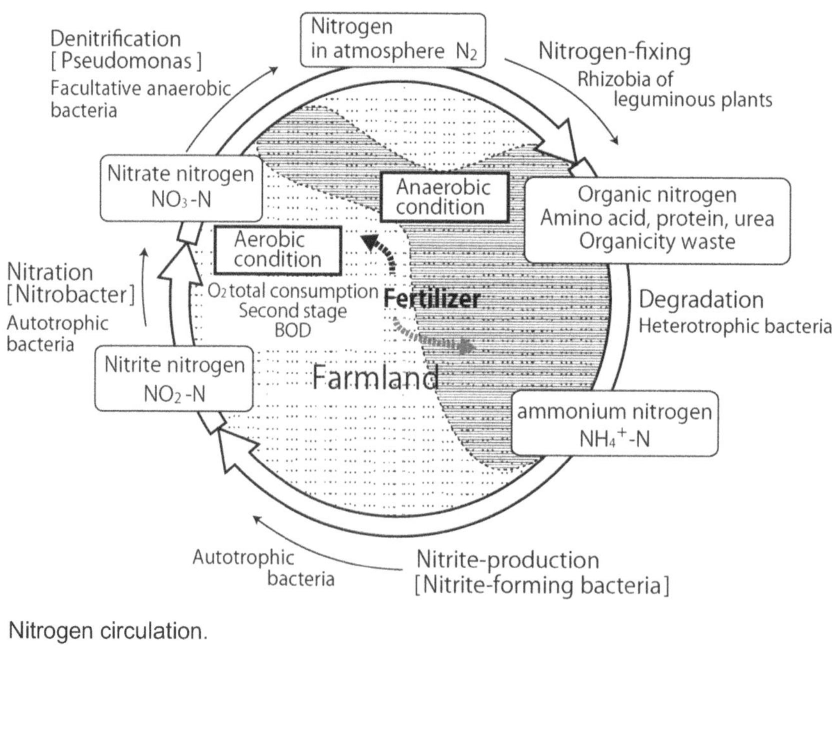

Figure 2.32 Nitrogen circulation.

who drink such water. High concentrations of nitrate nitrogen, which is assumed to originate from fertilizers, may also be detected in groundwater and other sources in Japan.

As an upper limit for the application of nitrogen fertilizers and composts, the nitrate nitrogen concentrations in groundwater are generally controlled to be 10 mg/ℓ or less. Anaerobic nitrogen removal is used as a process prior to water purification treatment when raw water containing high concentrations of nitrate ions are used in a few water supplies that use groundwater downstream of a large European river as a water source. In addition, in water areas where water is repeatedly used in the same water system, prior to the discharge of sewage and industrial wastewater with high nitrogen concentrations, aerobic nitrification and anaerobic nitrogen removal treatments are used to lower nitrogen concentrations. Many wastewater treatment plants in Europe such as in the Ruhr River valley are constantly operating such treatment systems.

In natural water systems, when nitrogen concentrations reach 10^0 mg/ℓ (1 mg/ℓ), algae are generated in water areas where water stagnates – in inner bays, lakes and marshes – and so-called eutrophication phenomena are manifested, causing various obstacles to water use.

In particular, phosphorus, which is required to be controlled alongside nitrogen as a source of eutrophication disorders in closed inner bays and lakes, needs to be kept below 10^{-1} mg/ℓ (0.1 mg/l), which is one order of magnitude lower than nitrogen. Phosphorus removal can be achieved to a certain extent by the activated sludge method or a modification thereof, but can be performed more efficiently by coagulation sedimentation treatment by complexation with aluminium salts and iron salt hydroxides. On the other hand, since phosphorus is an internationally limited non-renewable resource, it is important to cope with the depletion of phosphorus components in fertilizers, and studies are underway to recover phosphorus from sludge by the activated sludge method. How phosphorus is handled as a resource recycling system (such as by using phosphorus in thickening with sludge), as well as how to directly recover phosphorus, remain key issues.

2.3.2.4 Circulation of sulphur

As part of water-based organic recycling, sulphur recycling is a key issue to be kept in mind. Figure 2.33 depicts the circulation of sulphur around a water system.

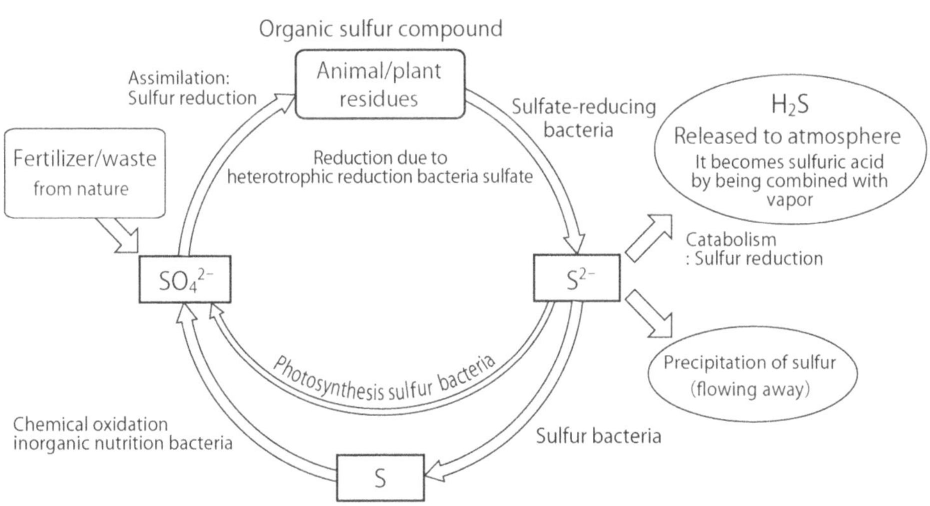

Figure 2.33 Sulfur circulation.

Organic matter including animal and plant residues in anaerobic conditions can become highly corrosive when producing sulfuric acid, due to the generation of sulphide hydrogens by sulfuric acid-reducing bacteria in the presence of water – which is a serious environmental concern. Such organic matter can also be reduced by sulphur bacteria to precipitate sulphur particles and can generate blue tides in polluted sea areas, causing a variety of damage.

FURTHER READING

Eckenfelder W. W. and O'Connor D. J. (1961). Biological Waste Treatment. Pergamon Press, New York, Oxford, London, Paris.

Ishidate M. (ed.) (1979). Living Environments and Carcinogenesis: Air, Water, and Food. Asakura Bookstore, Japan. (In Japanese).

Ito H. and Hirai N. (1999). Physiology Chart Basic Medicine Series. Medical Review Co., Ltd, Tokyo.

Schlesinger W. (1991). Biogeochemistry: An Analysis of Global Change. Academic Press, London.

Tambo N. (1983). Water Supply and Trihalomethane. Gihoudo Publishing, Tokyo. (In Japanese).

World Health Organization (WHO) (2008). WHO Guidelines for Drinking water Quality, 3rd edn. WHO, Geneva, Vol. 1.

REFERENCES

Ames B. N., Magaw R. and Gold L. S. (1987). Ranking possible carcinogenic hazards. *Science*, **236**(4799), 271–280.

Arakawa H. (1991). Mystery at 4°C: Exploring the essence of water. Hokkaido University Press. (In Japanese).

Calabrese E., Gilbert C. E. and Pastides H. (1989). 'Safe Drinking Water Act'. (USEPA) Lewis Publishers, Michigan, USA.

Carson R. (1962). Silent Spring. Houghton Mifflin, Boston USA.

Degremont and Lyonnaise des Eaux (eds) (1991). Water Treatment Handbook, 6th edn. Degremont Water and Environment, Paris, Vol. l.

Hishinuma N. and Kitamura S. (2005). Component and Function of the Human Body. Open University of Japan, Text book. (In Japanese).

Montgomery J. M., Consulting Engineers, Inc. (1985). Water Treatment Principle & Design. John Wiley & Sons, New York, Chichester, Brisbane, Toronto, Singapore.

Rook J. J. (1972). Production of potable water from a highly polluted river. *Water Treatment and Examination*, **21**(Part 3), 259.

Rook J. J. (1974). Formation of haloform during chlorination of natural water. *Water Treatment and Examination*, **23**(Part 2), 234.

Phelps E. (1944). 'Stream Sanitation'. John Wiley, New York.

Tambo N. (1980). Water Supply. (Civil Engineering Series, Vol. 88, Japan Society of Civil Engineers). Gihoudo Publishing, Tokyo. (In Japanese).

Tambo N. (1985). Evaluating the Flow and Material Balance in Water Environment. Japan Society of Civil Engineers, Paper 363. (In Japanese).

Tambo N. (2012). History and Technologies of Urban Water Metabolism Systems. Kashima Publications, Tokyo. (In Japanese).

Tambo N. and Ogasawara K. (1985). Water Purification Technology. Gihoudo Publishing, Tokyo. (In Japanese).

Tambo N., Kamei T. and Takahashi M. (1981). Behavior and properties of metabolic waste components from aerobic biochemical processes (1). *Journal Japan Sewerage Association*, **18**(210). (In Japanese).

Yasuhiro K. (1995). New Water Sciences. NHK Books 729. (In Japanese).

Chapter 3

Understanding water separation processes and systems

3.1 EVALUATION OF TREATABILITY

3.1.1 Water treatment system

As discussed in Chapter 1, Section 1.2.1 (The spectrum of water circulation), water quality must always be maintained in a sustainable state to ensure human health and the quality of the water environment. Water use is not the consumption of water itself, but the quality of the water. Therefore, when water is used or removed from the water cycle, as shown in Figure 1.6, it is necessary to convert each cycle into the expected next water quality level. The operation of water treatment is responsible for controlling the quality of water whilst in the process of transferring water. Typical examples are water treatment and sewage treatment.

Water from the first cycle, which is a major natural cycle, is taken to a higher quality and is used in the second cycle for urban household water and activity water. Urban use adds pollutants to water and degrades its quality. However, returning it to the first cycle directly leads to a deterioration of the water environment. Therefore, water must be returned after improving its quality. At the center of quality enhancement is the removal of unwanted impurities in the water, and separation of impurities from the water system and their transfer to the air (as a gas) or to a solid state in a suitable manner. In this way, water treatment is an operation that allows substances to be transferred to the air or solid process streams without being held in the water system, in order to enhance and maintain the relative value of the water system in the environment. In most cases, this operation is required to cross the cycle boundaries (Figure 3.1).

The basic principles of operation in environmental engineering are 'keeping of strict material balances of the subject components' and 'minimization of energy-consumption and utilization space when performing transport and phase change.' (Tambo & Kamei, 1975; Tambo, 2012; these principles are also basic principles for chemical and environmental engineers). Therefore, the design of water treatment systems should satisfy these principles.

Water treatment systems are built with appropriate combinations of different water treatment processes to bridge the water quality gap between the quality of the water used and the impurity concentrations of the raw water and wastewater. Water treatment is the transfer of impurities from the liquid phase (water) to the solid

Author(s): Norihito Tambo and Koichi Ogasawara
doi: 10.2166/9781789061307_0055

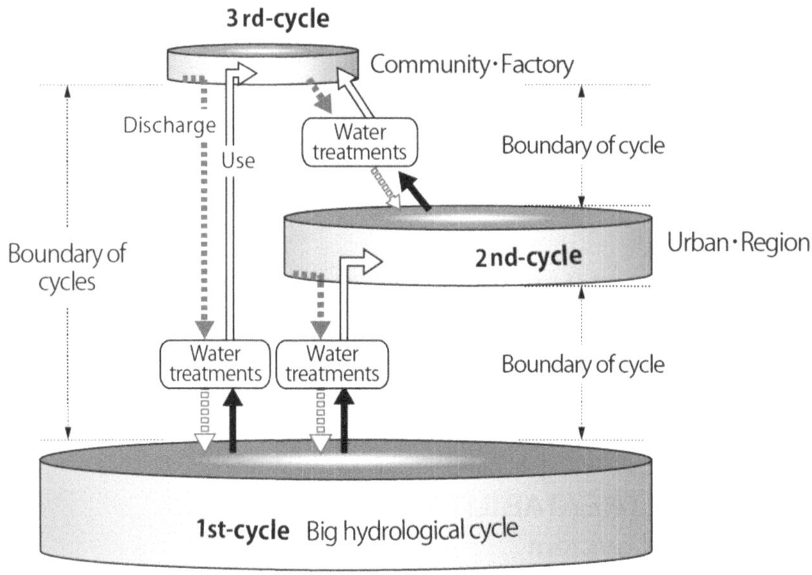

Figure 3.1 The spectrum of water cycles and water use.

and/or gas phases used, so systems are designed with as little load to the solid/gas phase as possible. A treatment process is a physical, chemical, physicochemical, or biochemical operation that can be handled as a unitary process based on certain principles of operation and is also called a unit operation/process.

3.1.2 Separation and adjustment process

The processes used for water treatment are classified according to the characteristics of the operation into 'separation processes (operations) for removal (transfer of impurities in water to a solid/gas system)' and 'conditioning processes for adjusting the presence of impurities such as detoxification operation' (Tambo, 1964, 1994 and 2012). The basis of water treatment is a separation operation, but pretreatment operations to increase impurity size are crucial for water treatment systems. This key item includes coagulation in a rapid filtration system (which will be described later) and the formation of biological flocs in an aeration tank in an activated sludge system. In addition, with low concentrations of impurities, water treatments such as disinfection and neutralization are also performed without removing impurities from the water, controlling harmful water qualities *in situ*.

In order to build many processes into a comprehensive system, each process must be characterized as clearly as possible, and the material balance, removal characteristics, connectivity with neighboring processes, required energy consumption, etc. of each process must be evaluated consistently as system components.

Figure 3.2 shows the operation units used for water treatment, grouped by separation and adjustment functions.

3.1.3 Mechanisms for assessing the treatability of water impurities

When attempting to evaluate the treatability (or natural variability) of impurities, linking the presence of impurities in water with the effectiveness of the process, three factors must be considered: the size, chemical reactivity and concentrations of any impurities.

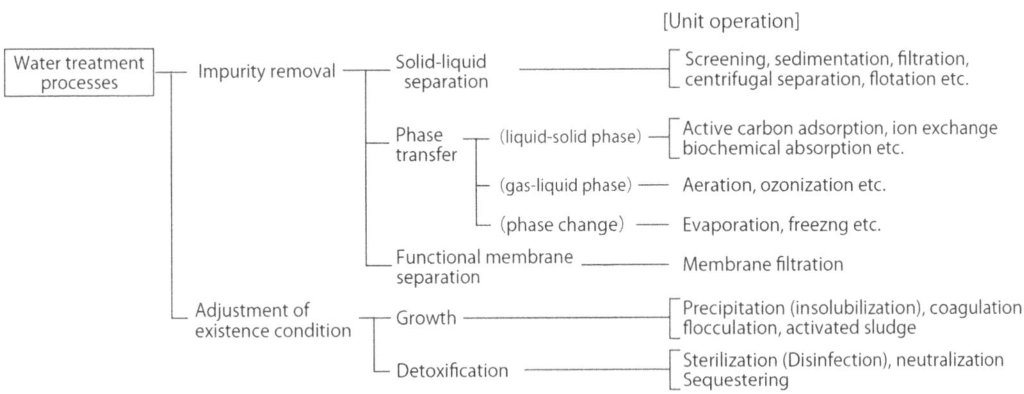

Figure 3.2 Classification of water treatment unit operations.

The first step in process selection is to consider how impurity size is distributed and how any impurities can be separated. As mentioned in earlier chapters, water molecules have a size of the order of 3 Å = 10^{-10} m = 10^{-7} mm. Water treatment is an operation that removes various impurities from water molecules of 3 Å as a continuum. Therefore, the difference between the size of the impurities and the size of the water molecules is the first issue. Among the impurities in water, sand particles and large suspended matter lie in the size range of mm to cm, and have a size difference of 1 million (10^6) times the size of water molecules of 3 Å, and the two can be separated easily by their difference in size. Processes that use gravitational/inertial forces, such as sedimentation and filtration, are used as drive forces (energies). The separation operation finds the difference between the substances (here, water and impurities) to separate them, and uses separation measures by focusing on that difference. Therefore, if the size is largely different, the difference in size becomes the first separation operation factor to focus on.

As the size of an impurity approaches the size of a water molecule, the separation process is required to be more precise, even though there are still considerable differences in size (of the order of 1000 to 10,000 times), by adding differences of physicochemical and chemical properties to the size differences, such is the case for colloidal and macro molecular substances, Since there is almost no difference in size among soluble impurities, precipitation and coagulation are promoted by focusing on the difference in chemical properties of the substances themselves, then operations are used to emphasize or widen the difference in size with water molecules, before the finally adjusted components are separated by size difference.

In lean systems unable to use size difference separation, a traditional (conventional) water treatment system uses the difference in chemical properties between water and other innocuous impurity components for separation, and /or to detoxify the components that cause the failure (dilute system), or introduce a transfer (adhesion or sorption) operation into a solid or gas state, thereby protecting the water system (Figure 3.3).

Membrane separation has been a widely available technology since the beginning of the 21st Century. In the case of micro-filtration (MF) and ultra-filtration (UF) membranes, solid-liquid separation is carried out by reducing the pore size of the membrane and stopping the passage of fine particles, but real progress in functional membrane separation has been made with nano filtration (NF) membranes, which can directly remove soluble organic/inorganic salts, and reverse osmosis (RO) membranes which have been used for desalination of seawater since the end of the 20th Century. This was a revolutionary development in separation technology, taking into account differences in physical size and in chemical properties, and

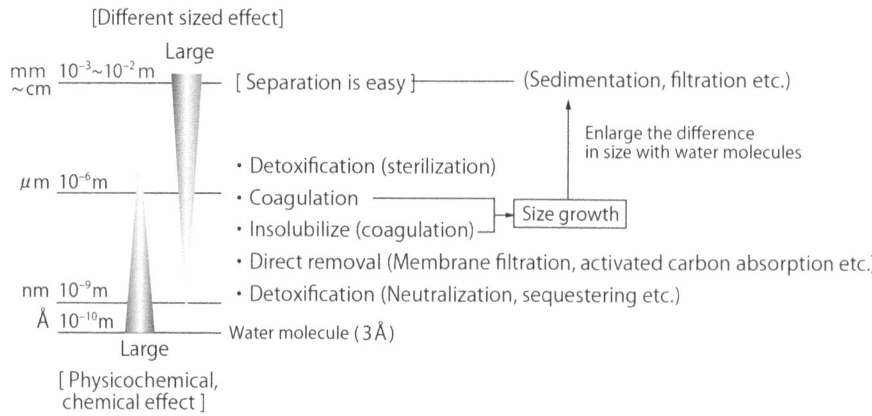

Figure 3.3 The concept of separation methods using differences in the properties of water and pollutants.

capable of direct application to impurities of the Å-order. Whereas water treatment technologies were, up to the 20th Century, ecological technologies introduced into human societies after learning global scientific mechanisms, membrane separation technologies are physiological separation methods for biological metabolism that allow animals and plants to continue their daily activities. Social infrastructure systems now have ultimate separation technologies at the same scale and accuracy as the metabolism of individual organisms. In principle, separating of impurities by size is an excellent method, but only when the size of any impurities and of water molecules are not equal; when they are, the separation of impurities by a membrane or physicochemical factor becomes important. Membrane separation is discussed in more detail later in this chapter and also in Chapter 8.

In order to position the relation between the presence of impurities and their separability comprehensively, two-dimensional representations of an impurity and the nature of that impurity are required. These status labels are commonly used as methods for classifying and characterizing substances and they are presented in a form that is easy to understand and organized as 'two-dimensional matrices' in a later section below.

In addition, the concentration of impurities that we want to treat is related to process selection. Depending on the size and chemical nature of the impurity being treated, the removal and conditioning process to be used will be chosen, but the level of concentrations of the impurity will be another key to process selection. Significant differences in impurity concentrations, such as g/l, mg/l and ng/l will result in different processes that can be employed. Taking into account the concentrations of the components of interest, the evaluation of treatability is likely to be expressed as a three-dimensional matrix system that takes into account the three factors of 'impurity size', 'impurity chemistry', as well as the third factor: 'impurity concentration' (Figure 3.4).

Generally, impurities for removal are not present alone but are present in an admixture with various other ingredients. The total concentration of impurities in the water to be treated varies from high concentrations, such as industrial wastewater and municipal sewage, to clear water, such as surface water and groundwater, used as raw water for water supply. As a whole, therefore, not only the concentration of the individual components of the impurities but also the total concentration of the impurities in the water to be treated have a major impact on the removal process selection of the individual components. It is essential to think about the system when selecting a process for water quality consisting of various ingredients and concentrations. In this book, the authors will discuss how to build a water treatment system by

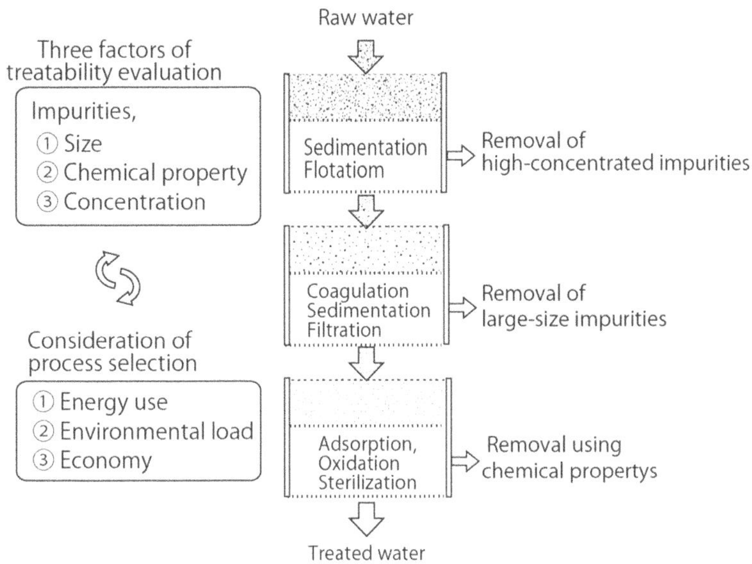

Figure 3.4 The three factors to be considered as part of treatability evaluation and process selection.

integrating various processes which are outlined in the following sections. In this section, we will proceed with the statement that by successively stacking the composition of comprehensive water treatment systems one after the other, from 'high-concentration ingredients' and 'high impurity size ingredients,' we will be able to economically and accurately select components to be treated and their appropriate treatment processes with 'minimum energy use and environmental load' (Tambo & Kamei, 1978; this is a central tenet of Tambo's treatment system operation which is described further in Section 3.5 below).

3.2 SEPARATION MECHANISM DEPENDING ON DIFFERENCES IN PHYSICAL PROPERTIES (PARTICLE SIZE)

3.2.1 Differences in physical properties and operation force

It is possible to effectively separate impurities from water due to differences in their physical properties:

(1) The size of the impurity is much larger than the size of the water molecule ($\fallingdotseq$ 3 Å), and the assembly of water molecules can be regarded as a continuum.

(2) The density of the impurities to be separated differs from that of water ($\fallingdotseq$ 1 g/cm^3) by an appropriate amount (either larger or smaller).

(3) In some cases, when the electromagnetic properties of impurities have characteristic features and values, they can also be used as physical property differences for separation.

In order to properly select and design the physical separation processes, it is crucial what kind of 'force' is used to drive the separation. In the case of natural water systems on the ground, the driving force of the separating operation is of course a gravitational force, and the differences in the size and the densities of water molecules and the impurities are of significance. The fact that 'stones do not float in water and trees do' is a key phenomenon in the natural world. Even with the same stones, if the size is small enough, such as with clay particles at about 3 μm or less, sedimentation becomes slow even if there is a

Table 3.1 Examples of settling velocity.

Gravitational Acceleration and Settling Velocity	Size µm	Density g/cm^3	Gravitational Acceleration G		
			Earth's Surface 1 G	Centrifugation 1000 G	Ultracentrifugation 20,000 G
			Settling Velocity cm/min		
Clay particle	5	2.5	0.12	120	–
Biological cell	10	1.08	0.026	26	520

By Stokes equation at Water temperature 20°C.

difference in density, and it becomes difficult to separate the clay particles as they are. This is because the sedimentation rate is extremely small because the size of the particles is small. Because the removal rate of the device is inefficient if the size of impurities is too small, physicochemical reactions are used to add a coagulation operation to agglomerate the impurities and perform a conditioning (growing) operation to increase the size of the particles so that sedimentation is easier, as described later in Section 6.1. Alternatively, centrifugal force is applied, using a high-speed rotating force such as a centrifuge or cyclone, and separation is performed using a force much larger than the normal gravitational force. Centrifuges can increase sedimentation speed by increasing the gravitational acceleration G on the ground to several thousand times by rotating at high speed. In addition, ultracentrifuges can be used to create a centrifugal force that is several tens of thousand times the gravitational acceleration G because of the extremely poor sedimentation potential of particles with a small size (such as bio-cells with a density of close to 1 g/cm^3) (Table 3.1). In addition, there is a dissolved air flotation method in which air bubbles are attached to particles with a small density difference to integrate the particles and air bubbles, and the agglomerated particles are floated and separated from water by the large introduced density difference. In addition, it has now become practical to use magnetic force as the separation force, such as in the case of magnetic separation by inflow agglomerates formed by coagulation of magnetic powder and particles with a magnetic field, followed by recovery and reuse of the magnetic powder.

3.2.2 Size of impurities in water

When common impurities present in natural water are arranged by size, they occur in the following order: 10–100 µm for silt and algae; 1 µm for clays and bacteria; 0.1–0.2 µm for viruses; and 0.001 µm in humic substances. Many pesticides, called trace organic pollutants, have a size of about 0.0001 µm. Figure 3.5 schematically shows that the removal target components are distributed over a wide range of 10^7 to 10^6 times the size differential.

Various impurities are present in water with a certain particle size distribution, and the general distribution region is shown in the top row of Figure 3.6. The valid removal processes that can be used for each impurity size are shown in the second row, whilst methods for measuring the size of impurities are described in the third row. When particle size is in the range of about 0.00x µm, it is generally expressed by the particle size itself, and various methods of particle size measurement have been devised. In recent years, computer-based measurement and recording methods have been developed. Automated and high-speed measurement has been promoted, and particle size distribution of particle groups can be measured with high accuracy in a short period of time.

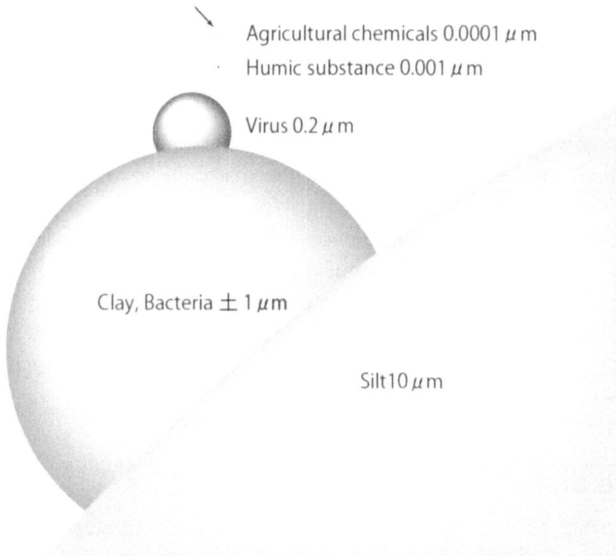

Figure 3.5 A comparison of the relative size of water impurities.

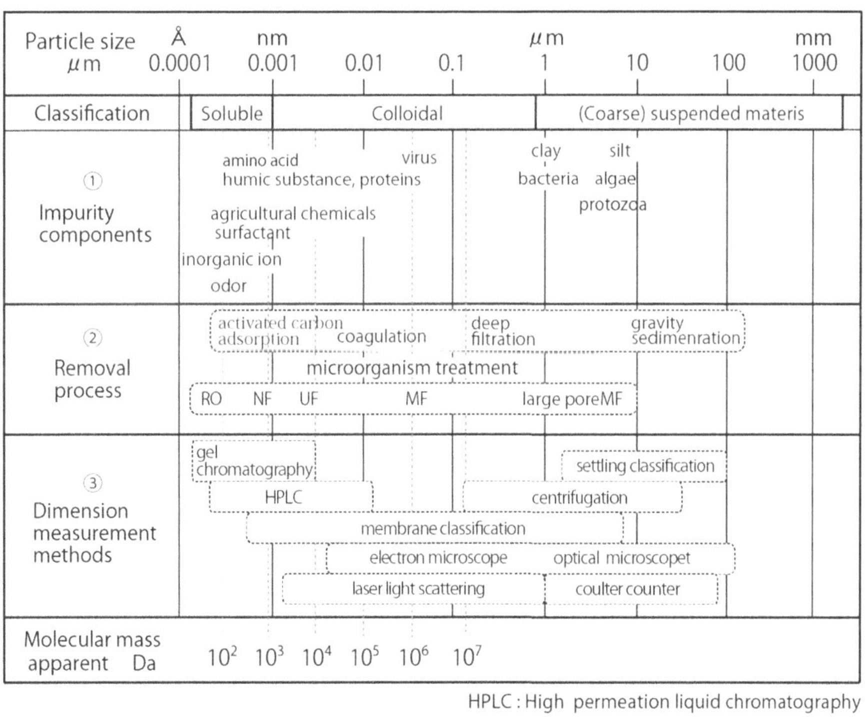

Figure 3.6 Particle size distribution and removal processes.

In particles smaller than 10^{-8} m (0.01 μm), rather than measuring particle size itself, liquid chromatography, gel chromatography and other methods (e.g. exclusion chromatography) are used to express the size of the object in terms of molecular weight (mass). Therefore, correspondence with the actual particle size is roughly apparent. Chromatography enables the measurement of the chemical/physicochemical properties of a substance by selecting various detectors (TOC-continuous measurement, ultraviolet/visible part/infrared absorption, radio absorption, etc.) in accordance with the distribution of the size (mass) of an impurity (removal substance) and the distribution of the chemical/physicochemical properties of the substance. As molecular weights decrease and the size difference between water molecules and impurity particles decreases, the difference in physicochemical/chemical properties begins to have greater significance than the difference in size properties. Separating impurities (solutes) close to the size of water molecules that are ultimately classified as soluble substances takes advantage of differences in chemical properties. In order to express this relationship, Tambo and others devised the concept of the 'Water Quality Conversion Matrix', which will be described below in Section 3.5.

3.2.3 Impurity size distributions and removal limits

The particle size of a group of impurities in water is usually distributed over a size range. Certain microorganisms and germs are distributed over a very small size region, and as far as their components are concerned, they can be treated as single-particle-size populations. However, in the presence of various particles/groups of microorganisms with different particle sizes, their size distributions must be considered if removal is considered using size as a property difference. If we are trying to remove such a group of particles by some sorting technique, we must calculate where and to what extent of the removal components with size distributions can be removed. 'Removal limits' are basic values when considering from where impurities can be removed. A removal limit is generally defined as 'the limiting removal size (cutoff level) at which particles larger than that are removed by 100%'.

Two functions are often used to describe particle size distributions in natural systems: normal and log-normal. In terms of particle size, the size representation of a distributed particle population varies depending on the method of measurement and the intended use. Even if measurement is made of a diameter, d, in one direction (a constant diameter), calculating other values is required, such as mean values and standard deviations from the distributions of the second-order surface area diameter, nd^2, in the case of a reaction system in which surface area is a problem, and the number, n, of each third-order volume diameter, nd^3, in the case of mass changing etc. Figure 3.7 shows the particle size distribution of quartz sand as primary, secondary, and tertiary size distributions.

In many cases, the distribution is distorted due to the presence of a large number of particles on the small diameter end of the particle size distribution which is accurately represented by a normal (Gaussian) distribution, in which the particle size distribution is bilaterally symmetrical across the mean value. It is sometimes expressed as an exponential function or Poisson distribution (Figure 3.8). Although it seems difficult to explicitly express various particle size distributions using mathematical distribution functions, many natural phenomena can be expressed by correcting the distortion (skew) from the normal distribution. Log-normal distributions are typical examples. Figure 3.9 shows a log-normal plot of the distribution of a one direction diameter measurement of quartz powder particles. The distribution is well represented by a straight line. (As shown in Figure 3.10, a one direction diameter is the diameter of the projected surface of a particle measured in the same direction, regardless of the direction of the particle or of any irregularity, and is a type of statistical diameter). A practical exponential equation of the form shown in Equation 3.1 shows the percentage of particles larger than the particle size d.

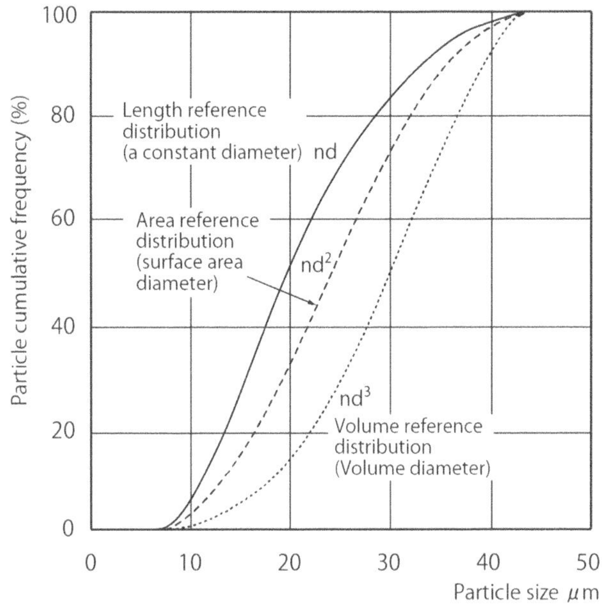

Figure 3.7 An example of a particle size cumulative frequency curve.

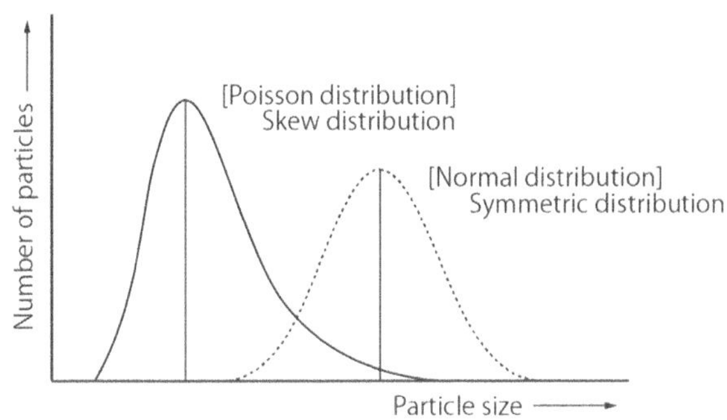

Figure 3.8 Normal and Poisson distribution curves.

Rosin-Rammler distribution equations are well known for expressing the distribution of soil particles:

$$P = 100\exp(-bd^n) \qquad \text{—}(3{\cdot}1)$$

P : Ratio of particles larger than a certain size (%)
b : Parameter (If value is large, the diameter is small.)
d : Particle size
n : Parameter (If value is large, the distribution is small.)

(3.1)

Equation 3.1. The Rosin-Rammler distribution equation

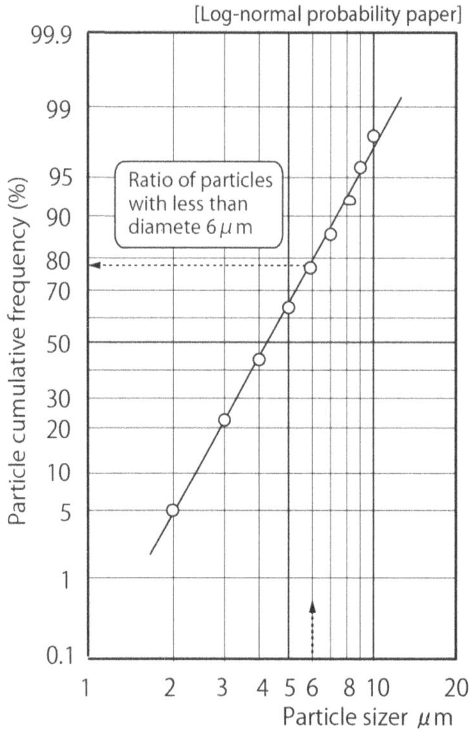

Figure 3.9 An example of lognormal distribution.

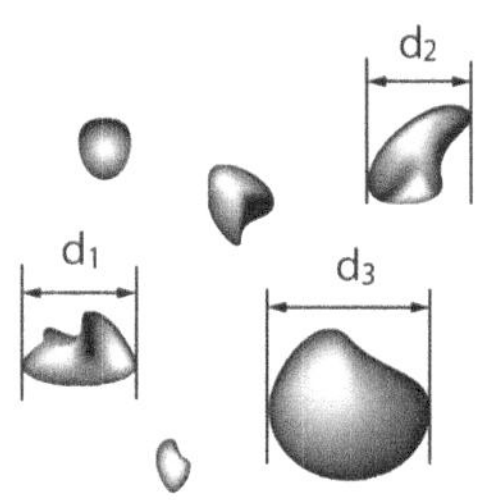

Figure 3.10 A one direction diameter, d.

But why do we devise a probability distribution, a logarithmic probability distribution, or a variety of specific representations to represent the size of the distribution? If we plot two points on a coordinate line, we can accurately recognize the situation. Likewise, a straight line defined by a single point and an angle can be recognized with the same accuracy. This means that a phenomenon is linearly related. For this reason, various functions are created to make the plot of a straight line which is easy to understand. This is called linearization. For this reason, we create various functions to make the plot linear. Linearized data can be simply summed and there is no sensory discomfort. By contrast, humans cannot evaluate nonlinear phenomena instantaneously and accurately – which is perhaps a basic reason why we use logarithms and other mathematical functional deformations.

When trying to remove impurities that are present with a certain size distribution, you must choose a separating tool and consider where to remove them. Size is thus a removal setting (a cut off level). In some processes, such as membrane separations, the removal limit can be set sharply, and the removal of sized impurities above the limit can be expected with certainty. Such membrane processes include the precision filtration (MF-membrane) method, the removal characteristics of which are expressed by molecular masses, and which has the ability to completely remove most particles with fractionated molecules of 3000 Da or more (Figure 3.11).

By contrast, in some operations such as sedimentation separation, impurities above the removal limit size can be removed almost completely, whilst for impurity components below the removal limit size, they can be expected to be removed at a certain ratio. The sedimentation surface load rate (overflow rate: W cm/min) described in Section 6.1 is the removal limiting size of particles, defined as the 'minimum particle sedimentation velocity capable of 100% removal in the pond' and can be calculated as the sedimentation velocity value $W_0 = (Q/A)$ cm/min, obtained by dividing the treatment flow rate Q cm^3/min by the sedimentation basin surface area A cm^2. However, even particles with a sedimentation velocity W smaller than W_0 are removed by the removal rate $r = W/W_0$ (the Hazen number), and a considerable part of particle size distributions are subject to removal. In these cases, both the particle size (sedimentation rate) distribution of the impurities and the overflow rate setting of the sedimentation basin are related to each other, with weight as an operation variable (Figure 3.12). In order to increase the sedimentation rate of particles and increase removal efficiencies, coagulation and flocculation processes are introduced in an attempt to shift the particle size distribution towards a larger size (described in detail in Section 4.1).

Separation processes such as deep filtration, superficial (cake) filtration, and membrane filtration, which have defined removal limits, are often used for more precise separation of components that have become smaller after pretreatment, and generally consume more energy per mass of separated components. On the other hand, pretreatment of sedimentation and crude filtration, etc., is often placed at the front of the

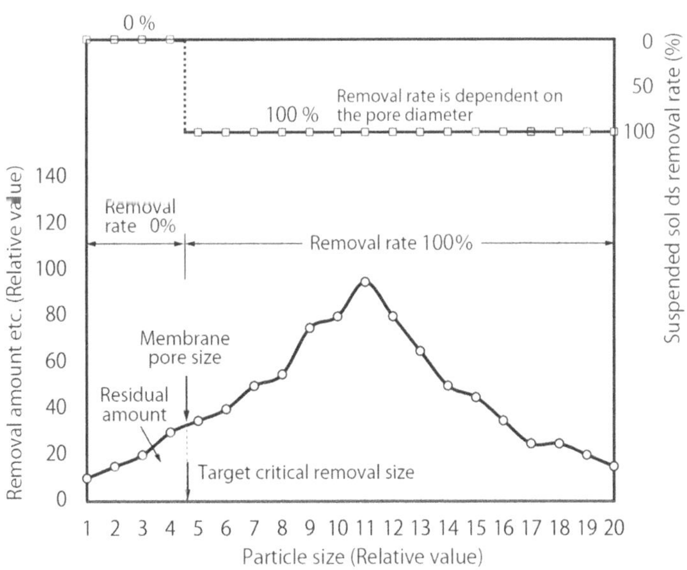

Figure 3.11 The concept of removal by membrane filtration.

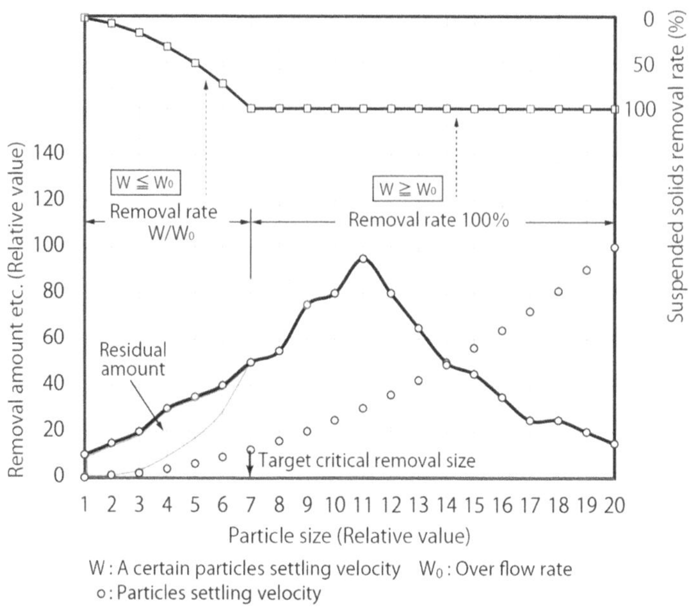

Figure 3.12 The concept of removal in a sedimentation basin.

system with a focus on quantitative removal (concentration reduction) of total raw water components rather than on a defined separation limit. This system has a hidden goal of achieving water with a low concentration of impurities at a low energy consumption rate.

3.2.4 Destabilization/separation of colloids by physicochemical mechanisms

After separation, which is considered a purely physical mechanism, the coalescence of physicochemical/electrochemical particles is also important for understanding water treatment processes.

The biggest condition for stabilizing colloidal particles is the presence of ions in water that are distributed near the surfaces of the particles, forming an electric double layer, as shown in Figure 3.13. Particle stability is a state in which particles repel each other without coalescence and are dispersed in water, in a stable state. When two colloidal particles approach each other, the electrical condition of the electrical double layer on each particle determines whether the particles repel each other and remain stably dispersed, or whether the electrical repulsion forces become small and unstable, causing the particles to attract each other and coalesce, causing the particle size to increase (coagulation) and separate from the water. The behavior of the particles due to the electrical properties of their surfaces is the basic segregation and regulation mechanism of the most widely used chemical or biochemical coagulation operations in modern water treatment systems. Based on the results of independent discussions by Derjaguin and Landau and by Verwey and Overbeek in the 1940s, the mechanisms are now explained by what is commonly referred to as DLVO theory (Kruyt, 1952). The coagulation operation of individual water treatment processes will be discussed in more detail in the relevant sections of Chapter 4, where we will also consider the interfacial electrochemical phenomena that form the basis for considering the stability and instability of colloidal particles.

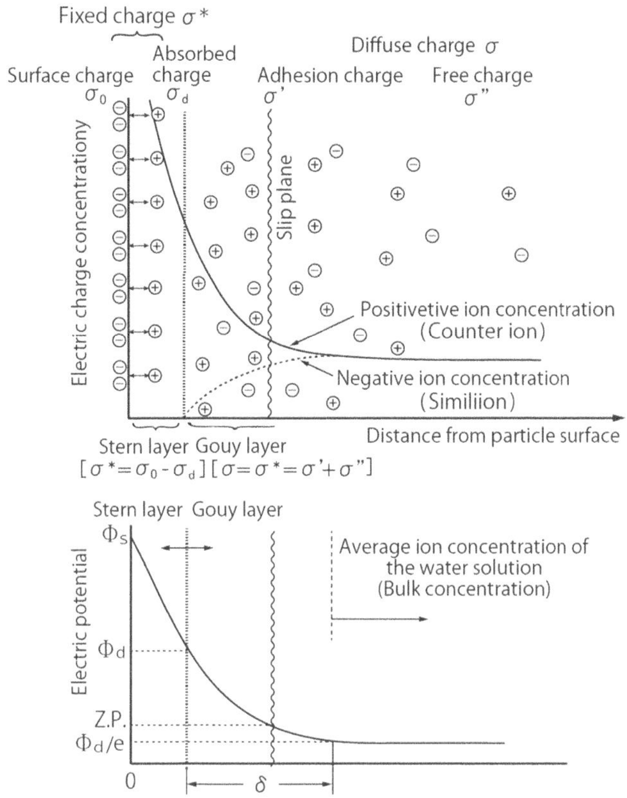

Figure 3.13 The Gouy-Chapman–Stern electric double layer model.

When charged particles are placed in water, the natural principle of making the entire system electrically neutral is applied on the solid surface, and opposite charged ions (counter ions) in the water are collected to neutralize the charge on the individual surface of the charged particles, and this is considered to constitute the electric double layer. Helmholtz considered a capacitor-like two-layer arrangement of surface charges and counter ions, and named it an electric double layer (Mysels, 1959). However, because the ions in water always diffuse through thermal motion, the oppositely charged (counter) ions are attracted to the solid surface charge and are present in high concentration near the particle surface, but they are diffused by thermal motion and become diluted by the average ion concentration of the aqueous solution as they move away from the solid surface. Therefore, it is believed that the counter-charged ions diffuse to form a diffused double layer that distributes in a dynamic equilibrium, rather than being fixed as in the Helmholtz double layer. In this case, the charged ions which have the same charge as the surface charge of the particle (similiion), are electrically neutral in an amount equivalent to that of the oppositely charged ions far enough from the surface of the particle, but diffuse in opposite directions to the oppositely charged ions in such a manner that they are repelled and become a no-charge on the surface of the particle. The concentration of oppositely charged (paired) ions near the particle varies depending on the size of the surface charge of the particle and can be much larger than the mean (bulk concentration) of the solution, but the charged ions can only vary from the mean (bulk concentration) of the solution to zero at the solid interface. It is therefore the oppositely charged (counter) ions that play a major role in destabilizing colloidal particles. These types of double layers are called Gouy-Chapman diffuse double layer models. However, as in the Helmholtz model, ignoring the presence of specifically arranged counter ions on the surface of the particle has been found to be inadequate to express the phenomena, and we need to consider the specific (selective) adsorption layer (the Stern layer) on the surface of the particle. This model is particularly useful in the presence of oppositely charged counter ions, either highly charged or polymeric, and is called the Gouy-Stern electric double layer model. Figure 3.13 is a conceptual diagram of the Gouy-Stern electric double layer model of negatively charged colloidal particles.

The lower diagram in Figure 3.13 shows the surface potential distribution of the double layers of electricity, generally expressed as four characteristics: the surface potential of the particle φ_s, the Gouy-Stern layer boundary potential φ_d, the zeta potential (ZP) and the thickness δ of the diffusion boundary layer. The thickness δ of the diffused boundary layer indicates the distance from the boundary layer until the Gouy potential reaches 1/e of boundary layer potential Φ_d/e. Among the indicators that are easy to measure is the ZP generated on the sliding surface between bulk water and static water that adheres to particles. It is widely used as a key potential representing the properties of the electric double layer, and various measurement methods have been proposed for various purposes, including commercially available automatic measurement.

The potential distribution in the electrical double layer changes as the ion concentration in water increases, as shown in Figure 3.14. As it increases, the potential and thickness of the diffusion layer decrease, and the potential of the Stern layer, which is in dynamic equilibrium with it, also decreases. When the ion concentration is increased as shown in (0) to (1) in Figure 3.14, the Stern layer potential approaches 0 at the limit, and the ZP also approaches 0. The state of ZP 0 is called the isoelectric point. The electric repulsive forces between the colloidal particles become inactive, and the interparticle

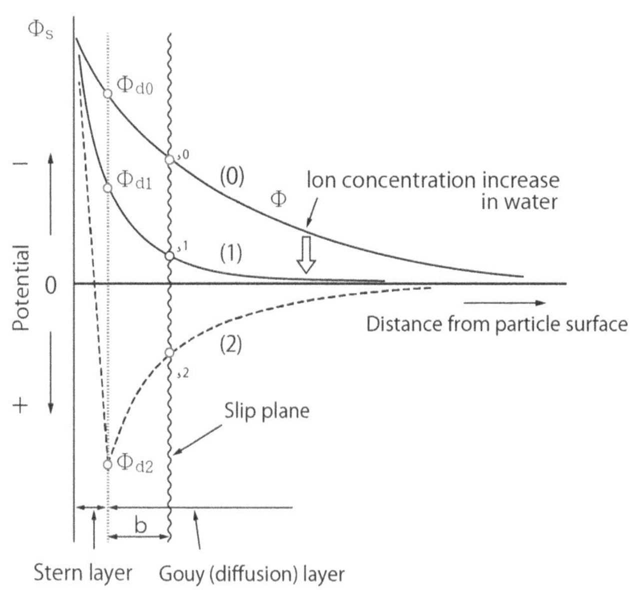

Figure 3.14 Distribution of interfacial potential and increase in ion concentration.

attraction (van der Waals forces) becomes easily concentrated. In this case, the counter ions play an overwhelming role, as discussed earlier, and the action of the charged ions is often negligible. The effects of oppositely charged ions on the potential decrease are stronger for polyvalent ions (see the Shultz-Hardy's Valence Law described in Chapter 4, Section 4.1.2 and Table 4.5).

In the case of polymeric ions, specific adsorption of counter ions to the Stern layer proceeds to the point of Φ_{d2} where the potential of the Stern layer is reversed, as shown at (2) in Figure 3.14. This charge reversal usually does not occur in a typical simple ion, and the ZP changes only to the iso electric point even when the simple ion concentration is increased.

The particles in an aqueous solution are each surrounded by an electric double layer as described above, and the solution as a whole is in an electrically neutral state. However, as the particles approach each other, the electric double layers overlap and repel each other. In Figure 3.15, which shows interactive forces, at (1) the electrical repulsion potential between particles can be seen; at (2) the attraction potential, based on London-van der Waals forces acting between particles is shown; and at (3) the energy-barrier (E_{max}) of the synthetic potential is identified, beyond which the kinetic energy of the particles makes it possible to determine whether or not the particles can reach the attached agglomeration region of the dominant attraction potential, which exhibits a large negative value. In other words, the possibility of agglomeration between particles depends on whether the height (E_{max}) of the curve of the synthetic potential energy (3) can be reduced to be less than the magnitude of the kinetic energy that drives the mutual approach of the particles in the case of colloidal particles.

In the case of colloidal particles, Brownian motion energy becomes the driving energy of collisions, and depends on whether the height (E_{max}) of the barrier curve of the synthetic potential energy can be lowered to below the kinetic energy of $E_K = 3KT/2$ (where K is Boltzmann's constant and T is absolute temperature). This is the 'removal (coagulation) limit condition' (Tambo, 1964); see Figure 3.16.

Individual operations will be discussed later. From the results of many experiments, The authors would like to propose in later discussions that this condition can be met in practice if ZP is brought to 10 mv or less in absolute terms. This is a 'critical operation condition' (Mysels, 1959; Tambo, 1964) that is widely accepted in water, wastewater and sludge treatments, and in natural coagulation reactions.

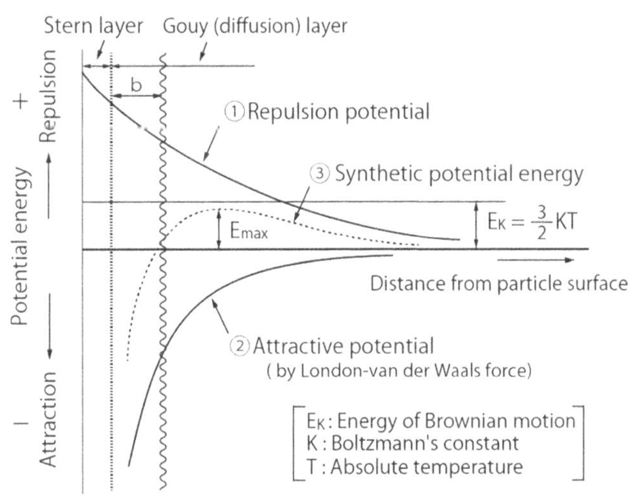

Figure 3.15 The energy generated in particle surface interactions.

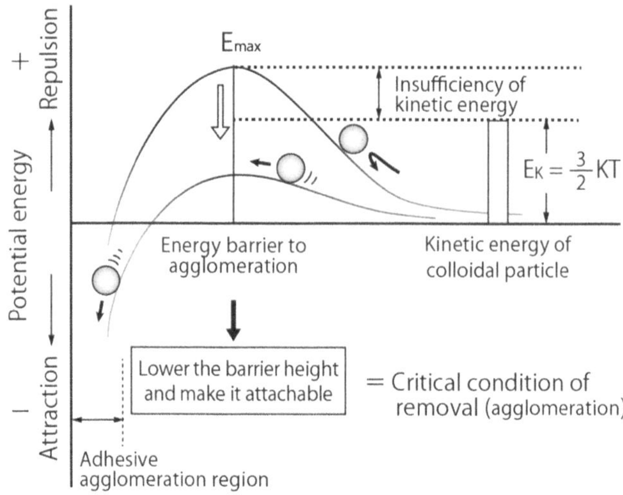

Figure 3.16 Particle agglomeration and the energy barrier.

3.2.5 Application limit size of intragranular diffusion-type treatment

Removal of impurities in water by solid-liquid phase transfer, such as activated carbon adsorption and ion exchange, occurs by removal of impurities in water by physicochemical reactions or chemical reactions at the inner surfaces of pores existing in a network inside the adsorbent solid phase. From the lower limit of colloidal size to the soluble area, the size of the substance to be removed (adsorption substance) spreads. In this area, impurity size is generally expressed in terms of molecular weight, measured by liquid chromatography or other methods. The removal limits and removal rates of impurities of this size are affected by the chemical nature of the impurities. If an impurity has a higher affinity for an adsorbent such as activated carbon than for water molecules, it will be adsorbed at the solid-state interface. As will be described in detail later in Chapter 7, adsorption and ion exchange processes proceed on the surfaces of the pores inside the activated carbon and the ion exchange agent. Therefore, impurities (adsorbed substances) must first diffuse into the pores and reach the adsorption and ion exchange sites (points).

Figure 3.17 shows a schematic diagram of the pore component of activated carbon. Macropores of the order of micrometers run three-dimensionally and horizontally inside the activated carbon, and micopores of about 10^{-2} to 10^{-4} μm are arranged in a network on the walls of the macropores. The inner surface area of the micropores ranges from 700 to 1400 m^2 per 1 g of activated carbon, accounting for most of the inner surface area of the micropores. Therefore, a polymer component such as a molecule having a mass of 1000 to 1500 Da or more cannot diffuse into the pores and, therefore, an effective adsorption process cannot be performed. Thus, the first limiting condition when considering the interfacial treatment of an intragranular diffusion type is that impurity particles not larger than the limit particle diameter (the limit diameter at which diffusion cannot be effectively performed) are not included in the object of this treatment. The coarser impurity particles removed from this treatment must be pretreated by coagulation followed by sedimentation and filtration (Tambo, 1964).

Since activated carbon adsorption is a physicochemical interfacial reaction, the adsorbate must diffuse and migrate through the pores before being adsorbed and fixed to the interface of the adsorbent particles (mostly the inner surfaces of the pores). The adsorption isotherm method (detailed in Section 7.1) is a static method for quantitatively expressing the extent to which such adsorption reactions proceed

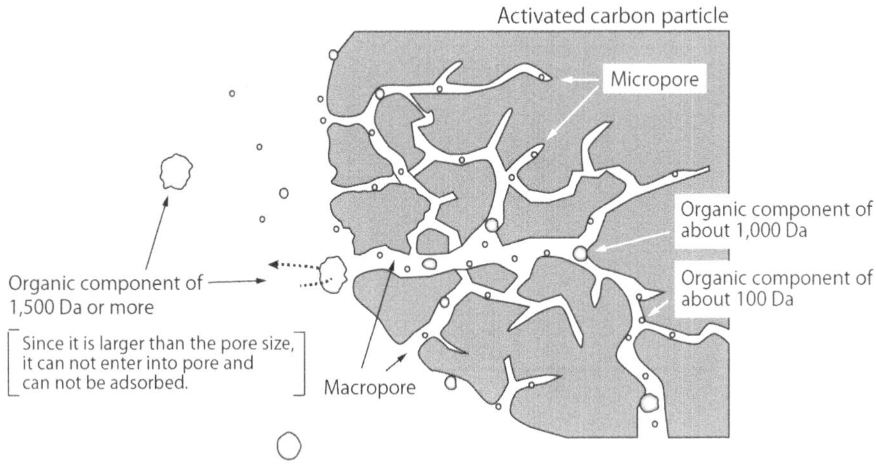

Figure 3.17 A schematic diagram of the pore components of activated carbon.

(adsorption equilibrium). To quantitatively express the ease of adsorption, a relationship needs to be formulated between the concentration of the removal substance in the liquid system (C mg/l) and the concentration of the substance to be adsorbed in activated carbon (when the amount of activated carbon is q mg/g of the substance to be adsorbed in the liquid system). Various relational expressions have been developed theoretically or experimentally. This relationship varies depending on the temperature condition and therefore the C-q relationship is shown at a constant temperature and is called the adsorption isotherm equation. One example is the Freundlich model, which provides Freundlich's adsorption isotherm equation (Equation 3.2).

$$q = a\,C^{1/n}$$

or

$$\log q = \log a + \frac{1}{n}\log C \qquad \text{—(3·2)} \tag{3.2}$$

q : Adsorbed substances amount (mg/g-act.carbon))
C : Equilibrium concentration of adsorbate in water (mg/l)
a,n : Experimental constant

Equation 3.2. The adsorption isotherm equation of Freundlich

As the size of removed components decreases to this point, the chemical properties associated with the substances have significant implications for the magnitude (q-C relation) of the extent of adsorption on the activated carbon interface (the inner surface). In terms of how the adsorbate is distributed between water and activated carbon surfaces, substances that have a high affinity with water (water solubility) and other adsorbed substances are less likely to be adsorbed – and adsorption is therefore less effective. It is generally known that, qualitatively, solubility in solution (in water) is inversely proportional to the adsorption quantity (Lundelius's rule). It is also known (following Traube's rule) that materials that are more hydrophobic are easier to adsorb – which may also be considered as an example of Lundelius's rule. It is known that the higher the molecular weight of an organic substance, the higher the equilibrium

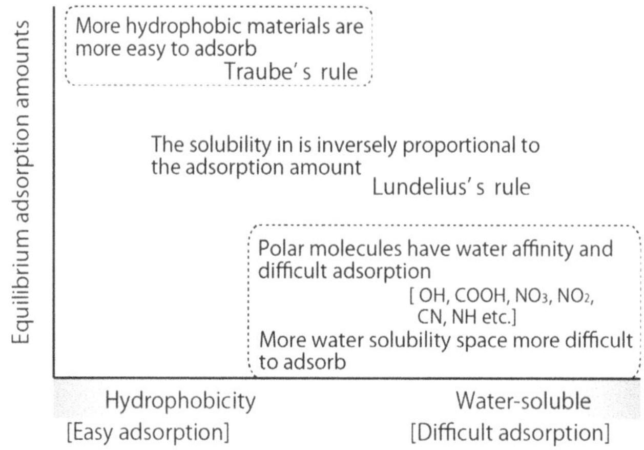

Figure 3.18 The relationship between the properties of the adsorbate and activated carbon adsorption.

adsorption amount, if it can diffuse in the pores. It should be noted that the same series of compounds that can be discussed in terms of equilibrium adsorption can also be discussed in terms of the size of the impurity. It is also known that the adsorption of ionized amorphous materials is best performed at the isoelectric point. In addition, substances with strong polarity (a polar solute) are attracted to other substances with strong polarity, but compounds with OH, COOH, NO_3, NO_2, CN and NH groups are attracted to water molecules (easily soluble in water) and do not appear to exhibit strong adsorptive properties, even if activated carbon has some polarity. In particular, compounds with a large number of sugars and OH groups, which have a strong polarity, are rarely adsorbed. Thus, activated carbon adsorption is strongly governed by physicochemical properties. The authors believe this can be referred to as physicochemical reaction control (Figure 3.18).

3.2.6 Ion exchange

Even in intragranular diffusion-type interfacial treatments with the same intra-pore network, it is the chemical reaction itself that controls the magnitude of the adsorption equilibrium of ion exchangers such as synthetic resins and natural zeolite particles. The electrostatic force of the functional group of an exchanger separates it into a cation exchanger (capable of exchanging and adsorption of cations) or an anion exchanger (capable of exchanging and adsorbing anions). The type of exchangeable ionic groups in the exchanger determines the type of adsorbate that can be exchanged and adsorbed, and the total number of ionic groups in the unit volume of the exchanger determines the amount of adsorption equilibrium. Positive ion exchange resins with acidic groups and alkaline negative ion exchange resins with functional groups such as amines are typical exchangers. The operation is discussed in more detail in Section 7.2 below.

The basis for any removal is chemistry, in which the functional groups of the ion-fixed layer on the surface of the exchanger (ion exchange groups) and the ions of the material under removal on the liquid side (bulk) constitute a diffused double layer. As described in the previous section, oppositely charged ions are distributed at high concentrations on the front surface of the fixed layer, resulting in ion exchange adsorption and removal from the water. As mentioned in the section above on the electrical double layer, it is known that the larger the charge, the more easily the exchanged ions are absorbed. In

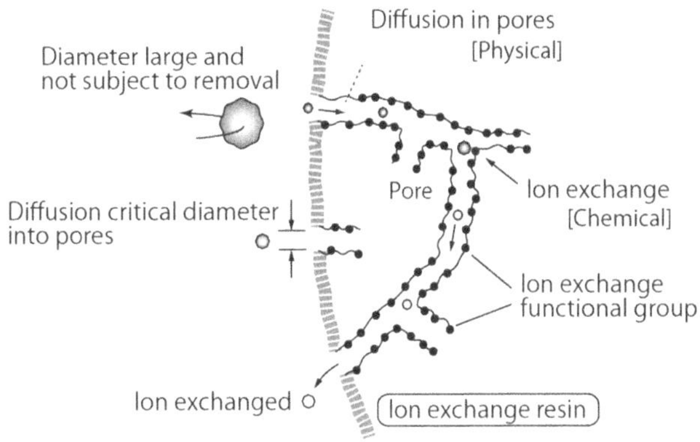

Figure 3.19 The concept of ion exchange.

dilute systems, the smaller the ionic radius, the more easily adsorbed the substance is, and other physical properties seem to affect it somewhat, including hydration. Changes in exchanged adsorption with temperature are negligible in normal temperature ranges and are not evaluated by isotherms as in activated carbon adsorption.

Since adsorption is by in-pore diffusion, as described above, components larger than the critical particle size that cannot enter the pores (due to the large particle size of the target ion to be removed) cannot be regarded as removal targets (Figure 3.19). If humic substances with a large molecular weight are to be removed by anion exchange resin (or activated carbon adsorption), pretreatment is performed by coagulation to remove components with a molecular weight of 1000 to 1500 Da or more, or a macro molecular (MR) ion exchange resin with a large pore size is used. In addition, pure cesium ions can be exchanged to the full volume in ion exchange reactions without problems, such as for the removal of cesium (Cs^{137}) in reactor wastewater using zeolites (natural cation exchangers). However, if there is a large molecular weight component that constitutes part of a complex, there will simply be a part that cannot be removed. The pretreatment operation needed to remove Cs adsorbed on large molecular weight complexes and fine particles will be described in de tail later.

3.3 SEPARATION MECHANISM DEPENDING ON CHEMICAL PROPERTY DIFFERENCES

As the size of the removal component gradually approaches the size of the water molecule, 3 Å, the main consideration for separating impurities from water is the chemical property difference. Even in the adsorption process described in the previous section, when the physicochemical mechanism is ion exchange adsorption, the process equilibrium was evaluated centering on the chemical equivalent. In this section, we will focus on the differences in the chemical properties between true solute molecules and water molecules in solution. We will discuss the basic points of how to separate solutes accurately and efficiently by making the differences prominent by progressing chemical reactions that are tailored to the purpose. Chemical reactions, unlike physical or physicochemical reactions, proceed either individually or independently. In most cases, the conditions under which a reaction proceeds are the features of each individual compound, although the reaction can be explained by a common mechanism. This is the characteristic difference between how chemistry and physics are handled.

3.3.1 Equilibrium and the Le Chatelier Principle

In most cases, a reaction proceeds reversibly, and the equilibrium state is determined by environmental conditions. In the case of water treatment, in addition to the redox potential and pH etc. of the indicators used to assess the state of the water environment (mentioned above), a variety of environmental factors serve to determine the mechanism and equilibrium of a reaction. In particular, the main factors that determine the final equilibrium state are the concentration relationships among the substances A, B, C, and D in the reaction, found, for example, in the treatment water. For example, assume that A and B react in the form of $aA + bB + \cdots \rightleftharpoons cC + dD + \cdots$ to form products such as C and D, where a, b, c, d, etc. are the concentrations (in moles) of the substances A, B, C, and D etc., respectively, involved in the reaction. In such a case, if any of the concentrations of components A, B, ··· increase, the equilibrium proceeds rightwards, and if any of the concentrations of components C, D, ···. increase, the equilibrium proceeds leftwards. The shift of equilibrium point in response to changes in reactant concentrations is known as the Le Chatelier Principle. The equilibrium of these reversible reactions is defined as the law of mass action, in Equation 3.3.

$$\frac{[C]^c[D]^d\cdots}{[A]^a[B]^b\cdots} = K \quad —(3\cdot3) \tag{3.3}$$

[A],[C] etc.: Concentration of reactant, product
In dilute systems expressed in mol / l
a,c etc.: Number of moles of substance
K : Equilibrium constant

Equation 3.3. The equilibrium equation of reversible reaction based on Le Chatelier's principle

Generally, concentration (activity) is used to calculate equilibrium by multiplying the molarity by the activity factor γ for concentration compensation. The equilibrium constant K is large or small, and the equilibrium point of a chemical reaction depends on the activity of various components. In water treatment, the presence of reactants produced by pretreatment is subject to separation and adjustment.

It has been mentioned above that the water solubility of a component is a central mechanism when water treatment separates solids and gases from water. Basically, equilibrium constants are defined as being either reactants (solids) or dissolved ions, i.e., solubility products, according to the law of mass action, and are used to calculate water solubility and precipitation of various components, as well as for hardness treatment, heavy metal removal, iron removal, manganese removal, phosphoric removal, coagulation, and other chemical reactions related to the basics of operation. Individual operations are discussed in more detail in Chapters 5 and 7. The phenomena related to oxidation-reduction potential (ORP) and phenomena related to pH mentioned above can basically be regarded as a phenotype based on the law of mass action.

3.3.2 Three basic rules of thermodynamics

There is one branch of thermodynamics that is the basis for describing the mechanisms of operation, chemical reactions and various power machines. The discussion here focuses on water treatment technology. In discussing the basic part of water quality change/conversion, which is a chemical reaction, the authors would like to briefly discuss the thermodynamics that form the basis of modern science. Starting with the performance of power machinery, there are many environmental technologies and engineering that must be considered through thermodynamics, including the progress of chemical reactions and the global energy budget. Water treatment is one of these, and the authors would like to expand the breadth and depth of understanding by paying a little attention to the basic sciences in this area.

Thermodynamics discusses the physical mechanisms by which the equilibrium condition of various operations are determined, in a branch classically described as 'equilibrium thermodynamics'. At the end of the eighteenth century, when steam engines were invented and emerged as a source of power, scientists began to think about the fueling efficiencies of the operation of empirically built engines. Research on the balance of heat and energy, including research on Carnot's virtual heat engine (the Carnot cycle), progressed. In the 1850s, thermodynamics as the basis of science even beyond mechanical thermodynamics, was established by Clausius. By that the concept of free energy by Helmholtz and the definition of Gibbs's chemical potential developed into a broad range of phenomena, including the equilibrium of chemical reactions. Further, the results of statistical dynamics developed by Boltzmann and Maxwell were incorporated into the evolution of mechanical thermodynamics into chemical thermodynamics, making thermodynamics the most important discipline of modern natural sciences and thermodynamic laws the most important laws of classical science, alongside Newton's law of inertia (dynamics) and Maxwell's law of electromagnetism.

Thermodynamics assumes three states, as shown in Figure 3.20: an isolated system, a closed system and an open system. Isolated systems are completely independent of the outside world, with neither energy nor material remaining in the system, and there is no change in the overall balance. Generally speaking, it is a virtual system, and there is no system corresponding to an actual neighboring environment. If we think that the whole universe is a system, we can say that it is an isolated system. This system has no reality but hypothetical. Closed systems are systems in which heat/energy enters and exits through the boundaries of the system, but substances remain within the boundaries and undergo qualitative changes through internal circulation. An open system is a system in which energy and materials are interrelated and enter and exit across a boundary and are also associated with other systems.

The first law of thermodynamics is the energy conservation law, which states that the total amount of heat and energy does not change in a closed space unless there is an input/output of substances, heat and energy from/to the outside. The second law explains the quality of energy by the law of entropy and is the most widely applied basic law of thermodynamics in which various definitional expressions are made. The third law of thermodynamics establishes an energy level criterion in which entropy becomes zero at the absolute temperature of T = 0 K (zero degrees Kelvin), there is no lower temperature, and the absolute criterion of the temperature is 0 K (Figure 3.21).

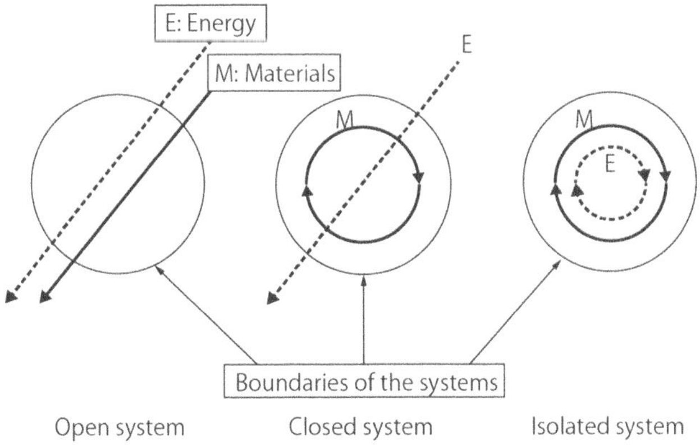

Figure 3.20 The three systems of the environment in thermodynamics.

The first law : Law of energy conservation
[The total amount of energy is preserved in closed space]

The second law : Law of entropy
[Explain the quality of energy]
[Heat flows from higher to lower direction]
(Entropy increases)

The third law : Law of energy level criteria definition
[Entropy is zero at absolute temperature zero (0 K)]
[There is no temperature lower than absolute zero (0K=-273.15℃)]

Figure 3.21 The three laws of thermodynamics.

3.3.3 Energy and enthalpy

According to the first law of thermodynamics, 'energy is not created or extinguished,' and heat or work entering and leaving a system's boundaries will result in a change commensurate with the total amount of energy in the system. Therefore, the relation expressing the first law is expressed by Equation 3.4.

$$dE = dq + dw \qquad \text{—(3·4)} \tag{3.4}$$

dE : Amount of change of internal energy
dq : Amount of change in the heat quantity taken in from environment
dw : Amount of change in work given by the environment to the system

Equation 3.4. The first law of thermodynamics

If the volume change before and after the reaction can be ignored, $dw \fallingdotseq 0$ can be assumed, and the amount of heat absorbed under constant volume is $dE \fallingdotseq dq$, which is equal to the amount of change in the internal energy. When heat enters the system, q takes a positive value (endothermic reaction) and when it exits, it takes a negative value (exothermic reaction).

Many water treatment operations are performed at atmospheric pressure, and reactions generally proceed at a constant pressure. Thermodynamic conditions in these environments are defined by the Enthalpy equation (shown in Equation 3.5).

$$H = E + PV \qquad \text{—(3·5)} \tag{3.5}$$

H : Enthalpy
E : Internal energy
P : Pressure of the system
(In the case of water treatment, constant atmospheric pressure)
V : Volume of system

Equation 3.5. The Enthalpy equation

Considering certain states 1 and 2, the amount of heat absorbed under constant pressure (change in enthalpy) between states 1 and 2 is given by Equation 3.6. To generalize this, the enthalpy change $\Delta H = q_p$ is obtained under constant temperatures and pressure P conditions. The change in enthalpy is the amount of heat exchanged under constant pressure. When ΔH is positive, an endothermic reaction occurs. When ΔH is negative, an exothermic reaction occurs.

$$E = E_2 - E_1 = q_p - P(V_2 - V_1) \quad \text{—}(3\cdot6)$$

$\downarrow$ q_p : Heat absorbed under constant pressure

The change amount ΔH of enthalpy is

$$H_2 - H_1 = (E_2 + PV_2) - (E_1 + PV_1) = q_p$$

Under constant temperature and constant pressure

$$\Delta H = q_p$$

[ΔH : When positive, endothermic reaction
When negative, exothermic reaction] (3.6)

Equation 3.6. The enthalpy change of certain states 1 and 2

It is difficult to measure the enthalpy of a system, and when considering the equilibrium state of a reaction, the change in enthalpy, ΔH, or the calorimetric value, q, that increases or decreases in the reaction is measured – or the difference between the sum of the standard enthalpies of the reactants and products of the ongoing chemical reaction is calculated from a database of standard enthalpies of known chemical substances. Standard enthalpy, cal/mol (J/mol), required to produce a variety of compounds (e.g., OOH_2O or $CaCO_3$) from a stable monomer under one standard atmospheric pressure (1 atm) is described in scientific databases. Gibbs Free Energy and Entropy Standard State Values, described in the next section, are also described in databases.

3.3.4 Entropy and free energy

The second law of thermodynamics can be explained by defining the entropy $S = q/T$ (in J/K or cal/K) that represents the state of a system. This means that the entropy of a system increases by q/T when the system, at absolute temperature T, is subjected to a heat quantity q. Work is being done only in systems that are not in equilibrium, and nothing happens in equilibrium. So, knowing the state of equilibrium and the distance from the current state allows you to know in what direction the system changes.

The total entropy change dS of a system is the sum of the entropy change d_iS inside the system and the entropy flow d_eS between the system and the environment, such that $dS = d_iS + d_eS$. If the environmental inflow is q, the inflow entropy is $d_eS = dq/T$. If the system is in equilibrium or a reversible state, $d_iS = 0$, whereas if the system is in a natural or irreversible state, $d_iS > 0$. Described in another way: 'entropy is constantly increasing when spontaneous changes occur in an independent system' (the Clausius inequality; Equation 3.7).

- Entropy
  $S = q/T$ Unit: J/K or cal/K
  q : Inflow energy from the environment
  T : Absolute temperature
- Total entropy change of the system
  $dS = d_iS + d_eS$
  d_iS : Entropy change inside the system
  d_eS : Entropy flow rate between system and environment
- Inflow entropy (increase) to the system
  $d_eS = dq/T$
  dq : Influence energy change

State in the system

System is in an equilibrium or reversible state $d_iS = 0$ —(3·7)
System is in a natural or irreversible state $d_iS > 0$

[Clausius inequality] Entropy always increases when spontaneous change occurs in one independent system.

(3.7)

Equation 3.7. The Clausius inequality

To illustrate the second law of thermodynamics by giving an example of the phenomenon, it follows that:

(1) Heat cannot be transferred from a low temperature object to a high temperature object without adding work from outside: a large amount of heat contained in low temperature water (in high entropy water) is not able to be naturally transferred to a small amount of water to generate high temperature water (low entropy water) – as explained by the Clausius principle (Figure 3.22). In order to collect low temperature heat and obtain high temperature heat, it is possible to transfer a large amount of the low temperature heat to a high temperature object only by adding work from the outside, like from a heat pump, etc.

(2) 'Permanent engines' that continue to move without being supplied energy from the outside cannot be created: heat generated by work can be recovered internally from a system and cannot continue to

① [The Clausius principle]
Heat cannot be transferred from a low-temperature object to a high-temperature object without adding work from outside.

② [The perpetual motion machine is impossible]
Moving continuously without an external supply of energy is impossible.

Figure 3.22 The second law of thermodynamics.

work as a heat source (Type 2 perpetual motion machine is impossible) (Figure 3.22). Although energy, E, such as heat quantity and mechanical work, and mass, M, follow the conservation law, entropy, S, is not preserved and, if left to stand, it increases toward the maximum. When the system moves toward equilibrium, the potential energy will be at its minimum and entropy will be at its maximum.

Common reaction systems, such as environmental engineering systems, use the notion of free energy to think about how reactions proceed spontaneously. The introduction of free energy, as described below, is considered as a function that explains the state change in which potential energy is minimized at the equilibrium point and entropy is maximized. As a quantitative measure of both energy and entropy, we define Gibbs free energy $G = H - TS = E + PV - TS$ (see Equation 3.8) in the isothermal and isobar process to determine the spontaneous progression of reactions, where H is the enthalpy (T: absolute temperature (K)) and S is the entropy (J/K). If we define a quantitative index of Helmholtz free energy $A = E - TS$ to evaluate the isothermal and constant volume process, we can write the relationship between Gibbs free energy and Helmholtz free energy as $G = A + PV$. In an environmental system, Gibbs free energy is used to evaluate the isothermal and isobar (constant pressure) process in many cases, since it is common for the reaction to proceed with P and T as variables in environmental systems (Equation 3.8).

Gibbs free energy : G

$$G = H - TS = E + PV - TS \quad \text{— (1)}$$

Helmholtz free energy : A

$$A = E - TS \quad \text{— (2)}$$

The relation between (1) and (2) $G = A + PV$ — (3·8)

H : Enthalpy
T : Absolute temperature (K)
S : Entropy (J/K)
E : Internal energy
P : Pressure
V : Volume of system

(3.8)

Equation 3.8. The free energy equation

Under isothermal and isobaric conditions (constant T and constant P), the change in the free energy of the reaction considered is $\Delta G = \Delta H - T\Delta S$, where $\Delta H = q - w + P\Delta V$, assuming a slow system change, $\Delta G = w_{max} + P\Delta V$, since it can be assumed that the energy donation from the environment is approximately equal to the maximum work w_{max} when the energy donation from the environment is considered to have little loss. The effective work $W_{effective}$ obtained by subtracting the part of $P\Delta V$ wasted by the volume change against pressure from the environmental work w_{max} determines ΔG. If the state of $\Delta G = W_{effective}$ is negative, the process will proceed; if the value is zero, the system will be in equilibrium; and if the value is positive, the process will proceed in the opposite direction (Equation 3.9). To evaluate the free energy, you can find the standard enthalpy of formation ΔH^0 (kJ/mol, cal/mol), standard entropy S^0 (Jmol: K^{-1}), and standard free energy of formation ΔG^0 (kJ/mol) at 1 atm at a standard state of 25°C (298.15 K) – which is the standard state of change – from databases obtained according to the states of various element compounds.

Changes in free energy ΔG under isothermal and isobaric stats,

$$\Delta G = \Delta H - T\Delta S \quad —(3\cdot9)$$

$$\Delta H = q - w + P\Delta V$$

Assuming a slow system change,

$$\Delta G = w_{max} + P\Delta V$$

ΔG is determined by $W_{effective}$

$W_{effective} = W_{max} - P\Delta V$ part wasted

$\Delta G = W_{effective}$

q : The energy donation from environment
w : the work in the system
W_{max} : the maximum work donation from environment
$W_{effective}$: The effective work

$\Delta G < 0$ The process proceeds.
$\Delta G = 0$ The process stops in equilibrium.
$\Delta G > 0$ The process proceeds in the reverse direction. (It can not happen naturally.)

(3.9)

Equation 3.9. The free energy and the proceed of process

Considering a reversible reaction, such as aA + bB ⇋ cC + dD, as an example of the change in free energy due to a chemical reaction, the amount ΔG of the change in free energy of the reaction can be calculated from the concentrations (activities) of the reactants A and B, and the products C and D [A.B. C.D], and the molar numbers a, b, c and d of the substances involved in the reaction, by Equation 3.10.

The change amount ΔG of the free energy in the reversible reaction (a A + b B ⇄ c C + d D) is

$$\Delta G = \Delta G^0 + RT \ln\left\{\frac{[C]^c[D]^d}{[A]^a[B]^b}\right\} \quad —(3\cdot10)$$

[A], [C] : The concentrations of the reactants and the products
a, c : The molar numbers of each substances
R : Gas constant (8.31J/K·mol = 9.11cal/K·mol)
T : Absolute temperature (K)
ΔG^0 : Standard free energy of formation
(Free energy change amount at standard state 298.15 K, 1 atm)

(3.10)

Equation 3.10. The change in free energy by the chemical reaction

At the equilibrium point of the response, $\Delta G = 0$, so that $\Delta G^0 = -RT \ln\{[C]^c[D]^d/[A]^a[B]^b\}_{equi,}$ and the equilibrium constants $K = \{[C]^c[D]^d/[A]^a[B]^b\}$, as mentioned earlier. This means that the ΔG^0 can be calculated as $\Delta G^0 = -RT \ln K$ if the equilibrium constants K can be measured.

Thermodynamic properties are interrelated with the equilibrium state of the reaction (Equation 3.11).

$$\Delta G^0 = -RT\ln\left\{\frac{[C]^c[D]^d}{[A]^a[B]^b}\right\}_{equilibrium} \quad \text{—(3·11)}$$

↑ Equilibrium constant K

$$\Delta G^0 = -RT\ln K \tag{3.11}$$

Equation 3.11. The Thermodynamic properties and the equilibrium state of reaction

3.3.5 The energy balance and thermal characteristics of the Earth

The Earth is the largest energy and material system that we are actually involved in. If we consider an energy system and a material circulation system using the Earth as a reaction vessel, it is easier to understand the Earth as a comprehensive system that also serves as the base for other systems, when we look at the Earth assuming boundaries for the convenience of various situations, such as economic systems, social systems, urban systems, environmental systems, natural ecosystems and engineering systems.

A total of 177,000 TW (terawatts) = 177 PW (petawatts) = 177×10^{15} W (Watts) of solar energy flow into the Earth. The same amount of energy flows out of the Earth into outer space. During the inflow and outflow, we can think of a closed system in which the nature of the system is changed (restored) while material continues to circulate and the order of the system is maintained. The driver of this cycle is the solar energy, which reaches the Earth at 177,000 TW. One Watt is the rate of work (=work/sec) denoted by $1\ W = 1\ J/s = 1\ N \cdot m/s = kg \cdot m^2/s^3$ (Equation 3.12).

$$1\ W = 1\ J/s = 1\ N\cdot m/s = 1\ kg\cdot m^2/s^3 \quad \text{—(3·12)}$$

W : watt

$$\text{Power} = \frac{\text{Work}}{\text{Time}} \tag{3.12}$$

Equation 3.12. The power

Figure 3.23 shows the approximate values of solar energy radiation temperature inflow to the Earth at 6,000 K, and the Earth energy radiation temperature re-radiated into outer space at 265 K. Energy quality expenditure (increased entropy, S), which decreases by this difference in temperature, is used to drive material circulation and quality change on the Earth, thereby maintaining the order of the Earth and providing a global example of the second law of thermodynamics. Entropy S is the quality of energy defined by the second law of thermodynamics and is defined by the value $S = Q/T$ obtained by dividing the amount of energy Q by the temperature T in situ. The smaller the value, the better the quality of energy. The global ecosystem is a closed system with drive at 177,000 TW. The inflow entropy S_{in} = 177,000 TW/6,000 K = 29.5 TW/K, and the outgoing entropy S_{out} = 177,000 TW/265 K = 667.9 TW/K. The entropy increasing rate ΔS = 638.4 TW/K, indicating that the Earth's ecosystem is sustained by that entropy difference (Figure 3.23).

Economic, modern social systems are driven by intensive use of 12 to 13 TW of high-quality (low entropy) energy such as fossil energy, atomic energy, etc. This is only a small fraction of

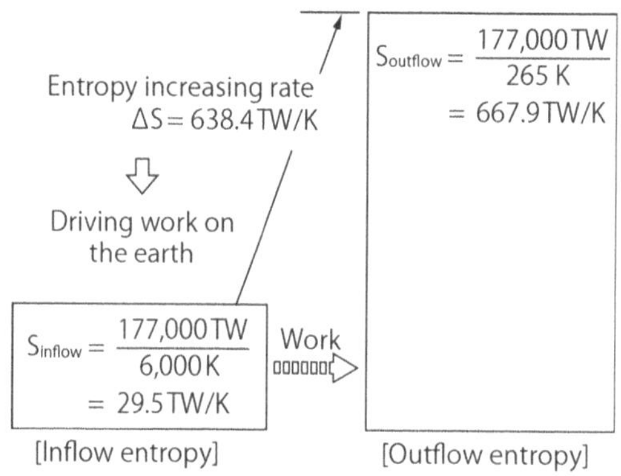

Figure 3.23 The increase of entropy on the Earth.

less than 1/100,000 of the global inflow of solar energy. However, the economic system of modern societies can be regarded as an open type in which both energy and materials can enter and exit through urban/industrial boundaries, and modern society has grown as an open type growth system in which the quantitative expansion of the system's balance of income and expenditure has economic value, without considering the strict boundary balance of that income and expenditure. However, even with this small increase in energy expenditure, greenhouse gases, which were slightly increased by releasing fossil energy (which has been stored in small increments over the order of one million years so far) during the short 350 years of the modern period, have pushed global surface temperatures from 290 K (17°C) by over 1.5 K (0.5%), leading modern industrial societies to the global climate change crisis.

Modern Earth is operated by combining the daily balance of solar energy with a very short-term release of geological energy (fossil/nuclear energy) stored for the order of one million years. In addition to solar energy, the Earth itself produces about 44 TW of geo-heat (from the radioactive decay of the Earth's crust). In addition, the tidal energies induced by the attraction of the sun, moon and other astronomical bodies are of the order of about 3 TW.

Approximately 150 TW of solar energy is fixed as biomass in forests and fields. In forests, about 50 to 100 TW, and in fields, about 1 to 2 TW, are recycled in the bio-cycle. The amount of biomass fixed in a farm field is only about 10 to 12 TW. Until the beginning of the modern era, social operation was driven by the daily solar-energy balance and the balance of the biomass-energy cycle, averaging over a few decades. Through the energy and industrial revolution, modern societies in the 1800s and beyond introduced the use of fossil energy (which is a short-term withdrawal of non-renewable resource stocks that had been stored over millions of years) into daily social systems operations, and humanity has expanded greatly within a limited short period of time – but it is an expansion which will, nevertheless, end after 350 years of modern economic behavior.

This system uses only 10 to 12 TW of high-quality fossil/nuclear energy. The absolute amount of energy consumed is less than 100,000 times smaller than that of solar energy. Intensive use of low-entropy fossil energies has grown global humanity from a population of 1 billion to 7 billion in 300 years. This is the energy significance of modern times. About 30% of solar energy is reflected by the atmosphere (6%), clouds (20%) and surface (4%), and does not enter the Earth. The ratio of reflected energy to solar energy is called Albedo reflectivity (Figure 3.24).

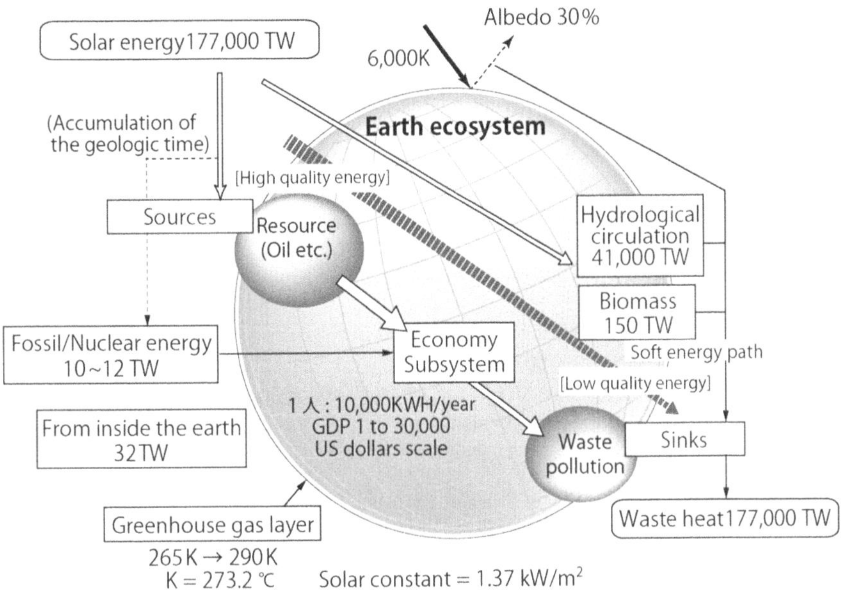

Figure 3.24 The Earth's ecosystem and economic system.

3.4 DETOXIFICATION AND IMPURITY CONCENTRATIONS

3.4.1 Indication of concentrations

There are various methods for expressing the concentration of impurities in water. 1 mg/l or 1 ppm is a practical unit often used in water quality standards, etc. Basically, the substance concentration of a liquid is expressed in units based on the mass, in mg/l. In thin aqueous solutions, 1 mg/l ≒ 1 ppm but, strictly speaking, 1 mg/l and 1 ppm are not the same. At higher concentrations, operation and assessment differ depending on which units are used. Consider the Yellow river's water in China, which has a very high concentration of 200,000 mg/l of suspended matter, and 178,094 ppm (Equation 3.13). Yellow river turbidity concentrations have been reported at Zhengzhou, with annual averages 40 g/l and 150 to

Table 3.2 Unit prefixes used in the decimal system.

Digit	Prefix	Symbol
10^{-3}	milli	m
10^{-2}	centi	c
10^{-1}	deci	d
10^{1}	deca	da
10^{2}	hecto	h
10^{3}	kilo	k

Note: $10^0 = 1$: reference.

200 g/l at flooding. At this high concentration, 200 g/l is no longer 200,000 ppm, but is 178.094 ppm.

① In the case of water containing 200 g/l of clay particles

When the density of clay particles is 2.6 g/cm^3
Volume of 200 g   200/2.6 = 77 cm^3
Mass of sample 1 l   {(1,000 − 77) × 1} + (77 × 2.6)= 1,123 g/l
200 g/l is   200/1,123 = 0.178094 = 178,094 ppm
*[ppm is Mass / mass or Volume / Volume etc. : Dimensionless]

② In the case of water containing 20 mg/l of clay particles (3.13)

Volume of 20 mg   0.02/2.6 = 0.0077 cm^3
Mass of sample 1 l   {(1,000 − 0.0077) × 1} + (0.0077 × 2.6) = 1000.01 g/l
20 mg/l is   0.02/1000.01 = 19.9998 ppm
20 ≒ 19.9998

*In the low concentration range, mg/l and ppm are often used without distinction.

Equation 3.13. The meaning of concentration unit

To cope with such high turbidity concentrations, Tambo and Wang developed the pellet-coagulation method as an ultra-high-concentration technique for separation treatment (described in Chapter 4), enabling the universal application of coagulation treatment to high concentration suspensions.

When referring to the relationship of quantities in terms of stoichiometry, the concentration indicator mol/l is generally used. As described above with relation to Le Chatelier's Law, when molecules react with each other, the number of molecules involved in the reaction is essential. Therefore, a molar concentration of mol/l or mg/l is used. One mole is defined as the mass (12 g for carbon) of the molecular weight (g), and there are 6.022×10^{23} molecules /mole. The figure Na ≒ 6.022×10^{23} is known as the Avogadro number (Avogadro constant) in a 1 mole mass, regardless of the molecular weight of the individual material. Mass molarity is the concentration that represents the number of molecules involved in a reaction.

In the SI unit system, only mass is defined in kilograms (1 Kg) of the base unit, and 1/1000 of this is expressed in grams (g). The basic units are the meter (1 m) for length, pascal (1 Pa) for pressure, and Joule (J) and wattage (W) for energy. In decimal notation, 1 km = 1000 m is used. Whatever the unit, decimal system unit prefixes are used, as detailed in Table 3.3.

Table 3.3 SI unit prefixes (power 3 multiples).

Digit	Prefix	Symbol	Digit	Prefix	Symbol
10^{-6}	micro	μ	10^{6}	Mega	M
10^{-9}	nano	n	10^{9}	Giga	G
10^{-12}	pico	P	10^{12}	Tera	T
10^{-15}	femto	f	10^{15}	Peta	P
10^{-18}	atto	a	10^{18}	Exa	E
10^{-21}	zepto	z	10^{21}	Zetta	Z
10^{-24}	yocto	y	10^{24}	Yotta	Y

3.4.2 Essential elements and trace elements

An element that must be ingested by an organism in order to maintain the organism is called an essential element. In addition to the four main elements of hydrogen (H), carbon (C), nitrogen (N) and oxygen (O) as bio-constituents, the eight elements are considered as essential minerals in nutrition: calcium (Ca), potassium (K), sodium (Na), magnesium (Mg), iron (Fe), phosphate (P), sulfur (S) and chlorine (Cl). Examples of trace essential elements, which are essential in trace amounts, include manganese (Mn), cobalt (Co), nickel (Ni), chromium (Cr), molybdenum (Mo), selenium (Se), copper (Cu), zinc (Zn), aluminium (Al), silicon (Si), vanadium (V), boron (B), fluorine (F), iodine (I) and arsenic (As) (see Table 3.4).

The function of essential elements is concentration dependent. An inadequate intake may cause deficiency, making health maintenance difficult, whilst excessive intake may cause poisoning symptoms and health impairment. A reasonable dose range may be useful for a drug or health supplement. The body's daily balance of ingestion, excretion and accumulation of elements from and to the environment is a delicate balance. If an element is clearly deficient, the element is given as a drug, and if it is excessive, treatment such as controlled ingestion is performed. Natural foods/supplements, which are promoted as being healthy, are difficult to evaluate in terms of the metabolism/accumulation of the individual as a starting point, and it is not easy to provide them without excessive or deficient metabolism. If the product is of a level that is used as a health food within the limits of a range that does not have a biological effect, it may not be very effective.

It is important to adopt an extremely low concentration of trace pollutants as a standard value for water quality in order to flatten the environmental value on which intake is based. In any event, the maintenance of trace element concentrations by water quality control is generally limited to over control, which is a possible cause of disease.

Clark numbers, estimated by the American Frank W. Clarke (Clarke & Washington, 1924) for the presence of global elements in the lithosphere (rock sphere), the hydrosphere up to 10 miles below sea level (≑16 km) and the aerosphere are helpful when considering the appearance of metallic contamination and trace element problems. The mass of elements in the lithosphere accounts for 93.06% of the total, but there is also 6.91% in the hydrosphere and 0.003% in the atmosphere. The amount (concentration) actually present in a water system is related to the amount of water solubility of each element (compound), but it is also useful to know the absolute amount of water in the system as basic knowledge. For further understanding, we can also divide elements into those that are used at high concentrations as resources, those that constitute part of the environment, and those elements that are valuable for functional materials, even if their abundance is small and can sometimes become poisonous (Table 3.5).

Table 3.4 Elements for organism maintenance.

Main Elements	Essential Elements	Essential Trace Elements	
Hydrogen H	Calcium Ca	Boron B	Nickel Ni
Carbon C	Potassium K	Fluorine F	Copper Cu
Nitrogen N	Sodium Na	Aluminum Al	Zinc Zn
Oxygen O	Magnesium Mg	Silicon Si	Arsenic As
	Iron Fe	Vanadium V	Selenium Se
	Phosphate P	Chromium Cr	Molybdenum Mo
	Sulfur S	Manganese Mn	Iodine I
	Chlorine Cl	Cobalt Co	

Table 3.5 Clarke numbers.

Rank	Element	Clarke Number(%)		Rank	Element	Clarke Number(%)	
1	Oxygen	49.5	Environmental element	26	Tungsten	0.006	
2	Silicon	25.8		27	lithium	0.006	
3	Aluminum	7.56		28	Cerium	0.0045	
4	Iron	4.7		29	Cobalt	0.004	
5	Calcium	3.39		30	Tin	0.004	
6	Sodium	2.63		31	Zinc	0.004	
7	Potassium	2.4		32	Yttrium	0.003	Rare element
8	Magnesium	1.93		33	Neodymium	0.0022	
9	Hydrogen	0.87		34	Niobium	0.002	
10	Titanium	0.46	Resource element	35	Lanthanum	0.0018	
11	Chlorine	0.19		36	Lead	0.0015	
12	Manganese	0.09		37	Molybdenum	0.0013	
13	Rin	0.08		38	Thorium	0.0012	
14	Carbon	0.08		39	Potassium	0.001	
15	Sulfur	0.06		40	Tantalum	0.001	
16	Nitrogen	0.03		41	Boron	0.001	
17	Fluorine	0.03		42	Cesium	0.0007	
18	Rubidium	0.03		43	Germanium	0.0007	
19	Barium	0.023		44	Samarium	0.0006	
20	Zirconium	0.02		45	Gadolinium	0.0006	
21	Chromium	0.02		46	Bromine	0.0006	
22	Strontium	0.02		47	Beryllium	0.0006	
23	Vanadium	0.015		48	Praseodymium	0.0005	
24	Nickel	0.01		49	Arsenic	0.0005	
25	Copper	0.01		50	Scandium	0.0005	

An element with an abundance of 1% or more is a principal element that constitutes part of the environment, and people are always living in an abundance of that element. It can be said that it is an environmental (basic) element. Elements less than 1% have historically been used as essential resources for civilization, after concentration as mineral resources. Tiny amounts of rare elements, as small as 0.002 to 0.0005%, are attracting attention as resources required for advanced technology in recent years, and are used as a featured resource for new materials using advanced concentrating and refining technologies.

Consider zinc as an example of a resource element. It has been used for over 2000 years as a raw material for brass, and in recent years it has been widely used in zinc-pulverized iron plate and as a battery and pigment. If the amount of zinc consumed from water or food exceeds a certain level, even if it is an essential trace element of the human body, it may cause health problems. The mean required daily dose for a Japanese person is said to be 8 mg for males and 6 mg for females. Concentrations exceeding an upper limit of 30 mg are not recommended. Insufficiency may cause symptoms of deficiency disorders, such as impaired taste, decreased spermatogenesis, amenorrhea, decreased immune function and delayed wound healing. This is an important element for enzymatic activity for the maintenance of the living body. However, it is known that high concentrations of these elements cause skin irritation, paralysis of the body, respiratory disorders and other diseases when ingestion exceeds 100 mg. Zinc damage is often caused by the inclusion of impurities such as lead and cadmium. Since water turbidity phenomena occur when concentration exceeds 3 mg/l, the standard value of drinking water is set at 1.0 mg/l or less for the control of water turbidity, and health impairment is not a basic value for the formulation of the standard. It is considered that the water quality standard is set at a level that does not cause coloration damage by iron and manganese. Although 0.3 mg/l of iron and 0.05 to 0.1 mg/l of manganese are considered to be an acceptable water quality to prevent coloration damage, the upper limit of intake as an essential element is assumed to be 50 mg/day (40 mg/day for women) and 11 mg/day, respectively, and much of it is consumed in food. Toxicity occurs at unusually higher concentration levels in natural environments and there are reports of neuropathy due to long-term ingestion of manganese due to industrial pollution.

Arsenic is a rare element that has similar physicochemical properties to phosphorus and is often treated as a poison, which makes it easier to treat it together with phosphorus as a bioresource element. Arsenic alone is not generally considered to be useful in living organisms (though it has been reported that arsenic is useful in very small amounts as a non-toxic organic compound), and it is treated as a poison suspected to be carcinogenic even in small amounts. The first reported toxicity of arsenic was in the case of black foot disease, which originated from Taiwanese well water reported in 1964, followed by soil at the Toroku Mine (Kyushu island, Japan), contamination poisoning with Morinaga milk, and widespread well water contamination in Bangladesh in recent years. A number of serious health disorders have been reported, beginning with skin coloration, leading to the development of bladder, lung, and skin cancer, and it is considered to be one of the typical cumulative toxicants. It has long been known as an acute/chronic poison, and used as an assassination poison and as 'Iwami Silver Mine' rodent removal (killing) agent.

Figure 3.25 summarizes the relationship between the utility of pollutants in water and the pattern and control of hazards. Refer to the WHO (2008) *Guidelines for Drinking Water Quality* for methods of occurrence and control of hazards to individual substances.

Substances with various concentration levels and characteristics are handled in environmental engineering and water treatment. When displaying water quality, it is important how many decimal places should be used and what value should be used for the unit. For example, dividing the amount of reagent used for titration in a BOD test yields a number of digits. However, in general, the third decimal place has little meaning. Rather than say that the BOD $\doteqdot$ 7.575 mg/l, it is better to write BOD $\doteqdot$ 7.58 mg/l etc. to indicate that the accuracy cannot be guaranteed at the third digit, or even to write 7.6 mg/l. Recently, some automatic instruments have very high accuracy, so a high number of digits may be used depending on the purpose, but it is generally better to think of this as a special case.

3.4.3 Disinfection/sterilization

In the previous section, the authors explained the concentration-dependent effect of a chemical substance, which may be an obstacle to water use or indicate a poison, which may be harmless in the low concentration range or as a minimum trace amount. With regard to carcinogens and radiation, the difficulty of clearly determining hazard thresholds was explained in Section 2.2. Along with these various chemical and radioactive materials, there are microbial infections caused by bacteria and viruses, which are among the large classic groups of water quality factors that cause danger to humans. The pandemics of acute

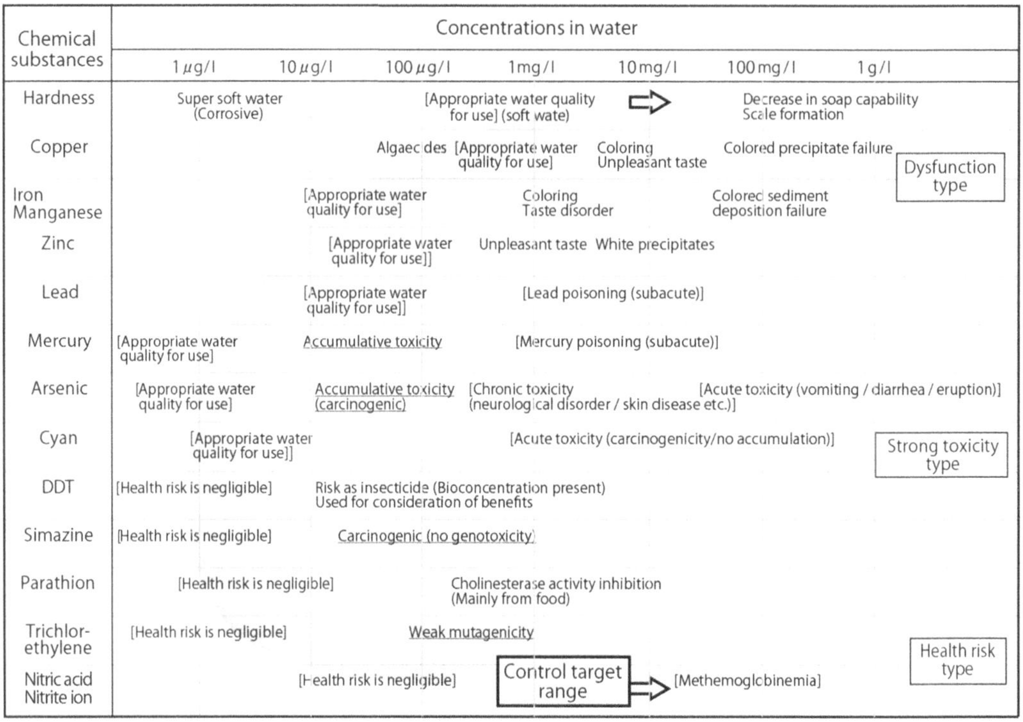

Figure 3.25 Examples of pollutants in water, occurrence patterns and control of hazards.

gastrointestinal infections caused by cholera, typhoid fever and other severe water-borne infections began on the Indian subcontinent with the expansion of the British Empire, and spread instantaneously to the West, the Orient, the Americas and Africa, and shook the world from the 17th to the 18th Century.

As outlined in Figure 3.6, a bacterium is a hydrophilic particle with a size of the order of 10^0 μm, and viruses are 10–100 nm (10^{-2} to 10^{-1} μm). The number of Escherichia coli bacteria in municipal sewage contaminated by humans and animals is of the order of 10^8 per liter, the amount of municipal rainwater runoff is of the order of 10^6 per liter, and the amount of river water used for raw water supply is of the order of 10^4 to 10^3 per liter. Microorganisms, such as bacteria, decrease due to the effects of the self-purification actions in rivers and lakes such as ultraviolet light, oligotrophic state and ingestion by other organisms. However, microbes grow rapidly in sewage at an appropriate temperature, resulting in a level of epidemiological health threat. The level demanded for drinking water is required to be 10^1 to 10^0/l or less when using human potable water.

If we approximate the size of a bacterium with a 1 μm diameter sphere and estimate the density to be as small as 1 g/cm^3, the weight of a single bacterium is about 0.52×10^{-12} g ($= 0.52 \times 10^{-9}$ mg). A table of correspondence between the number and mass concentrations of bacteria, using these assumptions, is given in Table 3.6.

Since the basic aim of water treatment is to remove undesirable impurities from a water system, a separation operation is used first. Sewage water, rainwater and river water containing high concentrations of bacteria are separated from the water system by solid-liquid separation processes such as aerobic/anaerobic microbial treatment, coagulation/sedimentation, and various filtration treatments,

Table 3.6 Number and mass concentrations of bacteria. (Assuming density 1 g/cm^3)

Number concentration (numbers/l)	10^8	10^7	10^6	10^5	10^4	10^3	10^2	10^1
Mass concentration	50	5	0.5	0.05	5	0.5	0.05	5
	μg/l				ng/l			pg/l

Assuming the size of bactera is 1 μm diameter sphere.

along with other components. As is discussed later, conventional water treatment processes based on geoscientific/ecological principles can achieve removal rates of approximately two orders of magnitude (2digit, 2 log) on a stand-alone basis. However, at higher concentrations of raw water and wastewater, there are bacteria etc. that remain unavoidable for removal. In addition, when the concentration is as low as 10^{-1} mg/l or less, conventional water treatment systems are not always effective. Therefore, removal is abandoned and treatment is carried out to kill microorganisms so that they cannot be toxic. Sterilization is performed at high temperatures until the bacteria and viruses are completely killed, just as sterilization of medical devices is performed at hospitals and other facilities.

In water treatment, high concentrations of microorganisms are handled in serial processes depending upon conditions. First, the separation process/system attempts to reduce the concentration of microorganisms by removal, thereby reducing the concentration of microorganisms by two to four orders of magnitude. After that, operations are performed so that the remaining bacteria of 10^3 to 10^2 per liter or less are killed and reduced by two to three orders of magnitude, by adding a disinfectant such as chlorine/ozone or irradiating UV rays so that the number of active bacteria remaining is 10 or less per liter. Thus, killing microorganisms and bacteria etc., to the extent that they cannot cause harm to the human body, is called disinfection or a disinfecting operation. Among the separation methods, ultra-filtration (UF) and nano-separation membranes (NF) treatment can almost completely remove bacteria and viruses etc., by solid-liquid separation.

As described above, disinfection is a typical operation of detoxification, in which bioactivity is controlled after solid-liquid separation of high-concentration bacteria in water and concentration reduction. Therefore, detoxified (killed) bacteria and other debris (which are referred to as carcasses) remain in the water. If the application of germicides, ultraviolet rays and radioactive radiation is forced in the presence of high concentrations of bacteria, secondary biological and environmental effects of the debris may occur. Occasionally, high concentrations of suspensions (viable bacteria or debris) may interfere with the bactericidal operation. We believe that any disinfection operation can only be carried out properly after removal of the debris to ensure low concentrations (Figure 3.26). The principles and operation of sterilization processes are not discussed in this book.

3.4.4 Detoxification treatment by neutralization and complex formation

As mentioned above, it is often possible to neutralize highly acidic/alkaline water and adjust it to a neutral water of pH 6 to 8 – the most common detoxification/conditioning process that does not involve removal processes. In addition, when iron or manganese is eluted in trace amounts from pipes or systems to cause coloration damage to water, phosphoric acid salts are added to form complexes of iron, manganese and phosphoric acid, thereby preventing the water from oxidizing metals and developing coloration. Alkalinity adjustment and metallic complexation, which have been widely used to control the water quality of boiler water systems, are examples of adjustment processes that do not involve removal processes.

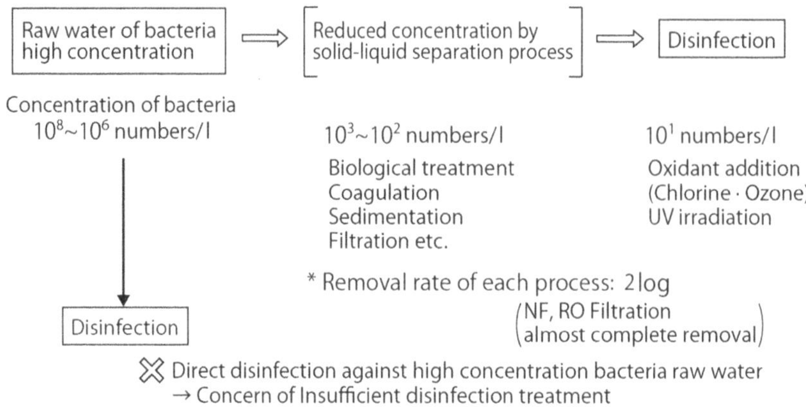

Figure 3.26 General principles of disinfection treatment.

3.5 EVALUATION OF TREATABILITY USING 'WATER QUALITY CONVERSION MATRICES'

3.5.1 Assessment of water quality conditions/treatability in water metabolic systems

In modern urbanized watersheds, water quality changes and flow rates throughout the water metabolizing systems, key information at the point of raw water intake and utilization spots, and the extent to which water quality is restored after utilization are all constantly correlated to keep the water quality and basin environment in a good state. Unfortunately, even today, the basics of building water systems by considering the use of water alongside the idea of water recovery has not been fully materialized, even in developed countries. Research and planning methods, such as assessing the potential for water quality recovery and building urban/regional water metabolism systems, are still not fully matured, even though the current author Tambo raised the issue in the mid-1970s and began research in this area nearly half a century ago. Both authors believe that excellent proposals for various advanced and traditional water technologies will only be effective if a basic agreement for water quality evaluation is comprehensively described and accepted.

One of the roughest ways to generalize water quality assessment of treatment is by classifying and treating water quality with emphasis on solid-liquid/gas separation characteristics, as shown in Figure 3.27 (Tambo, 1964). Professor J.J. Morgan of the California Institute of Technology, University of Florida, a colleague of Tambo in the United States, developed the figure into a detailed diagram in his book 'Aquatic Chemistry' (Stumm & Morgan, 1981), and similar diagrams now appear in many publications. As we have seen in this chapter, for common components involved in urban water metabolism, it is almost possible to judge the treatability (processability) of an impurity with a particle size of several micrometers to a few mm. If the molecular weight is small at 10^4 Da or less, the physicochemical/ chemical properties become increasingly significant. At the simplest molecular level, the chemical properties themselves become the dominant factor for treatability evaluation.

The simplest combination for assessing water quality using impurity size and chemical factors can be seen in Figure 3.28. This primitive but common method divides the chemical properties of water impurities into inorganic and organic substances, and size properties are divided into two groups such as suspensions and soluble matters, which together make up a four element matrix.

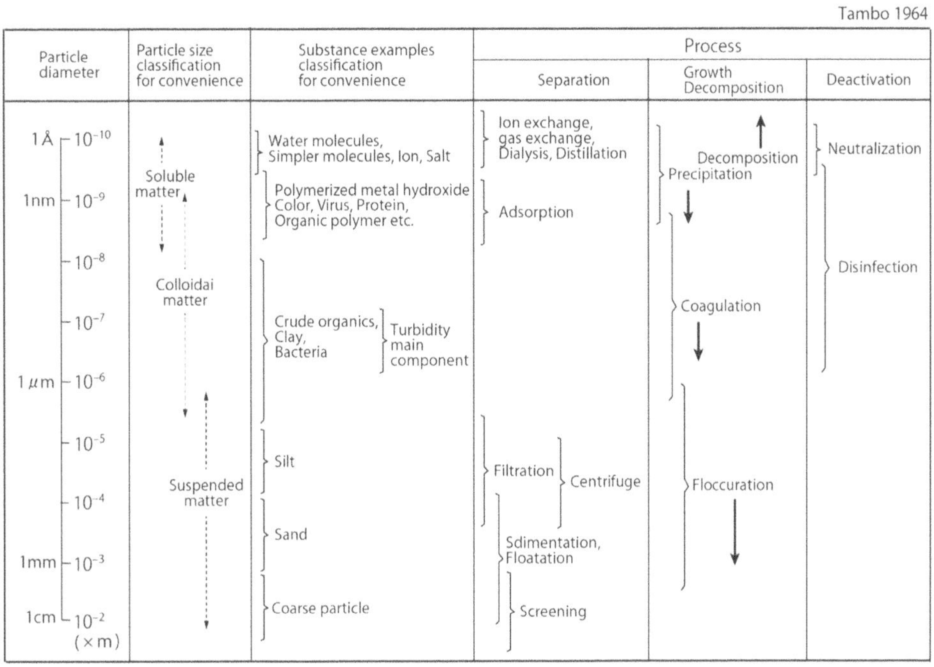

Figure 3.27 A process selection chart, by impurity size.

Chemical nature \ Impurity size	Suspended matters	Soluble matters
Inorganic matters	○	△
Organic matters	○	□

(Marks mean treatability etc.)

Figure 3.28 An example of component classification.

3.5.2 Construction of a matrix for water quality labeling

The following series of 'Water quality matrices' and 'Water quality conversion matrices' were devised to express the behavior of water quality components of an assumed artificial/natural system by a matrix with the minimum number of elements.

The columns of matrices are sorted by appropriate methods (mechanical/optical analyses, gel chromatography, etc.) into impurity size. The width of this class division for both types of matrices (quality and conversion matrices) are set to be equal taking into account the characteristic features of the size of impurities in water in the environment and the removal limit (cutoff level) of the water quality conversion operation, which will be described later.

In selecting the water quality factor groups that make up each row, we will:

(1) focus on water quality factors that are as comprehensive as possible;

(2) focus on ensuring clear material balances, and
(3) limit the number of water quality factors to a minimum number so that the number of matrix elements is the minimum that can be practically used.

Figure 3.29 is a standard matrix used by Tambo since the 1970s, and has been considered the minimum basic number of element divisions. The amount of material in each element (i rows and j columns) is expressed as the basic concentration $C_{ij,}$ or mass M_{ij} of the test method for each element.

Generally, increasing the number of water quality factors (the number of water quality test items) increases the times and costs of measurements and operation significantly. In addition, an increase in the number of operation factors causes significant difficulties in operating the system by clarifying the material and energy balance. Therefore, the basic operation of the system will focus on the more common main factors. Various special or trace factors can be operated by setting up a sub-matrix as required. The matrix/submatrix configuration used when assembling a treatment process into a system is a summary of the general configuration order of processing into a system from a high concentration component and from a large impurity component to smaller one. In many cases, we believe that both material and information systems are consistently satisfied.

3.5.3 Two-dimensional treatability evaluation by TOC/E260 ratio and molecular weight fraction

A water quality conversion matrix indicates how much the water quality element C_{ij} of a water quality display matrix (as described above) can be removed by the water quality conversion processes being considered, and the matrix element is displayed by the removal rate R_{ij}. Inorganic matter can be represented easily for individual components (or groups of components), as shown in Figure 3.29. The use of general organic indices such as BOD or COD has historically been used in the Streeter/Phelps/Eckenfelder system to

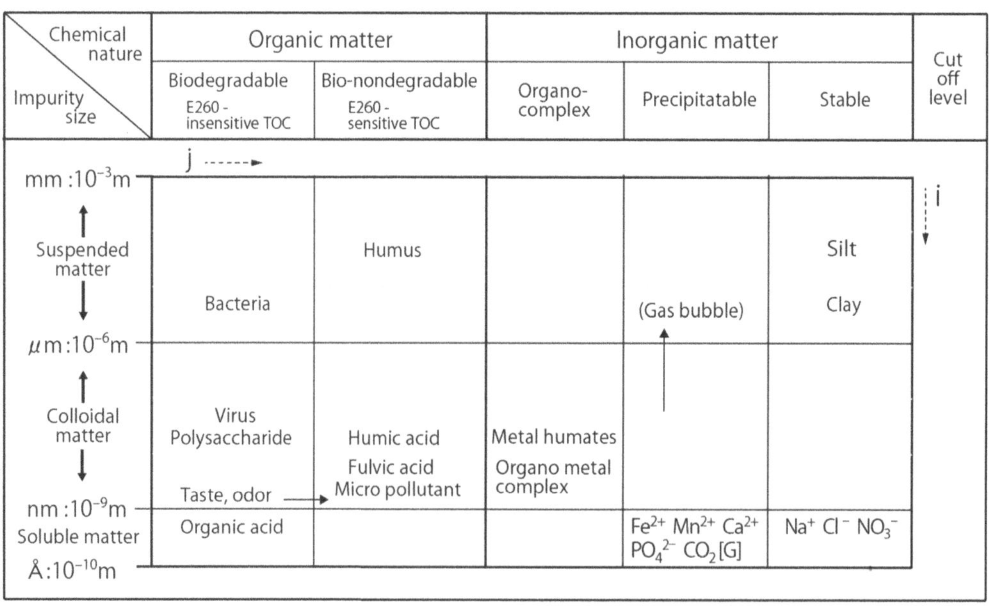

Figure 3.29 A standard water quality matrix.

describe the behavior of organic matter groups in urban water systems, but it was too comprehensive to evaluate the component behavior adequately. In addition, the method of analyzing and integrating individual organic matter or Groups of organic matter takes too much time and work for individual analysis, and the amount of organic matter identified by conventional analytical methods for sewage is only 50–70% or less of the total amount, making it difficult to obtain an accurate material balance of the system as a whole.

Therefore, Tambo and Kamei (1975,1976, 1978, 1980, 1989) and others proposed to use total organic carbon content (TOC mg/l) as the main quantitative index, and to evaluate the characteristics of organic matter by using ultraviolet absorbance at 260 nm and 220 nm as a sub-index, and to express matrix elements representing the behavior of organic matter by adding molecular weight (size) distributions by gel chromatography.

Figure 3.30 shows chromatograms of Sephadex G15 gels of raw sewage, coagulated and settled sewage, treated sewage by the activated sludge method, and additional activated carbon absorption effluent. The authors performed molecular weight fractionation chromatography in the early 1970s. The satisfactory material balance of the sample was obtained by two-stage extrusion of distilled water and 0.1N-NH_4OH. Sephadex G15 gels require two-step extrusion to have adsorption properties on some parts of organic matter. A sample 10 ml of 20-fold thickening sewage is placed on top of the gel column, then first inflow water (H_2O) and then NH_4OH are added to force the thickening sample into the gel column. Portions of polymer with a molecular weight of 1500 Da or more pass through the intragap between the gel particles in the column and flow out of the lower end of the gel column first (equivalent to the volume of voids filled with gel particles in a void volume). Organic matter of small molecular weight diffuses into the internal pores of the packed gel particles while flowing down the gel column, and returns to the extruded solution by re-diffusion. The flow out of the various components from the gel column is delayed: the smaller the molecular weight, the later it flows out of the column. Figure 3.30

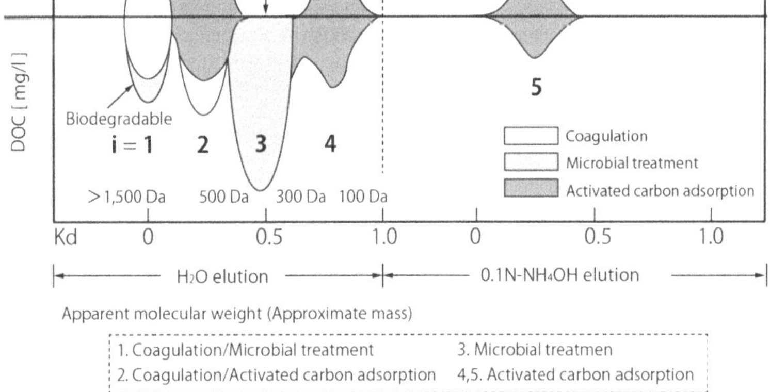

Figure 3.30 Removability by the activated sludge method.

shows the apparent molecular weight and size of components flowing out of the column. When water treatment systems are constructed, coagulation, separation/aerobic microbial decomposition and finally activated carbon adsorption are carried out in accordance with the principle that the sequence of the constituent processes is taken from the order of components having the largest molecular weight, and gel chromatograms before and after the respective treatment processes are obtained, which are synthesized and shown in the drawing. The chromatograms are displayed with the DOCs (dissolved organic carbons) of each image group on the lower side of the vertical axis with the UV absorbance at 260 nm: $E260_{(1\ cm\text{-}cell)}$. One fraction of the chromatogram is 10 ml, and the apparent molecular weight of the organic matter is estimated by examining the apparent molecular weight corresponding to the number of the fraction to be extruded in advance using a standard for which the molecular weight of polysaccharide etc. is known. The figure shows the size of the organic impurity, taking the apparent molecular weight Da as the horizontal axis. $K_d = 0$ means a large molecular component that does not enter the gel pore, and $K_d = 1$ means a small component that can diffuse inflow into the gel pore.

General organic matter in sewage is roughly classified into five visual groups: i = (1), » 1500 Da, i = (2) 1000 to 500 Da, i = (3) 500 to 300 Da, i = (4) around 150 Da, and i = (5) which is absorbed to the gel; i = 1, 2, 4, and 5 show absorbance at 260 nm, but i = 3 dos not show adsorption at 260 nm. DOCs that do not absorb at 260 nm in ultraviolet (E260 insensitive) are biodegradable organic matter that correspond to BOD substances. In contrast, DOCs (E260-sensitive) which exhibit ultraviolet absorption in Groups 1, 2, 4 and 5, are bio-nondegradable components. One group of 1500 Da or more macromolecules can be treated with coagulation, whereas the two groups of smaller components (around 1000 to 500 Da) do not provide sufficient coagulation treatment removal. In the smaller groups (Groups 4 and 5), components are gradually removed by activated carbon adsorption. Biologically persistent components are substances described as 'COD_{cr} – BOD' in conventional water quality notation. Biological components such as Group 3 (E260-insensitive DOCs) are not eligible for coagulation or adsorption (Table 3.7).

Taking the ratio of ultraviolet absorbance E260 (1 cm) to TOC mg/l as (TOC/E260), peaty groundwater without biodegradability shows values of 30 to 60. For sewage, this value is as high as 500, and decomposition proceeds by biological treatment until it reaches a value of about 50. Tambo first revealed this fact at the American Society of Chemistry Annual Conference Humus Symposium in 1987. In the 1990s, US professors Edzwald and Van Benschoten proposed E253(1 m)/TOC ratio as specific UV absorbance (SUVA) which is the reverse of the author's (TOC/E260) index (Edzwald & Van

Table 3.7 The fractionation of general organic matter in sewage and its characteristics.

Fractionation Classification	260 nm-Ultraviolet Absorption Characteristic	Treatment
Group 1 »1500 Da	E260 Absorbance -sensitive [Bio-nondegradable (COD_{cr}–BOD)]	Coagulation
Group 2 1000~c.500 Da		Coagulation Activated carbon adsorption
Group 4 c.150 Da		Activated carbon adsorption
Group 5 Gel adsorption matters		
Group 3 500~c.300 Da	E260 Absorbance -insensitive [Biodegradable (BOD)]	Biological treatment

Benschoten, 1997), and it was then widely used as an indicator of the presence of biologically persistent organics in natural waters in USA. Beginning in the 1970s, Tambo and others revealed that two-dimensional matrices using molecular weight distributions and TOC/E260 clearly demonstrate biological processability for natural organic matter, and Kamei obtained his Doctor of Engineering (1975) from Hokkaido University, as a result of work on this study.

The basic study used a 260 nm spectrophotometer, which gives the highest absorbance of humic substances without interfering with nitric and/or nitrous compounds, but subsequently an inexpensive 253 nm fixed wavelength absorber is commonly used in practice. As will be described later, using the TOC (mg/l)/E260 (1 cm) ratio used by the authors, the large numerical difference between biodegradable components with values of e.g. 30 to 50 in peaty groundwater and those as large as 500 in sewage water and biologically persistent components (such as humic substances) can be clearly understood. By contrast, SUVA = E260(1 m)/TOC (mg/l) has the disadvantage that the characteristic expression of organic matter becomes straightforward in terms of the absorbance of TOC units. However, based on unusual values such as the absorbance of a 1 m-cell, it has the disadvantage that it can only express humic substances in values of 5 or 6, and non-humic substances in 3 or less. Matrix-based evaluations using ultraviolet absorbance E260 and TOC mg/l ratios are described below.

In natural and urban water systems, humic substances, carbohydrates and other general organic components are the predominant ingredients. In order to clearly describe the behavior and treatability of general organic components, which are the quantitative main components of natural organic matter, we performed molecular weight fractionation by gel chromatography and examined treatability by dividing them into two components, with and without E260 absorption. However, the types of organic matter in water are enormous and present in very small quantities regardless of whether synthetic or natural. In order to identify trace chemical substances of pesticides etc., a mass spectrometer was used as a detector for liquid/gas chromatography. These trace organic substances are hidden behind the behavior of general organic substances, which are quantitative main components. Trace toxic metals also usually behave in the presence of general organic matter (Figure 3.31).

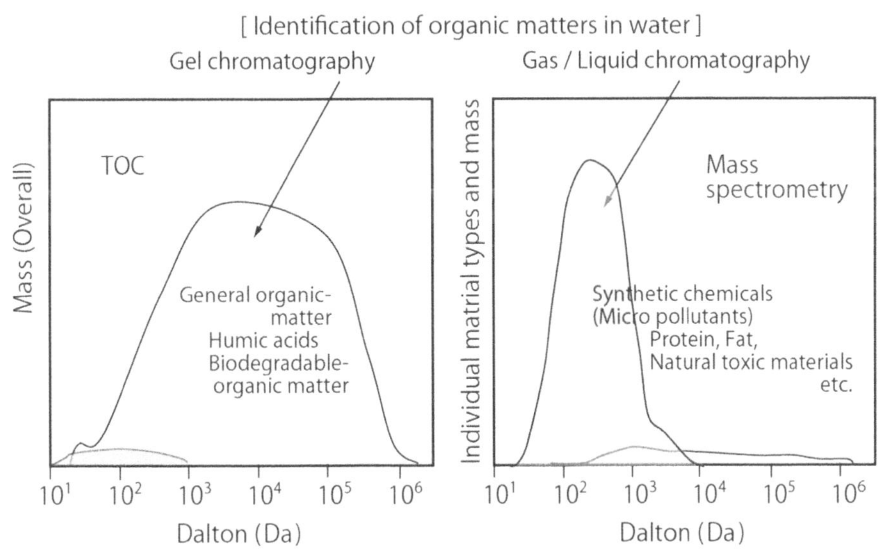

Figure 3.31 Various chromatographic results for the identification of organic matter in water.

3.5.4 Water quality conversion matrix

Figure 3.32 shows a two-dimensional water quality conversion matrix that describes the relationship between the presence of water impurities and the corresponding treatment processes for conventional water treatment systems (that simulate ecological/geochemical transformation processes). The conventional water treatment systems of microbial, coagulation, activated carbon adsorption and ion exchange treatments provide an overview of the types of water treatment systems that can cope with the removal of impurity components.

Figure 3.33 shows a matrix depicting the relationship between the nature (presence) of impurities in water and the types of membranes that are effective for separation, for functional membrane treatment that mimics physiological/biological separation mechanisms. Since the function of membrane separation is basically expressed in terms of separation limit size (molecular weight), one size (dimensional) representation of the treatability on size alone may be sufficient. Of course, secondary chemistries, such as affinity among membranes, water and impurities, affect their functionality slightly.

In a water quality conversion matrix, elements are cut out of the impurity size in the vertical axis. However, when water treatment systems are assembled, it is generally considered that arranging the processes so that removal proceeds from high concentration components and/or coarse components first reduces the energy consumption required. Therefore, the range of the target water quality elements is narrowed as the treatment stage progresses and, in reality, the matrix element division in the concentration axis direction is not very large. In the drawing, the size of particles moves in increments from 1 Å (10^{-10} m) near the size of water molecules to 1 mm (10^{-3} m), which is the size of sand/organic matter particles, which are considered to be the coarsest suspended matter in water.

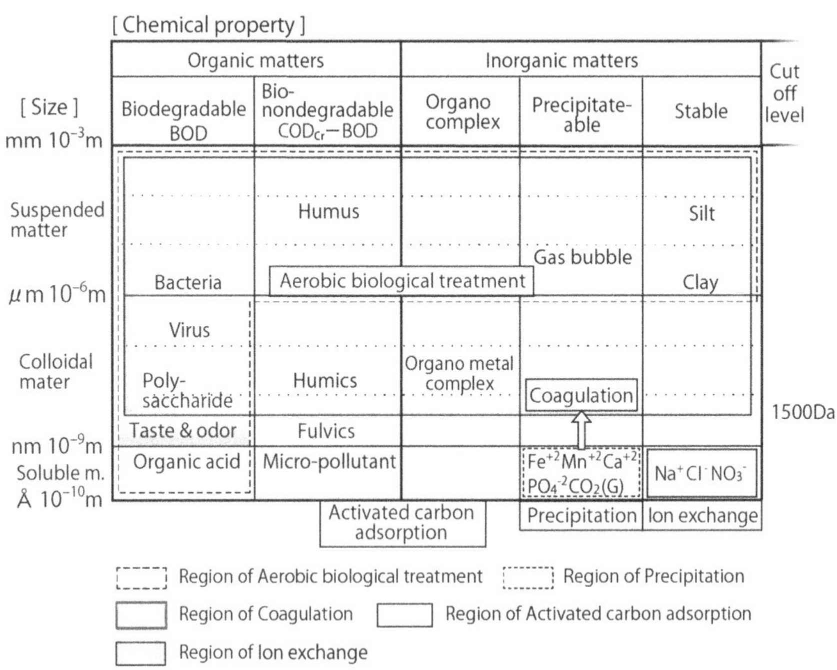

Figure 3.32 A water quality conversion matrix.

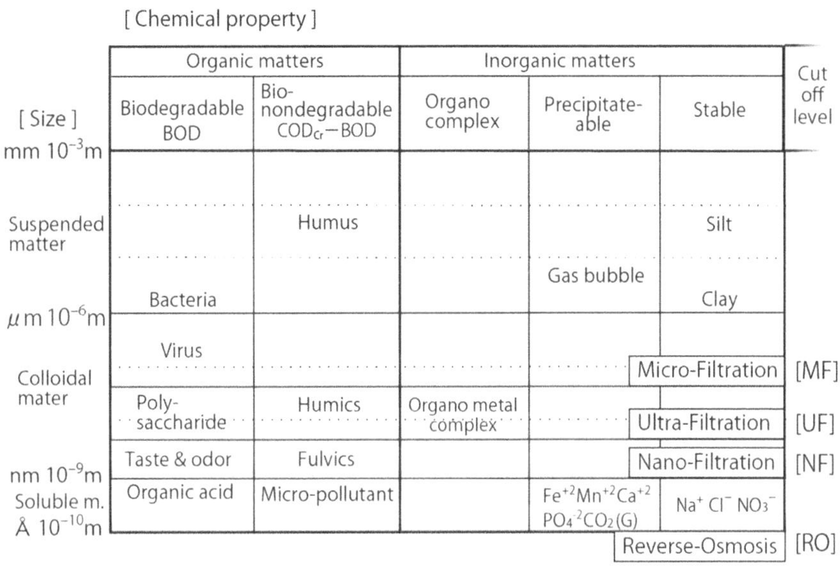

Figure 3.33 A water quality conversion matrix of functional membranes.

Classification of chemical properties of impurities is obviously divided into organic matter and inorganic substances, and the correspondence with historically applied water treatment/quality control indices is practically important when dividing these substances into organic and inorganic substances. For this reason, we take into account the quality conversion process, which is somewhat different from the classification of chemical substances. In current quality control of urban/regional water systems, where water environmental management is carried out under the OD (Oxygen deficit) concept and treatment of organic matter is carried out biochemically, assuming that the shortage of dissolved oxygen is the biggest factor impairing the healthy state of water systems, the relationship with organic matter indicators such as BOD and COD is also crucial.

When organic matter is classified using BOD and COD as indices, biodegradable organic matter is expressed as BOD, and the other bio-persistent organic matter is expressed as (COD − BOD) obtained by subtracting BOD, which is an index of biodegradable organic matter, from COD (chemical oxygen demand). The COD used here is the COD that represents the oxygen equivalent O_2 required to oxidize the total amount of organic matter. Therefore, it is necessary to ensure that the total amount of organic matter is oxidized and mineralized. This COD must be a di-chrome acid COD (COD_{cr}), which can be regarded as an internationally established rule.

For soluble inorganic components, it has become possible to analyze the abundance of individual elements and compounds simultaneously and control their behavior to a considerable extent. However, even if volatile organic substances having various components are analyzed/controlled by considering their behavior individually, it is extremely expensive, precise and requires enormous work, and is not easy on a daily basis, except when specific stable compounds (e.g., carcinogens) are individually targeted, even in view of recent advances in analytical techniques.

Therefore, in order to quantitatively evaluate the behavior of organic matter main components contained in water in natural environments and urban water metabolism systems, we have devised a new index to replace COD and BOD which were used in the past. It should provide a clearer picture of the material balance and be an indicator of the selection and operation of different water quality conversion processes. As a measure, the size of impurities was fractionated by gel chromatography or membrane classification, and total carbon concentration (TOC mg/l) was used as a main index to more directly express the amount of organic matter. The properties of the components were separated into biodegradable and bio-nondegradable properties to be evaluated/identified, and matrices were used to comprehensively control the organic components in the water cycles of natural and urban/regional systems. However, many types of biologically persistent and volatile organic substances (trace organic components) that are problematic in terms of carcinogenicity and toxicity must be individually identified and treated in the same manner as inorganic harmful substances. The development of liquid/gas chromatography mass spectrometry has enabled this.

In order to distinguish between biodegradable and bio-nondegradable organic substances (TOCs), we considered the treatment of organic substances in natural systems as pseudo-two-component systems by dividing soluble organic carbons (dissolved organic carbons; DOCs) into 'components that do not exhibit ultraviolet absorbance (E260 insensitive) and components that exhibit ultraviolet absorbance (E260 sensitive)' as described in Section 3.5.3, Figure 3.30. In addition, we have double bond/triple bond etc. in the body of the component, and unsaturated compounds exhibiting an ultraviolet absorbance of 260 nm (E260) would be a typical biologically persistent organic matter. In Tambo's original papers (Tambo & Kamei, 1975, 1978 etc.), absorbance at E260 nm was used as an index but, in general, absorbance at 253 nm (E253) using an inexpensive single-wavelength absorbance meter has been used by many researchers worldwide for economy. Wavelengths in this area are the wavelength range in which the absorption of organic matter can be measured most sensitively by minimizing the interference of the nitric compounds system. The behavior of the nitric compound system can be grasped easily by simultaneously measuring ultraviolet 220 nm absorbance (E220). Since the separation operation of the suspension system is generally easy, and ultraviolet absorbance can be measured only for the dissolving system, the TOC measurement is carried out using the DOC (soluble organic carbon) of the 0.45 μm-pass filtration sample when the abundance and behavior of the general organic matter are displayed to evaluate treatability.

The first group of components in such evaluations is a group of biodegradable organic substances such as carbohydrates, aliphatic organic acid, etc., which are quantitated as organic carbon (TOC) that does not exhibit E260 absorbance. Conventional indicators almost correspond to BOD components. The second group of ingredients is biologically persistent organic matter, which is quantitatively dominated by humic substances in a clear natural water system. Most of the major components of organic matter in natural and urban water metabolism systems are biodegradable organic matter in sewage and other wastewater, and biologically persistent components such as humic substances, which are decomposition components. Biologically persistent organic components are mostly organic matter that flow through soil systems into rivers as the end products of metabolism in which organisms are decomposed and organic matter that is contained in the effluent stabilized through sewage treatment. These organic matter are also defined as biologically persistent natural organic matter groups (natural organic matters, NOM). A group of organic substances with a heterogeneous and complex component, also referred to as 'coloured substances' are a brown group of organic substances. These are often referred to collectively as humic species, which have molecular masses ranging from 200 to 20,000 Da, with mean compositions of less than 50% carbon, 40% oxygen, and 5% (or less) hydrogen-and nitrogen (Figure 3.34).

Figure 3.35 compares the infrared absorption spectra of microbial metabolic wastes generated and accumulated by repeated microbial degradation by repeatedly adding glucose (sugar) to a water tank seeded with a small amount of aerobic microorganisms, and humus in natural coloured water from peatland in Hokkaido, Japan. It can be seen that the substance in the glucose metabolizing water and the

Major organic matter groups in water systems

[The first group] Biodegradable matter (E260-Insensitive TOC)

Carbohydrates, Aliphatic organic acids etc.
BOD matters

[The second group] Bio-nondegradable matter (E260-Sensitive TOC)

humus etc.

- Origin of natural systems → Metabolic end products of organisms
- Origin of sewage systems → Metabolic end products of sewage organic matter treatment

These organic matter are also defined NOM (Natural organic matter)
Coloring water (Dark brown, 200~20,000Da)
Trihalomethanes formation precursor

Figure 3.34 Organic matter groups in natural and urban water systems.

substance in the peaty-coloured water are almost the same. In such aerobic microbial degradation, about 3 to 5% of biodegradable organic matter in inflow raw wastewater is produced as biologically persistent TOCs (NOMs) of metabolic wastes. Therefore, even if the aerobic sewage treatment plant is adequately treated, 3 to 5% of the TOC of the raw wastewater remains in the effluent liquid system in the form of persistent humus as the metabolic wastes, and the removal of standard TOCs is limited to 95–97%. The amount of persistent effluent mass produced during anaerobic microbial decomposition (fermentation) seems to be as low as 0.5%, one order of magnitude lower (Tambo *et al.*, 1981) (Table 3.8). These components exhibit good absorbance at 260 nm ultraviolet (E260-sensitive) and can be characterized as components with an ultraviolet absorbance ratio of 30 to 50 (DOC/E260) measured in a 1 cm quartz cell to TOC mg/l. It appears that high-molecular

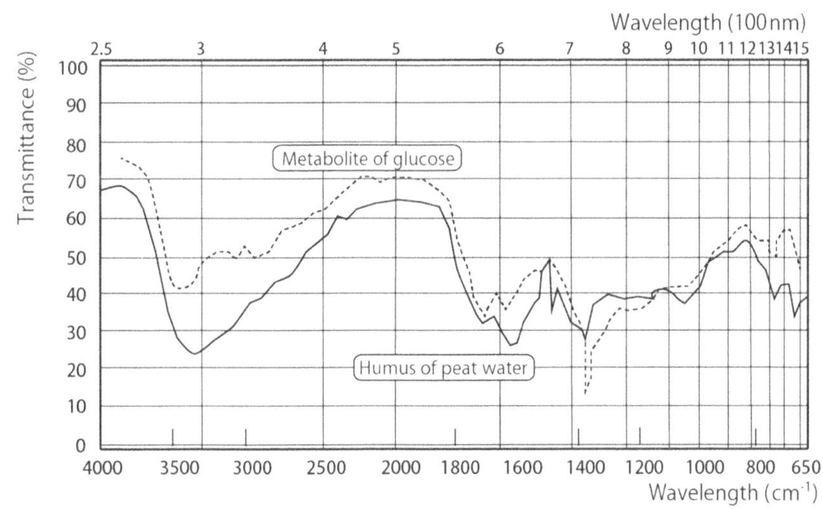

Figure 3.35 The infrared spectra of peatland water organic matter and glucose aerobic degradation metabolite.

Table 3.8 Rate of production of microbial decomposition (TOC).

	Production Rate of Metabolic Waste Matter TOC to Biodegradable TOC
Aeroblic treatment	3~5%
Anaerobic treatment	About 0.5%

*TOC removal ratio of aerobic treatment is 95 to 97% upper limit

components (e.g., humic acids) with molecular masses of >1500 Da show DOC/E260 ≑ 30 and low-molecular components (e.g., fulvic acids) with <1000 Da show DOC/E260 ≑ 50.

Take an organic wastewater with an index of (DOC/E260) = 500. Tracking the larger white arrow in Figure 3.36 shows that 0.1 = 10% of the total organic matter as DOC mg/l is E260 sensitive biologically persistent organics (i.e. humic acid). The remaining 90% of organic matter is therefore a component that can be further degraded by aerobic microorganisms. Inflow sewage is close to this value. Conversely, if (DOC/E260) is observed to be close to 30 to 50, this means that there is no further improvement in water quality, even if further microbial treatment or river self-purification is expected, since it only contains biologically persistent humic substances.

In this way, humic substances making up several percent of the major organic component in wastewater (10 to 100 mg/l), which is treated as a BOD component in urban/regional water metabolizing systems, become the second main component of biologically persistent organic matter. The second organic component is present in clear water (effluent of forest lands/wetlands/lake water) subjected to microbial

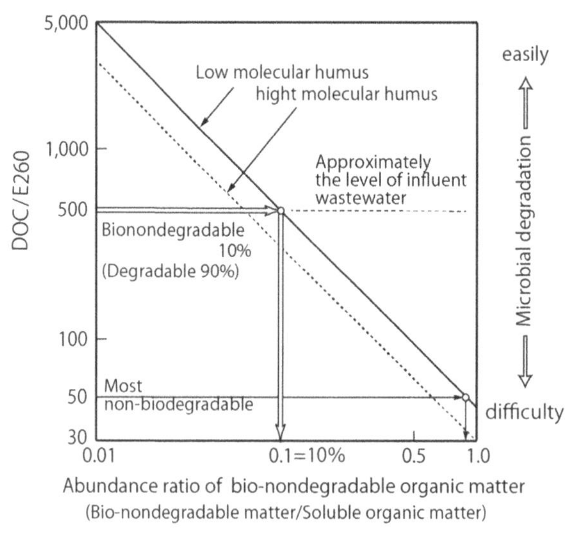

Figure 3.36 shows the proportions of biodegradable and persistent components in a sample.

decomposition in the order of several mg/l to several tens μg/l, and reacts with chlorine added for disinfection in water purification systems to produce trihalomethanes such as chloroform. It is a possible cause ingredient that has become a pioneer in the incidence of trace organic contaminations. When considering the behavior and treatment of trace contaminants (volatile micro-organic pollutants, etc. of the order of μg/l to ng/l), such as flavors, smells, pesticides and other carcinogens, the presence/behavior of these humic substances should be considered as the base organic (main) ingredient that is constantly prevalent in natural clean water. Competition between the humic substances and trace contaminants for activated carbon adsorption is discussed below in Section 7.1.5, and it is a significant factor in regulating the adsorption of pesticides, etc.

In addition, the 'organic complex' shown in the third column of the water quality conversion matrix (see Figure 3.32 above) forms a strong complex of such humic substances and polyvalent metal ions, and behaves as if it were made up of humic substances itself in the natural system/treatment system. Iron and manganese in groundwater from peatland are stably present as organic complexes, and they cannot be precipitated (insolubilized) by aeration/chlorine oxidation and removed by sedimentation/filtration, unlike simple metal ions.

In Figure 3.30, the E260-sensitive DOCs and E260 values on the polymer side in Group 1 are distributed symmetrically on the horizontal axis with the ratio of DOC to E260 staying around 30. On the smaller molecule side of Groups 1 or 2, the ratio of DOC to E260 is 50. Portions removed by coagulation in Groups 1 and 2 and those removed by activated carbon adsorption in Groups 2 and 4 are both considered humic substances (NOMs) with sensitivity to ultraviolet absorbance E260. In contrast, DOCs fractionated in Group 3 are microbially readily degradable components which are E260-insensitive, and which are almost completely removed by aerobic microbial treatment. In Group 1, the DOC portion that exceeds 30 times the E260 value is a biodegradable logical polymer that can be estimated from Figure 3.36. It is a slowly decomposing polymer but can be degraded by long-term microbial treatment. Group 5 is removed by adsorption if molecular weight is not excessive.

Compared to natural components, the humic components that flow out from forests and other areas during normal times are mainly from Groups 2, 4, and 5, and the DOC/260 is 50. Waters in natural lakes and marshes with long surface retention time are also supplemented by polymer Group 1 humics (which is humic acid, to be thought of as polymerized humic substances in long-term storage). In peatland and wetland water, the main component is polymerized, and the value of DOC/E260 approaches 30. As rain water precipitation increases the overflow of wetlands/swamps and surface spillage from the A_1 stratum where surface humus has accumulated, polymer humic constituents increase even in river water, and the overall removal rate of TOCs by coagulation treatment improves. When municipal sewage is added, a large amount of biodegradable organic substances such as carbohydrate that do not express ultraviolet absorbance E260 are added to Group 1 and Group 3. The DOC/E260 values for all samples are 300 to 500 (Figure 3.37). Microbial treatment works effectively against such values of water. These findings and the operation of the typical process system for water treatment described in the previous section will now be discussed in connection with the use of the water conversion matrices shown in Figures 3.32 and 3.33 (above).

3.5.5 Details of the water quality conversion matrix operation

In conventional types of water treatment, aerobic microbial treatment separates those components surrounded by dotted lines in Figure 3.32 from a water system and draws impurities out as sludge into a solid system. Suspended components (diameters mm- μm) located in the first three rows of the matrix are physically or physico-chemically captured (by biosorption or bio-flocculation) by activated sludge

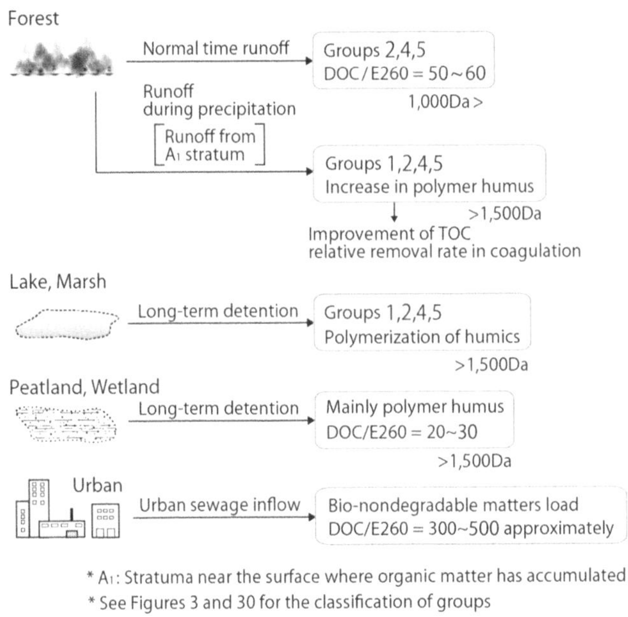

Figure 3.37 Humic matter discharge.

flocs or existing biomass on a biofilter bed, whether inorganic or organic, and solidified and separated. Among the biodegradable components in the first column, the dissolved components in rows 4 to 7 are also absorbed/adsorbed by the biomass to solidify them, and are also decomposed by biochemical reactions to grow the biomass into a solid system. By contrast, in ordinary sewage treatment plants where microbial treatment is carried out, of the biologically persistent components and inorganic components in columns 2 to 5, the components in rows 5 to 7 cannot be removed and are discharged into the effluent water of the sewage treatment plant, and removal of the components in row 4 is insufficient Table 3.9.

Table 3.9 Operation of a water quality conversion matrix (1). (See Figure 3.32 Water quality conversion matrix).

Applicable Range	Characteristic	Operation	Process
Enclosed by dotted line	—	Aerobic microbial treatment	Impurities are sludged and removed from the water system (Excluding some)
Columns 1–5: rows 1–3	Particle size mm-μm Inorganic, organic matter	Biosorption Bioflocculation	Solidification Separation
Column 1: rows 4–7	Soluble biodegradable matter	Bioabsorption Biooadsorption Growth	Degradation Solid-liquid separation
Columns 2–5: rows 5–7	Bio-nondegradable, Inorganic matter	Sewage treatment	Cannot be treated and discharged

There may be complaints that the river water quality is not good enough even if the sewage is processed and the treatment plant is in operation, and in extreme cases the sewage effluent may even be said to be polluting the river. The reason for this is that the removal of components below the colloidal components in columns 2 to 5, and in particular in rows 5 to 7, is often inadequate, so that residents may feel problems below the discharge point where a large amount of sewage is collected, treated and discharged.

Rapid filtration systems with coagulation treatment as a basic function have been the main form of water treatment to date. This process has been widely used around the world, despite it removing most of the substances larger than colloidal components in the area enclosed by a dotted line in Figure 3.32 (above). In most modern coagulation treatments, aluminium salts or iron salts are added to raw water and hydrolyzed to form positively charged colloids of aluminium/iron hydrate complexes that neutralize the usually negatively charged colloidal particles (clays, humus, bacteria, algae, parasitic eggs, etc.) in water to eliminate any electrical repulsion between colloidal impurity particles. After that, the particles collide with each other by agitation, crosslink-agglomerate to a floc of the order of millimeters, increase particle size, separate by sedimentation/filtration, and transfer to a solid state. Aluminium salts (such as aluminium sulfate) added to water hydrolyze to form positively charged colloidal particles such as $Al_8(OH)_{20}{}^{4+}$, and some of them undergo hydration to form neutral aluminium hydrate particles, such as $Al(OH)_3(H_2O)_3$, which work together to promote coagulation and flocculation.

The various negatively charged colloidal particles, which are removal targets, have a wide variety of size, including clay particles (micrometers), humic acid particles (10,000 to 1000 Da) and fluvic acid particles (500 to 100 Da). The high-charge hydrate complex of aluminium has a molecular mass of about 600 Da, and aluminium hydrate complexes are larger than low molecular weight fulvic acid. As a result, the aluminium coagulate must form an adsorption-type aggregate by forming a nucleus, so that a large amount of aluminium is required and operation is not efficiently carried out (Figure 3.38).

Low-molecular-weight humic constituents (NOMs) are not adoptable for coagulation, but are components that should be removed by activated carbon adsorption, as shown in Figure 3.30. In addition, trace contaminants with low molecular weights (micro-pollutants, volatile organic matter) are likewise activated carbon adsorbents, corresponding to two columns, with six to seven rows of matrices. Therefore, the authors think that the low-molecular-side limit of coagulation treatment will be of the order of 1500 Da, which is slightly larger than the 600 Da molecular weight of effectively acting aluminium polymers such as $Al_8OH_{20}{}^{4+}$. As for clay components, the presence of both charge neutralization (by a small amount of highly charged aluminium polymers) and a suitable amount of coagulated neutral aluminium hydrate as a cross-linker are required, and operations proceed easily to near neutral pH which satisfies these conditions. Higher-molecular weight humic substances (humic acid), having a smaller particle size and a larger specific surface area than clay with much more specific negative charges, require the presence of a large amount of positively charged aluminium polymer for charge neutralization. In addition, neutral aluminium hydrate coagulated as a cross-linking substance is

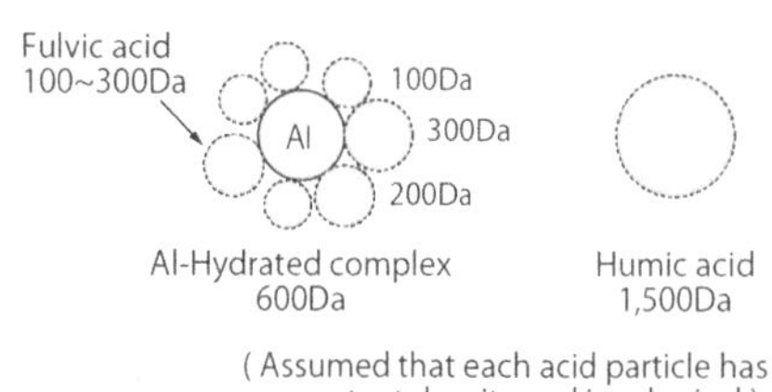

Figure 3.38 The coagulation pattern of fulvic acids and the size of humic acid.

required to maintain the agglomeration (flocculation) of the charge neutralized colloid at the same time. This condition is met in the weakly acidic range of pH 6.0 to 6.5. Coagulation requires appropriate pH control depending on the type of coagulant and the material under coagulation. For more information, see Section 4.1. Based on these findings, we have confirmed from a number of measured examples that the smallest size of particles that can be treated effectively with coagulation is about 1500 Da, which is slightly larger than the molecular weight (600–800 Da) of highly charged aluminium polymers as coagulant.

It is extremely useful for coagulation operation if highly charged aluminium polymers can be stably prepared in advance. Dr Shigeo Ban, Daimei Chemical Co. Japan, succeeded in stably constructing a mononuclear complex $[Al_2(OH)_n\ Cl_{6-m}]_m$ (where $1 \leqq n \leqq 5$, $m \leqq 10$) of poly-aluminium chloride (Pac) as early as the 1960s. This was one of the biggest inventions in the postwar period, not only for the Japanese but also for water purification engineers worldwide. Since he opened the patents to be used widely by anyone else, the use of poly-aluminium coagulants spread through Japan at once, replacing conventional aluminium sulphate for almost all water/wastewater treatment up to today. When the International Water Association (IWSA) World Conference took place in Kyoto, Japan in 1978, the author (Tambo: Vice Chairman of the IWSA Scientific and Technical Council at that time) introduced this Japanese technology and handed samples to Mr F. Fissangee (then Chairman of IWSA Scientific and Technical Council, and also working at Lyonnaise des Eaux, France). As a result, the Japanese inventions spread worldwide and PACl became a common asset in the world through numerous follow-up tests.

The use of positively charged synthetic organic polymeric polymers in coagulants is known to be useful for highly concentrated thickening, but the molecular weight of such polymers can range from thousands to tens of thousands, so that the lower limit of the size of negatively charged particles that can be agglomerated is also beyond the size of polymeric coagulant levels of tens of thousands of Da. This coagulant is valid only for coarse particles with a size >1 μm, such as clay or coarser turbidity organic soil (humus), livestock feces, and coagulation colloidal clumps.

Coagulation treatment is barely effective against the hydrophilic substances in the first column of the matrix (E260 insensitive TOCs) and is barely effective against components in the first column, rows 1 to 4, which contain larger impurities. However, it is almost ineffective against the components in the fifth row or higher. Polysaccharides, organic acid, alcohols, etc. present in the matrix element cannot be subjected to coagulation treatment and must be subjected to microbial treatment. This matrix element (low-molecular-weight E260 insensitive TOCs), surrounded by the dotted line in Figure 3.32, is an area where microbial treatment is effective, such as for sewage. In low-concentration systems, slow sand filters are appropriate. Microorganism coagulation sometimes fails to respond properly to the hydrophilic colloidal components in rows 4 and 5 of Column 1 (E260 insensitive TOCs). In particular, the polysaccharides in row 5 may remain in the treatment water as an intermediate product of microbial metabolism in significant quantities. They are difficult to remove even by physicochemical coagulation, and theirs hydrophilicity (E260-insensitive TOCs) and colloidal domain size are also difficult to remove by activated carbon adsorption. Therefore, polysaccharides cannot be expected to diffuse into the intrapore to be removed, and leak into the distribution system and become a source of regrowth of microbial regeneration in the pipes. In the Netherlands, the greatest concern for water quality control is how to remove assimilative organic carbons (AOCs) in a treatment system to prevent the regrowth of microorganisms in pipes without chlorination. This is the most cumbersome ingredient that leaks from the activated carbon adsorption system in water and wastewater advanced treatment systems (Table 3.10).

As described above, it is understood from the matrix in Figure 3.32 that the group of impurities in water that can be treated as standard by the use of ordinary water and sewage treatment consists of organic matter

Table 3.10 Operation of a water quality conversion matrix (2). (See Figure 3.32 Water quality conversion matrix).

Applicable Range	Characteristic	Operation	Process
Inside blurring range	Particle size larger than the colloid matters	Coagulation	Almost removable
Column 1: rows 1–4	Particle size: mm ~0.1 μm	Coagulation	Row 4 is barely possible
Column 1: rows 1–5 and below	Polysaccharide organic acid etc. Low molecule E260-insensitive TOC	Biological treatment (low concentration system: slow filtration)	Coagulation impossible
Column 1: rows 4–5	Hydrophilic colloids	Sewage treatment	Treatment impossible sometimes
Column 1: row 5	Microbial metabolic intermediate products	Treatment difficult	Leak into the water distribution system Regrowth of microorganisms

(BOD: E260 insensitive TOC) and biologically persistent organic matter (E260 sensitive TOC) in the first column, and of inorganic matter in the second to fifth columns, except for items in rows 6 and 7.

The ingredients in rows 6 and 7 of the second and subsequent columns remain in the water system without being removed by normal treatment of the urban water metabolism system. Among the inorganic substances which are constituents in rows 6 and 7 of the fourth column are metal ions and phosphoric acid ions. These are insolubilized (precipitated) by chemical reactions to form colloidal size, which is then agglomerated by coagulation operation and separated by sedimentation/filtration. This treatment is widely used for hardness treatment of groundwater and heavy metal treatment. It is called a precipitation operation in water treatment engineering, but sometimes it has been called 'special treatment' in urban water systems. However, this is the main treatment usually used in industrial wastewater treatment and groundwater-derived treatment plants.

The organic complexes in column 3 are complexes of biologically persistent organics in column 2, rows 5 and 6, and minerals such as heavy metals in column 4. These ingredients are removable in a process similar to that of E260-sensitive humic substances, in some cases using strong potassium permanganate/ozone as oxidizing agents to separate organo-iron/manganese into metallic hydroxides. E260-sensitive TOCs (biologically persistent organic matter), such as low-molecular-weight humus (low-molecular-weight NOMs, fulvic acids), and trace organic contaminants (micro organic pollutants: volatile organics) are also included in the matrix elements in column 2, rows 6 and 7. Figure 3.18 (above) shows the operation of the activated carbon adsorption-compatible constituents.

Water treatment depends on the components that can be treated, depending on the length of the operation time (retention time). Even with biodegradable low-molecular-weight components, odor and taste components take some time to be degraded and mineralized in water. They are biodegradable but not easily degradable. Even if they are not persistent, they are considered to be slow-degrading components. Therefore, it is necessary to fix them by adsorption and decompose them slowly on the inner surface of the activated carbon. Polysaccharides can be treated in the same manner, and although not much is required in terms of quantity, they can only be adsorbed on the surfaces of particles and continuously regenerated to enable long-term activated carbon filtration tower operation, in an operation called a biological activated carbon adsorption tower, which is also used for ammonia-based nitrification in the pores. These various activated carbon processes are now used widely in a number of advanced treatment trains.

The last thing that remains is the stable inorganic ions in column 5, row 7. The most abundant and typical matter are the Na- and Cl-ions in seawater and brine. NO_3 ions and others accumulated in groundwater are also included in the group. Water molecules and inorganic ions have similar size, and they are hydrated ions which are the most difficult components to separate. As in the desalination of seawater by the evaporation method, water is brought into evaporative conditions in air, and is then condensed and recovered. The separation operation is performed by moving the water in a phase change. This is the most energy-intensive operation. It may be the matrix element that is the most difficult to treat. At the end of the 20th Century, RO/NF membranes were put into practical use, and this matrix element has finally become an item to be removed with acceptable energy and time consumption. (Table 3.11).

The above-mentioned water quality conversion matrices are described with the removal of impurity components in mind. On the other hand, for dilute systems, treatment such as detoxification of microorganisms by a germicidal operation and deactivation (sequestration) of metallic ions using chelating agents is often carried out, but the essence of water treatment is considered to be the removal of impurities from the water system to a solid-air system.

Figure 3.33 shows water treatment matrices for separation function membrane treatment. Compared to conventional water treatment systems, the separation elements are determined by the impurities/ molecular sizes and the cut-off level of separation function membranes used in removal. In principle, the separation matrices are straightforward. A one-dimensional representation is shown in Figure 3.39.

As can be seen from Figure 3.39, water separation membranes exhibit characteristic features in relation to the removal of impurities which are smaller than colloids. Functional separation membranes have lower treatment fluxes and generally have higher energy consumption rates than conventional methods. Much lower molecular weight substance removal can lead to much higher energy consumption rates for energy. Therefore, it is advisable to remove high concentrations of coarse suspensions in advance by conventional sedimentation/filtration, to cope with low molecular weight components. In addition, even when colloidal removal is performed, membrane filtration can easily be made more efficient by increasing the impurity size by a coagulation treatment.

Table 3.11 Operation of a water quality conversion matrix (3). (See Figure 3.32 Water quality conversion matrix).

Applicable Range	Characteristic	Operation	Process
Column 2: rows 6 to 7	Soluble bio-nondegradable organic matter	Adsorption	Activated carbon treatment
Column 3: rows 5 to 6	Bio-nondegradable organic matters Heavy metals complex of 4 column	Adsorption Metal oxidation	Activated carbon treatment Solid–liquid separation
Column 4: rows 6 to 7	Polyvalent metal ions phosphate ion etc.	Precipitation →Growth	Solid–liquid separation
Column 5: row 7	Inorganic ions	Evaporation Membrane filtration	RO/NF membrane filtration

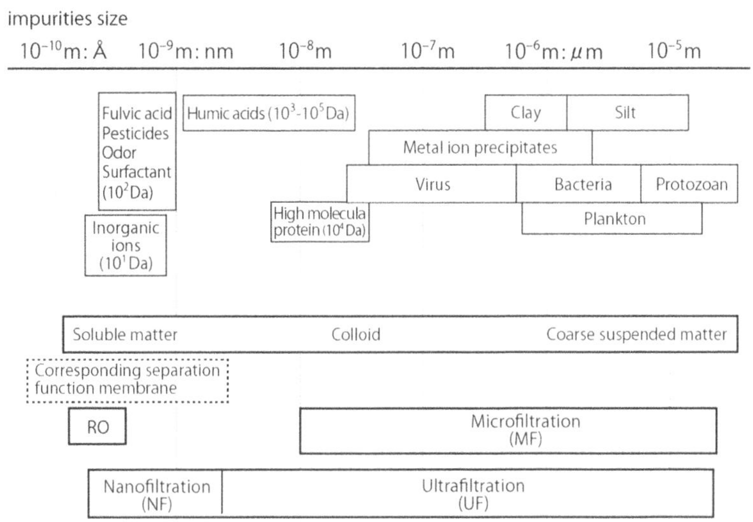

Figure 3.39 Size of water impurities and corresponding separation membranes.

3.5.6 Process selection considering impurity concentration

The third item that must be considered, after the size and chemical nature of any impurities, is their concentration level and its effect on the process. Water treatment must cope with extremely thin pollutant waters (such as 1 to 2 parasite eggs/m^3) to high turbidity water (such as of up to 10^2 g/l of suspended matters in the Yellow River). Figure 3.40 shows a two-dimensional matrix of impurity size and impurity concentration for typical water treatment separation processes. The combination of

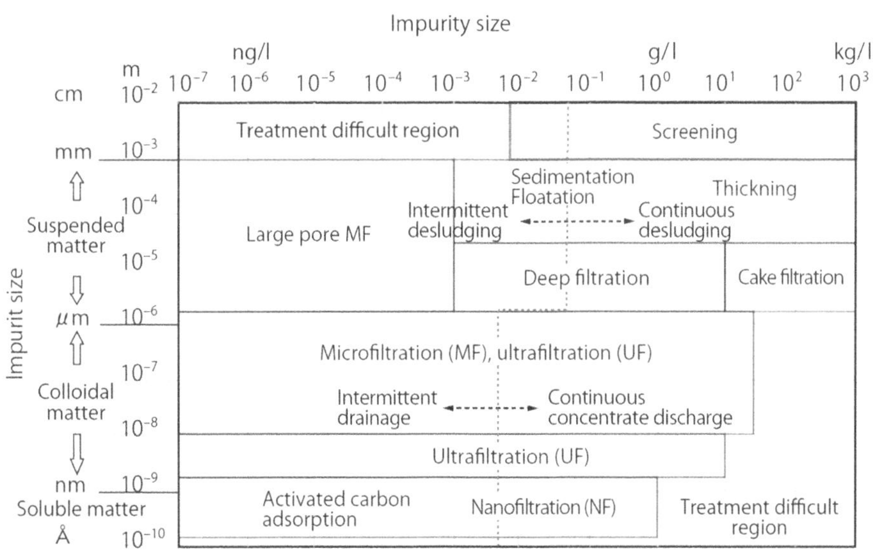

Figure 3.40 A two-dimensional matrix of impurity size and concentration for separation processes.

Figures 3.32 and 3.33 provide three-dimensional quality conversion matrices of impurity size, impurity properties and impurity concentrations.

The elimination of soluble metabolic components (NOMs) and biologically persistent trace organic components in rows 5, 6, and 7 of the water conversion matrices is the only way to cope with water treatment in today's high density societies. Therefore, it is considered a particularly important treatment.

3.5.7 Removal of general organics in detail

In the previous section, the authors proposed the quantitative design, operation and evaluation of organic substances in natural and urban water metabolizing systems by a pseudo-two-component system consisting of 'E260 sensitive' and 'E260 insensitive' organic carbons (TOCs). Organic matter that does not express E260 may be considered to be biodegradable organics that are categorized by the conventional index BOD. E260-sensitive TOCs are biologically persistent organic substances called color or humic substances, which are expressed as (COD–BOD) in conventional indices. As shown in Figure 3.36, humic substances with a molecular weight exceeding 1500 Da show values such as $TOC/E260_{(1\ cm)} \leqq 20$ to 30. As noted above, such humic substances are observed in surface waters that are stagnant for a long time in wetlands and reservoirs, and they flow into rivers only after strong rainfall. In addition, when groundwater passes through the soil layer, humic substances in river water during normal times and in low water periods are mainly composed of low-molecular components, which are a lost macromolecule humic portion from soil particle adsorption, etc. and are recharged into rivers with other low molecular weight substances as intermediate ground water runoff. These low molecular weight substances with UV 260 nm absorbances of less than 1000 Da in mid-stream water have a $TOC/E260_{1\ cm} \doteqdot 50$ to 60. They also appear to be the major components of younger humus, such as in the microorganism metabolic wastewater and treated glucose mentioned earlier. The removal of low-molecular-weight E260-insensitive organic components, which are the main components of this younger humus, requires treatment such as activated carbon adsorption. The potential for trihalomethane (THM) formation to be reduced by switching pre-chlorination treatment to intermediate chlorination treatment in rapid sand filtration systems may be considered effective for the removal of humic substances (E260-sensitive components) in this macromolecular part by coagulation and sedimentation. The humic substances of these macromolecules are thought to be caused by the polymerization of low-molecular-weight components in natural reservoirs and /or marshy ponds.

Figure 3.41 shows the TOC/E260 ratio by molecular weight of Ishikari peaty groundwater (Hokkaido). The TOC/E260 ratio shows the characteristic value of 20 in the area of molecular weight 1500 Da or more, and the value of 50 in the area of 1500 Da or less. The V_e/V_0 in the horizontal axis is the measurement value used to estimate the molecular weight of the extruded liquid in high performance liquid chromatography divided by the volume of the void channel volume of the gel column.

Figure 3.42 shows the removal rates of E260-sensitive TOCs (humic substances) in water treated with aluminium coagulation. The humics of water with molecular masses of 2,500 Da or more are almost completely removed by coagulation, whereas removal is not expected for water with molecular masses of 1000 Da or less. This means that activated carbon adsorption is required for the removal of humics in smaller molecules.

The results of aluminium coagulation of E260-insensitive TOCs (BOD component) are shown in Figure 3.43, again in terms of molecular mass and TOC removal rate. The removal of coagulation is much lower than that of humic substances, and removal at molecular masses of 2500 Da or less cannot be expected. The removal rate is only around 50 to 60% even for polymer moieties with molecular masses of 5,000 to 10,000 Da.

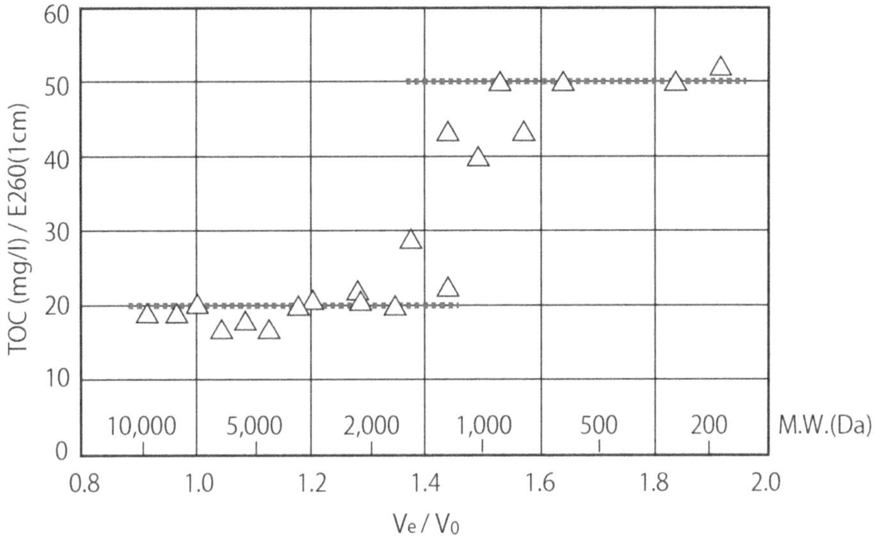

Figure 3.41 The characteristics of the high and low molecular components in peatland water.

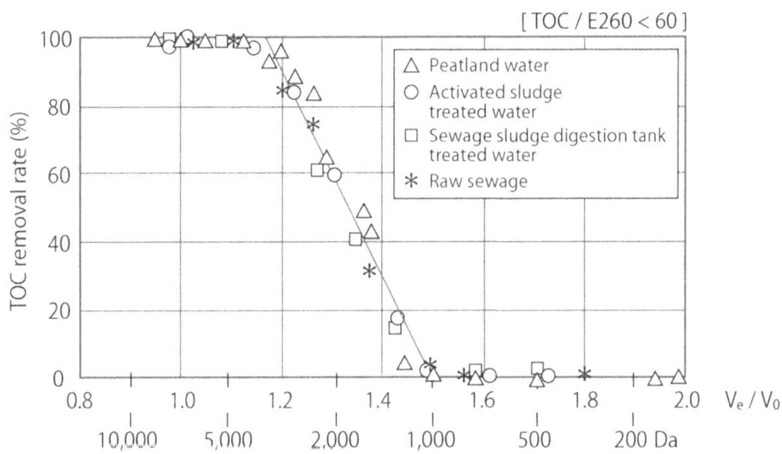

Figure 3.42 The removability of bio-nondegradable general organic matter by coagulation.

Removal of E260 insensitive TOCs (BOD components) by aerobic microbial treatment is shown in Figure 3.44, and a sufficient removal can be expected with molecular weight of 2000 Da or less. However, if the molecular weight increases to 5000 to 10,000 Da, it takes time for microbial degradation to occur and sufficient removal cannot be obtained within a limited treatment time. Even if biodegradability is present, it should be considered slow degradability. We have often experienced in many advanced treatment systems that TOCs less than 1 mg/l have remained in treated effluent, after coagulation sedimentation/activated carbon adsorption for sewage regeneration or treatment using a less clean downstream raw river water. This may be attributed to the presence of E260-insensitive organic matters (probably hydrophilic polymeric polysaccharides) that cannot be removed in the series systems of Figures 3.43 and 3.44.

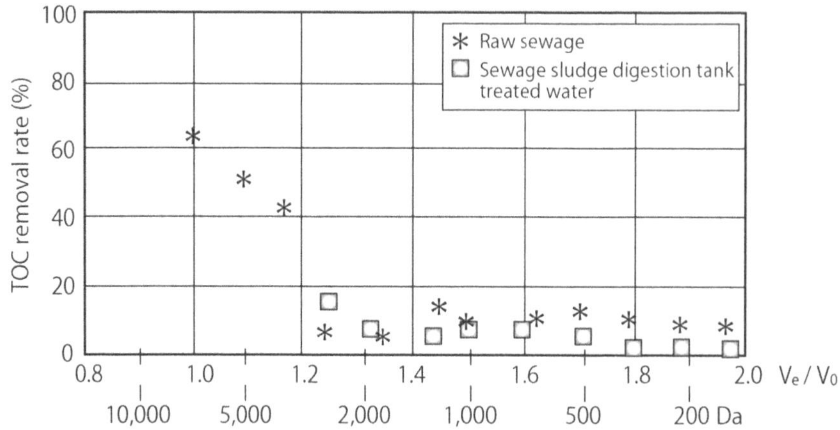

Figure 3.43 The removability of biodegradable general organic matter by coagulation.

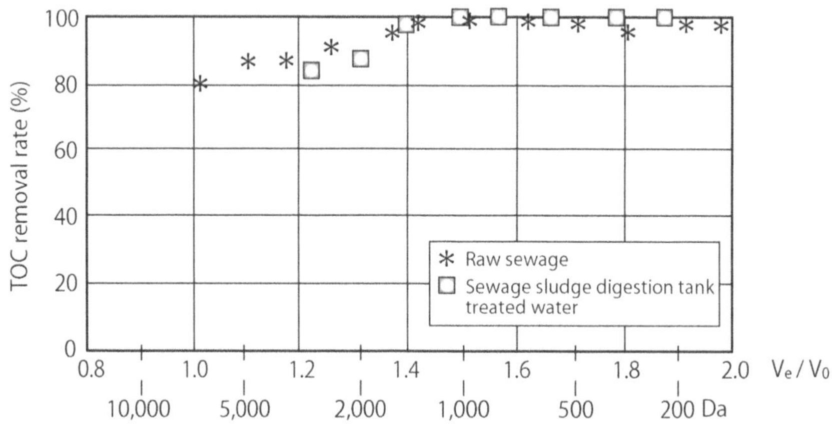

Figure 3.44 Removability of biodegradable general organic matter by aerobic microbial treatment.

3.5.8 Removal rate estimation formula for general organic matter in a pseudo-two-component system

As described above, we have found that a two-dimensional water quality conversion matrix can be used to express the treatment efficiency of wastewater in a somewhat uniform manner. Especially, when attempting to handle a wide variety of organic substances in the second-cycle urban water metabolic system (water supply and sewage), it is clear that it is effective to treat general organic substances as a 'pseudo-binary system' in which both the material balance and the treatability can be displayed beyond the BOD and COD of conventional expression systems, and to estimate the behavior by molecular weight. Table 3.12 summarizes this characterization in a list which is an evaluation of the treatability by two-dimensional elements of E260-sensitive and -insensitive organic matter (TOCs) and their molecular weight.

Table 3.12 Treatability by two-dimensional elements of E260-sensitive and E260-insensitive organics.

Main indicators [Material balance indicator: TOC
Humus characterization indicator: E260]

Pseudo 2 Component Classification of Soluble Organic Matter	**Relationships with the Conventional Indicators**	
	Oxygen Demand (Conventional Indicator)	**Organic Carbon Content TOC (New Indicator)**
E260 sensitive matter	COD_{cr}–BOD	TOC (Humus)*
E260 insensitive matter	BOD	TOC (Biodegradable general organic matter)
Characteristics of the indicators	Indirect/complementary evaluation of organics balance	Direct evaluation of material balance

*Separation of humic substances TOC/E260
High molecular side ≒ 20~30 (>1500 Da)
Low molecule side ≒ 50 (<1000 Da)
[Carbon equivalent to humic material] = [Total organic carbon content] – [Carbon content of biodegradable component]

The water quality conversion matrix shown in Figure 3.32 (earlier) shows how the matrix elements can be effectively processed to removal. Equation 3.14 shows the experimental estimates of the maximum removal rate, expressed by the fractionation ($i = 1, 2, 3$ etc.: see Figure 3.30) of the pseudo-two components of E260-sensitive and -insensitive TOCs.

Removal rate calculation formula

$$R_i = \frac{A_i\,(\text{E260sen.TOC}) + B_i\,(\text{E260insen.TOC})}{\text{TOC}} \quad \text{---}(3\cdot 14)$$

R_i : Removal rate of fraction i component
A_i : E260 expression component concentration
B_i : E260 non-expressinon component concentration

Removal rate [Aerobic microbial treatment system] (3.14)

	i = 1 (Da > 1,500)	i = 2 (Da < 1,500)
A_i	0 %	0 (+3 ~ +5 %)
B_i	95 %	100 %

Removal rate [Chemical coagulation treatment system]

	i = 1 (Da > 1,500)	i = 2 (800 < Da < 1,500)	i = 3 (Da < 800)
A_i	100 %	100~0 % (Transition region)	0 %
B_i	20~5 %	0 %	0 %

i = 1, 2 etc. are the fraction numbers in Fig.3.30

Equation 3.14. The Estimation of removal rate on the matrix

The maximum removal rate (which is considered to be reached at final equilibrium) is obtained by multiple attempts of batch-type activated sludge microbial treatment and coagulation treatment of a large number of fractionated samples by gel chromatography, to determine the extent to which the components can be removed by these treatments. In the operation and evaluation of actual water systems, samples of raw water and raw wastewater are taken beforehand, and samples are characterized by molecular weight by liquid chromatography to calculate TOCs and ultraviolet E260 distributions for each group of images. By estimating the equilibrium removal rate using Equation 3.14, it is possible to quantitatively determine the extent to which removal can be expected by aerobic microbial and coagulation treatments. If the results of the current treatment do not reach these values in the examination of actual treatment plants and natural water test systems, it can be understood that the treatment is not yet sufficient. By devising the operation of the treatment plant, we can expect that the removal rate will be improved to the highest practical level estimated by this equation. Many demonstration studies have shown that the accuracy of this formula is very high and is useful in the calculation of processing limits.

3.6 CONSTRUCTION OF A WATER TREATMENT SYSTEM

3.6.1 Combining processes in a system

Water treatment systems that correspond a wide variety of water pollutants can be constructed by appropriately combining water treatment processes corresponding to individual areas of impurity (matrix elements). Generally, treatment processes are arranged in descending order of impurity size.

Consider the typical activated sludge sewage treatment system shown in Figure 3.45. As indicated by the impurity concentration and the size treatment matrix, inflow sewage is first passed through a screen to remove the coarsest size of rubbish, sand grains are settled in the grit chamber, and then silt and clay particles are removed as much as possible by the next primary sedimentation tank. After return sludge is added to the reaction tank (an aeration tank), a microbial reaction is carried out, and then the activated sludge is then separated by a final sedimentation tank, and chlorine is added to the supernatant water to disinfect the discharge.

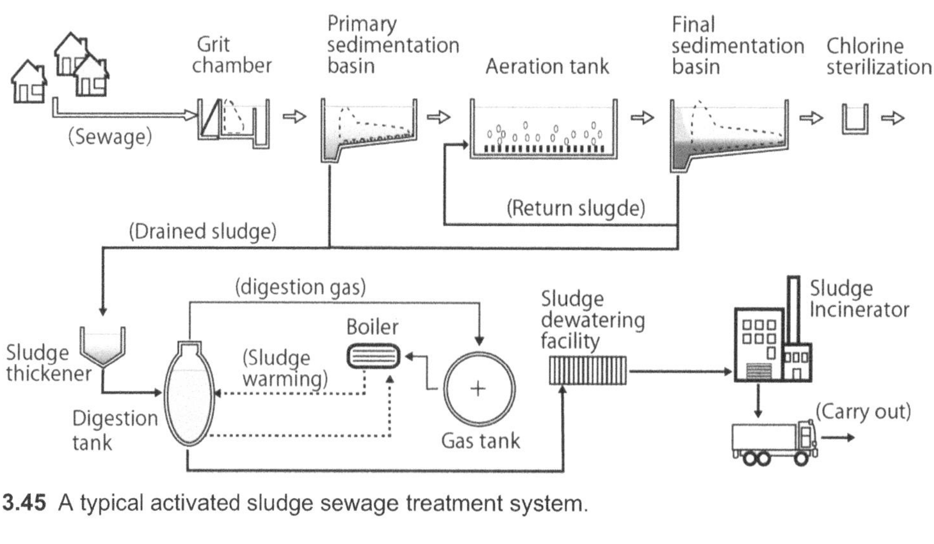

Figure 3.45 A typical activated sludge sewage treatment system.

The separated high-concentration activated (excess) sludge is transferred to a sludge treatment system, thickening proceeds in a thickening bed to increase the concentration, and then it is introduced into a digester to stabilize/reduce the volume by anaerobic reactions. The produced sludge is dewatered by cake filtration, etc., and transferred to a solid system for disposal.

High quality low molecular weight and low concentration treatment processes can be added to recycle treated sewage, such as precision filtration, ultra-filtration, nano-filtration, and activated carbon adsorption. Furthermore, sewage which has been treated by membranes and activated carbon adsorption and which has reached extremely low concentrations is also subjected to advanced treatments such as chlorination, ozonation, UV disinfection and other detoxification processes that do not involve removal of components. In sewage treatment, we have learned to refer to sedimentation treatment as primary treatment and microbial treatment as secondary treatment. Therefore, we refer to subsequent treatments as tertiary treatment. However, since coagulation/filtration and activated carbon treatments as well as membrane treatments are successively followed, the term 'tertiary treatment' is insufficient and these treatments are better described as 'advanced' or 'higher-order' treatments.

However, other types of treatment trains (systems) are also conceivable, apart from this traditional sewage treatment system configuration. Inflow sewage can be subjected to plane sedimentation, followed by coagulation/sedimentation treatment to remove higher molecular weight components, and the remainder become low molecular components. However, if a high molecular weight component, in particular a biologically persistent or slowly degradable component, is separated in the sewage, the remaining low molecular weight biodegradable component can be processed and separated in a short time by biological filtration or a trickling filter bed. In this case, the microbial treatment is the third step. High-concentration organic (polymer) matter separated by coagulation at the second stage is transferred to an anaerobic digested sludge treatment process after thickening, and can be treated with energy/phosphoric resource recovery (Figure 3.46). This method requires less energy consumption than conventional activated sludge methods. In the mid-1970s, whilst the authors were researching this system at Hokkaido University, three Nordic countries were developing a similar system together.

As described in the coagulation section below, high-speed solid-liquid separations using a polymer coagulant in a pellet-fluidized bed separation of raw sewage water, followed by fixed-bed aerobic biological treatment, can be processed in as short as 1 to 2 hours. Today, if an aerobic membrane microorganism treatment, in a membrane bio-reactor (MBR), was followed by an immersion-type MF/UF membrane coagulation operation, it would enable the creation of state-of-the-art water, energy and resource recovery systems within a range of recycling.

3.6.2 Sewage reuse

In the previous section, we mentioned a hybrid process in which high efficiency and good operability can be achieved by the coalescence process. In a somewhat broader sense, by considering the appropriate combination of coagulation, aerobic microbial treatment and anaerobic microbial treatment, we believe that the assembly of organic wastewater and sewage treatment systems – which have traditionally been standardized as primary treatment, secondary treatment and tertiary (higher-order) treatment – will differ significantly. Figure 3.47 is a conceptual diagram of a hybrid sewage treatment system, such as the HYPRO process that the authors have worked on since the 1970s and which was also developed jointly by three Nordic countries (Sweden, Norway and Denmark) during the same period (Tambo, 2012).

By pre-coagulating sedimentation treatment, aerobic biological treatment can be effectively carried out only for the most appropriate components, and sludge treatment can also effectively and energy effectively set up an anaerobic process with a high concentration.

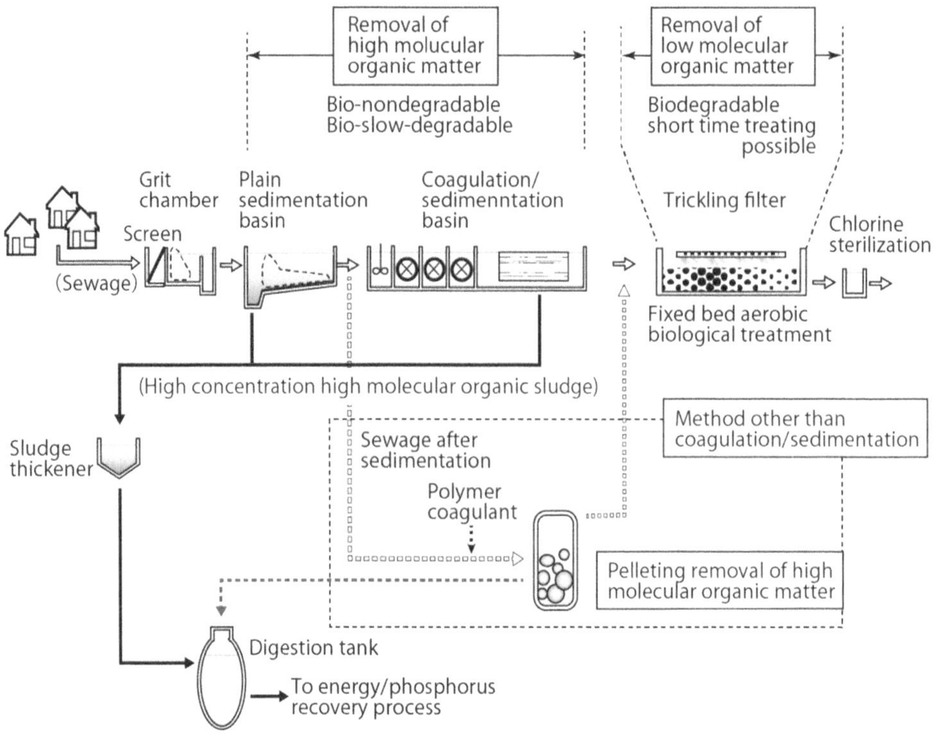

Figure 3.46 Another new sewage treatment system.

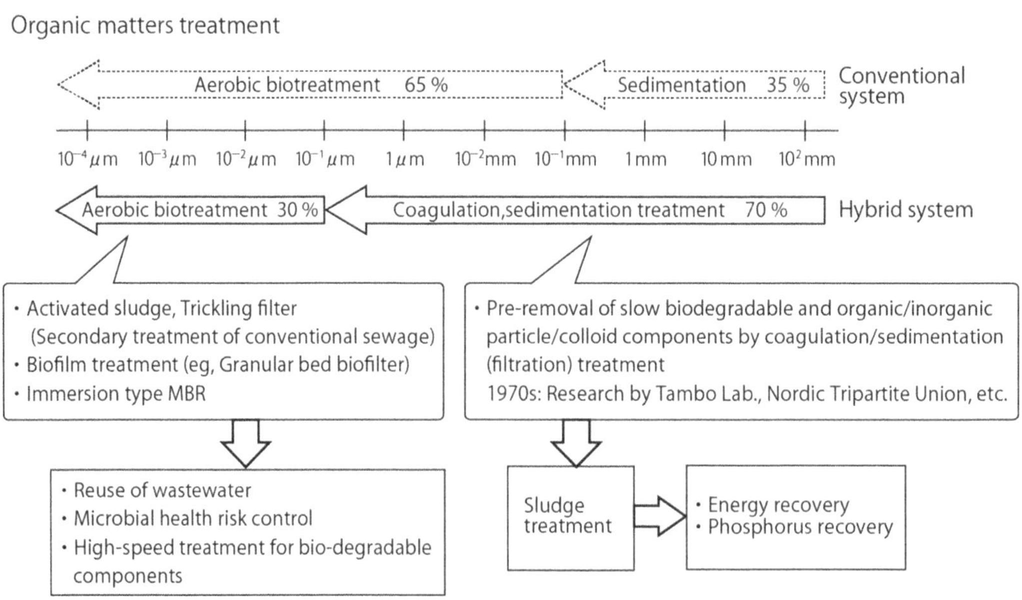

Figure 3.47 A conceptual diagram of a hybrid sewage treatment system.

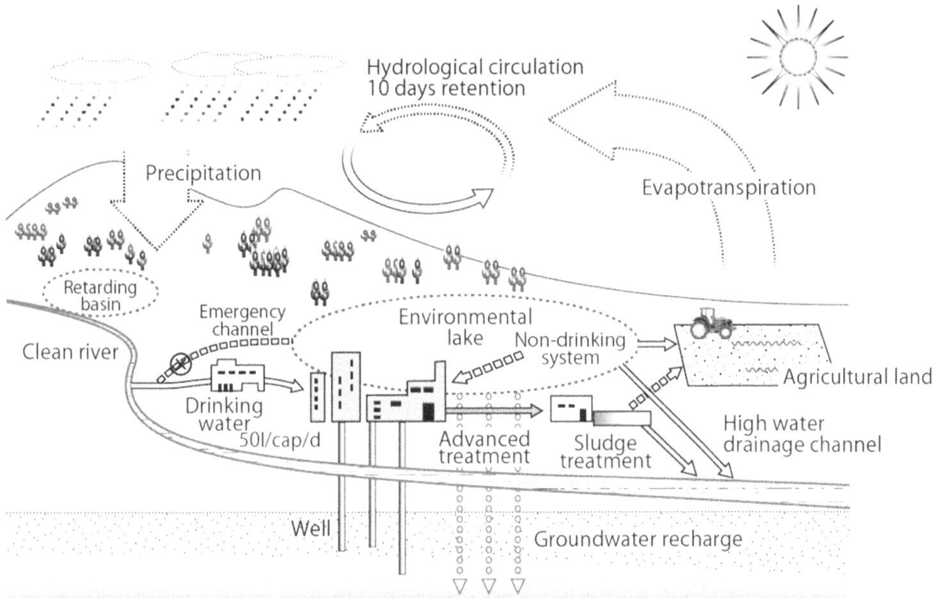

Figure 3.48 A water district.

The authors believe that such a complex water treatment system will become the core water treatment system to establish a water environment district when autonomously establishing water metabolism systems in urban areas, as illustrated in Figures 1.9 (above) and 3.48, which have been proposed by Tambo since 1976.

A variety of process configurations can be considered, one of which is the combination of pellet agglomeration (described below in Sections 4.3 and 5.2) and aerobic biological filter beds (combined processes). The results of a series of studies on the environmental lake downstream of the Sapporo metropolitan area are described as examples.

Figure 3.49 is a combined sewage treatment system (CBT) in which the agglomeration separation process by pelleting is applied first for macro colloid suspensions. The BOD load of small (soluble) molecules is removed by the following aerobic biological filter bed, and final impurity separation is finished by an MF membrane. (Shimizu *et al.*, 1993a, 1993b).

In this process, an up-flow granulation (pelleting) process (described in Section 4.3 below) is used for the first stage treatment of the inflow sewage, involving pellet floc formation (using granule precipitation/pellet flocculation). Poly-aluminium chloride (PACl) is used as the main coagulation (flocculating) agent, and weakly anionic polymers are added to the lower part of the up-flow pellet fluidized bed as floc-forming aids and gently stirred by paddles (at 3 rpm), to make organic pellets with a superficial velocity of 200 m/day. This operation removes suspended organic matter in the inflow sewage, greatly reduces the burden of suspended organic matter (BOD) on the next aerobic biological treatment, and enables high-speed biological treatment. In sewage flocculation treatment, flocs formed only by metal coagulants are fine and fragile because most of the suspended components are organic substances. A small amount of a weakly anionic polymeric floc-forming aid is added to promote cross-linked flocs. This makes it possible to produce organic pellets with a high effective density.

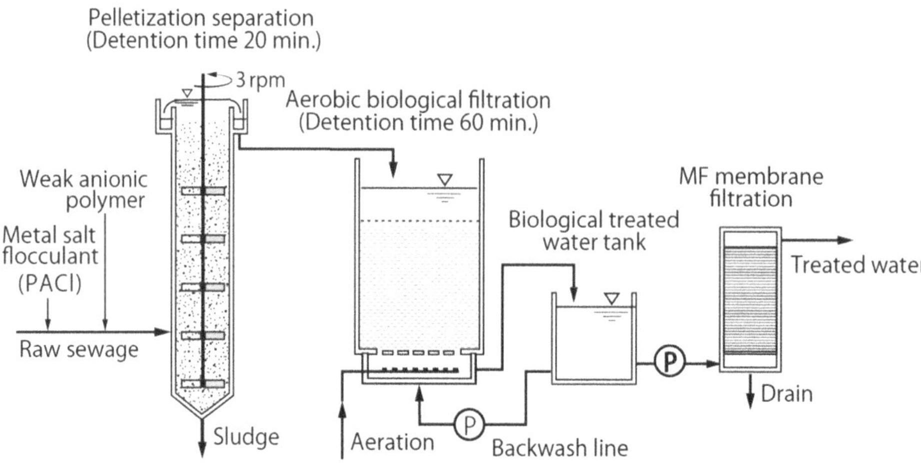

Figure 3.49 A schematic diagram of a combined sewage treatment system using pelletization separation, biological filtration and membrane filtration.

Pelleting sewage in the process shown in Figure 3.50 does not lead to the development of a uniform-density pellet agglomeration of the 'one by one' type achieved using metastable regions (and as shown in Figure 4.124 below). This is thought to be due to the fact that the polymer cross-linked floc is largely dependent on the process of compaction by the syneresis effect as agitation is continued.

In a combined palletization separation and membrane filtration (CBT) study conducted in a pilot plant installed at the Sewage Treatment Plant in Sapporo City (1990–1992):

(1) The turbidity of the treated water to be used as the blanket effluent was about 10 to 15 TU (turbidity units). Throughout the plant operation, the turbidity of the effluent after the pellet layer (with a thickness of about 1 m) was always stabilized at a level of 10 TU or less.
(2) Micro flocs produced in the lower part of the palletization bed were captured by the pellet blanket bed (pellet group concentration: 1000 to 3000 mg/l suspended in the ascending stream, which had a superficial flow rate of 200 m/day) to form pellet particles, and the separated water with low turbidity overflowed from the upper part of the bed.

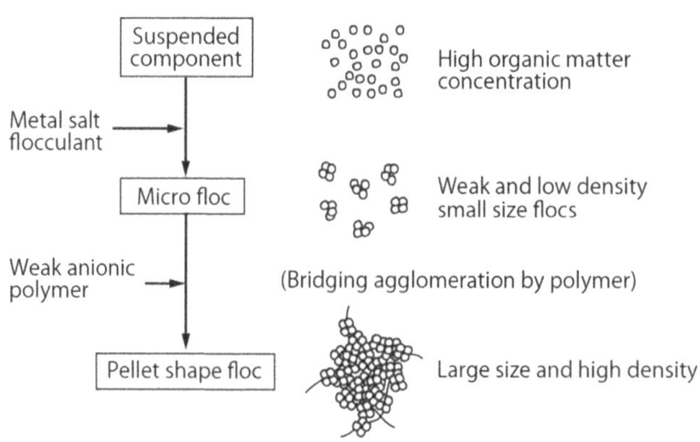

Figure 3.50 Sewage pellet floc formation by syneresis.

(3) The pellet blanket layer gradually increased in layer thickness as the agglomeration progresses. Sensors detected when the layer thickness reached 2.03 m, and the drain valve at the bottom of the pellet separator opened, so that the pellet blanket layer thickness was controlled to 2.02 m or less. Waste sludge (pellet sludge) was easily dewatered and sent to an anaerobic digester, where it was mineralized and gasified at low energy-consumption rates. Sludge production increased by 30% at 1.32 kg-SS (suspended solids)/inflow 1 kg-SS.
(4) The granulated and separated overflow treated water became organic wastewater of low concentration and low molecular weight solutes, and an aerobic organism filter bed, which is easy to operate and consumes little energy, could be used to treat most of the soluble BOD load in a short period of time – about one hour in superficial tower residence time.
(5) A coagulation process removed phosphate in the inflow water, and the remaining phosphorus was ingested as a nutrient salt for fixed bed aerobic treatment, resulting in a very low concentration of phosphorus that was transferred to the treated water.
(6) In the experiment, MF membrane separation was placed in the final stage, and the treated water was examined to see whether it could be used for general purposes other than drinking, and whether it could be used as raw water for environmental lake and groundwater recharge in the water environment district.

The operating conditions from the study are shown in Table 3.15. The average quality of treated water is shown in Table 3.16 for each process step. The DO level of the aerobic filter bed was 3.0 to 4.5 mg-O_2/l, the soluble BOD was about 25% of the total BOD, the mean BOD value of biotreated water was 4.7 mg/l (96.7% removal), and the mean BOD value of MF treated water was 1.9 mg/l (98.6% removal). For ortho-P, the mean value of pellet layer separated water was 0.104 mg/l (removal rate: 94.2%), the mean value of biological treatment water was 0.012 mg/l (removal rate: 99.3%), and the mean value of MF-membrane treated water was 0.01 mg/l.

Table 3.13 Operational conditions for advanced process experiments with pellet flocculation and biological filter membranes.

	Operation Conditions
Pelletization separation	Superficial velocity 200 to 300 m/day Aggregate condition PACl 10~20 mg as Al_2O_3/l Weak anion polymer 1.6 mg/l
Aerobic biological filter bed	Downflow aerobic biological filter bed: 2 towers in series Filter material: Size 5 to 20 mm Lightweight aggregate Filter bed thickness: 1200 mm + 1000 mm 1990.12.20~1991.2.15 Lv 20 m/day 1991.2.15~1992.2.26 Lv 50 m/day
MF membrane filtration	Polypropylene porous hollow fiber membrane Cut off dimension: 0.2 μm Operation method: Dead end filtration Backwashing method: High pressure gas backwashing Operating flux: about 2.5 m/day Chemical washing: 0.5% H_2O_2 + 2% NaOH alkaline washing 4% citric acid washing

Table 3.14 Results from advanced process experiments undertaken with pellet separation, biological filter bed and membrane filtration. 1991.2.1 ~ 1991.10.31

	Raw Water (Sewage)	Pelletization Treated Water	Biological Filter Bed Treated Water	Membrane Treated Water
PH	7.04 (6.8~8.2)	6.90 (6.5~7.3)	7.07 (6.4~7.5)	6.98 (6.7~7.3)
Turbidity	100.1 (69.7~153)	9.6 (1.9~23)	1.43 (0.3~5.8)	0.03 (0~0.6)
SS mg/l	146.1 (75.9~244)	13.8 (4~31)	1.92 (0.29~17.5)	–
BOD	139.5 (86.7~193)	29.8 (13.8~69.5)	4.66 (0.8~12.2)	1.94 (0.6~4.3)
COD_{cr}	283.8 (166~512)	73.9 (36.8~163)	36.5 (15.2~67.5)	25.0 (3.0~44.8)
Ort-P*	1.80 (0.65~3.0)	0.104 (0.01~0.34)	0.012 (0~0.03)	0.05 (0~0.02)

*For Biological and membrane treated water, the detection limit of 0.005 mg/l or less is expressed as 0.
() Indicates maximum and minimum values.

Figures 3.51 and 3.52 show the effectiveness of the pelletization process for removing BOD and COD_{cr}. Figure 3.51 shows that the effluent from the pellet separation tank has a gradient in the plot of close to 0.9, and most of the suspended BOD and as much as 10% of the dissolved BOD were removed. By contrast, the gradient of 0.65 in the plot in Figure 3.52 seems to remove most of the suspensions and 35% of the solubility of the COD_{Cr} was also removed.

Table 3.15 Influent and treated water quality at each stage of the CWT process, in the warm temperature water period. (Normal temperature period)

	Inflow Sewage	AFPB Outflow Water	Contact Aeration Tank Outflow Water	Sand Filtration Outflow Water
SS	220	4.10	2.10	–
Total COD_{cr}	228	32.5	17.1	16.5
Soluble COD_{cr}	62.1	21.0	13.2	13.0
Total BOD	158	15.0	8.12	6.50
Soluble BOD	54.1	13.2	6.51	4.80
TN	33.1	8.12	6.10	5.85
Soluble TN	26.5	6.90	0.22	0.25
NH_4^+	25.4	6.60	0.18	0.15
NO_3^-–N	0.10	0.04	5.77	5.85
TP	3.50	0.15	0.10	0.06
PO_4^{3-}–P	1.60	0.10	0.05	0.02

Operating temperature: 15°C or more, Total detention time HRT: 4 hours Circulation ratio: 3.

Table 3.16 Influent and treated water quality at each stage of the CWT process, in the low water temperature period. (Low water temperature period)

	Inflow Sewage	AFPB Outflow Water	Contact Aeration Tank Outflow Water	Sand Filtration Outflow Water
SS	181	10.0	4.20	—
Total COD_{cr}	285	34.8	18.1	16.2
Soluble COD_{cr}	60.9	22.7	15.5	12.9
Total BOD	210	19.8	9.10	8.12
Soluble BOD	61.6	16.1	6.90	5.15
TN	30.2	7.92	5.51	4.60
Soluble TN	20.7	4.60	4.05	4.20
NH_4^+	19.6	4.90	0.10	0.08
NO_3^-–N	0.21	0.10	4.00	4.20
TP	3.10	0.15	0.11	0.08
PO_4^{3-}–P	0.74	0.02	0.03	0.02

Operating temperature: 10~12°C, Total detention time HRT: 7.5 hours, Circulation ration: 3.

Figure 3.53 shows the relation between the BOD volume load and treated water quality in the aerobic filter bed. In most cases, the BOD volume load is lower than or equal to 1 kg · BOD/m^3/day. As a result, the BOD of the treated water in the aerobic biological filter bed is lower than or equal to 10 mg/l. The relationship shown in Figure 3.53 can be regarded as almost the same as that of ordinary sewage treatment, and it is considered that, as a result, the inflow load of the next aerobic biological filter bed is

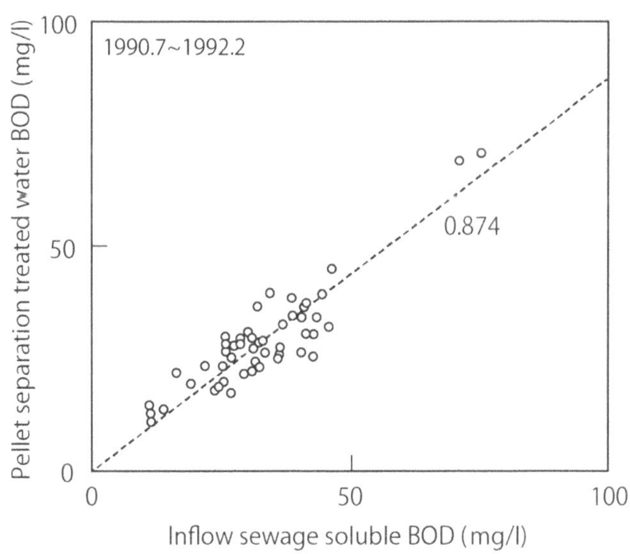

Figure 3.51 Effectiveness of the pelletization process for removing BOD.

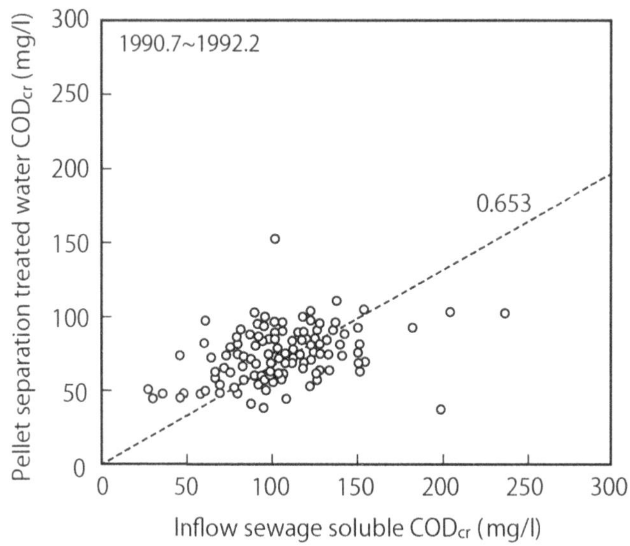

Figure 3.52 Effectiveness of the pelletization process for removing COD_{cr}.

reduced greatly at all times because of the removed suspended BOD, etc. by the preceding granulation and coagulation separation.

But what are the operating conditions of the palletization separation tank, which is essential for this combined processing system? The separation operation was considered to be related to both the concentration of organic matter and the concentration of inorganic matter, and the amount of poly-aluminium chloride was variably added to the raw water (adjusted by adding 100 to 750 mg/l of kaolin to the influent sewage). The results of the experiments, plotting the relation between the dosing

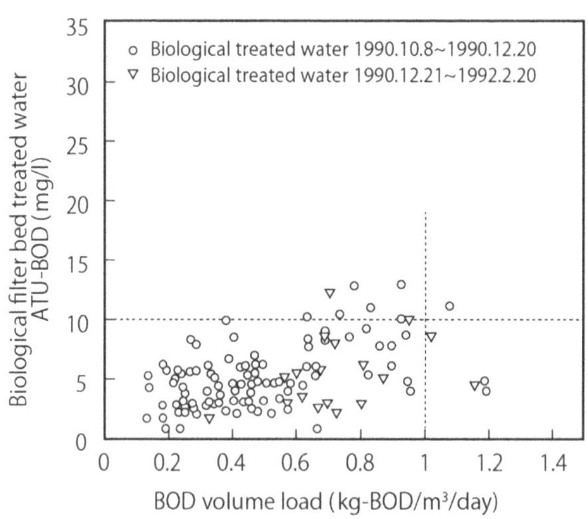

Figure 3.53 The BOD volume load and treated water quality in an aerobic filter bed.

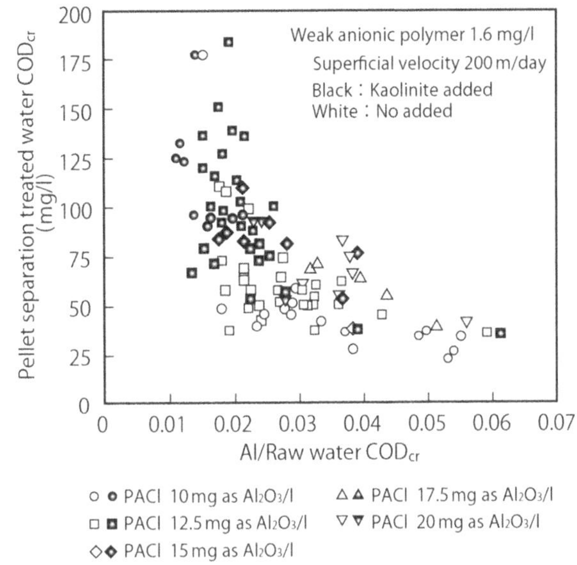

Figure 3.54 The effects of kaolinite addition on sewage water treated by pelletization separation.

rate of aluminium (PACl) to the COD_{cr} and the granulation separation treatment water are shown in Figure 3.54. The addition rate of weak anionic polymer as a crosslinking aid is fixed at 1.6 mg/l. As can be seen from the figure, the removal rate of the palletization separation tank is governed only by COD concentration, and it is hardly affected by the magnitude of the inorganic (kaolin) concentration. In the palletization and separation treatment of raw water with a high concentration of organic matter (such as sewage water), the amount of inorganic coagulant is mostly used for charge neutralization of organic matter (microbial cell surfaces), and the dosing condition of coagulant seems to be governed by the concentration of organic matter (COD_{cr} concentration in this case). This is a phenomenon similar to that of organic colloids in terms of color intensity and conditions governing clay-based multicomponent aggregation, as described in Chapter 5, although there is a difference between suspended organic matter and soluble organic matter.

3.6.3 Combined waste treatment assuming denitrification/dephosphorization

Following combined waste treatment (CWT) system research in which a pelletization separation and coagulation system was installed for the reuse of sewage water, etc., the composition of a CWT to stably promote nitrification, denitrification and dephosphorization was examined (Shimizu *et al.*, 1993a, 1993b, 1994).

The basic difference from the preceding set up is that the aerobic catalytic filtration layer of the second stage is used as a nitrification reaction vessel for nitrification of ammonia, etc.; furthermore, he treated liquid is circulated into the lower part of the palletization separation vessel of the first stage and the pellet fluidized bed is used as an anaerobic denitrification process.

The conventional system configurations for nitrification and denitrification by biological treatment are two- and three-tank systems for anaerobic and aerobic processes. Figure 3.55 shows a flow chart of a

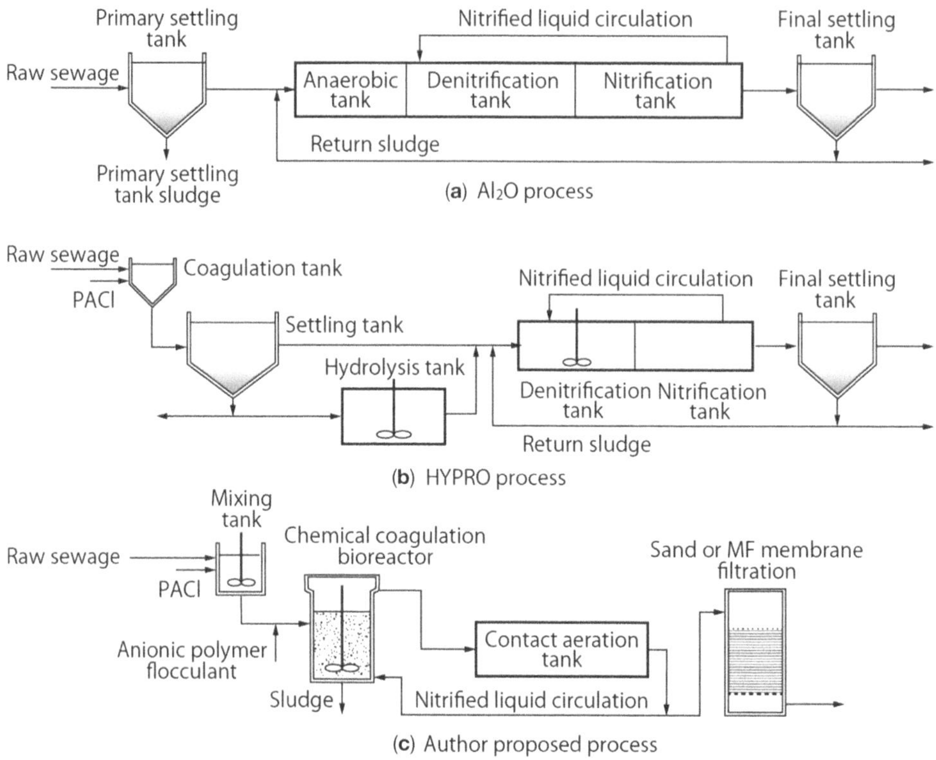

Figure 3.55 Various nitrification and denitrification systems.

conventional three-tank system (two-anaerobic, one-aerobic three-tank system: the A_2O method) and the HYPRO system, and the composition of a further nitrification and denitrification CWT, to be explained here below.

The aerobic biological treatment process consumes a large amount of power in the aeration tank, and the CWT method aims to reduce aeration energy. It is desirable to remove phosphorus and supply a substrate (carbon source) in an anaerobic denitrification reaction system, as well as maintain a high removal rate in the final treatment stage. In addition, the sludge is easily separated, and methane gas is recovered at high efficiency in the sludge treatment process.

The Al_2O method shown in Figure 3.55(a) is one of the most common biological dephosphorylation and denitrification methods currently used. However, since much of the organic carbon is removed as a suspended component in the primary sedimentation tank, the substrate necessary to drive the denitrification reaction in the latter part is insufficient, and a carbon source such as alcohol is added. To compensate for these drawbacks, the HYPRO process was developed jointly by three Nordic countries (Figure 3.55(b)) (Andersson & Rosen, 1990; Henze & Harremoes, 1990; Hoffmann & Klute, 1990; Jorgensen, 1990; Karlson & Smith, 1991). As shown in the figure, in this method, poly-aluminium chloride is added to inflow sewage water to remove suspensions and phosphorus compounds by coagulation precipitation, followed by aerobic and anaerobic nitrification and denitrification to remove nitrogen. In order to make the denitrification process proceed completely through continuous experiments, components (organic acids, etc.) produced by hydrolyzing sludge obtained by coagulation and precipitation are used as hydrogen donors for the denitrification reaction. However, many reaction operations are required, including those involved in coagulation and in sedimentation tanks, anaerobic hydrolysis tanks, anaerobic denitrification tanks, and in aerobic nitrification tanks, which all complicate the processing system.

The CWT system proposed by the authors for the removal of suspended components and denitrification/dephosphorylation is a simple configuration consisting of an anaerobic granulation (pellet) separator and an aerobic aeration tank, as shown in Figure 3.55(c), with sand filtration or MF membranes for final separation. The structure of the anaerobic organic pellet separation tank and the following aerobic aeration tank are shown in Figure 3.56. A system following this configuration operated at the Sewage Treatment Plant in Sapporo City, from 1992–1994. (Shimizu *et al.*, 1993a, b).

Using raw sewage water from the Sewage Treatment Plant in Sapporo City as raw water, 10 mg/l of an Al_2O_3 equivalent of a flocculant (PACl) was quantitatively injected into the influent water. Rapid agitation was applied to generate a micro-floc suspension. At the line end, which flows into the lower part of the anaerobic granulation separation (pellet flow) tank, a weak anionic polymer floc-forming aid (2 mg/l) was added and mixed to uniformly disperse so that the floc mass could be stably suspended in the fluidized bed. Oxygen was not supplied to the pellet fluidized bed, and the fluidized pellet bed was basically maintained in an anaerobic condition (ORP potential of -300 to -450 mV). To help pellet formation and particle size retention, the bed was stirred slowly with a paddle, at 1 rpm. The fluidized bed seems to have had a characteristic microbial ecosystem in which microaerophilic *Beggiatoa sp.* and

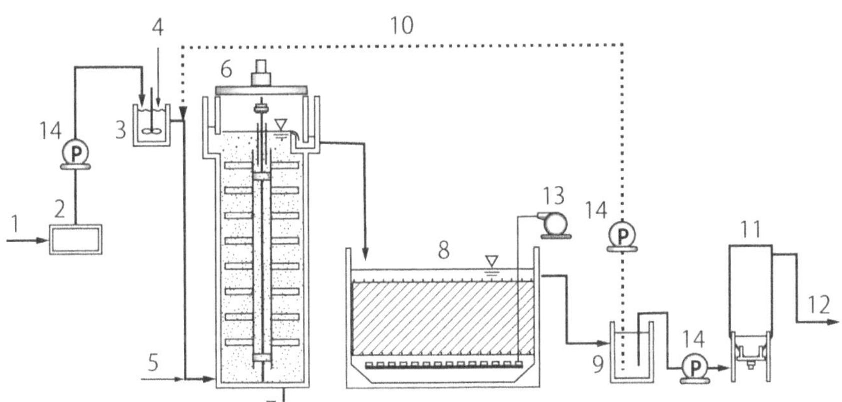

Shimizu, Tambo et al., 1994,1995

1. Sewage inflow 2. Screen 3. Rapid mixing tank 4. Flocculant (PACl) dosage
5. Anionic polymeric flocculation aid
6. Anaerobic fluidized pellet bed bioreactor (AFPB)
Capacity 100l,Inside diameter 25 cm, Height 205 cm
7. Sludge (pellet) 8. Contact aeration tank
(50l × 2 tanks in series, Honeycomb tube filled aeration tank)
9. Aerobic outflow storage tank 10. Aerobic outflow circulation line
11. Sand filter (or MF membrane filter)
12. Final treated water 13. Air compressor 14. Pump

Figure 3.56 A CWT system for removal of suspended components and for denitrification and dephosphorylation.

denitrifying bacteria, sulfate-reducing bacteria, and methane-fermentation bacteria were present together. Before the pellet fluidized bed became steady, it required at least several hours of operation. Granulated sludge (bio-pellets) were extracted by a drain valve at the bottom of the anerobic fluidized pellet bed (AFPB) and sludge treated. Since the waste mud was wet granulated, it could be easily dehydrated using a belt press, etc. (Tambo *et al.*, 1993).

The layer thickness of the fluidized bed section of the AFPB was kept constant by an on-off control connecting photosensors installed in the solid-liquid separating section at the upper part with drainage valves. From the top of the anaerobic granulation separation (pellet flow) tank, the separated water (anaerobic) with a low SS concentration flowed out to the next contact aeration tank. In aerobic catalytic aeration tanks, mainly an oxidative decomposition of soluble BOD components and nitrification of ammonia were promoted. The AFPB aerobic aeration tank in the previous stage removed most of the SS and COD loads, and the inflow load to contact aeration tank was low. Therefore, the SRT of the biofilm in the contact aeration tank was 20 days or more, and high concentrations of nitrifying bacteria could be retained, and high nitrifying activity was exhibited. The nitrified liquid produced by the aerobic reaction was returned to the inflow of the anaerobic granulation and separation tank (AFPB) by the circulation lines, and the anaerobic denitrification reaction proceeded in the pellet fluidized bed. Since the effluent from the catalytic aeration tank contained the separated suspended components, the final treated water was disposed of or recycled through sand filtration or MF membrane filtration.

The results of the continuous operation, over a long period of time (two and a half years) are summarized in Table 3.15 (warm water period) and Table 3.16 (low temperature water period), which divide the results into a warm water period with a water temperature of 15°C or higher and a low temperature water period of 10 to 12°C, and examine the differences in removal characteristics.

Table 3.15 shows the results of operation with a total hydraulic residence time (HRT) of 4 hours (2 hours in the AFPB bioreactor; 2 hours in the catalytic aeration tank) and a circulation ratio (circulating water volume/influent sewage volume) of 3 during the period when water temperature was 15°C or higher. More than 95% of the 220 mg/l SS components and 3.5 g/l total phosphorus (TP) in the raw water were removed as pelleted sludge in the AFPB. The removal rate of the SS type COD is 85%. As can be seen in Table 3.16, even if the water temperature drops to around 10°C, the rate of removal of these pollutants in the AFPB bioreactor does not decrease.

At a water temperature of 15°C or more, denitrification of the AFPB bioreactor and nitrification of the catalytic aeration tank are almost completely achieved with a total residence time of 4 hours, and the total nitrogen (TN) removal rate is 80% or more. However, when the water temperature is 10 to 12°C, $NO_3{-}N$ and $NH_4^+{-}N$ remain in both reactors, reducing the nitrogen removal rate to 40–60%. Therefore, if the total HRT is extended to 7.5 hours (AFPB bioreactor: 3 hours; catalytic aeration tank: 4.5 hours), the nitrification rate and denitrification rate both increase, the TN removal rate recovers to around 80%, and the TN concentration in the final treated water becomes 4.6 mg/l, as shown in Table 3.16.

The granulated sludge (pellets) in the AFPB bioreactor not only physico-chemically constitute a high-density agglomeration but are also composed of microorganisms such as various bacteria. As a result of observation of the bacteria at normal water temperature and at low water temperature using a scanning electron microscope, a large number of cocci with a size of about 1 μm were found to be present together with filamentous fungi. No filamentous fungi were found in the winter sludge. Figure 3.57 compares the number of denitrifying bacteria in the aeration tank of the conventional activated sludge method with the number of denitrifying bacteria in the AFPB bioreactor. The number of bacteria in the AFPB bioreactor was of the order of tens of times higher than the number of bacteria in the conventional treatment plant. The sludge retention time (SRT) of the bioreactor was estimated to be about 2.2–8.1 days, with a mean retention time of 2 hours. Since the maximal specific growth rate of

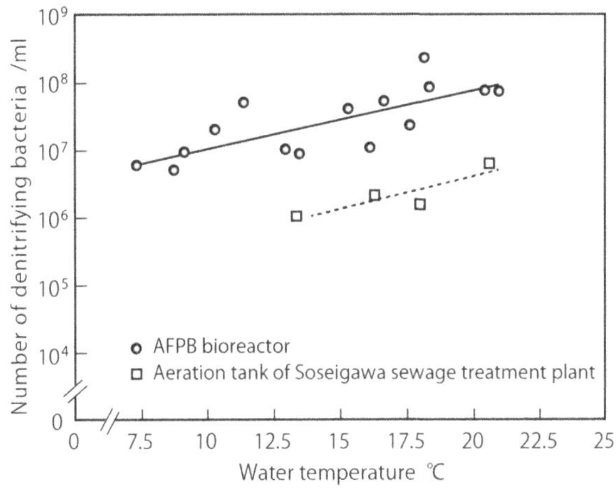

Figure 3.57 Comparison of the distribution of numbers of denitrifying bacteria.

denitrifying bacteria under anaerobic conditions is $\mu_m \doteqdot 0.2$ to 0.4 hr, the SRT of the AFPB bioreactor was much greater than $1/\mu_m$. Therefore, sufficient residence time for the growth of denitrifying bacteria was ensured (no wash out), so that the concentration of denitrifying bacteria could be maintained at a high level. In addition, the amount of substrate BOD necessary for the denitrification reaction was constantly replenished sufficiently, which resulted in a stable treatment process.

3.6.4 Process/system configuration: kinetics and dynamics

This chapter has focused on equilibrium and interaction, to understand and evaluate the development of different water treatment processes and systems, primarily by combining the effects of their use on water quality. However, in order to design and operate real world treatment systems and urban water-metabolizing systems, kinetic expressions are essential, and we need to be familiar with dynamics in order to respond to changing environmental and social conditions. The science that deals with these types of processes and systems in the field of engineering have evolved significantly since the 1960s, primarily in the field of chemical engineering – and the new knowledge and methods involved are now essential in many other fields of engineering, and sometimes even in the field of social sciences. Science and technology in the 20th Century exceeded that of the 19th Century through systematic development. From the prosperity of petrochemicals to more detailed biotechnologies and large-scale engineering such as nuclear power, discussions now extend to the boundaries of the earth and its ecosystem. However, the study of systems created in the 20th Century has not yet developed sufficiently for us in the 21st Century. The authors believe that this academic study will be completed in an era of systematics, in which problems such as how boundary conditions are summarized, which have been treated as differential equations, will be developed into integral equations.

When Tambo worked as a researcher in the Department of Chemistry in the University of Florida (Gainesville), USA, more than a half-century ago (1962–1963), he encountered three books in a Florida university bookstore that heavily affected his later life: 'Chemical reaction engineering' (Levenspiel, 1962), 'Transport phenomena' (Byron Bird *et al.*, 1962) and 'Physicochemical hydrodynamics' (Levich, 1962). Each was a new book just published that year, which later became creative classics as the foundation of chemical engineering worldwide. These were innovative descriptions of each process, providing a basic description and a path to future development of a comprehensive system.

Levenspiel's book represents a well-patterned mathematical expression of the progress of the phenomenon and shows the path to the system synthesis of the process. The book by Bird *et al.* describes the momentum (fluidics)/heat/mass transport using the same basic equations, taking the science of fluid/heat/material transport from the same viewpoint, and represents transport phenomena that have been developed in various fields of engineering in a unified manner. Levich was a young vice president of the Soviet Academy of Sciences. The founder of micro-hydrodynamics, which links chemical and physicochemical reactions with fluid behavior, presented basic insights into integrating transport and reaction processes and synthesizing problems into larger systems beyond the long-accustomed hydrodynamics. The original was written in Russian and the English version was a typewriter produced copy. There were a few errors, but I have not forgotten my amazement when I first had a look at it.

FURTHER READING

Degremont and Lyonnaise des Eaux. (1991). Water Treatment Handbook, 6th edn. Degremont water and environment. Degremont, Paris, Vols. **1 & 2**.

Edwards M. (1997). Predicting DOC removal during chemical coagulation. *Journal of the American Water Works Association*, **85**(5), 78–89.

Goda K. (1985). Basics of Water Quality Engineering. Maruzen Book, Tokyo. (In Japanese).

Kitahara F. and Watanabe M. (1972). Interfacial Electric Phenomena-Basics, Measurements and Applications. Kyoritsu Publishing, Tokyo. (In Japanese).

Montgomery J. M., Consulting Engineers, Inc. (1985). Water Treatment Principle & Design. John Wiley & Sons, New York, Chichester, Brisbane, Toronto, Singapore.

Orr C., Jr. and Dallavalle J. M. (1959). Fine Particle Measurement-Size, Surface and Pore Volume. The Macmillan Co., New York.

Sawyer C. N. and McCarty P. L. (1978). Chemistry for Sanitary Engineers, 3rd edn. McGraw-Hill Co., New York, St. Louis, San Francisco, Dusseldorf, London, Mexico, Panama, Sydney, Toronto.

Shimizu H. (1985). Ion Exchange Resin Technologies and Applications. Kyoritsu Publishers, Tokyo (In Japanese).

Shimizu T., Tambo N., Kudo K. and Ozawa G. (1993). Advanced sewage treatment by granulation precipitation and biological treatment process. *Journal of Japan Sewage Works Association*, **29**(339), 42–52. (In Japanese).

Suzuki T., Tambo N. and Ozawa G. (1993). A new treatment system with fluidized pellet bed separator. *Water Science and Technology*, **27**(11), 185–192.

Tambo N. (1980). Water Supply. (Civil Engineering Series vol.88, Japan Society of Civil Engineers). Gihoudo Publishing, Tokyo. (In Japanese).

Tambo N. (1983). Water Supply and Trihalomethane. Gihoudo Publishing, Tokyo. (In Japanese).

Tambo N. (1985) Evaluation of the Flow and Material Balance in Water Environment. Japan Society of Civil Engineers, Paper 363. (In Japanese).

Tambo N. and Kamei T. (1980). Water quality conversion, matrix of aerobic biological processes. *Journal of Water Pollution Control Federation*, **52**(5).

Tambo N. and Kamei T. (1989). Evaluation of Extent of Humic Substance Removal by Coagulation. Advances in chemistry series No.219. American Chemical Society, Washington DC.

Tambo N. and Ogasawara K. (1985). Water Purification Technology ('Josui no gijyutsu'). Gihoudo Publishing, Tokyo. (In Japanese).

Tambo N., Wackernagel M. and Rees W. E. (1996). Our ecological foot print: Reducing human impact on the Earth. New Society Publishers, Gabriola Island, Canada (1996).

Weber W. J., Jr. (1971). Physicochemical processes for water quality control. Wiley-Interscience, New York, London, Sydney, Toronto.

REFERENCES

Andersson B. and Rosen B. (1990). Upgrading for nitrogen removal some full-scale experience from Sweden. *Water Science & Technology*, **22**(7–8), 93–104.

Byron Bird R., Stewart W. E. and Lightfoot E. N. (1962). Transport Phenomena. John Wiley and Sons, Inc., New York & London.

Clarke F. W. and Washington H. S. (1924). The composition of the Earth's crust. Professional Paper. United States Geological Survey (USGS), **127**. USGS, Washington.

Edzwald J. K. and Van Benschoten J. E. (1997). Aluminium Coagulation of Natural Organic Matter. In: Chemical Water and Wastewater Treatment, H. H. Hahn and R. Klute (eds). Springer, Berlin Heidelberg New York, pp. 341–359.

Henze M. and Harremoës P. (1990). Chemical-Biological Nutrient Removal — The HYPRO Concept. In: Chemical Water and Wastewater Treatment, H. H. Hahn and R. Klute (eds). Springer, Berlin, Heidelberg, pp. 499–510.

Hoffmann E. and Klute R. (1990). Improving the denitrification potential in biological wastewater treatment by dosing carbon from sludge hydrolysis. In: Chemical Water and Wastewater Treatment, H. H. Hahn and R. Klute (eds). Springer, Berlin, Heidelberg, pp. 53–56.

Jorgensen P. E. (1990). Biological hydrolysis of sludge from primary precipitation in chemical water and wastewater treatment. In: Chemical Water and Wastewater Treatment, H. H. Hahn and R. Kute (eds). Springer, Berlin, Heidelberg, pp. 511–520.

Karlson L. and Smith G. (1991). Pre-precipitation facilitates nitrogen removal without tank expansion. *Water Science & Technology*, **23**(4–6), 811–817.

Kruyt H. R. (1949 & 1952). Colloid Science. Elsevier publishing Co., Amsterdom, Houston, New York, vol. I & II.

Levenspiel O. (1962). Chemical Reaction Engineering. John Wiley and Sons, Inc., New York & London.

Levich V. G. (1962). Physicochemical Hydrodynamics. Prentice Hall, Englewood Cliff, N.J.

Mysels K. J. (1959). Introduction to Colloid Chemistry. Interscience publishers, Inc, New York, London.

Shimizu T., Tambo N., Kudo K. and Hamguchi T. (1993a). High efficiency sewage wastewater treatment system consisting of fluidized pellet separator, aerobic bioreactor and anaerobic digester. *Journal of Water and Wastewater*, **35**, 425–432. (In Japanese).

Shimizu T., Tambo N., Kudo K., Nakabayashi A., Hamaguchi T. and Suga U. (1993b). An advanced wastewater treatment system consisting of anaerobic fluidized bed bioreactor and an aerobic bioprocess, Proceedings of Envir. Sanit. Eng. Res.Vol.**30**, pp. 187–196.

Shimizu T., Tambo N., Kudo K. and Ozawa G. (1994). An aerobic fluidized pellet bed bioreactor process for simultaneous removal of organic, nitrogenous and phosphorous substances. *Water Research*, **28**(9), 1943–1952.

Stumm W. and Morgan J. J. (1981). Aquatic Chemistry, 2nd edn. John Wiley & Sons, New York, Chichester, Brisbane, Toronto.

Tambo N. (1964). 'A Fundamental Investigation of Coagulation and Flocculation'. Doctor of Engineering Thesis, Hokkaido University, Japan.

Tambo N. (1994). Water Metabolism and Water Environment Transition in Modern Urban Cities. Proceedings of an International Symposium on the 80th Anniversary of the Japan Society of Civil Engineers. Japan Society of Civil Engineers, Tokyo (in Japanese).

Tambo N. (2012). History and Technologies of Urban Water Metabolism Systems. Kashima Publication, Tokyo. (In Japanese).

Tambo N. and Kamei T. (1975). Water quality matrix and water treatment processes. *Chemical engineering (kagaku kougaku)*, **39**(10), 544–549. (In Japanese).

Tambo N. and Kamei T. (1976). Assessment of urban water metabolism by matrices. *Journal of Japan Water Works Association*, **39**(502), 2–24. (In Japanese).

Tambo N. and Kamei T. (1978). Treatability evaluation of general organic matter. Matrix conception and its application for a regional water and waste water system. *Water Research*, **12**, 931–950.

Tambo N., Kamei T. and Takahashi M. (1981). Behavior and Properties of Metabolic Waste Components from Aerobic Biochemical Processes. *Journal of Japan Sewage Works Association*, **18**(210).

Tambo N., Kihara Y. and Ohba A. (1993). A new sludge treatment system using fluidized bed wet granulation and filtration method. *Journal of Japan Sewerage Works Association*, **30**(360), 42–48.

World Health Organization (WHO) (2008). WHO Guidelines for Drinking water Quality, 3rd edn. WHO, Geneva, Vol. **1**.

Chapter 4

Particles and particle separation basics

4.1 COAGULATION MECHANISMS OF COLLOIDS

4.1.1 Nature of colloids

4.1.1.1 Coagulation: Growth process of colloids for separation

Classical water treatment processes (geochemical and ecological processes) do not have a direct means of solid-liquid separation for impurities suspended in water with diameters ranging from 1 μm to less than 10 nm. Water resources in the ground include particles of the order of micrometers such as clay (which originates from soil systems and has a main particle size distribution in the region of 10^{-3} to 10^{-2} mm and which is a turbidity component of water) and coarse colloidal particles, such as microbial algae and protozoa.

These are the biggest ingredients that impair water clarity. But there is also a group of micro colloids with particle size distributions in the region of 10^{-6} mm (nm) derived from natural humic substances which are called color components, and which have long been an obstacle to water use. In addition, a diverse group of microorganisms are present in the micrometer to nanometer size, including those that are pathogenic even at low concentrations (such as bacteria and viruses), and which are among the biggest obstacles to water safety.

These particle size components are the major components addressed in water treatment, as described in Chapter 2 for colloidal properties. In Chapter 3, in Figures 3.6 and 3.27, impurity size and removal processes are linked. In addition, in the matrix labeling section of Chapter 3, the authors explained that in evaluating treatability (as shown in Figure 3.32) by incorporating physicochemical coagulation treatment and aerobic microbial treatment into a solid-liquid separating process, it is possible to cope with the removal of a wide range of impurities.

The functional membrane separation process, which emerged at the end of the 20th Century, enables direct separation of a wide range of impurities, ranging from soluble mineral salts to crude colloids, by membrane permeation, and enables the construction of a water treatment system train without the need for size growth treatments of impurities (such as coagulation). Membrane separation is a physiological

Author(s): Norihito Tambo and Koichi Ogasawara
doi: 10.2166/9781789061307_0129

and biological separation method that is different from mimics of classical geochemical and ecological separation methods. Membrane separation process focusing on the fact that the separation function of individual living organisms such as animals and plants is composed mainly of membranes. However, even if direct separation is feasible, it has become increasingly clear that it is desirable to perform appropriate coagulation treatment as a pretreatment in order to drive the separation continuously and effectively at a low energy-consumption rate.

4.1.1.2 Aggregation and dispersion of colloidal particles

Colloids have dimensions of 10^{-5} to 10^{-3} mm, as described above. It is possible to classify colloids from the shape of particles and their aggregate states:

(1) Granular colloids – those that are colloids in three dimensions;
(2) Filiform (thread) colloids – filaments with only two colloidal dimensions; and
(3) Membrane colloids – with colloidal dimensions in one dimension only

From a distributed perspective:

(1) Dispersion colloids are micro-multiphase systems consisting of a dispersion medium (water, etc.) and a dispersed phase;
(2) Associated (micelle) colloids are solute molecules assembled into colloidal particles (soap is an alkali salt of higher fatty acids with a molecular mass of about 200 to 300 Da, which collects about 50 to 100 molecules in water to form micellular colloids); and
(3) Molecular colloids are true solutions of high molecular weight polymers (proteins, starches, etc.) in solution, etc.

Water-soluble micellar colloids and molecular colloids are more stable than water-dispersible colloids. The reason is that micellar colloids and molecular colloids are true solutions that have an affinity for water, whereas dispersed colloids have a low affinity for water (dispersed by water molecules). Therefore, from the point of view of affinity for water, the former are called hydrophilic colloids and the latter hydrophobic colloids.

Dispersed colloids account for the majority of the colloids we target, but they can be divided into eight categories, as shown in Table 4.1, depending on the combination of the dispersant and the dispersed phase.

Table 4.1 Classification of colloidal suspensions and accumulation systems.

Dispersion Medium	Dispersed Phase	Dispersion System	
		Suspension System	**Accumulation System**
Gas phase	Gas	Fog } Aerosol	—
	Solid	Smoke } Aerosol	Powder, Xerogel
Liquid phase	Gas	Bubble	Froth
	Liquid	Emulsion	Cream
	Solid	Suspension Sol - Especially in the colloidal size	Gel { Colloidal precipitate Jelly }
Solid phase	Gas	Solid colloid	Solid colloid
	Liquid	Solid colloid	Solid colloid
	Solid	Solid colloid	Solid colloid

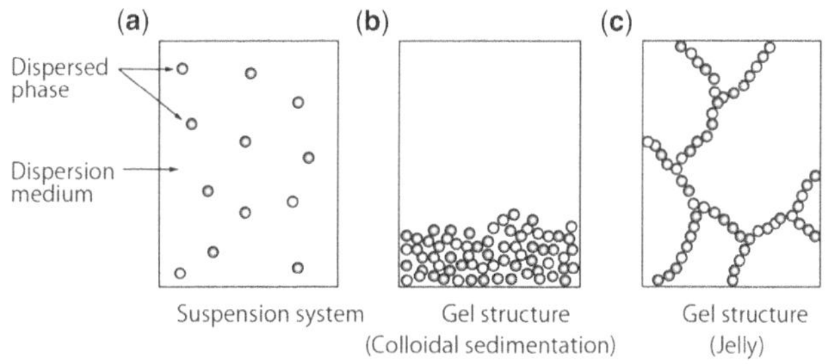

Figure 4.1 The structure of a colloidal dispersion system.

Colloids in which a gas is used as a dispersion medium and for a dispersed phase do not exist because gas molecules are uniformly mixed. Dispersion colloids are divided into a suspension system in which the particles constituting the dispersed phase are present independently in the dispersion medium, and in a collection system in which the particles float or deposit, or form a bonded state, as shown in Figure 4.1.

4.1.1.3 Movement of colloids

4.1.1.3.1 Brownian motion

Colloidal particles exhibit mechanical (kinetic) behavior that is characteristic in the dimension. A typical behavior is Brownian motion. The medium (water) molecules surrounding the colloidal particles constantly undergo random thermal motion and collide with the colloidal particles from all directions. When particle size is large, the collisions of water molecules from the respective directions cancel each other out and are averaged, and the particles do not move. However, the collision of water molecules with much smaller colloidal particles is uneven, and therefore the impact of the water molecules on the colloid particles fluctuates, and the impact force of the water molecules colliding by thermal random motion in succession causes a 'random' motion of colloid particles. This phenomenon has been termed Brownian motion (movement) since Robert Brown observed it in 1827 in the random motion of particulates spilled from pollens ruptured in water.

The kinetic energy E_{kb} of the motion of individual Brownian particles is given by Equation 4.1.

$$E_{kb} = \frac{1}{2} mV'^2 = \frac{1}{12} \rho \pi d^3 V'^2 = \frac{3}{2} KT \qquad \text{—(4·1)}$$

m : Particle mass
V': Square velocity fluctuation of the particle = $(v'^2)^{1/2}$
v' : Moment velocity fluctuation of the particle
ρ : Particle density
d : Particle size
K : Boltzmann's constant
= 1.381×10^{-15} erg/℃
= 1.381×10^{-23} m^2 kg sec^{-2} K^{-1}
T : Absolute temperature

(4.1)

Equation 4.1 The Kinetic energy of Brownian movement particles.

As can be seen from Equation 4.1, the Brownian kinetic energy E_{Kb} of a single colloidal particle is constant regardless of the size of the particle. Take a case where water temperature is 25°C as an example: E_{kb} (25°C) = 6.175×10^{-14} [erg] or [g/cm^2/sec^2], which is a constant value.

Table 4.2 Particle size and squared fluctuation rate.

Particle Size d	**10^{-7} cm (1 nm)**	**10^{-6} cm (10 nm)**	**10^{-5} cm (100 nm)**	**10^{-4} cm (1 µm)**	**10^{-3} cm * (1 µm)**
Square fluctuation rate V'^2	9430 cm/sec	296 cm/sec	9.43 cm/sec	0.296 cm/sec	0.00943 cm/sec

*Outside of colloid size

Assuming that $\rho = 2.65$ for clays, the squared fluctuation rate of various size soil particles is shown in Table 4.2.

Figure 4.2 shows the Brownian motion of particles, plotting their positions on the XY plane at regular time intervals. The position of each interval is connected by a straight line but, in reality, a zigzag motion is carried out during this period as well. The number n_ℓ of times colloidal particles are present between ℓ and $(\ell + \Delta\ell)$ during time t after n measurements is given by Equation 4.2, which is similar to the Maxwell velocity equation in molecular kinetics.

$$\frac{n_l}{n} = 2ae^{-al^2} l \Delta l \quad \text{—(4·2)} \tag{4.2}$$

n : Number of measurements
n_l : Number of colloidal particles are present in the l and (l+Δl) between within t time.
a : Particle size

Equation 4.2. he Brownian motion of particles.

Einstein (1905) and von Smoluchowski (1906) both independently revealed that Brownian motion caused colloidal particles to move in one direction. The mean squared distance Δ is given by Equation 4.3.

$$\bar{\Delta}^2 = \frac{RT}{N}\frac{t}{3\pi\eta r} \quad \text{—(4·3)} \tag{4.3}$$

$\bar{\Delta}^2$: Average square value of the distance Δ that colloidal particles move.
R : Gas constant
N : Avogadro's number 6.03×10^{23}
T : Absolute temperature (K)
η : Viscosity of the dispersant (g/cm·sec)
r : Particle radius (cm)
t : observing time

Equation 4.3. The Smolchowski equation for Brownian motion.

4.1.1.3.2 Diffusion by Brownian motion

Diffusion is the movement of colloidal particles from high to low concentrations, and Brownian motion is the only driving force in a stationary body of water. The diffusion coefficient of Brownian motion shown by

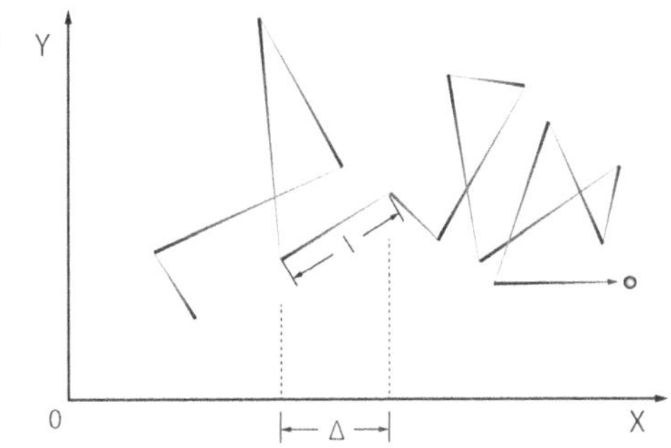

Figure 4.2 Particles demonstrating Brownian motion.

Einstein is given by Equation 4.4 using Equation 4.3.

$$D = \frac{\overline{\Delta x^2}}{2t} = \frac{KT}{6\pi \eta r} \qquad \text{—(4·4)} \qquad (4.4)$$

D : Coefficient defined by Fick equation a=D(dc/dx)
a : Number of moved colloidal particles in unit cross-sectional area during unit time
c : Colloid concentration
dc/dx : Concentration gradient

Equation 4.4. The diffusion coefficient of Brownian motion.

Even if a colloidal particle suspension is settled at static condition, the sedimentation velocity w is lower than the diffusion coefficient D, so that all the particles will not settle completely and will exist with a concentration gradient balanced with the concentration gradient of the diffusion transport. Solving the equation for the equilibrium condition in Equation 4.5, the equilibrium concentration of the hydrostatic sedimentation cylinder particles in the depth direction is shown as in Equation 4.6.

$$D\frac{dc}{dz} - wc = 0 \qquad \text{—(4·5)}$$

$$\ln\frac{c}{c_0} = \frac{2\pi r^3(\rho_c - \rho_0)N_g}{3RT} Z \qquad \text{—(4·6)} \qquad (4.5\text{–}4.6)$$

D : Diffusion coefficient
c_0 : Colloid concentration at the bottom of settling tube
c : Colloid concentration at Z_0 from the bottom
w : Colloidal particle settling velocity
r : Particle radius
ρ_c, ρ_0 : suspension density of colloid concentration c_0 and c

Equation 4.5, 4.6. The Equilibrium particle concentration distribution in the depth direction of hydrostatic settling tube.

As we have seen, Brownian motion and diffusion are the same molecular motion phenomena, but when we look at the number of colloidal particles present at a given moment in a colloidal suspension in a given space, the values differ depending on the time (instant) observed. This phenomenon is called 'fluctuation'. This is the third expression of colloidal molecular motion.

4.1.1.3.3 Viscosity of the colloidal suspension

The viscosity μ of a liquid is defined by the Newtonian law of Equation 4.7.

$$\tau = \mu \frac{du}{dx} \qquad \text{—(4·7)} \tag{4.7}$$

τ : Local shear stress
u : velocity (x direction only)
μ : Viscosity
du/dx : Velocity gradient

Equation 4.7. Liquid viscosity.

A fluid with viscosity μ independent of the velocity gradient is referred to as a Newtonian fluid. Various colloidal suspensions have a unique viscosity, and for a dilute suspension of colloidal particles of a given size in μ_{sp}, the Einstein Equation 4.8 is used.

$$\mu_{sp} = \mu_0(1 + 2.5\Phi) \qquad \text{—(4·8)} \tag{4.8}$$

μ_{sp} : Viscosity of colloidal suspension
μ_0: Viscosity of suspension medium (water)
Φ: Volume fraction of colloidal particles
(This equation applies to $\Phi < 0.003$)

Equation 4.8. The Einstein equation for viscosity of colloidal suspensions.

4.1.1.4 Optical properties of colloids

4.1.1.4.1 Turbidity

In colloidal suspensions, the liquid appears cloudy because the suspended particles scatter light. Assuming that the intensity of the incident light is I_0, the intensity of the transmitted light is I_t, the turbidity of the colloidal suspension is τ, and the thickness of the colloidal suspension is ℓ, Equation 4.9, Lambert's Law, holds.

$$\frac{I_t}{I_0} = e^{-t^{*}l} \qquad \text{—(4·9)} \tag{4.9}$$

I_0 : Intensity of incident light
I_t : Intensity of transmitted light
t^* : Turbidity
l : Colloidal suspension thickness

Equation 4.9. Lambert's law.

Equation 4.10, which is similar to Beer's Law, holds true for turbidity and suspensions.

$$t^* = k^* N \quad —(4 \cdot 10) \tag{4.10}$$

k^* : Turbidity corresponding to turbidity per a particle
N : Number of particles in unit volume of suspension

Equation 4.10. Beer's law.

Since it is more convenient to use density C [g/cm^3] than to use the number N experimentally, let $\tau = KC$ be defined as turbidity (specific turbidity). In general, turbidity in water treatment is defined as turbidity that exhibits intensity in units of suspension of 1 mg/l of kaolinite.

4.1.1.4.2 Tyndall effect

The Tyndall effect is a phenomenon in which a strong light beam is applied to a colloidal dispersing system, and the optical path appears to shine when viewed from a direction perpendicular to the direction of the light beam. The intensity I (the induced vibration of a type of electrodipole) of scattered light when the particle size is much smaller than the wavelength of light was shown by Lord Reyleigh (see Equation 4.11).

Equation 4.11 shows that shorter wavelengths of light are less likely to be scattered, while larger size particles are more likely to be scattered. If the particles are larger than the wavelength of light, a large number of dipoles are induced in a particle, and the scattered light generated interferes with each other, thereby disabling the relationship shown in Equation 4.11.

$$I = K' \frac{NV^2}{\lambda^4} \sin^2 \alpha \quad —(4 \cdot 11) \tag{4.11}$$

N : Number of particles in unit volume of suspension
V : Particle volume
α : Angle between the incident and scattered light
K' : Constant
λ : Wavelength of light

Equation 4.11. The Scattered light intensity due to Tyndall effect.

4.1.1.4.3 Colloidal color

In general, the color of an object is caused by absorption of a portion of visible light, but in colloidal suspensions (solutions), coloration is also caused by scattering as well as absorption. We can measure the minimum concentration of a liquid layer of a given thickness that is found to be barely colored, and consider the inverse of the minimum concentration as the coloring power. When Cu^{2+} is 1, Fe^{3+} is about 3, but an iron hydroxide colloid has a coloring power of 500 and a gold colloid has a coloring power of 20,000, and colloidal solutions (suspensions) make a remarkable change in the appearance of water depending on the colloidal components.

The brown color produced by humic substances (which are produced when plants are decomposed by microorganisms and which are universally present in natural waters) is called the 'color of water'. Small fluvic acid molecules are about 300 Da in size and are soluble categories, whereas polymeric humic acids are color colloids with a size of less than 1500 Da. Color colloids are components of all parts of a water resource system, including aerobic treated sewage, anaerobic digested sewage supernatants and wetlands and closed lake waters, and effluents from the A_I surface soil layer (accumulated fallen leaves and grasses on top ground surface with rich natural organics) of forests. A brown color of 1 mg of platinum and 0.5 mg of cobalt in 1 litre of water is evaluated as a color of 1 degree. It is a universal indicator of metabolic emissions in ecosystems that are quite different in different hues. Although it is not harmful in itself, it reacts with disinfecting chlorine to form a raw material for chlorinated organic compounds such as trihalomethane.

4.1.1.5 Colloidal charge

Most of the colloidal particles suspended in natural water are loaded with negative charge and the electrical double layers shown in Figure 3.13 in Chapter 3 are on the surface, where the oppositely charged particles (counter ions) are predominantly present as charge groups. The zeta potential (ZP) developed on the sliding surface between the water molecules attached to the particles and the bulk water mass is commonly used as an indicator of a representative potential as a measurable potential. In many cases, natural colloids such as clay and humic material have a ZP of −20 to −25 mV, and the charged particles repel each other in the group of suspended particles to form stable colloidal dispersions. When colloidal particles and ions etc. with opposite charges are added to such a system to neutralize the charge in the electrical double layer on the suspended particle surface, the ZP drops and approaches zero, as shown in Figure 3.14. The ZP zero state is called an iso-electric point, and it is considered that the electrical mutual (repulsion) action between suspended particles disappears. As discussed in Figure 3.13, charge neutralization proceeds only to the iso-electric point if the oppositely charged particle is a simple counter ion. However, if the oppositely charged particle is a polymeric substance, specific adsorption to the Stern layer proceeds, and the ZP of the colloidal particle reverses, resulting in a positively charged suspended particle. The presence or absence of this phenomenon also helps determine whether oppositely charged particles are macromolecules or not.

Colloids or sols in water are charged either positively or negatively. As described above, the positive and negative charges are derived from the nature of the suspended particles themselves, but the following three cases are considered to be encountered in water treatment:

(1) In chemically inert substances such as water bubbles and paraffinic particles, there is a weak negative charge caused by selective adsorption of OH^- ions.
(2) In examples which include a defect in the crystal lattice on the surface of minerals (a failure of the internal equilibrium at the fracture plane), such as in clay particles, a negative charge is generated when heterologous atoms enter the crystal, such as the replacement of Si^{4+} with an Al^{3+}, as shown in Figure 4.3.
(3) In metal hydroxides and metal oxides, OH^- and H^+ ions form potential-determining ions that change the charge from positive to negative regions depending on the pH. Such compounds are called ampholytes. The charge of these oxides and hydroxides becomes positive at low pH and negative at high pH.

In the last case, the colloidal particles' charge is caused by the divergence of complex ions. For example, if they are from an ionizing functional group, such as a hydroxyl group (—OH—), a carboxylic group (—COOH), a sulfonic group (—S2OH), or a phosphoric acid group (—PO2), the colloidal charge

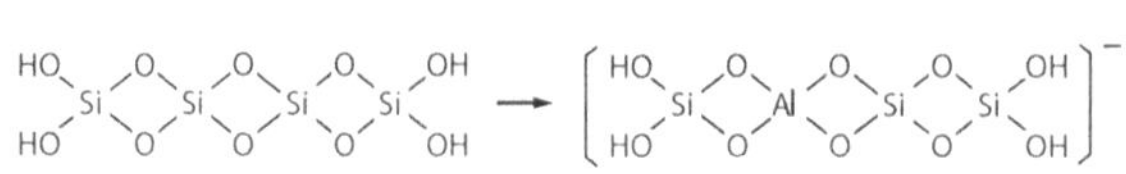

Figure 4.3 An example of a crystal of heterologous form.

Table 4.3 Examples of protein charge and pH.

Low pH side	Intermediate pH	High pH side
$R^{COOH}_{NH_3^+}$	$R^{COO^-}_{NH_3^+}$	$R^{COO^-}_{NH_2}$
Positive charge	Isoelectric point	Negative charge

depends on the degree of ionization (dissociation) and the pH of the system. Because the surface charges of proteins and their aggregates (microorganisms in water) have an amino group (NH2) that dissociates on the acidic side to give a positive charge, and a carboxyl group (—COOH) that dissociates on the alkaline side to give a load, the positive and negative charges vary with high and low pH, as shown in Table 4.3.

In natural colored water, carboxyl and hydroxyl groups are dissociated to form a negatively charged colloidal solution.

If the colloidal particles' charge is high, the ratio of the amount of negative charge adsorbed to the Stern layer becomes large, the change in the potential of the Stern layer becomes rapid, and the total potential drop (typically by the ZP) is mainly governed by the change in the Stern layer potential Φ_d.

The ZP is often measured by electrophoresis and is determined by Equation 4.12 from the rate at which charged particles are attracted to oppositely charged poles under a certain potential gradient, and rewritten to a practical unit by Equation 4.13.

$$Z.P. = \frac{4\pi U \mu}{DH} \quad \text{—}(4\cdot 12)$$

Z.P. : Zeta potential (e.s.u)
μ : Viscosity coefficient of water (g/cm·sec)
U : Particle mobility (cm/sec)
D : Water dielectric constant (≒80)
H : Electric field strength (e.s.u)

Practical unit system :

$$Z.p. = \frac{4\pi\mu}{D}\,\frac{l}{t}\,\frac{A}{i}\,\frac{1}{R_s} \quad \text{—}(4\cdot 13) \qquad (4.12\text{–}4.13)$$

Z.P. : Zeta potential (mV)
l : Particle travel distance during time t (cm)
t : Time to move distance l (sec)
A : Cross sectional area of electrophoresis cell (cm^2)
i : Direct current flowing through the cell (mA)
R_s : specific resistance of the suspension (Ohm-cm)

Equation 4.12, 4.13. Zeta potential

In practice, mobility M is often used instead of the ZP to show relatively charged states (Equation 4.14).

$$M = \frac{U}{H} = \frac{lA}{t\,i\,R_s} \quad \text{—}(4\cdot 14) \qquad (4.14)$$

M : Mobility
(μ/m/sec)/(v/cm)

Equation 4.14. Mobility.

4.1.2 Stability of colloid suspensions

4.1.2.1 Stability and instability of colloids

Two factors determine the state of colloidal particles suspended in water. The first is the stabilizing element that tends to keep colloids dispersed, and the second is the unstable element that tends to coagulate them (Table 4.4).

There are two phenomena that act as stabilizing elements, one of which is the hydration of the surfaces of particles. In hydrophilic colloids, a layer of water molecules adsorbed on the surfaces of the particles is sandwiched and prevents direct contact of the colloidal particles.

The other phenomenon is an electric double layer. The electric repulsion between the charged particles prevents the particles from approaching each other.

The most important instability factors are:

(1) The Brownian motion of colloidal particles and turbulence fluctuations in slightly grown particulate masses. Both of these give particles an opportunity to collide with each other. They also overcome the electrical repulsion forces typically present on the particle surface and provide the kinetic energy necessary to bring the particles directly together.

(2) In addition, and to be described below, the London van der Waals force acting between particles as the short-range physico-chemical force that maintains bonding when particles approach each other. Other chemical bonds include ionic bonds, covalent bonds, and hydrogen bonds.

In the case of hydrophilic colloids, removing the adsorbed water layer on the hydrated surface creates an opportunity for the particles to directly associate with each other by Brownian motion etc., and the particles adhere to each other by their own physical and chemical bonding forces (London van der Waals forces and hydrogen bonding forces etc.) and agglomerate.

For colloidal suspensions that are electrically stable, electrolytes are added to reduce electrical repulsion and create a situation where London van der Waals forces are dominant.

Colloidal particles that are stably present are referred to as primary particles, and aggregated particles are referred to as secondary particles, and the states of individual particles are sometimes classified.

In hydrophobic colloidal solutions (sols), when an electrolyte is added, oppositely charged ions (counter ions) are adsorbed on the colloidal surfaces as described in the previous section, and the potential of the diffusing double layer is lowered, resulting in coagulation with ease by intermolecular force to form secondary particles. Thus, the minimal level of electrolyte required to produce coagulation in hydrophobic colloidal particles is called the coagulation value. This is the primary function of a coagulant, which will be described later. Comparing the coagulation values of various electrolytes, the higher the valence of oppositely charged ions, the stronger the coagulation force. Therefore, the coagulation value decreases rapidly with a smaller valence. For monovalent ions: 10^1 to 10^2 m mol/l are needed; for divalent ions: 10^0 to 10^{-1} m mol/l; and for trivalent ions: 1 to 10^{-2} m mol/l are needed

Table 4.4 The stabilizing and destabilizing factors affecting colloids.

Stabilizing elements	① Hydration of the particle surface	Contact disturbance
	② Electric double layer	Electrical repulsion
Unstabilizing elements	① Brownian movement	Collision kinetic energy
	Turbulent fluctuation	〃
	② London-van der Waals' forces	Physical absorption
	Chemical bonding force	Chemical bond

Table 4.5 Coagulation values for different electrolyte ions.

	Monovalent Ion	Divalent Ion	Trivalent Ion
Coagulation value	$10^1 \sim 10^2$ m mol/l	$10^0 \sim 10^{-1}$ m mol/l	$10^0 \sim 10^{-2}$ m mol/l

(Table 4.5). The coagulation value decreases with the ionic value in a geometric series. This phenomenon is called the Hardy Schulze law.

Because hydrophilic colloids are hydrated, coagulation does not proceed easily with the addition of electrolytes alone. This is because the dispersion and stabilization of colloidal particles is caused not only by the repulsion of surface charges but also by solvation (hydration). Dehydration must be used in combination with advanced coagulation. Generally, even if a dewatering process using alcohols or similar is not specifically considered, a large amount of electrolyte can be added to neutralize charges and at the same time induce dehydration phenomena, thereby generating coagulation. This process is called 'salting out'.

The effectiveness of salts on hydrophilic sols depends on the extent of their dewatering. The order indicating the strength of the salting out force is called the liquid separation order (Lyotropic series) or the Hoffmeister series. For monovalent cations with the same anion, $Cs^+ > R_b{}^+ > NH_4{}^+ > Na^+ > L^+$ decreases in that order. For monovalent anions with the same cation, it is known that $F^- > IO_3^- > H_2PO_4{}^- > BrO_3{}^- > Cl^- > ClO_3{}^- > Br^- > NO_3{}^- > ClO_4{}^- > I^- > CNS^-$ (Table 4.6).

Hydrophobic colloids or hydrophilic colloids adsorbed on the surfaces of hydrophobic particles stabilize and make coagulation difficult. These colloidal particles are called protective colloids.

4.1.2.2 Potential coagulation barriers and driving energy inducing collisions

Suspended particles come closer together and collide or agglomerate with each other, leading to the progression of coagulation. In discussing such agglomeration of particles, we must first consider what is shown schematically in Figure 3.15 of Chapter 3, as follows:

(1) The potential for electrical repulsion (electrical double layer interaction) between the primary particles;
(2) The reciprocal attraction potential based on London-van der Waals forces acting between particles; and
(3) Whether both particles can make direct contact beyond the energy-barrier Emax created by the synthetic potential.

This energy barrier is a barrier that must inevitably be crossed in order for two particles to come together directly and is considered to be the activation energy of coagulation reactions. In order for both particles to exceed the synthetic energy barrier E_{max}, the particles must have enough kinetic energy to exceed (pass over) the barrier. For a colloidal size with smaller particles, the energy is given by Brownian motion, whilst for grown coagulated agglomerates of a larger size, the kinetic energy of turbulent fluctuations of the

Table 4.6 The Lyotropic series.

Monovalent cation with the same anion	$Cs^+ > Rb^+ > NH_4{}^+ > Na^+ > Li^+$
Monovalent anion with the same cation	$F^- > IO_3{}^- > H_2PO_4{}^- > BrO_3{}^- > Cl^- > ClO_3{}^- > Br^- > NO_3{}^- > ClO_4{}^- > I^- > CNS^-$

micro-turbulent eddies becomes the drive force. The former could be referred to as 'Brownian motion coagulation' and the latter as 'turbulent coagulation'.

As can be seen from Equation 4.1, the magnitude of the kinetic energies of the approaching particles is constant for Brownian motion coagulation and is $3KT/2 \fallingdotseq 6.175 \times 10^{-14}$ erg (or gcm^2 per sec^2 or 10^{-7} J) at 25°C. Depending on the temperature, the energy of the drive in the colloidal regions where the particles undergo Brownian motion coagulation is in the order of 10^{-14} erg or 10^{-20} J, and coagulation will not proceed unless the energy barrier for inter-association is lowered to near the iso-electric point, as shown in Figure 3.16. For this purpose, an agent, called a coagulant, is added to the electrical double layer and generates counter-ions. The functions and properties of a coagulant are described in more detail in the following sections.

As shown in Table 4.1 above, when particle size increases, the possibility of collision is not brought about by Brownian motion alone and coagulation does not proceed. As will be described later, the repulsion potential increases as particle size increases, and it becomes difficult to exceed the barrier of the repulsion energy of the order of 10^{-7} J with Brownian motion. For this reason, a driving force for the coagulation action to proceed must be sought. This driving force is expected to be effective for the turbulence fluctuation rate of the micro-water eddy mass under forced agitation.

The scale of the vortex which produces turbulent fluctuation velocities acting on the coagulation of the particles is very small and has a size of almost the same order as the diameter of the suspended particles, which produces coagulation. For such small vortex flow phenomena, we can apply the concept of locally isotropic turbulence proposed by A.N. Kolmogorov (see: Levich, 1962), and the characteristic nature of the turbulence can be estimated from its similarity hypothesis.

4.1.2.3 Kolmogorov's theory of local isotropic turbulence

We now outline Kolmogorov's theory of local isotropic turbulence to help understand subsequent issues. When a flow becomes turbulent, vortices of different size overlap the mean flow, and the flow is defined as the combination of vortices, as shown in Figure 4.4. The nature of these individual vortices is defined by their velocity and the distance over which the change in velocity is defined (meaningful). This

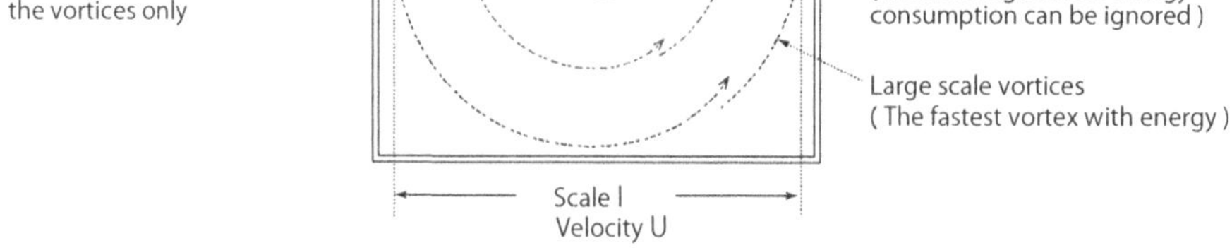

Figure 4.4 Schematic structure of turbulent vortices.

distance is called the turbulence scale ℓ, and the fastest vortex velocity on this scale is the difference ΔU between the mean flow velocities of two points separated by the turbulence scale ℓ. For example, considering a mixing pond with parallel flow paths, the largest turbulent vortex has a scale corresponding to the depth of the channel, and the velocity of this vortex is as large as the maximum flow velocity in the channel. If the Reynolds number of a vortex is defined as $\Delta U\ell/v$, the largest vortex has the same value as the Reynolds number of a water channel. There are many vortices in a turbulent flow with a smaller size λ and a lower velocity V_λ. Although the number of such small vortices is very large, the share of kinetic energy in the total flow is small. However, it is these small size vortices that act effectively on coagulation.

As the size λ of a vortex becomes smaller, the velocity V_λ of the vortex becomes smaller, and the Reynolds number $\lambda \cdot V_\lambda/v$ of the vortex becomes smaller and smaller. In large scale vortices (having a large vortex Reynolds number), viscosity has little effect on vortex flow and is not associated with energy dissipation (consumption). However, the superposition of such vortices produces small vortices, and further superposition produces much smaller vortices. Then, the vortex Reynolds number rapidly decreases with the decreasing size of the vortex, and the effect of energy consumption due to viscosity cannot be ignored. A small vortex of scale $\lambda = \lambda_0$, such as a vortex Reynolds number $Re(\lambda_0) = \lambda_0 \cdot V_{\lambda 0}/v = 1$, shows that viscous dissipation is not negligible due to the characteristics of Reynolds number. Therefore, kinetic energies are consumed (lost as heat) by viscosity in small vortex regions of $\lambda < \lambda_0$.

Thus, the energy imparted to the fluid by external forces creates large vortices, which in turn create small vortices that transfer kinetic energy to smaller vortices. Energy consumption can be ignored until the size of the vortex reaches the λ_0, where the vortex Reynolds number is 1 (unity). When the vortex size is smaller than λ_0, the effects of viscosity are not to be ignored. A vortex smaller than size λ_0 will always be accompanied by viscous energy consumption. In this way, size λ_0 vortex groups can be divided into two major groups: large inertial vortex regions that can ignore viscous effects, and viscous vortex regions that consume energy (Table 4.7). The scale λ_0 of the vortex dividing these two regions is called the microscale of the turbulence (where the vortex Reynolds number = 1).

In the inertia sub range area, the effects of viscosity can be ignored, so there are three elements defining the disturbance: the unit-time energy consumption (agitation strength) ε_0 (erg/cm^3s); the turbulence (vortex) scale λ (cm); and the fluid density ρ (g/cm^3). When a dimensional analysis is performed on these combinations, the equation representing the velocity of the vortex in the inertial domain is $V_\lambda = \alpha\,(\varepsilon_0\,\lambda/\rho)^{1/3}$, where α is the proportionality constant.

In a viscous sub range where the size of the vortex is smaller than the microscale λ_0, the viscosity of the agitation energy is not negligible, and in the viscous region, the dynamic viscosity coefficient v (cm^2/sec) of the fluid is added to the element considered in the inertial region. Dimensionless analysis in the same manner yields $V_\lambda = \beta\,(\varepsilon_0/\mu)^{1/2}\lambda$ where β is the proportionality constant.

Table 4.7 Vortex regions and energy consumption.

Eddy: Large	Re : Large	Inertia sub range	Viscous action negligible
	Eddy Re = 1	[Re attribute = Inertial force /viscous force]	
Eddy: Small	Re : Small	Viscous sub range	Energy consumption by viscosity

Substituting V_λ for $Re = (\lambda V_\lambda/\nu) = 1$ yields the microscale of the isotropic turbulence, $\lambda_0 = (\nu^3\rho/\varepsilon_0)^{1/4}$ (see Equation 4.15).

Vortex velocity equation

In Inertia sub range $V_\lambda = \alpha\left(\frac{\varepsilon_0 \lambda}{\rho}\right)^{\frac{1}{3}}$

In Viscous sub range $V_\lambda = \beta\left(\frac{\varepsilon_0}{\mu}\right)^{\frac{1}{2}}\lambda$

Substituting V_λ into $Re = \lambda V_\lambda/\nu = 1$

Microscale of turbulence λ_0 $\lambda_0 = \left(\frac{\nu^3\rho}{\varepsilon_0}\right)^{\frac{1}{4}}$ —(4·15)

V_λ: Vortex velocity of scale λ (cm/sec)
ε_0: Energy dissipation rate (m^2/sec^3)
ρ: Water density (g/cm^3)
ν: Kinematic viscosity (cm^2/sec)
μ: Viscosity (g/cm·sec)
α, β: Proportional constant

(4.15)

Equation 4.15. Vortex velocity and turbulence scale.

4.1.2.4 *Critical size for coagulation and flocculation to proceed*

Generally speaking, the area where vortices are available for coagulation under rapid agitation is considered to be a viscous area. This is because particles on a vortex with a scale much larger than the particle diameter rotate only by the movement of the vortex (imagine that two beans on a record on a turntable rotate without changing their relative positions), and the suspended particles cannot approach and collide. In order for the suspended particles to collide closely with each other, it is necessary to consider a turbulent transport in which particles riding on a small turbulent vortex with a size close to the diameter of the suspended particles are conveyed to adjacent vortexes in succession and come close to each other randomly. The velocity of turbulent fluctuations V' (the root mean square fluctuations) available for the transport of the suspended particles is then the, $V' \propto (\varepsilon_0/\mu)^{1/2}d$. Where d is a size similar to the diameters of the coagulating suspended particles.

The kinetic energy E_{kt} of suspended particles with turbulent flow fluctuations is shown in Equation 4.16.

$$E_{kt} = \frac{1}{2}mV'^2 = \frac{1}{12}\rho\pi d^3 V'^2 = \frac{1}{12}\gamma_s\pi\left(\frac{\varepsilon_0}{\nu}\right)d^5$$

$$= \frac{8}{3}\gamma_s\pi\left(\frac{\varepsilon_0}{\nu}\right)a^5 \quad \text{—(4·16)}$$

m : Spherical suspended particle mass ν : Water dynamic viscosity
γ_s: Specific gravity of suspended particle ε_0: Energy dissipation rate
d,a : Suspended particle diameter and radius

(4.16)

Equation 4.16. Kinetic energy of suspended particles with turbulent flow fluctuations.

Generally speaking, the intensity ε_0 of rapid agitation during coagulation treatment is considered to have a magnitude of about 10^3 erg·cm/s $= 10^{-4}$ J. From Equation 4.16, it can be inferred that practical drive energies are $E_{kt} = 9.38 \times 10^5 \times a^5$ erg.

Figure 4.5 shows how particle size and drive energies of suspended particles change from a Brownian motion coagulation region to a turbulent motion coagulation region using rapid agitation strength ε_0 as a parameter. Estimating the particle size at which the energy of Brownian motion is equal to the energy of turbulent diffusion transport, the radius $a = 1.45 \times 10^{-4}$ cm appears to be the branching point between the two, under the condition of agitation strength $\varepsilon_0 = 10^3$ erg/cm^{3}s. Beyond the particle diameter 1.5 μm, turbulent diffusion transport rapidly increases in proportion to the fifth power of the particle size, and rapidly becomes the dominant drive force.

When charge neutralization of the electric double layer proceeds until the absence of an energy barrier (zero ZP, the isoelectric point), the it is not necessary to count the magnitude of the drive energy, but the degree to which turbulent diffusion is effective beyond Brownian motion to increase the collisional velocity becomes the center of the problem.

As we have seen, the two main types of random motion (Brownian and turbulence fluctuations) provide the energy used to drive the relative collisions of particles. The first criterion is whether coagulation can proceed by exceeding the maximum synthetic potential barrier E_{max}, which indicates the magnitude of the interaction between two particles, resulting from the synthesis of the repulsion potential of the electrical double layer and attractive potential of the London van der Waals force at the particle interface, as shown in Figures 3.15 and 3.16. In most cases, the electrical double layer has various potentials of a certain magnitude (in many cases, it is typically represented at the ZP), and it is possible to discuss in detail whether or not the electric double layers can mutually repel each other and collision-agglomeration proceed under a certain potential condition. In the next section, we discuss the mechanisms and explain the process of proposing ± 10 mV as the practical critical coagulation potential (critical ZP) for the process.

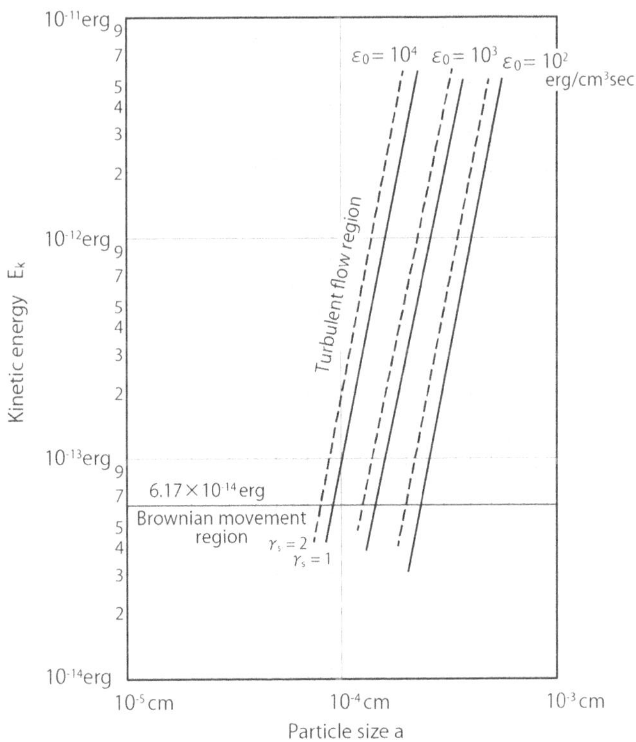

Figure 4.5 Driving energies and particle sizes in coagulation.

4.1.3 Critical zeta potential for coagulation

When the surface potential of the particles reaches the iso-electric point, coagulation can naturally occur. In the very many cases measured by the authors, coagulation occurs with reducing ZP to some level. Now we will discuss whether a practical critical ZP can exist or not when the interface potential is reduced to a certain level.

For the critical coagulation of colloidal particles in water treatment, Riddick (1961) describes fine particles such as natural colored particles to be coagulated about ZP $= \pm 5$ mV and coarse suspended particles to be coagulated about ZP $= \pm 10$ to 20 mV, empirically and qualitatively. However, no theoretical study has been conducted. It has also been reported that it is impossible to determine a certain critical ZP under widely varying external conditions. The author (Tambo, 1964a, b, c) experimentally and theoretically clarified that there is a critical ZP which can be used as an index of coagulation treatment in practice, and clarified that critical ZP can be used even for coagulation behavior and silica sol (such as classic Baylis silica) formation for water treatment as well.

A discussion of the interaction of two homogeneous suspended particles, or coagulation, begins with a quantitative study of the energy balance and predominance of the kinetic energy of Brownian motion or turbulent fluctuations (eddies) and the repulsive energy of the electric double layer when they interact, as shown in Figure 3.15. The DLVO theory (of Derjaguin, Landau, Verwey and Overbeek), which discusses the relation between dispersions and coagulation between two particles, quantifies the handling of problems.

The electrical energy distribution of particle repulsion is determined by multiplying the potential distribution in Figure 3.14 by the effective charge of the oppositely charged particle. This relationship is shown in Equation 4.17 when the surface potential of two particles of equal nature is not very high and the suspended particle size is relatively large compared to the thickness of the diffused double layer.

As shown in Figure 3.14, the potential distribution of the electrical double layer from the point of the actual process is based on the assumption that the opposite charged particles adsorbed on the Stern layer cover the surface of the particle and the particle is covered by the opposite charged particles. If the diffusion layer starts from the boundary potential Φ_d of the Gouy-Stern layer, it is easier to deal with the issues by considering the interference between the electric double layer. (Where Φ_d represents electric potential (electrostatic system of units (ESU)) at the interface between the Stern and Gouy layers).

$$E_r = \frac{a d \Phi_d^2}{2} \ln\{1 + \exp(-\kappa H)\} \qquad \kappa a \gg 1 \qquad (4\cdot17)$$

$$\kappa = \sqrt{\frac{4\pi e^2 \Sigma n_i Z_i^2}{DKT}} \qquad (4\cdot18)$$

E_r : Repulsive energy between two particles (10^{-7} or 10^7erg)
D : Water dielectric constant (≒ 80)
a : Particle radius (cm)
Φ_d : Interface potential of Gouy-Stern layer (e.s.u.)
$1/\kappa$: Length of the ionic atmosphere (cm)
H : Distance between two suspended particles
e : Elementary charge 4.803×10^{-10} (e.s.u.)
n_i : Number of i- ons in unit volume water (particle/cm³)
Z_i : Valency of i -ion
K : Boltzmann's constant
T : Absolute temperature

(4.17–4.18)

Equation 4.17, 4.18. Repulsive energy between two particles.

The potential energy E_a (London van der Waals force) at which two suspended particles attract each other is given by Equation 4.19.

$$E_a = -\frac{A}{6}\left[\frac{2a^2}{H^2+4aH} + \frac{2a^2}{(H+2a)^2} + \ln\frac{H^2+4aH}{(H+2a)^2}\right] \quad \text{— (4·19)}$$

If the distance H between two particles is very small then,

$$E_a \fallingdotseq -\frac{a}{12}\frac{A}{H} \quad \text{— (4·20)}$$

E_a : Attractive energy (erg or J)
A : London-van der Waals force constant

(4.19–4.20)

Equation 4.19, 4.20. Mutual attractive energy between two suspended particles.

Here, the London van der Waals constants for A are estimated to be around $A = 10^{-12}$ erg, although it is difficult to obtain them directly in an independent manner. E_a is approximated by Equation 4.20 when H is very small.

Adding Equations 4.17 and 4.19 or 4.20 together yields Equations 4.21 or 4.22 as the combined interaction potential energy equation, as is shown in Figure 3.15.

$$E_i = \frac{ad\Phi^2_d}{2}\ln\{1+\exp(-\kappa H)\} - \frac{A}{6}\left[\frac{2a^2}{H^2+4aH} + \frac{2a^2}{(H+2a)^2} + \ln\frac{H^2+4aH}{(H+2a)^2}\right] \quad \text{— (4·21)}$$

$$E_i = \frac{ad\Phi^2_d}{2}\ln\{1+\exp(-\kappa H)\} - \frac{a}{12}\frac{A}{H} \quad \text{— (4·22)}$$

(4.21–4.22)

Equation 4.21, 4.22. The combined interaction potential energy distribution equation.

As can be seen from Equation 4.21, the magnitude of the interaction energy increases proportionally to particle size a, whereas the drive energy is constant at $E_{kb} = 6.175 \times 10^{-14}$ erg (water temperature: 25°C) in the Brownian drive force region. Therefore, the height of the energy barrier increases proportionally as the particle size of the suspension increases. In the Brownian motion region, the larger the particle size, the more resistant it is to the coagulation progress. On the other hand, in the area dominated by locally isotropic turbulence, as shown in Figure 4.5, the drive force increases rapidly in proportion to the fifth power of the size of the active viscous vortex (the diameter of the micro-vortex corresponds to particle size). Thus, as the agglomerate particle size increases, the drive force increases by the fifth power of the particle size, rapidly exceeding the energy barrier growth, leading to easy progression of the coagulation reaction. This means that the increase in drive force as the agglomerate size increases rapidly outweighs the increase in the energetic barrier, leading to the progress of coagulation reactions. Therefore, the particle size that transit from the Brownian kinetic drive region to the turbulent fluctuation drive region, may be considered to be the most important factor for coagulation to proceed (critical size to promote coagulation).

The critical particle size when transiting from a Brownian movement drive region to a turbulent eddy drive region is shown in Figure 4.5. When the agitation energy intensity is $\varepsilon_0 = 10^4$ erg, the critical particle radius $a \fallingdotseq 10^{-4}$ cm = 1 μm is taken. When the agitation intensity is as low as $\varepsilon_0 = 10^2$ erg, the critical particle radius is as large as $a \fallingdotseq 2$ μm and it is difficult to smoothly connect the Brownian motion range to the turbulent flow range. It is important to increase the intensity of rapid agitation (energy dissipation rate) ε_0 to a level of 10^3 erg/cm^3 or more.

In order to discuss the critical moment which proceeds to coagulation, with respect to the coagulation energy barrier (activation energy for coagulation) and driving energy, the author Tambo selected particles at the critical diameter mentioned above, 1.5 μm, in the following discussions.

The resultant mutual potential curves in the vicinity of the particles as shown in Figures 4.6 show that when the Gouy-Stern interface potential is as low as $\Phi_d < 10$ mV, the attraction potential is superior in normal natural water (κ is 10^5 to 10^6 cm^{-1}) and the mutual coagulation proceeds unimpeded, as shown in Figures 4.6(a–c). However, when the interface potential Φ_d exceeds 15 mV, an energy barrier E_{max} develops, as shown in Figure 4.6d, and the magnitude of the kinetic energy E_k of suspended particles caused by Brownian motion or turbulence fluctuations must exceed the energy barrier height E_{max} during particle interaction, as shown in Equation 4.23. The kappa values of $\kappa = 10^{5.5}$ to $10^{6.85}$, which were used as parameters for calculating mutual energies, were calculated by Tambo from water quality analysis data of various water sources in the United States.

$$E_{max} < E_{\kappa} = \frac{1}{2} mV'^2 \quad \text{—(4·23)} \tag{4.23}$$

Equation 4.23. Energy barrier potential.

Figure 4.7 shows the critical line of the coagulation phenomena in which the height of the energy barrier in Figures 4.6(d–j) is approximately 6.17×10^{-14} erg (by interpolation), with the $1/\kappa$ of the thickness of the diffused double layer (for $\kappa = 10^{5.5}$ to $10^{6.5}$ cm^{-1} considered for natural water levels) in the vertical axis and the Gouy-Stern interface layer potential Φ_d in the horizontal axis.

It can be seen that the thickness $1/\kappa$ of the ionic atmosphere exhibiting the critical coagulation condition decreases rapidly (the κ value increases) as the interface potential Φ_d increases. Actually, since the Stern-Gouy layer boundary potential cannot be directly measured, the relationship with ZP, which is the only interface potential that can be actually measured, is used in practice. One of the functions that express the potential distributions in the diffuse double layer around the sphere is the Debye–Hückel (Equation 4.24).

$$\text{Z.P.} = \frac{\Phi_d a e^{-kb}}{a+b} \fallingdotseq \Phi_d e^{-kb} \quad a \gg b \quad \text{—(4·24)} \tag{4.24}$$

Equation 4.24. The Deby-Huckel equation.

Here, b is considered to have a substantially constant value at the interface between the Stern-Gouy layer, where Φ_d is defined, and the slip plane (shear plane) where ZP is defined. However, b is not precisely known and is considered to be in the order of 10 to 100 Å. We used b = 65 Å (the reason is described later) as a plausible value estimated from many experiments conducted by Tambo. Therefore, if we rewrite the critical coagulation ZP diagram, which expresses the horizontal axis of Figure 4.7 in terms of the relationship between κ and ZP, using the relationship shown in Equation 4.24, we find the relationship shown in Figure 4.8. From this diagram, if the region is on the left side of the critical coagulation ZP line, coagulation will progress, and dispersion will persist in the region on the right side and coagulation will not occur. The critical coagulation ZP follows a line of approximately 11 to 12 mV, and the critical value drops to 9 to 10 mV when the κ value reaches a higher water quality range of $10^{6.3}$. Incidentally, this is the condition of raw water from the New Orleans water supply at the mouth of the Mississippi

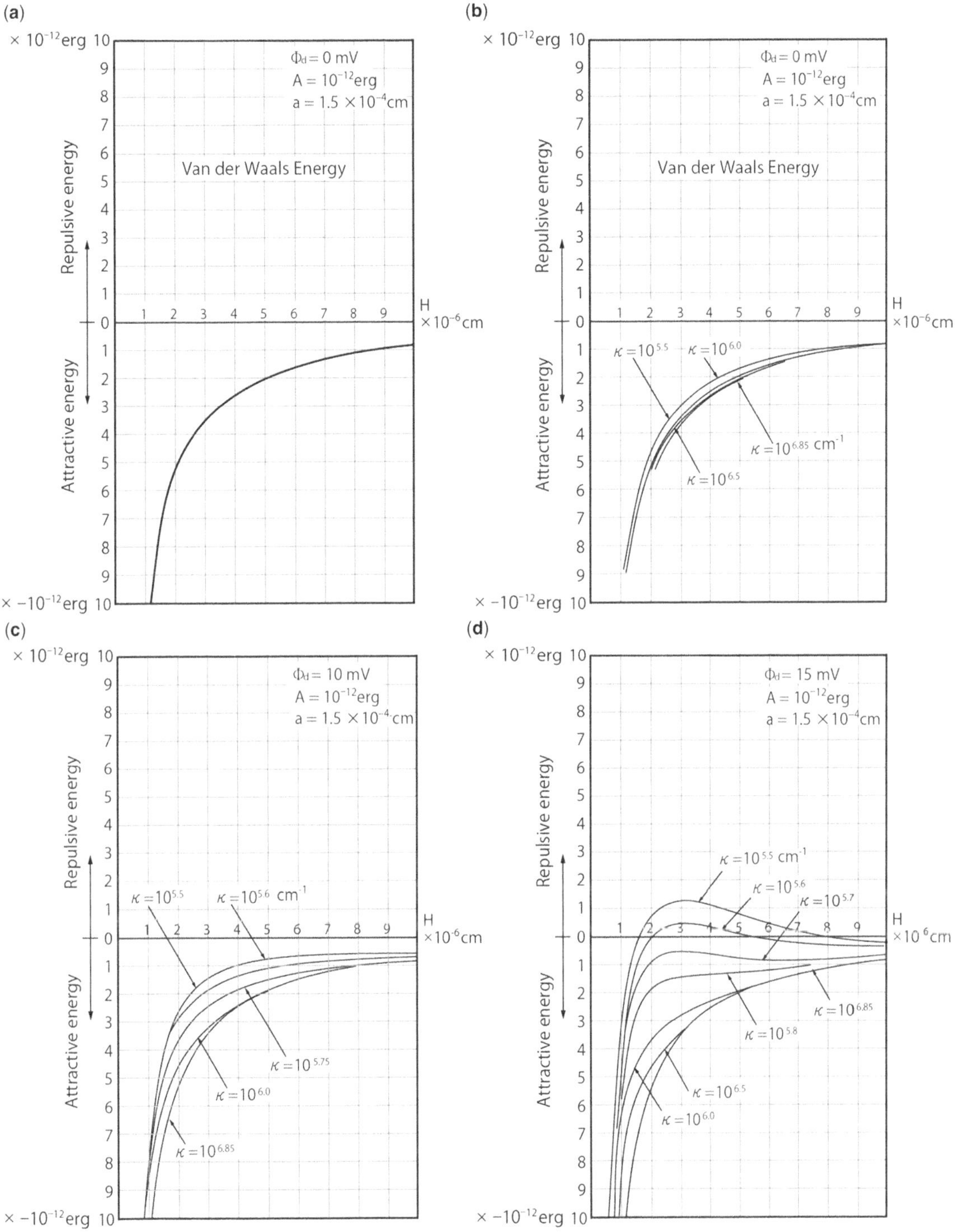

Figure 4.6 (**a–d**). Mutual potential curves in the vicinity of the particles.

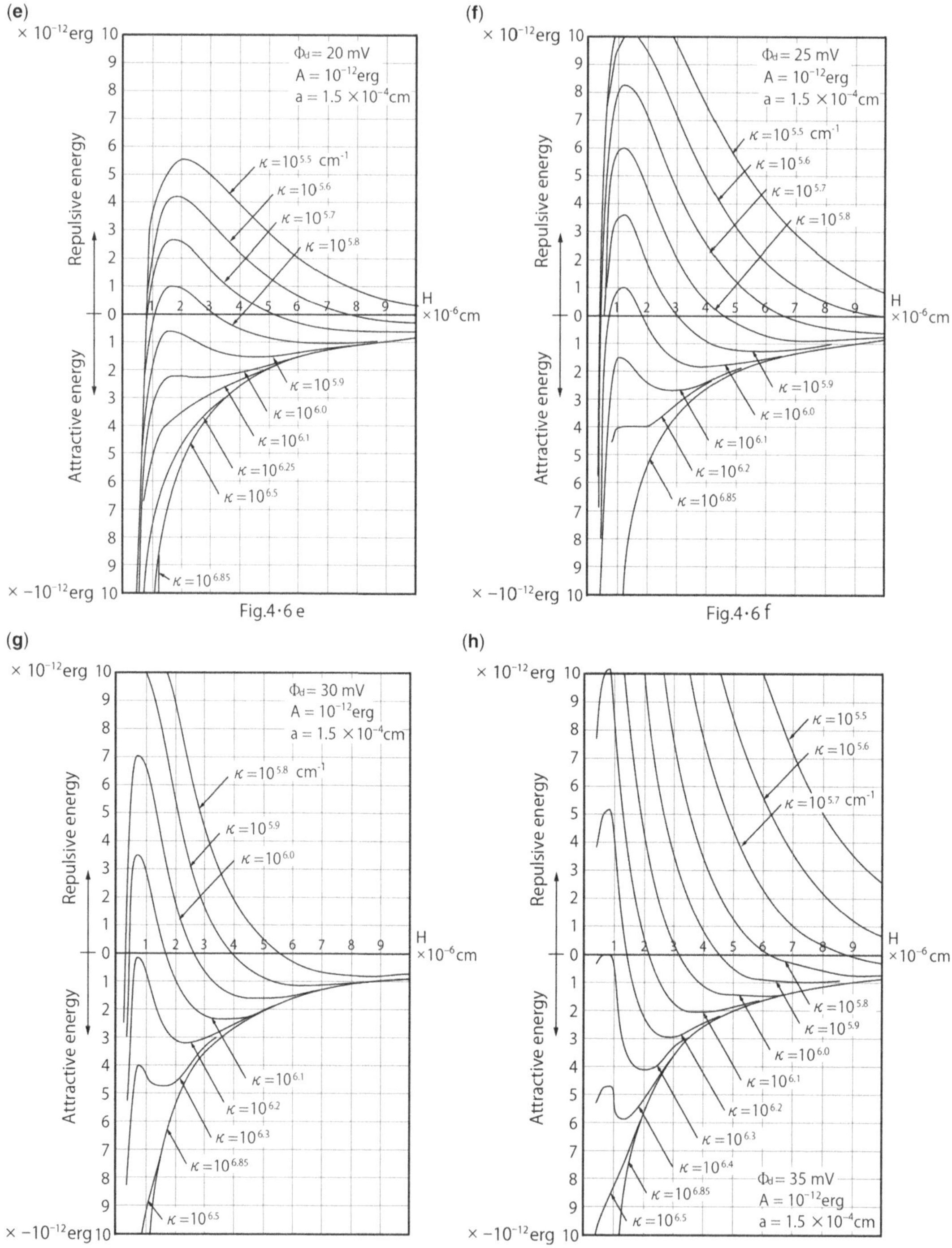

Figure 4.6 (**e**–**h**). (*Continued*)

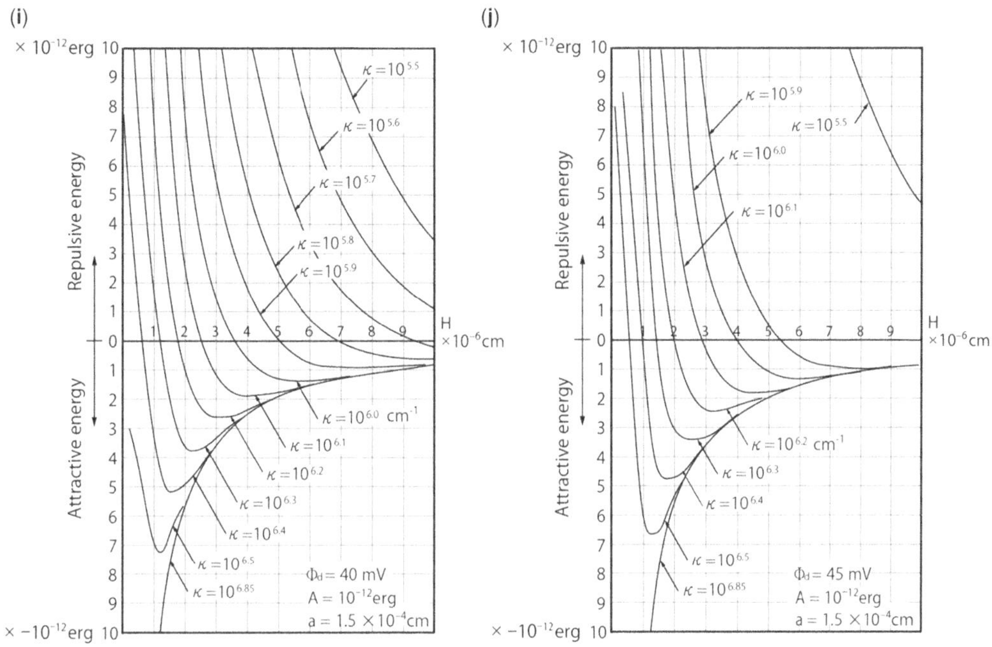

Figure 4.6 (**i–j**). (*Continued*)

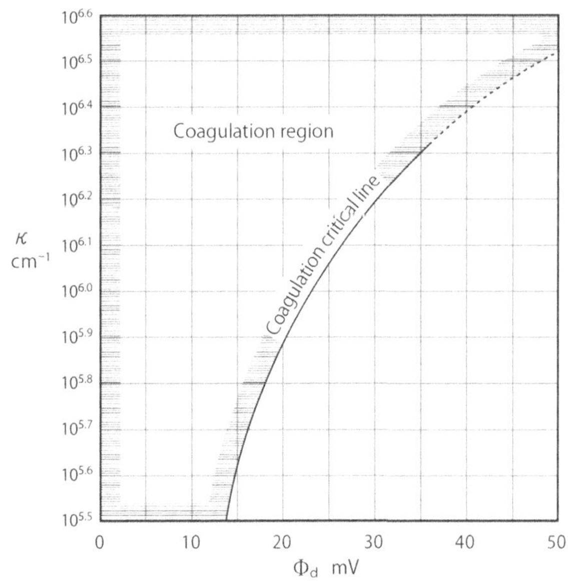

Figure 4.7 The critical line in coagulation phenomena.

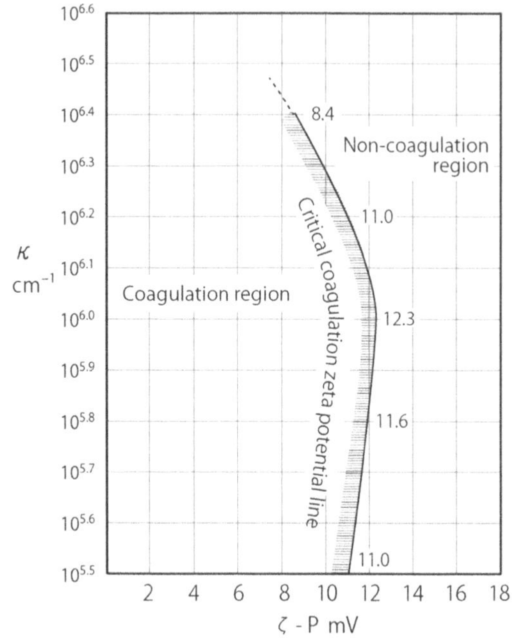

Figure 4.8 The critical coagulation zeta potential (ZP) line.

River. In other water work's raw water in US and Japanese rivers, it is only necessary to carry out a coagulation process using 10 to 11 mV as the critical coagulation ZP. Tambo proposes that the practical critical coagulation ZP should be set to ± 10 mV for general purposes.

4.1.4 Coagulation rate

If suspended particles reach the iso-electric point, there is no mutual energy barrier problem, and the extent to which the opportunities for particle-to-particle association differ between Brownian motion and turbulence fluctuations. The velocity gradient coagulation, in which particles on the high flow velocity flow line of a laminar flow field (e.g., Poiseuille flow) catch up with and coalesce with the particles on the neighboring low flow velocity flow line, occurs in a fixed bed flow field, etc., but is not used in a typical coagulation-flocculation process, so here we consider the relative reaction (collision) ratio of Brownian motion to isotropic turbulence fluctuations in an agitating system.

In Brownian motion, the number of times a particle collides within a unit of time per unit volume of suspension is given by Equation 4.25.

$$N_{br} = 8\pi D_{br} d n^2 \qquad \text{—(4·25)} \tag{4.25}$$

N_{br} : number of times the particles collide per unit time per unit suspension volume
D_{br} : Diffusion coefficient by Brownian movement
d : Suspended particle size
n : Number of particles in unit suspension volume

Equation 4.25. The number of times the particles collide due to Brownian movement.

The number of times particles collide within a unit volume of suspension due to turbulence fluctuations is given by Levich (1962) as in Equation 4.26, under the local isotropic turbulence condition defined by Kolmogorov.

$$N_{turb} = 12\pi\beta\sqrt{\frac{\varepsilon_0}{\mu}}\,d^3 n^2 \quad \text{—(4·26)} \tag{4.26}$$

N_{turb} : Number of collisions between micro-particles per unit time per unit suspension volume
β : Proportional constant (approximately 1)
ε_0 : Energy consumption rate that occurs in unit time and unit volume
μ : Viscosity
d : Particle size
n : Number of particles in unit suspension volume

Equation 4.26. The number of collisions between micro-particles due to turbulent fluctuations.

From Equations 4.25 and 4.26, the ratio of the Brownian and turbulent coagulation velocities (the number of collisions of particles in times) is given by Equation 4.27.

$$\frac{N_{tub}}{N_{br}} = \frac{12\pi\beta\sqrt{\frac{\varepsilon_0}{\mu}}\,d^3 n^2}{8\pi D_{br} d n^2} \fallingdotseq \frac{3}{2}\frac{d^2}{D_{br}}\sqrt{\frac{\varepsilon_0}{\mu}} \quad \text{—(4·27)} \tag{4.27}$$

Equation 4.27. Ratio of Brownian and turbulent coagulation velocities.

The D_{br} of the diffusivity of the particles due to Brownian motion is given by the Einstein equation in Equation 4.28 (Einstein's diffusivity).

$$D_{br} = \frac{KT}{3\pi\mu d} \quad \text{—(4·28)} \tag{4.28}$$

D_{br} : Diffusivity by Brownian movement
K : Boltzmann's constant
T : Absolute temperature
μ : Viscosity

Equation 4.28. Einstein's diffusivity

Considering the case where water temperature is 25°C, Equation 4.29 is obtained.

Water temperature = 25° C

$$T = 298^\circ$$
$$K = 1.38\times10^{-16}\,\text{erg/°C}$$
$$\mu = 0.0089\,\text{g/sec/cm}$$
$$D_{br}(25°C) = 4.91\times10^{-13}/d \quad \text{—(4·29)} \tag{4.29}$$

Equation 4.29. Diffusivity at water temperature 25°C.

The turbulent agitation strength ε_0 is expressed by Equation 4.30 because of the vortex characteristics of large inertial regions.

$$\varepsilon_0 \fallingdotseq \frac{\rho(\Delta U)^3}{l} \quad \text{—(4·30)} \tag{4.30}$$

ε_0: Energy consumption rate of the turbulent agitation (erg/cm³·sec)
ΔU: Average velocity difference of distance at eddy scale l (cm) (cm/sec)
ρ: Water density (g/cm³)

Equation 4.30. The turbulent agitation strength

Therefore, when the water temperature is 25°C, the transition particle size from the region where Brownian motion predominates to the region where turbulence fluctuations predominate is given by Equation 4.31.

Substituting Eq.5 · 29 into Eq.5 · 27,

$$\frac{N_{tub}}{N_{br}} = 3.24\sqrt{\varepsilon_0}\, d^3 \times 10^{13}$$

at Water temperature 25℃

Transition particle size ($N_{tur} = N_{br}$) d is

$$d = (3.09 \times 10^{-14} \times \varepsilon_0^{-\frac{1}{2}})^{\frac{1}{3}} \quad \text{—(4·31)} \tag{4.31}$$

Equation 4.31. Transition particle size between Brownian movement.

Since the rapid (turbulent) agitation energy ε_0 in a water treatment plant is considered to be approximately 10^3erg/cm^3 sec, the transition diameter d in terms of the collision rate in that water treatment system is of the order of submicrons, i.e., approximately 2–3 × 10^{-5} cm.

In view of the discussion of both the transient diameter of these impingement rates and the critical coagulation ZP, we can see the importance of increasing the agitation intensity so that the vortex diameter at which the local isotropic turbulence intensity works effectively is as close as possible to a submicron. This means that in a typical jar test (described later), which is widely used to set coagulant injection conditions, rapid agitation is performed at an agitation strength of 10^3 erg/cm^{3}sec, such as at 200 rpm. Of course, the actual agitation strength in the chemical mixing pond should also be of the order of at least 10^3 erg/cm^{3}sec.

The lack of intensity of rapid agitation in many Japanese and European water supply and industrial water purification plants is a common defect. In addition to addressing energy barrier problems, some optimal transport times are required for particles to collide with each other. Since the number of collisions per unit time is proportional to $(\varepsilon_0)^{1/2} = G$, the velocity gradient value defined by Camp and Stein (1943), the duration of agitation must be increased inversely proportional to the G value for weak agitation. The specific design of the agitation process will be described later in Section 4.2 for the process of flocculation. It seems that many practical treatments of this issue have not progressed with sufficient theoretical understanding.

4.1.5 Interparticle binding force and cross-linking coagulation

The DLVO theory and Tambo's theory of critical coagulation potential, which have been discussed on the basis of the energy-barriers to interactions due to the interference of ionic atmospheres (electric double layers) between particles, suggest that colloidal particles can associate with each other due to agglomeration. However, even after the particles have met, the subsequent solid-liquid separating process often requires reinforcement of chemical and electrical bonding forces beyond London van der Waals forces to maintain agglomeration against disturbances. Cross-linking substances that exert chemical bonding forces, such as ionic bonds or hydrogen bonds, are required to generate these bonding forces.

There are two possible situations in which such chemical bonding forces are required:

(1) Even if both particles have uniform surface-charge distributions with charge neutralization, the colloidal particles cannot exhibit bonding force to maintain/grow agglomerates, and there is a third cross-linking material (bridge) that can exert additional chemical bonding force between the colloidal particles (Figure 4.9). For example, in the coagulation of kaolinite clays, even if the ZP is neutralized and reaches the iso-electric point, it is difficult for simple ions alone to maintain agglomeration against turbulent flow fluctuation energies that increase in proportion to the fifth power of the diameter as the particle size (agglomeration) increases under turbulent flow conditions, and positively charged polymerized ions are required to co-adsorb both particles and to cross-link and strengthen the bond between the particles.

(2) In the other case, even if the electrical double layer of the two particles reaches the vicinity of the iso-electric point and is ready for direct contact, the surface of the colloidal particles leaves a heterogeneous charged surface area and they can be attracted to each other. For example, even after a component of proteins became a $^{COO-}R^{NH4+}$ and the iso-electric point was generated, the functional groups of COO^- and NH^{4+} were distributed heterogeneously on the colloidal

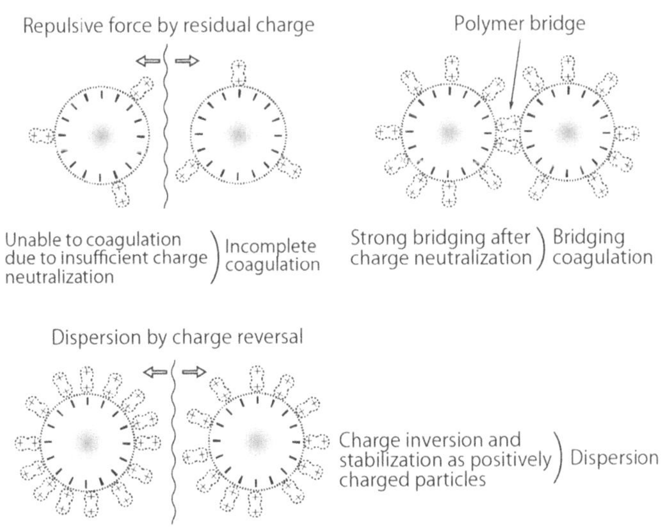

Figure 4.9 The various actions of a polymer coagulant.

surface; or with montmorillonite-based clay, in which charge neutralization of the electric double layer as a whole progresses and the electric charge of individual clay particles is distributed locally on the edges and surfaces of the clay particles.

In the case of microbial flocs, it is thought that the former condition exists loosely.

In order for primary particles to steadily associate and agglomerate with secondary particles, it is required that two conditions are satisfied for the basis of stable operation:

(1) Neutralization of appropriate repulsive charges; and
(2) An appropriate binding (bridging) force.

Examples of chemical coagulants that satisfy these two requirements include inorganic and organic polymer coagulants. Because polymeric substances are strongly adsorbing, when added to suspensions, almost all of the suspended matter collects on the surface of the colloidal particles until the surface reaches adsorption saturation. Therefore, most of it is concentrated in the Stern layer. In these cases, aggregation is most likely to occur when the positively charged polymeric coagulant covers 50% of the surfaces of negatively charged suspended particles and the total reciprocal potential of the particles is sufficiently reduced. If the amount of oppositely charged polymer added is too much, the oppositely charged polymer covers the surface of the colloidal particle, increasing the opposite repulsion force and redispersing the colloidal particle. This phenomenon is called deflocculation, as shown in Figure 4.9.

However, we must note that polymer coagulation has a definitive application failure condition and cannot be universally used. The particle size of a polymer at the time of coagulation operation must be small enough for the material to be coagulated. This is why large polymers cannot be used for small organic colloids, and aluminium and ferrous salts with a molecular weight of at most 600 to 700 Da are used. The coagulation limits mentioned in the water treatment matrices in Chapter 3 are around 1500 Da in molecular weight, which is thought to be a relatively workable size for an aluminium or iron coagulant. Briefly, a polymer coagulant with a molecular mass of 10,000 Da does not adsorb on humic particles with a molecular mass of 1500 Da. Conversely, performance as an adsorbent (see Chapter 3, Section 5) cannot be expected at all for polymer colloids.

Further details are provided in the colored water coagulation section below.

4.1.6 Coagulation agents

The chemicals used for coagulation treatment are as follows:

(1) Coagulants. These neutralize the surface charge (electric double layer) of colloidal particles and the like, destabilizing the repulsive potential barrier Er between particles, enabling collisions between particles. Coagulants have the function of exerting a cross-linking action between two particles to strengthen the bond and to make it possible to proceed to floc formation.
(2) Coagulation aids. These are chemicals that control the pH of a suspension system and increase the pH buffering capacity of water in order to make a coagulant work effectively.
(3) Flocculation aids. These are agents that exhibit cross-linking action between particles and strengthen the binding of flocs. (Although coagulation aids and flocculation aids may be referred to together as a coagulation adjuvant, they are chemicals that are expected to work differently in terms of their function).

Metallic salts, such as aluminium or ferrous salts[1], which hydrolyze easily in water to form polymerized polyvalent cations, are used primarily as coagulants. Cationic organic polymers have also been widely used in sludge treatment systems for the coagulation of high-concentration coarse particle size slurries in recent years. However, they have too large a molecular mass to be used for double-layer charge neutralization of fine particles, and the concentration of corresponding molecules is insufficient for dilute systems, which limits their general use in water treatment systems.

The amount of a coagulant that generates the oppositely charged particles required to neutralize the charge of the electric double layers drops sharply when the charge value of the oppositely charged (counter) ions increases. This phenomenon is known as the Hardy Schulze law, as mentioned earlier. The coagulant required to neutralize the charge in the double layer decreases by about one order of magnitude as the ionic number of the oppositely charged layer increases from 1, 2, 3, 4.

Aluminium and iron salts have been widely used as a coagulant because they form hydrolyzed tetravalent ions that have a large charge neutralizing capacity in the weakly acidic pH range where the actual processes are carried out, as will be described later, Tables 4.8 shows examples of commonly used coagulants.

For floc-forming reactions of aluminium coagulants, classic water treatment textbooks explain that the reaction of aluminium salts and calcium bicarbonate proceed in the form of Equation 4.32. Aluminium salts, which are acid, consume the alkali $Ca(HCO_3)_2$ in water to form aluminium hydroxide $Al(OH)_3$ precipitates, release CO_2 to reduce the pH of the water, and the formed aluminium hydroxide precipitates (flocs) by encapsulating colloids and suspended matter in water for separation. But actual coagulation mechanisms cannot be achieved with the neutral aluminium hydroxide $Al(OH)_3$, as described in Equation 4.32.

$$Al_2(SO_4)_3 \cdot 18H_2O + 3Ca(HCO_3)_2 \rightarrow 2Al(OH)_3 + 3CaSO_4 + 6CO_2 + 18H_2O \quad \text{—}(4\cdot 32) \tag{4.32}$$

Equation 4.32. The reaction of aluminium salts and alkalinity.

Aluminium salts are ampholytes that hydrolyze easily and are as weakly acidic as acetic acid. On the acidic side, they exist as positive aluminium ions, but as pH changes to the alkaline side, they become negative aluminum acid ions (aluminate). The process is shown schematically in Figure 4.10. Matijevic's and other pioneering studies have shown that the addition of alkalis to acidic aluminium species to raise the pH results in the binding of OH^- ions to aluminium, resulting in the formation of various aquo-complexes, such as $Al_6(H_2O)_5(OH)^{2+}$, $Al_6(OH)_{15}{}^{3+}$, and tetravalent positively charged polymer $Al_8(OH)_{20}{}^{4+}$, which maximizes charge neutralizing capacity, as shown in Figure 4.10. Figure 4.11 shows the chemical structure of tetravalent positively charged polymeric $Al_8(OH)_{20}{}^{4+}$ that maximizes charge neutralizing capacity.

---

[1]As noted earlier, the mechanisms by which aluminium and ferrous salts (explained by the critical coagulation potentials and other arguments based on DLVO's theories etc.) emerged in the early 1960s when Tambo was working under Professor A.P. Black at the University of Florida, USA. Young colleagues of the same era, including Professor J.J. Morgan (CALTECH), Professor C. O'Melia (Johns Hopkins University), Professor Packham (Imperial College London), contributed a great deal on these topics. Prof A. Amirthrajah (Georgia Tech) later drew from Tambo's theories to establish his excellent work in the United States in the 1980s. In Professor Black's group, many friends including Professor R.F. Christman (University of North Carolina) and Professor F.B. Berkner (University of Maryland) have been colleagues for many years. This was the result of almost a quarter century of research by our research staff and graduate students such as Professor Y. Watanabe (Professor Emeritus Hokkaido University/now Chuo University), Professor K. Kimura (Hookaido University, Specialist Group Chair of Particle Separation at IWA: the Group has continued its activities since being established in London (c.1970) by Professor K. Ives (University College London), Professor Bernhalt (Archen Institute of Technology), Professor C. O'Melia, Tambo, etc., and Professor W. Stumm Harvard University, who led the group).

Table 4.8 Examples of coagulants.

Name of Substances	Chemical Symbols
Aluminium sulphate	$Al_2(SO_4)_3 \cdot 18H_2O$
Sodium aluminate	$Na_2Al_5SO_4$
Ferrous sulfate	$Fe\,SO_4 \cdot 7H_2O$
Ferric sulfate	$Fe(SO_4)_3$
Ferric cloride	$FeCl_3$
Chlorinated copper	$FeCl_2 \cdot Fe_2(SO_4)_3$
Calcium hydroxide	$Ca(OH)_2$

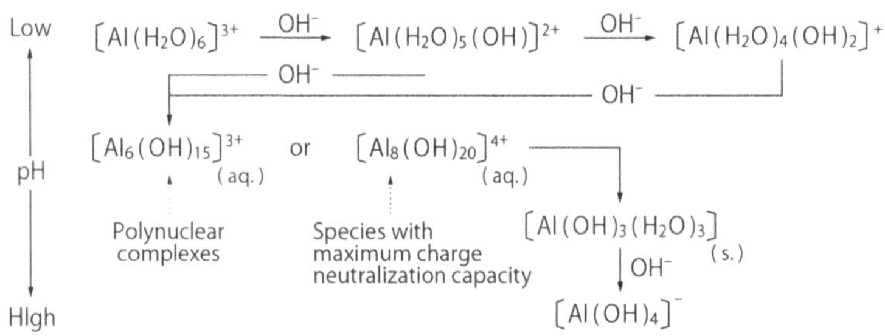

Figure 4.10 pH and various aluminium species.

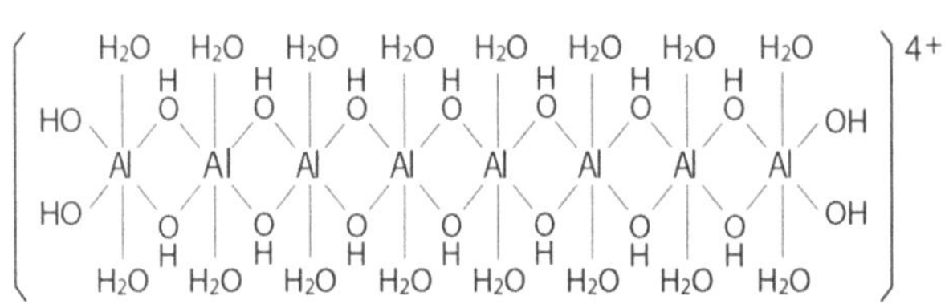

Figure 4.11 The chemical structure of tetravalent positively charged polymeric $Al_8(OH)_{20}{}^{4+}$.

As shown in Figure 4.12, the NaOH titration curves for aluminium solutions show a high buffering capacity between pH4 and pH5, indicating that hydrolysis proceeds rapidly in this area. Aluminium combines almost entirely with added OH^- ions near pH 4.5 to 5.0, indicating that hydrolytic polymerizations proceed to form polynuclear complexes, such as $Al_8(OH)_{20}{}^{4+}$ or $Al_7(OH)_{17}{}^{4+}$, around pH 4.5. In an actual aluminium solution, they are not present as a single species but as a plurality of aquo-complexes which are mixed in each pH range. Aluminium-polynuclear complexes are said to have pentavalent ions at different temperatures and concentrations due to aging proceed, but the authors have not confirmed them.

Figure 4.13 shows the results of the electrophoretic transport tube method produced by the authors which can measure the mobility of both soluble and colloidal aluminium. For the measurement method, refer to Tambo and Ito (1977; p. 38).

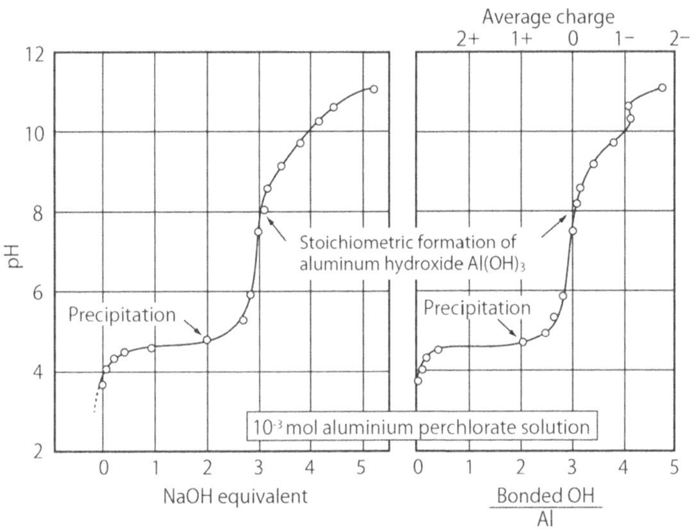

Figure 4.12 Titration curves for aluminium solutions.

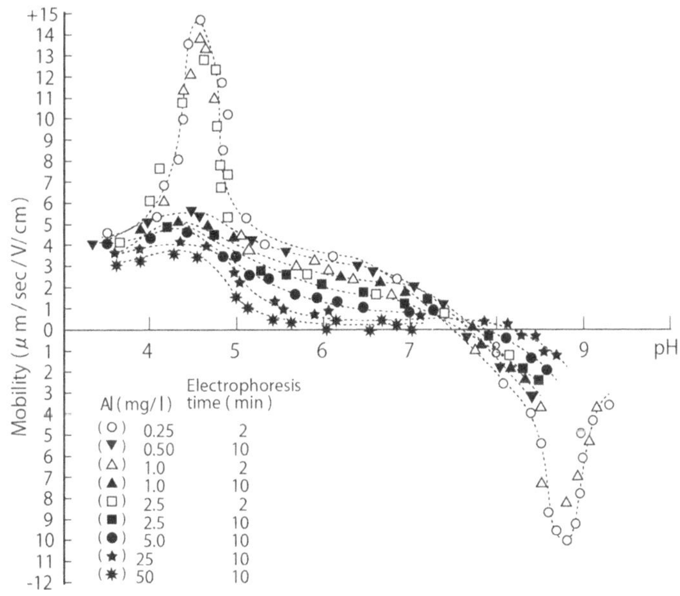

Figure 4.13 The mobility of aluminium hydroxide and pH.

On the acidic side, a high positive mobility is exhibited. However, as the pH increases, electrophoretic mobility decreases, reaching the iso-electric point (electrophoretic mobility 0) near pH 7.5; on the alkaline side, negative electrophoretic mobility is exhibited and the value decreases greatly. The features of this method are to measure the electrophoretic mobility (ZP) of soluble aluminium aquo-complexes, which cannot be measured by microscopic electrophoresis. From the relationship between Equations 4.13 and

4.14, we can proceed with the discussion by replacing electrophoretic mobility with ZP. For water at 25°C, the ZP and the degree of electrophoresis are related to Equation 4.33, and the electrophoresis potential is approximately 12.8 mV per electrophoretic mobility. The critical coagulation ZP proposed by Tambo is considered to be less than 1 degree of electrophoretic mobility.

$$\text{Z.P.(mV)} = 12.8 \times \text{M}(\mu\text{m/sec/V/cm}) \quad \text{—(4·33)} \qquad \text{Z.P.}: \zeta\text{-potential}, \quad \text{M}: \text{Mobility} \tag{4.33}$$

Equation 4.33. ζ - potential and mobility.

A migration time of 2 minutes is the shortest time for measuring the degree of migration by our method, and it is impossible to identify the aluminium species that occur in a much shorter time. In addition, if the aluminium concentration is high, the reaction proceeds at the same reaction time. Therefore, in Figure 4.13, when the aluminium concentration is high, the absolute value of the mobility is small in both the positive and negative regions (as seen in the case of measurement with a long electrophoresis time in a state in which the reaction progresses). Therefore, in order to estimate the ZP (electrophoretic mobility) of an aluminium polynuclear complex in a state as close as possible to a rapid mixing tank in an actual coagulation treatment, it is best to refer to the white circles and squares in Figure 4.13 which represent approximately 2 minutes of electrophoresis time.

There is a prominent high mobility region at pH 4.5. When pH 5.5 is exceeded, the mobility decreases rapidly, the positive mobility decreases gradually with a plateau, reaching the iso-electric point at pH 7.5, then increases negative value rapidly, reaching a negative maximum at pH 9, and then decreases again.

Figure 4.14 shows the particle size distribution of solutes and colloids with an Al concentration of 0.8 mg/l. as determined by classification with 5 to 10 μm filter paper, in order to estimate the size of the aluminium species exhibiting the mobility shown in Figure 4.13. Unfortunately, we were unable to identify these species, but the work of Professor Singley and colleagues at the University of Florida, calculated the ratios of 10^{-4} molar (2.7 mg/l) aluminium solutions in various pH ranges based on the

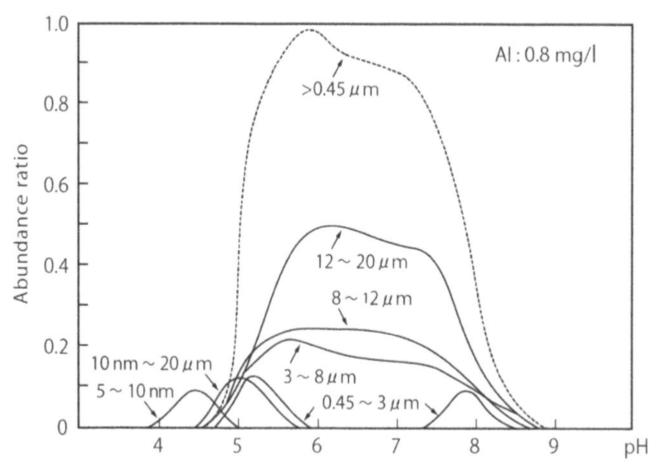

Figure 4.14 The pH and particle size distribution of aluminium hydroxides.

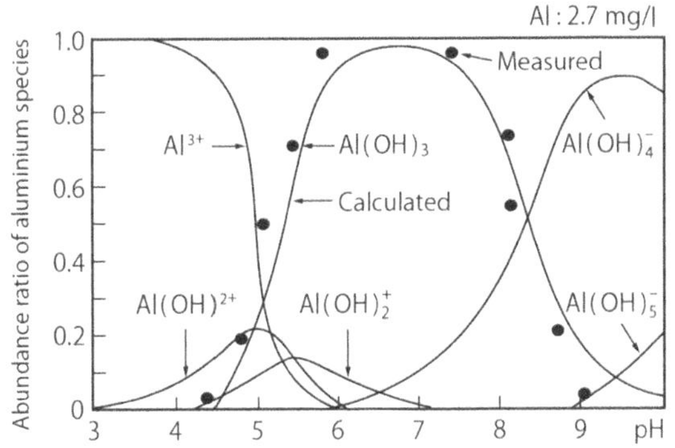

Figure 4.15 The pH and presence ratio of aluminium species.

assumption of six types of aluminium. The results are shown in Figure 4.15, and they appear to be well correlated with the results of our particle size distribution measurements.

It is estimated from the titration curves in Figure 4.12 that the aluminium monomeric Al^{3+} shown in Figure 4.10 is a divalent aquo-complex. As shown in Figures 4.10 and 4.11, the aluminium monomer is further polymerized to form a +4 valent polymeric polynuclear aquo-complex with a size of about 5 nm to 10 nm in the region of pH 4.5. In the pH range of 5 to 8, all particles are in suspension with a particle size exceeding 0.45 μm and become insoluble as neutral aluminium hydroxide precipitates, as shown in $Al(OH)_3$, resulting in the formation of so-called aluminium flocs.

This is part of the explanation of the classical mechanisms of action described earlier in the book concerning coagulants. Later, in discussing the coagulation phenomena of turbidity and color colloids, we mention that the combination of prominent positively charged polynuclear complex species around pH 4.5 and the presence of many low-charge aluminium hydroxides extensively present in weakly acidic regions dominates various coagulation phenomena. The range of insoluble aluminium hydroxide in the neutral pH range, like other components, is greatly affected by aluminium concentration. Figure 4.16

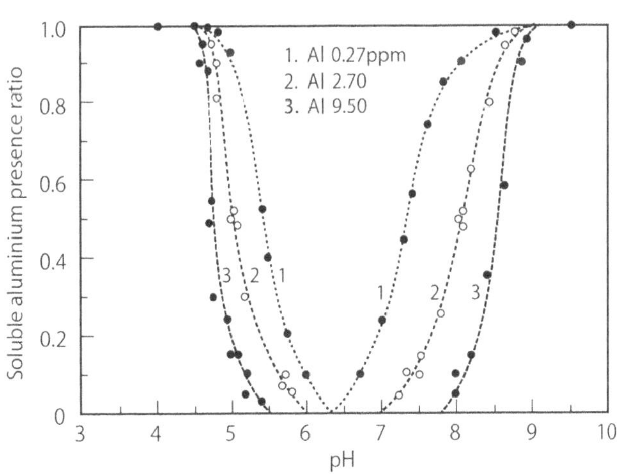

Figure 4.16 The soluble aluminium presence ratio in a water-aluminium system.

shows that insoluble aluminium flocs are present in a wider pH range at higher concentrations. The key to the design of turbidity and color colloid coagulation treatment described below is how to use aluminium hydroxide, which has a low charge neutralization capacity, and polynuclear complexes, which have a high charge neutralization capacity, near pH 4.5.

In the case of iron salts, hydrolysis proceeds remarkably in pH range 3 to 4, and an aquo-polynuclear complex such as $[Fe_2(H_2O)_8OH_2]^{4+}$ is considered to be the main charge neutralizing agent. Hydrolytic polymerization appears to begin at a pH lower than that of aluminium. Figure 4.17 shows the state of the presence of various types of aquo-ions of aluminium and iron shown by Stumm and O'Meria (1968), and the state of the coagulation can be understood in comparison with the state of the presence of various types of aquo-complexes by comparing such data as are described in this chapter 4 by Tambo.

Aluminium sulfate Al $(SO_4)_3 18H_2O$ (alum), which has been used for water treatment since the 18th Century as a typical aluminium-based coagulant, has a molecular mass of 186 Da, and its properties are described above. As a coagulant of aluminium aquo-complexes, PACl (polychlorinated aluminium polymer) was a major Japanese invention developed by Dr Shigeo Ban in the 1970s (Ban *et al.*, 1968) and patents were opened to the public so that it became the mainstream inorganic coagulant used around the world. The molecular mass is assumed to be about 600 to 850 Da.

The effectiveness of a coagulation process by hydrolyzed polymerization of metallic salts and charge neutralization/cross-linking of negatively charged suspended particles is influenced by how the mixing of fluids can be controlled as the chemical reaction progresses, since the mixing/dispersion of added coagulant and water mass (suspension) is critical because the aluminium concentrations and the reaction rate of hydrolyzed polymerization proceed closely and are delicately intertwined.

Aluminium polychloride is one of the greatest achievements of Japanese water technology which has delivered a stable, high-performance coagulant worldwide. The authors highly appreciate the achievements of Dr Shigeo Ban, who was Vice-President of the Daimei Chemical Co., Ltd, and who disclosed a wide range patented manufacturing methods both in Japan and overseas.

Using aluminium chloride or nitrate as a raw material, hydrolyzed polymerization as shown in Figure 4.10 is carried out at high concentrations, and the pH is raised to a level just below (OH/Al) = 2 just before the coagulation initiation point as shown in Figure 4.12 to promote polymerization of soluble aquo-polynuclear complexes. It seems that commercially available PACl is produced with a ratio of

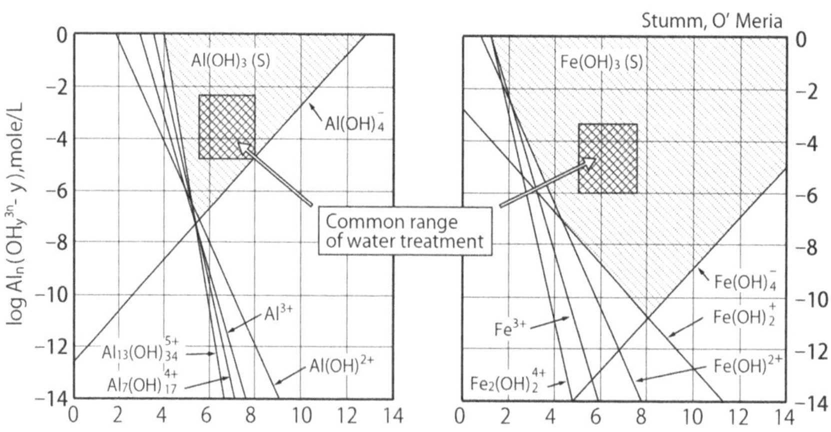

Figure 4.17 The extent of the presence of various types of aluminium and iron aquo-complexes.

about (OH/Al) = 1.5 (basicity 50%). Coagulation (precipitation) begins when the titration curve is close to 2.0 (66% basicity), so it seems difficult to exceed 70% basicity even under well-controlled conditions. As the basicity increases, the average positive charge content of the system decreases. However, as the degree of polymerization of each of the aquo-complexes increases and the charge value of each of the aquo-complexes increases, the charge neutralization capacity increases by an order of magnitude according to the Hardy Schulze law, and the average charge neutralization capacity of the system increases as a whole.

The general chemical formula for PACl is shown in Equation 4.34. Concentrations are made of almost 5%, or of 10–11% solutions, of an Al_2O_3 solution. Since it exists as a stable aquo-complex at these concentrations, as a general rule the raw solution is added and mixed to raw water at the start of use. Dilution changes the equilibrium and leads to turbidity and deactivation.

$$[Al_2(OH)_n Cl_{6-n}]_m \quad \text{—}(4\cdot 34) \qquad 1 < n < 5 \qquad m \leqq 10 \tag{4.34}$$

Equation 4.34. The general chemical formula of PACl

In various papers and reports, the amount of coagulant added is expressed as x ppm or mg/l of sulfuric acid or of PACl. However, in some alums (aluminium sulfate compounds), the water of crystallization may be $18H_2O$ or $14H_2O$ and the component content of PACl differs depending on the manufacturing method. Therefore, in technical reports and papers, it is required to express the added concentration in mg/l of Al by aluminium sulfate in order to make the relative evaluations correct. Aluminium nitrate aquo-complexes are also made in large quantities. Aluminium sulfate compounds (alums) at 10 ppm (water of crystallization $18H_2O$) have an Al concentration of 0.81 ppm, and polychlorinated aluminium at 10 ppm has an Al concentration of about 0.25 ppm. However, the values vary depending on the products.

In the production of PACl, hydrolysis proceeds under controlled conditions with a high aluminium concentration of several percent. However, in a rapidly agitated vessel (or jar tester, as described below), hydrolyzed polymerization proceeds heterogeneously at low aluminium concentrations of the order of ppm, with concentration distributions that are not negligible (due to poor mixing). The properties and uniformity of the products differ greatly between PACl production, which undergoes homogeneous hydrolyzed polymerization at very high concentrations, and the aquo-polynuclear complex, which is formed from aluminium sulfate added to a rapidly agitated tank. In other words, if we imagine the conventional method of adding aluminium sulfate (alum) as a 100 m running race, we can imagine a situation in which children run at different speeds at the start of the race (hydrolysis of aluminium) and later surge towards the finishing line at different speeds (collision and charge neutralization with suspensions). In a similar race with PACl, hydrolysis is carried out uniformly at a high concentration first at 60 to 70 m in the chemical reaction vessel (coagulant production of the order of 60 to 70% of the base), followed by charge neutralization of colloids in a rapid mixing pond for the remaining 30 to 40 m (with the later part in a water treatment chamber).

High concentrations of hydrolyzed polymerizations have enabled highly charged aquo-complexes to be homogeneously formed, and aluminium flocs can be stably formed even at low coagulant loadings in a coagulation chamber (flash mixer). In terms of floc properties, which will be described later, fractal floc structure can be observed. However, in the 1960s, PACl became available, enabling flocs to be produced in a wide range of aluminium content (dosing ratios). The authors believe that the appearance of PACl compensated for the inadequate design of rapid mixing tanks in water purification plants (with their danger of a 'short-circuit flow' due to insufficient agitation strength and duration, and insufficient number of agitation stages).

When aluminium sulfate (alum) is added directly to a rapid mixing tank, and the hydrolyzed polymerization of aluminium proceeds in raw water whilst the charge neutralization reaction of the suspension by the polymer produced is carried out simultaneously in parallel, the two-stage reaction proceeds in parallel in one reaction tank with a short time lag. With the advent of PACl, the authors believe that a clear-cut agitation process design can be achieved in which the formation of an aquo-aluminium complex as a coagulant is preceded by a separate reaction, and then the charge neutralization reaction and flocculation cross-linking reaction of the suspensions can proceed physiochemically in series and with certainty.

4.1.7 Coagulation patterns of suspensions by metallic salts

For the coagulation of colloidal particles suspended in water, appropriate positively charged aluminium aquo-complexes should be selected to neutralize the surface-charge of the suspensions (colloids) to enable their coagulation. Subsequent solid-liquid separation, such as by sedimentation or sand filtration, must further take into account the application of cross-linking forces between the coagulated particles to maintain an appropriate floc size against external forces. Flocculation aids (polymeric substances) may be added, but they must coexist with coagulated particles of aluminium (neutral aluminium hydroxide). As mentioned earlier, under certain conditions of dissolution, various aluminium species are present in a mixed state. Therefore, the pH of the raw water immediately after rapid hydrolysis is used as an indicator for the process condition, and appropriate utilization of these mixed aluminium (or iron) species is planned.

In the case of added aluminium (dosage), there are groups of aluminium aquo-complexes with a maximum negative charge neutralizing ability in the region of pH 4.5 to 4.7. $Al_8(OH)_{20}^{4+}$ is considered to be the main species. Sometimes species such as $Al_{13}(OH)_{24}^{5+}$ are thought to exist but the authors have not clearly confirmed the effects of pentavalent polymers. As pH rises, the total amount of positive charge decreases, and the charge of the compounds decreases to 2+. In combination with the Hardy Schulze law, the charge neutralizing capacity decreases markedly, and in the neutral region around pH 7, the charge neutralizing ability of the complexes almost disappears. In the pH range around pH 4.5, polynuclear complexes of aluminium hydroxide are present as soluble polymers because they repel each other with their own high charges. When the pH reaches pH 6 or more, the charge amount of the single substance becomes small, and the charge neutralizing coagulation of the clay-based suspensions with a small total amount of negative charge can be achieved, but it is not possible to proceed with the coagulation of color micro-colloids with a large total amount of negative charge. In the neutral pH range near pH 8 where the iso-electric point of aluminium hydroxide is observed, the aluminium polymers do not require the presence of negative colloids to perform charge neutralized coagulation and, as shown in Figure 4.16, they coagulate by themselves as aluminium hydroxide flocs. Needless to say, there is no charge neutralizing capability, and the negative colloids in the raw water remain dispersed and do not have a coagulation effect. However, when charge neutralized coagulation particles coexist, they contribute to flocculation as cross-linking substances.

4.1.7.1 Coagulation patterns of coarse suspensions (clays)

In this section, we describe clays, which appear as both typical coarse suspensions (as colloids) and in typical fine colloid suspensions in natural peaty water (as color colloids). In a natural system, many colloids appear to exist stably with a ZP of about −25 mV. Human red blood cells appear to have these same values and to be stably dispersed in blood vessels.

The results shown here are from experiments conducted by the author Tambo in 1962–1963 at the University of Florida, USA. The test water used was a suspension of 51.4 ppm of Kaolinite 4 (kaolin) in distilled and/or ion-exchanged water with 50 ppm of $NaHCO_3$ as alkalinity. The water temperature was 25°C, rapid mixing was continued for 10 minutes with 120 rpm by a jar tester, followed by slow stirring (floc formation: flocculation) for 20 minutes at 40 rpm. After a sedimentation operation for 30 minutes, the supernatant water turbidity was measured.

The ZP of the floc was measured by microscopic electrophoresis (Brigg-Cell method) after the floc was rapidly agitated and destroyed.

This study was basically concerned with coagulation and flocculation, assuming a sedimentation process. Micro flocculation, involving both direct filtration and membrane filtration, is discussed later in Chapters 6 and 8. The coagulation condition was adjusted adding 0.2N HCl or NaOH to the suspension as soon as possible after adding coagulant so that the coagulation pH was controlled to the expected values. In the absence of a coagulant, the ZP of kaolin 4 ranges from −30 to −35 mV at pH 5 and above, with a slight increase in absolute values with increasing pH. Conversely, if the pH was lowered to around 3, the ZP decreases to around −13 mV due to the effects of H^+ ions, and the ZP approaches the critical coagulation potential, as Tambo suggested.

Figures 4.18 to 4.21 show the relation between turbidity removal and ZP at various pH when the dosing rate of aluminium sulfate (alum) was changed to 10 ppm, 20 ppm, 40 ppm and 80 ppm. Figure 4.22 shows how the ZP changes as the amount of aluminium added increases at this constant pH.

Table 4.9 summarize the features of coagulation by pH and aluminium species. Results at the various pH ranges were as follows:

(1) In the region of pH < 4 or pH ≑ 4.0; The ZP (of coagulant added suspensions) was always in the negative region and, even at saturation, ZP only reaches the iso-electric point and the potential reversal does not occur. In this region, aluminium exists mainly in trivalent simple ions and, as

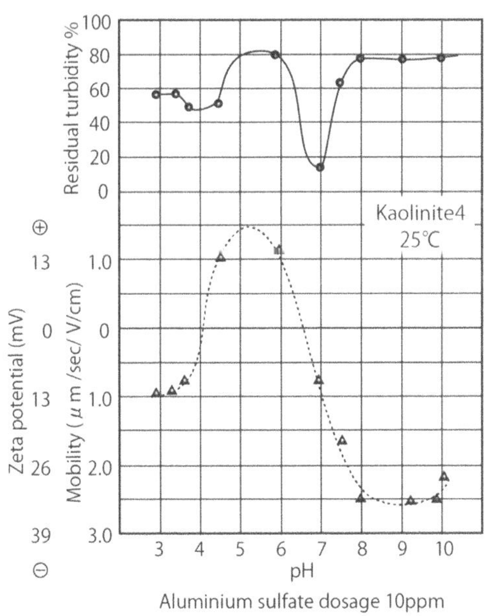

Figure 4.18 Aluminium sulfate dosage (10 ppm), turbidity removal and zeta potential.

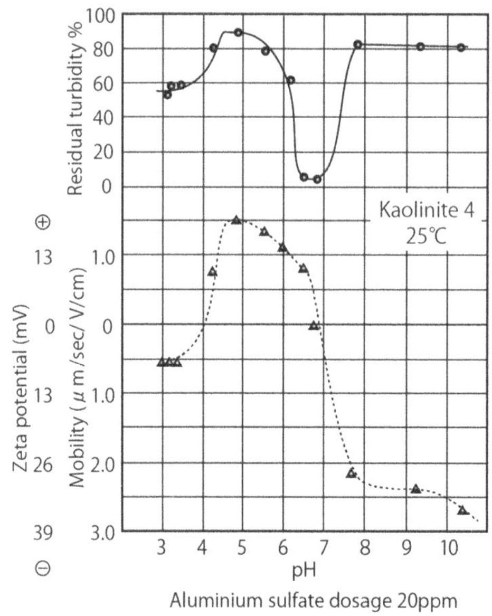

Figure 4.19 Aluminium sulfate dosage (20 ppm), turbidity removal and zeta potential.

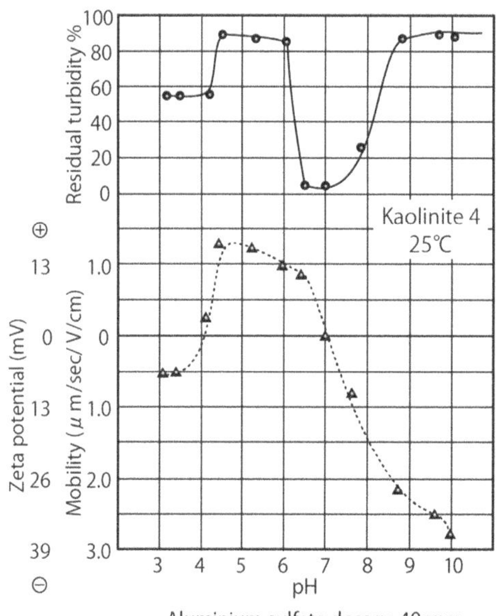

Figure 4.20 Aluminium sulfate dosage (40 ppm), turbidity removal and zeta potential.

described in the description of the diffused double layer shown schematically in Figure 3.14, the reversal of potential does not occur with simple ions but only occurs when polymer ions are adsorbed to the Stern layer. No definite flocculation was observed even at the iso-electric point.

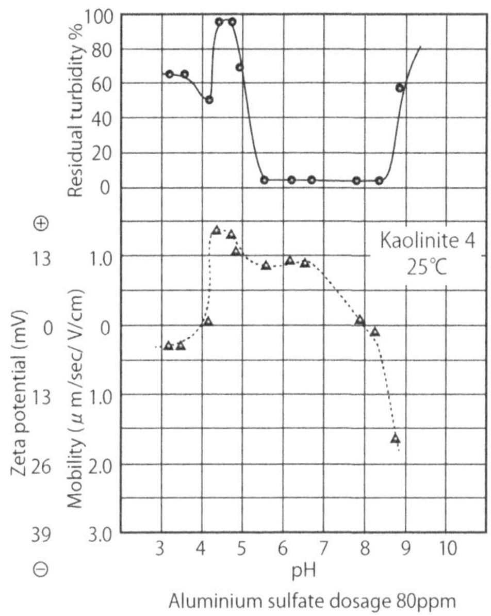

Figure 4.21 Aluminium sulfate dosage (80 ppm), turbidity removal and zeta potential.

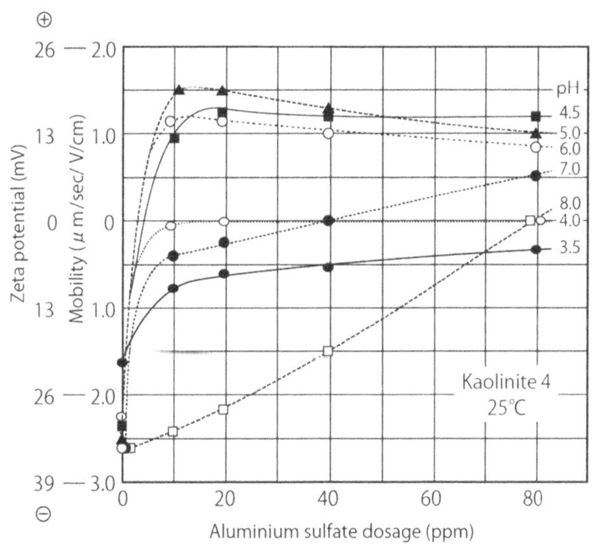

Figure 4.22 The relationship between aluminium sulfate dosage and zeta potential.

This was due to the absence of aluminium precipitates that maintain floc bonding and there was only the formation of loose agglomerates due to London van der Waals forces (charge neutralization alone). A jar test does not show flocculation (floc formation) because it cannot withstand shear failure during the flocculation agitation available for sedimentation. In recent pre-treatment

Table 4.9 Coagulation features by pH and aluminium species.

pH Range		**Characteristics of Coagulation by Aluminium Species**
pH $\leqq 4$		It can be neutralized, but can not be bridging, and flocculatlon difficult. Main specie is Al^{3+}.
pH4 ~ 5		Inappropriate range for coagulation
	pH4.5 ~ 4.7	Aluminium aquo complexes with maximum charge neutralization capacity Mutual repulsion due to the high charge of the complex itself Main specie is $Al_8(OH)_{20}{}^{4+}$.
	pH5 near	Mutual repulsion due to the charge of the complex itself
pH5 ~ 6.5		Aluminium dosage small : Dispersion by the charge inversion of clay particles " large : Charge neutralization ability is reduced and the repulsive force is reduced by the formation of low charge aquo complex Flocculatlon in bridging by neutral aluminium hydroxide coagulation
	pH6 or more	Possible of neutralize and coagulation clay having small total negative charge amount Difficult of coagulation of color having large total negative charge amount
pH6.5 ~ 8.5		Neutral aluminium hydroxide is formed, Flocculation proceeds smoothly by the neutralization and bridging. Main specie is $Al(OH)_3$
	pH8 near	Isoelectric range of aluminum hydroxide Aluminium itself is insolubilized as aluminium hydroxide (coagulation) No charge neutralization capacity Act as bridging agent
pH8.5 or more		Only negative charge aluminium species are present and disperse due to their inability to neutralize charge $Al(OH)_4{}^-$ etc.

coagulation for membrane filtration (MF, UF, etc.), a new study is needed on whether London-van der Waals agglomeration, which is based only on charge neutralization, is effective or not.

(2) In the region $4 < pH < 5$; Between pH 4.5 and 4.7, the ZP increases rapidly with increasing pH, peaks near pH 4.8 to 4.9, and then begins to decrease with increasing pH. The turbidity of the supernatant water (residual) in this area is higher than that of the kaolin suspension prior to the addition of coagulant, and the added aluminium is considered to be a highly charged aquo-complex that adsorbs on kaolin and disperses above the critical coagulation ZP, and in addition increases the turbidity by the dispersed suspension of colloidal aluminium. Excessive injection of aluminium (charge reversal state critical coagulation ZP) is an inappropriate coagulation pH-range that occurs even at low dosages.

(3) In the region $5 < pH < 6.5$; The ZP gradually decreases with increasing pH and forms a plateau. At low loadings of 10~40 ppm, the highly charged polymers reverse the charge of the clay suspension, and the critical coagulation potential is exceeded, causing the suspension to become charge reversed and dispersed and no coagulation occur. However, at 80 ppm as shown in Figures 4.15, 4.16 and 4.17, hydrolyzed polymerizations from higher concentrations proceed to form lower charge aquo-complexes, which, in turn, reduce the charge neutralizing capacity, clearly develop plateaus, and broaden the range within which the system fits within the critical coagulation ZP.

At the same time, neutral aluminium hydroxide coagulum is present and cross-linking between charge-neutralized agglomerates occurs, and good flocs are formed by slow (floc-forming) agitation.

(4) Area near neutral pH of $6.5 < pH < 8.5$; As the pH increases, the ZP rapidly decreases, and the charge neutralizing coagulation proceeds from the positive critical potential to the negative critical potential through the iso-electric point. In this region, aluminium forms a neutral aluminium hydroxide precipitates, which, in addition to charge neutralization, is the region where cross-linking occurs properly and where coarse suspension coagulation and flocculation are most common.

(5) In the region $8.5 < pH$; Only negatively charged aluminium species, with no charge neutralizing ability, return to a state of dispersion with a charge approximately equal to the charge of the clay particles.

Kaolinite particles in water do not coagulate even if their ZP is below the critical coagulation potential of 10 to 12 mV, and some neutral insoluble aluminium hydroxide is required for agglomeration. It was observed that the presence of $Al_2(SO_4)_3 18H_2O$ of about 10 ppm, i.e., Al of about 0.81 ppm (less than 1 ppm) was required for floc formation.

Plotting the kaolin concentration obtained in the experiment and the amount of aluminium added until reaching the iso-electric point, Figure 4.23 shows that approximately two-thirds of the added aluminium is present in the diffused layer. In addition to the ionic composition of the water under test, it can be estimated, by simplification, that $Al_2(OH)_{20}{}^{4+}$ ions with a molecular weight of 556 can be added to the diffused layer by about 1.39 ppm by adding 10 ppm of aluminium sulfate. From the experimental results obtained in this manner from $10^{5.94}$ ppm aluminium concentration (10 ppm aluminium sulfate added) to $10^{6.697}$ ppm (80 ppm added) in the diffused layer, the inverse κ cm^{-1} of the ionic atmosphere thickness of the Debye–Hückel equation (Equation 4.24) is obtained as the first step of determining the value of b. Next, in Equation 4.24, we substituted the experimental critical coagulation potential and the values of the Gouy-Stern layer boundary potential Φ_d and the ion atmosphere to obtain b values. We obtained

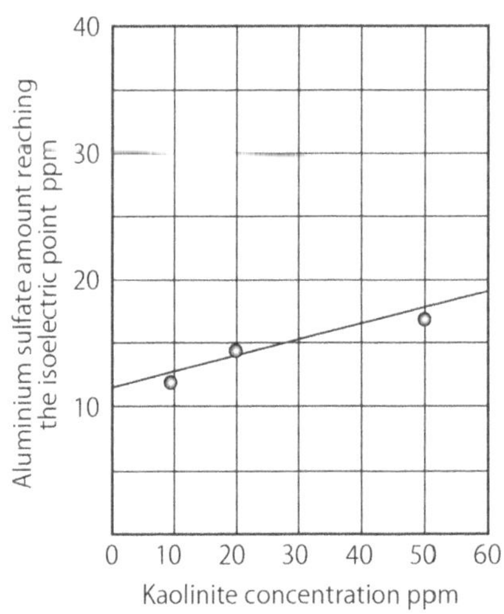

Figure 4.23 Kaolinite concentration and quantity of aluminium added before reaching the iso-electric point.

approximately constant b values of about 6.5×10^{-7} from every four calculated results from b = 6.52×10^{-7} cm (10 ppm aluminium sulfate) to 6.34×10^{-7} cm (80 ppm).

Kaolinite particles in water do not coagulate even if their ZP is below the coagulation critical potential of 10 to 12 mV (e.g., in conditions near pH 4), and neutral-insoluble aluminium hydroxide must be present in certain amounts for coagulation to make flocs. $Al_2(SO_4)_3 18H_2O$ should be present at about 10 ppm (as Al about 0.81 ppm) (less than 1 ppm) as precipitates and the bonding (bridging) force of the precipitates should work to maintain the agglomeration against the external breaking force at the time of flocculation agitation. This is observed to be necessary for flocculation in jar tests. This phenomenon has been relatively well known among water treatment engineers since W. F. Langelier (1921) presented the binder alum hypothesis but the mechanism was explained by the author's doctoral thesis (Tambo, 1964a, b, c). Professor Amirtharajah's research followed Tambo's (Amirtharajah, 1981), and established the present concept shown in Figure 4.24.

Suspended kaolinite-based clays allow direct collisions between particles when the ZP falls below the coagulation critical potential. According to DLVO's theories, particles are attracted to each other by London van der Waals forces to form clumps. The agglomerate size is small and London van der Waals force is insufficient to maintain agglomeration against flocculation agitation (turbulent fluctuation velocity) intended to grow agglomerates to a few millimeters in size for effective sedimentation. This can be achieved with the presence of cross-links that are formed by co-adsorption of aluminium precipitates adsorbed on the both Stern layer bed of collision particles. This cross-linked coagulation produced by the co-adsorption of aluminium coagulant (when the size of precipitated aluminium reaches about 100 nm) extends the cross-link (co-adsorption) to the outer side of the energetic barrier and coagulation proceeds

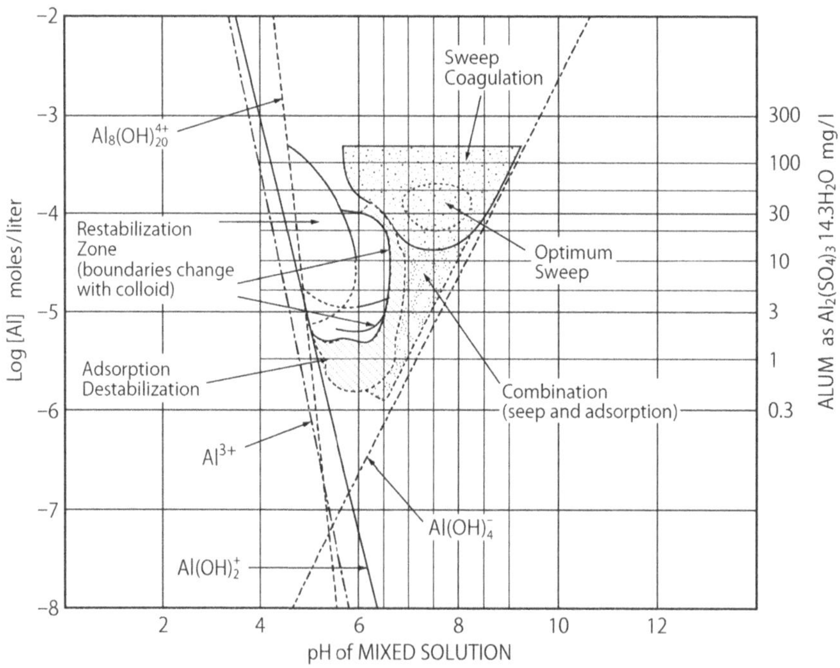

Figure 4.24 Characteristic range chart for aluminium sulfate coagulation (Amirtharajah, 1981).

even at a potential higher than the critical coagulation ZP shown by Tambo (1964a, b, c). and Amirtharajah (1981) designated the coagulation (flocculation) of this area as 'sweep coagulation'.

As mentioned above, in the coagulation process, two separate stages are commonly referred to:

(1) Charge neutralization and agglomeration by London van der Waals forces; and
(2) The flocculation (growth) of London van der Waals agglomerates against external forces by the cross-linking (co-adsorption) action of insolubilized aluminium (precipitates).

Although jar testing can be used to evaluate these two steps, there are few theoretical and experimental studies on whether the coagulation of membrane processing that does not require sedimentation or coagulation processing for direct sand filtration of low turbidity systems need only charge-neutralization for London-van der Waals agglomeration, or to what extent co-adsorption of the aquo-complex is required. Progress in research in this area is desirable.

In the case of a suspension of bentonite (about 90% montmorillonite), when charge neutralization progresses even in the region of pH 4 to pH 5 where there is no cross-linked aluminium (as in the case of a kaolinite-based suspension), both charge neutralization and cross-linking occur, coagulation flocculation proceeds, and sedimentation takes place. Figures 4.25 and 4.26 show experimental jar test results of both kaolinite and montmorionite-based clay suspensions from Professor Packham of Imperial College London. The results of a jar test of 50 ppm kaolinite-based suspension show pH in the horizontal axis and the amount of aluminium sulfate required to reduce residual turbidity by half are measured along the vertical axis. For kaolinite suspensions, removal is effective only when aluminium co-exists in the form of both an aquo-aluminium complex and insoluble hydroxide, similar to the experimental results described earlier by the authors. In these conditions, an insoluble aluminium hydroxide presents less than 1 ppm aluminium (whilst aluminium sulfate presents at about 10 ppm), and the suspended particles are charged at a coagulation critical potential of 10 to 12 mV or less. These conditions are coagulation floc-forming conditions. In kaolinite-based suspensions, the first requirement is the presence of 10 ppm of insoluble alum (about 1 ppm of Al), in which the ZP of the particles falls

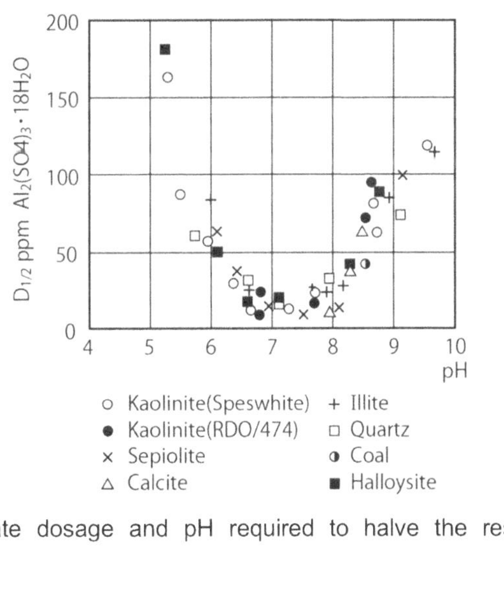

Figure 4.25 Aluminium sulfate dosage and pH required to halve the residual turbidity of kaolinite suspensions.

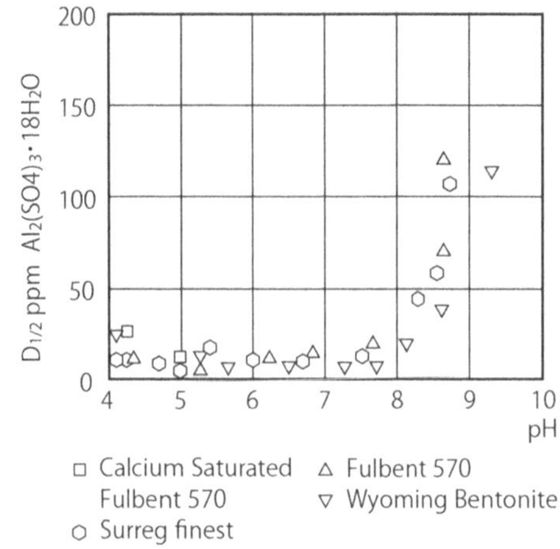

Figure 4.26 Aluminium sulfate dosage and pH required to halve the residual turbidity of montmorillonite suspensions.

within the critical 10 mV, obtained in a great number of coagulation and flocculation tests results at Hokkaido University over a period of more than 50 years.

For montmorillonite-based suspensions, charge neutralization proceed in almost the same manner as for kaolinite, but removal effectively proceeds in a wide acidic range up to pH 4.0. Coagulation appears to occur whenever ZP reaches a critical coagulation potential (10 to 12 mV or less). The reason for this is explained as follows. Montmorillonite (the main component of bentonite) is a flake-like mineral whose ZP as a whole is present in natural water as negatively charged particles of the order of −20 to −25 mV. However, if we look at the state conditions for individual particles microscopically, the upper and lower planar portions of its surface area are negatively charged, and the edge portions of the flakes are partially positively charged (Iler, 1955), behaving as negatively charged particles as a whole (values obtained by electrophoresis). Therefore, in the coagulation of bentonite, a coulombic force is applied between the positive-charged edges and the planar portions of the charges to strengthen the bond between the charged neutralizing particles (London van der Waals aggregates) and to form a floc with a three-dimensional stereoscopic component without insoluble aluminium for cross-linking. The injection of bentonite-based high-concentration slurries for the excavation of wells utilizes the structural viscosity of clay slurries with such component-to-particle bonding capabilities.

4.1.7.2 Coagulation treatment of natural colored water

The general nature of natural organic colored waters (humic substances) seems to be as follows. Natural water, which contains a large amount of organic colored substances (as humic substances), is widely distributed in marshes, low wetlands and peaty areas around the world. Color in wastewater treated with microbial treatments also contain humic-like substances, and pulp wastewater (lignin) is also found in many rivers. Although there are few reports of immediate health impairment from the drinking of brown water due to the presence of these colored substances, it is aesthetically unpleasing and often accompanied by unusual odors and tastes, and therefore removal of color is required when being used as drinking water. In industrial water, particularly in the papermaking and food industries, color removal is essential because of the high requirement that these products are not discolored by the water.

One example of a direct health-threatening consequence of the drinking of natural colored waters (see Section 3.5) is Kaschin-Beck disease, first reported to a Japan Society of Civil Engineers Health Group in the 1960s by Russian military physicians (N. Kaschin and E. V. Beck) in the North-East Chinese region (the former Manchurian Region). Ingestion of high-color peaty groundwater is said to cause symptoms such as poor limb development and shorter fingers, but pathological studies do not appear to have definitive conclusions. In another example, Professor Hanya and others at Tokyo Metropolitan University issued a warning about the use of the polluted Tama river for raw water for the Tokyo Water Supply in 1955, and the Tokyo Metropolitan Government stopped its use. However, it seems that the scientific conclusions regarding the danger were not proven. In another case, Dr Rook (1972, 1974) of the Amsterdam Water Works downstream on the Rhine River, identified trihalomethane formation caused by the chlorination of river water containing colored (humic) substances, and then linked it to the problems of various trace chlorine compounds and ozone oxides by water disinfection processes. Since the 1970s, control of natural colored substances in water has become a major issue for water quality control.

The strength of natural brown color is measured by comparison of the standard platinum cobalt solution series and the test colored water by the naked eye. Brown humic acids and standard platinum-cobalt solutions are based on the fact that they exhibit almost the same hue under white light, and do not necessarily show consistent absorption curves even when absorbance ratios are determined by photocolorimetry at characteristic wavelengths. Therefore, it should be understood that this measure, if used, quantitatively relates the ambiguity of naked vision comparisons with certain criteria.

For color strength measurements, the authors used the 430 nm absorptions and pH 8.4 proposed by Professor A. P. Black (University of Florida) and Professor R. F. Christman (University of North Carolina, Chapel Hill) to study the treatment of water in the marsh areas of the eastern states of the United States at the University of Florida (1962–1963). The results are described in the following section.

When studying brown water from peat land in Hokkaido in Japan, the author Tambo used 390 nm and pH7.0 as colorimetric conditions. The latter is also used for research on Kraft Pulp (KP) wastewater, which has similar conditions to those of natural organic colored water. Professor Y. Nasu, a colleague at Hokkaido University, Japan, used two wavelengths (370 nm and 410 nm) simultaneously. The Japanese water supply standard recommends 390 nm. As shown in Figure 3.34, the absorption curve for organic colored water shows a rapid rise in absorbance with increasing wavelength below 400 nm (invisible region) and a plateau state with maximum absorbance in the range of 400 nm to 550 nm. Therefore, the authors believe that many researchers have selected wavelengths around 400 nm, which can be measured most sensitively and which are less susceptible to disturbance of suspensions by long wavelengths. Despite the popularity of basic colorimetry using the naked eye under white light, it should be remembered that photoelectric colorimetry is now available.

In Chapter 3, the authors explained that natural organics (DOC), such as cyclic compounds with absorbance at E260 nm (E260 sensitive), and compounds with unsaturated bonds are derived from humic substances and used as natural coloring ingredients in our lab. The ratio (DOC/E260) is defined as the ratio of biodegradable organic matter to bio-non-degradable organic matter (by DOC), and the ratio of (DOC mg/l/$E260_{(1\ cm)}$) is used to quantify the presence of biodegradable and bio-nondegradable organic substances in natural and urban water systems. In terms of microbial treatment and natural purification, the authors determined that the potential/impossibility of biodegradability can be quantitatively represented by the ratio of DOC/E260.

Using DOC/E260 ≑ 30 for macromolecular parts with masses of 1500 Da or more and DOC/E260 ≑ 50 for small molecular moieties with molecular masses of 1000 Da or less (which have been shown to be biologically persistent) (see Section 3.5, Figure 3.36) shows how much biodegradation can proceed by measuring the (DOC/E260) ratio of ordinary test water. In municipal wastewater (DOC/E260) ratios are as high as 500 to 1,000, but it is estimated that 90–95% of DOC can be degraded by biological treatment. If the DOC/E260 is 100 in lake water, 99.5% is biologically persistent humic substances (colored substances), indicating that microbial treatment is almost ineffective. This study was compiled into a number of papers around 1975 in my laboratory, reported to the *Journal of Water Pollution Control Federation* (1980), and became widely known internationally. Having compiled the work of evaluating treatability using gel for the evaluation of taxonomy in water and wastewater treatment processes (Tambo & Kamei, 1980), Tambo and Dr Kamei (an Assistant Professor in Tambo's group in Hokkaido University) established a method of natural water treatability taxonomy which is now widely used in the form of 'water quality conversion (evaluation) matrices'. Care must be exercised if the method is used for general purposes because of existing protein interference.

The authors published this idea in international (Tambo & Kamei, 1978a, b, c) and domestic journals (Tambo *et al.*, 1981a, b, 1982c) and were invited to make a presentation to the American Chemical Society's Annual Meeting in Denver in 1987. Afterwards, Professor J. K. Edzwald *et al.* (1985), at the University of Massachusetts, published an index characterizing natural organic matter in almost the same way as our method, called specific UV adsorption (SUVA). Initially, SUVA seemed to be an expression of the relation between TOC and UV (254 nm). We selected 260 nm – the most sensitive level on a spectrometer – but Edzwald selected a wavelength of 254 nm, for which an inexpensive ultraviolet spectrophotometer with a single wavelength could be used. At the same time, by making the absorbance cell length 100 cm, one order of magnitude can be obtained by inverting our index, which seems to have increased simplicity. This, on the other hand, reduces the accuracy of the estimation of the abundance ratio of biodegradable/biologically persistent organic matter, and because Edzwald did not consider any molecular weight distribution, his procedure cannot be used as an index for selecting processes such as coagulation, adsorption and biological degradations.

The humic substances of the brown colored component mentioned above, which make up a large part of the organic matter of natural origin (NOM: Natural organic matter), are aggregates composed of many molecules extracted from the humus state of plants. Although the components expressed vary, they are basically polymer components of aromatics. These are brown amorphous materials with functional groups such as COOH, $-OCH_3$, $-NH_2$, $-NH_3$ and $>C=O$. One ingredient that does not precipitate even under the condition of $pH < 1$ is defined as fulvic acid. The components are distributed on the relatively low molecular weight side. When pH is set to 1 or less, the polymer-side component that undergoes precipitation (coagulation) is called humic acid. Portions that do not dissolve at all are called humus (see Table 4.10).

Since the charged particle state is soluble (as fine as colloidal size), it cannot be measured by microscopic electrophoresis, and the electrophoretic transport tube method must be used. The mobility of the colored components of Ishikari peaty area groundwater in Hokkaido, Japan shown in Figure 4.43 is about -1 (am/sec/V/cm) mobility $\doteqdot -13$ mV ZP at around pH 3. The negatively charged value increases to around $pH = 6$, and the negative maximum value reaches about -4 (am/sec/V/cm) $\doteqdot -50$ mV. After

Table 4.10 Classification of humic substances.

Fulvic acid	in $pH < 1$: Low molecular weight components that do not precipitate
Humic acid	in $pH < 1$: High molecular weight components to precipitate
Humus	Insoluble soil regardless of pH

that, the negatively charged value decreases to a maximum value of −3 (am/sec/V/cm) ≑ −40 mV near pH7, and the negative mobility again decreases to −7 (am/sec/V/cm) ≑ −90 mV near pH 10. The detailed reason is not clear, but it appears that particles with a high negativity are stably dispersed in natural water in the entire pH range. The reason why humic acid coagulates at pH 2 to 1 or less is considered to be that the mobility becomes −1 (am/sec/V/cm) ≑ −13 mV or less and coagulation starts. Depending on the ionic strength of the solution, coagulation occurs below pH 2.

According to Professor Christman (personal communication), 90% of the organic colored component in the samples that he collected in the eastern US marshes are fulvic acid and 10% are humic acid. The size has been shown to be a 3.5 to 10 me in colloid dimension and his measurements have confirmed that it is a 10 me colloid or less. (Molecular mass could not be measured at that time.) On the other hand, the gel chromatogram of the Ishikari peat area (Hokkaido) water shows that the colored ingredients are divided into two groups: about 20% of the ingredients with a molecular weight of 1400 Da and about 80% of the ingredients with a molecular weight of about 1000 Da, as shown in Figure 4.27.

It is considered that organic colored water in the natural water system has been extracted into the water system by rainwater from decomposing plant residues over a long period of time. Wetlands are generally considered to have accumulated rainwater at a shallow depth on impermeable clay or soil beds on which leaves and plant residues accumulate, and humic substances are produced from aerobic decomposition and eluted by the rainwater to accumulate brown colored peaty water. According to our study, the amount of humic substances produced as metabolic wastes from the aerobic microbial decomposition of organic matter is about 10 times higher than the amount produced by anaerobic decomposition, and reaches as much as 5% of the decomposed TOCs (Tambo *et al.*, 1981a). This is the reason why the colored components of wetland waters have a high color concentration. In ponds with long-term retention time or voids on soils, humic substances are present in polymeric forms, which are thought to be due to the progress of certain polymerizations. Such polymeric components (humic acid components: DOC/E260 = 30 components) are retained in the soil layer when colored groundwater passes through the soil. As a result, groundwater inflow to rivers as an intermediate outflow mainly contains

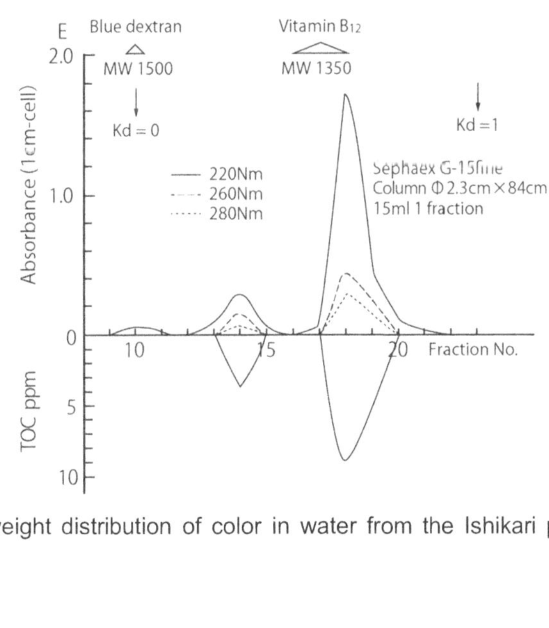

Figure 4.27 Molecular weight distribution of color in water from the Ishikari peat area. (Sephadex G15 Gel-chromatogram).

components of low molecular weight (fulvic acid components: DOC/E260 = 50). Therefore, the natural colored organic components of river water during normal times are mainly composed of components with a molecular mass of 1000 Da or less. During flooding, humic acid colorants, such as those of 1500 Da molecular weight, appear in river water because of the high concentration of human in the A_1 layer of forest soil surface (which had been stored in the soil surface layer) which are spilled to rivers due to surface runoff (Table 4.11). As mentioned earlier in the section on water quality matrices, coagulation sedimentation can remove only the polymer molecules. The low-molecular-weight colored component is removed by activated carbon adsorption. This phenomenon is the key to trihalomethane control, which will be described later.

Peaty land and low wetland (swamps) have a water impermeable layer below the water level, and it has poor water drainage, forming shallow pools of water by accumulation of rainwater. As a result, suspensions are separated by sedimentation in the static water, generally in the form of a transparent colored water, in which humic substances generated by aerobic decomposition of accumulated vegetable residues are extracted and present in rainwater. Therefore, the previous measurement methods and removal by coagulation etc. will be discussed first with respect to transparent natural organic colored waters. However, the impermeable layer that constitutes the bottom of the wetland is often not perfect. Since the aquifer below the impermeable layer is generally anaerobic (without oxygen), iron, manganese and other substances are dissolved in groundwater in the form of Fe^{2+} and Mn^{2+}. When such groundwater rises from a break in the impermeable layer of the swamp bottom and mixes with the aerobically colored water mass, iron and manganese associate with humic molecules to form protective colloids and remain in the reduction state. Iron and manganese stably exist as so-called humic-acid iron. In the Ishikari peat land groundwater, for which we measured polarities by electrophoretic mobility and ion-exchange resin, iron was present as negative colloids, and was also present in a form protected by humic substances. Iron could not be precipitated by conventional aeration, chlorination or ozonation, and it could only be coagulated as iron hydroxide by strong oxidation with potassium permanganate. On the other hand, iron can be removed with aluminium coagulation in a protected form while the iron remains as negatively charged color colloids. However, the authors have not yet succeeded in artificially producing such protective color colloids of iron and manganese ions by humic substances in the laboratory.

When clear, naturally colored water flows out from wetlands to rivers due to strong rainfall, turbidity components are mixed, and both turbidity and color components are present in the river water system. The algorithms and process characteristics of color component measurement and treatment when coexisting with turbidity are addressed later in this chapter. As shown in Figure 4.28, the authors use the fact that kaolin suspensions exhibit constant absorption over a wide wavelength range of 200 nm to 1000 nm to express the standard absorbance of 260 nm (or 254 nm) of soluble peaty-colored water by correcting and separating the turbidity estimated from the absorbance of 860 nm in the near-infrared region.

Table 4.11 The natural colored organic matter in river waters.

during Normal	Component of low molecular weight component of 1000 Da or less is main. High molecular weight components are retained in the soil and low molecular weight components flow out.
during High water	Contains high molecular weight components over 1500 Da Humic matters on the soil surface flow out by surface runoff Coagulation and Sedimentation

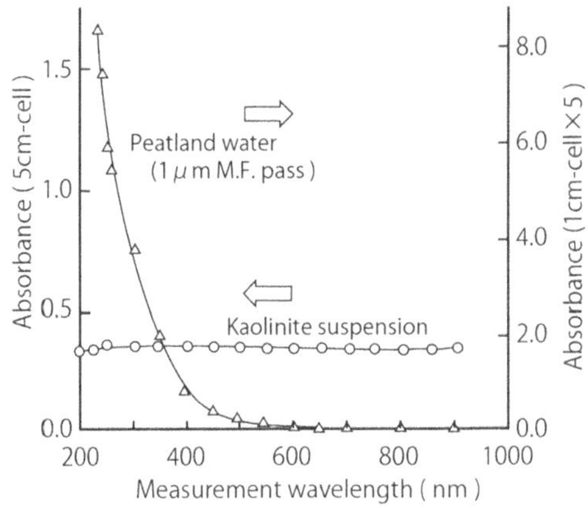

Figure 4.28 Color and turbidity absorbance, and wavelength.

4.1.7.3 Examples of colored water coagulation in the Eastern United States

The author Tambo was at the University Florida to study water treatment from 1962–1963. This section explains the results of experiments conducted at Professor A.P. Black's colored waters laboratory in the eastern USA. The water sampling sites and water quality are outlined in Table 4.12. We can estimate the condition of raw water in the early 1900s in cities in the Eastern United States that developed a rapid filtration method in a short period of time, with one of the main aims being dealing with colored waters that could not be treated by slow sand filtration. The sample waters in the US were typically high-colored natural waters exhibiting a weakly acidic pH, and were similar to the peaty groundwater in Hokkaido, Japan (described further below). Samples contained little turbidity. Professor R.F. Christman, a colleague in the laboratory, analyzed the content of humic substances and found that humic acid was about 10% and fulvic acid about 90%.

In the experiments, as in the coagulation test described above, agitation/sedimentation separation by a standard jar test and measurement of electrophoretic mobility of micro-flocs by a microscopic method using a Briggs cell were carried out for all samples. A 1-litre sample was taken in a jar, to which was

Table 4.12 An outline of colored water quality in the Eastern United States

Water No.	**1**	**2**	**3**	**4**	**5**	**6**
Water sampling sites	Emerson Brook, Tewksbury, Messa.	Great Dismal Swamp, Norfolk, Va.	A Stream HartFord Conn.	A Stream near Miami, Fla.	A Stream near Miami, Fla.	Newnan's Creek, Gainesville, Fla.
Color (Color Unit)	70	434	240	292	312	94
Alkalinity (ppm as $CaCO_3$)	10.0	4.0	2.0	0	0	22.0
Total Hardness (ppm as $CaCO_3$)	30.0	10.0	8.0	8.0	8.0	18.5
pH	6.7	4.8	4.8	5.45	5.4	7.15

added a variable amount of 0.1 N HCl or saturated lime water $Ca(OH)_2$ so that the pH became the specified pH after adding coagulant; a specified amount of aluminium sulfate $Al_2(SO_4)_3 18H_2O$ was added while rapid stirring at 120 rpm for 2 minutes and floc-forming gentle stirring at 40 rpm for 20 minutes were performed, before allowing to stand for 30 minutes, collecting 100 ml of supernatant water, and filtering out the supernatant water using qualitative filter paper. The pH of the filtrate was then adjusted to 8.4 according to the recommended methods of Black and Chrisman, and the color measured using a photo colorimeter at 430 nm versus platinum-cobalt standards. The ZP of the floc remaining in the jar was measured by a Brig cell, and the process pH was confirmed at the same time.

Figures 4.29 to 4.34 show the changes in color removal and ZP of the resulting flocs at different pHs when 100~150 ppm of aluminium sulfate was added to colored water in the eastern states of the United States. Figures 4.35 to 4.39 show the relation between color removal and the ZP of floc with the change in pH when aluminium sulfate was added from 30 to 200 ppm to the colored water of a small river flowing from Lake Newnan in the suburb of Gainesville, at the University of Florida. Figure 4.40 summarizes the experimental results of the Lake Newnan samples (Figures 4.35 to 4.39) and shows how the ZP of flocs formed (in the horizontal axis) by increasing the amount of aluminium added (in the vertical axis), using pH as a parameter.

The experiments revealed that, when aluminium sulfate was added and various natural organic colored waters were treated with coagulation, the following general coagulation patterns resulted:

(1) Region pH < 4.0; The ZP was not reversed by approaching the iso-electric point even in the area where the added amount of aluminium sulfate reached the maximum value of 200 ppm, and color removal did not occur even at the iso-electric point. It is probable that the oppositely charged ion was a simple aluminium ion, so there was neither a cross-linking effect nor a charge reversal effect, and the floc could not be formed.

(2) Region 4.0 < pH < 4.5; As the pH increased, the ZP increased rapidly and charge reversal also occurred. Maximum ZP was observed at around pH 4.5 to 4.7. This is presumed to be due to the maximum action of the aluminium polynuclear complex (probably tetravalent) with the

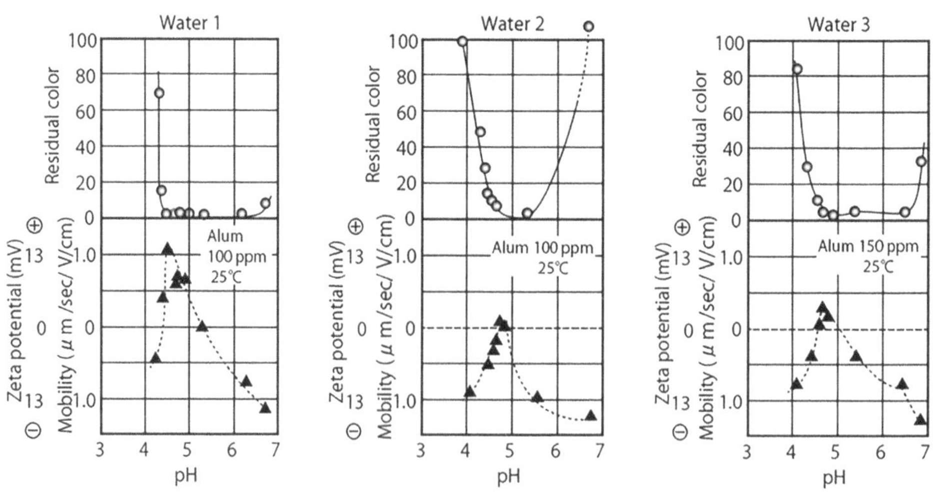

Fig.4.29 Fig.4.30 Fig.4.31

Figures 4.29–4.31 Jar tests for colored water in swamps in eastern states of the USA.

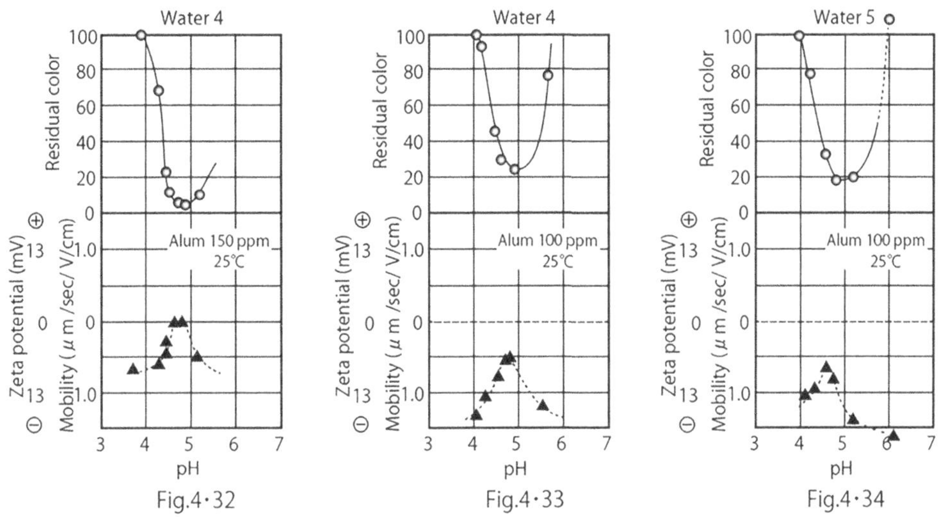

Figures 4.32–4.34 Jar tests for colored water in swamps in eastern states of the USA.

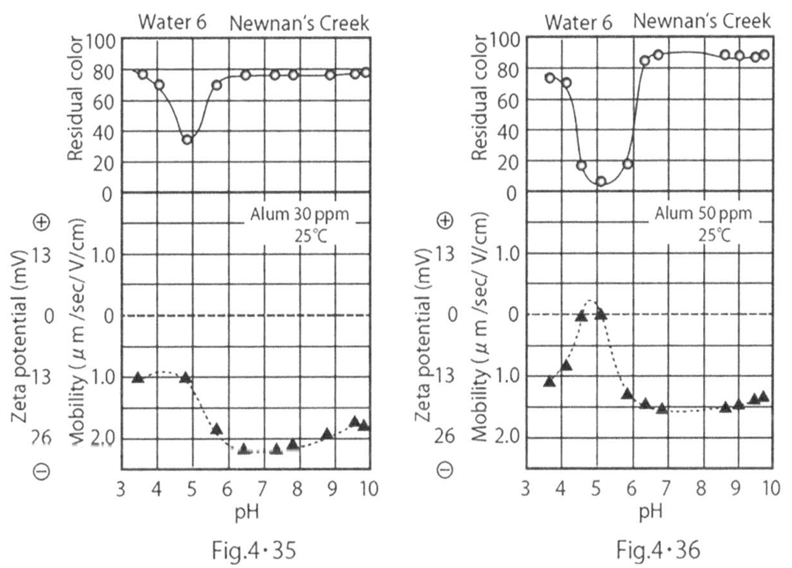

Figures 4.35–4.36 Jar tests for colored water taken from Lake Newnan Creek, USA.

maximum charge neutralizing ability. No coagulation was observed even at the iso-electric point but it is considered that there were no aluminium precipitates for cross-linking.

(3) Region 4.5 to 4.7 < pH < 6; The area with the best color removal appeared in the pH-region immediately after the highest ZP was observed. In this region, when the ZP was rapidly decreased and the ZP reversed, a good color removal was observed in a wide region between +10 mV and −10 mV. It may be considered that the absolute ZP was within the critical

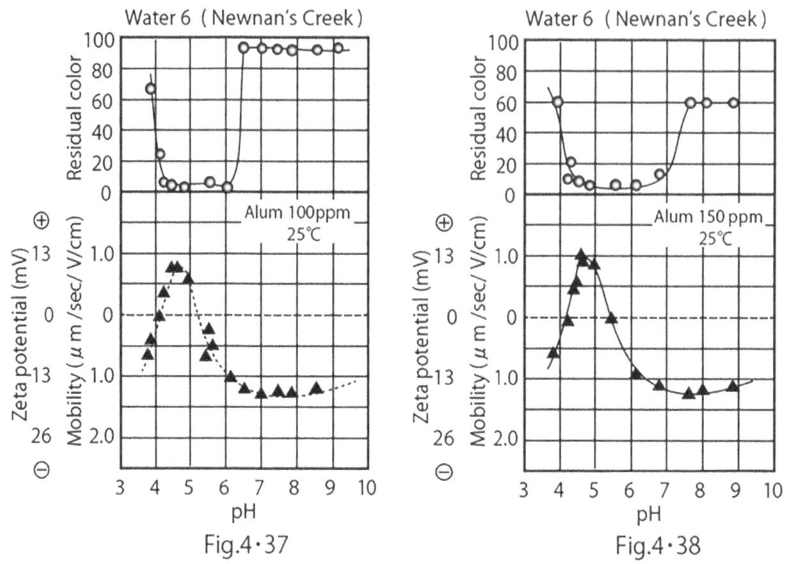

Figures 4.37–4.38 Jar tests for colored water taken from Lake Newnan Creek, USA.

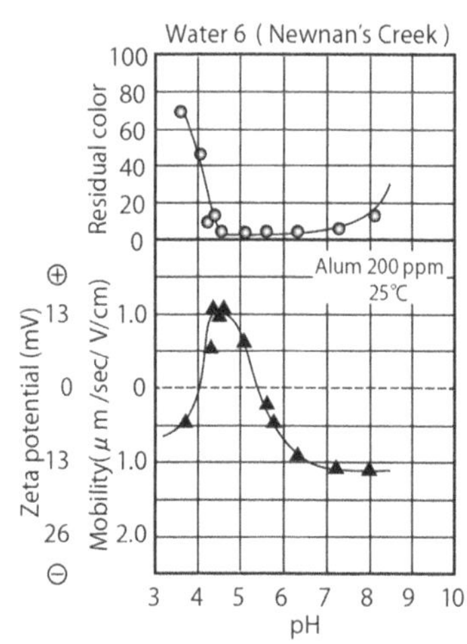

Figure 4.39 Jar test for colored water taken from Lake Newnan Creek, USA.

coagulation potential of 10~12 mV and the neutral coagulum of alum (precipitate) – working as a crosslinking binder for flocculation – was present at about 10 parts per million (a little less than 1 ppm as Al). Two requirements for floc formation – critical charge neutralization and binding precipitate (such as 1 ppm of Al) – are both satisfied.

(4) Region $6.0 < pH < 8.5$; The change in ZP was slow, and the ZP of aluminium precipitate decreased from +3 to +2 (μm/sec/V/cm) (see Figure 4.13), and the mobility of color particles changed slowly from −5 to −3 and then back to −5 around pH 7, as shown in Figure 4.43.

Measurements, results and discussions of the mechanism will be addressed below.

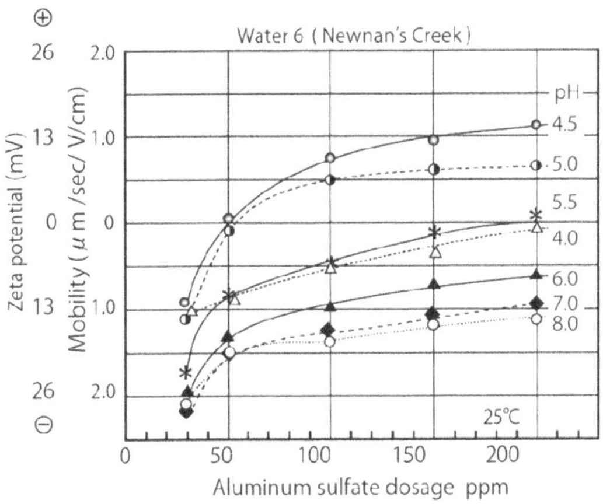

Figure 4.40 Aluminium dosage and floc zeta potential for colored water taken from Lake Newnan Creek, USA.

Micro colloids, such as color colloids, usually have a particle size of less than or equal to nanometers, and coarse colloids, such as clays, are present with a particle size of 10^0 micrometers. Considering the same amount (volume) of particles, a micro colloid has a specific surface area as large as 10^6 times that of a coarse colloid. Therefore, if there is no extreme difference in surface charge density (in fact, mobility μm/sec/V/cm differs at most by only one order of magnitude), the first requirement of the process is that the total charge amount of a micro colloid is several orders of magnitude larger than that of a coarse colloid, and the charge neutralization of a negatively charged micro colloid of a color is attempted by using the pH range in which the maximum charge neutralization capability of the aluminium coagulant is shown, and the potential of the aluminium and color colloid micro flocs is brought within the author Tambo's critical coagulation potential range of 10 to 12 mV as a prime requirement. In addition, bound precipitated aluminium (cross-linkers) must be present as 0.8 ppm aluminium (10 ppm aluminium sulfate) to grow agglomerates to the required size under floc-forming agitation strength. In many experimental examples of colored water, the added amount of aluminium sulfate required for charge neutralization is several tens to several hundred ppm, and then, the required additional crosslinking is sufficient.

Table 4.13 Summarizes the pH and coagulation pattern of natural organic colorants.

pH range	Characteristics of Coagulation by Al Species
pH<4	Even if Al 200 ppm dosage, it does not coagulate only by approaching the isoelectric point. It is a simple ion type (Al^{3+}) and can be neutralized by its charge, but it is difficult to flocculate without bridging.
pH4 ~ 4.5	Moves to charge neutralization but does not coagulate. Aluminium aquo complex action with maximum charge neutralization ability. No precipitation of aluminium for bridging. Main species $Al_8(OH)_{20}{}^{4+}$
pH4.5 ~ 4.7<6	Optimal range for color removal. ζ - potential is in the critical coagulation range. Presence of neutral precipitate of aluminium for bridging.
pH6 ~ 8.5	The ζ - potential is outside the critical coagulation range and coagulation is difficult.

Contrary to the coagulation condition of the coarse suspension system described above, for natural colored water it is considered to be appropriate that the process condition is set to a slightly higher pH, with a pH of 4.5 at which the aluminium aquo-complex has the highest charge neutralizing capacity, and the ZP of the coagulation particles is brought into the critical potential of $+10\text{ mV} \sim -10\text{ mV}$; the amount of precipitated aluminium is slightly increased in due course (5 to 10% of the added amount) to increase pH to pH 4.7 to 6.0.

4.1.7.4 Mechanism of agglomeration of aluminium aquo-complexes and color colloids

The previous section, based on the results of a study conducted by the author at the University of Florida, clarified the pattern in which the color component of natural water interacts with the aluminium aquo-complex etc., which is expressed at various pH values and agglomerates to form flocs.

Sometime after the original experiments, the authors (at the Department of Environmental Engineering, Hokkaido University) became able to measure the mobility of color colloids in the soluble region by using the electrophoretic transport tube method, which was created by the authors after returning to Hokkaido from the US, so that the above-mentioned coagulation patterns can be explained by understanding the mechanism of coagulation by superimposing measurement results on the particle size composition distribution of the aluminium/color colloid agglomerates formed through the soluble/suspending region. Here, solubility and insolubility are expressed separately by the passage and non-passage of 0.45 μm filter paper. A particle size of 0.45 μm approximates to the worst-case particle size condition for coagulation proposed earlier by Tambo. From the viewpoint of interfacial electrochemistry, it is possible to consider an insoluble particle as a coagulated particle.

The authors first investigated the situation in which only aluminium was added to the water, to provide basic data. While stirring with a magnetic stirrer, aluminium sulfate was added to distilled water to a specified concentration, and hydrochloric acid or sodium hydroxide was then added dropwise to achieve the specified pH. Aluminium was analyzed using a pipetted sample at pH 2.0, and samples analyzed by the JISK0102 oxychloro-extraction method were colorimetrically determined as monomeric ions. The amount of aluminium particles present in each filter is shown in Figure 4.14 (above), from the amount of aluminium in the filtrate, which is separated by pore size by membrane filters with pore sizes of 12, 8, 3, 0.45, 0.10 μm, 10, and 5 nm. The mobility of aluminium measured by the Hitohoff electrophoretic transport tube method is shown in Figure 4.13 above. Compared to Figure 4.14, it can be seen that there are soluble aluminium polynuclear complexes with the largest charge neutralizing capacity at around pH 4.5, and that as pH increases, the amount of positive charge decreases sharply and the charge neutralizing ability decreases, resulting in the appearance of insoluble precipitates. This relation explains that the proper combination of the presence of an aluminium aquo-polynuclear complex of pH 4.5 with a high charge neutralizing capacity and the presence of aluminium hydroxide precipitates formed in the neutral pH range as flocculation binder are key to the design of coagulation process for both clay-based suspensions and color-based micro colloidal solutions. Figure 4.41 shows an example of the presence of aluminium, which provides an overview of this relationship (Tambo & Ito, 1977).

The original test water used in the coagulation test was well water from peaty land in the suburb of Sapporo City, Hokkaido, and it was clear colored water with little turbidity. The well water was soft water with pH6.0, color unit 80, TOC20 mg /l, and electric conductivity 73.7 μs/cm.

The jar test was repeated at various levels of added aluminium, and changing the pH for a constant amount of aluminium dosing. The raw water was filtrated with 0.45-μm filter paper, the color was adjusted to 20 degrees, and the test performed. The supernatant was filtrated after coagulation sedimentation with 12 μm filter paper to determine the color and aluminium concentration. The 12 μm

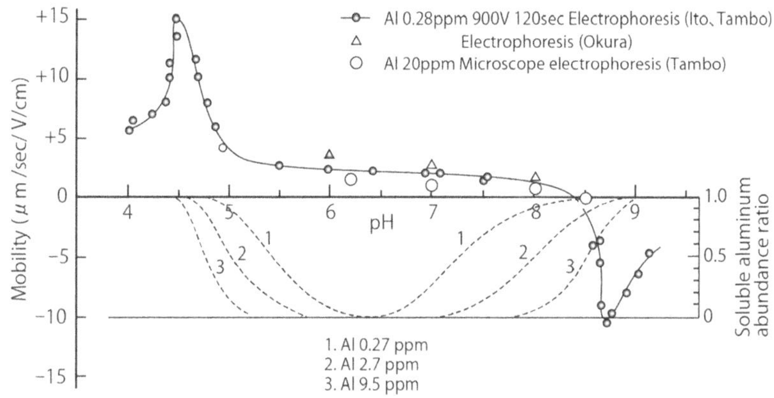

Figure 4.41 Aluminium mobility and the soluble aluminium abundance ratio.

filter paper mimics direct filtration separations. Figure 4.42 shows the color and aluminium concentrations before and after filtration with a 12 μm filter paper. The solid line shows the residual color and aluminium concentration when the aluminium sulfate content was changed to 10 ppm, 30 ppm and 60 ppm (0.8, 2.4, and 4.8 ppm as Al). The dashed line shows the aluminium residual when the jar test was performed with aluminium coagulant only. Figure 4.43 shows the color component of the test water (color colloids) measured by the electrophoresis transport tube method and the change in aluminium and color mobility due to the pH of the component which was passed through the 12 μm filter paper. Characteristics of the mobility of color colloids depending on pH have been described above.

If the color and aluminium mobility curves agree (coincide), all humic particles and added aluminium coagulant aquo-complexes can be considered to exist as complexes or agglomerates of humic substances and aluminium. These conditions occur in the pH range of 4.7 to 7.0. On the other hand, in the range of

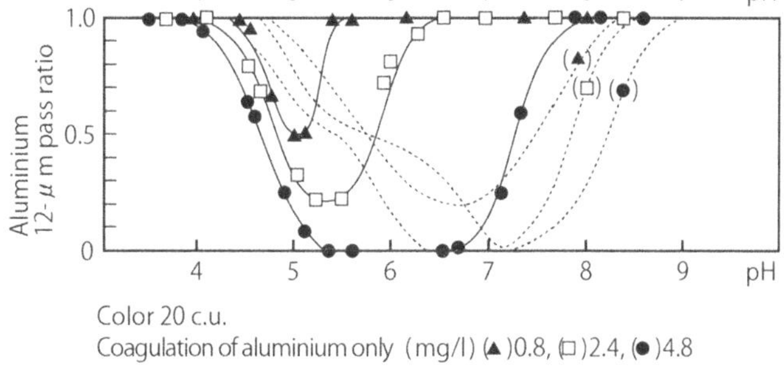

Figure 4.42 A jar test for color removal.

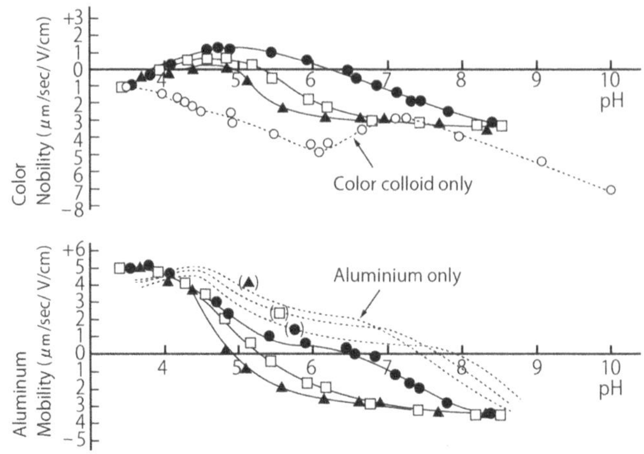

Figure 4.43 The mobility of color and aluminium flocs.

pH < 4.5 to 4.7, the mobility of color and aluminium differ greatly, and aluminium polynuclear complexes (probably +4) having high solubility (polymers having a molecular weight of the order of 10^2) have a high charge neutralizing capacity. However, as explained in Figure 4.23, it is considered that the dispersed aluminium part exhibits a high mobility due to the separation of the portion that adsorbs to the Stern layer, and acts together with the color colloids and the portion that exists in the bulk water and in the Guoy diffusion layer. As shown in Figure 4.13 above, hydrolyzing polymerization proceeds rapidly in the region where pH approaches 5.0. The positive charge of aluminium decreases, reducing the charge neutralization capacity. The positive charge of aluminium becomes a positively charged polymer with a diameter as large as 1 μm, which is co-adsorbed by color colloids to form cross-links, enabling flocculation to be promoted. Under these conditions, color-aluminium complexes within the critical coagulation potential grow into agglomerates and become flocculated.

Figure 4.44 shows the distribution of particle sizes of aluminium and color components in each pH range measured by direct filtration, and shows the results that directly support the above-mentioned states.

4.1.7.5 Structure of aluminium flocs produced

Figure 4.45 shows a schematic representation of the structure of the coagulated agglomerates formed by clay suspensions, humic (color) substances, and by combining clay and colored colloids with aluminium, aquo-complexes and/or hydrated precipitates. There are three basic types of coagulation:

(1) Cross-linking coagulation (bridging coagulation); Clay-based suspensions have particle sizes of the order of 1 micrometer, whereas color colloids are considered to have a particle size of about 10 nanometers. Therefore, the specific surface area of clay-based suspensions differs by as much as 10^4. For this reason, as described above, in order to neutralize a large amount of surface charge of micro-colloids such as color components, we select a pH condition in which an aquo-polynuclear complex ($Al8(OH)204^+$, etc.) – having the largest charge neutralization capacity – is significantly present, and perform a process centering on charge neutralization. For clay-based turbidity, on the other hand, some amount of cross-linking (bridging) of aluminium precipitate is required, primarily at a pH near neutral, with charge neutralization within the critical potential range.

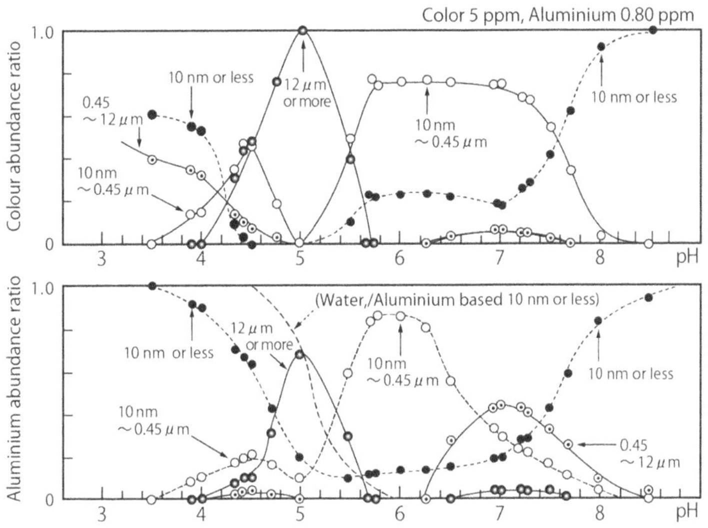

Figure 4.44 Particle size distribution of aluminium and color in each pH range.

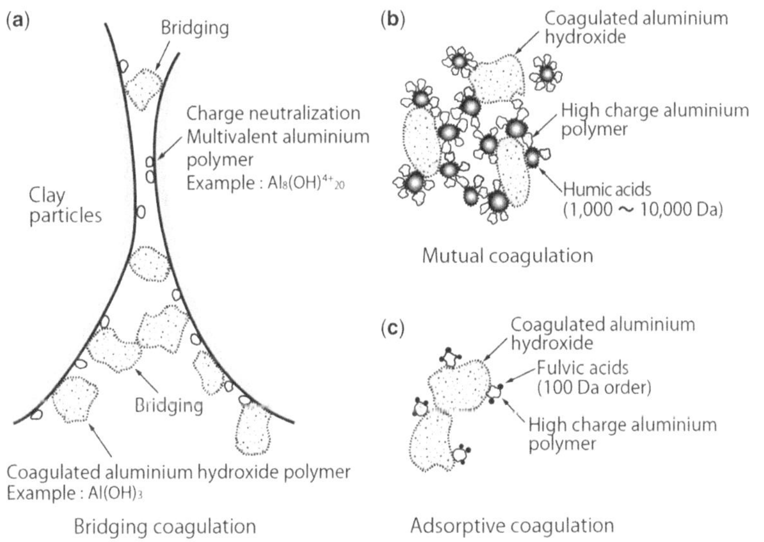

Figure 4.45 Schematic representations of agglomerate structures.

As a result, clay-based coagulation (agglomerate) is neutralized by a high-charge poly-nuclear complex at a low injection level under conditions in which a minimally required amount of coagulated aluminium hydroxide (1 ppm or more as Al up to about 100 turbidity) is generated, and the clay particles are connected to each other by the cross-linking action of precipitated coagulants to form flocs, so that they fall within the critical coagulation potential, as shown in Figure 4.45(a). This is called cross-linking (bridging) coagulation.

(2) Mutual coagulation; Colored components such as humic acid with a molecular weight of 1500 Da or more are coagulation process compatible components (see Chapter 3, Water Quality Conversion Matrix) but the coagulant used must have a high charge-neutralizing capacity because the size of these components means they also have a relatively large specific surface area compared to clay-based suspended particles. Aluminium species appearing in the pH4.5 to 4.7 range have molecular masses a little less than 1000 Da, and are polymers slightly smaller than the color components. As a result, they form micro flocs of color and aluminium in the form of a mutual coagulation that charge-neutralizes in pairs with each other. However, under these conditions alone, it is difficult to grow flocs to a size that can be separated by sedimentation. Therefore, an appropriate amount of coagulated (precipitated) aluminium hydroxide (of a size of the order of micrometers) is required as a floc-forming agent for bonding. Polymeric polymers other than aluminium (less charged) can be substituted. The flocs of the color particles thus formed are complex agglomerates of aluminium and color particles that are in coagulation with each other, and are binded by agglomerates of precipitated aluminium hydroxide, being 10–20-times larger than the coagulated aluminium and color particles.

(3) Adsorptive coagulation; Color components with molecular masses less than 1000 Da (e.g. fulvic acid) have a smaller size than highly charged aluminium polynuclear complexes and, therefore, when associated with aluminium species, coagulation occurs in such a way that the color components adsorb around the aluminium particles. This is a negative protective colloid in which color particles are adsorbed on the Stern layer of a positively charged aluminium polynuclear complex to form agglomerates. These negative colloids (protective colloids) are further cross-linked by the coagulum of aluminium hydroxide to form.

Highly charged aluminium particles and aluminium hydroxide precipitates have a limited specific surface area which is required to adsorb color components, and are therefore not useful as adsorbents because of their low adsorption area. As described in the section on Water Quality Matrices in Chapter 3, these low-molecular-weight colored components are treated with porous adsorbents, such as activated carbon, which has a large adsorption specific surface area.

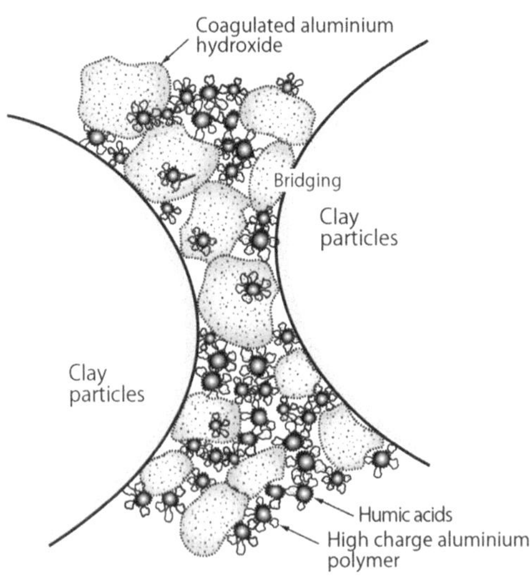

Figure 4.46 A schematic structural model of clay and a color complex floc.

4.1.7.6 Multicomponent coagulation

The presence of both color components (humic substances) and clay-colloidal components is seen in many rivers. The coagulation pattern of raw water in which both color and turbidity components are mixed at problematic levels proceeds in a pattern very similar to that of successful color removal near pH 5 of weak acids. The reasons for this are:

(1) A mutual coagulation and complex formation of aluminium and color particles first occurs;
(2) Micro flocs of formed aluminium and color colloids cross-link the charge-neutralized clay particles; and
(3) Secondary agglomeration is used to form a composite floc of clay and color colloids.

A schematic component of a clay-color composite floc is illustrated in Figure 5.46 (Tambo, 1965d).

In order to confirm the mechanism by which such flocs were formed, a synthetic raw water with a turbidity of 50 degrees (made up of kaolin particles, of which 50% were 12 μm or more, and 50% were 3–12 μm) and a colored peaty water of 20 degree were used; 2.4 mg/l of aluminium sulfate (as Al) was added as a coagulant and a jar test was carried out by the usual method. As a result, the relationship between pH and the supernatant water concentration of each component, and the pH and mobility of each component were plotted using turbidity, color intensity and aluminium concentration as indices, as shown in Figures 4.47 and 4.48, and in Table 4.14 (Itoh *et al.*, 1981).

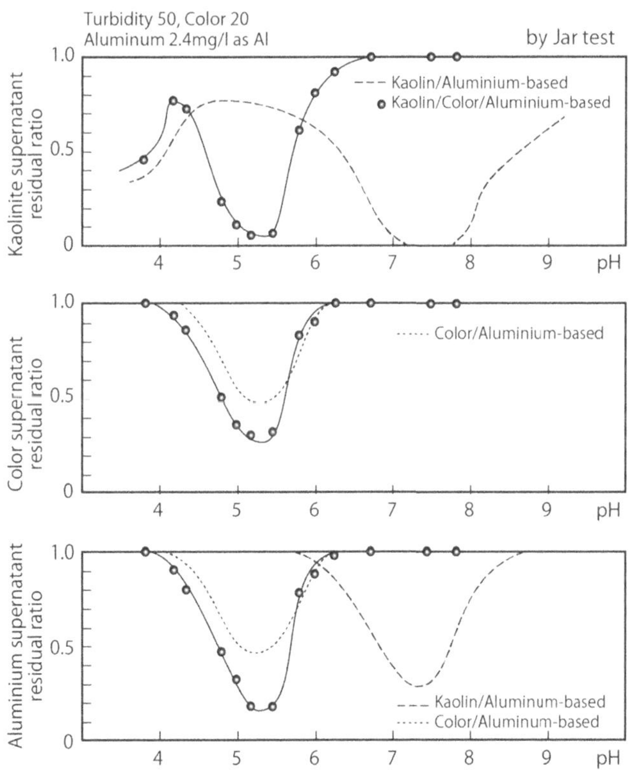

Figure 4.47 Concentration of color, turbidity, and of aluminium and pH in supernatant water.

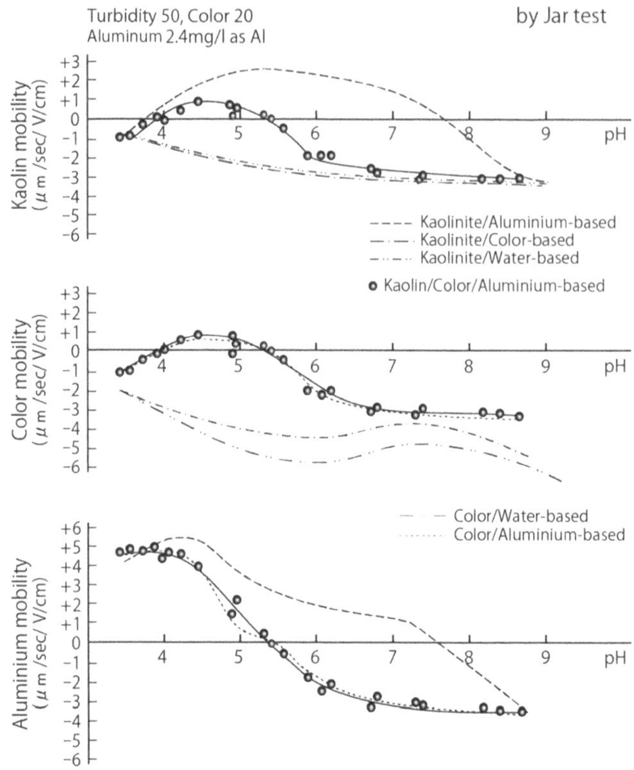

Figure 4.48 Mobility of kaolinite, color and aluminium.

It is clear that:

(1) Results of the jar test in Figure 4.47 show that the removal patterns of color and turbidity are identical, and that the coagulation patterns of the color components alone are almost identical with an optimal coagulation range around pH 5.0.

Table 4.14 The flocculation pattern for color and turbidity at high concentration.

Suspended solids can be removed, but part of the color remains Low molecular weight fulvic acids remain	
near pH4	Aluminium is present as a monomer. There is no bridging effect and no coagulation.
near pH5	Optimal coagulation range The optimal coagulation range for suspended solids only is approximately pH 7.5, but the removal pattern is the same as the case of the color only.
pH6 or more	There is no color charge neutralization ability of aluminium Although some of the color particles proceed to mutual coagulation and adsorptive coagulation, no ability to make good flocs

(2) The aluminium loading completely eliminates turbidity, but the color is slightly better removed in the presence of turbidity than in the color-aluminium system, though it is not completely removed. It is probable that low molecular weight fluvic acids and the like remain.
(3) Although the optimal removal range of the kaolin suspension alone was around pH 7.5, its properties disappeared completely in the mixed system and were governed only by the patterns of the color system.
(4) This pattern shows that the coagulation occurs under the condition that the absolute value falls within the mobility 1 (μm/sec/V/cm) proposed by Tambo, as seen in Figure 4.48.
(5) On the acidic side, such as at pH 4, a simple aluminium ion does not effectively react with the color/clay colloids, and coagulation does not occur. If aluminium is distributed in monomeric form in the diffused layer, there is no cross-linking component near the iso-electric point, and no floc is formed.
(6) Also, in the neutral to weakly alkaline range above pH 6, aluminium does not have the ability to neutralize the charge of the color component, and only a small fraction of the color component proceeds towards color-aluminium cross coagulation or to form an adsorbed protective colloid in which the color colloid covers the surfaces of the micro-aluminium hydroxide particles, but does not have the ability to produce a good floc and sufficiently advance the removal.

From the changes in kaolin mobility shown in Figure 4.48, it seems that the surface of clay is covered with aluminium-colored colloidal coagulation particles, and the charge changes at the same level as the load of the protective colloid. However, the authors could not directly observe the charge.

In order to prevent eutrophication, the removal of phosphorus in wastewater is an important coagulation treatment. For this purpose, it is necessary to move it from the neutral to the weakly acidic aluminium insoluble region as in the case of color components, to have the maximum ability to form aquo-complexes with phosphorus, and to have at least 1 mg/l of bound aluminium (insoluble cross-linker).

4.1.7.7 Jar test

Unfortunately, despite the theory of coagulation and its process factors having been described in detail in the previous section, it is difficult to determine process conditions using theory alone. Although the process of agglomeration has been understood and indexed by evaluating the critical coagulation potential, it is not possible to evaluate the cross-linking element for advancing to floc structure by itself. This requires batch coagulation-flocculation tests that simulate the real process on a bench-scale, and jar tests have been used for this for many years.

A jar test is performed on a desktop unit that can drive a number of agitators, called jar testers, as shown in Figure 4.49, with various agitation strengths (at least two steps). Coagulation, flocculation, and sedimentation of the test water is undertaken in several beakers with individual coagulation (chemical) conditions to select the optimal condition for obtaining the relatively best sedimentation water (supernatant water).

About 1 litre of test water is placed in a beaker or a rectangular reaction vessel. It is then rapidly stirred at about 150 to 200 rpm for about 5 to 10 minutes, then slowly stirred at 40 to 50 rpm for about 10 to 20 minutes, and allowed to stand for 10 to 30 minutes to measure the quality of the supernatant water. Direct scaling up to the prototype is extremely difficult and the stirring conditions do not need to be very agitated, except in extreme circumstances. However, as shown in Figure 4.5, to test the condition of rapid agitation requires a rapid agitation strength exceeding the Brownian motion coagulation region and needs to be in the turbulence fluctuation drive region, since the initial stage (exceeding the worst particle size condition of 1 μm in coagulation) does not proceed smoothly unless the energy-consumption rate ε_0

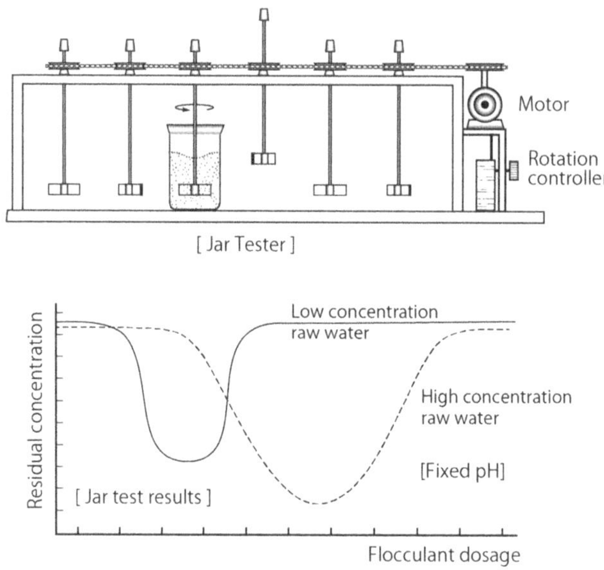

Figure 4.49 A jar test.

of agitation is 10^4 to 10^3 erg/cm^{3}sec or more. According to Equation 4.30, for a 12 cm diameter beaker, the energy dissipation rate is of the order of 10^5 erg/cm^{3}sec or more, even when the rotating water flow rate is considered. For 40 rpm mixing in the beaker, the agitation strength is about 10^3 erg/cm^{3}sec. For particles with a large particle size, turbulent flow flocculation will proceed under adequate conditions.

As mentioned earlier, coagulants based on metallic salts, such as those of aluminium, react sensitively to pH during the process. Therefore, in order to determine the appropriate coagulation condition for a given raw water, that condition must be determined by at least the combination of the coagulant loading and the process pH. Generally, as shown in Figure 4.50, a quantitative injection on 'constant dosage with variable pH method' is used, in which the injection amount of the coagulant to all beakers is kept constant, pH is changed from beaker to beaker by adding HCI or NaOH of about 0.1 to 0.01 normal, and the same process is then repeated by changing the amount of coagulant added. Chemicals need to be added to beakers side-by-side and simultaneously, and commercial jar testers used to equip a small tube system are used to achieve this.

As shown in Figure 4.49, jar tests with varying amounts of coagulant can only be performed if the optimal coagulation pH range is well known empirically. For example, in the moderate concentration range (about 10~100 degrees) of clay turbidity, the neutral region around pH 7 is understood as the optimal pH range. If colored water or turbid water containing high color is being investigated, the weakly acidic region around pH 5 is understood to be the optimal pH range, as described in the previous section. However, it is often the case that, in manual operation, it is not easy to change the coagulant level at a constant pH, but rather an operation is used in which the buffering capacity is increased by adding an alkalinity such as bicarbonate. Changing the coagulant volume to a strictly constant pH requires automated control, which results in an expensive jar test. The authors believe this is a useful method only in studies in which small changes are monitored empirically.

As shown in Figure 4.50, we can find out how much coagulant can be increased to achieve the highest removal rate by performing jar tests using constant chemical injection with variable pH. In many cases, such

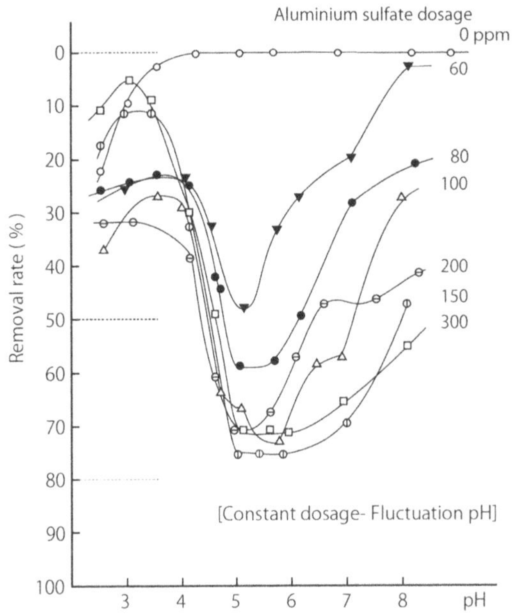

Figure 4.50 A flocculation graph for peat water (DOC 32 ppm).

as for organic color removal, the removal rate is not improved even if the amount of coagulant is increased. We call this the marginal removal rate, and when we use the phrase 'removal by coagulation' in our laboratory in Hokkaido University, including in previous studies of water quality conversion matrices, we refer to this as a numerical value after calculating the marginal removal rate. The search for optimal conditions for coagulation phenomena is considered to be a two-dimensional plane, i.e., reaction pH and the amount of coagulant added are considered together.

The commonly used jar test simulates the processes of coagulation, flocculation and sedimentation. As mentioned earlier, rapid filtration systems using coagulation and direct filtration processes are recommended under low turbidity conditions, and the coagulation of membrane pretreatment is becoming common due to the widespread use of MF/UF membranes; we will explain these in detail in the membrane treatment section below. However, if we take into account the agitation strength and the duration of agitation, we can evaluate the results immediately after rapid agitation by filtering with a filter paper of about 10 to 12 μm.

4.1.8 Flocculation aids

As mentioned in the preceding section, van der Waals agglomerates alone have insufficient bonding strength, and in order to proceed to the process of flocculation and sedimentation, a certain precipitation of aluminium hydroxide needs to be added as a bonding agent (binder). The design of high-speed filtration with polymer adjuvants is also used to increase the adhesion strength of filter particles to sand grains, even when micro floc agglomerates such as used in direct filtration are used (as described later, under filtration bed coagulation).

In Japan, the use of synthetic polymeric coagulant and flocculation aids is restricted (because of a fear of leakage of monomers containing a small amount of polymeric coagulation aids) and the flexibility of the processing design is therefore perhaps greatly impaired.

As described above, when using a polymer coagulant, the coagulant size at the time of coagulation needs to be sufficiently smaller than that of the components (suspended particles), and therefore, a metal aquo-complex which is highly charged and has a low molecular weight (1000 Da or less) has been widely used as the main coagulant. Synthetic polymer coagulants with molecular masses ranging from 10^5 to 10^6 Da can effectively neutralize the charge and grow agglomerates of clay-based coarse suspensions of a size of the order of micrometers or micro flocs agglomerated with a metallic coagulant. Polymer coagulants and flocculation aids are widely used in high-concentration systems, such as in sludge treatment, because they increase not only the strength but also the density of the produced flocs. Our own pellet coagulation theories are explained in a later section of this chapter, and the presence of polymeric flocculation aids is essential to their establishment.

Polymer coagulants include both natural and organic synthetic polymers. In natural systems, sodium alginate (extracted from seaweed), chitosan (extracted from crab shells), activated silica, etc. are all used as negatively charged polymers as flocculation aids. Activated silicic acid has been widely used as a mineral polymer aid since J. R. Baylis of the Chicago Water Works, USA, proposed it (Baylis, 1937), and it has recently been used as a main ingredient of the silica-iron coagulant developed by Japanese water engineers. The authors were involved in the development study of this new silica-iron coagulant at Hokkaido University in the 1980s, after elucidating the mechanism of formation of Baylis silica at the University of Florida (1962). This development is described in the next section.

4.1.8.1 Synthetic polymeric coagulation aids

Synthetic polymeric coagulant /auxiliaries are made by polymerization of monomers, and both monomolecular polymerization and copolymerization of several monomers are used. In any event, some unreacted monomers remain after polymerization and their toxicity is a problem. Different molecular weights, charge densities and polymeric chain morphologies (linear, side-chain) are produced. The type of charge can be cationic, neutral (nonionic) or anionic. Only cationic polymers are used as a coagulant, but neutral or anionic charges act as flocculation aids (polymeric cross-linkers) or filtration aids. An example of typical components is shown in Figure 4.51.

Polymer cross-linking is basically a co-adsorption phenomenon of polymer chains with respect to suspended particles. It can be described as a type of Stern layer adsorption of the electrical double layer. One end of the polymer chain adsorbed on a particle extends to the tip of the Guoy layer, and cross-linking occurs in the form of co-adsorption to the Stern layer of neighboring particles. This is a very useful coagulant for high-concentration suspensions. Lamer and Healy (1963) described the co-adsorption patterns of polymer chains, as shown in Figure 4.52. We can see a more detailed description of the process leading to more schematic stabilization and redistribution of the phenomena in Figure 4.9 above.

4.1.8.2 Activated silicic acid (Baylis sol)

A study by the author (Tambo) under Professor A.P. Black at University of Florida in 1962 (Tambo, 1966b) was the first theoretical work to verify the production process (using the Baylis method) of activated silica, which was used widely as a mineral coagulation adjunct at that time.

Activated silica is negatively charged polymerized colloidal particles of silicic acid, which can be added at the order of ppm as an adjuvant to coagulants such as aluminium to increase the strength and density of the produced flocs and accelerate/stabilize coagulation and flocculation reactions. It is widely used because there is no concern about toxicity. It also enables high-speed filtration as a filtration aid, but because of its high bonding strength with the filter media, the loss of water head increases rapidly in a normal filter

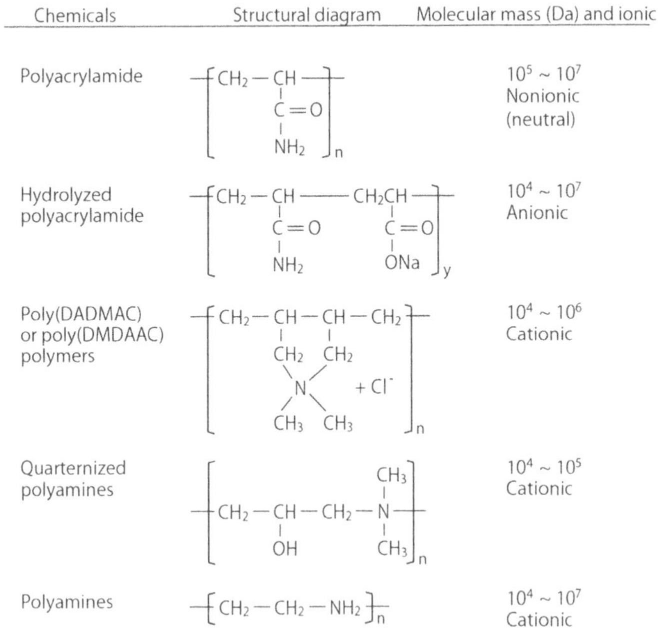

Figure 4.51 Structure of polymer coagulants and coagulant aids.

bed particle size configuration. However, the use of coarse filter media enables high-speed filtration of 300 to 400 m/day.

The production method requires the alkalinity of water to be neutralized with acid to promote the formation of polymerized colloidal silica particles, and to stably store the particles in the form of dispersed colloids for a long period of time so that the polymerized dispersed colloidal particles do not further bind and generate a gel. (see Figure 4.1. Three forms of colloidal dispersion).

Various acids are used for neutralizing alkalinity, including sulfuric acid, hydrochloric acid, carbon dioxide and ammonium sulfate, but sulfuric acid, hydrochloric acid and carbon dioxide are commonly used in Japan as well as overseas.

Historically, use of activated silicic acid began when it was noted that the water in Lake Michigan, a water source for the City of Chicago water supply where J. R. Baylis worked, showed better coagulation reactions to aluminium sulfate than synthetic water (distilled water plus salt) with similar turbidity and pH values. It was found that this difference occurs when sodium silicate was converted into sodium oxide and sodium bicarbonate by carbon dioxide in the atmosphere. On the other hand, fresh sodium silicate alone was not effective. Baylis investigated whether the coagulation reaction was improved by adding sodium silicate (water glass) neutralized with acid, and after repeated trial and error, followed the production method of activated silicic acid known as the Baylis method (Baylis, 1937).

In active silicic acid production:

(1) Water glass diluted to the appropriate concentration is neutralized to a certain extent to form a silica sol by an appropriate acid (with the choice of activator).
(2) The silica sol produced are aged for a predetermined time to grow the sol particle size.
(3) The resulting sol is diluted to a predetermined level so as not to gel, thereby allowing its preservability.

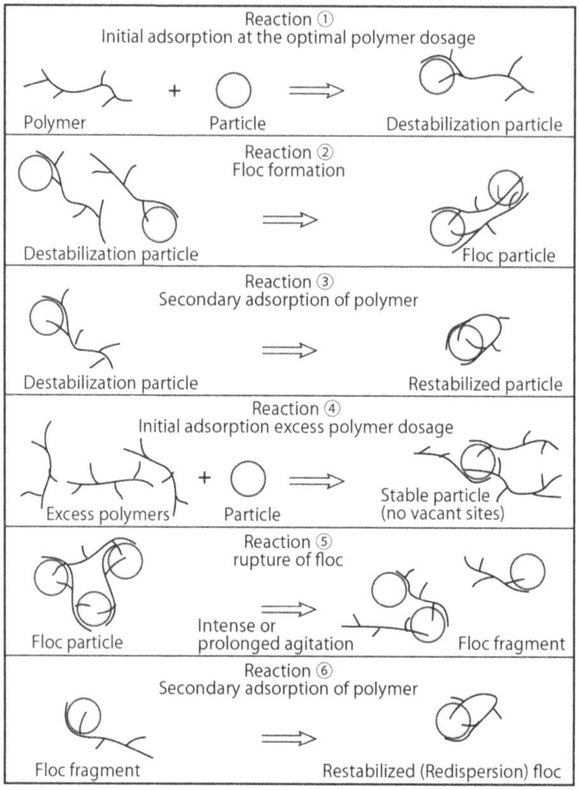

Figure 4.52 Bridging between polymers and colloidal particles (after O'Melia); see Lamer & Healy, 1963.

In Baylis's method:

(1) Dilutions of silicic acid soda (water glass) are used to make 1.5% solutions of SiO_2 (8000 ppm as $CaCO_3$ alkalinity).

(2) Commercially available concentrated sulfuric acid (94.5% or more) is diluted about 5 times to make 20% dilute sulfuric acid, which is added to 1.5% solution of SiO_2, proceeding next with neutralization of alkalinity while stirring to adjust the alkalinity to approximately 1200 ppm as $CaCO_3$ (80% neutralization). This produces a silica sol. The pH in this case is 9.0.

(3) These resulting silica sols are aged (particle size growth due to the progress of polymerization of the sol) after standing for about 2 hours.

(4) A three-fold dilution to a 0.5% SiO_2 level ensures that the sol does not further bind, gelate, and lose activity. It can stably act as an active silica sol for a long time.

Baylis probably discovered the coagulation promoting effect of silicic acid colloids by repeated trial and error, and established a sulfuric acid activation method called Baylis Sol, but no detailed theories have been reported. Therefore, the author (Tambo) investigated the extent to which the Baylis condition can be explained by a logical judgment, concluding that Baylis seems to have found the optimal conditions in the early days of 1937 – a surprising achievement and effort.

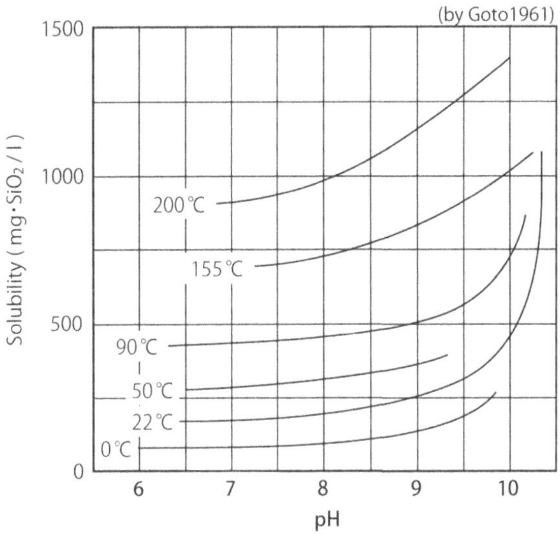

Figure 4.53 The solubility of silica (Goto, 1960).

The procedures and conclusions of Baylis's work will now be discussed, followed by a discussion of Tambo's examination at the University of Florida in 1962 and at Hokkaido University 1946 after he returned to Japan.

Silicic acids dissolved in water are present as monomeric silicic acid Si $(OH)_4$, or as silicic acid ions (e.g., $HSiO_3^-$ and SiO_3^{2-}) when the pH is sufficiently high. As shown in Figure 4.53, the solubility of silicic acid is almost constant around neutral pH, and varies greatly with temperature. Lowering the pH of a system containing a large amount of soluble silicic acid decreases the dissolution of the silicic acid ion monomer Si $(OH)_6^{2-}$ beyond solubility and the silicic ion polymerizes into polymer ions. The composition of polymeric ions (trimers) is shown in Figure 4.54.

The polymer thus formed increases its degree of polymerization (molecular weight) by aging (curing) as amorphous silicic acid colloids. If left intact, the colloidal particles (sols) solidify (gel) into a gelatinous form due to secondary coagulation and they lose their effectiveness as a coagulation aid. Therefore, dilution is performed to a low level that does not cause gelation, to maintain activity.

However, it is extremely difficult to conduct experiments by combining all these conditions, and the author (Tambo) had to search for optimal conditions based on the following physicochemical conditions.

First, among the standard products (JIS No.1 to No.3 silicic acids) which have different proportions of Na_2O and SiO_2 (the main components of water glass), JIS No.3 silicic acid was selected since it has the lowest amount of Na_2O (9–10%) which requires less sulfuric acid for neutralization.

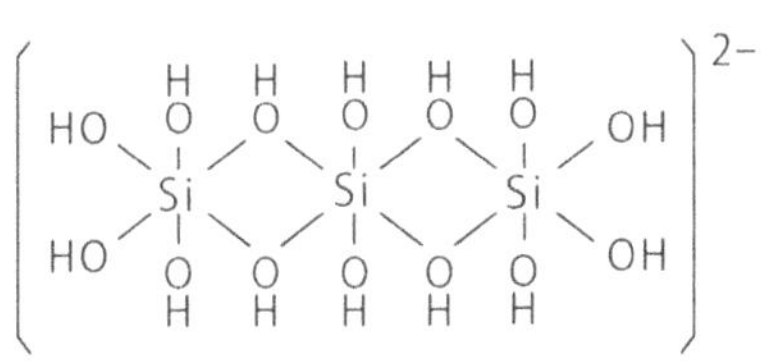

Figure 4.54 Polymeric silicate ions (trimers).

As shown in Figure 4.53, the neutralization condition of activation is such that the solubility hardly changes in the region of normal temperature below pH 9. Therefore, pH 8.5 is set as the target neutralization value with a view to a margin of operation.

The extent to which the concentration of the SiO_2 can be increased is related to the productivity of the silica polymer. However, at high concentration, the gelation time after neutralization is too short to allow stable ripening of the colloidal sol. Figure 4.55 shows the time–pH relationships between the time to gelation after neutralizing sodium Na_2O and SiO_2water glass) in a ratio of 1:3.3 with various concentrations of sulfuric acid. As shown in this Figure, the shortest gelation time appears in the neutral to slightly alkaline region. When neutralized to pH 8.5, trying to achieve a maturation time of 2 hours means that the polymerization-maturation can proceed prior to the dilution of the 2% SiO_2 to a non-gelling level. If the concentration is reduced to 1.5%, the time to gelation can be delayed for several months, and this can be used well in practice. In this way, the pH at which gelation occurs greatly influences the control of the active silicic acid, and therefore, the setting of the activation pH must be strictly determined.

As described above, when the pH is higher than the pH at which the polymerization proceeds, the OH required for the polymerization is sufficiently supplied, and the formation of colloidal silicic acid can be promoted satisfactorily for the following reasons. As shown in Figure 4.56, the electro kinetic potential of colloidal silica (sol) formed with a 0.75 ppm SiO_2 has values of −20 to −30 mv and the colloids repel each other. Because of the colloidal silica's higher absolute potential than the critical coagulation potential proposed by Tambo, it is thought that polymerizations proceed as individual particles without coagulation. As shown in Figure 4.56, the critical coagulation ZP is 13 mV (mobility of 1 μm/sec/V/cm), which is approximately pH 6.0. If we extrapolate the dotted line that minimizes the

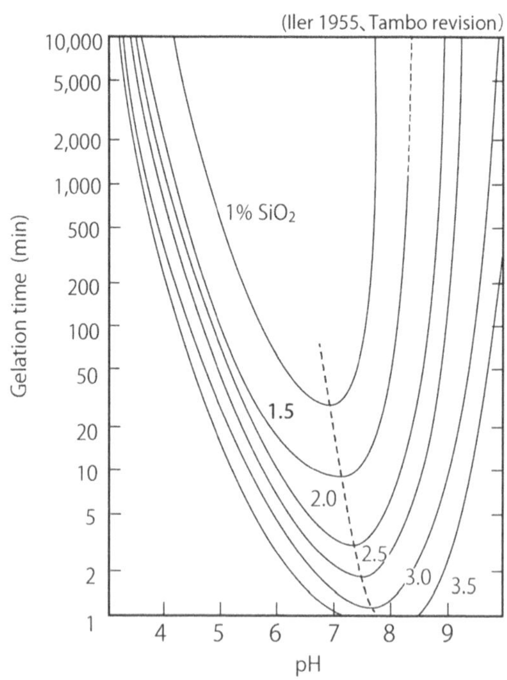

Figure 4.55 The relationship between gelation time, pH and SiO_2 concentration. (after Iler, 1955).

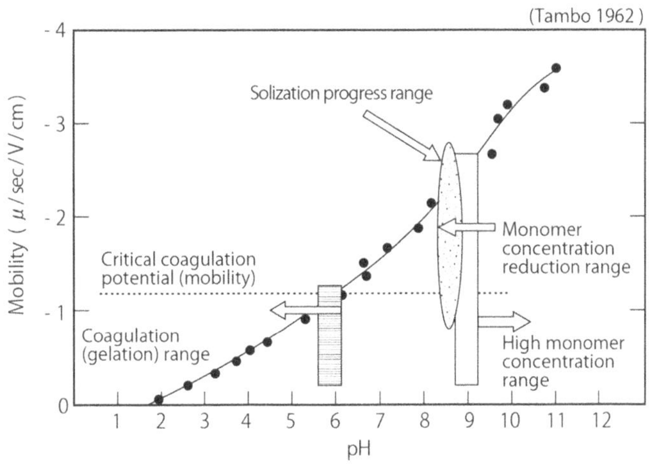

Figure 4.56 The mobility of silica colloids and pH.

gelation time of the concentrations in Figure 4.55 to the position of the SiO_2 concentration of 0.75%, the pH that gives the minimum gelation time can be estimated as approximately 6.5 hr. Expression of coagulation at the critical coagulation point leads to gelation, and the mechanism of the critical coagulation potential is considered to exist as a universal mechanism.

The size of the polymeric silica sol increases (polymerizes) by aging for about 2 hours. Figure 4.57 shows the results of measuring the dissolution rate of the colloid produced by NaF using the Goto method (Goto, 1958) under the conditions of addition of hydrochloric acid of 0.12 N. This shows the ratio of the dissolution rate of each maturation time of the polymerization in the vertical axis and maturation time (up to 6 hrs) in the

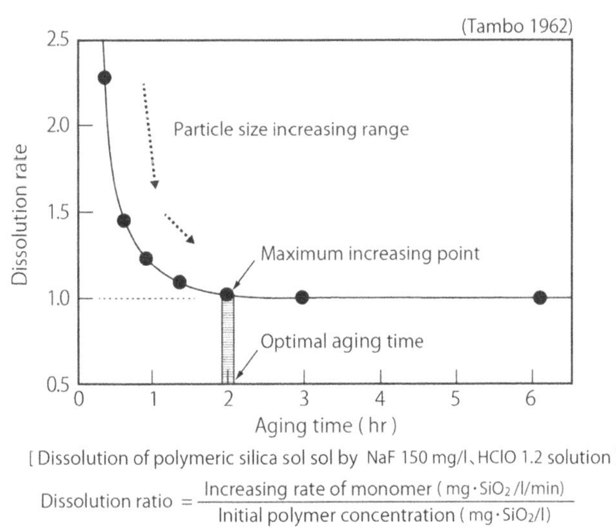

Figure 4.57 Silica sol dissolution rate.

horizontal axis. A high dissolution rate means that the size of colloidal particles formed is low, and a value of 1 means that aging beyond that time is meaningless.

As can be seen from Figure 4.57, the minimum necessary aging time is two hours, as Baylis suggested. From the 1960s to the 1980s, Baylis silica sol produced in this way was widely used in coagulation-flocculation sedimentation, high-speed filtration and water softening systems as inorganic floc-forming (flocculation) aids in Japan and the United States. In particular, in Japan, polymeric synthetic auxiliaries were regulated and not used for water purification. Therefore, they were widely used to improve sedimentation efficiency at low turbidity and low temperature conditions to stably maintain water purification treatment without causing sand filter clogging in amounts of 1 ppm (SiO_2) or less.

4.1.8.3 Polysilica-ferric coagulants

In the 1980s, a number of metal silicate coagulants were proposed by Japanese water supply engineers such as T. Hasegawa and others. The innovative development of silica sol utilization began. As shown in Figure 4.56, water glass (sodium silicate) is basically stably present on the strongly alkaline side. Neutralization with acid causes it to approach a neutral or weakly alkaline pH, thereby rapidly decreasing solubility, polymerizing to form a sol, and subsequently gelling to form solids. Baylis found that the polymer sol was stably present in practical use. A study by the author (Tambo, 1966b) at the University of Florida described the mechanism of critical coagulation potential regulation as above. Similar to the alkaline side, the acidic side also exhibits high solubility of silica. However, because the original raw material was alkaline water glass, we did not attempt to produce silica sol by shifting the system from the acidic side to the alkaline side. The author does not know the motivation for considering the formation of a sol from the acidic side, but it is considered to be the groundbreaking step for later developments. Figure 4.58, like Figure 4.55, is a schematic representation taken from Iler (1955; p. 45), which I have referred to since the 1960s for the relation between gelation time and pH of SiO_2, and shows that there is the greatest stability (solubility) around pH 2. As can be seen from Figures 4.55 and 4.58, silica is stably present as a solution or sol when the pH of the water glass is lowered to 4.0 or less.

The authors started research work to clarify the mechanism and characterize Hasegawa silica sol and propose a new metal-silica complex coagulant. When JIS No.3 water glass is processed to lower the pH

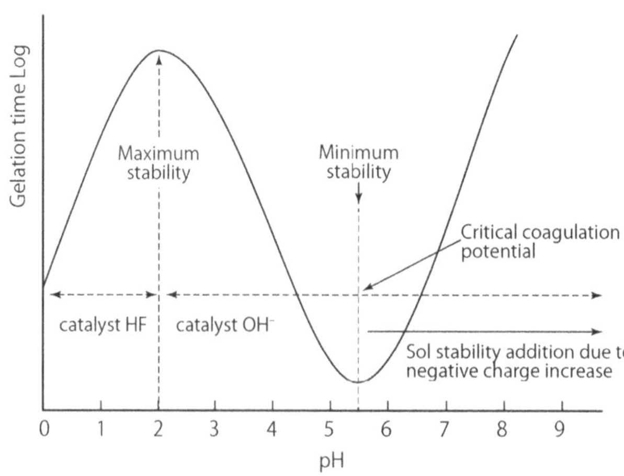

Figure 4.58 The relationship between gelation time and pH of SiO_2. (Taken from Iler, 1955).

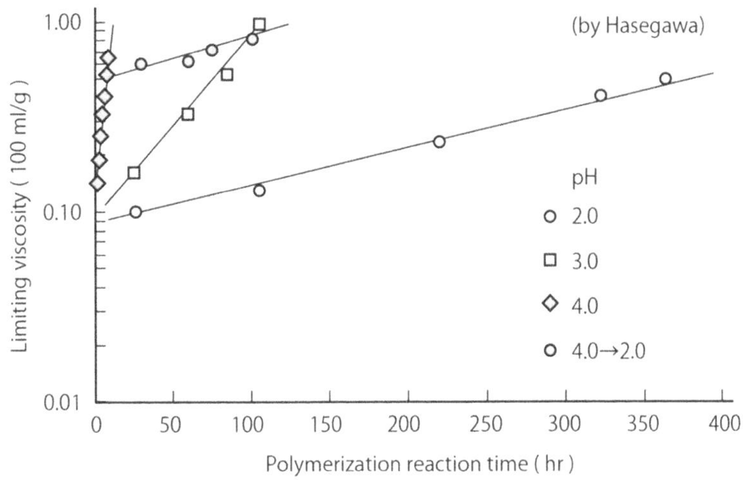

Figure 4.59 The time-course of the polymerization reaction and increase in limiting viscosity.

to about pH = 4.0 using sulfuric acid etc., polymerization proceeds and a sol is formed. However, the sol forming condition (polymerization of monomers) is most effective when the critical coagulation potential at which gelation progresses rapidly is slightly exceeded. Thus, as shown in Figure 4.58, the polymerization proceeds for several hours at a pH around 4.0. The catalyst for polymerization is an OH group. Ultimate viscosity is used as an index to control the danger of deactivation due to aging and gelation. Figure 4.59 shows the relationship between the time-course of the polymerization reaction and the increase in ultimate viscosity (limiting viscosity) for each pH. Figure 4.60 shows the relationship between ultimate viscosity and the molecular weight of the polymerized silicic acid sol. On both logarithmic graphs, we can see that the degree of polymerization increases with a positive linear relationship.

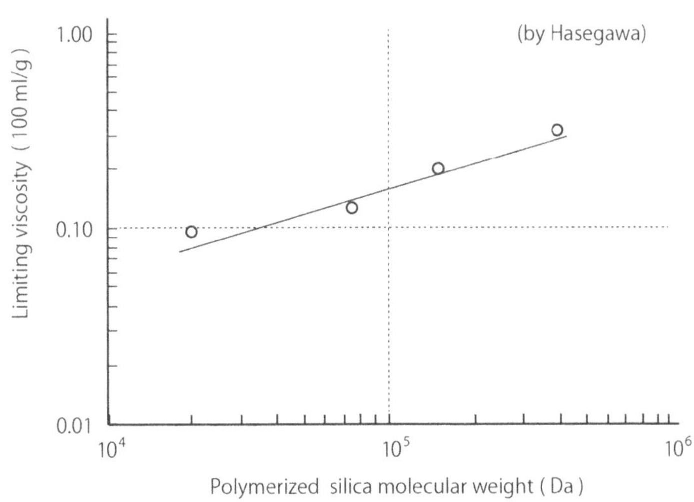

Figure 4.60 Limiting viscosity and the molecular weight of polymerized silicic acid sol.

Referring to both Figures, it can be seen that the degree of polymerization rises rapidly near pH 4. However, if the pH is lowered to 1.5 to 2.0 there is a slight time margin just before gelation, the polymerization becomes a maximum stability condition and the progress of polymerization becomes slow, and the stable sol state can be maintained for a long time. As shown in Figure 4.61, the acid-side sol (Hasegawa sol) produced in this way is a macromolecular colloidal particle with a molecular mass distribution ranging from 10^5 to 10^6 Da. Compared to the molecular mass of a Baylis sol of only 10^4 Da, it is an order of magnitude larger, and the cross-linking capacity is greatly superior. The reason why such a large molecular weight can be obtained has not been clearly demonstrated experimentally, but it is inferred that the silica sol proceeds to polymerization at a critical coagulation potential of around pH 4.0, that the coagulation reaction occurs partially, and depolymerization and repolymerization proceed by lowering the pH to 2.0. For Baylis sol, the need for more acid to lower pH from pH 4 to pH 2 is a little more of an issue, but the biggest advantage of a metal-silica composite coagulant is that it can more easily form metal salts and composites.

Amongst the properties of complex metal-silica coagulants, aluminium and iron exist as simple metal cations under strong acid conditions (pH 1.5 to 2.0) (see Figure 4.17), and aluminium and iron can be superposed to form molecule-to-molecule pairs in the form of a completely uniformly dispersed solution and a stably uniformly dispersed silica sol adjacent to each other. It is enough to simply mix the iron or aluminium solution with a silica sol. Any ratio may be used but, for water treatment, Si:Fe ratios ranging from 1:4 to 3:1 seem to be mixed. When iron or aluminium ions hydrolyze and polymerize to form positively charged multinuclear aquo-complexes, negatively charged colloids are formed in water and charge neutralizing coagulant reactions proceed, with negatively charged silica sols as crosslinkers always present in pairs in the immediate vicinity of the coagulant. Therefore, in the most energy-intensive aspect of the coagulation floc-forming process in which polymer adjuvants are added and uniformly mixed, it is possible to overlook such a deficiency as floc breakage. Floc component densities are also high because of the uniform floc configuration.

Figure 4.62 shows an example of a comparison of the sedimentation properties of the flocs formed by PACl and PSAl, indicating that the effect of silica sol is significant and the sedimentation properties are improved. In addition, the electrophoresis of polysilica iron (PSI) containing soluble regions is determined by the author's electrophoretic transport tube method described above. Figure 4.63 shows the

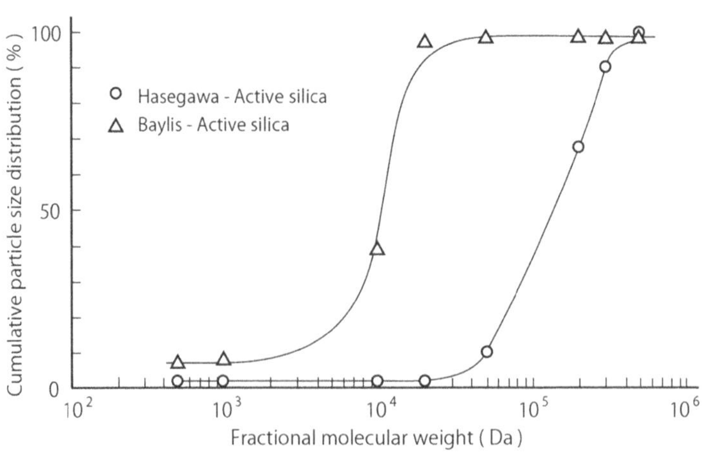

Figure 4.61 The fractional molecular weight distribution characteristics of silica sol.

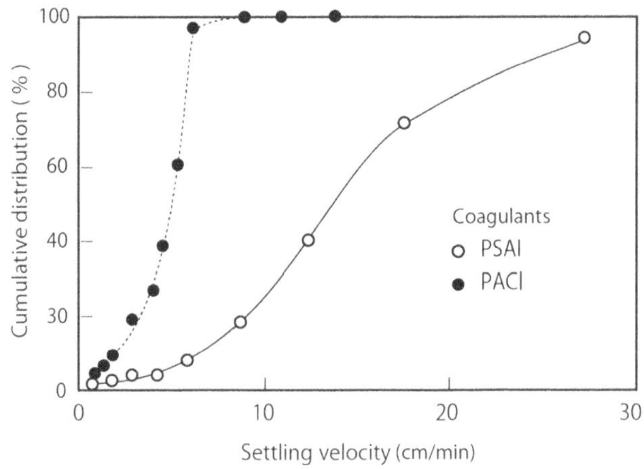

Figure 4.62 Comparison of floc settling velocities.

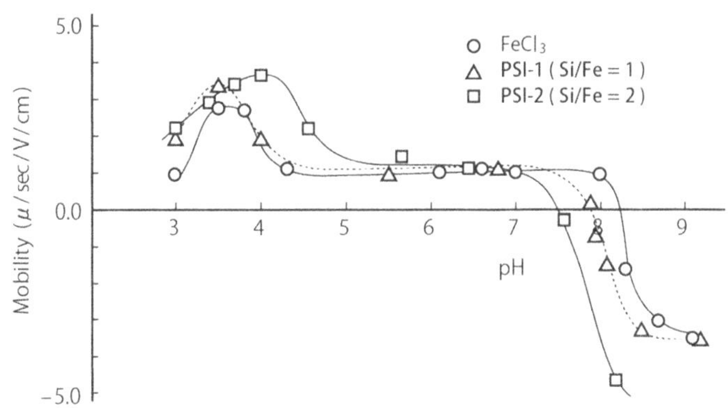

Figure 4.63 Changes in the mobility of iron chloride and PSI, by pH.

results of the electrophoresis of PSI containing soluble regions. Basically, it has the same coagulation characteristics (the same electrophoresis patterns) as iron salts. As shown in Figures 4.64 (clay suspension) and 4.65 (humic color component), when mobility is 1 μm/sec/V/cm or less (under Tambo's critical coagulation potential theory), coagulation proceeds regardless of the positive or negative charge. PSI improves the physical properties of flocs, but there is no magic to the occurrence of charge neutralizing coagulation reactions. It is only necessary to follow the conventional process, and it can continue to function as a traditional coagulant.

Aluminium and iron are the most common metal ions that coexist with Hasegawa sol. In recent years, Alzheimer's disease has been suspected of being related to the accumulation of aluminium in the brain. Efforts to minimize the amount of coagulant aluminium remaining in drinking water have been required. PACl, in which colloidal silica is permanently present with an aluminium aquo-complex, has the advantage that the risk of aluminium escaping is much less than the risk of coagulation with sulfuric acid bands or PACl alone. In addition, the ferrous salt-coexisting PSiFe is not related to Alzheimer disease. Suidou-Kikou Co., Ltd, markets a mixed coagulant of iron and silica sol, simply called PSI. Iron is an

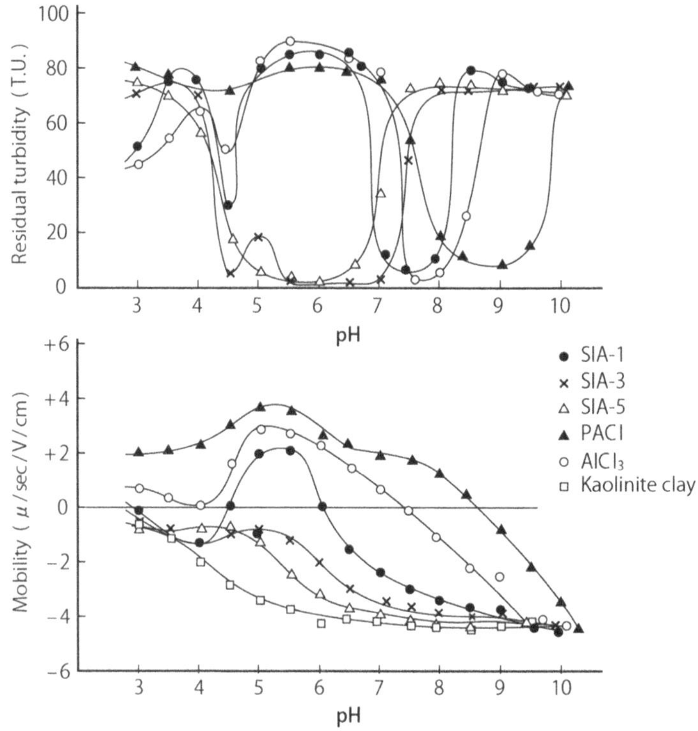

Figure 4.64 Residual turbidity, mobility and pH in kaolinite flocs.

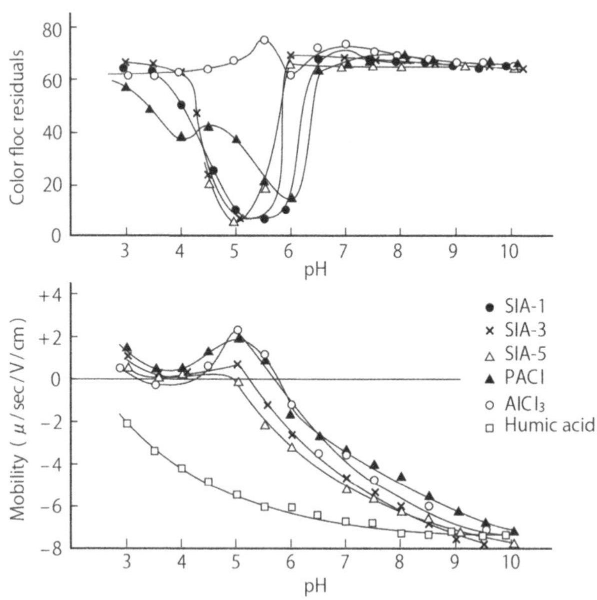

Figure 4.65 Color floc residuals, mobility and pH.

older coagulant but it has a disadvantage in that even trace amounts of iron residual in treated water produces color, and high-purity iron agents are relatively expensive; high-purity aluminium has been used for a long time, even when made by conventional methods. As with residual aluminium, micro-coloration problems can be solved easy in a PSiFe mixing coagulant where silica sols are present at all times.

4.1.9 Flash mixers (rapid mixing)

In order to proceed to coagulation, a flash mixer is often used. At first, coagulant, alkaline and flocculation aids must be added to the raw water in a timely manner, and in a suitable order and manner to achieve as rapid a homogeneous distribution as possible to produce a homogeneous coagulation reaction. At this stage, chemical reactions such as hydrolysis, charge neutralization and interfacial electrical reactions are completed, and very small micro flocs of the order of 10^0 to 10^1 μm are formed. This process is called flash mixing (rapid mixing).

Chemicals are usually added in the order of alkaline adjusters, flocculation aids and then coagulant. The former may be agitated at the required strength and duration to uniformly completely dissolve or disperse, and chemicals are often added and mixed in an inflow channel/tube. In order to avoid any strong short-circuit flow, injection is made in a turbulent channel system as close as possible to the extrusion flow, as shown by the tracer curves in Figure 6.25 (Chapter 6), taking into account the retention time required for mixing. The addition of coagulant requires the creation of locally isotropic turbulence with a predetermined strength ε_0 of 10^3 to 10^4 erg/cm^{3}sec. Therefore, a reliable forced agitation method is required. Since a stirred tank is a completely mixed flow, the generation of a short-circuit flow can appear strongly in a single tank, and the uniform reaction is prevented; therefore it is necessary to arrange at least a number of mixing tanks in series. Three tanks in series are desirable.

In Japan, rapid mixing tanks have hardly ever been converted into multiple tanks in series. This is a major difference from the design philosophy of the United States. The big reason for this difference is that Japan and the United States differ in the amount and kind of water-borne impurities to be treated. In the coagulation of micro-colloids, such as color colloids, which must have the greatest charge neutralizing capacity, the homogeneity/reliability of the reactions in a rapid mixing tank require a very delicate agitation process, whereas if coarse suspensions, such as clay colloids, are of interest, a small amount of coagulant will undergo cross-linking coagulation, and sometimes because of their low dispersity, a concentrated mass of coagulant is present locally, resulting in a relatively large particle size of the metal hydroxide produced, which have an adverse effect on subsequent flocculation.

The injection concentrations of the coagulant used are also related to the mixing intensities and retention time distributions. Low concentrations (e.g., 1.0 to 0.1% of alum) may be used in mixing ponds with low agitation intensities for micro colloid coagulation, where charge neutralizing capacity is of greatest concern. For coagulation of crude suspensions such as clays, high concentrations (10%) are used. Generally, when handling coarse suspensions, problems seem to be less pronounced in the case of micro-suspensions such as colloids. The design guideline for Japanese water supply may have been inappropriate in this respect but, as mentioned earlier, the invention of PACl was a major invention which compensated for the roughness of the rapid agitation.

The agitation systems of rapid mixing apparatus (such as ponds) include an up-and-down channel, a mechanical agitation tank, a static mixer, and a low lift pump etc., and any of them can be used for adding an alkaline auxiliary agent or a floc forming auxiliary agent. However, the coagulant requires a series configuration of a number of mechanical agitation tanks with retention times of about 2 to 5 minutes. A few minutes of retention time is desired to ensure micro flocs are formed. Mixing coagulant with a hydraulic jump is inadequate because of the lack of ample retention time to form micro flocs.

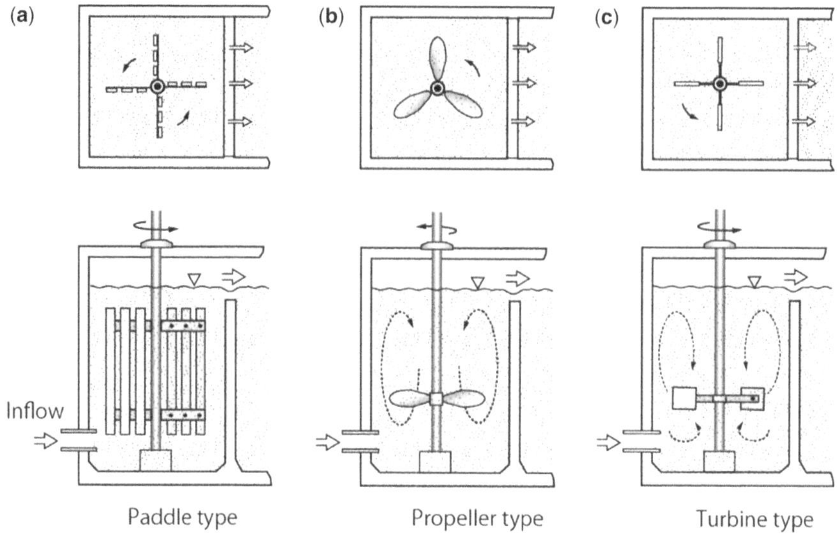

Figure 4.66 Various types of rapid mixing tank.

Microflow construction requires consideration of the agitation intensity G as well as agitation time T; the GT value will be discussed further in the next section.

Mixing ponds equipped with mechanical stirrers have multiple tanks, and a single tank configuration is shown in Figure 4.66. Usually, the ratio of the depth of the stirring tank to the length of the flat side is 1:1 to 3:1. A paddle-type agitator is used for small agitation tanks with a volume of up to 2 m^3, but for larger tanks up to 30 m^3 a propeller or turbine agitator are used. The center of the agitation shaft is offset from the center of the tank to reduce co-rotation of the water flow and increase power consumption to enhance agitation effectiveness.

4.2 THE FLOC FORMING PROCESS

4.2.1 Historical evolution of design theories: T.R Camp's G-value and GT-value

As shown in Figures 3.27 and 3.32 in Chapter 3, a water treatment system uses processes to grow the particle size of impurities to a point where the separation process shown in Figure 3.40 can be undertaken with a solid-liquid separation process that minimizes process energy and cost. The reason for this is that water, a heavy material with a weight of 1 ton (1000 kg)/m^3, must be treated every day – from at least 100 m^3 to, in many cases, as much as 1 million m^3 in a single plant, requiring the use of an extremely low energy consumption separation process. The oldest solid-liquid process used by humankind since the beginning of civilization was sedimentation treatment. In order to separate water with less energy, it is necessary to modify the nature of the impurities to be separated to a state that differs greatly from the state of the water molecules.

This process, in which dissolved impurities are insolubilized by chemical or biochemical reactions and formed into colloids or coarse suspensions, is called precipitation. On the other hand, it is through coagulation treatment (mentioned above) that agglomeration is promoted by physicochemical processes (charge neutralization, van der Waals agglomeration, etc.). Coagulation causes impurities to become a size that can be separated by filter paper or sand filter layers. In addition, flocculation and sedimentation allow colloidal particles, coarse suspensions, and van der Waals agglomerates to grow into larger agglomerates that can be easily separated from water with less energy-intensive uses of physical drive forces such as gravitational and centrifugal forces.

When the author (Tambo) first began to study coagulation treatment in the 1960s, a slow-stirring processor called a flocculator had not yet been introduced into the sedimentation systems of water purification plants. In fact, the processes involved in the rapid filtration method used up to the 1950s did not have flocculators. The old Moiwa rapid sand filtration plant in Sapporo City, which Tambo used as a site for experiments during his university student period (1953 to 1958) was one of Japan's earliest rapid sand filtration purification plants. Indeed, it was one of the famous water purification plants built in 1937 under the guidance of Professor Y. Kuratsuka of Hokkaido University. The water, with coagulant added, was inflowed from the rapid mixing tank to the sedimentation basins immediately.

The first floc formation theory (G-Value, GT-Value Theory), developed by T. R. Camp, was published by the Journal of the US Society of Civil Engineers (Camp, 1953). Tambo graduated from the Faculty of Engineering at Hokkaido University in 1955. At that time, flocculation research was the state-of-the-art research for water treatment, and researchers in Japan and abroad were competing for results. Looking back at his earlier career and after a half century of research, the author Tambo wishes to convey that research to the readers of today.

In his paper, 'Study of Agitation in Coagulation Operation' submitted to the *Journal of Japanese Water Works Association* (Tambo, 1959), the author confirmed the significance of Camp's GT value by experiments. A model flocculator and a sedimentation pond (overflow rate $Q/A = 0.42$ cm/min) were used to determine how the clarification of the water after sedimentation improves as the stirring progresses. Figure 4.67 shows the results of coagulation of a Kaolin suspension with a turbidity of 50

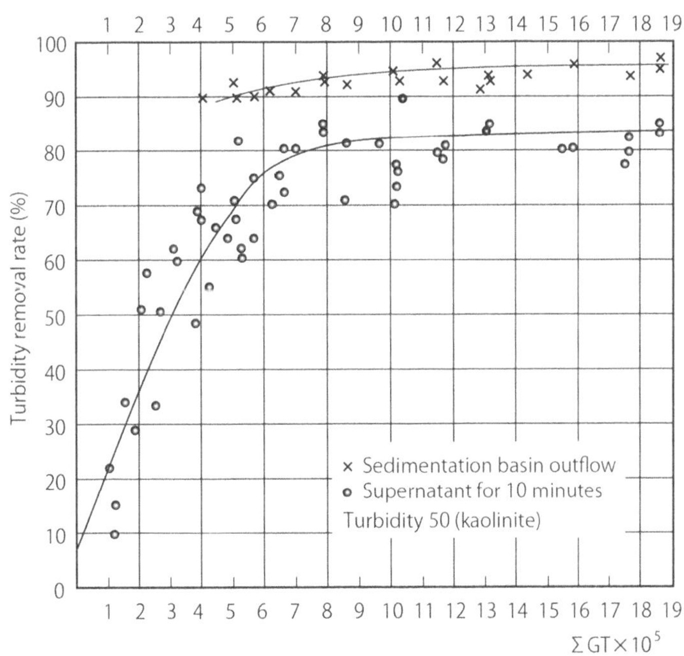

Figure 4.67 Turbidity removal rates with respect to cumulative GT values (Tambo, 1959).

TU and floc formation using various agitation intensities (G values) in a flocculator with a horizontal shaft paddle mixing flow. The cumulative GT values were put on the horizontal axis, and the turbidity of the collected sample's supernatant for 10 minutes settling on the vertical axis. It is probable that this was the first time in Japan that the effects of GT values were demonstrated in concrete experimental results. The author thinks this test was the first demonstration that the absorption of micro flocs into large flocs was determined by the dimensionless product of the G value (power consumption rate of agitation: sec^{-1}) and agitation time T (sec).

The graph shows that the turbidity removal rate increases with increasing (dimensionless) cumulative GT value to a certain value (about 6 to 8 × 10^5 in experiments) in a highly correlated manner, indicating that either the agitation strength G (1/sec) or the agitation time T (sec) does not have independent significance. This series of studies explained the effects of suspensions on set GTs at high agitation intensities and low turbidity with breakup of flocs, and led to further extended research by the authors.

Based on a first coagulation paper published with Stein (Camp & Stein, 1943), Camp's 'Velocity gradient and internal work in fluid motion' (Camp, 1953) found that the progression of flocculation at a point in fluids is proportional to the velocity gradient of the flowing water at that point and the concentrations of coagulation particles. However, his 1953 paper, which developed the design index of floc-forming ponds (flocculators) and proposed G and GT values, does not mention why it is related to the concentrations of particles that were in his original paper with Stein in 1943. In that paper, Camp and Stein state that the velocity gradient (gradient of fluid velocity, G) of a point is proportional to the power expenditure (power loss per unit volume) due to viscous shear of a unit volume of fluid. When we discussed coagulation drive energy in the previous section, the structure of the water flow in the stirred tank, explained by introducing Kolmogorov's locally isotropic turbulence theory, is essentially turbulent, as shown in Figure 4.4.

4.2.2 Amendment of Camp's G-value and GT-value theory

The energy consumption rate ε (power consumption rate W), essential for agglomeration, is defined by the relationship between the maximum vortex velocity V_l (the velocity difference ΔU between two points on a vessel, size ℓ) and the viscosity coefficient of water, as shown in Equation 4.35.

$$G = \left(\frac{W}{\mu}\right)^{\frac{1}{2}} \quad \text{—(4·35)}$$

G: Mean square variation velocity gradient in mixing tank (sec^{-1})

W: Unit time power consumption of mixing tank unit volume ($g \cdot 980\ cm^2/sec^2/sec \cdot cm^3 \cdot sec = erg/cm^3 \cdot sec$)

μ: Viscosity of fluid ($g/cm \cdot sec$)

(4.35)

Equation 4.35 The mean square variation velocity of turbulent mixing tank.

Camp's G-value is a representation of collisions in a laminar flow area combined with a velocity gradient indicating energy consumption due to molecular viscosity. There are some descriptions of the mean square variation velocity of a turbulent stirred tank, but it does not appear to be quantitatively indicated.

As described in Equation 4.30 above, Kolmogorov's theory allows the energy consumption rate ε_0 of a volume of fluid to be estimated from the physical values of the macroscopic agitator, such as the maximum vortex size ℓ and the maximum velocity ΔU of the agitator. Camp expresses such a physical condition as the velocity gradient $G = \Delta U/\ell$ (1/sec), and using the terms of the isotropic turbulence theory expression of Kolmogorov flow, Equation 4.36 is obtained.

$$G \propto \left(\frac{\varepsilon_0}{\mu}\right)^{\frac{1}{2}} \quad (\mathrm{sec}^{-1}) \qquad —(4\cdot 36) \tag{4.36}$$

ε_0 : Turbulent agitation energy consumption rate (erg/cm^3·sec)

Equation 4.36 The mean square variation velocity gradient of Kolmogoroff flow.

For example, suppose floc formation in a uniform-flow channel with a depth of 10 cm and a flow velocity of 100 cm/sec. The Reynolds number in this case is $Re = hV/\mu = 10 \times 100/0.01 = 10^6$, and the channel is perfectly turbulent. In Camp's terms, the size of the largest vortex in this channel is the depth h, the fastest water flow in the channel (near-surface velocity) creates the strongest vortex, the vortex size $l = h$, and the turbulence spectrum begins with a vortex with a vortex velocity ΔU (highest velocity in the channel). This also applies to Kolmogorov's isotropic turbulence. In this case, Camp's G value and the G value for Kolmogorov flow are shown in Table 4.15, where K is the conversion factor of the G (ε_0) value expressed as Camp's G value and the turbulent energy consumption rate ε_0. In this case, $K \doteqdot 1/30$. To be precise, the total energy consumption rate or the power energy consumption rate is calculated in detail by actual measurement or calculation, and the conversion factor is obtained.

The G-value theories of Camp and Stein were proposed based on the Stokes equation described under laminar flow conditions from the solution of the potential flow in which the transport of momentum is made by molecular motion. However, as mentioned earlier, the actual flocculation agitation occurs in the turbulent region, so the transport of momentum by convection (vortex) is excellent. Mixing by turbulent diffusion in such high Reynolds regions is caused by momentum transport by turbulent diffusion of the micro-water mass and is governed by so-called turbulent viscosity which is morphologically similar to viscosity (molecular).

The time-averaged (squared) instantaneous fluctuation values of the combined flow velocities of u′v′, u′w′, v′w′, etc. of the fluctuation velocities u′, v′, w′ in 3 directions (x, y, z) is taken into account in addition to the molecular viscosity, the apparent stress due to turbulent flow fluctuation (the latter is overwhelmingly larger: Reynolds stress), and the basic equations of Navier-Stokes are solved. However, Reynolds' stresses cannot be calculated without knowing the mechanisms of turbulence fluctuations and, because they are nonlinear phenomena, we need to define the turbulence component and perform complicated calculations with numerical solutions, and there is no common solution.

Therefore, the author decided to address the issue by defining the turbulence transport structure prevailing in Kolmogorov's isotropic turbulence. Veniamin G. Levich of the Russian Academy of

Table 4.15 Examples of G value calculations for Camp and Kolmogorov flow.

G value in the case of uniform flow channel with water depth 10 cm and flow velocity 100 cm/sec	
Camp(lamina)	$G = \Delta U/l = 100/10 = 10\ \mathrm{sec}^{-1}$
Kolmogorov flow	$G = K\ (1 \times 100^3/10)^{1/2} = K \times 10^{2.5}\ \mathrm{sec}^{-1} = K \times 316\ \mathrm{sec}^{-1}$
K: Conversion factor of G (ε_0) value expressed by Camp's G value and turbulent energy consumption rate ε_0	

Science explicitly described the theories, and his book was published in English as '*Physicochemical Hydrodynamics*' (Levich, 1962). The author used his ideas in many subsequent studies. A fundamental piece of literature on quality conversion engineering and chemical engineering is the book, '*Turbulence*', by J. O. Heinz, (1959), from which much more sophisticated theories can be obtained.

Fortunately, the G values theories which began with the 1943 original paper of Camp and Stein were that the form of power consumption (energy consumption) in the region where only molecular viscosity dominates could be used apparently in the turbulent region because of the approximation (almost the same shape) of the viscosity coefficient and the turbulence viscosity coefficient in energy dissipation formula. It should be noted, however, that they are totally different phenomena and cannot be interchanged in more developed coagulation theories.

The authors' own flocculation theory, developed since 1965 at Hokkaido University, are now described in the reference list: see Tambo *et al.*'s many publications from 1964, 1970–72, 1983 and 1991). Another important thing that must be taken into account is adding suspensions volume concentrations, C_v, to Camp's G-value and GT values, as they appear in many textbooks and technical books worldwide In the 1960s, it became common practice for rapid filtration systems to install a flocculator that provided slow agitation with a retention time of 20~40 minutes after introducing a coagulant into a rapid mixing pond. By contrast, in the activated sludge method, after the inflow of return sludge into the aeration tank, microbial communities are introduced directly into a sedimentation pond for separation. The reason for this is that the activated sludge method causes the inflow of particles as high as 3000 mg/l into the final sedimentation tank, whereas in the sedimentation basin of a water purification plant, the inflow of the suspended matter concentration is 10 mg/l at times of low turbidity and 10^1 to 10^2 mg/l at other times, and the concentration (precisely the volume concentration) of the fine particle group (micro-flocs obtained by coagulation) is small, so that the particles cannot collide with each other and grow to a floc with a large diameter that can settle within a predetermined time; therefore, the chance of collision joining must be advanced by continuous constant slow agitation (GT value of a predetermined size). In the case of low turbidity (10 turbidity units (TU) or less), the microflocs, which are only subjected to coagulation by rapid agitation, can be led to a sand filter bed as they are, and can be separated into solid and liquid by direct filtration. Suspensions as high as 50 mg/l must be grown to flocs with a size of overflow rate at least 95% in following sedimentation ponds. In the case of highly concentrated sludge slurries using the activated sludge method, particles are grown by collision in a short time (such as several tens of seconds) until the turbulent jet at the inflow of the sedimentation pond is attenuated. Needless to say, by adding slow agitation for a shorter period of time, growth can be promoted to a more homogeneous floc group. The size of the C_v is another key operation factor.

In the first Camp and Stein paper (Camp & Stein, 1943) which regarded the velocity gradient value (G value) as the strength index of coagulation, particle agglomeration progress in proportion to the 'number of suspended particles' was also discussed, but in the following paper by Camp himself, only the G value and the GT value were proposed and the number of particles was not discussed (Camp, 1953). In the same period, Tambo (1964a, b, c), Hudson (1965), Fair & Gemmell (1964), Swift & Friedlander (1964), Cockerham & Himmelblau (1974) and others all developed kinetic studies of flocculation. Of course, for the kinetic assessment of the floc-forming process it is essential to rely on material balance considerations, and the concentration (number concentration or volume concentration) of suspended components is a fundamental consideration for material balance conservation. The fundamental difference between our work at Hokkaido University, Japan and these studies is that the other authors dealt with the kinetic process without detailed discussion of the physical properties of flocs or floc groups, and defined the velocity index (operation variables) of the floc-forming process; hence they can only present simplified models that describe the floc-forming process.

As a kinetic equation, the author Tambo developed a study starting from 'Versuch einer mathematischen Theorie der Koagulationskinetik kolloider Lösungen' (Smoluchowski, 1917), a classic paper in the field of colloidal sciences describing the temporal changes (kinetics) in suspended particulate size distributions. Smoluchowski's paper is a population balance model describing the process by which colloidal particles (uniform size) coalesce by Brownian motion to increase particle size distributions, and mentions the clustering nature of floc particles, but does not discuss floc breakage. Tambo and his collaborators developed a population balance model that describes the process of agglomeration progression under locally isotropic turbulence, considering the structural densities (fractal properties) of flocs and their growth breakdown, starting from the Smoluchowski model, from a turbulent floc-forming pond (flocculator) to a dual phase flocculation procedure containing microflocs and a dissolved air in flotation method.

4.2.3 Floc densities

The flocculation process cannot be theoretically described unless the sedimentation rate of coagulation, flocculation, and growing particles can be explicitly calculated and evaluated prior to the design of the separating process, as described in the later section on sedimentation (Chapter 6, Section 6.1). What we need is the density and size (sedimentation rate) of the floc produced. Densities and growth-limit size are also governed by the agitation strength. Therefore, the authors first investigated how the density and sedimentation rate of the produced floc were determined, and then examined the agitation strength and the critical growth of floc diameter (floc strength).

Harmer E. Babitt and James J. Doland's *Water Supply Engineering* (1955) is one of the few so-called original books that the author Tambo bought in his under graduate school period. Perhaps it was the literature of the last days of the era when sanitary engineering was a small field of civil engineering. When visiting Soviet Russia in 1970 and spending some time with Professor Mintz, the filtration specialist in Moscow, Tambo was surprised that the Soviet Union had not prepared their own textbook and was using a full translation of Babitt & Doland. At almost the same time, Gorden Maskew Fair & John Charles Geyer's *Water Supply and Waste-Water Disposal* John Wiley Sons, Inc. (1956) came out. This became a standard textbook and a turning point for the major growth of sanitary and environmental engineering from the fields of civil engineering to environmental engineering. Tambo was strongly influenced by it and has long loved his copy of the old edition, complete with Professor Fair's signature on the book which he signed when he visited our laboratory at Hokkaido University. Tambo heard the then Dr Fair's lecture at the AWWA's general meeting in Denver (1963), and remembers clearly that he was introduced as 'the Moses in the field of water science and technology' for half a century. Over the years, there has been a great deal of change in our understanding of colloidal coagulation, the physical properties of flocs, and the floc-forming process. As we understand the work of coagulants, we have introduced the theories of DLVO, etc. as described in the section on coagulation earlier in this book.

4.2.3.1 Floc density measurement (structural density and fractal structure)

Regarding the physical properties of floc, it was said that Fe salt coagulation could produce floc heavier than aluminium salt coagulation, but there was no reference in any of the books that had been quantitatively evaluated. So the author simply considered finding out how dense aluminium floc is. Since floc should contain a large amount of void water in the course of coagulation, the author assumed that floc density might be structural, as defined by the content of void water.

Therefore, the authors began by developing a method for measuring floc density. Floc is an extremely fragile particle containing a large amount of water in the voids of hydrolyzed aluminium precipitate bound together and in an intertwined state. Therefore, measurements should be taken without direct contact, while the floc remains suspended in the water. In addition, as described in Equations 6.13 and 6.14 in Section 6 below, flocs collide with each other during sedimentation and exhibit a precipitation coagulation that converts particle sizes and shapes. Therefore, in order to see the properties of flocs as a single particle, it is necessary to measure flocs in the form of free particles in the form of a dilute suspension as much as possible. To meet these requirements, the authors installed (1) a transparent hydrostatic sedimentation tube under a rectangular jar tester (a floc-forming tank, as shown in Figure 4.68), and (2) a close-up photography device to capture the type and size of floc particles that settle in the tube from the outside and track the sedimentation rate at the same time. The equipment was operated as follows:

(1) The stirred tank was filled with a 3-litre suspension (50 ppm kaolinite), adding a predetermined quantity of aluminium sulfate or PACl, adjusting the pH to a predetermined pH condition at a NaOH of 0.025 N, and performing coagulation flocculation at a predetermined stirring strength and duration (5 minutes at 100 rpm and 20 minutes at 40 rpm). In this case, temperature convection occur in the sedimentation tube when there is a difference between room temperature and water temperature, so the suspension was allowed to reach room temperature sufficiently before starting the experiment. In addition, there is was a fear that the internal water temperature of the floc particles would rise due to heat convection in the sedimentation (measurement) tube and heat radiation to the floc particles. In addition to photographic illumination during the sedimentation test, careful attention was paid to heat radiation from the measurer himself/herself. For this reason, illumination was from a weak fluorescent lamp, and adiabatic

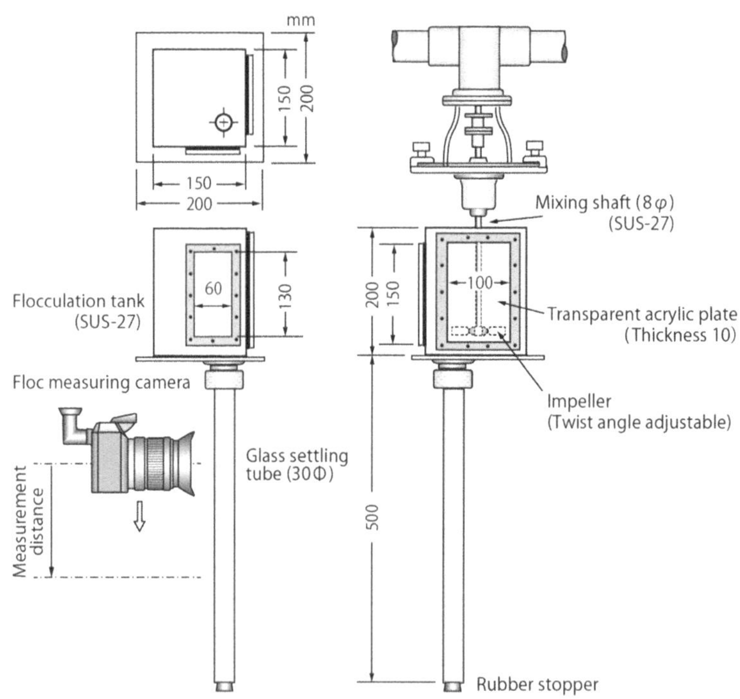

Figure 4.68 A floc settling velocity (density) measuring device.

shielding was provided between the instrument and the person measuring, thereby minimizing thermal disturbance.

(2) When the floc was formed after the slow agitation was completed, the small iron cover above the sedimentation tube attached to the bottom of the floc-forming tank was gently opened for a short period of time, several flocs were dropped into the settling tube, and the lid closed again. The measurable time of one floc formation stirring was about 10 minutes, and the sedimentation rate and particle size of several or more flocs were measured by repeating dropping several times.

(3) The floc sedimentation rate was measured using a single-lens reflex camera with a close-up bellows attached, placed on a tripod positioned about 20 cm from the sedimentation tube, lighting with a flash, and taking a picture to estimate the size and shape of the particles as the floc dropped into the tube within the field of view of the camera. At the same time, the floc particles were tracked while the tripod elevator was moved (rolled down) by 5 cm, and the sedimentation speed was calculated by measuring the time required for the floc particles to settle by 5 cm.

(4) After the film was developed, it was projected onto a screen and the correct shape and size (2D diameter) were recorded.

The precipitated particles settle at a constant rate at a terminal sedimentation rate. Based on the estimated Reynolds number, we can define the relationship between the particle size and sedimentation rate using the Stokes formula shown in Figure 6.2 and Table 6.2 (see Chapter 6, Section 6.1 below). Here, the resistivity coefficient C_D is a coefficient that varies depending on the shape of the particles, and it is assumed from observation that the coefficient is approximately the midpoint between a sphere and a cube, and that the coefficient takes approximately the value of a regular octahedron, and the calculation is performed using the sphericity coefficient $\varphi = 0.847$ shown in Table 6.3. From the relation between the resistivity coefficient C_D and the Reynolds number Re for non-spherical particles shown in Figure 6.5 below, the resistivity coefficient for the Stokes range for particles with $\varphi = 0.847$ is estimated to be $C_D \doteqdot 45/Re$. Substituting this relation into the Stokes equation gives the sedimentation rate of floc particles in Equation 4.37, and the sedimentation rate w (cm/sec) of floc in the sedimentation tube gives the floc density (structural) ρ_f (g/cm^3), where d_f is the floc diameter (cm) and ρ_w is the water density (g/cm^3).

$$w = \left[\frac{4g}{3}\cdot\frac{R_e}{45}\left(\frac{\rho_f - \rho_w}{\rho_w}\right)d_f\right]^{\frac{1}{2}} \doteqdot \frac{g}{34\mu}(\rho_f - \rho_w)\,d_f^2 \qquad \text{—}(4\cdot37) \tag{4.37}$$

w : Floc settling velocity of in the settling tube (cm/sec)
d_f : Floc sizer (cm)
ρ_w : Water density (g/cm^3)
ρ_f : Flock density (g/cm^3)
$\rho_f - \rho_w = \rho_e$ (Flock effective density)

Equation 4.37 The floc settling velocity equation.

4.2.3.2 *Effective density*

The authors defined the effective density (buoyant density) ρ_e of flocs as $\rho_e = \rho_f - \rho_w$, where ρ_f is floc density and ρ_w is the density of water, and expressed all experimental results carefully using the effective density as the main index, rather than the density of the floc itself. For example, if there are floc particles with a floc density ρ_f of 1.01 and 1.02 (g/cm^3), the density ρ_f difference is only $(1.02 - 1.01)/1.01 = 1\%$. However, when expressed by the difference in effective density ρ_e, it becomes $(0.02 - 0.01)/0.01 = 100\%$, which can clearly describe the result and express the physical properties of the flocs. At the same time, as can be seen from Equation 4.37, the sedimentation velocity w of floc particles is linearly related to the effective-density ρ_e.

In order to confirm this relation, the following additional experiments were conducted.

First, aluminium sulfate was added to a suspension (of well water: water temperature 11°C, pH 6.9, M-alkalinity, 1.97 meq/l) adjusted to a kaolin concentration of 50 ppm. The results obtained showed that the effective density ρ_e was widely distributed in the range of 2×10^{-2} to 10^{-3} (g/cm^3), and did not quantitatively answer the first question of how dense the floc was. Therefore, based on another assumption – that the floc density is not a property of the particles constituting the floc but is due to structural elements – the author hypothesized that the floc density is governed by floc size, which is considered to be the main component to control effective floc density, and plotted the relation between floc effective density and floc size. As shown in Figure 4.69, the authors obtained a clear linear relation.

Having confirmed the linear relationship, the authors proceeded with a study of how this linear relationship is governed by floc factors (e.g., coagulation condition, type of suspensions, etc.).

4.2.3.3 The floc density function

As shown in Figure 4.69, effective floc density ρ_e and floc particle size d_f show good linear regressions in both logarithmic plots. As we will see later, the relationship between all floc diameters and effective densities is basically linear in both logarithmic plots. The regression equation that satisfies this relationship is considered to have the form shown in Equation 4.38. Tambo & Watanabe (1967) named

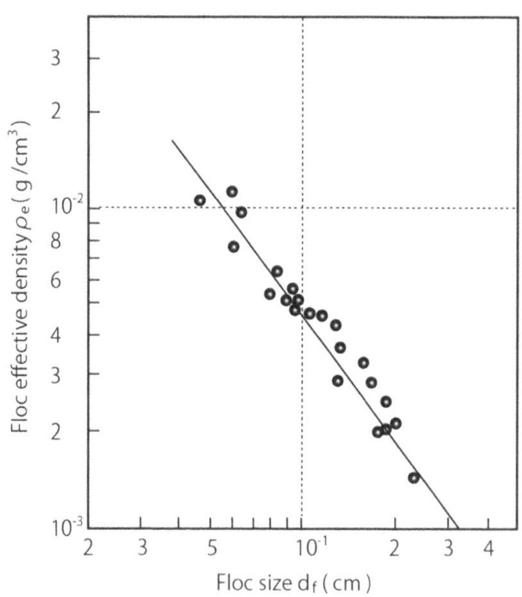

Figure 4.69 Floc size d_f and floc effective density ρ_e.

this regression equation the 'Floc density function'.

$$\rho_e = \rho_f - \rho_w = \frac{a}{(d_f/1)^{K_\rho}} \quad \text{—(4·38)}$$

ρ_e : Floc effective density (g/cm³)
ρ_f : Floc density (g/cm³)
ρ_w : Water density (g/cm³)
d_f : Floc size (cm)
a,K_ρ : Constant determined by coagulation conditions
a : Effective density ρ_e assumed assumed that $d_f = 1$ cm
K_ρ : Slope of a straight line on both logarithmic graphs (-)
1 : Numerical value to adjust dimension (cm)

(4.38)

Equation 4.38 The Floc density function.

where ρ_e, ρ, and ρ_w are the effective density of flocs, the density of flocs, and the density of water (g/cm^3), respectively; d_f: is the floc diameter (cm); a and K_ρ are constants determined by the coagulation condition, with K_ρ being a numerical value (dimensionless) indicating the slope of a straight line on both logarithmic graphs and a the effective-density ρ_e (g/cm^3) assuming that $d_f = 1$ cm.

Therefore, in order to arrange the dimensions of the equation, the floc diameter d_f is divided by the diameter of 1 cm, then the right-hand denominator is made dimensionless, the equation is unified and the experimental equation is generalized to a characteristic equation.

The following experimental values are summarized, and the constants a and K_ρ are calculated by the least square method for each series of experiments, and the characteristic values are used to express floc densities.

4.2.3.4 Coagulant dosage (ALT ratio) and floc densities

First, the authors examined how the floc density changes by varying the kaolin concentration and the amounts of coagulant added (rate of dosage). A 20 to 1000 mg/l kaolinite suspension containing 50 mg/l of $NaHCO_3$ as alkalinity was prepared in the neutral pH range. Aluminium sulfate and poly aluminium chloride (PACl) were added as coagulant. The added aluminium concentrations (dosage as Al) were changed largely in the range of 0.5 to 100 mg/l to obtain characteristics of density functions.

Figures 4.70 to 4.73 and their characteristic values are summarized in Figure 4.74, which show the change in floc structural density caused by the amount of coagulant addition rate (dosages) with the ALT ratio (Aluminium Turbidity Ratio) proposed by Tambo *et al.* (1969) and Tambo and Watanabe (1979d) as follows:

ALT ratio = (concentration of added aluminium)/(concentration of suspensions) [dimensionless]

A diagram showing the experimental results can be represented in a unified manner by a dimensionless ALT ratio.

As is clear from these experimental results, the amount of coagulant added greatly influences the effective density of flocs, and clay flocs formed at a given ALT ratio have a floc density determined by the ALT ratio, regardless of the type of aluminium coagulant and the individual concentration levels of turbidity. Figure 4.74 shows the relationship. It is clear, therefore, that:

(1) The higher the ALT ratio (the percentage of the aluminium injection rate), the lower the Floc densities of the same particle size.
(2) When the ALT ratio becomes smaller, the reduction rate of the floc density becomes smaller as the particle diameter increases, and when the ALT ratio becomes about 1/50 to 1/100, the index value

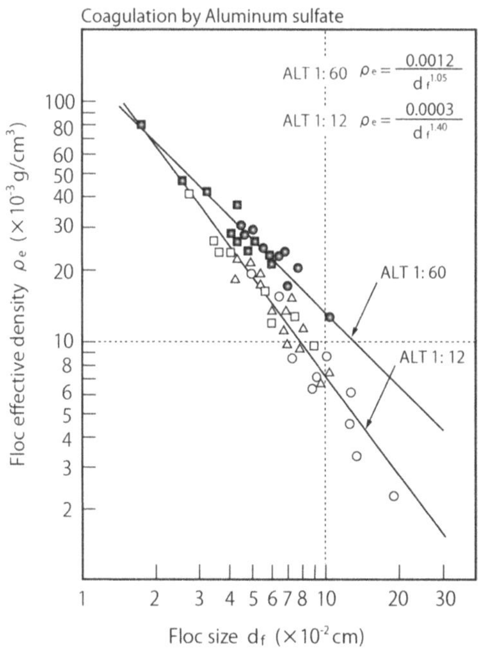

Figure 4.70 Relationship between floc effective density and size with different ALT-ratio.

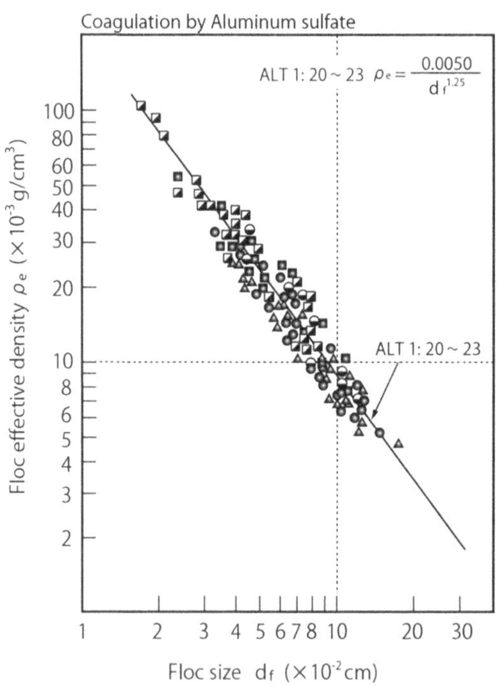

Figure 4.71 Relationship between floc effective density and size with different ALT-ratio.

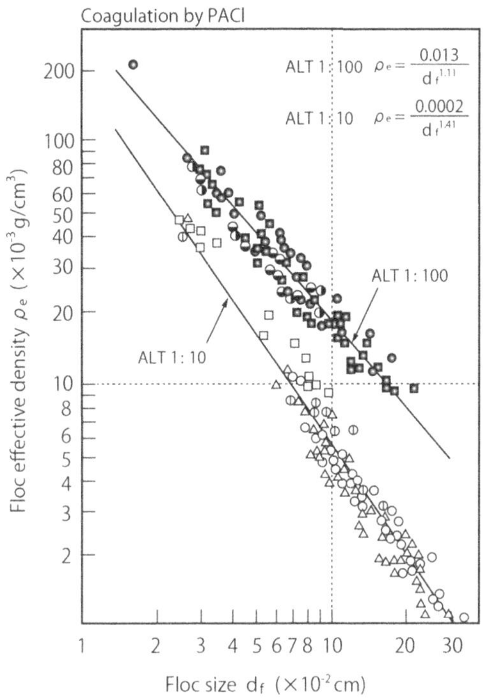

Figure 4.72 Relationship between floc effective density and size.

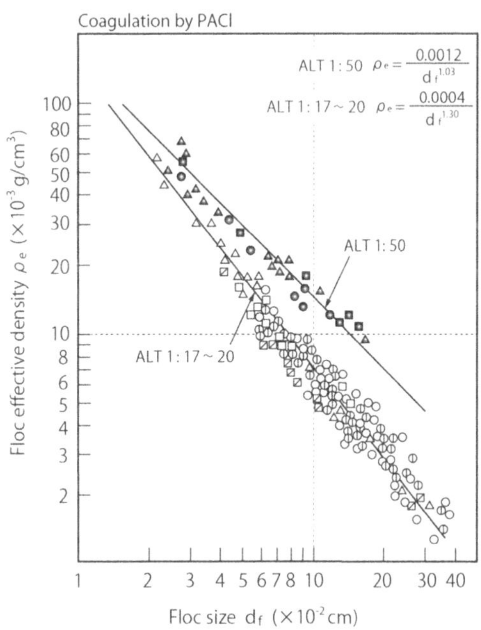

Figure 4.73 Relationship between floc effective density and size. with different ALT-ratio.

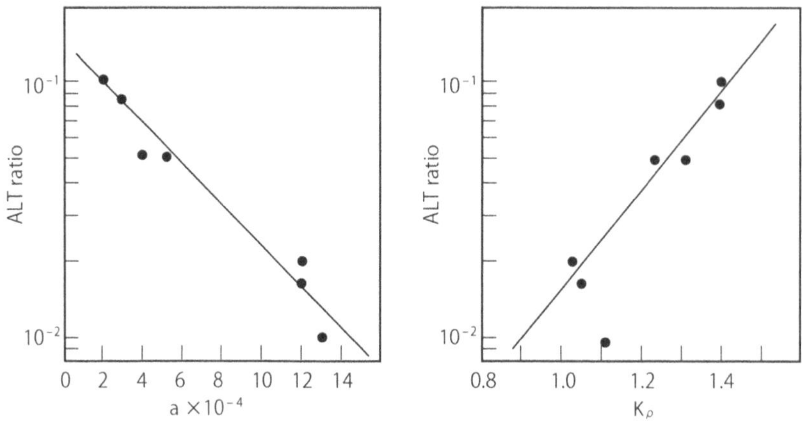

Figure 4.74 The characteristic values a and K_ρ for floc density function and ALT ratio.

K_ρ of the floc density function becomes almost 1, and the settling velocity of the floc increases in proportion to the first power of the particle diameter.

(3) Since $K_\rho \doteqdot 1.3$ to 1.4 in the range of an ALT ratio of 1/20 to 1/10, which is close to the ordinary value of the aluminium sulfate coagulation, the settling rate of floc changes in proportion to the 0.7 to 0.6 power of the particle diameter.

(4) For the same ALT ratio, the floc caused by aluminium sulfate and the floc caused by PACl are expressed by the same density function, without differing from each other.

(5) The reason why the floc precipitation by PACl is considered to be good is that PACl coagulation can progress floc formation at a much lower ALT ratio, and floc density becomes larger than that of aluminium sulfate floc particles of the same diameter which require a high ALT ratio and, at the same ALT ratio, a larger floc is formed than with aluminium sulfate.

4.2.3.5 Coagulation pH and floc densities

Figure 4.75 shows the results of experiments conducted at pH ranging from 6.5 to 8.0 to see how coagulation pH influences the effective densities of flocs.

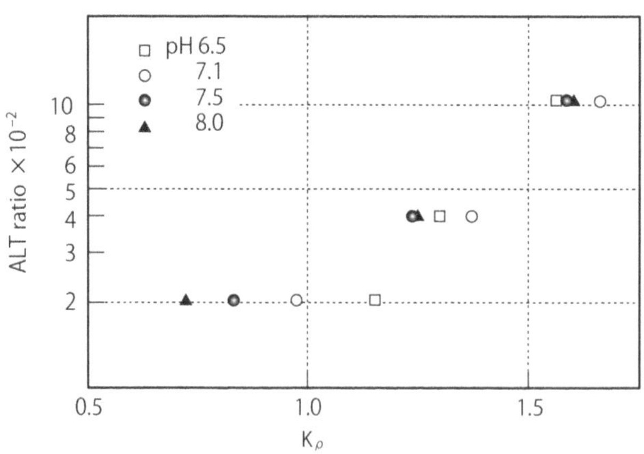

Figure 4.75 The ALT ratio and characteristic value K_ρ with pH as index.

Differences in pH in the high injection range with ALT ratios of 4×10^{-2} or higher have little effect on floc density. However, in areas with low injection rates, such as 2×10^{-2}, K_ρ decreases with increasing pH, reducing the effective density of flocs and forming swollen flocs.

4.2.3.6 Agitation intensity and floc density

Figure 4.76 shows the relationship between floc-forming agitation intensity and the effective density of floc particles. There appears to be no significant change in the floc density function over this normal range of agitation intensity variation. If the agitation strength is further increased, fracture of the floc proceeds and floc diameter decreases, resulting in a higher density, but the density function itself remains almost unchanged. This explains the basic relation between floc structure and density, which is governed by size.

4.2.3.7 Effect of flocculation aids on floc density

Figure 4.77 shows how the effective density of flocs change with the addition of various floc-forming (flocculation) aids. Even with these flocculation aids, the K_ρ values are virtually unchanged (there is no change in floc structure) but because of the increased bonding strength, larger flocs can be made with the same agitation strength, increasing the sedimentation rate.

4.2.3.8 Effect of alkalinity on floc density

The results of flocculation coagulation experiments using a 50 mg/l kaolin suspension based on distilled water and City water are shown below.

An alkaline solution ($NaHCO_3$) of 50 to 200 mg/l was added to distilled water and to Sapporo City water to form a coagulation floc. The relation between the effective floc density and the floc diameter was obtained as shown in Figures 4.78 and 4.79. City water has an alkalinity of 20 mg/l (as $CaCO_3$). The results showed that the alkalinity of different raw waters had little effect on the floc density function in the normal range.

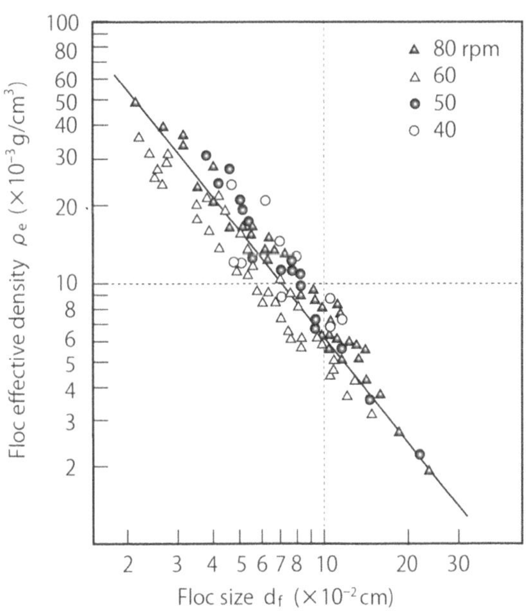

Figure 4.76 Floc-forming agitation intensity and floc effective density.

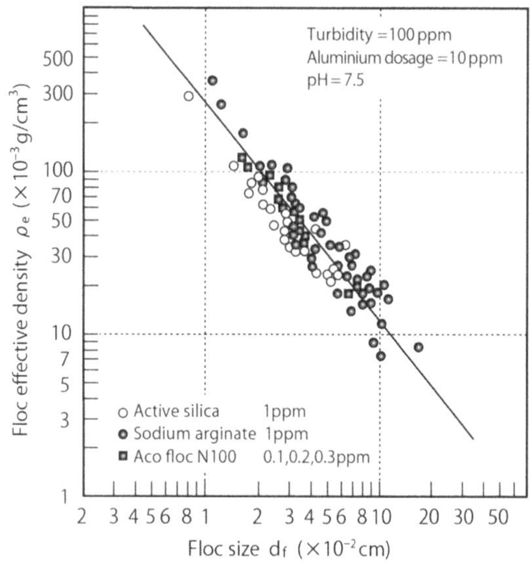

Figure 4.77 The effect of flocculant aids on floc density.

4.2.3.9 *Density functions of the various flocs*

Figure 4.80 shows the relation between the effective density and floc diameter of microbial sewage flocs produced by the activated sludge process in a normal state and in a bulking state. Because of the large size of floc-forming elementary particles, which are bacteria and microorganisms, the effective density of

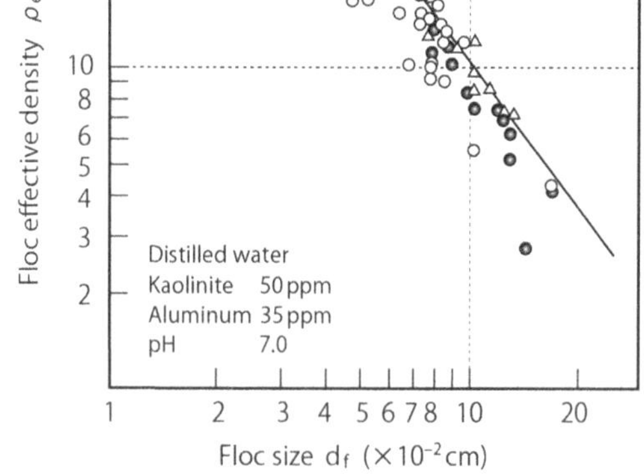

Figure 4.78 Alkalinity and floc density.

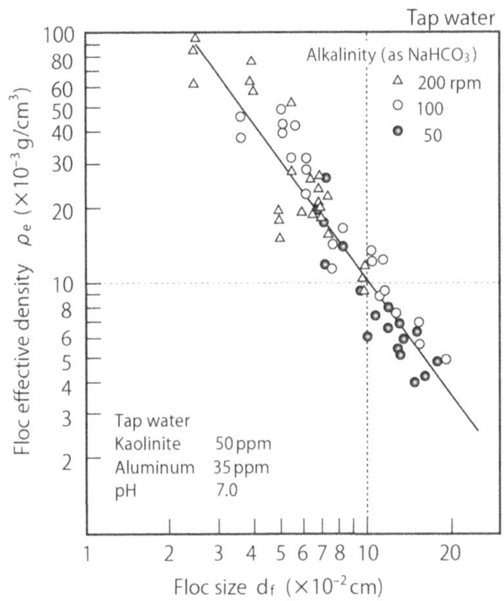

Figure 4.79 Alkalinity and clay floc effective density.

normal activated sludge flocs appears to be an order of magnitude higher than the effective density of aluminium flocs. Bulking sludge has an effective density similar to that of colored water flocs. Figure 4.81 shows the relation between the effective density of kaolin floc and the floc diameter after coagulation with ferric sulfate. Iron-salt flocs seem to be almost the same as aluminium-based coagulation flocs.

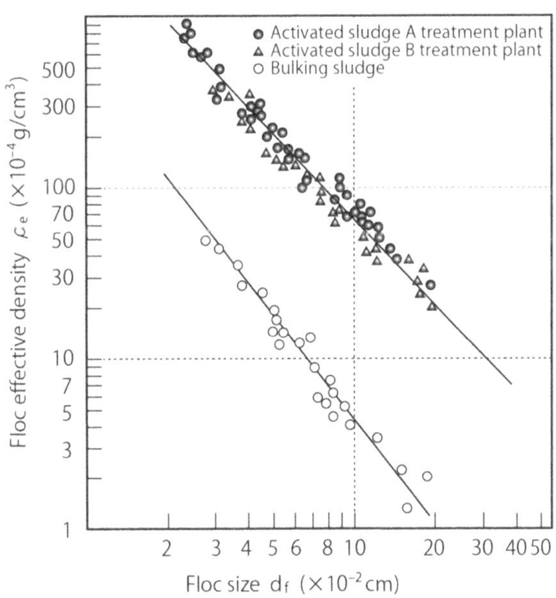

Figure 4.80 The effecitve density and floc diameter of microbial sewage flocs.

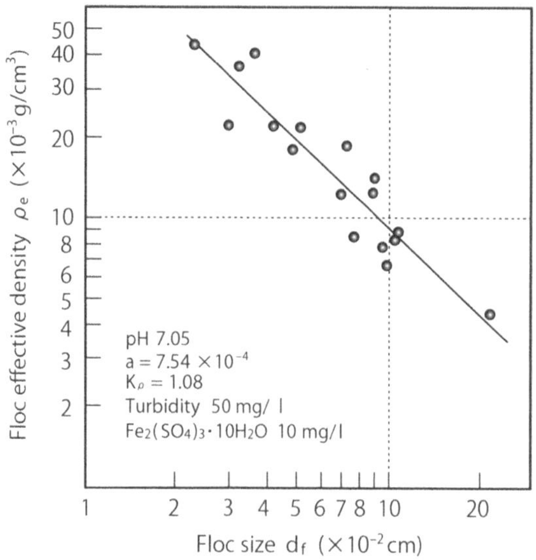

Figure 4.81 The effective density fo kaolinite flocs after coagulation with ferric sulfate.

The density of color colloids and clay coagulation flocs is clearly different, and the effective density of color flocs is only about one quarter of that of clay flocs. Since the size of the original elementary particles forming the floc differs (the color colloid particles are small), the number of agglomerates of the elementary particles required to form floc of the same size is large, so that the void ratio is large and the effective density is small. On the other hand, the effective densities do not differ depending on the type of coagulant (see Figures 4.82 and 4.83).

Table 4.16 summarizes the relationship between floc density and the major factors that may affect floc formation.

4.2.4 Numerical model (simulation) of floc density function and fractal characteristics

4.2.4.1 Simulation of floc structure

In the above experiments using aluminium sulfate and poly-aluminium chloride as the coagulant, it was found that the K_ρ of the floc density function (Equation 4.38) was uniquely determined by the ratio of aluminium dosage to turbidity (ALT ratio). When the ALT ratio is 1/50 to 1/100, the $K_\rho \doteqdot 1.0$, and the $K_\rho \doteqdot 1.3$ when the ALT ratio is as high as 1/20 to 1/30. Then, in the normal coagulation flocculation process, $K_\rho \doteqdot 1.0$ to 1.4.

Focusing on the fact that the floculation process (collision agglomeration process) can be treated as a kind of random phenomena, Vold (1963) first attempted to clarify the floc structure by simulating the collision process on computer, using random numbers. The Vold model is a model in which the flocs grow by repeating collisions, one by one, on the flocs as the early particles move from one position to the next position defined by a random number. However, the normal floc formation process is not modelled properly by Vold, as the flocs (agglomerated particles), in which a plurality of initial particles are aggregated, collide with each other. Therefore, the flocs formed by the Vold simulation model are more dense than the flocs actually formed.

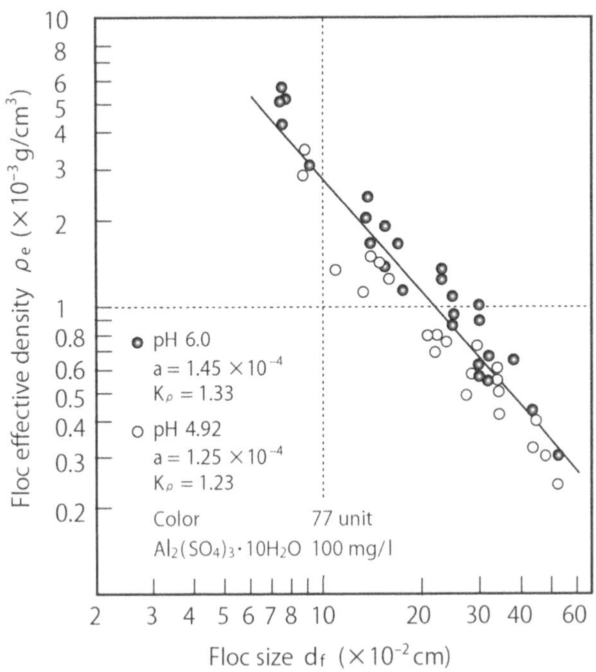

Figure 4.82 The effective density of color flocs after coagulation with aluminium sulfate.

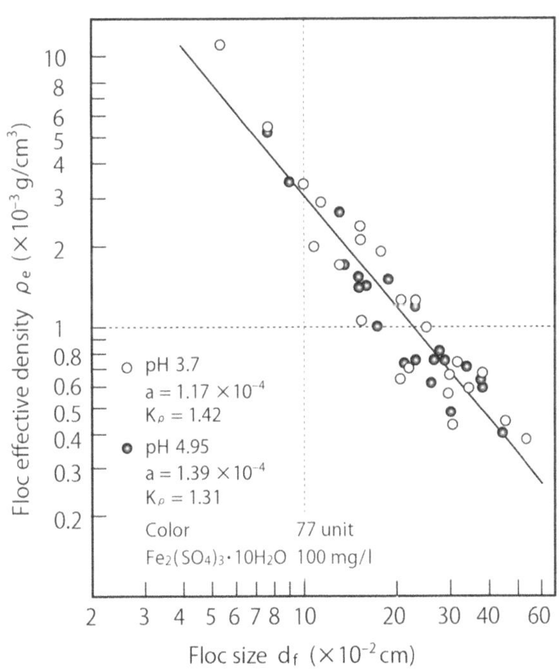

Figure 4.83 The effective density of color flocs after coagulation with iron sulfate.

Table 4.16 Floc density and the major factors which can affect floc formation.

Factor	**State of Floc Density Change**
ALT ratio	ALT ratio large: Density is small if size is same. small: Even if particle size increases, the density decrease rate is small.
PACl coagulant	Aggregation and flocculation proceed even at low ALT ratio. If the size is the same, ALT ratio Is smaller than Al and density is large.
pH	High ALT ratio range: Almost no influence on density. Low ALT ratio range: As pH increases, density decreases and bulking.
Mixing Intensity	Normal level: No major change. Mixing intensity large: Divide into smaller diameter, higher density.
Flocculant aid	No change in floc structure. Bind strength is increased, size can be increased. Size increases and the settling velocity increases.
Alkalinity	Almost no influence.
Iron coagulant	Similar to aluminium.
Color particle	Because of the small diameter, flocculation may require a large amount of particles. Since the number of particles is large, porosity is large and density is small. No difference due to coagulant.

For this purpose, the following simulation model was considered by the authors in order to bring the simulated floc formation process closer to the actual process:

(1) Simulation model operation is performed in two dimensions and converted into density functions (three dimensions).
(2) The initial particles having a unit diameter are uniformly distributed on a plane.
(3) The position of the initial particle is displayed in coordinates (X, Y).
(4) The position of the moving destination is specified by a random number, and the collision is unified.
(5) The particles consolidated by collision have a particle size (bidirectional average diameter: collision radius) and a center of gravity as a new aggregate.
(6) Particle selection is performed by a uniform random number weighted by the size of particle.
(7) The frequency of collisions is calculated.
(8) The entire coalesced particle is regarded as a floc, and the number i of contained initial particles and the floc mean diameter d_f are obtained.

By repeating this process, the relation between the diameters d_f of the progressively growing flocs and the number of clusters i is obtained.

The results of various calculations are summarized as shown in Figure 4.84, where the relative diameter d_f/d_1 of floc relative to the initial particle is taken in the horizontal axis, and the initial particle number i contained in floc is taken in the vertical axis.

From the diagram, the relation of $I = \alpha\ (d_f\ /d_1)^{\beta}$ is established, and the slope β of the regression line is approximately 1.4. Therefore, when the two-dimensional model is converted into a three- dimensional model, $\beta \doteqdot 2.1$, and the coefficient $\alpha \doteqdot 1$. Therefore, the effective-density ρ_{ei} of a floc containing i initial

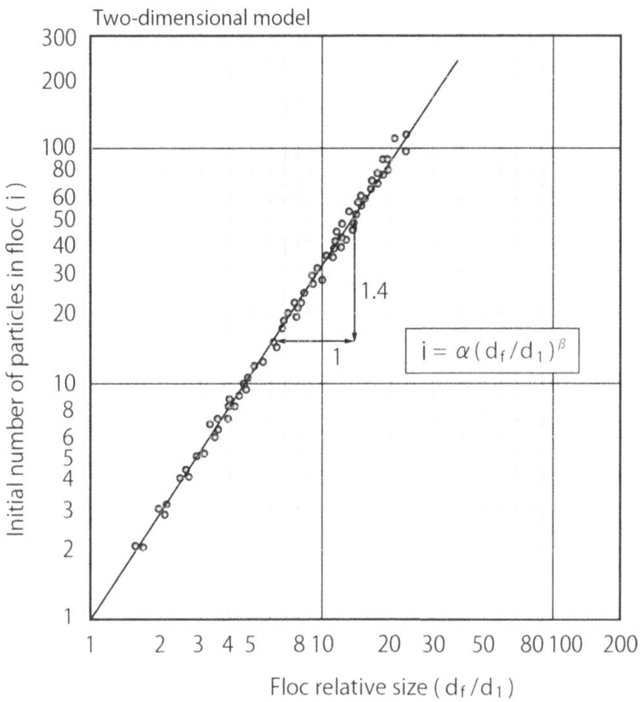

Figure 4.84 Floc relative sizes and the initial number of particles in the flocs.

particles is given by Equation 4.39.

$$\rho_{ei} = \alpha(\rho_s - \rho_w)\, d_1^{3-\beta} d_f^{\beta-3} = K^* d_f^{\beta-3} \fallingdotseq (\rho_s - \rho_w)\, d_1^{0.9} d_f^{-0.9} \quad \text{---}(4\cdot 39) \tag{4.39}$$

ρ_{ei} : Floc effective density containing i initial particles (g/cm^3)
d_1 : Initial particle size (cm)
d_f : Floc size made of i initial particles (cm)
ρ_s : Density of initial particles (clay) (g/cm^3)
ρ_w : Water density (g/cm^3)
α、β : Constant (-)
$K^* = \alpha(\rho_s - \rho_w)\, d_1^{3-\beta}$

Equation 4.39 The effective density of Floc containing i initial particles.

In the case of turbidity, when clay particles are the primary particles, $\rho_s = 2.65$ g/cm^3, $d_1 = 2$ to 3×10^{-4} cm, therefore, the constants K_ρ and a in Equation 5.38 are calculated as 0.9 and (1.1 to 1.5) $\times 10^{-3}$, respectively.

The density function constants a and K_ρ shown in the previous experimental data (Figures 4.72 and 4.73, etc.) are largely dependent on the ALT ratio, and the change is summarized in Figure 4.75 for the suspended matter in laminar states. The model floc produced by the simulation is considered to be an initial (elementary) particle of the turbidity itself, so this model floc corresponds to the case where the ALT ratio is zero. Considering this point, the values of a and K_ρ shown in Figure 4.75 (which shows the

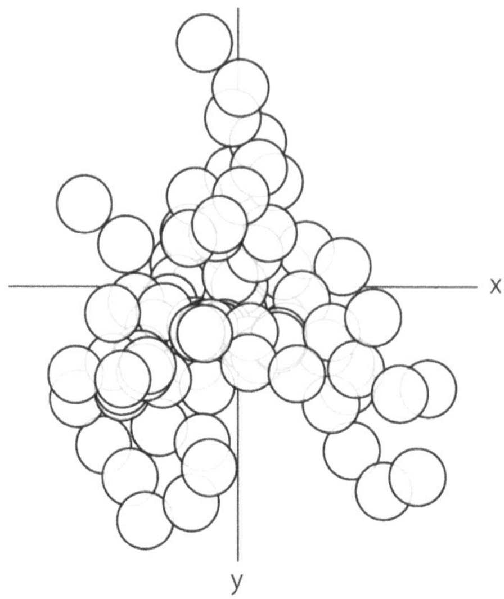

Figure 4.85 Vold's floc model.

experimental results) and those of the model floc show a good fit, and it is clear that the structure of floc is as assumed by Tambo and Watanabe (1967, 1979d). Figure 4.85 shows the results of Vold (1963) and Figure 4.86 shows the results of the model floc of Tambo and Watanabe (1967, 1979d, e).

Our floc model, which considers the bonding of higher-order aggregates, makes it possible to reproduce a real floc structure, whereas the Vold model (which combines early-stage particles one by one) produces a much higher-density floc than real floc.

4.2.4.2 A fractal physics perspective

Professor Matsushita, a specialist on fractal physics at Chuo University,Tokyo, review our work in the Journal of the Japanese Society of Physics (Matsushita, 2006), as follows. In the early 1980s, fractal

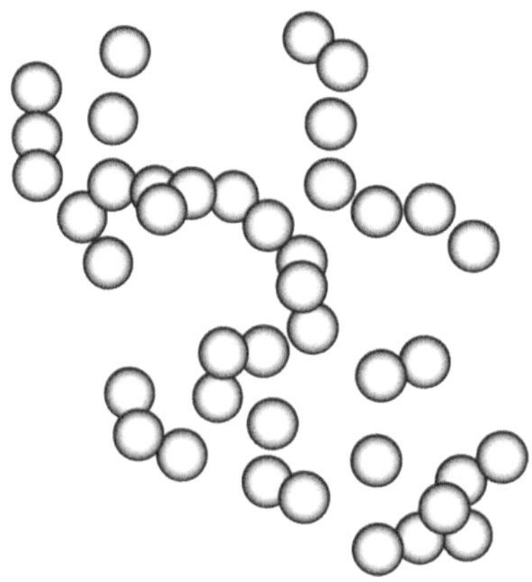

Figure 4.86 Tambo and Watanabe's floc model.

physics represented the self-similarity first proposed by Mandelbrot (1977) and widely discussed by him (Mandelbrot, 1982). Colloidal coagulation was also modeled as fractal agglomerates. Witten & Sander (1981) defined an agglomeration in which elementary particles collide together one by one in Brownian motion as a diffusion-controlled agglomeration (a diffusion-limited aggregation: DLA). In contrast, coagulation phenomena in which small agglomerates collectively grow were also described, in Meakin (1983) and M. Kolb *et al.* (1983) who presented models called cluster coagulations (cluster-cluster aggregation: CCA).

Basing their work on colloidal gold coagulation, Weitz and Oliveria (1984) tested the CCA models. The study of floc densities functions by Tambo and colleagues was the first study on fractal structure conducted prior to these studies, presented 10 years prior to the presentation of fractal concepts. The coefficient K_ρ of the floc density function is a self-similar fractal value representing floc effective density as the structural density of the agglomerates, and is a model index representing fractal agglomeration characteristics that can be directly converted into the fractal dimension $D = 3 - K_\rho$. For the structure of the colloidal coagulation to which the CCA model corresponds, the fractal dimension for reaction-rate-limited agglomeration is $D \doteqdot 2.00$, and the fractal dimension for diffusion-rate-limited agglomeration is $D \doteqdot 1.75$, based on the results of large-scale computer simulations.

The results of Tambo and colleagues' experiments are shown in Figure 4.74 and the fractal dimension D is calculated as follows: $K_\rho = 1.4$ for a high dosage system with an ALT ratio of 10^{-1}, and the fractal dimension is $D \doteqdot 3 - 1.4 \doteqdot 1.6$. This is typical of diffusion limited CCA. On the other hand, a low injection rate system with an ALT ratio of 10^{-2} has a $K_\rho \doteqdot 1.0$ and the fractal dimension $D \doteqdot 3 - 1.0 \doteqdot 2.0$, which accurately matches the reaction-rate-limiting CCA-type agglomeration.

As described above, the density of flocs formed by a coagulation process has fractal characteristics at a density determined by the agglomeration structure, and the structure is determined by the constant K_ρ of the floc density function or the fractal dimension D. Practically describing this phenomenon, the sedimentation velocity w of the particles in the Stokes area is proportional to the square of the diameters of the particles in $w = (g/18)\rho_e d^2$, as shown in Table 6.2 in Chapter 6 (below), but by introduction of the density function, the settling velocity $w = (g/18)(a/d^{K\rho})d^2 = Ad^{(2-K\rho)}$. Here, A is a constant. As described above, when the high-concentration aluminium injection rate ALT ratio is 10^{-1}, $K_\rho \doteqdot 1.4$. When the low-concentration dosage rate ALT-ratio is 10^{-2}, $K_\rho \doteqdot 1.0$. Therefore, a sedimentation velocity $w = Ad^1$ (low-concentration injection rate) or $w = Ad^{0.6}$ (high concentration injection rate) in the Stokes area is obtained.

For dense low aluminium dosage flocs, the sedimentation rate only increases proportionally to particle size, and it only doubles at twice the diameter. For swollen high dosage flocs, the sedimentation rate only increases by a factor of 1.52 as much as doubling the diameter, and the sedimentation rate improves slowly with increasing particle size. From this, it can be seen that knowing how to make a dense floc and increase the sedimentation rate is important in terms of both the solid-liquid separation rate and the amount of sludge generated. With the advent of poly-aluminium coagulant (AlCl), it is important to be able to make a low ALT ratio floc.

4.2.4.3 Background of the innovation

This experimental study was conducted as a series of B. Eng. graduate research in the Department of Sanitary Engineering, Hokkaido University in the 1960s. Yoshimasa Watanabe, a 1967 graduate (later Professor of Miyazaki University, Hokkaido University, succeeding to the author's Chair) and S. Shimizu, a 1968 graduate (later President of Nissui-Con; Japan Water Works Consultant) both took part carefully, day and night. In 1968, PACl was fortunately available for the first time and, with it, a much lower aluminium dosage rate could be used which had not been possible with alum (until 1967).The low ALT ratio experiment was able to be undertaken in practice for the first time during Shimizu's days, drastically changing the coagulation process in lower ALT ratios. Coagulation flocculation experiments in the wide range of dosage including much lower dosage rates (which generate much higher density flocs) were carried out. As a result, the concept of the ALT ratio was established. Later, 'Physical Characteristics of Flocs I: The Floc Density and Aluminium Floc'(Tambo & Watanabe, 1979d) was selected as a TOP 10 article of the 40 years from 1967 to 2007 in the same IWA-based journal, *Water Research*. Reports in Japanese were published in 'Studies on Aluminium Floc Densities (I), (II), and (III)' prior to the above in the *Journal of Japan Water Works Association* by Tambo, Watanabe and Shimizu (Tambo & Watanabe, 1967, 1971; Tambo *et al.*, 1968).

It is said that English reports were first given by A. L. Lagvankar and R. S. Gemmell of North Western University under the title 'A size-density rerationship for flocs' in the Journal of the American Water Works Association (Lagvankar & Gemmell, 1968) in almost the same period as the authors' study. They measured the buoyant density of the floc by dropping the flocs, one by one, in a pipette in sucrose (sugar-dissolved) water of differing concentrations (densities) to put the flocs in a neutral state (not sinking, not floating). However, this method could not measure the density in a non-disturbed state (without touching the floc) as a prerequisite for the floc measurement mentioned above. In addition, with this method it is hard to obtain qualitative data close to the floc density function by measuring a large number of flocs continuously and stably with precision, as in the case with the method proposed by Tambo *et al.* They were not able to develop measurement results with versatility and reproducibility in engineering or in floc formation theory.

4.2.5 Floc strength (Maximum growth size)

In the early stages of primary particles agglomerating to form flocs, flocs continue to grow if only the coagulation condition is met. However, when a certain size is reached, the external force applied to the floc mass exceeds the binding force of the floc, causing breakage. The size of the floc bond, which can resist external forces that cause breakage, determines the greatest growth floc diameter that can grow under certain coagulation conditions. It is qualitatively known that, for flocs produced under the same coagulation conditions, the larger the agitation strength, the smaller the growable floc diameter.

By quantifying these phenomena, Tambo and Hozumi (1970, 1979a) first clarified how to measure the binding strength of flocs that cannot be measured directly, using the largest growing floc diameter under a certain agitation intensity (localy isotropic turbulent flow conditions) as an index. The destruction of flocs was not mentioned in the Smoluchowski paper; Prker *et al.* (1971, 1972) reported similarly on destruction in the same period as Tambo but there seem to be no reports before us. Later, Professor J. Gregory and colleagues at Univesity College London posted a review paper to Water Research Vol. 39, No. 14, Sept. (2005). The authors had introduced a series of papers from 1967 onwards in the UK, including Dr Bach's paper, which were developed on the basis of Tambo *et al.*'s original paper in *Water Research* (1970 and 1979a), but Professor Gregory seems to have overlooked previous papers submitted to *Water Research*. Of course, there was no reason to know that the original paper appeared in the *Journal of the Japan Water Works Association* (1968), in Japanese, more than ten years before the contribution to *Water Research* (1979a).

4.2.5.1 Theories to find the largest growable floc diameter $d_{f\ max}$

Floc formation proceeds under turbulent stirring conditions, as described above. The size of the vortex that breaks the floc is a micro-vortex nearly close to the size of the floc. The mechanism by which turbulent energies are transmitted to the vortex up to this size may be considered to be locally isotropic turbulence. Therefore, we believe that the forces due to the locally isotropic turbulent vortices (whether torsional or shear-type breaks) exceed the binding forces of the flocs, which are described in the Kolmogorov theory (Heinz, 1959; Bachelor, 1958; Levich, 1962) in which the large vortices generated by the agitator are consumed in a unit-volume of fluid by way of a micro vortex in the viscus subrange by the intensity ε_0 (erg/cm^{3}sec).

The floc bonding force B_e (g·cm/sec^2) with porosity e (Porosity, dimensionless) is expressed as $B_e \sim d_f^2(1-e)\sigma$ by multiplying the actual cross-sectional area of the floc fracture A_e (cm^2) $= d_f^2(1-e)$ by the bonding strength σ (g/cmsec^2) of a substantial part of the floc particles.

The void fraction e is organized in the form of the relation between the floc density function (Equation 4.38) and the volume V, the agitation strength of the density ρ_f, and the floc fracture diameter (greatest growing diameter d_{max}). When the void fraction e is replaced with the relation of the floc particle structure (the water fraction V_e of the void and the substantial fraction $(1-e)V$ of the floc) by the relation $\rho_w V_e + \rho_0(1-e)V = \rho_f V$ and using the floc density function property K_ρ, the binding force B_e of the floc of the void fraction e is shown in Equation 4.40.

$$B_e \sim \sigma d_f^{\,2-K_\rho} \quad \text{—(4·40)} \tag{4.40}$$

B_e: Bonding force of floc (g/cm·sec^2)
σ: Bonding strength of substantial part of floc particles (g/cm·sec^2)

Equation 4.40 The bonding force of floc of porosity e.

On the other hand, the maximum dynamic external force (differential dynamic pressure strength Δf_{df} (g/cmsec^2)) due to locally isotropic turbulence eddies acting on the floc with a diameter d_f is expressed as the square mean difference of the turbulence velocity between the two points of the distance d_f at both ends of the floc, as in Equation 4.41.

$$\Delta f_{d_f} \propto \rho_w \frac{\left|\overline{v_1' - v_2'}\right|^2}{2} \quad \text{—(4·41)} \tag{4.41}$$

ρ_w: Water density (g/cm^3)
v_1', v_2': Each turbulence velocity between the two points of the distance d_f (cm/sec)
$\left|\overline{v_1' - v_2'}\right|^2$: Time-averaged absolute velocity of turbulence fluctuations between two points

Equation 4.41 The maximum dynamic external force due to localy isotropic turbulent vortexes.

According to Kormogoroff's locally isotropic turbulence theory, the time-averaged absolute velocity of turbulence fluctuations between two points is considered to be equal to the velocity $v_\lambda = d$ of the turbulence vortex on a scale $\lambda = d$ between two points separated by a distance d (the diameter of the floc). This is illustrated by the inertial vortex region in Equation 4.42 and the viscous vortex region in Equation 4.43

(Levich, 1962). The turbulence vortex microscale is given in Equation 4.44.

$$v_{\lambda=d} = \alpha \left(\frac{\varepsilon_0 d}{\rho_w} \right)^{\frac{1}{3}} \quad d \gg \lambda_0 \text{ (Inertial region)} \quad \text{—}(4\cdot42)$$

$$v_{\lambda=d} = \beta \sqrt{\frac{\varepsilon_0}{\mu}}\, d \quad d \ll \lambda_0 \text{ (Viscous region)} \quad \text{—}(4\cdot43)$$

$$\lambda_0 = \left(\frac{\nu^3 \rho_w}{\varepsilon_0} \right)^{\frac{1}{4}} \quad \text{(Turbulence vortex microscale)} \quad \text{—}(4\cdot44) \qquad (4.42\text{–}4.44)$$

v_λ: Turbulent vortex velocity of scale λ (cm/sec)
ε_0: Energy consumption rate (erg/cm^3·sec)
ν: Water kinematic (cm^2/sec)
μ: Water viscosity (g/cm·sec)
$\alpha \fallingdotseq 1$, $\beta \fallingdotseq 1/4$ or $(1/15)^{1/2}$

Equation 4.42 - 4.44 The Turbulence vortex velocity and Turbulence vortex microscale.

Substituting Equations 4.42 and 4.43 into Equation 4.41, assuming that the force acting is proportional to the surface area of the floc, the total destructive force on the floc ΔF is expressed as Equations 4.45 and 4.46.

$$\Delta F \sim \rho_w^{1/3} \varepsilon_0^{2/3} d^{8/3} \quad \text{(Inertial region)} \quad \text{—}(4\cdot45)$$

$$\Delta F \sim \rho_w \left(\frac{\varepsilon_0}{\mu} \right) d^4 \quad \text{(Viscous region)} \quad \text{—}(4\cdot46) \qquad (4.45\text{–}4.46)$$

Equatio 4.45, 4·46 The total destructive force on flocs.

The fracture limit (d_{max}) of the floc is considered to be determined by the condition that the bonding strength of the floc is equal to the turbulent fracture force. Therefore, the largest floc diameter $d_{f\ max}$ corresponds to $\Delta F = B_e$, and under the condition that Equation 4.40 is equal to Equation 4.45 or Equation 4.46, Equations 4.47 and 4.48 are obtained.

$$d_{fmax} = k^* \sigma^{-(2/3+K_\rho)} \cdot \rho_w^{-(1+K_\rho)} \cdot \varepsilon_0^{-(4/3+K_\rho)} \quad d \gg \lambda_0 \text{ (Inertial region)} \quad \text{—}(4\cdot47)$$

$$d_{fmax} = k^\# \left(\frac{\sigma}{\rho_w} \right)^{1/(2+K_\rho)} \cdot \left(\frac{\mu}{\varepsilon_0} \right)^{1/(2+K_\rho)} \quad d \ll \lambda_0 \text{ (Viscous region)} \quad \text{—}(4\cdot48) \qquad (4.47\text{–}4.48)$$

d_{fmax} : $\Delta F = B_e$ Equivalent diameter
k^*, $k^\#$: Coefficient (each with a dimension)

Equation 4.47,4.48 The largest floc diameter.

Assuming the values of the floc density function characteristic value $K_\rho = 1.0$ (at low ALT ratio of 10^{-2}) to 1.4 (high ALT ratio of 10^{-1}), the effective energy consumption rate ε_0 under turbulent agitation conditions is proportional to the third power of the rotational speed Nr(rpm) of the agitator wing of the floc forming tank. Thus, Equations 4.49 and 4.50 are used.

$$d_{fmax} = k^{*}\varepsilon_0^{-1/(4/3+K_\rho)} \fallingdotseq k^{*}\varepsilon_0^{-(0.43\sim0.37)} \sim Nr^{-(1.3\sim1.1)} \quad d \gg \lambda_0 \text{ (Inertial region)} \quad \text{—} (4\cdot49)$$

$$d_{fmax} = k^{\#}\varepsilon_0^{-1/(2+K_\rho)} \fallingdotseq k^{\#}\varepsilon_0^{-(0.33\sim0.29)} \sim Nr^{-(1.0\sim0.88)} \quad d \ll \lambda_0 \text{ (Viscous region)} \quad \text{—} (4\cdot50)$$

K_ρ: Floc density function characteristic value
Nr : Blade rotation speed (rps)
ε_0 : Eenergy consumption rate ($erg/cm^3 \cdot sec$)
μ : Water viscosity ($g/cm \cdot sec$)
λ_0 : Microscale of local isotropic turbulence (Kormogoroff scale , cm)
$= (\nu^3 \rho_w / \varepsilon_0)^{1/4}$ ν : Water kinematic viscosity (cm^2/s)
ρ_w : Water density (g/cm^3)
k^{*} : Coefficient indicating floc strength,
Function of exponent K of floc density function

(4.49–4.50)

Equation 4.·49, 4.50 The rotational speed of agitator blade and the largest floc diameter

4.2.5.2 Experimental evaluation of the relationship between maximum growth diameter d_{max} and effective agitation strength ε_0

Using a floc-forming tank with a paddle-shaped agitator blade of 10 × 10 × 20 cm as shown in Figure 4.87, a coagulation suspension in a square column bath rotating at 1 rpm around a horizontal rotating shaft was mixed with paddles to facilitate large-diameter floc formation. Operation was performed to prevent

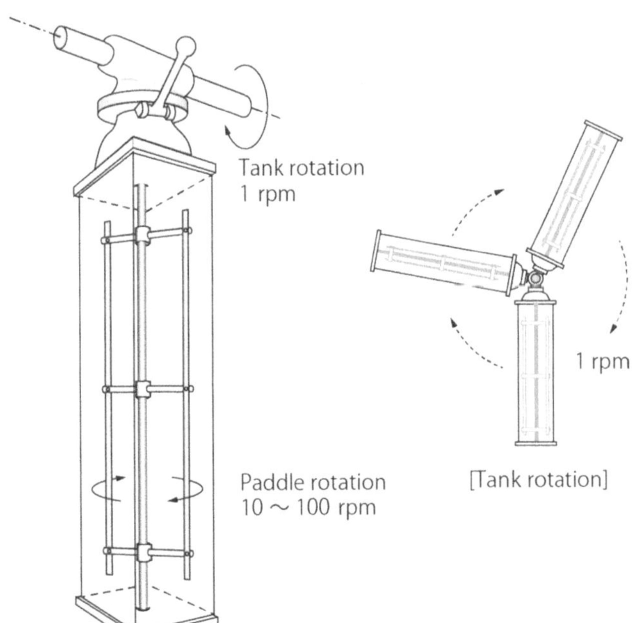

Figure 4.87 A floc forming tank for measurement of maximum floc size growth.

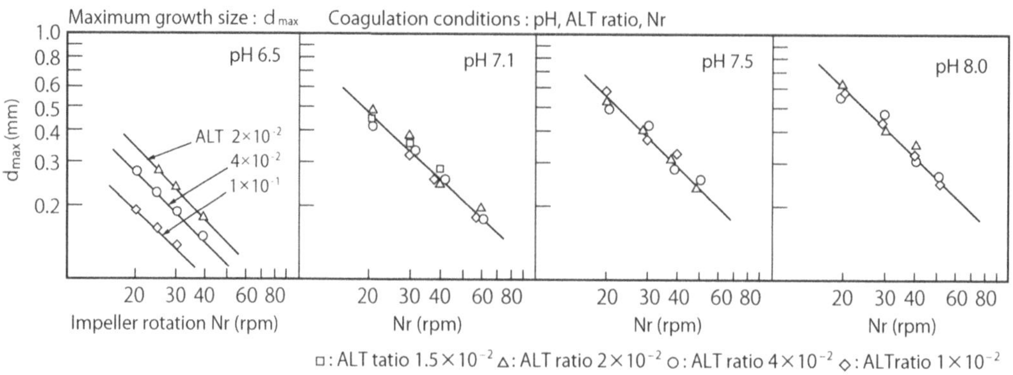

Figure 4.88 Maximum floc size and coagulation conditions for clay aluminium floc.

large-diameter floc from settling in the agitator bottom even when floc grows, and the phenomenon that progressed in the actual floc-forming tank was modeled.

Using $Al_2(S0_4)_3 18H_2O$ (an alum) for the coagulation of kaolin clay suspension, the entire agitator vessel was rotated at a speed of 1 rpm around a horizontal shaft with rapid agitation by the paddle at 100 rpm for 5 minutes and slow agitation at a predetermined intensity for 30 minutes. The diameters of 20 to 30 flocs grown were photographed, and the maximum floc diameter d_{max} was estimated.

Figures 4.88 and 4.89 plot the relationship between the agitation rotational speed Nr[rpm] and the maximum floc diameter d_{max}[mm] when forming clay aluminium floc on logarithmic graphs. The slope regresses to approximately −1.0 to −1.2, and the predicted state can be confirmed by Equations 4.49

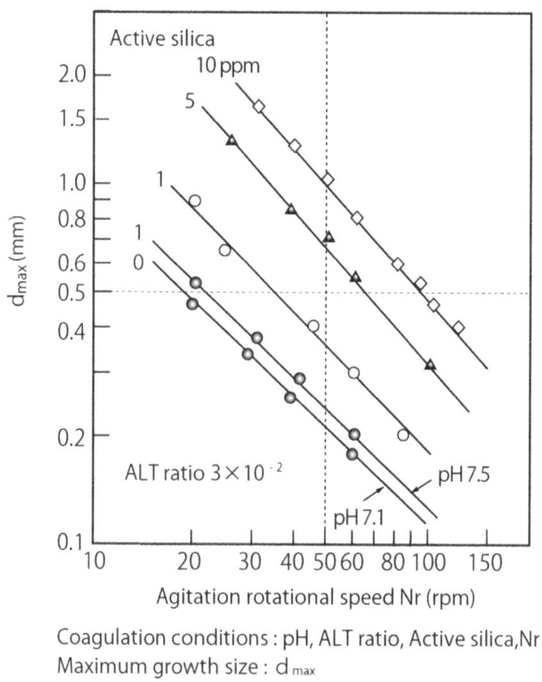

Figure 4.89 Maximum growth size and coagulation conditions for clay aluminium floc.

and 4.50. As can be seen in Figure 4.88, as the pH increases from 6.5 to 8.0, the maximum floc diameter that grows under the same agitation intensity increases with pH and the floc intensity increases. At pH 6.5, the maximum floc diameter changes slightly depending on the ALT ratio, and the diameter becomes larger at the low ALT ratio, which seems to result in stronger flocs. This phenomenon is remarkable in the sense that the adhesion strength was secured in the filtration of direct filtration or low turbidity raw water coagulation, and the loss of water head is unlikely to occur. In the neutral pH range above pH 7.1, the change in ALT ratio does not appear to be significant. It is considered that the amount of precipitated aluminium particles is sufficient in the neutral region.

Figure 4.89 shows that adding Baylis-activated silicic acid as a coagulation adjuvant greatly improves the largest floc growth diameter. This is very useful in coagulation sedimentation due to the increased sedimentation rate, but it is well known that large filters and high filtration rates need to be accommodated because of the rapid clogging of the filtration media gap with the addition of activated silicic acid. However, as a prerequisite for fast filtration, the floc strength issue has not yet been quantified. It is a classic problem, but is a basic problem worth consideration.

In order to generalize the power process of flocculation, the agitation strength needs to be expressed as a common index at the effective-energy dissipation rate ε_0 [erg/cm^{3}sec]. To estimate ε_0, Tambo and others devised the following:

(1) A torque meter was attached to the agitating shaft of a vertical floc forming tank, and the total energy consumption (dissipation) rate ε^* was directly obtained by Equation 4.51. The locality of the energy consumption distributions of the agitating tank was corrected to assume that the effective energy consumption rate $\varepsilon_0 \doteqdot (0.1 \text{ to } 0.2) \times \varepsilon^*$ [erg/cm^{3}sec] of the bulk portion where floc formation was proceeding effectively.

(2) Using the contact flocculation rate theory of Tambo and colleagues – see, for example, Tambo and Hozumi (1979b) – Equation 4.52 is based on the time-decay (first-order response) gradient of the microfloc (turbidity) of supernatant obtained by mixing microfloc with the highest concentration floc group of clay particles coagulated by aluminium to form a contact coagulation-flocculation, followed by static sedimentation at a predetermined time interval. The effective-energy-consumption rate ε_0 [erg/cm^3 sec] was estimated by using the equation considering the sorptive collision of microflows with respect to the maximally growing floc group.

(3) In addition, the validity of the results was examined by comparing the two equations.

$$\varepsilon_* = \frac{T \cdot g_c \cdot \omega}{V} \qquad (4\cdot51)$$

$$\frac{dn}{dt} = \frac{-1}{5\sqrt{15}}\sqrt{\frac{\varepsilon_0}{\mu}} \cdot V_f \cdot n \qquad (4\cdot52)$$

ε_* : Total energy consumption rate (erg/cm^3·sec)
T : Mixing torque (g_{weight}·cm)
g_c : Gravity conversion factor of Newton (980gcm/g_{weight}·sec^2)
ω : Angular velocity of impeller (1/sec = 2π Nr)
Nr : Rotation rate of impeller (rps)
V : Flocculator volume (cm^3)
n : Micro floc number concentration (1/cm^3) or concentration (mg/l)
V_f : Maximum growth floc volume concentration (-)

(4.51–4.52)

Equation 4.51, 4.52 The effective energy consumption rate and flocculation rate.

Table 4.17 Estimation of the effective energy consumption rate ε_0 of a floc-forming tank.

Nr (rpm) Rotation Rate	ε^*(erg/cm^3·sec) Total Consumption Rate	$\varepsilon_0 = (0.1\sim0.20\}\ \varepsilon^*$ Estimated from the Total Consumption Rate	ε_0 (erg/cm^3·sec) Estimated from the Contact Coagulation
10	3.4×10^{-1}	$(3.4 \sim 6.8) \times 10^{-2}$	—
20	2.4×10^{0}	$(2.4 \sim 4.8) \times 10^{-1}$	3.8×10^{-1}
30	7.7×10^{0}	$(7.7 \sim 1.5) \times 10^{-1}$	1.5×10^{0}
35	1.3×10^{1}	$(1.3 \sim 2.6) \times 10^{0}$	3.4×10^{0}
40	1.9×10^{1}	$(1.9 \sim 3.8) \times 10^{0}$	3.4×10^{0}
45	2.6×10^{1}	$(2.6 \sim 5.2) \times 10^{0}$	—
50	3.5×10^{1}	$(3.5 \sim 7.0) \times 10^{0}$	—

Table 4.17 shows the effective energy consumption rate ε_0 estimated by the above two methods. Approximate values were obtained by two methods of estimating the ε_0. In such a paddle-type floc-forming stirred tank, an approximate value of $\varepsilon_0 \doteqdot (0.1 \text{ to } 0.20)\ \varepsilon^*$ appears to be established.

The relation between the agitation strength with the effective energy consumption rate ε_0 and the corresponding maximum floc growth diameter d_{am} (fracture diameter) is shown as a straight line on both logarithmic graphs (plots), as shown in Figure 4.90. The diagram shows the straight line of the microscale λ_0 of the turbulent vortex. The fracture area of the aluminium clay floc is located near the top and bottom of the microscale, and is located at a delicate position that spans both the inertial and viscous subrange areas. We can determine the exponent to be around −0.33 in Equations 4.49 and 4.50 in practical experiments.

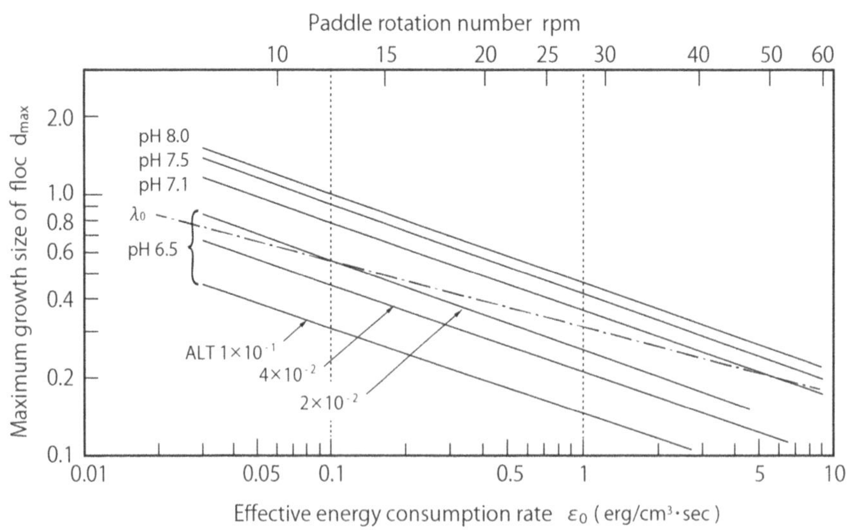

Figure 4.90 The maximum floc growth size and effective energy consumption rate for a clay aluminium floc.

4.2.5.3 *Estimation of floc strength*

It is more practical to use the largest growth diameter d_{max} under a given agitation condition than to directly determine the bonding strength σ of floc particles (this is discussed further below). The relation between d_{max} and the bonding strength of flocs in the viscous range is shown in Equation 4.48. Since there is no way to theoretically know the coefficient K* in this equation, in practical use, a floc group generated under a certain standard condition is used as a standard, and plots showing the floc bonding force showing the relation between the maximum growth diameter d_{max-1} and the maximum growth grain diameter d_{max-i} of the floc group are made, as shown in Figure 4.91. From Equation 4.48, the ratio of the coupling strength σ_1 to σ_i for the maximum growth diameter d_{max-i} of the floc group compared with the maximum floc growth diameter d_{max-1} under the coagulation state considered can be demonstrated, as shown in Equation 4.53.

$$\frac{\sigma_i}{\sigma_1} \doteqdot \left(\frac{d_{max-i}}{d_{max-1}}\right)^{2+K_\rho} \quad \text{—(4·53)} \tag{4.53}$$

σ_i: Bonding strength of the maximum growth diameter d_{max-i} of comparison flocs

σ_1: Bonding strength of the maximum growth diameter d_{max-1} of standard flocs

Equation 4.53 The maximum growth size and bonding strength of standard and comparison flocs.

Under neutral pH conditions shown in Figures 4.88 and 4.89, aluminium sulfate (alum) bands with an ALT ratio of 2×10^{-2} generates standard floc ($K_\rho \doteqdot 1.07$), and increases the σ_i / σ_1 ratio rapidly 6-fold when the ALT ratio is increased 5 times as 1×10^{-1}, but the floc density becomes small and swells to $K_\rho \doteqdot 1.4$. The addition of 1 ppm of activated silica (Baylis sol) at an ALT ratio of 3×10^{-2} at pH 7.5 increases the floc strength ratio by 3.9 times, but K_ρ remained at 1.15 times and held a dense floc state. In contrast, attempting aluminium-based coagulation with the addition of a polymeric coagulation adjuvant (Accofloc: neutral) yields a greatly increased maximum floc diameter d_{max} as shown in

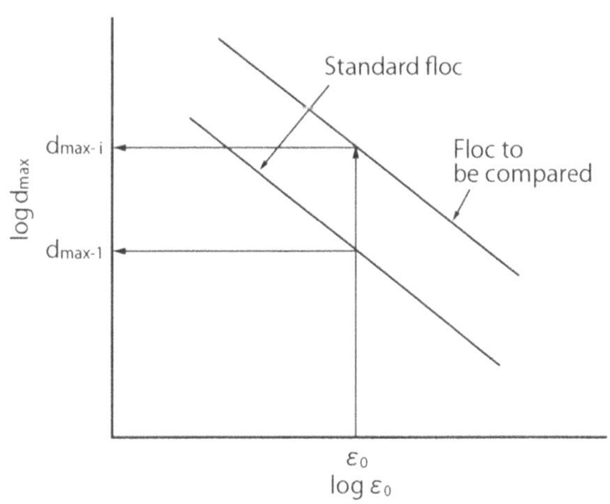

Figure 4.91 Floc strength comparison.

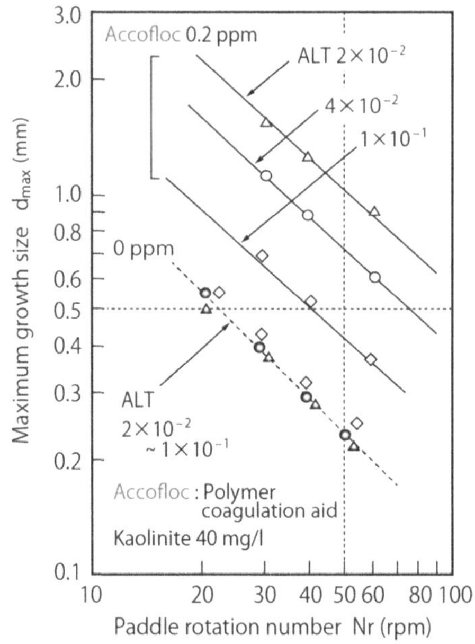

Figure 4.92 The increased maximum size of aluminium floc with the addition of Accofloc.

Figure 4.92. For the same agitation strength, adding 0.2 ppm of Accofloc to ALT ratio 2×10^{-2} of alum greatly improves $d_{max\text{-}i}/d_{max\text{-}1} \doteqdot 5$. If we estimate the relative bonding strength using Equation 4.53, we can assume that the relative bonding strength will not swell flocs, such as $K_\rho \doteqdot 1.0$, even if a polymer auxiliary is added as described in the previous section. Therefore, we can give a large bonding strength, such as $(\sigma_i / \sigma_1) \doteqdot 5^{(2 + K\rho)} \doteqdot 5^3 = 125$ times. However, various attempts have been made to measure the bond strength as an absolute value, but there are still no examples of reliable measurements.

4.2.6 Theoretical description of the flocculation process

4.2.6.1 Basic formula for collision

Smoluchowski (1917) presented Equation 4.54, where a number n of suspended particles with diameter d in a unit volume collide within a unit time under laminar condition as N_{grad}.

Laminar conditions

$$N_{grad} = \frac{4}{3} G_{grad} d^3 n^2 \qquad \text{—(4·54)} \qquad (4.54)$$

N_{grad} : Number of particle collisions per unit time
G_{grad} : Average velocity gradient
d : Suspended particle size
n : Number of particles in unit volume

Equation 4.54 The Smoluchowski's particle collision equation.

Camp (1953) proposed the mean velocity gradient G_{grad} value (sec^{-1}) as a characteristic value indicating the collision condition of flocculation ponds, and presented Equation 4.55, which was obtained from the energy consumption rate (power consumption) ε ($erg/cm^3 sec$).

$$G_{grad} = \left(\frac{\varepsilon}{\mu}\right)^{\frac{1}{2}} \qquad \text{—(4·55)} \tag{4.55}$$

ε : Energy consumption rate
(Power consumption rate erg /cm³sec)
G_{grad} : Average velocity gradient (1/sec)

Equation 4.55 The Camp's average velocity gradient equation.

However, in an actual flocculator, the energy consumption rate ε_{grad} (velocity gradient G_{grad}) produced by molecular motion under laminar conditions and the energy consumption rate ε_{turb} (turbulent velocity gradient G_{turb}) under turbulent viscosity produced by microfluidic momentum transport differ greatly, and turbulent conditions almost always proceed in practice. As we saw in the previous section, from a rough estimate of G_{turb} in Equation 4.55 from the laminar velocity gradient and the energy consumption under turbulent conditions in the water flow in a uniform flow channel, G_{grad}: G_{turb} ≑ 1:30. This is a big difference. If we take Camp's G value expression, $G_{turb} = (\varepsilon_{turb}/\mu)^{1/2}$, we see that Camp and Stein's original paper operated at a distance from reality, with only morphological approximations.

Under turbulent flow conditions, Levich (1962) gave the number of collisions per unit time of N_{turb} [cm^{-3}] of particles in a unit volume suspension, such as in Equation 4.26 above, based on Kolmogorov's locally isotropic turbulent flow conditions.

Since most flocculation proceeds under turbulent conditions, a comparison of the rate of collision under the laminar velocity in Equation 4.54 for Smoluchowski, and Camp and Stein with Equation 4.26 for under the locally isotropic turbulence theory of Levich yields a factor of $4/3:12\pi\beta$, if we assume that β ≑ 1 and the factor (ratio) is 1.33:38 ≑ 1:29 (which is a good fit to my previous rough calculations of uniform flow in parallel channels). It can be seen from the basic equations that, even for the same G-value, there are differences in the frequency of collisions between laminar and turbulent systems by a factor of 30.

4.2.6.2 Particle size growth

When the initial particle group satisfying the coagulation condition of uniform size (of particle diameter d_1) is introduced into the floc forming tank and agitation is added, the collision between the initial particles d_1 and each other produces a double particle (a two-fold particle) d_2; a three-fold particle d_3 is produced by collision of a d_1 particle and a d_2 particle, and d_4 particles are produced by collision of d_2 and d_2; and a 5 fold particle d_5 is produced by collision of a d_2 and a d_3, so that higher-order bound i-fold particles d_i are produced one after another. The number of collisions of an i-fold particle d_i and a j-fold particle d_j that occur in a unit time in a unit volume of fluid is shown in Equation 4.56, based on Equation 4.26.

$$\begin{aligned} N_{ij} &= 12\pi\beta\sqrt{\frac{\varepsilon_0}{\mu}}\left(\frac{1}{2}d_i+\frac{1}{2}d_j\right)^3 n_i n_j \\ &= K d_1^3 (i^{1/3}+j^{1/3})^3 n_i n_j \qquad \text{—(4·56)} \end{aligned} \tag{4.56}$$

d_1 : Initial particle size (cm)
d_i : i-fold particle size ($=d_1 i^{1/3}$)
d_j : j-fold particle size ($=d_1 j^{1/3}$)
n_i, n_j : Number concentration of i-times particles, j-fold particles (1/cm³)
K : Floc particle collision rate constant
$= 3/2\sqrt{15}\pi\sqrt{\varepsilon_0/\mu}$ (cm³/sec)
ε_0 : Effective energy consumption rate of the flocculation tan (erg/cm³·sec)

Equation 4.56 The Number of collisions between i-fold particle and j-fold particles.

Flocs of various size are formed in the floc-forming tank by the progress of the collisions, but the upper limit of agglomeration of the primary particles to S-fold particles ($d_{max} \leqq d_s$) which can be maximally grown by the agitation strength given is obtained from the author's equation described in the last paragraph. Thus, the two-way action of collision-to-collision and destruction results in the presence of S kinds (sizes) of flocculent particles in the stirred tank with a certain size distributions at final equilibrium.

By considering Tambo and Watanabe's floc density functions, as floc agglomeration progresses, the density of the floc decreases rapidly (as the collision radii increase correspondingly), thereby accelerating the collision frequency (the floc growth rate). If the d_i of the diameter of the i-fold particles is corrected by increasing the amount of void water contained in the floc using the constant K_ρ of the density function, the collision equation for floc formation, which takes floc density function into account, is corrected as in Equation 4.58 (Tambo & Watanabe's collision equation).

i-fold particle size using density function

$$d_i = i^{1/(3-K_\rho)} d_1 \qquad \text{—(4·57)}$$

Tambo-Watanabe collision equation

$$N_{ij} = K d_1^3 \left(i^{1/(3-K_\rho)} + j^{1/(3-K_\rho)}\right)^3 n_i n_j \qquad \text{—(4·58)}$$

(4.57–4.58)

Equation 4.57, 4.58 The i-fold particle size and Collision equation for floc formation.

4.2.6.3 Growth and particle size distribution estimation of floc groups

The increase or decrease for a time t of R-fold floc particles can be written as shown in Equation 4.59. The expression holds for the sum of the term for the production of the R-fold particle by collision of the i-fold particle with the R-i = j-fold particle, and the other term for the R-fold particle by collision with the i-fold particle to be no longer the R-fold particle. The term describes the extinction of the R-fold particle, and the sum of all the i-fold particles that may be present is obtained.

$$\frac{dn_R}{dt} = \frac{1}{2} K d_1^3 \sum_{i=1}^{R-1} \left[i^{1/(3-K_\rho)} + (R-i)^{1/(3-K_\rho)}\right]^3 n_i n_{R-i} - K d_1^3 \sum_{i=1}^{x} \left(i^{1/(3-K_\rho)} + R^{1/(3-K_\rho)}\right)^3 n_i n_R \qquad \text{—(4·59)}$$

$$\left(\sum_{i=1}^{x} i n_i = n_0 \quad , n_1 = n_0\right)$$

(4.59)

The first item: This term indicates a production of [i+j = R]-fold particles by the collision of i-fold particles and [R – i = j]-fold particles.

The second item: This term indicates R-fold particle reduction by collision of R-fold particles and i-fold particles

Equation 4.59 The Increase or decrease of R-fold floc particles.

Equation 4.59 shows that all collisions effectively go to unity. Furthermore, considering the growth limitation that there are no flocs above the S-fold particles that have reached the largest growth particle size d_{max}, we need to correct the integration area as shown in Equation 4.60.

$$\frac{dn_R}{dt} = \frac{1}{2} K d_1^3 \sum_{i=1}^{R-1} \alpha_R [i^{1/(3-K_\rho)} + (R-i)^{1/(3-K_\rho)}]^3 n_i n_{R-i} - K d_1^3 n_R \sum_{i=1}^{S-R} \alpha_{R+i} (i^{1/(3-K_\rho)} + R^{1/(3-K_\rho)})^3 n_i \quad \text{—(4·60)} \tag{4.60}$$

Equation 4.60 The Increase or decrease of floc particles considering growth limit S-fold particles.

Equations 4.61 and 4.62 show the dimensionless relationship between the number of particles n and the floc-forming rate coefficient K, so that Equation 4.60 can be represented as Equation 4.63.

$$N_i = \frac{n_i}{n_0}, \quad \sum_{i=1}^{R} i N_i = 1 \quad \text{—(4·61)}$$

Dimensionless flocculation time : m

$$m = K d_1^3 n_0 t \fallingdotseq 1.22\sqrt{\frac{\varepsilon_0}{\mu}}\, d_1^3 n_0 t \quad \text{—(4·62)}$$

$$\frac{dN_R}{dm} = \frac{1}{2} \sum_{i=1}^{R-1} \alpha_R [i^{1/(3-K_\rho)} + (R-i)^{1/(3-K_\rho)}]^3 N_i N_{R-i} - N_R \sum_{i=1}^{S-R} \alpha_{R+i} (i^{1/(3-K_\rho)} + R^{1/(3-K_\rho)})^3 N_i \quad \text{—(4·63)}$$

(4.61–4.63)

Equation 4.61 - 4.63 The Dimensionless of Increase or decrease of floc particles.

Figure 4.93 shows an example of solving Equation 4.63 with a numerical solution using the Runge-Kutta method, using the simultaneous equations of order $S = 50$. In this case, we proceed with the calculations

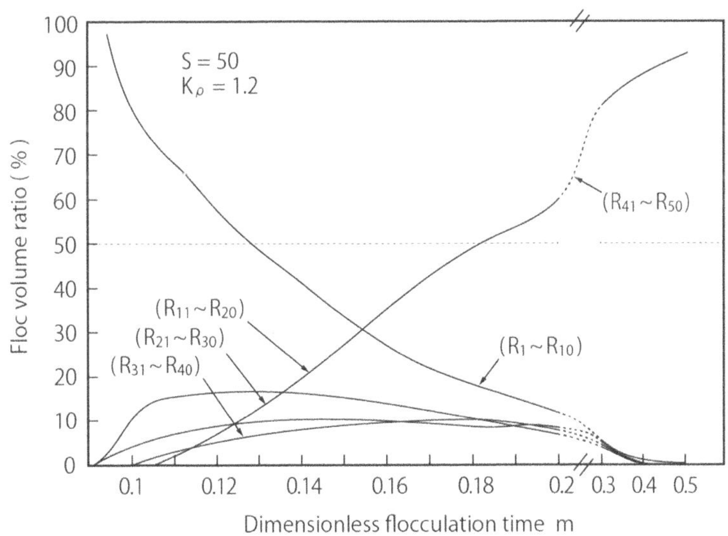

Figure 4.93 Dimensionless flocculation time.

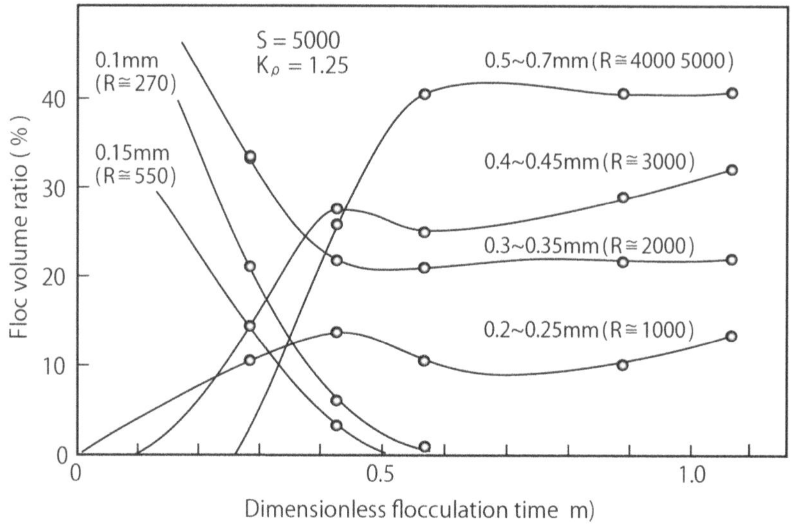

Figure 4.94 Actual results of changes in the number of particles in each size group.

assuming that $\alpha_R = 1$. However, the actual results of the changes in the number of particles in each size group in the floc-forming tank are as shown in Figure 4.94, and it is inferred that this difference does not take into account the changes in the collision probability α, and it is possible to assume that this difference does not take into account the changes in the number of particles in each group in the floc-forming tank.

For the collision coefficient α in Brownian motion, see Figures 4.95 and 4.96 (taken from Kruyt (1960)) for the solution to Smoluchowski (1917), but there are no descriptions of the collision coefficient α under isotropic turbulent conditions.

The probability of collision under turbulent flow conditions in an actual floc-forming vessel must be expressed taking into account the number of floc breakdown clumps S.

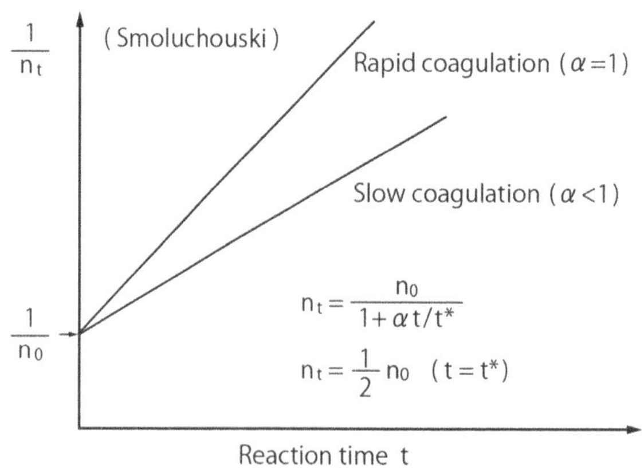

Figure 4.95 The number of particles decreases in rapid and slow coagulation (Kruyt (1960)).

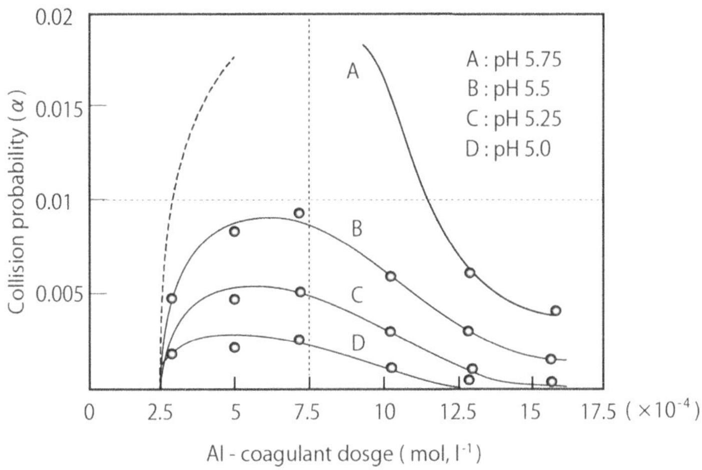

Figure 4.96 Collision probability and coagulant dosage (Kruyt (1960)).

Since there are no particles larger than an S-fold particle, the probability of collision with an S-fold particle is 0. The α_0 is a number determined by the degree of coverage of the coagulant particles in the suspension. An R-fold particle between a 1-fold (initial) particle and an S-fold particle will have a value in between. Therefore, the authors thought that the collision probability of R-fold particles can be expressed in the form of $\alpha_R = f(\alpha_0, S, R)$. Based on experimental observations such as in Figure 4.94, the authors thought that the collision probability can be expressed qualitatively in the form of Equation 4.64.

Trial and error application of various experimental results (particle size distributions, sedimentation cumulative removal rates) yielded n ≑ 6 as the best fit.

$$\alpha_R = \alpha_0\left(1 - \frac{R}{S+1}\right)^n \qquad \text{---}(4\cdot64) \tag{4.64}$$

α_R : Collision probability of R-fold particles

α_0 : Collision probability of initial particles

Equation 4.64 The collision probability equation.

Substituting the collision probability of Equation 4.64 into Equation 4.63 yields an equation (floc formation equation) for the dimensionless floc growth process shown in Equation 4.65.

$$\frac{dN_R}{dm} = \frac{1}{2}\sum_{i=1}^{R-1}\alpha_0\left(1 - \frac{R}{S+1}\right)^6 [i^{1/(3-K_\rho)} + (R-i)^{1/(3-K_\rho)}]^3 N_i N_{R-i} - N_R\sum_{i=1}^{S-R}\alpha_0\left(1 - \frac{R+i}{S+1}\right)^6 (i^{1/(3-K_\rho)} + R^{1/(3-K_\rho)})^3 N_i \qquad \text{---}(4\cdot65) \tag{4.65}$$

Equation 4.65 The dimensionless floc growth eqution.

4.2.6.4 Four parameters defining the flocculation process: m, S, K_p, and α_0

From Equation 4.65, we can quantify all processes if we know the four parameters of m, S, K_ρ, and α_0.

The m-value is a dimensionless flocculation (floc-forming) time, defined as in Equation 4.62, and is an index proportional to the 'GC_0T' value when expressed in the Camp G-value way, where C_0 is the (dimensionless) volume concentration of the early floc particles. This indicates that the progress of the flocculation reaction is defined by the GC_0T value, and the Camp proposed GT value is inadequate.

The S-value is given by Equation 4.66, which shows the number of aggregates (maxima) of the initial particle d_1 to the maximum particle d_{max} that can grow under ε_0 or G, where the maximum d_{max} of floc that can grow under the given stirring conditions is given by Equations 4.49 and 4.50, and Figure 4.90 (Tambo & Hozumi, 1979b).

$$S = S_m^{(3-K\rho)/3}, \quad S_m = \left(\frac{d_{max}}{d_1}\right)^3 \qquad \text{—}(4\cdot66) \qquad (4.66)$$

S : Number of aggregates of initial particles at maximum particle size
d_{max} : Maximum particle size
d_1 : Initial particle size

Equation 4.66 The number of aggregates of initial particles at maximum particle size.

The K_ρ value is the characteristic value of the above-mentioned floc density function (Tambo & Watanabe, 1979d, e), which is defined by Equation 4.38, and with aluminium clay floc it is given as shown in Figure 4.74.

The parameter α_0 is the probability of collisions of the primary particles and is likely to be governed by the (adsorptive) coverage of coagulant on the surfaces of the coagulation particles, such as clays. If we compare the measured ZP of clay aluminium floc with the results of numerical experiments (simulations), the coating rate may be considered to be about 20–30% under normal coagulation conditions, and it may be considered that $\alpha_0 \fallingdotseq 0.3$. We believe that the α_0 values of color floc and microbial floc should be further investigated.

4.2.6.5 Normalization of floc size distributions

Figure 4.97 shows an example of Equation 4.65 using the Runge-Kutta method to find numerical solutions for various S and K_ρ values. This is a cumulative particle size diagram in which the number of aggregates (R) of the primary particles is plotted in the horizontal axis, and the cumulative volume percentage is plotted in the vertical axis. The size growth of floc particles at a stage where α_0 m is small in the early stage of floc formation is remarkable, but when formation time progresses and the α_0 m enters a range of large values, the growth of the distribution slows down and finally reaches the maximum growth distribution (ultimate equilibrium floc size distribution) where the growth does not exceed a certain distribution.

The number of agglomerates $R_{50\text{-ult}}$ when the cumulative volume ratio of the floc group reaches 50% of the maximum growth distribution takes the value of the number of agglomerates $R_{50\text{-ult}} \fallingdotseq 42$ despite various coagulation conditions; the maximum floc particle size distribution under various coagulation conditions is standardized by $R^* = R/R_{50\text{-ult}}$ (normalization) using $R_{50\text{-ult}}$ as the denominator, and when the number of agglomerates R^* is converted into the floc volume ratio by considering the floc density function, it is ascribed to a single cumulative maximum growth floc volume distribution curve, standardized as shown in Figure 4.98 regardless of the difference between the maximum number of agglomerates S_m and the index K_ρ of the density function. This indicates that there are 'self- preserving characteristics' in the floc

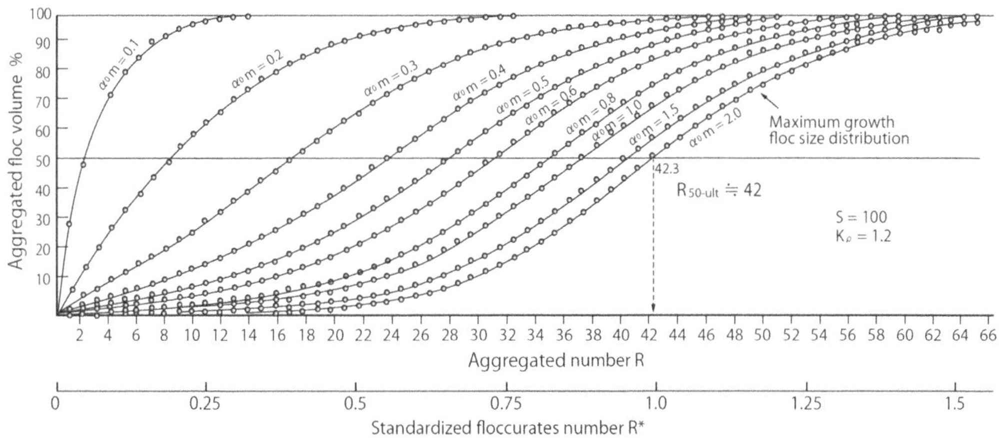

Figure 4.97 The flocculation time α_0 m and aggregated floc volume (particle size) distribution.

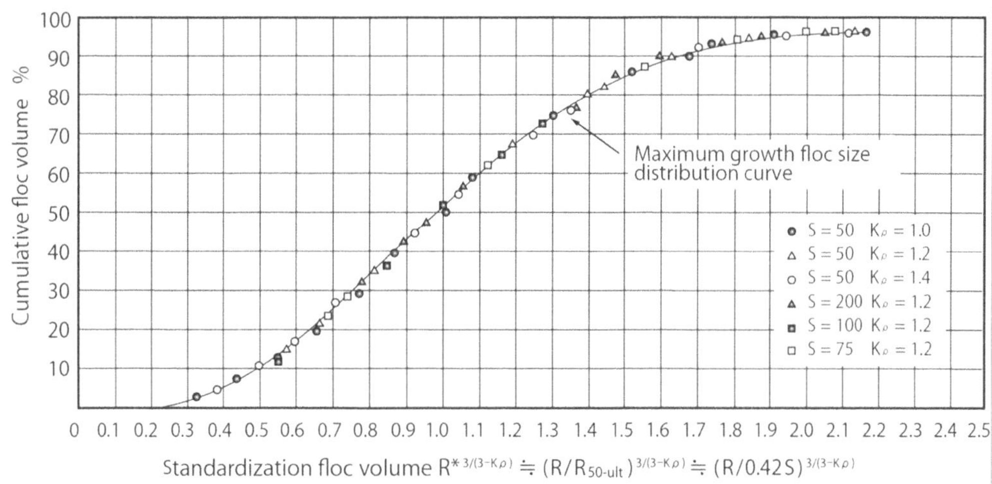

Figure 4.98 A standardized cumulative maximum growth floc volume distribution curve.

size distributions at the final stage of the floc growing process. Swift and Friedlander (1964) reported that the coagulation of hydrated sols in Brownian motion has a self-equilibrium.

In addition to obtaining such a single standardized maximum (ultimate) growth floc volume distribution curve, we also attempted dimensionless standardization of the distribution of the $R^*_{50} = X$ other than the floc group maximum growth curve $R^*_{50} = 1$, and the results of the numerical calculations performed by the authors are shown in Figure 4.99. The results indicate that all standardized cumulative floc volume curves in the floc formation process using R^*_{50} as a parameter are represented by the same cumulative curve regardless of the K_ρ and S_m of the coagulation and flocculation conditions.

R^*_{50} is the normalized clump count R^* when the cumulative floc volume reaches 50%. The dimensionless floc-forming time α_0 m $\propto GC_0T$ to reach the volume distribution of the floc group shown in R^*_{50} and is a function of the maximum growth cluster S_m of floc. The results of simulation

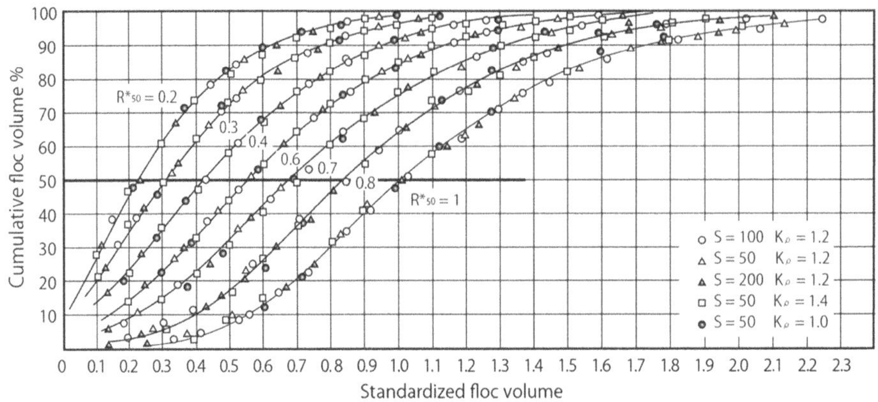

Figure 4.99 Standardized cumulative floc volume curves.

calculations of floc-forming processes under various conditions are shown in Figure 4.100, and the relation of Equation 4.67 has been found.

$$m \sim S_m^{-\gamma} \quad \text{—(4·67)} \tag{4.67}$$

m : Dimensionless flocculation time

S_m : Maximum growth aggregated floc number

Equation 4.67 Relationship between m and S_m

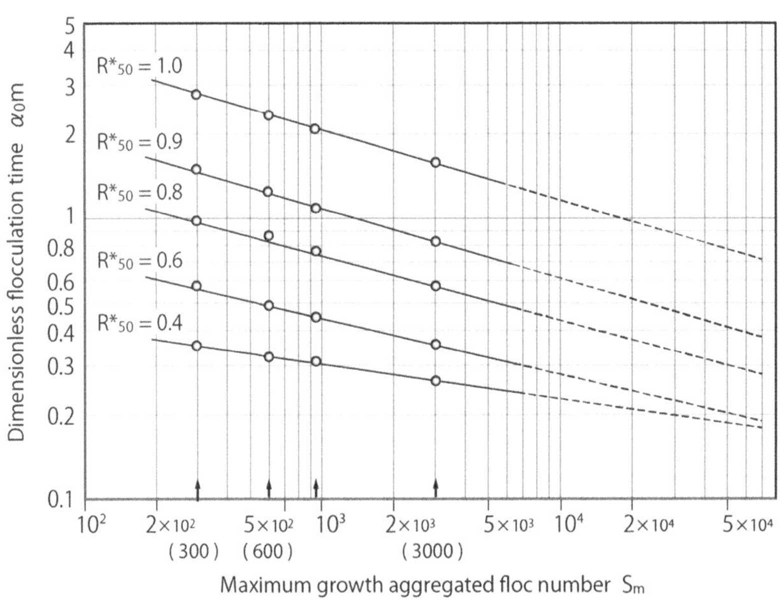

Figure 4.100 Dimensionless flocculation time and maximum growth aggregated floc number.

4.2.6.6 *Experimental verification of the theory*

The flocculation theory described above was tested by batch floc formation experiments with a model flocculator. The square water tank shown in Figure 4.101 was equipped with a large test device with a horizontal shaft paddle-type stirrer. In order to check the progress of floc formation, water was sampled at intervals without disturbance while the system was operating, guided into the transparent cell, and the floc particle group was photographed to obtain the particle size distribution. The agitation strength was determined by the same agitation tank strength evaluations as in Figure 4.87 and Table 4.17 (Tambo & Hozumi, 1979b). The relation between effective agitation strength ε_0 and Nr^3 was clearly observed. The effective agitation intensity ε_0 ranged from 10^{-2} to 2×10^0 (erg/cm^{3}sec) when the paddle agitation rotational speed Nr is between 3 and 20 rpm.

The experimental results are presented here in terms of the cumulative sedimentation rate of an idealized sedimentation pond rather than cumulative particle size distribution (see Chapter 6, Figure 6.7 for the sedimentation behavior of ideal sedimentation ponds and Figure 6.9 for cumulative sedimentation rate distribution curves). Figure 4.102 shows the cumulative sedimentation removal rate curves for floc groups grown maximally with standard dimensionless sedimentation times θ* in the horizontal axis and cumulative removal rate q (%) in the vertical axis.

Where the dimensionless sedimentation time $\theta = (t_i/t_1) \times 100$ (%), t_1 is the time (sec) required to settle the early (one-fold) particle down to the sedimentation pond bottom and t_i is the time (sec) required to settle the i-fold particle. When the cumulative sedimentation curves at all maximum growth times are expressed as standardized sedimentation times $\theta^* = K^*\theta$ multiplied by the correction factor K*, the cumulative sedimentation curves of one standardized maximum growth floc group are obtained.

Figure 4.103 shows the values of K* coefficients for standardization, which are determined by the maximum (initial particle) agglomeration factor S_m for floc particles, obtained from the results of floc-forming experiments on clay-aluminium flocs. Numerical experiments during the previous theoretical results and discussions have confirmed that the same auto-equilibrium distributions exist experimentally in the maximum growth floc group, even if the batch floc forming experimental results are represented by sedimentation curves in the same manner as in the combinations of Figures 4.98 and

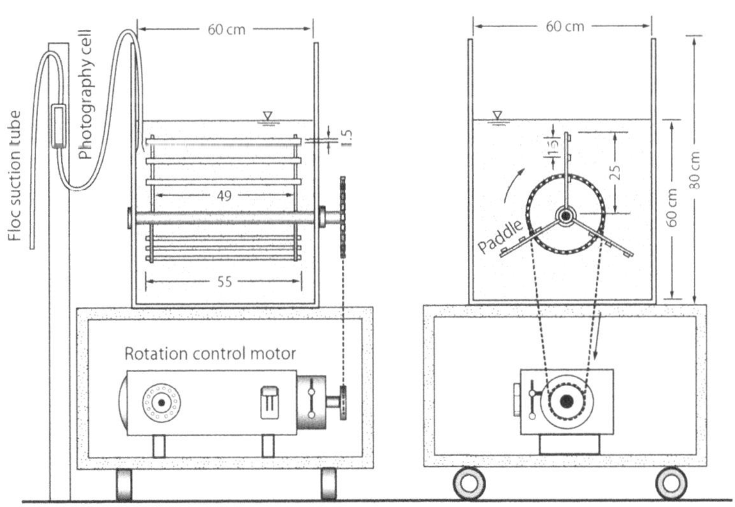

Figure 4.101 A batch paddle-type flocculator test device.

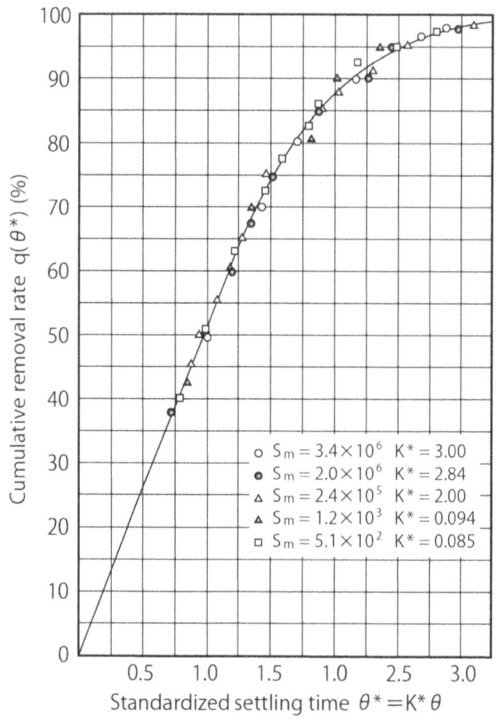

Figure 4.102 A cumulative removal rate curve for maximum growth floc groups.

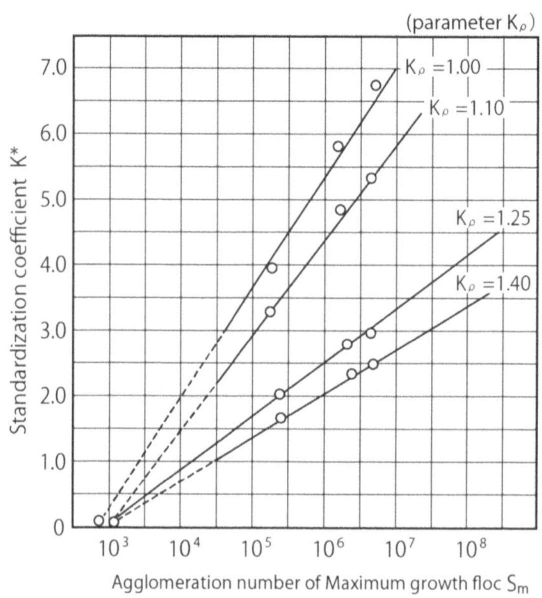

Figure 4.103 Maximum growth floc agglomeration and standarization coefficient.

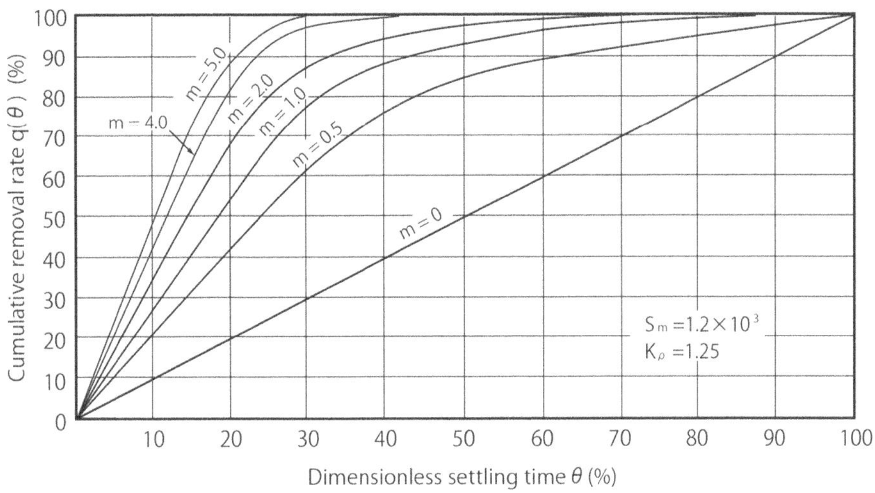

Figure 4.104 Cumulative settling curve.

4.100, and by Equation 4.67 which provides a single standard maximum growth curve. Figure 4.102 shows one standard removal rate cumulative curve for a normalized time sedimentation in which $\theta^* = K^*\ \theta$ at which the maximum growth equilibrium particle size distribution has been reached. The relation between the correction factor K^* for standardization and the maximum (initial particle) agglomeration factor S_m for floc particles is shown in Figure 4.103.

Figure 4.104 shows the results of photographic measurements of how the curve of the cumulative removal rate $q(\theta)$ (%) progresses toward the maximum growth curve by taking the dimensionless settling time θ (%) on the horizontal axis and using various agitation durations (dimensionless agitation duration indicator m). At initial state $m = 0$, the cumulative sedimentation curve is straight but, with increasing stirring time m, the clot progresses to the maximum growth curve (as shown in Figure 4.102). According to Figure 4.104, the maximum growth equilibrium distribution is almost reached with a dimensionless stirring time of $m = 5.0$.

Figure 4.104 shows the relationship between the cumulative removal rate $q(\theta)$% and the dimensionless flocculation time $\theta = t_i/t_1$. When the relationship between $\theta^* = K^*\theta$ is introduced and the cumulative settling removal rate $q(\theta^*)$ is obtained by standardization, the various dimensionless cumulative precipitation curves are attributed to a single curve as shown in Figure 4.102. Since it was found that the standardized dimensionless cumulative settling curve of the maximum growth floc group can be represented by the same curve under all coagulation floc forming conditions, the floc distribution during growth of other agitation times m is also standardized as $\theta^* = K^*\theta$ so that the maximum growth equilibrium distribution becomes the same curve. Whether standardization could be carried out using the same procedures in these cases as in the case of maximum growth is now examined.

Therefore, the sedimentation curve was rewritten as shown in Figures 4.105, 4.106, and 4.107 by expanding and contracting the lateral axis by K^* in units of standardized sedimentation time $\theta^* = K^*\ \theta$, which corresponds to the sedimentation time θ of the cumulative sedimentation curve when the removal rate of the cumulative sedimentation curve (curve m) reaches 50%. In that case, the standard agitation time (parameter) mK^* corresponding to the dimensionless floc forming time m is calculated, and the standardized dimensionless agitation duration $q^*_{50.max} = 1$ is defined as the maximum growth floc group.

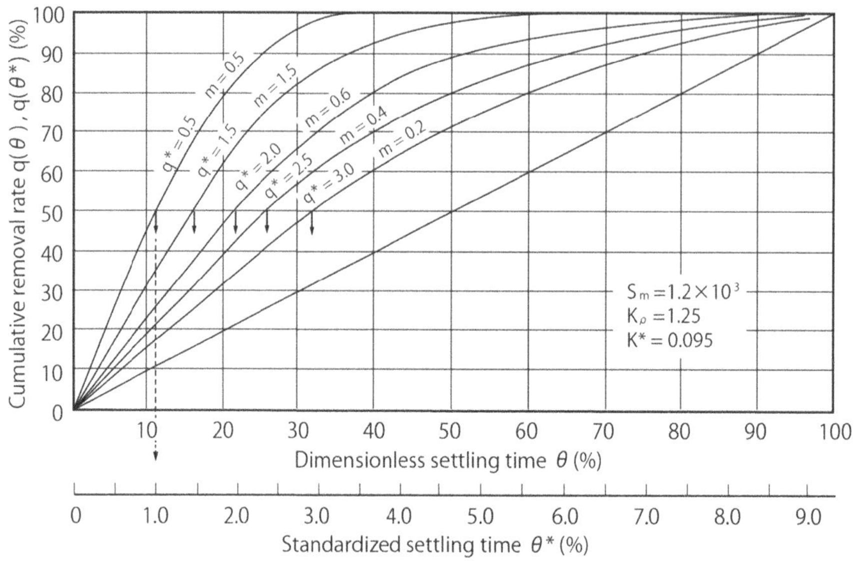

Figure 4.105 A standardized cumulative settling curve (measured value).

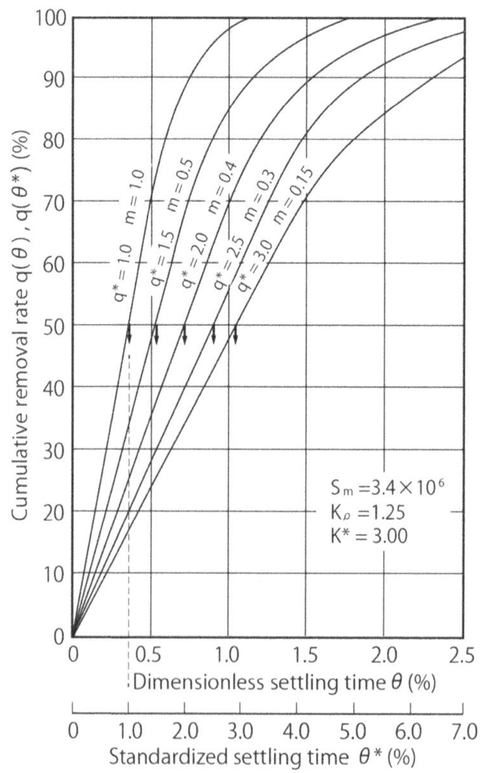

Figure 4.106 A standardized cumulative settling curve.

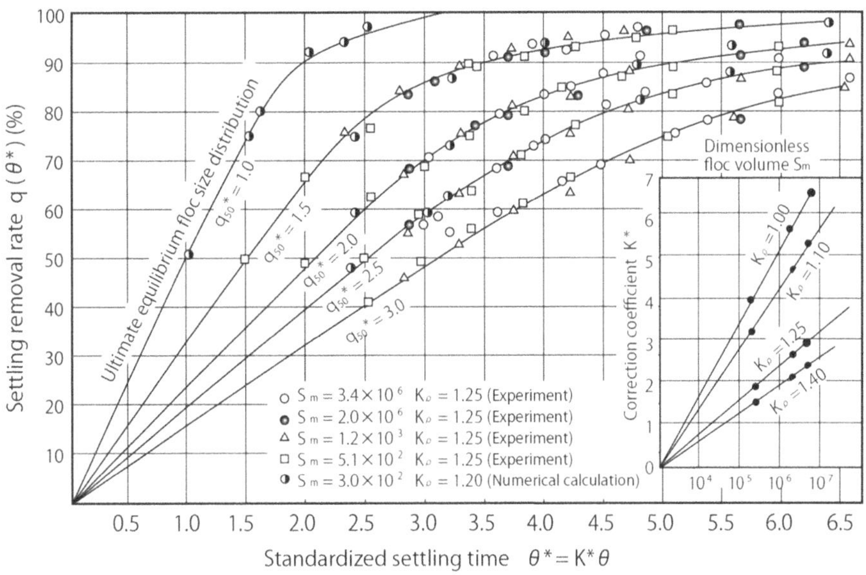

Figure 4.107 A dimensionless general purpose flocculation operation diagram (M-curve).

Figure 4.107 shows a standardized cumulative sedimentation curve [$\theta^* = K^*\theta$ vs $q(\theta)$ or $q(\theta^*)$] obtained under various coagulation conditions. The cumulative sedimentation curve $q^*_{50\max} = 1$ for the largest growing floc group was used as a ruler, and the lateral axis of each curve group was synthesized by expansion and contraction at standardized sedimentation times θ^*. In a similar way to how the largest growing particle group is attributed to a single self-equilibrium cumulative curve, the cumulative sedimentation curve group during growth has the same θ^* with a 50% removal rate (the cumulative sedimentation curve group was standardized with q* as a dimensionless stirring time) is also standardized with q* as a parameter. In this case, the normalization factor K* (measured value) is shown

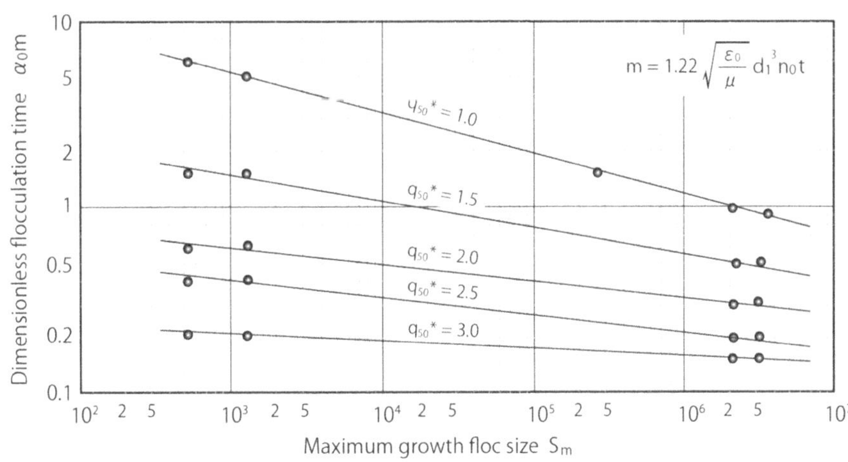

Figure 4.108 Flocculation time m and maximum growth floc size S_m ($\alpha_0 = 0.3$).

in Figure 4.107 using the index K_ρ of the density function as a parameter, as a function of the aggregation factor S_m of the initial particles to the maximum growth floc. In this case, the relation between the dimensionless floc formation time m and the maximum growth rate of the floc (initial particle agglomeration number) S_m is shown in Figure 4.108 based on actual results. From the maximum growth rate of the floc (agglomeration number S_m), it is possible to determine the dimensionless floc formation agitation duration m required to form a floc group of a distribution q*.

4.2.6.7 Design of a flocculator using a flocculation process control diagram

The theoretical examination and experimental validation described above enable the theoretical design of the floc formation process using the following procedures: Figure 4.107 (A dimensionless general purpose flocculation operation diagram), Figure 4.108 (Flocculation time m vs. maximum growth floc size S_m), Figure 4.109 or Figure 4.90 (Floc formation agitation intensity G or ε_0 vs. maximum grown floc size d) and Figure 4.74 or Figure 4.75 (Floc density function characteristics: Index diagram).

The flocculation process can be described by the following four dimensionless indices: m, S_m, α_0, and K_ρ.

(1) The m value of the dimensionless index is obtained by the progression index of the dimensionless floc formation degree using Equation 4.62 (originally shown above):

$$m = 1.22\sqrt{\frac{\varepsilon_0}{\mu}}\, d_1^3 n_0 t$$

$$\fallingdotseq 1.22\sqrt{\frac{0.1\,\varepsilon}{\mu}} \cdot \left(\frac{6}{\pi}\right) C_0 t \fallingdotseq 0.723\, G C_0 t \quad -(4 \cdot 62) \qquad (4.62,\ \text{again})$$

$$\text{or } G C_0 t \fallingdotseq 1.38\, m$$

Equation 4.62 (Reprint) Dimensionless flocculation time (m).

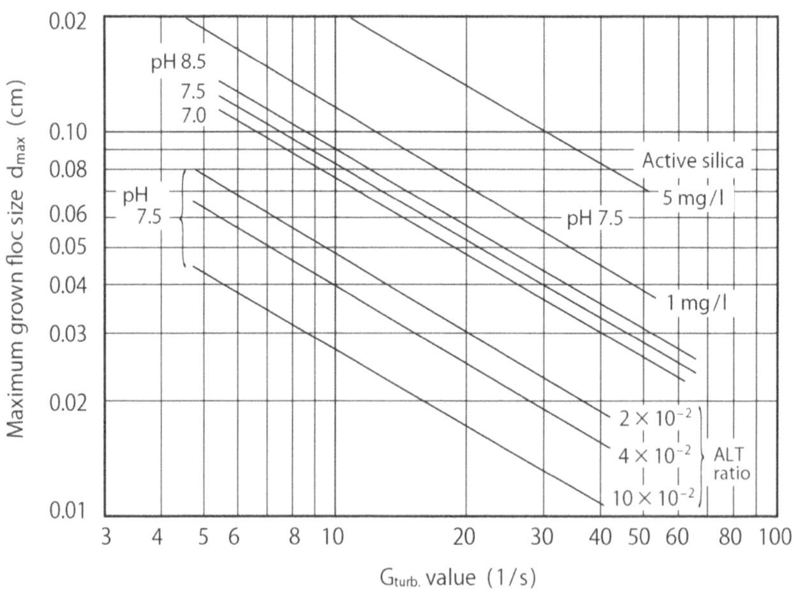

Figure 4.109 Agitation intensity ($G_{turb.}$value) and maximum grown floc size (d_{max}).

(1) The dimensionless index S, S_m is expressed as Equation 4.66 by the number of aggregates of initial particles or the volume ratio of initial particles to the maximum floc that can be grown by the flocculator.
(2) The maximum growth floc diameter d_{max} is expressed by Equation 4.49 or Equation 4.50.

In many cases, assuming that the floc size is in a viscous range smaller than the microscale λ_0 of locally isotropic turbulence, it seems reasonable to consider Equation 4.50. However, as shown in Figure 4.90 above, a floc of high bond strength produced using a coagulation adjuvant or polymer coagulant may be in an inertial range.

As mentioned above, Camp's proposed G and GT values are inadequate to describe the flocculation process.

We need to proceed with the design by using a diagram of the GC_0T value expressed in the form of the GT value which is accustomed to using agitation intensities or correctly using the m-value of the dimensionless floc-forming progress taking into account the C_0 of the volume density of the primary floc group proposed by Tambo and Watanabe, or the $G_{turb.}$value $= (\varepsilon_0/\mu)^{1/2}$ defined under turbulent flow. Assuming that about 10% of the total energy consumption rate is the effective energy consumption rate $\varepsilon_0 \doteqdot 0.1\varepsilon^*$, the process chart explores how far floc formation should proceed using the dimensionless standard value m or the $G_{turb}C_0T$ value in order to determine the agitation time T necessary for the floc group to reach maximal growth equilibrium.

The necessary dimensionless floc-forming agitation time index m(ult) and the $G_{turb}C_0T$(ult) value may vary depending on how many times the initial particle joins the maximally grown floc (S_m value) in the flocculator. Normal flocculation involves operation to agglomerate a few micrometers of microfloc into a few hundred micrometers of well-grown floc. This corresponds to calculating the agglomeration of $S_m \doteqdot 10^6$. In this case, it can be estimated that the required $G_{turb}C_0T$ value will be about 1.7 and the m value will be about 1.2, based on Figure 4.110, which calculates the relation between the m-value or the $G_{turb}C_0T$ value and the S_m value required to reach the maximum growth level.

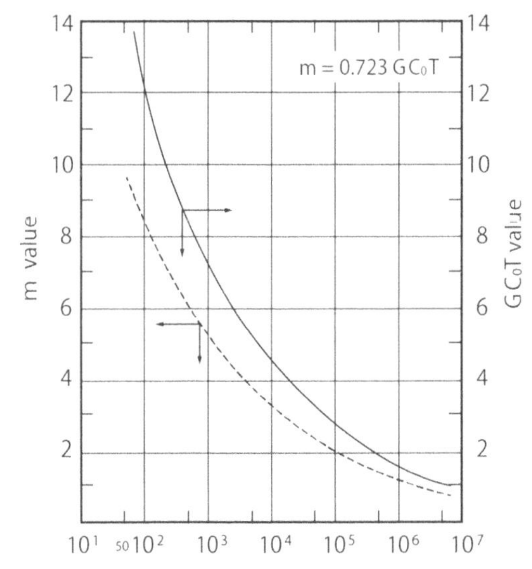

Figure 4.110 Dimensionless flocculation time and maximum growth aggregated floc number.

For low turbidity raw water with turbidity up to about 20 T.U., it is reasonable to use a direct filtration method without a sedimentation pond (or only through the pond) that minimizes the ALT ratio and suppresses increased head losses in the sand filter bed and turbidity breakthrough over time. Direct filtration also requires that the filtration be performed with floc appropriate to the process. The ALT ratio is preferable as small as 0.005 to 0.0025 to produce a floc as small as 20 to 30 μm. Such a floc-forming process is called micro-flocculation. Referring to Figures 4.108 and 4.109, Figure 4.110 shows that the dimensionless agitation time $G_{turb}C_0T$ value required to reach the maximal growth-equilibrium profile requires a value of about 7 for $S_m \doteqdot 10^3$.

For example, if $S_m = 1.2 \times 10^3$ in Figure 4.105, the m-value that leads to the maximum growth is about 0.2, and if $S_m = 3.4 \times 10^6$ in Figure 4.106, the m-value is about 1.

Using Figure 4.110, the $G_{turb}C_0T$ or m values that lead to maximal growth are known, and then the volume concentration C_0 or $d_1^3n_0$ of the initial floc group inflow from the rapid mixing tank must be known in order to determine the actual agitation duration T (the Retention Time of floc formation values) for the design.

There are several different ways to do this:

(1) Calculate the diameter d_1 (assuming a uniform diameter) and the number density n_0 (assuming a uniform distribution) of the initial floc (rapidly mixing pond outflow microflocs). Recently, instruments that automatically measure particle size and count are available at reasonable prices. If they can be maintained, they are useful not only for daily operation management but also for basic design data.

(2) A simple estimation of the C_0 from previous findings, with the examples of measurements, is also helpful on site.

Photographic measurements were performed on a group of particles in which 1 mg/l of aluminium was added to a 25 mg/l clay suspension for rapid mixing. The mean particle size $d_1 \doteqdot 4 \times 10^{-3}$ mm, and the number density at that time was $n_0 \doteqdot 1.1 \times 10^6$ (cm^{-3}). Therefore, the volume concentration of the initial floc particles $C_0 = 3.4 \times 10^{-5} \doteqdot 0.003\%$, and the relation between the turbidity C_T (mg/l) and the volume concentration C_0 (dimensionless) is shown as $C_0 \doteqdot 1.35 \times 10^{-6}C_T$. From Figure 4.109, the effective agitation strength $G_{turb} = (\varepsilon_0/\mu)^{1/2}$ giving a required maximum growth diameter of floc $d_{max} \doteqdot 0.5$ mm which can be obtained at a rate of $20 s^{-1}$. The maximum agglomeration ratio $S_m = (d_{max}/d_1)^3 \doteqdot (500/4)^3 = 1.7 \times 10^6$. From Figure 4.110, the dimensionless flocculation time required to reach maximum growth equilibrium is $G_{turb}C_0T \doteqdot 1.0$, and from Equation 4.62, $T \doteqdot 1.0/G_{turb}C_0 = 10^6/\ (20 \times 34) \doteqdot 1470\ s \doteqdot 24$ min.

Camp used a GT-value of about $20 \times 1470 \doteqdot 30{,}000$. When similar calculations are performed at concentrations as low as 5 mg/l, $T = 10^6/(20 \times 7) \doteqdot 7140\ s \doteqdot 120$ min, and the GT value is as large as $20 \times 7140 \doteqdot 142{,}800$, which is a 4.5-fold difference between the two, and the GT value cannot be regarded as a practical and reliable process/design index.

Since Camp did not consider the floc volume concentration C_v, he proposed a wide range of recommended GT values, ranging from 23,000 to 210,000, which has long been considered an indicator of floc formation, but it cannot be a meaningful design criterion. As proposed here and elsewhere by Tambo, by introducing the initial volume concentration C_0 of the suspensions, it is possible to define the definitive dimensionless design index $G_{turb}C_0T$ as described above and to clearly define the design procedures.

(3) A third approach is to determine the C_0 from the volume ratios of clay and aluminium. In the neutral pH range, insoluble aluminium hydroxide accounts for most of the floc volume. It has been observed that 1 mg/l of aluminium hydroxide insolubilized in the neutral region occupies about

55 times the volume of 1 mg/l of clay (Matsui & Tambo, 1990). The procedures are that the volume of 1 mg of clay with the density of 2.6 g/cm^3 is about 4×10^{-4} cm^3, therefore, the volume of 1 mg/l of insoluble aluminium hydroxide is considered to occupy about 50 times the volume of 2×10^{-2} cm^3. Therefore, in application with the previous coagulation condition, in which 25 mg/l of clay suspension is added with 1 mg/l of aluminium, the volume concentration of the original floc $C_0 \doteqdot 3 \times 10^{-3} = 0.003\%$ by adding the volume concentration of aluminium and clay, respectively. This volume ratio is the same as the actually measured volume ratio in the microphotograph in the previous section, and it is recognized that two estimation methods are certainly used. This third estimation method can be used in general as a method for calculating the C_0 value necessary for calculating the index GC_0T value in the design of the coagulation and floc-forming process of a wide range of clays with differing ALT ratios.

(4) The initial volume fraction of coagulated colloid flocs with aluminium can be considered to be approximately equal to the initial volume fraction of aluminium only precipitated coagulum in the neutral pH region. However, when colored aluminium flocs are produced in the weakly acidic region, the volume fraction is much smaller than in the neutral region.

(5) Consider the method of estimating the volume density C_0 of an early floc by the addition method for a micro-flocculation process such as direct filtration. In such cases, if the microfloc diameter is assumed to grow from 3 μm to 30 μm, $S_m = 10^3$. The dimensionless agitation time GC_0T required for flocs to reach their maximal growth (equilibrium) distribution is approximately 5, from Figure 4.110. If we extrapolate $d_{max} \doteqdot 30\,\mu m$ from Figure 4.109, the design agitation intensity $G_{turb} \doteqdot 5 \times 10^2$ would be set.

Direct filtration involves coagulation of 10 mg/l of a low turbidity clay suspension at an ALT ratio as low as 0.005, followed by filtration. The initial floc volume density C_0 is estimated to be approximately 5×10^{-6} by the estimation method described above. Therefore, the duration of agitation required for micro-flocculation (the required retention time of the floc-forming tank), $T \doteqdot 5/GC_0 \doteqdot 5/(5 \times 10^2)(5 \times 10^{-6}) \doteqdot 2000\ s \doteqdot 33\ min$.

For floc formation under conditions of extremely low ALT ratio even in the formation of microflocs, an agitation time close to that of normal floc formation is required. Direct filtration often causes problems with bacterial and ultrafine particulate leaks. This indicates that even if the turbidity is low, sufficient floc formation time is required to obtain a steady reasonable operational result. On the other hand, if long agitation is applied at a high G value, interface deterioration of the floc may occur, but quantitative evaluation of this is not yet sufficient, and it requires further research.

Floc size distribution (cumulative precipitate velocity distribution) should be known when flocculation is discontinued prior to maximal equilibration.

In cases where floc growth distribution is described as shown in Figure 4.107, in which the curve $q^* > 1$, we can calculate (Using Figure 4.108) the normalized sedimentation time versus percent sedimentation removal by using the generalized dimensionless floc formation curves in Figure 4.107, using the maximum-growth-number S_m and the dimensionless floc formation duration $m = G_{turb}C_0t$ (where t is the agitation duration) and using the q^* corresponding to the intersection of the two values as parameters.

Many floc-forming ponds (flocculators) are composed of multiple stages of agitating tanks. Initially, when the size of the floc is small, operation is recommended at high agitation strength (high G value) to increase the collision speed without worrying about breakage, and when the floc becomes large, agitation strength is decreased (low G value) and tapered floc forming (tapered flocculation) is performed to avoid breakage. Stepwise agitation from strong ($G \doteqdot 30\ s^{-1}$) to moderate ($G \doteqdot 10 s^{-1}$) to weak ($G \doteqdot 3{\sim}5 s^{-1}$) has been designed. It is difficult to process all of them using the author's method, but it requires sequential handling, such as the GC_0T_1, the GC_1T_2, and the GC_2T_3. The second and subsequent stages of the floc-forming tank do not have C_0, but as indicated by floc density functions, the number of aggregates of particles in one floc increases, the structural density decreases, and the C_1 or C_2 increases accordingly, resulting in an increase in the apparent volume. As a result, the agglomeration rate increases.

4.2.7 Contact flocculation

Contact flocculation is a process to rapidly reduce microflocs (inflow suspensions) in a system in which high-diameter flocs are suspended, and by adhering microflocs onto large-diameter flocs (grown flocs). A floc-forming process that is commonly used in floc blanket type sedimentation ponds (Figure 6.51) or slurry-circulating type high-speed sedimentation ponds (Figure 6.52), as described in Chapter 6, has similar phenomena in floc-forming ponds with contact floc-forming and partial back-mixing with return sludge in the activated sludge process.

4.2.7.1 Contact flocculation in a turbulent stirring tank

For the separation behavior of a contact high-speed coagulation sedimentation pond, the fluidized bed separation process is described in Equations 6.44 to 6.47 and by Figure 6.59 for the upstream sedimentation pond in Chapter 6. Here, contact floc formation under turbulent stirring conditions is described. The flocculation process (as contact coagulation sedimentation) in an accelerator sedimentation pond is typical.

Introducing microflocs where high concentrations of grown flocs are in suspension promotes the collisional attraction of microflocs to the grown flocs in a short period of time. For the group of large flocs with a high concentration that has reached the maximum grown floc particle size distribution as shown in Figure 4.107, defined above, and for the group of maximum grown flocs whose distribution can be specified, the collision of microflocs can be theorized.

In a contact high-speed sedimentation pond, a large GC_0T value with the highest-grown floc group volume can be maintained at all times, so that the agitation strength G and the duration T can be reduced by an amount corresponding to the large C_0. By reducing G, the maximum growth floc diameter d_{max} can be large, as shown in Figures 4.19, 4.49 and 4.50, and the overflow rate of the subsequent sedimentation separation devices can be large, enabling high-speed separation. Regarding the duration of agitation, the required time after discounting the amount of G reduced is calculated from the required GCT value. This is the process design of high-speed coagulation sedimentation ponds (see Figure 4.111).

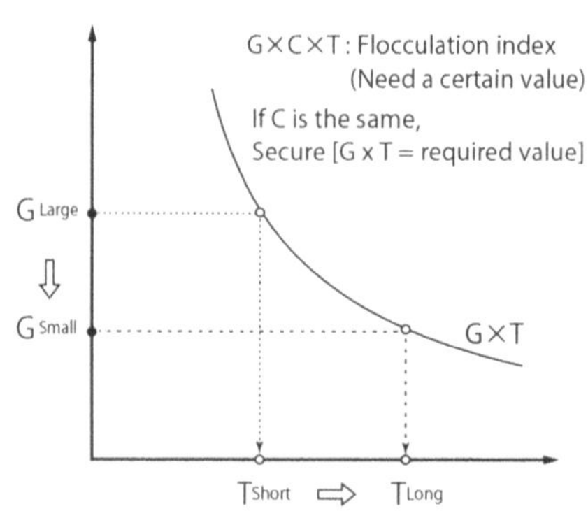

Figure 4.111 The operation design principle of high-speed coagulation sedimentation ponds.

Practically, the largest growing particle size d_{max} of floc under turbulent agitation conditions in contact floc-forming ponds ranges from (3 to 5) × 10^{-1} mm. On the other hand, the diameter d of the micro-floc particles coming into inflow with the contact agitation pond is about 5 to 20 μm, (i.e. 5 to 20 × 10^{-3} mm). The number of collisions, F, between the grown flocs and the microflocs in a unit time, unit volume fluid is shown in Equation 4.68 by Tambo and Hozumi (1979b), based on the Levich equation (1962).

$$F = \frac{12}{\sqrt{15}} \pi \sqrt{\frac{\varepsilon_0}{\mu}} \left(\frac{D}{2} + \frac{d}{2}\right)^3 Nn \quad \text{---(4·68)} \qquad (4.68)$$

D,d : Each size of grown flocs and microflocs (cm)
N,n : Each particle concentration of grown flocs and microflocs (cm^{-3})
ε_0 : Effective energy consumption rate ($erg/cm^3 \cdot sec$)
μ : Viscosity (g/cm·sec)

Equation 4.68 The collision numbers between grown flocs and microflocs.

When these extremely different particle size groups are agitated:

(1) Growth flocs have reached the maximum growth equilibrium floc particle size distribution, and growth between grown flocs is not expected.
(2) It is considered that the collision diameter of the microflocs is much smaller than the collision radius (D + d)/2 of the grown floc and the microfloc because of the micro collision diameter, and is almost negligible.
(3) For the reason that the sizes of the two particle sizes D and d are extremely different, d ≪ D, the collision radius can be approximated as (D + d) ≑ D and the volume concentration within a certain time of the floc group is able to assume $V_f \propto ND^3 \fallingdotseq Const.$
(4) Number of collisions F between the group of maximally grown flocs and the microflocs generate effective agglomeration of two group flocs with the rate of P*F at which multiplied by the collision agglomeration probability P* for the contact flocculation of the microflocs.

Considering these conditions, the number concentration of microflocs d is reduced as Equation 4.69 derived from Equation 4.52. as the basic contact flocculation formula.

$$\frac{dn}{dt} = -\frac{3\pi}{2\sqrt{15}} P^* \sqrt{\frac{\varepsilon_0}{\mu}} ND^3 n \fallingdotseq -\frac{9}{\sqrt{15}} P^* \sqrt{\frac{\varepsilon_0}{\mu}} V_f n$$

$$= -K_{turb}\, n \quad \text{---(4·69)} \qquad (4.69)$$

V_f : Volume of grown flocs (mother floc) group = $(\pi/6)ND^3$
K_{turb} : Contact flocculation reaction coefficient (Turbulent mixing) = $9\sqrt{15} P^* \sqrt{\varepsilon_0/\mu}\, V_f$

Equation 4.69 The Contact flocculation basic equation.

Equation 4.52 showed that the contact floc-forming process is the first-order response for the number of microflocs n. The K_{turb}: rate constants of the first-order reactions are proportional to the product of $V_f(\varepsilon_0/\mu)^{1/2}$ and, as a result, the degree of flocculation is governed by the GCT-value. Equation 4.52 is integrated under the condition that the number of particles $n = n_0$ at the initial time $t = t_0$, and the residual rate n/n_0 of the number of microflocs n achieved at time t is obtained. Equation 4.70 shows the removal rate of microflocs $r = 1 - (n/n_0)$

$$\frac{n}{n_0} = \exp\left(-\frac{9}{\sqrt{15}} P^* V_f t\right) = \exp(-K_{turb} t) \quad \text{— (4·70)} \tag{4.70}$$

n : Number of microflocs at contact time t
n_0 : Number of initial microflocs

Equation 4.70 The removal rate of microflocs at time t.

The number of microflocs n is proportional to the turbidity of the microfloc group. If the turbidity T^* of the supernatant that has been briefly allowed to settle by undisturbed sampling in the floc-forming process into a test tube is measured, and the ratio of the turbidity at the time t of sampling T_t to the initial turbidity T_0^* at $t = 0$ is determined, it is easy to estimate the progress of the reaction and its constants from the relation of $T^*/T_0^* \propto n/n_0$.

Figure 4.112 shows the test results plotted for a batch floc forming tank, in a process as follows:

(1) The kaolin flocs coagulated with aluminium sulfate (alum) were stirred continuously for a long time (40 min) to form flocs that reached the maximal growth-equilibrium distributions;
(2) Particle size distributions were obtained by photographic measurements, and the volume density V_f of the grown floc group was calculated;
(3) In a separate jar tester, microflocs produced under the same coagulation condition were added to a batch floc-forming vessel in which large flocs with maximal growth-equilibrium distributions were suspended;
(4) Water was sampled at regular intervals for a certain period of time, and the turbidity of the supernatant after settling for a short period of time (residual microfloc concentrations) was measured;
(5) The contact floc forming agitation time t was taken in the horizontal axis, and the turbidity T^* indicated by the microflocs was taken in the vertical axis.

As can be seen from Figure 4.112, the concentration of microflocs T^* decreases in proportion to the agitation time t, and the linear decay relationship of Equation 4.52 or Equation 4.69 is established. Figure 4.113 shows that the reaction constants K_{turb} of the first-order decay equations described above are proportional to the volume density V_f of the grown flocs.

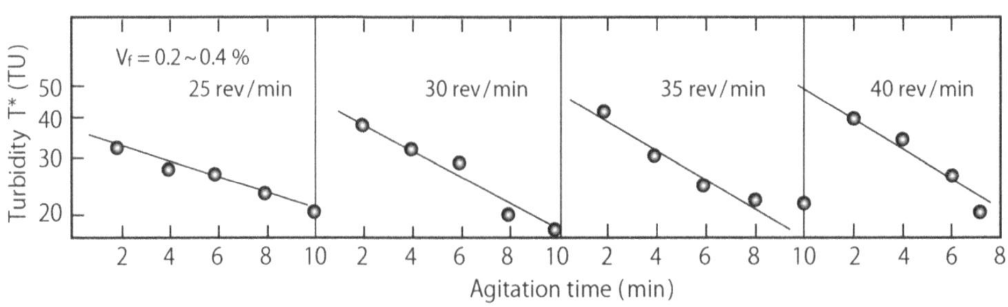

Figure 4.112 Agitation time and decrease in microfloc concentration.

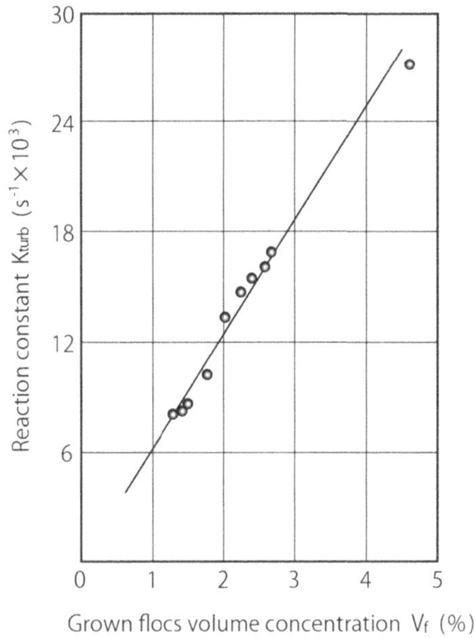

Figure 4.113 Reaction constant K_{turb} and grown flocs volume concentration V_f.

Figure 4.114 shows that K_{turb}/V_f is linearly related to the $N_r^{3/2}$ of the stirring blade rotation rate. Since the energy consumption rate $\varepsilon_0 \propto \varepsilon^*$ under the isotropic turbulent condition is proportional to the stirring blade rotation speed Nr^3, the relation of Equation 4.52 can be proved.

Next, the magnitude of the collisional agglomeration coefficient P*between the group of growing flocs and the microfloc particles is a matter of concern. With reference to the results of the operations of

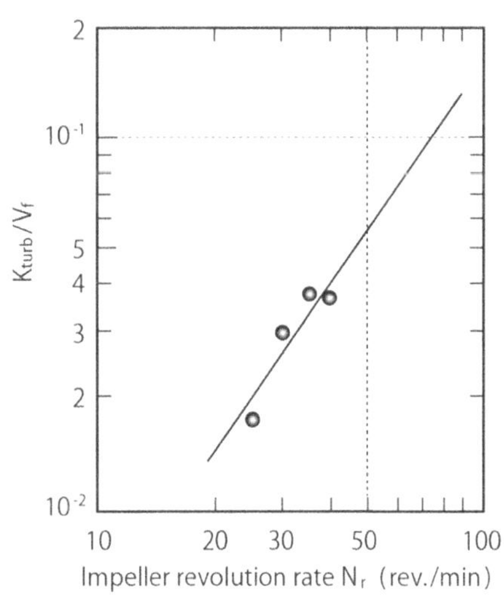

Figure 4.114 The stirring blade rotation rate and contact flocculation collision coefficient under turbulent flow.

Equations 4.64 and 4.65, the establishment of a collision between the maximum growing particle group and the microflocs can be described as in Equation 4.71 as has been discussed in the preceding section.

$$\bar{P} = \alpha_0 \sum_R X_R \left(1 - \frac{R}{S+1}\right)^6 \qquad \text{—(4·71)} \tag{4.71}$$

α_0 : Initial collision agglomeration probability (Dimensionless, In most cases around 1/3)
R : Initial particles agglomeration number to growth flocs (-)
S : Number of initial particles agglomeration to maximum grown flocs under certain mixing conditions (-)
X_R : R-times particles existence ratio in grown flocs group at ultimate (maximum grown) equilibrium distribution

Equation 4.71 The collision probability of Maximum grown particles and microflocs.

Comparison with experimental data from a batch contact flocculation process using the equipment of Figure 4.87, which was conducted by computer simulation on the group of grown (mother) flocs that reached the final equilibrium distribution, reveals the probability of collision at the time of the contact floc formation process under turbulent agitation, it is estimated that $P^* \doteqdot (2.8 \text{ to } 1.9) \times 10^{-2} \doteqdot 1/35 \text{ to } 1/53$.

In this case, it is sufficient to assume 10 to 20% of the total energy consumption rate ε^* measured by a torque meter, and to calculate the effective agitation energy $\varepsilon_0 \doteqdot 0.15\varepsilon^*$, and to round out the obtained results; a relatively small value such as $P^* \doteqdot 1/45 \doteqdot 0.022$ is assumed as the collision agglomeration probability between the mother floc (group of maximally grown particles) and the inflow microflocs in the turbulent contact floc-forming process.

When the above numerical values and the equation of microfloc concentration decreases in the turbulent flow contact floc forming process are summarized, the practical equation is given by Equation 4.72.

$$\frac{dn}{dt} = -\frac{1}{5\sqrt{15}}\sqrt{\frac{\varepsilon_0}{\mu}}\, V_f n = -K^*_{turb} n \qquad \text{—(4·72)} \tag{4.72}$$

Equation 4.72 The Equation of microfloc concentration decrease in the turbulent flow contact floc forming process.

4.2.7.2 Contact flocculation in floc blanket type fluidized beds

In Figure 6.59 in Chapter 6 below, we show that grown flocs are supported by a rising water stream to form a fluidized bed, and that microflocs which inflow from the bottom to the floating area (the floc blanket) are trapped by the grown flocs constituting the floc blanket to thus reduce the concentration, as formulated by Equations 6.46 or 6.47. Here, the authors proceed one step further, and experimentally determine the degree of collision agglomeration probability $P^{\#}$ of the microflocs coming into an inflow with large-diameter fluidized-bed constituent flocs (floc blanket particles) to determine the practical expression of contact removal in the blanket.

In an upflow floc blanket sedimentation pond, large-diameter floc groups that have grown are suspended by the upflow to form a fluidized bed (thickness: 1 to 2 m). Microfloc groups that have finished coagulation are entrained in the upflow from the lower part of the blanket to flow into the fluidized bed, collide with the large-diameter mother flocs of the fluidized bed, and are agglomerated into removal.

The process, in which the inflow microflocs contact the large (mother) flocs that make up the fluidized bed and collide and agglomerate together to reduce the concentration of microflocs, can be represented by Equation 4.73, assuming the following series of conditions:

(1) Large floc of diameter D, having a sedimentation rate commensurate with the ascending flow rate V_s, float at rest.
(2) Microflocs of diameter d inflow the blanket at an ascending flow rate V_s. The diameter of the microflocs is much smaller than that of the grown mother flocs, expressed as $d \ll D$, and the sedimentation rate of the fine particles is negligibly small compared to the sedimentation rate V_s of the large diameter mother floc particles.
(3) Collisions of two types of particles of diameters D and d can be defined as approaching flow velocity V_s and collision radius $(D + d) \doteqdot D$; the contribution of the fine floc particles is ignored.
(4) Collisions between microflocs are neglected because they have little effect on removal.

Consider the collision agglomeration probability $P^{\#}$ of a microfloc d on particles of diameter D.

The decreasing rate dn/dt of the concentration n of microflocs due to the flow time t of the microflocs moving upward from the bottom of the blanket is as shown in Equation 4.73.

$$\frac{dn}{dt} = -\frac{\pi}{4} P^{\#} V_s (D+d)^2 N n = -\frac{\pi}{4} P^{\#} V_s D^2 N n \quad d \ll D \quad —(4\cdot 73) \tag{4.73}$$

Equation 4.73 The decreasing rate of the concentration of microflocs due to flow time t.

This equation was given by Camp and Stein (1943), Camp (1953). The basic structure is the same as the coagulation equation proposed in Equation 6.13 in Chapter 6 below.

Equation 4.73 is expressed as Equation 4.74 by rewriting it to an equation indicating removal with flowing distance z from the bottom edge (inflow) of the floc blanket instead of flow time t:

$$\frac{dn}{dz} = -\frac{\pi}{4} P^{\#} D^2 N n = -\frac{3}{2} P^{\#} \frac{V_f}{D} n = -K^{\#} n \quad —(4\cdot 74)$$
$$K^{\#} = \frac{\pi}{4} P^{\#} D^2 N = \frac{3}{2} P^{\#} \frac{V_f}{D} \tag{4.74}$$

Equation 4.74 The removal rate of microfloc concentration with flow distance.

Integrating Equation 4.74 with $z = 0$ and $n = n_0$ gives n/n_0 of the removal index of the floc blanket of thickness z as in Equation 4.75, which shows that the exponential of the thickness of the floc blanket increases the removal rate of the blanket. Since the response factor at this time is proportional to V_f/D, adopting a low-speed fluidized bed is key to satisfy the condition of increasing the density of the growing floc group forming the blanket and keeping particles to a small size, and to seeking a high removal rate. Practically, a low ascending flow rate of 3 to 10 cm/min (overflow rate) appears to be obtained.

$$X = 1 - \left(\frac{n}{n_0}\right) = -\left(\frac{T}{T_0}\right) = 1 - e^{K^{\#}z} \quad \text{—}(4 \cdot 75) \qquad (4.75)$$

X : Ratio at which microflocs are coagulated and removed by mother floc

T_0, T : Microfloc concentration of the blanket inflow and height z

$K^{\#}$: Rate constant $[= -(3/2)\ P^{\#}(V_f/D)]$

Equatin 4.75 The removal rate of the microfloc concentration with flow distance.

Therefore, in order to confirm whether the relationships described in Equations 4.73 to 4.75 actually hold, and to accurately describe the structure of removal, we need to confirm the magnitude of the collision-agglomeration probability $P^{\#}$ in the blanket layers of the growing floc and the microflocs. To this purpose, a pilot plant as shown in Figure 4.115 was used for the study.

In the experiments, flocs were separated by coagulated kaolinite-based clay suspensions with aluminium sulfate (alum) at neutral pH. Microflocs that were brought by inflow to the blanket with a diameter $d = 2 \times 10^{-2}$ to 8×10^{-2} mm, which had undergone rapid agitation from the bottom of the blanket. The depth z of the floc blanket was changted by adjusting the height of the overflow weir, and T_z of the outflow turbidity from the blanket at different depths z was measured. The variable depth floc blanket operation measured the removal of microflocs by changing the raw water turbidity, the rate of dosing, and the upward flow rate and continuing the process for a long time in each case, and then confirming that the mother floc volume density V_f and the diameter D of the blanket were steady. Figure 4.116 shows experimental examples of raw water with turbidity $T = 20$ mg/l with different overflow rates and blanket thickness $z = L$.

For an ascending flow rate of 3 cm/min, how the treated turbidity varies with the blanket thickness can be summarized in semi-logarithmic plots as shown in Figure 4.117, which shows that the first-order response of suspended microfloc concentration n shown in Equation 4.75 promotes the inhibition of the blanket. The removal of microflocs F and G agglomerated to $(2{\sim}4) \times 10^{-2}$ mm is markedly improved compared to

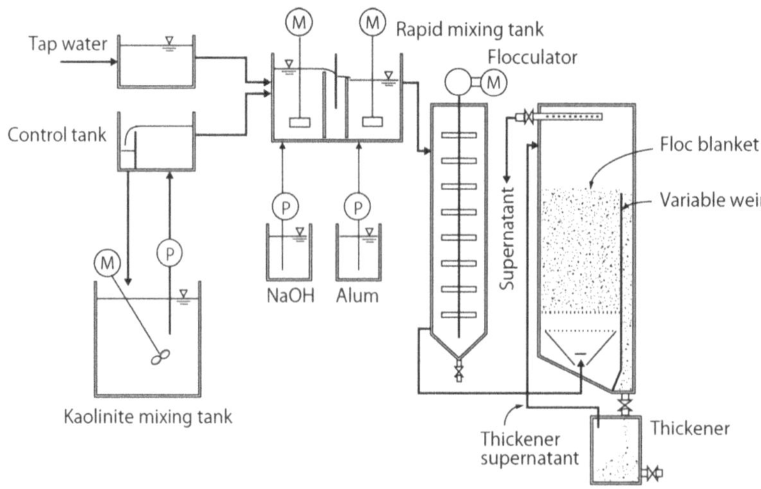

Figure 4.115 A floc blanket upward flow contact flocculation sedimentation pilot plant.

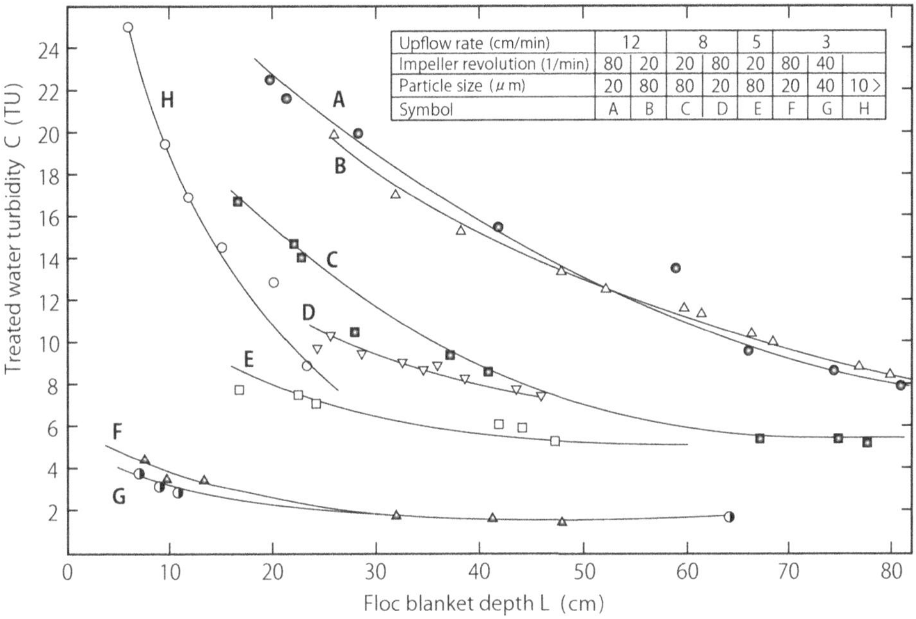

Figure 4.116 Floc blanket depth and treated water turbidity.

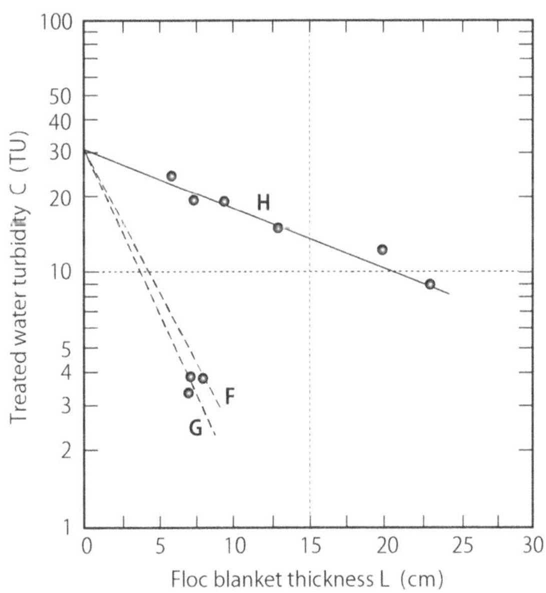

Figure 4.117 Semi-logarithmic plots for floc blanket thickness and treated water turbidity.

the case of H, which is considered to have a small d $\doteqdot 10^{-3}$ mm only, and it is worth adding GC_0T values of a certain size even when microflocs are formed. This suggests that to grow microflocs with a diameter of $d = 10^{-3}$ to 10^{-2} mm, it is also essential, even for a contact coagulation sedimentation pond, to provide agitation time T to achieve the greatest growth equilibrium in a rapid mixing tank with a large agitation strength. Of course, this is also significant for normal flocculation, which suggests that, in hydraulic jump and simple flow-impingement mixing systems, the time for the microflocs produced during rapid agitation to reach maximum-growth equilibrium at that strength and dimensionless agitation duration GC_0T (in particular, agitation duration T) are inadequate.

Rapid agitation should also be considered as a zero stage flocculation process when designing the process. The collisional probabilities of large-diameter particles D and small-diameter particles d, which exist independently is assumed to be 3(d/D) for potential flow. Assuming that the mother floc diameter $D \doteqdot 1.5$ mm, the microfloc diameter is $d \doteqdot 4 \times 10^{-2}$ mm (in the case of rapidly agitating and growing particles F and G) and 4×10^{-3} mm (in the case of mixing only H), and the collision probability is estimated by assuming the potential flow according to the Fuchs equation. The probability is 0.08 for microfloc-forming agitation and 0.008 for mixing only and about 10 times for mixing only. The difference in between two kind operation is about 10 times. Since we assume joint probability $\alpha_0 \doteqdot 1/3$, the joint probability $P^{\#}$ is the product of the two.

$P^{\#} \doteqdot 0.08/3 \doteqdot 0.027$ in the case of F or G with microfloc-forming agitation, and $P^{\#} \doteqdot 0.008/3 \doteqdot 0.0027$ in the case of H with mixing only. On the other hand, the probabilities of collisions estimated from experimental data are $P^{\#} \doteqdot 0.1$ (for F and G) and $P^{\#} \doteqdot 0.015$ (for H), which are several times larger than the estimated values assuming Fucks' potential flow. This may be thought of as increasing the probability of collision due to turbulent flow in the floc blanket. It is considered that the collision probability in a filtration pond with a low Reynolds count is higher than the potential flow assumption, which is equivalent to the case where the collision probability in a pond with a low Reynolds number is higher than the potential flow assumption.

If rapid agitation is continued to near the maximal floc-growth equilibrium profile (GCT), the authors believe that a probability $P^{\#} \doteqdot 0.1$ for collision of floc blankets can be assumed.[2]

---

[2]To date, agitation intensity has been expressed as (i) effective agitation energy-consumption rate ε_0 (erg/cm^{3}s) and (ii) total agitation rate ε^* (erg/cm^{3}s). In addition, in their original paper, Camp and Stein (1943) gave G values (s^{-1}) and demonstrated that "the progress of flocculation at a point of fluid is proportional to the velocity gradient of the flow water (stream) at that point", and defined the laminar flow velocity gradient G_{grad}, as shown in Equation 4.35, based on the energy consumption rate due to molecular viscosity under laminar flow conditions. However, as mentioned earlier, the actual floc process is performed under turbulent conditions. Under turbulent flow conditions, as Levich shows from Kolmogorov's locally isotropic turbulence theory, the turbulent energy-consumption rate ε is given by Equation 4.36. Since Equations 4.35 and 4.36 are morphologically identical, and if the laminar viscosity coefficient μ and the turbulent viscosity coefficient $(\mu + \nu)$ are used, the velocity gradient G_{turb} is expressed as Equations 4.30 and 4.36, and since the dimensions are the same, the term Camp's velocity gradient has been extended (although it is essentially misused). The authors explained earlier that the G_{grad} and G_{turb} of laminar and turbulent impact velocities differ by as much as 1:30 for the same rate of energy-consumption in a floc formation process. Laminar flow flocculation is seen in the interstices of the granular bed filtration, but it cannot be an efficient operation method from the viewpoint of energy-efficiency. When calculating the floc forming rate from the hydrological energy loss (consumption) rate (Power loss of the agitator or loss head of the water channel), calculate the G-value from the total energy consumption rate ε^*. Since the author's theory is based on the effective energy consumption rate ε_0, the effective energy consumption rate for flocculation is about 10% of the total energy consumption rate, such that $\varepsilon_0 \doteqdot 0.1\varepsilon^*$. Therefore, the effective energy consumption rate is: $G_{effect} = (\varepsilon_0/\mu)^{1/2} \doteqdot (0.1\varepsilon^*/\mu)^{1/2}$ and, as part of the author's process design diagram, the theory can be used with an effective energy efficiency value of 0.31 times the G value (gloss) obtained from mechanical power consumption. NB: 1 kg·m/sec = 1/75hp, 1hp = 75 kg·m/sec = 0.736kW, 1erg/cm^3 = 1.36×10^{-3} hp = 1×10^{-4} kW/m^3.

4.2.8 Flash mixer and floccculator

4.2.8.1 Flash mixer (Rapid mixing pond)

The first step in the coagulation process is to add coagulant, alkaline and flocculation aids in the appropriate order and in a timely manner to the raw water, to ensure that they are uniformly dispersed and hydrolyzed and bound uniformly as quickly as possible. At this stage, hydrolysis of the coagulant and interfacial physicochemical reactions with suspended colloidal particles are completed, and extremely small elemental coagulation particles (microflocs) are formed. This operation is called rapid mixing.

Chemicals are usually added in the order of alkaline aid, then coagulant. The flocculation aid is added either prior to or after adding the coagulant, whichever is most effective. For crude colloids, such as clay, it may be added before or after, whilst for fine colloids, such as color colloids, it may be added later.

The agitation method of the mixing varies, including an up-and-down bypass water channel, inlet pipes, static mixers, mechanical blade agitation, low lift pump, and hydraulic jump. When adding coagulant, it is recommended to use a series multistage stirred tank with a retention time of at least 2 to 3 minutes, and if possible 10 min or more for the zero stage flocculator. This is because by having multiple (2 to 3) stages, leakage of unreacted parts due to a short-circuit flow, which is unavoidable in the case of a single tank, can be suppressed. The use of a low-lift pump and a hydraulic jump has the advantage of achieving uniform mixing in a short period of time (instantaneously) but, on the other hand, they cannot grow microflocs of a certain size (10~30 μm) and, as described in the previous section, they then start from a small d_1 of the initial particle size condition of the floc-forming operation which requires long-term agitation under the low agitation strength of the floc-forming pond.

Inlet pipes and static mixers exhibit excellent mixing characteristics without short-circuit flow. It is also advantageous to be able to maintain a constant retention time and to start from a larger initial particle size d_1 of floc formation. Rapid mixing is an operation that can be thought of as a zero stage floc-forming pond without the need to discuss the breakage of flocs. The injection of chemicals into a short bypass channel is undesirable because it takes longer to disperse uniformly throughout the channel, resulting in uneven hydrolysis reactions. In the coagulation of crude colloids, such as clay suspensions, the selection of the necessary agitation condition is relatively less stringent, since the main crosslinking reaction of the crude particles occurs by hydrolyzing metal hydroxides. However, in the case of fine colloids, such as color colloids, where charge neutralization is a critical dominating process, an agitation process is required to satisfy both the rapid uniform dispersion of the coagulant and the retention time for contact bonding (which takes a certain amount of time) and design conditions therefore need to be rigorously discussed.

As shown in Figure 4.66, mixing ponds equipped with a mechanical stirrer usually have a square tank with a depth-to-width ratio of 1:1 to 3:1, and require at least two tanks in series to avoid non uniformity of reaction progress due to short-circuit flow. A paddle is suitable for a small agitator tank with a volume up to 2 m^3, and a propeller for up to 3 cm^3, or a turbine is commonly used. For turbines, the ratio of the agitator blade diameter to the mixing pond side length is usually 1:3 to 1:4, and for a propeller, the ratio is larger. Reducing the circulation of water and wings in the tank has also been attempted, as has enhancing the effectiveness of agitation by decentering the rotating shaft of the agitator from the center of the tank. As a required agitation strength for a mixing pond, a total energy consumption rate of about $\varepsilon^* \doteqdot 10^3$ (erg/cm^3s) is considered.

4.2.8.2 Flocculator (floc-growth operation)

As described above, the floc formation rate is proportional to G C_v, and the degree of floc formation is determined by the GC_vT value. For a given water quality (suspensions), the key to design conditions is how to determine the agitation strength, G, and duration, T, of the floc-forming pond. As the agitation strength ε_0 increases, the floc forming speed becomes big but, on the other hand, the floc breaks prematurely, reducing the maximum-growth S_m, and making it impossible to produce flocs with adequate sedimentation speed. Therefore, in the early stages of floc formation, the floc formation rate is increased by providing high agitation strength, and gradually decreasing the agitation strength to keep the floc diameter large as the floc grows (tapered flocculation) is widely carried out.

The agitation intensity is usually in the range of $\varepsilon_0 \doteqdot 0.1$ (effective) and $\varepsilon^* \doteqdot 10^0$ to 10^{-2} (erg/cm^{3}s) (gross), and the agitation intensity is reduced in three stages. Camp proposed changing the G*(gross G) values to three levels of 90, 50 and 20 (s^{-1}). Although the agitation duration T is generally about 30 minutes, we explained above that it takes a long time for low turbidity raw water and a short time for a high turbidity raw water, based on the definition of the m-value or $G_{tur}C_vT$ of the non-dimensional rate indicator (Equation 4.62).

Commonly used flocculators include bypass, vortex and paddle flocculators, as shown in Figure 4.118.

The flow pattern required for flocculators should have a low turbulence agitation intensity that allows large flocs to grow, and should ensure a large-scale circulation flow is produced across the entire cross section of the floc-forming tank in order that large-grown flocs do not sink to the bottom of the pond. It is necessary to satisfy both of these two conditions at the same time.

From this point of view, an up-and-down water-channel or paddle flocculator with a horizontal shaft is preferable. The model flocculator shown in Figure 4.87, which was used in the above-mentioned experiments to determine the largest floc growth diameter, is an experimental device at Hokkaido University, designed to allow floc formation to proceed without sedimentation of large grown flocs even under the weak agitation strength of an agitation paddle for making large flocs by rotating the agitation tank itself around the shaft at 1 rpm, and repeating the vertical turnover. Commercially available jar testers can study coagulation reactions, but the study of flocculation sedimentation is only a qualitative assessment.

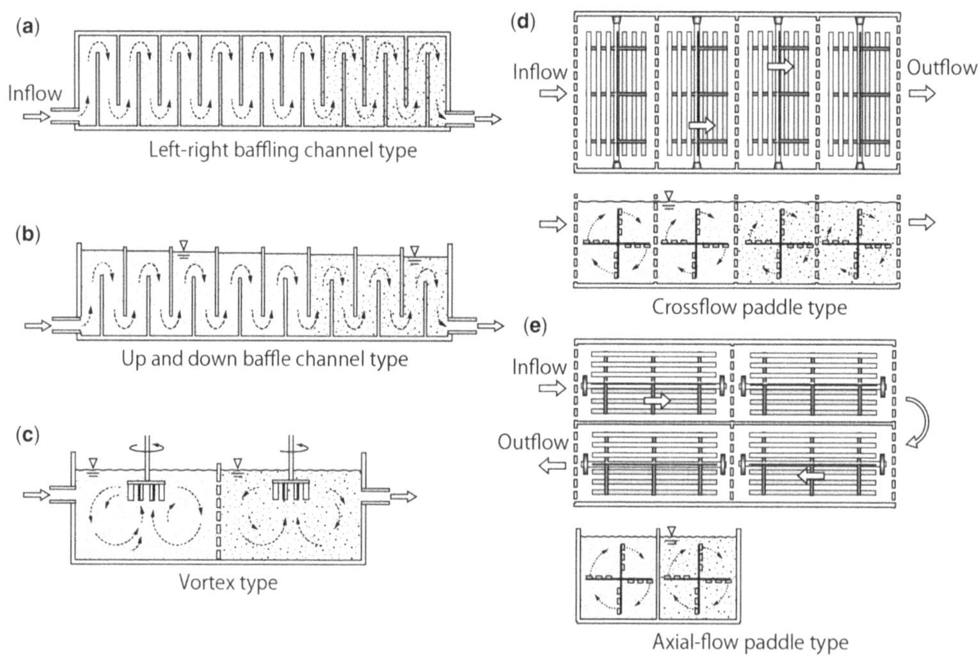

Figure 4.118 Examples of flocculators.

Horizontal shaft paddle-type agitators require submerged bearings. To avoid excessive gravitational downward loading, choose the material of the stirring blade to balance the buoyancy and gravitational force to prevent a knitting load from being applied to the bearing. In many water treatment plants where prechlorination was carried out, manganese and other metal precipitated in the underwater bearing, which resulted in malfunction. Instead of center bearing support, some have attempted to provide a stable drive by placing drive rollers on the rotating circumference of the flocculator shaft end. A vortex flocculator has difficulty in achieving both the generation of total tank convection and the formation of localy isotropic turbulence with a low agitation strength. It is difficult to design a flocculator with a stable and accurate formation rate for the same reasons that a jar tester has difficulty. Many water treatment plants in developing countries have floc-forming tanks with vertical agitators that are easy to manufacture, but it is difficult to accurately create the two functions, and this is not recommended from a functional standpoint.

A flocculator must have a gentle convection across its entire cross section and must be considered to cut short circuit flow. In a paddle-type system, water is flowed in the shaft direction or, in the case of a right-angle flow, an agitating tank is installed in multiple stages to suppress a short-circuit flow by rectifying walls. The up-and-down bypass flow type is the most resistant to short-circuit flow, and the sedimentation of growing flocs due to weak agitation and switching can be avoided, and excellent effects can be expected. On the other hand, it is difficult to cope with the fluctuation of the treatment flow rate, but is possible if multiple channels are considered. On the other hand, a vortex type is not recommended from the viewpoint of suppressing short-circuit flow.

Equations 4.76 and 4.77 show the relation between the gross agitation strength index G* (s^{-1}) of the flocculator, and the total energy-dissipation rate ε^* (erg/cm^3s) or the agitation power consumption P (g_w/scm)or bypass channel loss head h_f (cm).

$$G^* = \sqrt{\frac{\varepsilon *}{\mu}} = \sqrt{\frac{Pg_c}{\mu V}} \quad \text{(Mechanical mixing flocculator)} \quad —(4 \cdot 76)$$

$$= \sqrt{\frac{\gamma h_f g_c}{\mu T}} \quad \text{(Baffle channel type flocculator)} \quad —(4 \cdot 77) \qquad (4.76–4.77)$$

V : Flocculator volume (m^3)
T : Detention time (s)
γ : Unit volume weight of water (gw/cm^3)
g_c : Gravity conversion factor ($980gm \cdot cm/gw \cdot s^2$)

Equation 4.76, 4.77 The agitation strength index and total energy-dissipation rate.

The approximate value of total or gross ε^* for a horizontal shaft paddle flocculator can be estimated by Equation 4.78.

$$\varepsilon * = \frac{\gamma C_d[2\pi(1-K_r)n]^3}{2V\rho g_c} \sum_i A_i r_i^3 \quad —(4 \cdot 78) \qquad (4.78)$$

A_i : Perpendicular area to direction of motion of impeller (cm^2)
r_i : Distance between rotary shaft and impeller (cm)
C_d : Impeller drag coefficient (≒ 1.3~1.5)
n : Impeller revolution rate (rpm)
ρ : Water density (g/cm^3)
K_r : Co-rotation coefficient of water (-)
$K_r n$ indicates number of revolutions of water

Equation 4.78 The total energy dissipation rate.

4.2.8.3 Chemical injection device

Most chemicals used for water treatment have high solubility, so it is often possible to prepare and inject the stock solution as a concentrated aqueous solution. Low-solubility chemicals, such as lime, may be injected under constant agitation as slurries, or may be added directly, sometimes in powder form. When chemical injection devices are classified according to the nature of the chemicals at the time of injection, they are divided into wet and dry types. Since the environment in tropical countries and in Japan is generally hot and humid, it is better not to use dry systems that are susceptible to air quality (in areas other than the northern island Hokkaido and areas with relatively low humidity).

In addition, depending on the method of setting the injection quantity, quantitative injection, proportional injection of the throughput, and automated injection programs by creating operation algorithms linked to the flow rate and the main water quality factors are performed. In water treatment, the turbidity of the raw water fluctuates by more than 1:100. Therefore, process personnel (now likely to be automatically controlled injection systems) cannot cope with sudden fluctuations in turbidity and may be forced to discontinue operation. On the other hand, sewage and wastewater treatment are usually not very variable.

In addition to accuracy, the performance requirements for injection machines include ease of handling, flow rate and water quality fluctuation response, and resistance to chemical corrosion, wear resistance, earthquake resistance, etc.

4.3 PELLET FLOCCULATION (METASTABLE STATE AGGLOMERATION)

This section discusses the solid-liquid separation method developed by Hokkaido University (in Tambo's laboratory) as a new water purification method for treatment of high turbidity raw water, and as a core technology of a new treatment system (by coagulation and aerobic biological treatment) of organic sewage wastewater, for dephosphorization and treatment in a short detention time.

4.3.1 High-speed separation of highly turbid water by a pellet fluidized bed

In many Asian, African, and South American countries, water purification treatment is required using river water with raw water turbidity ranging from 10^2 to 10^4 mg/l. Taking the Chinese Yellow River water as an example, in the case at Zhengzhou City (in the lower Yellow river basin), the water always contains 1 g/l = 10^3 mg/l of suspensions and, in the case of flooding, 100 g/l = 10^5 mg/l is greatly exceeded. Rivers in Indonesia and Malaysia also commonly have high-turbidity water of 10^4 mg/l, and large rivers in India and Brazil, for example, normally have high-turbidity water of 10 to 10^3 mg. It is not easy to treat these high turbidity waters with conventional coagulation, flocculation and sedimentation.

In our experience, water intake is first subjected to simple sedimentation in a pre-sedimentation reservoir for a long period of time (e.g., 10 days in the Yellow River downstream at Keifeng city; several hours static sedimentation with polymer coagulation at Lanzhou upstream; or a simple polymer coagulation sedimentation for daily static sedimentation in Indonesian Slavaya) before being treated with conventional rapid filtration systems. Conventional high-speed coagulation sedimentation ponds (slurry-recycling or floc-blanket types) cannot discharge sludge in time, and the functions of sedimentation ponds are impaired (e.g. at Azisaba in Ethiopia and Slabaya in Indonesia).

Tambo *et al.* (1993a, 1993b) investigated the Yellow River main stream water and investigated a new treatment method using a pellet fluidized bed with an organic polymer coagulant. As a result, they clarified the mechanism and design method of a new method for the treatment of high turbidity water, which can be separated by sedimentation in a pellet fluidized bed with a retention time of several minutes, and the supernatant water can immediately be brought into inflow with a rapid filtration pond, whilst the water content of the produced sludge (pellets) is reduced to 70% by pelletizing, and can be easily transferred for a dewatering process (Tambo & Wong, 1993a, 1993b).

4.3.1.1 Concepts of coagulation pelleting by fluidized bed

In conventional coagulation flocculation processes, suspended particulates (primary particles) destabilized by the addition of coagulants bond by random collision to form fractal structure flocs which are swollen agglomerates. This is the random floc formation process described in detail in the previous section. Because of the fractal nature of random flocs, as flocs grow, the percentage of void water entrained in the flocs increases and the floc density decreases with increasing diameter. Even if the floc diameter increases, the sedimentation rate does not increase sharply in proportion to the square of the floc diameter, as Stokes' Law shows, and the sedimentation rate does not increase significantly by a factor of $d^{(2-K\rho)}$, which is shown by Tambo and Watanabe's floc density functions. Since K_ρ is in the order of 1 to 1.4, the sedimentation rate only increases from the first power of the diameter d to the power of 0.6, and the dewatering of swollen flocs produced is also not easy, as mentioned above.

The fluidized bed granulation process for water treatment proposed by Dr A. Graveland of the Rotterdam Water Works (the Netherlands) has been successfully put into practical use (Graveland, 1980; Graveland *et al.*, 1983). This kinetic theory and application will be detailed in Section 5.2 in Chapter 5 (below); the authors believe that this process can be more widely understood in the context of a 'metastable state region treatment'.

When metal ions are used as examples, the solution in the supersaturated region precipitates (insolubilizes) dissolved components easily. However, in the metastable region around the saturated region, as shown schematically in Figure 4.119, the solution is subject to insolubilization from the standpoint of solution equilibrium, but a certain aging time is required before precipitation can be observed. However, when a solid component having the same properties as the coagulum is introduced into a solution in a metastable region, the dissolved component of the liquid phase precipitates rapidly on the surface of the introduced solid, and the coagulum formed exhibits excellent separation characteristics in that the coagulum is often dense. This process is commonly referred to as crystallization.

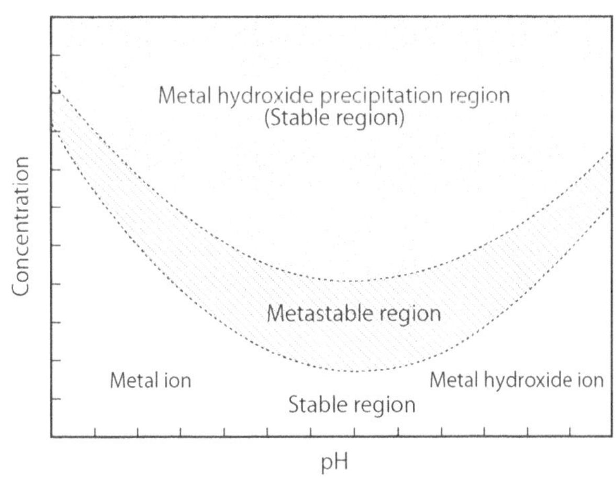

Figure 4.119 A schematic representation of the precipitation pattern of metal ions.

The authors thought that the concept of metastable regions as described above could be extended to coagulation flocculation of clay-based suspensions. Clay particles are coarse colloids with a center of particle size distribution around 1 μm. In natural water, clay particles are stably dispersed by mutual repulsion with a negatively charged surface potential (zeta potential) of about −20 to −30 mV.

In order to agglomerate such particles, the clay particles are charge-neutralized by polyvalent positive ions or polymers generated by adding coagulant, thereby reducing the electric repulsion between the clay particles and destabilizing them. This is the coagulation process described in detail in the preceding section. In the stable coagulation region, coagulant concentrations and pH range can also cross-link precipitated hydroxides, causing random flocs to grow into fractals.

On the other hand, in small areas with slightly less coagulant or slightly acidic or alkaline pH-biased conditions, there is a metastable coagulation range in which colloidal coagulation does not rapidly form random agglomerates, as in the case of metal-ion precipitation. This may be considered a quasi-precipitation zone in the case of aluminium hydroxide. If the initial coagulation condition can be set to the metastable region (i.e., so that it is not easy for individual particles to form agglomerated flocs and conditions are such that there are not a large number of voids in the floc agglomerates), the same metastable conditions will exist when the microparticles are mixed with grown solid phases just after polymers are added and mixed to enhance binding strength and to allow the attachment of microparticles in the metastable region, and allow adherence to the grown particle surfaces, thereby allowing the deposition to proceed one by one. This results in the growth of concentric spherical agglomerates (pellets) that do not contain higher-order voids due to floc fractal agglomerates but only have void fractions between the initial particles. The density of such pellet particles is independent of the particle size, resulting in uniform high density spheres with only initial porosity. Figure 4.120 schematically illustrates the features of the process of forming random flocs (fractal agglomerates) and pellets that grow on existing particles one by one.

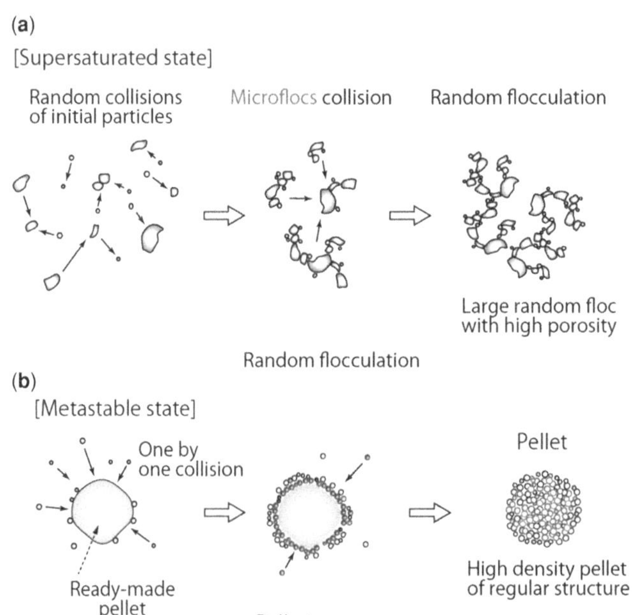

Figure 4.120 Features in the process of forming random flocs and pellets.

4.3.1.2 Pellet fluidized bed separator

Consider realizing the concept described above with the pelletized fluidized bed process shown in Figure 4.121. When clay particles (initial particles) neutralized with aluminium coagulant are continuously introduced into a fluidized bed in which preformed agglomerates are suspended, the initial particles tend to impinge on large agglomerates (pellets) one after the other. When a polymer coagulant is added at an appropriate time to strengthen the bonding strength between the initial particles and the preformed agglomerates (pellet particles constituting the fluidized bed), the initial particles adhere to each other, and the high-density pellet particles grow more concentrically. By adding appropriate agitation to the fluidized bed at this time, the initial particles which adhere to the surface of the large agglomerates (suspended pellets) one after another are prevented from growing irregularly on the pellet surface, and the initial particles can be grown to a shape close to a sphere.

A pellet fluidized bed separation system was investigated, consisting of three parts:

(1) A chemical mixing tanks for producing microflocs;
(2) A pellet fluidized bed with stirring blade (upflow); and
(3) A solid-liquid separation tube.

The system used in the basic experiment was an acrylic resin with an inner diameter of 10 cm for both the fluidized bed and the solid-liquid separation cylinder, and the fluidized bed thickness was set to 145 cm. The excess pellets were allowed to overflow from the upper surface of the fluidized bed to the separation column in a density flow to keep the fluidized bed at a constant thickness. Ascending velocity of the fluidized bed (superficial velocity) was set at a high level, typically at 30 to 70 cm/min (10 to 20 times that of a normal high-speed sedimentation pond).

The basic study proceeded with a raw water of suspension 3000 mg/l of kaolinite (d = 0.55 μm) from Sapporo City Water, assuming silt clay suspension systems such as those from the Yellow River and

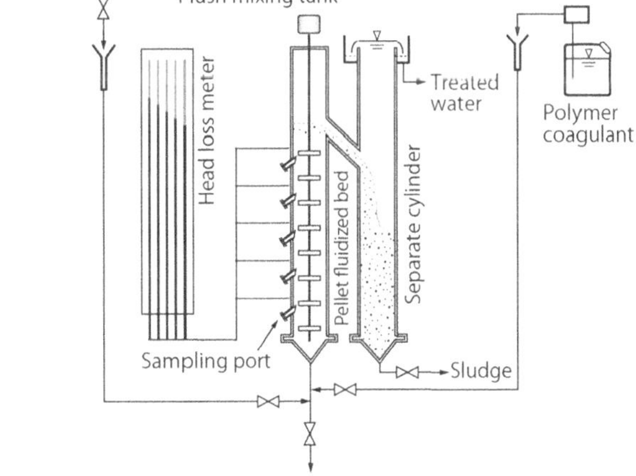

Figure 4.121 A flow chart for a pelletized fluidized bed operation.

Southeast Asia. As a metal coagulant, poly-aluminium chloride (10% as Al_2O_3) was used, with polyacrylamide-based Aquo-floc N-100PWG (MW1600 × 10^4 Da, nonionic) as an organic polymer flocculation aid; Aquo floc A-100PWA (MW1700 × 10^4 Da, weak anionic) and Aquo-floc A-110PWG (MW15600 × 10^4Da, anionic) were used to strengthen the bonds. The injection rates of the aluminium coagulant and polymer floc-forming aids were expressed in terms of ALT ratio (aluminium injection per unit turbidity mass) and PT ratio (polymer solids per unit turbidity mass), respectively, and clearly represent the differences between the conventional types.

4.3.1.3 Aging of the pelletized fluidized bed and reaching steady state

The first step in the high-speed treatment of high turbidity water by pellet fluidized bed is to stably produce large-diameter, high-speed sedimentation particle groups as nuclei for pellets in the fluidized bed. By introducing high turbidity raw water from the bottom of the fluidized bed columns and adding aluminium coagulant and floc-forming polymeric aids, random flocs are first formed. The resulting random flocs have a fractal structure, and although the void fraction is high, the floc strength is high due to the high polymer floc forming aid, resulting in large-diameter flocs. Therefore, operation is started at a upflow speed of about 3–5 cm/min, which is the same as that of conventional contact high-speed sedimentation ponds, basically in the area of operation with an overflow rate where large-diameter random flocs are collectively settled.

As operation continues, the random flocs that make up the fluidized bed become densified by reducing the porosity by the action of syneresis (liquid separation due to gel shrinkage), assisted by the binding force of the polymer auxiliaries, and the core particles of the pellets are formed maturationally. Therefore, it is possible to gradually increase the superficial velocity of the fluidized bed in accordance with the progress of aging (increase in the density of individual random agglomerated particles in the fluidized bed).

In essence, pellet coagulation treatment by a metastable zone process as proposed by the authors can theoretically be established only when the action of syneresis in high-speed fluidized bed separation is clearly positioned as a preparation mechanism in the transition stage leading to the expression of the third mechanism described below.

For a raw water turbidity of 3000 mg/l, the first fluidized bed void velocity of about 3–5 cm/min is gradually increased. After about 30 min, the velocity of the pellet fluidized bed separator is likely to reach its target processing velocity, such as 30 cm/min. Photographic observations confirm that, over time and as the fluidized bed thickness increases, the bed constituent particles gradually change from a random floc to spherical pelletized particles by synerisis.

As the particles comprising the fluidized bed become denser (shown in Figures 6.54 in Chapter 6 below) the concentration of the particles comprising the fluidized bed increases and the head loss of the fluidized bed increases. Figure 4.122 shows the relationship between the height of the fluidized bed and head loss. It is observed that even if the fluidized bed reaches the target speed of 30 cm/min, the thickness of each layer of the head loss increases until a few hours later, and the steady condition in which no change in the density of the entire layer (145 cm) is observed lasts up to 8 hours. As shown in Figure 4.123, there is no significant change in the particle size distribution during this period. The increase in the water head loss shown in Figure 4.122 is due to the slow syneresis effect which increases pellet density, and the progression of surface stacking (china marble-like) due to the accumulation of microflocs on existing pellets in a one by one process (the density of voids is the primarily the accumulation of microfloc voids; as shown in Figure 6.80 in Chapter 6; the size distribution of cubes is the largest, depending on the arrangement, with a size distribution between 0.476 and 0.2595 – the smaller rhombohedral values), which is the result of the progression of surface stacking being replaced by random flocs.

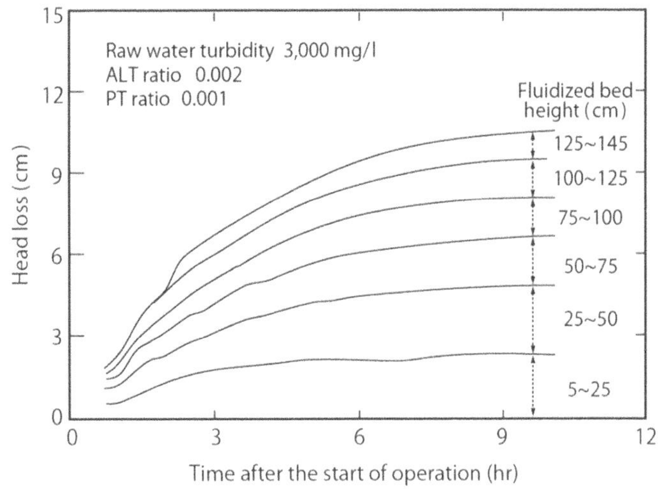

Figure 4.122 Relationship between the height of a fluidized bed and head loss.

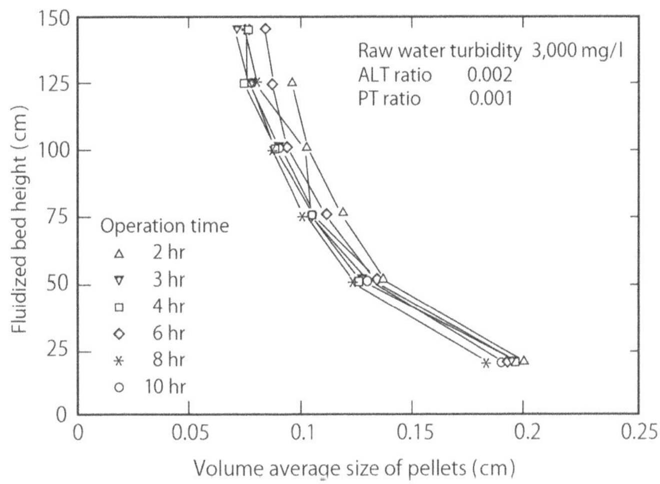

Figure 4.123 Changes in pellet size distribution.

4.3.1.4 Laminated adhesion of microflocs

The critical process for this is the adherence of the microflocs, one by one, to the surfaces of the mother pellets in a layered form. The key to this process is how to position and inject the polymers. As shown in Table 4.18, the addition of polymers to the raw water followed by the addition of polyaluminium chloride to form rapid agitation and floc does not produce effective pellets because random floc is first formed, and the simultaneous injection of polymers in the chemical mixing pond is not good. As a result, by direct injection of a polymer as a binder to the bottom of the fluidized bed immediately before the inflow of the microflocs, the microflocs may be individually bonded to the polymer without randomly agglomerating with each other, and the microflocs may bond only on the mother pellets. The

Table 4.18 The effects of polymer coagulant dosage.

Dosage Position or Method	Treated Water Turbidity (mg/l)	Effective Density of Pellets (g/cm^3)
Just below fluidized bed	<5	0.17
Rapid agitation before entering fluidized bed	25	0.11
Mixing tank	25	0.08
Via mixing tank and flocculator	Pellet fluidized bed is not formed.	

authors have considered creating a china marble type of particulate lamination without fractal properties and proceeding with the random floc generation of microflocs.

In this case, since the concentration of the fluidized bed pellets with a maximum floc particle concentration would govern the maximum velocity of the agglomerates in light of Tambo's GC_0T theory, it would be reasonable to perform floc formation in a short time in a fluidized bed by actively utilizing the C_v (solid concentration) of the fluidized bed.

4.3.1.5 Concentric spherical growth of pellets

Even if microfloc transport adheres to the surface of preformed (mother) pellets in the one by one mode, subsequent microflocs will adhere to the surface of the preformed (mother) pellets, creating a dendritic or hillock-like irregular surface that prevents the formation of concentric spherical pellets of the desired shape. The fluidized bed is then continuously agitated with a rotating paddle to provide agitation to dislodge irregularly growing microfloc bonds on the pellet surface. If the strength is too strong, the pellet body will be broken, but if the strength is insufficient, high-density pellets of good shape will not be produced.

Based on the results of further experiments, the author thinks it would be better to take the G^*_{turb} values about $30s^{-1}$. Even if the same G value is given, a highly concentrated pellet layer is agitated, so it is better to increase the number of agitating blades so that local excessive strength does not appear in the vicinity of the agitating blades. In this experimental study, 2 cm wide stirring blades were installed at 8 cm intervals and stirred at 36 rpm. Since we were unable to fully conduct scale-up research, we need to obtain more practical data.

4.3.1.6 Effective density of pellets

Figure 4.124 shows the results, using the floc density functions of Tambo and Watanabe, to compare the effect of densifying random floc by the syneresis effect by adding a polymer floc forming aid and the density of pellets formed by the one by one adhering stratification type of pellet fluidized bed process. The syneresis action of random flocs with a d $\doteqdot$ 0.1 cm, similar to that of the one by one adhering stratified pellets, requires agitation durations of 2 hours or longer, and it seems unlikely that the syneresis action of random flocs with larger diameters could be increased to densities equivalent to those of pellets.

4.3.1.7 Effect of upflow rate of fluidized bed

When the void column up flow rate is 30 cm/min, the single particle sedimentation rate of the pellets overflowing from the fluidized bed to the separation column is 1 cm/sec or more. A separation tower with an overflow rate of 0.5 cm/sec = 30 cm/min can perform solid-liquid separation and sludge

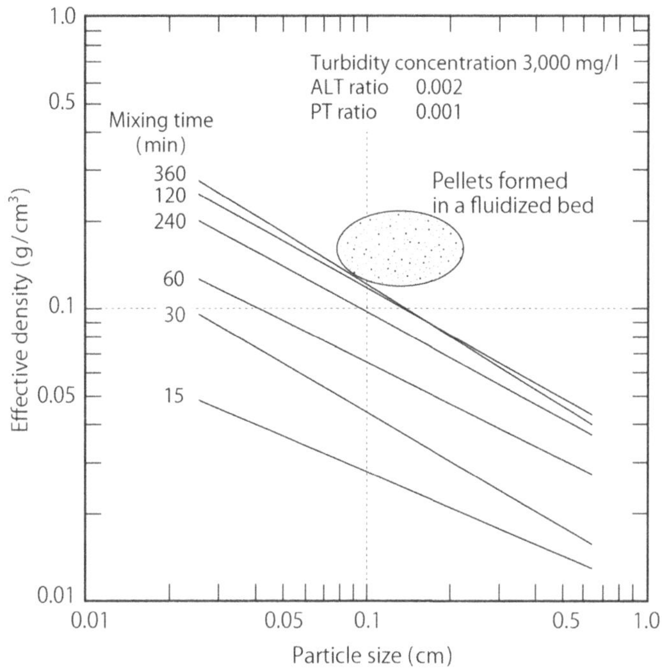

Figure 4.124 The change of floc density function shape in a Jar test.

discharge with a margin. Figures 4.125 and 4.126 show the results of the examination of the treatment effects of operation at higher upflow velocity.

At an ascending flow rate of 30 cm/min, a treated water turbidity of 2 mg/l can be obtained with a fluidized bed thickness of 145 cm. As shown in Figure 4.126, increasing the up flow rate increases the

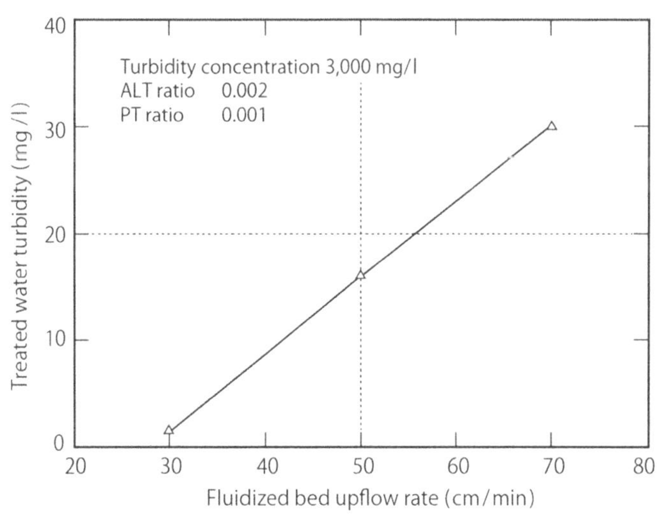

Figure 4.125 Superficial upflow rate and treated water turbidity.

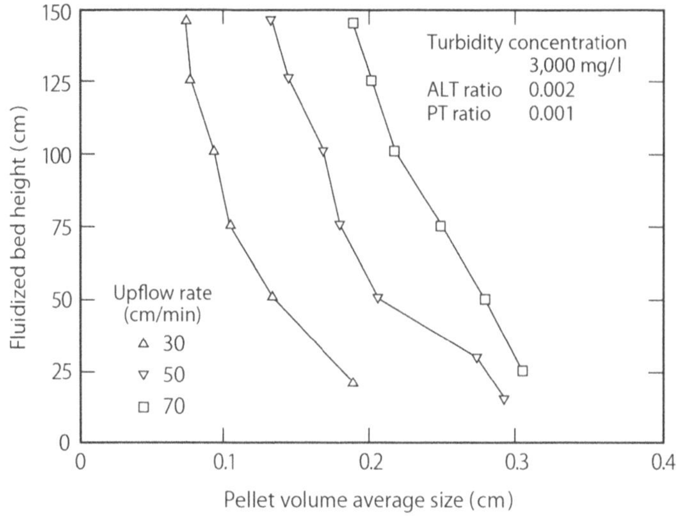

Figure 4.126 Superficial upflow rate and pellet size distribution.

pellet particle size of the fluidized bed, and the fluidized bed is formed stably in a self-equilibrium manner. This is an important finding on the stability of this pellet fluidized bed separation system. However, the treated water turbidity increases in proportion to the increase in the up flow velocity, as shown in Figure 4.125, because the convex microfloc agglomerates scraped off by the fluidized bed agitation to control the irregular formation length cannot stop flowing out from the fluidized bed. The majority of the suspended solids flowing out are agglomerates of irregularly formed and long microflocs detached from the surface of the pellets, which are observed to have a relatively high sedimentation rate. We believe that the solid-liquid separation can be improved by enlarging the upper cross-sectional area of the separation tower or by installing inclined tubes to greatly reduce the overflow rate, and ultra-high-speed operation with a void upflow rate of 50 to 70 cm/min is also feasible.

4.3.1.8 Aluminium dosing condition to set pellet coagulation

As noted above, both the ALT ratio and the polymer and turbidity (PT) ratio govern the properties of the resulting pellets, but the first step in pellet fluidization is how economically to make aluminium clay microflocs that are most suitable for separation. As shown in Figure 4.8 above, Tambo has proposed a practical critical coagulation ZP of about ± 10 mv in order to reduce the ZP of particulates in ordinary coagulation sedimentation. If nonionic or anionic polymers are used as flocculation aids, the negative surface charge of clay particles must be dropped to the critical potential beforehand.

The lowest possible ALT ratio should be used to create a high density floc. Figure 4.127 shows the relation between ALT ratio and ZP when the ZP of a clay-aluminium microfloc was measured using a Electrophoresis Apparatus Mark II (manufactured by Rank Brother Corporation) to obtain a turbidity level of 3000 mg/l.

The ZP at which a clear supernatant is obtained by a jar test is around -14 mV. In the normal water purification process, the ALT ratio is about 0.006 (Figure 4.127; point a) when a target value of -10mv is considered, which is 50% higher than the ALT ratio of 0.004 (Figure 4.127; point b) obtained by a strict jar test. On the other hand, treatment with a highly turbid raw water of 3000 mg/l in a pelletized

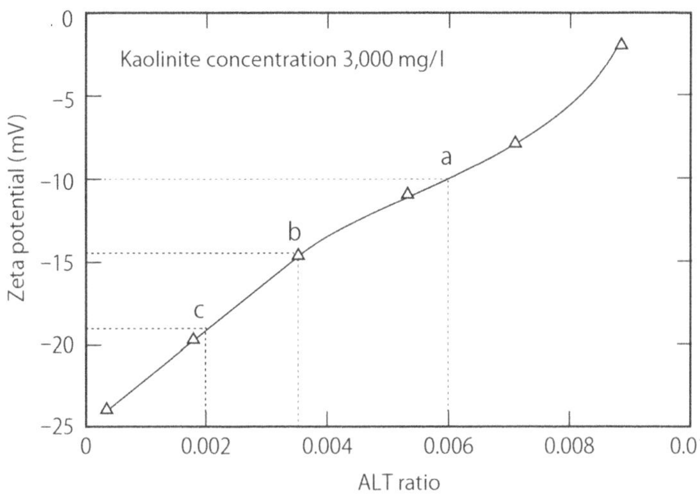

Figure 4.127 Microfloc zeta potential (ZP) and ALT ratio.

fluidized bed requires an ALT ratio of about 0.002 (Figure 4.127; point c), and the ZP at this time is about −20 mv. When a similar process was determined for kaolin concentrations of 300, 1000 and 10,000 mg/l, the relationship with the ALT ratio, which gave a predetermined ZP, was obtained as shown in Figure 4.128. In that Figure, the ALT ratio (1) is the ALT ratio required to reach the practical standard coagulation

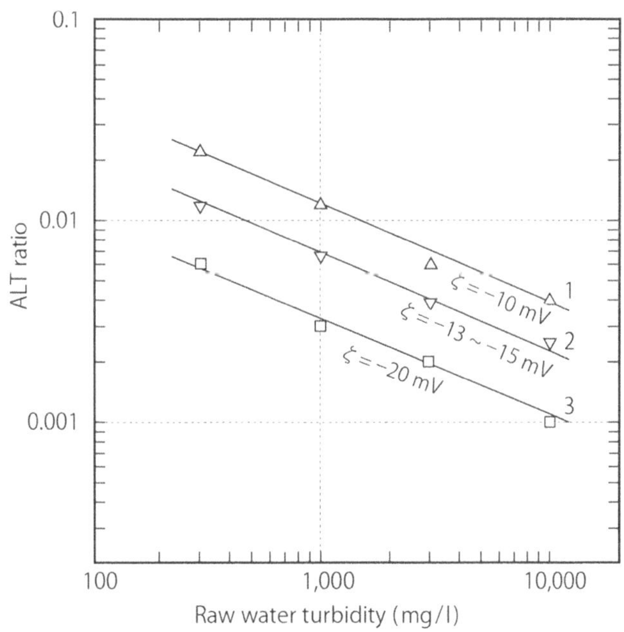

Figure 4.128 Raw water turbidity and ALT ratio, according to charge neutralization.

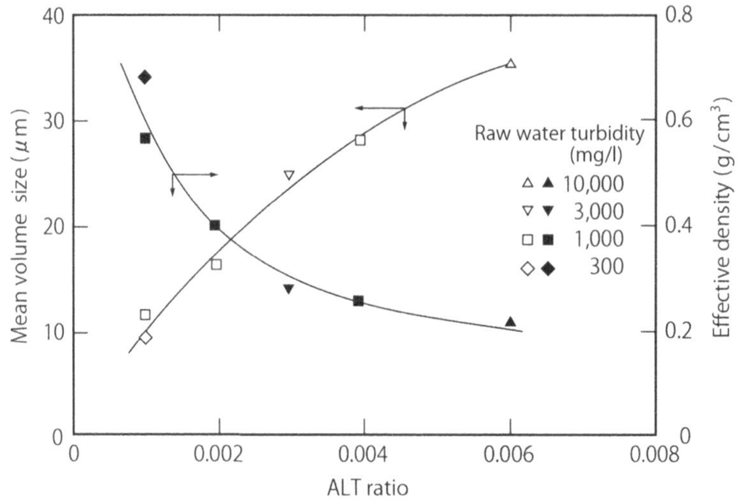

Figure 4.129 ALT ratio, particle size distribution and effective density of pellets.

treatment index of ZP −10mv; the ALT ratio (2) is the ALT ratio required to reach the coagulation ZP of −13 to −15 mv; (3) indicates the ALT ratio required to reach ZP −20 mv, which is considered to be the minimum requirement for pellet fluidized bed treatment.

In both cases, the relationship between the raw water turbidity and the required ALT ratio is a linear relationship on log-log plots. Note that the level of reduction in ZP required for pellet particle formation is −20 mv, which is about half the required ALT ratio for random flocs. The reasons for this are explained below. There are differences in the processes that can be clearly distinguished between the coagulation floc process (in which charge neutralization and cross-linking are carried out only by hydrolyzed poly-aluminium) and operation (in which charge neutralization is carried out by polyvalent polymerization ions of aluminium and cross-linking is carried out by sharing the cross-linking with the polymeric coagulant).

In addition, the diameter of the microflocs used governs the density of the pellets that are produced to form the pellet fluidized bed. The microscopic flocs produced at different ALT ratios are collected immediately after rapid agitation, and the relationship between the mean volume examined for particle size distribution by microscopy and the effective density of the pellets produced is shown in Figure 4.129. As the ALT ratio increases, the microfloc diameter increases, the porosity increases slightly, and the microfloc density decreases. The density in the Figure is a numerical value estimated by the floc density function. As a result, the density of pellets formed by concentrically stacking microflocs is also governed by the structural density of the original microflocs. Since the GCT value is valid and the agitation time is constant, the volume concentration C_v value, which is determined by the injection rate of coagulant and the concentration of raw water turbidity, determines the size and density of microflocs. Therefore, it is important to choose the smallest ALT ratio for pellet coagulation with the aim of achieving a critical ALT ratio of −20mv.

4.3.1.9 Investigation of injection conditions for organic polymer flocculation aids

Nonionic or anionic organic polymers are believed to adsorb on the surfaces of coarse suspended particles or microflocs that are charge neutralized with aluminium and form co-adsorption cross-links between microflocs under the same conditions. The ability of a polymer to cross-link is governed by the adsorption rate of the polymer (the ratio of the amount of adsorption to the amount of saturated adsorption). Langmuir isothermal adsorption lines can be used for the case assuming that polymeric molecules adsorb uniformly. The organic flocculation aid polymers used here have a small basic unit chain structure, so it is thought that one end of the elongated molecule is adsorbed on the surfaces of the fine suspended particles and the other end extends into the solution. Given that the number of basic units adsorbed is a constant percentage of the total chain, it is likely that the adsorption rate can be expressed in terms of the amount of adsorption of the entire polymer.

Therefore, we measured polymer concentrations (TOCs) residual in the solutions under various injection conditions, and investigated the adsorptive states of the polymer on the microparticles. The action of nonionic and anionic polymers remains almost unchanged, and this study focuses on the nonionic polymer Aquofloc N-100PWG (MW1600 $\times$ 10^4 Da). Polymers with molecular masses as high as 10^7 Da have chains as long as 10^{-4} to 10^{-5} mm in length, and charge-neutralizing cross-linking by aluminium polymers with molecular masses as high as 600 to 1000 Da can lead to long bonds outside the repulsive barriers in question. At charge neutralizations of the critical coagulation potential at which coagulation proceeds with polymerized metallic salts of up to -20 mv (close to the ± 13 mv proposed by Tambo), random agglomeration between microflocs is difficult and coagulation between microflocs does not proceed. The addition of polymers make cross-linking (co-adsorption) by the long polymer chain which outreaching distance of the maximum energy barrier of the electric double layer, and pellet agglomeration can proceed even at higher ZPs, such as 20 mV (higher than Tambo's critical coagulation ZP of 10–12 mV).

A polymer was added after rapid stirring. After 15 minutes of flocculation stirring, the injection rate of the polymer (PT ratio) was determined as a function of the percentage of polymer residual in the supernatant water (considered to be the equilibrium adsorption value), and the results are shown in Figure 4.130. The amount of polymer residual in water increases linearly from around the PT ratio of over 0.001, and the percentage of active polymer adsorbed decreases sharply. Figure 4.131 shows this relationship as an adsorption isotherm. The shape of this adsorption isotherm is typical of polymer adsorption, exhibiting a

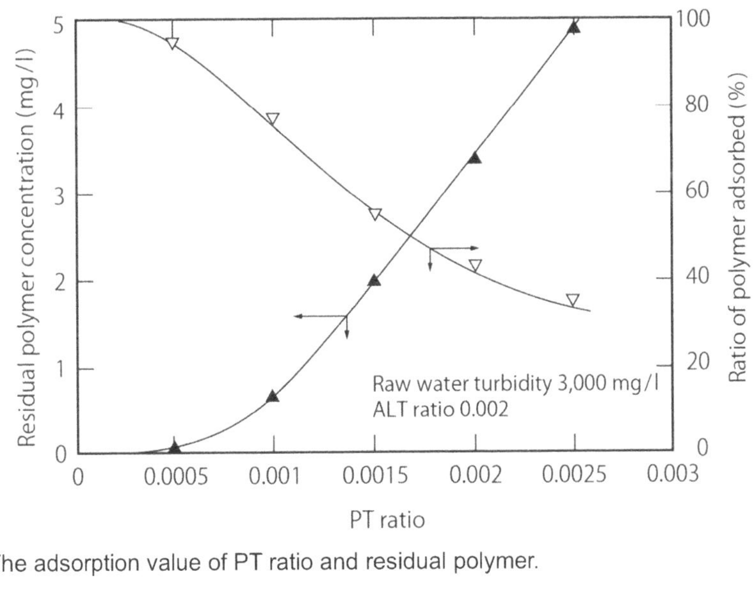

Figure 4.130 The adsorption value of PT ratio and residual polymer.

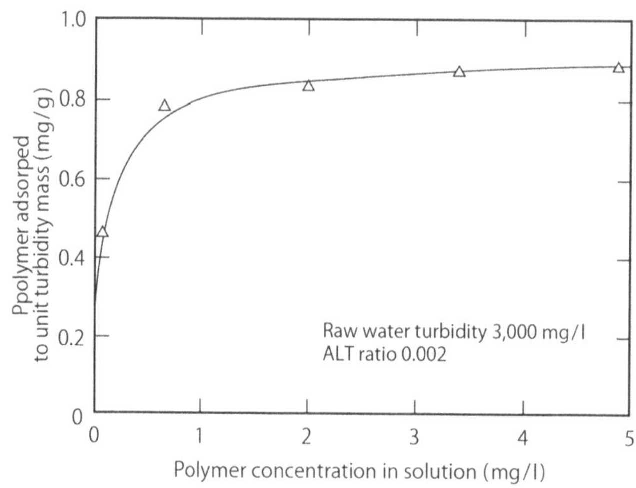

Figure 4.131 The adsorption isotherm of residual polymer.

rapid increase in adsorption rate at low concentrations and a higher affinity of particles and polymer. The plateau is reached in the high concentration range, and the amount of adsorption increases slightly. For a turbidity of 3000 mg/l, a minimum PT ratio of 0.001, which produces good pellets in a fluidized bed, corresponds to a position where about 80% of the polymer adsorbs on the fine suspended particles and just before the adsorption isotherm reaches a plateau.

In the raw water of kaolin concentrations of 300, 1000 and 10,000 mg/l, the injection rate of poly aluminium chloride (PACl) was selected to be 0.006, 0.003 and 0.001 in the ALT ratio so that the ZP becomes −20 mv. As a result of experiments conducted at various PT ratios, the polymer was injected so that the PT ratio was about 1/2 of the ALT ratio. As shown in Figure 4.132, the polymer adsorption rate was about 80%, and this value was presumed to be a practical and appropriate injection condition.

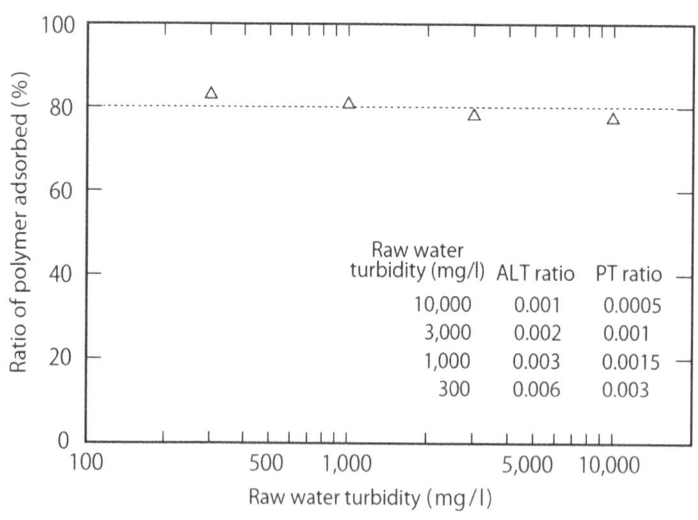

Figure 4.132 The polymer adsorption rate under optimal dosage conditions.

4.3.1.10 Injection procedures for poly-aluminium chloride (PACl) and crosslinking organic polymers

To determine the procedure for the appropriate addition of chemicals with two functions of coagulant and crosslinking agents:

(1) First, the aluminium dosing ratio (ALT ratio) was determined so that the ZP of the suspended particles would be -20mv;
(2) Second, the polymeric flocculation aids injection ratio was determined, so that the PT ratio is 1/2 of the ALT ratio (PT = ALT/2). Since a pelletized fluidized bed is formed by the paired effects of the chemicals, it is considered that this ratio does not change even if the ALT ratio changes. Other combinations of coagulants and polymers need to be developed, and new experimental studies are needed.

4.3.1.11 Operation results under optimal injection conditions: Conclusions

In accordance with the optimal operation procedures considered above, a pellet fluidized bed separation process with void upflow velocity of 30 cm/min for four high-level suspensions with kaolin concentrations of 300, 1000, 3000, and 10,000 mg/l resulted in treated water turbidity of 1 mg/l or less in the range of raw water turbidity from 300 to 3000 mg/l, increasing up to 10 mg/l up to 10,000 mg/l. An overview of the effective densities of the pellet particles formed is shown in Figure 4.124. If we look at each coagulation condition in detail, as shown in Figure 4.133, the ALT ratio to turbidity decreases significantly as turbidity increases, and the effective densities of the pellets produced also increase.

As described above, the pellet fluidized bed separation process is a 'separation process for high suspensions' that is highly effective in terms of operation time, required chemical amount, and the increased density of the generated agglomerated particles. The authors believe that pellet fluidized bed separation should be operated separately from any low-turbidity coagulation floc formation treatment.

In addition, if the pellet sediment (sludge) flowing over the separation tube in each experiment was collected and placed on a 1 mm wire mesh, the water content immediately dropped to around 70%. Therefore, the pelletized fluidized bed system eliminates the need for a sludge thickening process. By simply discharge sludge on a running filter cloth, sludge with a low water content can be produced quickly and easily, and can be processed by dewatering immediately. Again, as shown in Figure 4.133, the higher the turbidity, the lower the water content of sludge can be produced, and the features of the high suspension response separating process extends to sludge treatment.

To summarize the features of this pelletized fluidized bed process:

(1) A high turbidity raw water of 10^3 to 10^4 mg/l can be used in a short retention time of several minutes to complete coagulation (pelleting) and solid-liquid separation to obtain a clear treated water. The effective densities of the pellet particles formed and separated by this method are more than an order of magnitude higher than the random flocs produced by conventional coagulation floc forming methods, a high sedimentation separation rate is stably obtained, the water content of the separated sludge is low, and dewatering filtration is possible without a thickening process. It can be put into practical use in small and medium-sized facilities as an efficient new technology for countries that are struggling to respond to high turbidity water.
(2) The principle of the generation of pellets in the fluidized bed separation process is as follows: raw water suspensions neutralized to a ZP of about −20 mv are kept in a state in which random coagulation is difficult to proceed, and polymers for cross-linking are added immediately before inflowing into the fluidized bed. With the help of the polymers' long chains, concentric

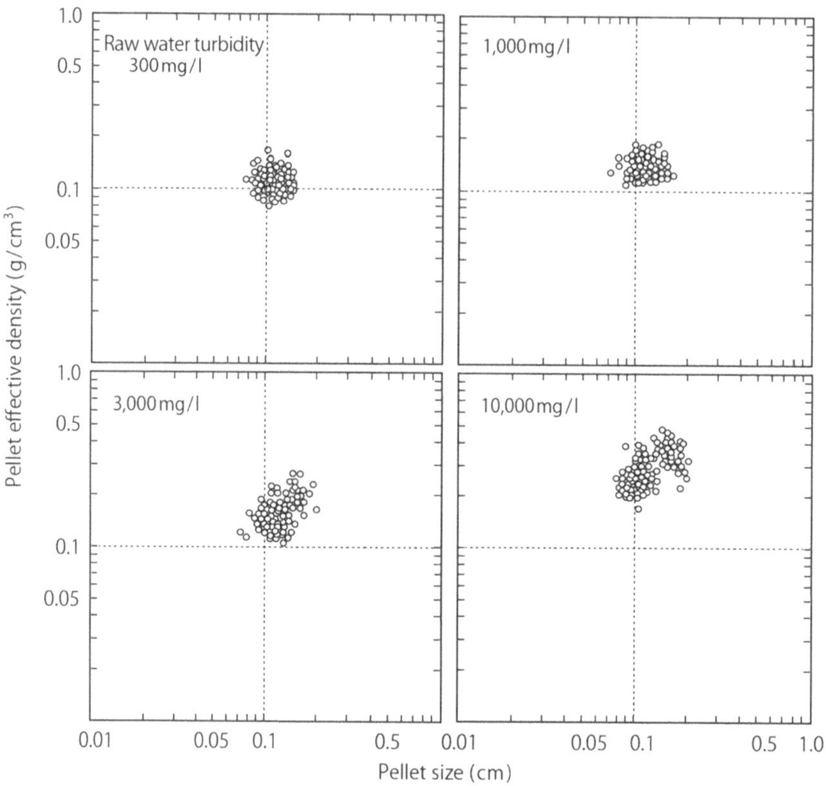

Figure 4.133 Raw water turbidity and pellet effective density.

spherical pellet growth (china marble growth) is carried out in such a way that co-adsorption cross-linking of the polymer is promoted, in a one by one fine particle deposition between the partly charge-neutralized turbidity particles and the pellet surfaces, across the electric double layer superimposed energy barrier in a potential profile in which random coagulation cannot proceed at −20 mv.

(3) After the addition of the polymer, the microflocs are randomly agglomerated at the pellet surface and a gentle stirring is always added to the fluidized bed to suppress the presence of random agglomerates so that irregular growth does not occur.

(4) The two types of coagulant, aluminium and organic polymers, in the coagulation of pellets in a fluidized bed have different roles. The former is responsible for partially neutralizing the charge until just before the random agglomeration occurs, whilst the latter is responsible for adhering fine particles (which cannot be subjected to random coagulation at a charge of ZP = −20 mv) to the pellets surfaces by co-adsorption phenomena in favor of the cross-linking lengths of polymer chains.

(5) The lack of a small potential drop of 5 to 7 mv between partially charge neutralized microparticles (which is insufficient to reach the charge neutralization level of −13 mv required for random coagulation, as Tambo described) is accompanied by the co-adsorption of long adsorption chains between partially charge neutralized elementary suspended particles, thereby both reducing the charge neutralization and providing bond strength.

(6) It is important to make a random floc and set the appropriate stable operation condition based on an accurate mechanistic understanding, taking into consideration that the coagulation produced by polymerization and syneresis action can produce similar results over time, but that the processes differ completely.

We took part in a study, started from 1983 to 1985, when invited by the Xi'an Architecture and Technology University in China. The study included lectures, technology exchange and joint research carried out in the Yellow River Basin. Based on the limited Chinese technologies and materials at that time (in the 1980s), we treated the high-turbidity water from the Yellow River using using organic polymers called 'Chinese coagulant No.3'.

4.3.2 Treatment of high turbidity water and colored water by a pellet fluidized bed

The pellet fluidized bed treatment process can also be applied to highly turbid water in Southeast Asia, Africa, and South America. Unlike the Yellow River water, however, in these southern regions, in parallel, we must also consider the treatment of colored water from dense forests derived from thick vegetation and high concentrations of humic substances discharged from wetlands. We therefore now examine how this pellet fluidized bed separating process can be applied to the treatment of highly turbid water with high color colloids.

The size of humic components is three orders of magnitude smaller than that of clay-based suspensions, ranging from 10^3 to 10^2 Da in terms of molecular weight (see Chapter 3; Water Matrix, Figures 3.37 and 3.38). The required dosing rate of aluminium and other mineral coagulants in the removal of such color by coagulation sedimentation is much higher than that of clays, and the flocs formed are swollen, the solid-liquid separation rate is low, and the dewatering of the separated sludge is poor. It would be desirable if the pellet fluidized bed separation process could be applied, with its advanced separation features. The starting point here is to investigate the microflocs of colored particles that differ in their properties, so as to understand how the process of stacking particles to form one by one on preformed (mother) pellets can be achieved.

The nature of the primary particulates (size, densities, and coagulability etc.) affects the separability of the produced pellets. The density of humic substances is only 1.2 g/cm^3, which is about an order of magnitude less than the density of clay particles in water. In addition, if fractal agglomeration has progressed in the microfloc forming stage and the structural density has greatly decreased, even if densification is attempted by syneresis, the composition of pellets cannot be achieved only by the color colloids components. However, if clay and color colloids are co-agglomerated by adding a mineral coagulant of aluminium, for example, in the presence of a considerable amount of high-density clay particles compared to the color components, if a high-density fine floc (initial particle) can be produced, it is expected that the generation of a pellet fluidized bed will proceed. Therefore, assuming the raw water contains the two components, we conducted a study using the experimental equipment shown in Figure 4.121 above.

4.3.2.1 Experimental description of the system

Kaolin and Kraft Pulp (KP) diluted black liquor filtration were added to city water, using a 0.45-μm membrane filter. A raw water test was prepared with a specified concentration of turbidity and color. The main component of KP black liquor is lignin, which has a coagulation behavior similar to that of natural colored water. Color is measured with reference to pH 9.0. As an inorganic coagulant, polyalanine chloride (PACl: 10% as Al_2O_3) was used and as an organic polymer (floc-forming crosslinking aid) a nonionic polymer (Aquofloc N-100PWG) was used.

The procedure was as follows.

(1) A standard test with a jar tester was performed by the constant dosage and variable pH method, and the appropriate pH was selected for coagulation (Figure 4.134). At the optimum pH5.0 determined as a result, the amount of aluminium coagulant was changed and the required amount of aluminium added was examined. The state of charge of the resulting fine coagulation particles was measured by microscopic electrophoresis (manufactured by Particle Micro Electrophoresis Apparatus Mark II: Rank Brothers). The addition of organic polymers did not improve the color removal with jar testers.

(2) The fluidized bed process follows the same conditions as those for forming clay-based pellets. The void up flow velocity was 30 cm/min. and the fluidized bed agitation speed was 36 rpm (G ≑ 30 sec^{-1}). The vertical particle size distributions of the fluidized bed pellets, the head loss of the fluidized bed, the treated water turbidity and color, and the water content of the separation-tube overflow pellets were measured over time. The single particle sedimentation velocity and effective density of individual pellets were measured by removing water at sampling points.

As described above in the section on color and clay-based coagulation (see Figures 4.46, 4.47, and 4.48, etc.), in the coagulation of color colloids, at around pH 4.5 to 5.0 the charge neutralization capacity of the aluminium hydrolyzed polymer is the highest and optimal for the operation of charge neutralization. However, in order to form flocs and separate by sedimentation, a cross-linking material is required which gives the binding force to withstand the agitation to the microfloc agglomerates in the turbulent flocculator, and for that the coexistence of the hydrated aluminium precipitate is required. In the actual treatment system, a region with a pH 5.5 or slightly lower than 6.0, where both the optimal pH and the crosslinked substance are coexist each other, is set, and the charge neutralization and the floc crosslinking bond conditions are satisfied. If an organic polymer can be added to the micro-color floc as a binder, the two functions of charge neutralization and cross-linking strengthening can be separated.

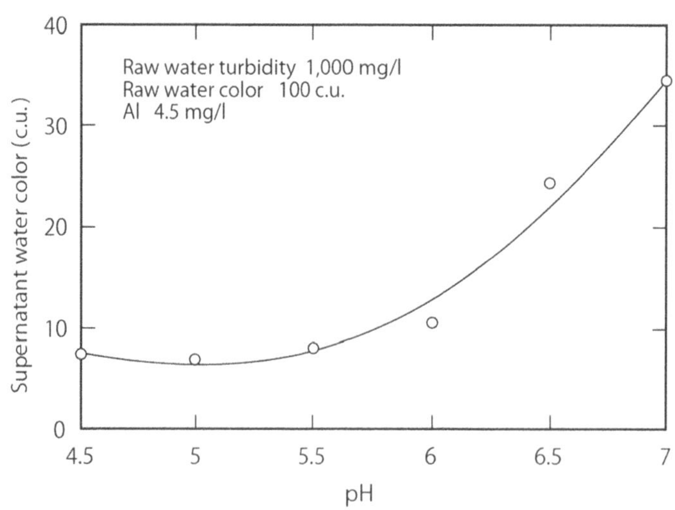

Figure 4.134 Changes in supernatant water color by coagulation pH.

Aluminium coagulation can be used to maximize the charge neutralization capacity at pH 5 or less, and the cross-linking strengthening action can be expected of the polymer.

Aluminium neutralization and cross-linking work together for the following reasons if turbidity and color colloids are present at the same time:

(1) Aluminium added to water undergoes hydrolysis within a few seconds to form tetravalent aluminium aquo complexes, which then undergo a complexation reaction with color colloids (humic substances), and charge neutralization is performed to form microflocs of color components.
(2) Because of the extremely high number concentration, the high charge density and high negative mobility of color colloids compared to clay particles, most of the aluminium aquo complexes react with color colloids.
(3) Some of the aluminium aquo complexes are also distributed to the clay particles, but the ratio of the aluminium aquo complexes that perform charge neutralization reactions with clay particles is low because the clay particles have a large particle size and relatively very low number concentration.
(4) In this way, charge-neutralized color colloids undergo collisional bonding due to Brownian motion and micro-turbulence fluctuations, grow to microflocs, grow to a size capable of cross-linking with clay particles, and form clay-color co-agglomerate flocs (see Figure 4.46 above).

The charge of a color aluminium complex adsorbed onto the surface of clay particles can be estimated by measuring the electrophoretic mobility of the color co-agglomerate microflocs by microscopic electrophoresis.

As a result of the experiment, under the condition of pH 5.0, the electrophoretic mobility of kaolin was −1.7 (μm/s)/(V/cm), and color colloid showed a larger negativity value of −2.3 (μm/s)/(V/cm). Figure 4.135 shows that the aluminium dosage rate required to reach the isoelectric point of the two-component system is measured by a combination of turbidity and color. The aluminium injection rate required for charge neutralization is dominated primarily by color in the mechanisms described above, even in the presence of high turbidity, such as 10^2 to 10^3 mg/l.

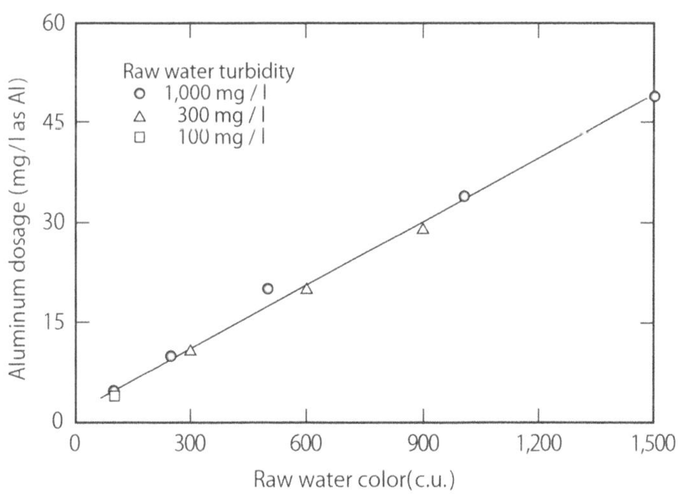

Figure 4.135 The aluminium dosage required to reach the isoelectric point and raw water color.

Table 4.19 Composition of raw water quality and fluidized bed situation.

Kaolinite Concentration (mg/l)	**1000**						**300**			**100**
Color (c.u.)	100	200	500	1000	1500	2000	300	600	900	100
Fluidized bed situation	O	O	O	O	Δ	×	O	Δ	×	O

O: Good pellet, Δ: Massive floc, X: Random floc.

The results of the two-component processing experiments with a fluidized bed are presented in Table 4.19, using color and kaolin suspensions, with aluminium additions (pH 5.0) and minimal polymer additions (set for optimum color removal using a void up flow velocity of 30 cm/min and a Jar test).

Table 4.19 shows that, in a wide range of suspensions from 100 to 1000 mg/l, a satisfactory pellet fluidized bed formation, suspension removal and concomitant satisfactory color removal up to 1:1 ratio of color to turbidity can be seen. As is the case for color, for suspensions there are coagulation difficulties with low molecular weight components of less than 10%, and the removal of aluminium coagulation is not possible as described in the previous section. At a kaolin concentration of 1000 mg/l and with color of 1500 CU and a ratio of 1:1.5, agglomerate particles were formed, but not good spherical pellets (Δ). This is thought to be the composition of high-density agglomerates due to the syneresis of high molecular weight organic polymers. When the color is increased to a higher ratio, only random flocs can be formed, and pelletization is impossible (indicated by X), and even if the amount of organic polymer added for crosslinking is increased, pellet formation is not observed.

Figure 4.136 shows how fluidized bed separation proceeds after beginning, in terms of the color removal rate. The color of the separated supernatant water improves over time and reaches a maximal removal rate of about 93–95%. When the separated water was filtered off at 0.45 μm, it was found that all the color particles were coagulated and micro-flocculated. The problem is that the solid-liquid separation of the fluidized bed takes about 4 hours for low color and about 0.5 to 1 hour for high color levels, to reach a final stable state. The authors think that the generated color microflocs may adhere to the pellets, resulting in a difference in the time required for effective surface layer formation. This can be considered as a process of shifting from high-density agglomeration by syneresis to stratified growth of one by one color-kaolin coagulation microflocs to generate concentric China marble type pellets.

Because the color components are fine particles with molecular masses around 10^3 Da, they must grow to color microflocs of a larger size enable cross-linking with the polymer. Given the presence of minimal structural densities, we believe that color-clay co-microflocs are deposited on the mother pellet and pellet particles are formed. During color micro-flocculation, component particles with voids appear in the early stage, but as the color coexistence ratios increase during the early color colloidal coagulation, structural densities decrease, and the effective densities of the resulting pellets also decrease, as shown in Figure 4.137.

Although higher turbidity systems exhibit higher pellet densities, the effective density of the resulting pellets is 0.002 g/cm^3 for 1 mm-sized particles, which is 2.0 to 2.5 times higher than for random flocs, even at moderate concentrations of raw water with turbidity of 100 mg/l. In highly turbid systems (1000 mg/l), the effective density of the colored clay pellets is 0.1 g/cm^3, which is 10 to 15 times the effective density of the random floc.

The water content of the sludge discharged from the separating tube outside the system is placed on a 1 mm mesh to allow easy dewatering of the sludge. The water content was measured as shown in Figure 4.138. Up to a 1:1 color/turbidity ratio, such simple dewatering separation allows the water content of sludge to be

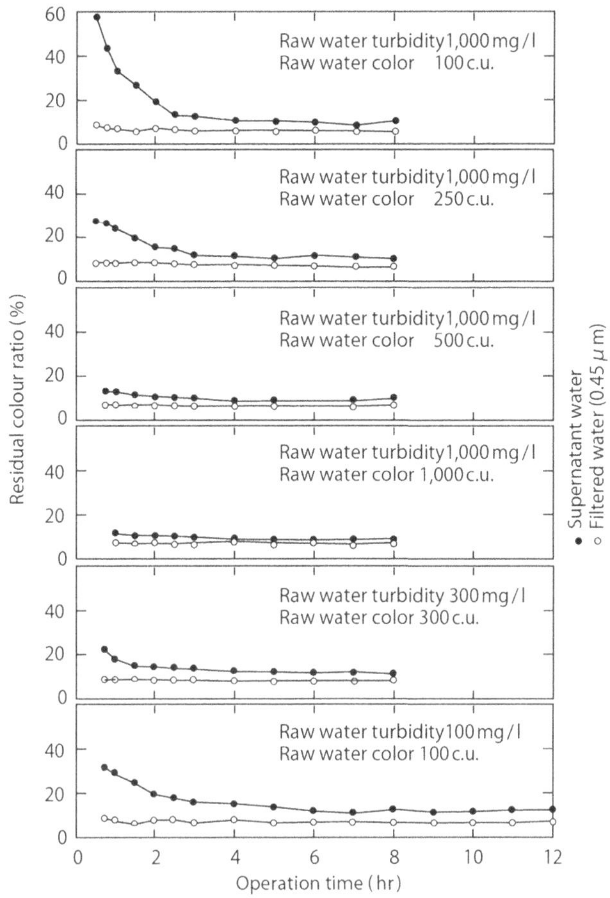

Figure 4.136 Color removal by fluidized bed operation.

immediately reduced to 90% or less and can be subjected to dewatering equipment without a sludge thickening process.

The authors believe that the pellet fluidized bed treatment is a definitively effective treatment method for high-concentration suspensions in binary turbid and color systems.

4.3.2.2 Results and discussions of the treatment mechanism

As described above, the agglomeration of pellets is a process in which the initial particles (particles which are in a condition that they can be agglomerated by charge neutralization and collision) are adhered one by one to the surfaces of the pellet particles (mother pellets) produced in advance, and the initial particles (particles which can be agglomerated by collision) are laminated (one by one type adhesion, china marble type laminated sphere formation). If we assume that the initial particles are uniform spheres, the porosity of the initial particle agglomerates that make up an ideal pellet is around 0.4, as in the case of filling a finite space with uniform spheres (see Chapter 6, Figure 6.80 below). When treating a clay suspension with a turbidity of 1000 mg/l, the average particle size of the clay aluminium microfloc

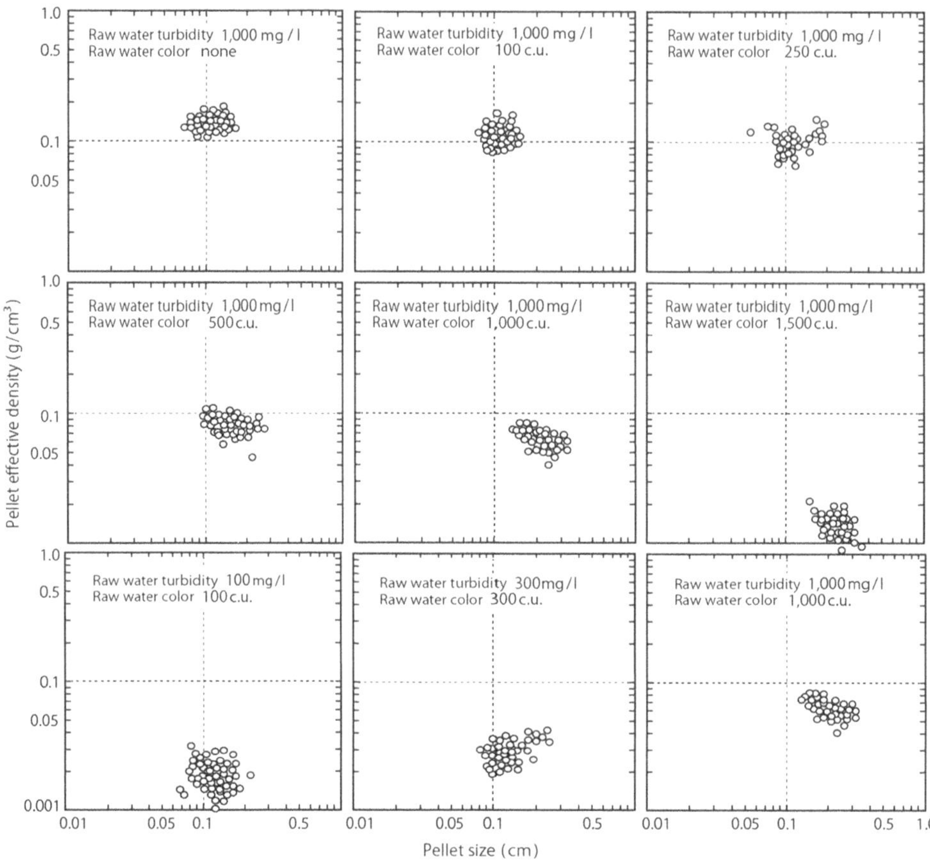

Figure 4.137 Effective density of fluidized bed pellets generated by a color-turbidity coexistence system.

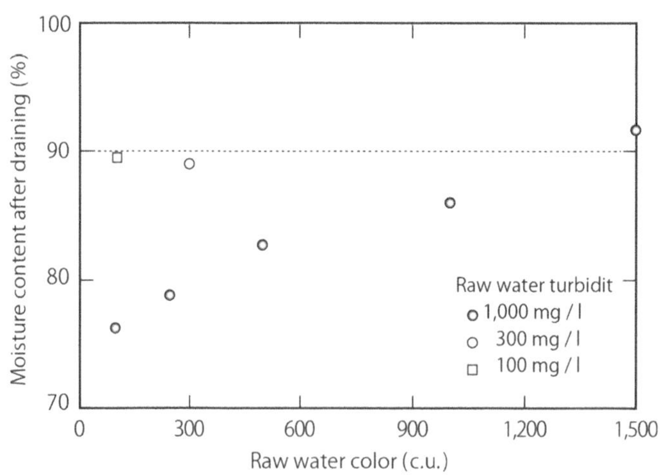

Figure 4.138 Pellet water content after draining through 1 mm mesh.

formed in a chemical mixing pond is about 25 μm. If the effective density is calculated using Tambo and Watanabe's floc density function, that is 0.28 g/cm^3. The measured effective density of pellets formed in a fluidized bed treated with 1000 mg/l of turbidity was as high as 0.139 g/cm^3. The void fraction of the aggregate of the initial particles that make up the pellet is estimated to be about 0.5. This value slightly exceeds the ideal packed porosity of the sphere of 0.40, but the actual microfloc is not a sphere, so the porosity of the pellet formed by the initial particulate build up in the one by one type should be only slightly higher. The estimated porosity of 0.5 seems to prove that the clay microflocs make up the pellet in almost the closest possible packing.

In order for pellet agglomeration to proceed, the presence of an organic polymer crosslinker is essential to strengthen the bond between the microflocs and the existing (mother) pellets. The organic polymers used have molecular masses of the order of 10^4 Da and linear chain extensions of the order of several tens of micrometers. In water, the effective length is considered to be of the order of several micrometers. Considering the diameters of the mother pellet and the microfloc, if the polymer chain length is between several tens of micrometers and several micrometers as described above, the bond between the two can take three forms, as shown in Figure 4.139.

When the diameter of the microfloc is sufficiently larger than that of the polymer molecule, the polymer takes the form shown in (a), and the space occupied by the polymer is minimized to form a high-density pellet. When the microfloc diameter is close to or much smaller than the chain length of the polymer, it takes the form of (b) or (c) and the space occupied by the polymer links (or 'bridges') is relatively large, and it is not possible to construct high-density pellet particles. If the diameter of the clay particle alone or the clay particle coagulation is several tens of micrometers, a good pellet will be formed. If the particle is a fine floc of several micrometers in diameter with only color colloids or excellent color components, a good pellet will be formed. However, if the particle is a fine floc of several micrometers in diameter with only color colloids components, the particle will be in the state shown in (c) and no pellet will be formed.

As a result of these studies, the following process is required to treat a two-component system of color and turbidity in a pelletized fluidized bed:

(1) Even if the color is as high as 100 to 1000 color units (CU), treatment with a pellet fluidized bed with a void upflow speed as high as 30 cm/min can be stabilized if the turbidity is present in concentrations of 100 (mg/l) or higher.
(2) For the raw water described above, removal of most of the color components (molecular masses greater than 1500 Da) and turbidity components that can be coagulated is possible.

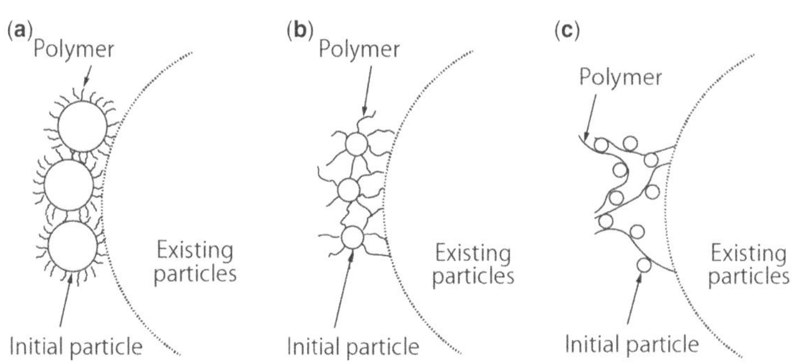

Figure 4.139 Relative initial particle size and forms of polymer linking.

(3) The effective densities of the two-component pellets produced under coagulation conditions capable of pelletizing are of the order of 0.02 to 0.1 g/cm^3 and can be thickened and dewatered easily.
(4) The coagulation condition of the color turbidity two-component system is almost the same as the coagulation of only color, and a high removal rate can be achieved by adding aluminium until reaching the isoelectric point or the critical coagulation potential: −10 mv at pH 5.0. The effect of coexisting turbidity is hardly affected with respect to charge neutralization.

River water with such high color and high turbidity conditions is widely distributed in Southeast Asia, Africa, South America and in China, and it is desirable to carry out scaled-up research for larger facilities.

4.4 COAGULATION AND FLOCCULATION FOR FLOTATION (AIR BUBBLE-SOLID DIPLOPHASE COAGULATION)

4.4.1 Agglomeration of air bubbles and solid colloids for dissolved air flotation

4.4.1.1 Outline of air flotation method

Flotation treatment is a unit process widely used to separate solid and liquid particles from the liquid (water) phase which is further described in Section 6.2, Chapter 6, below.

When the density of solid particles or different types of liquid particles suspended in water is lower than that of water, the low density of these particles causes a phenomenon opposite to that of sedimentation, and the suspended particles begin to move upward due to the buoyancy inherent in them. These solid or liquid particles float and collect on the surface of the water, where they are collected and removed out of the system. For suspensions that differ in density from the suspending medium (water) and separate (by ascending/descending) at a very small rate, and which are difficult to separate by simple flotation or sedimentation, an operation is often performed to increase the buoyancy (density difference) by attaching bubbles (such as air bubbles) to the suspending medium to increase the separation rate by flotation. Dissolved air flotation, in which microbubbles are deposited on and separated from less dense solid suspended particles (lumps), has been widely used in water, sewage and industrial wastewater treatment, and the technologies are sophisticated. Removal of algae caused by eutrophication in lakes and reservoirs (where low turbidity and high color water must be treated by removing both algae and microturbidity as a pretreatment for desalination and seawater use) is widely practiced. Coagulation flocs of metallic salts found during the recycling of sewage treatment are typical suspended particles that are difficult to settle, and are a typical application of the dissolved air flotation method.

The foam separating method (froth flotation) is one form of the flotation method which has been widely used in the refining process of mine engineering. This process is basically conducted mainly by gas-solid adsorption, and the mechanism will be further explained in Section 6.2.

Three types of pressurized air flotation methods are most widely used in the treatment system of dissolved air flotation (see Chapter 6, Figure 6.70) but they can perhaps more simply be reduced to just two: the raw water pressurization method and the circulation water pressurization method:

(1) *Raw water pressurization method.* If the substances for removal which are suspended in the raw water are coagulation particles, the flocs can be destroyed and dispersed when the treated water passes through the pressurized pump and the vacuum valve. For this reason, this method can be used only for floc particles that have a high recoagulation ability, or when coagulation is advanced by injecting a chemical after depressurization of the chemical injection point.

(2) *Circulation water pressurization method.* This method was considered to avoid the drawbacks of the raw water pressurization method. A portion of the treated water that does not contain suspensions is subject to pressurized air uptake. Immediately before the inflow of the flotation tank, the pressurized water containing a large amount of air is mixed with the raw water immediately after the pressure is reduced and the fine bubbles are precipitated, and the bubbles become attached and contained with the particles to be removed. In this method, the raw water containing floc does not pass through the pressurized pump or the pressure reduce valve, and the removal rate does not deteriorate due to the breakdown of coagulation particles (floc). However, in the circulation method, the treatment flow rate of the flotation tank increases due to the addition of circulation water to the raw water. This increases the size of the tank needed and increases the amount of power required. Generally, the circulation flow rate is set to 20 to 30% or less of the original flow rate, and the air dissolution pressure is increased, and design is made to remove the need for an increase in the size of the tank. This method is the most widely used for achieving high overall treatment efficiency.

In this section, we elucidate the kinetic process of gas-solid mixing and adhesion in the floc-forming zone for the circulation water pressure flotation processes shown in Figure 4.70(c).

The Water Research Centre (Medmenham Laboratory and Stevenage Laboratory in the United Kingdom) conducted vigorous practical studies on circulation water pressure flotation processes in the late 1970s. In the Proceedings of the International Conference held in Felixstowe in Suffolk (1976 and 1977), R. Zabel and R. A. Hyde compiled the very valuable results of the conference (Melbourn & Zabel, 1977). In the 1990s J. K. Edzwald *et al.* (1990) proposed a modified model of the Camp-type for an air bubble deposited floating model (a single collector collision model) in the laminar environment. Tambo, Fukushi and colleagues conducted a comprehensive kinetic modeling study in the early 1980s, and reported a series of papers in the Journal of the Japan Water Works Association (in Japanese), (1986a, b). They pioneered the elucidation of the mechanism of mass deposition under actual turbulent conditions. Their research was presented at a World Congress III of Chemical Engineering in Tokyo (1986) and was recognized in the field of chemical engineering as a kinetic model for air solid clusters running under turbulence.

The kinetic model was published later in Water Science and Technology (Fukushi *et al.*, 1995), after a report by Tambo at the IWA International Specialized Conference on the Flotation Process held in Florida, Orlando in 1994, where it was discussed by specialist groups. Today, it is recognized as a standard model describing the kinetic process of microbubble deposition on suspended solids and flotation segregation for growing flocs under turbulent conditions in ponds.

The model of Edzwald *et al.* (1990) can be thought of as a solution under specific conditions that do not exist in the real pressure flotation process and where the floc diameter (d_f) is sufficiently small ($d_f \ll d_a$) relative to the bubble diameter (d_a); see Fukushi *et al.*(1995).

As shown in Figure 4.140, the dissolved air flotation system consists of a rapid mixing tank for adding and mixing chemicals for coagulation, a floc forming tank (multiphase flocculation), and a flotation tank with a turbulent mixing zone at the inflow end to allow microbubbles to adhere to the growing floc. The air bubbles are completely attached to the floc, and the air-solid (floc) aggregate with a higher flotation velocity floats as floated sludge (flotation residue) on the surface of the flotation tank. The aggregate is scraped off from the surface of the flotation tank, and clear treated water is removed from the lower layer of the separation tank. A portion of the treated water is sent to an air dissolution tank that produces a circulation water containing a predetermined quantity of dissolved air, and at the same time a predetermined pressure of air is sent from the compressor. In addition, floated sludge scraping system, pressure pump, pressure reducing valves, etc. are required.

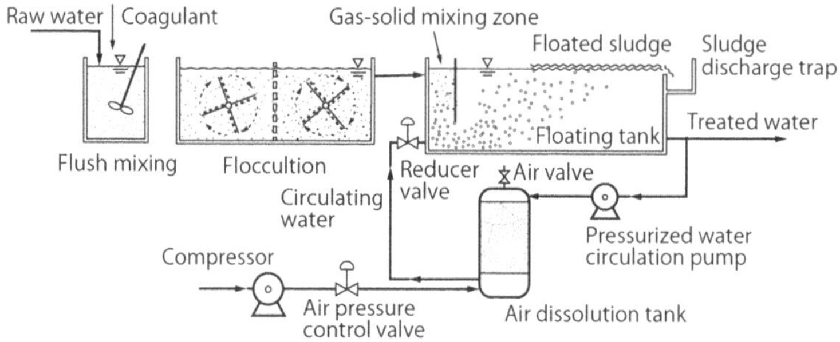

Figure 4.140 A circulating water pressure flotation process.

Preparation for operation begins with the creation of pressurized water with a predetermined quantity of dissolved air. Initially, tap (city water supply) water is used, and when the process is established, it is switched to treated water. The air dissolution tank is filled with pressurized water (part of treated water) injecting compressed air by compressor until the pressurized water reaches a predetermined air level (saturation).

The main constituents of air are Nitrogen and Oxygen, which are poorly soluble in water. The solubility of each constituent increases in proportion to the pressure (partial pressure) of each gas according to Henry's Law (Equation 6.80) below. Figure 6.71 shows the added pressure and the amount of saturated air dissolved. The basic structure of the various dissolution tanks currently in use is shown in Figure 6.72.

Next, a predetermined quantity of coagulant and coagulation adjuvants is added to the raw water in a rapid mixing tank, and the produced microflocs are guided to the next floc-forming tank to form a group of flocs that grow to a predetermined size. These coagulation and floc forming processes are described in detail in Chapter 4.

By allowing the flocs to gently inflow the air-solid (floc) mixing zones that promote microbubble deposition in the flotation tank, floc breakages are avoided. Air is dissolved into the pressurized water circulating from the air dissolution tank at a pressure of about 4 kPa (kg/cm^2) to saturation, and the pressure is instantaneously reduced by a pressure reduce valve to atmospheric pressure (about 1.3 kPa), to which the water in the flotation tank is added to generate microbubbles in the mixing zone. At this time, the water becomes cloudy with fine bubbles.

The results of photographic measurements of the generated bubble diameter distribution are shown in Figure 6.74. The average diameter is about 60 μm, and the measurement result of the bubble diameter distribution after adhering to kaolin floc is almost the same, and the size distribution of the microbubble diameter hardly changes either immediately after depressurization or after adhering. This is considered to be because the size distributions of the microbubbles do not change and aggregate in the gas-solid mixing zones, and the microbubbles collide with and adhere to the inflow growing floc groups while stably maintaining the dispersed state at the time of generation. The reason for this is that the ZP distributions of microbubbles generated in city water measured by the flotation potential method are as shown in Figure 6.76. The ZP distributions of microbubbles have large negative surface charges, such as −150 mv which is much higher than −10 mv (coagulation critical potential levels shown by Tambo) in the neutral pH range, and the collision of bubbles is disturbed by their repulsion of each other, and they continue to float stably in the water as dispersed microbubbles.

As shown in Figure 6.141, the water flow in the air-solid mixing zone is a mixed flow drive to the inflow vacuum jet through inflow pipes. The water has a flow velocity $v \doteqdot 10^2$ cm/sec and an order of cross-sectional dimension $L \doteqdot 10^1$ to 10^2 cm. In terms of the mean flow velocity of inflow water, when $v \doteqdot 30\sim50$ cm/sec, $Re \doteqdot (30 \text{ to } 50) \times (10 \text{ to } 100)/0.01 \gg 4{,}000$, and the air bubble-solid mixing zone is

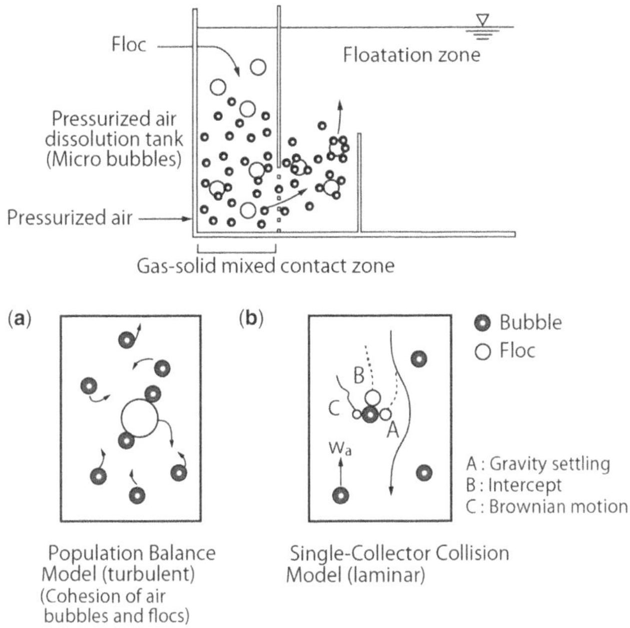

Figure 4.141 The collision and attachment of air bubbles and flocs in the gas-solid mixed contact zone.

under turbulent flow conditions. The collisional attachment of micro-bubbles to flocs must be treated as a kinetic phenomenon under locally isotropic turbulence, similar to the floc-forming process described by Tambo and Watanabe above. Therefore, it is not a laminar basin as Edzwald *et al.* (1990) thought, and is not a phenomenon of bubble deposition proceeding in the form of a single bubble colliding with a floc under laminar flow conditions, or the Brownian motion of the bubble causing it to collide with the floc. Details of the kinetics of collisions between flocs and bubbles under isotropic turbulence are described in the following sections.

For reference, Figures 4.141(a) and 4.141(b) schematically show the population balance (kinetic) model under turbulent flow (assuming locally isotropic turbulence) considered by Tambo, Fukushi, and Matsui (Fukushi *et al.*, 1995), and the single-collector collision model in which the deposition of bubbles progresses in different settling and floating velocities for floc particles and bubbles under laminar flow conditions (which may be hydrostatic).

4.4.1.2 Adhesion of bubbles to floc particles and flotation process

A batch flotation tester (Hokkaido University Type; see Figure 6.73) was used to conduct a batch flotation test similar to the jar test for coagulation floc formation (Tambo *et al.*, 1984). The following suspension components with different characteristics were tested to investigate the differences compared to coagulation sedimentation:

(1) aluminium hydroxide coagulum floc only;
(2) Clay aluminium floc;
(3) Color aluminium floc; and
(4) Bubble adhesion and floating separation of aluminium hydroxide floc.

4.4.1.2.1 Aluminium hydroxide coagulum floc only

Figure 4.142 shows how aluminium only flocs are separated by bubbling with microbubbles in relation to pH, aluminium removal rate, and floc mobility.

Aluminium sulfate was added to Sapporo City Water to have a predetermined concentration, and the pH was adjusted to a predetermined pH with NaOH or HCl immediately, and a floc-forming jar test and bubble introduction flotation treatment were carried out. The mobility of aluminium hydroxide was measured by micro-electro phoresies (Rank Bros. Mark II).

Removal by sedimentation appears well in the range of mobility from +1 to −1 μm/s/V/cm, indicating the separation limit of critical coagulation potential by Tambo. However, in the removal of dissolved air flotation, the separation area was limited to +1 to 0 μm/s/V/cm, and good removal was observed only in the area where aluminium flocs are positively charged. In the region where the hydroxide is negatively charged, flotation separation cannot be expected. This means that micro-bubbles with a negative charge as high as −100 mV near neutral pH cannot adhere to aluminium hydroxide in the negatively charged region, as shown in Figure 6.76 (in Chapter 6) and, therefore, air bubble-solid adhering agglomerates do not proceed and the dissolved air flotation method cannot be applied.

4.4.1.2.2 Flotation separation of clay aluminium flocs

Comparative tests of sedimentation and flotation treatment in Sapporo city water, with a suspension 50 mg/l of kaolin, were performed in the same manner as in the preceding paragraph. The results are shown in Figure 4.143.

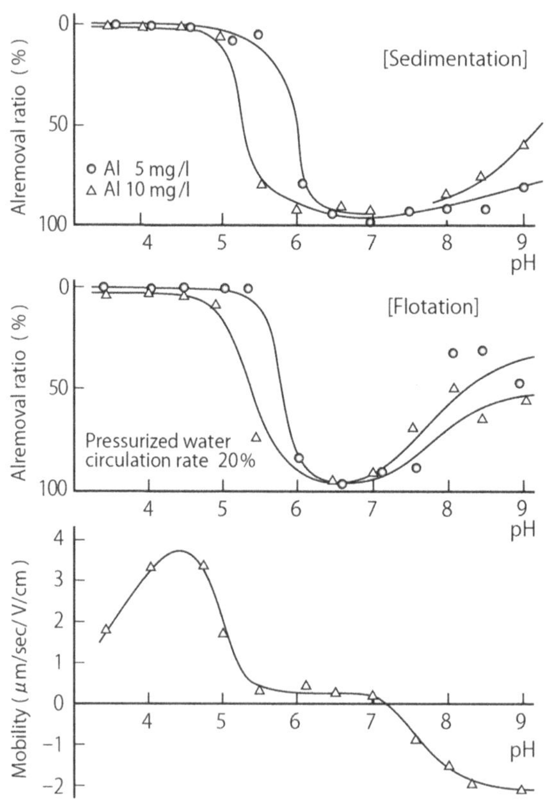

Figure 4.142 The flotation of aluminium flocs.

The bottom Figure in Figure 4.143 shows the mobility of kaolin floc measured by electrophoresis. Kaolin particles have a negative mobility (potential) of around −2 μm/s/V/cm. The coagulation and sedimentation removal are consistent with the coagulation condition in which the ZP (mobility) falls within the critical coagulation potential shown by Tambo when the coagulation floc potential is present in the range of +1 to −1 μm/s/V/cm mobility, in the neutral to weakly acidic pH range. However, even if the mobility is −1 μm/s/V/cm or less towards the weak alkaline side, the removal rate of separation of flotation will rapidly deteriorate when the pH is 7.2 to 7.3. This is probably because, as shown in Figure 4.143, the potential of aluminium hydroxide becomes negative, and it is impossible to adhere to negatively charged micro-bubbles. Needless to say, negatively charged micro-bubbles cannot adhere to negatively charged kaolin particles, indicating that air bubble-solid adhesion to flocs proceeds by the mediation of a positively charged precipitation coagulant such as aluminium.

4.4.1.2.3 Flotation separation of color aluminium flocs (KP black liquor)

Figure 4.144 shows the results of a batch-type sedimentation and flotation test similar to the one described in the previous section, using dilute KP black liquor as the test water. Color removal indicates the ultraviolet absorbance at 390 nm.

As with clay suspensions, sedimentation separations proceed effectively when the mobility of coagulation flocs is in the critical coagulation potential range of −1 to +1 μm/s/V/cm. As discussed

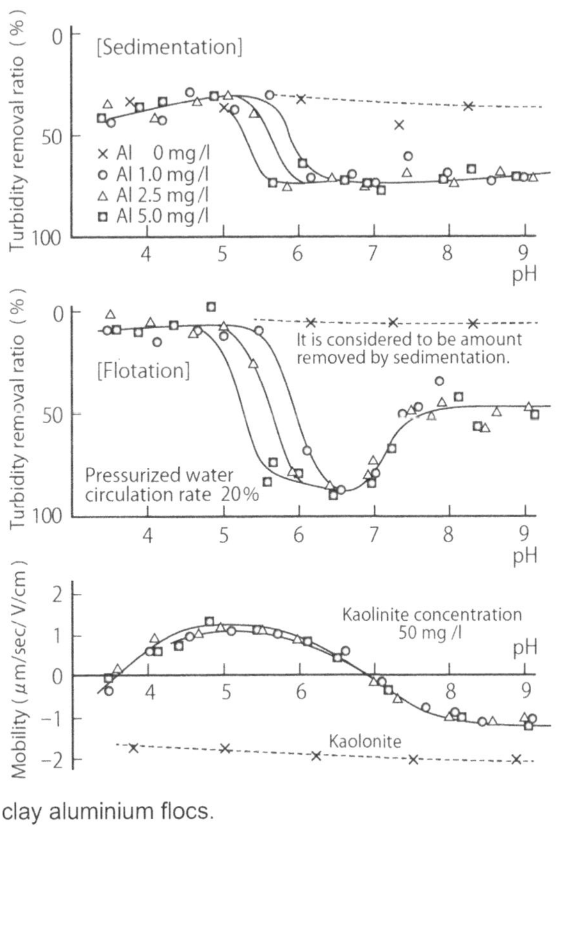

Figure 4.143 Flotation of clay aluminium flocs.

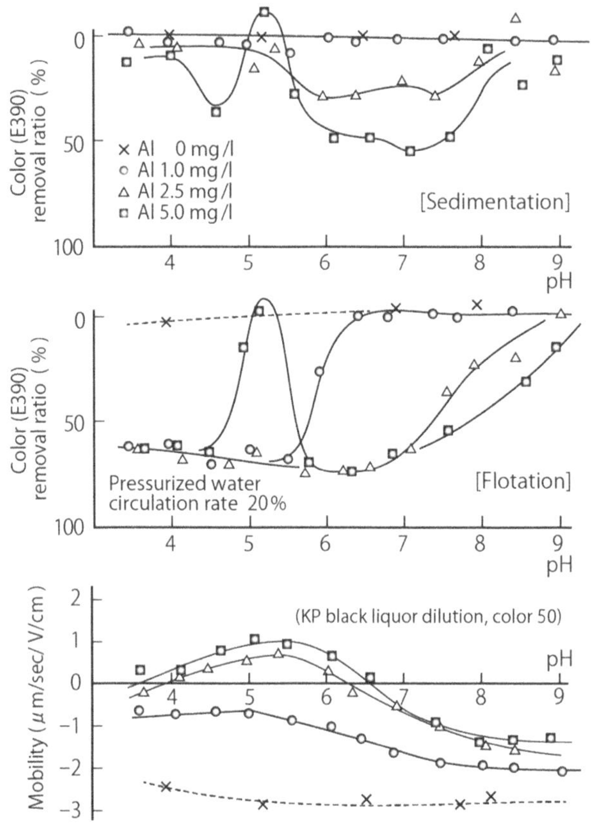

Figure 4.144 Color clay aluminium flocs.

above in the coagulation section of color colloids, color colloids have mobility ranging from −2 to −3 μm/s/V/cm, with a slightly more negative potential than clay particles. However, since the particle size of color colloids differs by more than 10^{-2} times from that of clays, the specific surface area of color colloids is 10^4 times larger and the surface charge is 10^4 times greater at the same concentration.

For color colloids, the coagulation condition was selected to be around pH 5.0, which maximizes the charge neutralization capacity of aluminium. For clay-based colloids, the crosslinking capacity of a certain amount (about 1 mg/l of Al) which is effective for maintaining flocs is required by suppressing excessive charge neutralization capacity. The authors believe that the neutral pH range satisfies the appropriate coagulation condition. See Figures 4.13, 4.16, 4.18 to 22, Tables 4.9 (clay-based), Figures 4.29 to 4.39 (color-colloid-based), Figures 4.45 and 4.46 (Coagulation patterns).

Thus, as shown in Figure 4.144, the diluted KP black liquor undergoes charge reversal due to aluminium overload at pH 5, bringing about the absence of removal with both sedimentation and flotation. Coagulation sedimentation and flotation proceed within the range from the weakly acidic region to the positively charged region of aluminium, within the range of the critical coagulation potential. However, in the case of flotation, as in the case of the clay-based system described above, air bubbles cannot adhere to the floc above pH 7.2, where negatively charged aluminium hydroxide is formed, and the flotation process does not proceed. As for the color/clay system, we mentioned earlier that the coagulation condition is governed by color content.

However, like the flotation of color colloids, the flotation process is not possible in areas where aluminium has a negative charge.

Importantly, we must be aware that the dissolved air flotation method, as compared to coagulation sedimentation, has a theoretical limitation that negatively charged micro-bubbles do not adhere to the alkaline side where metallic hydroxide changes to negative, and operation cannot be performed.

4.4.2 Kinetics of air bubble-solid agglomerate formation in dissolved air flotation

4.4.2.1 Collision formula for bubble and floc groups

As discussed in detail in the previous section on floc formation, the number of collisions between microparticles occurring in a unit volume of fluid due to turbulent fluctuations in micro-water masses, $N_{turb.}$, can be expressed by Equation 4.26 when the phenomenon is considered to be a diffusion rate-limiting phenomenon in the viscous sub-region (following Levich's locally isotropic turbulence theory). Equations 4.42 to 4.44 determine whether the phenomenon should be discussed in the inertial or viscous sub-region of locally isotropic turbulence (Tambo & Fukushi, 1986a,b).

In the air-solid mixing contact zones of a flotation tank, as shown in Figures 4.140 and 4.141, if agitation is mainly carried out by jets of pressurized air-soluble water and the kinetic energy of the pressurized water jet is all dissipated during mixing, the total energy dissipation rate (erg/cm^3 sec or 10^{-7} W/cm^3) was considered to be about 10^{-1} to 10^2 erg/cm^{3}sec in our experiments. Therefore, the microscale λ_0 of the turbulent vortex is considered to be of the order of 500 μm to 1 mm, and the diffusion-limited response dominated by the viscous subrange domain can be assumed. Therefore, Equation 4.26 is considered to be the basic equation.

For floc formation, we need only consider the collision of floc particles with each other. However, in the contact mixing zones of a flotation tank, there are three types of collision combinations, i.e. bubble-bubble, bubble-floc and floc-floc. The ratio of the number of collisions in each case is roughly calculated by Equation 4.26, referring to experimental data such as the bubble diameter $d_a = 62.8 \times 10^{-4}$ cm, the bubble number $n_a = 5.86 \times 10^4$ cm^{-3} (applied pressure 4 kg/cm^2, 15°C, circulation ratio 0.1), the kaolin coagulation floc diameter $d_f = 400 \times 10^{-4}$ cm, and number of floc $n_f = 2 \times 10^2$ cm^{-3} (ALT ratio 0,05, kaolin 50 mg/l, pH 7.25). Assuming these data, the relative number of collisions would be $N_{air\text{-}air}$: $N_{air\text{-}floc}$: $N_{floc\text{-}floc} \doteqdot 330: 57: 1$.

As mentioned earlier, bubbles with the highest collision probability and the coalescence of bubbles cannot exist because they are hindered by the high negative surface potential as high as −100 mV or more (Tambo *et al.*, 1985). In addition, the frequency of the floc-to-floc collision (growth) is one order of magnitude lower than that of bubble collision, and is practically negligible in the discussion of the kinetic process of air bubble-solid deposition. Therefore, the kinetics of attaching microbubbles to a group of grown flocs should be discussed in the study of the flotation enhancement operation in the mixing zone of a tank.

4.4.2.2 Kinetics of the bubble deposition process

The formulation of the bubble deposition process is expressed in terms of how the number of flocs (to which i bubbles are attached) changes over contact mixing times in the air-solid mixing contact zones (Tambo & Fukushi, 1986b).

Consider the following assumptions for the formulation:

(1) Agitation intensity of the gas-solid mixed contact zones may be expressed as the mean effective agitation intensity (energy dissipation factor) ε_0 (erg/cm^{3}sec, W/cm^{-3}).
(2) Bubble diameter d_a is constant, and the mean diameter is used.

(3) Impingement diameter ($d_f + d_a$) of bubbles and flocs will be constant regardless of the number of bubbles attached.
(4) Once attached to the floc, bubbles do not detach.
(5) There are a largest number of attached bubbles, number m, depending on the diameter of the floc d_f.
(6) In the operation of an air floc agglomeration system, the number of microbubbles in a unit volume of water is unchanged (low concentration suspension treatment). This means that the number of bubbles used for flotation (the number of attached bubbles) is small compared to the total number of bubbles deposited. That means the air-solid ratio, which is a significant operation index for high-concentration suspensions, is not discussed in the case of the low concentration clarifying flotation process.
(7) During the collision of the floc particles with the bubbles, there is a collision-attachment factor α, and only a certain percentage of the collided bubbles adhere.

The process by which N_{fi} of the number of flocs of diameter d_{fi} with i bubbles attached changes with time can be written as shown in Equation 4.79, based on Equation 4.26. Equation 4.80 is obtained by dividing the equation for decreasing the number density (N_{f0}) in the early stage when no air bubbles are attached. Also, the rate of reduction in the concentration of the number of bubbles in the water is calculated as in Equation 4.81.

$$\frac{dN_{fi}}{dt} = \frac{3}{2}\pi\beta\left(\frac{\varepsilon_0}{\mu}\right)^{\frac{1}{2}} n_a(d_a+d_f)^3\left(\alpha_{fi-1}\cdot N_{fi-1} - \alpha_{fi}\cdot N_{fi}\right) \quad (i=1-m_f) \quad \text{—}(4\cdot79)$$

$$\frac{dN_{f0}}{dt} = \frac{3}{2}\pi\beta\left(\frac{\varepsilon_0}{\mu}\right)^{\frac{1}{2}} n_a(d_a+d_f)^3\left(-\alpha_{f0}\cdot N_{f0}\right) \quad (i=0) \quad \text{—}(4\cdot80)$$

$$\frac{dN_a}{dt} = -\int_0^{\infty}\left\{\frac{3}{2}\pi\beta\left(\frac{\varepsilon_0}{\mu}\right)^{\frac{1}{2}} n_a(d_a+d_f)^3\sum_{i=0}^{m_f-1}\alpha_{fi}\cdot N_{fi}\right\}dd_f \quad \text{—}(4\cdot81)$$

$$m_f = \pi\alpha_0\left(\frac{d_f}{d_a}\right)^2 \quad \text{—}(4\cdot82)$$

$$\alpha_{fi} = \alpha_{f0}\left(1-\frac{i}{m_f}\right) = \alpha_0\left(1-\frac{i}{m_f}\right) \quad \text{—}(4\cdot83)$$

$$\frac{d\bar{i}_f}{dt} = \frac{3}{2}\pi\beta\left(\frac{\varepsilon_0}{\mu}\right)^{\frac{1}{2}} n_a(d_a+d_f)^3\alpha_0\left(1-\frac{\bar{i}_f}{m_f}\right) \quad \text{—}(4\cdot84) \qquad (4.79\text{–}4.84)$$

N_{fi} : Number of flocs in unit volume with i bubbles attached (cm^{-3})
t : Detention (contact) time in gas-solid mixing zone (sec), $\beta \fallingdotseq 1/\sqrt{15}$ (-)
ε_0 : Effective energy dissipation rate (Consumption rate, $erg/cm^3\cdot sec$)
μ : Water viscosity ($g/cm\cdot sec$), d_a : Bubble size (cm), d_f : Floc size (cm)
d_{fi} : Size of i bubbles attached ($\fallingdotseq d_f$, Ignore change in size due to bubble attached.
n_a : Number concentration of bubbles (cm^{-3})
m_f : Maximum number of bubbles attached to d_f size floc (-)
α_{fi} : Collision cohesion (attach) probability of bubbles being transported. to the floc where i air bubbles are attached. (-) Decreased by the number of bubbles covering floc surface.
α_0 : Initial collision cohesion probability (-)

Equation 4.79 - 4.84 The gas attachment and the floc number concentration.

Because the bubbles are strongly negatively charged (as described Section 6.1.2 in Chapter 6), they must adhere to the positively charged aluminium hydroxide on the floc surface. If the presence ratio of the aluminium coagulant effective for adhesion of floc surfaces is α_0, α_0 can be assumed to be the initial collision adhesion probability (coefficient), and the maximum microbubble number m_f that can be attached to a floc of diameter d_f can be determined as in Equation 4.82. Many experimental results from Tambo and colleagues suggest that $\alpha_0 \doteqdot 0.3$ to 0.4. As the adhesion of bubbles progresses and the percentage of the surface covered by negatively charged bubbles increases, the α_{fi} of the probability of collisional adhesion of subsequent bubbles to flocs with i bubbles attached to the surface gradually decreases, as shown in Equation 4.83. When the maximum number of adhered air bubbles $i = m_f$ is reached, the air-solid adhesion phenomena end. The mean bubble i_f adhering to the floc of diameter d_f increases with the duration of agitation as shown in Equation 4.84.

4.4.2.3 Dimensionless formula and practical solutions of the kinetic equations

The kinetic equations for the air bubble-solid deposition process described above can be made dimensionless by introducing the following dimensionless variables. Prof. K. Fukushi (Hachinoe Institute of Technology, Japan) in his doctoral thesis (1986, Hokkaido University), performed a Laplace transformation of the original system to obtain an analytical solution (Tambo & Fukushi, 1986b).

The dimensionless equation for calculating the change in the N_{Fi} of the number of i bubbles adhering to the floc F is calculated using the mean value of the distributions of the bubble diameter d_a and the floc diameter F, as in Equation 4.85.

$$N_{Fi} = {}_{m_F}C_i \exp\{-(1+F)^3\theta\}[\exp\{(1+F)^3\theta/m_F\}-1]^i \quad (i = 0 - m_F) \quad —(4\cdot85)$$

$_{m_F}C_i$: Combination of m_F and i (Mathematical combination)
F : Floc size ($= d_f/d_a$)
N_{Fi} : Number concentration of F size floc with i bubbles ($= n_{fi}/n_f$)
N_a : Number concentration of non-attached bubbles ($= n_a/n_0$)
m_f : Maximum number of bubbles attached to F size floc ($= \alpha_0 F^2$)
θ : Normalized mixing diffusion time ($= d\theta/dt = N_a$)
n_a : Average number concentration of floating bubbles

(4.85)

Equation 4.85 The floc number concentration of size F with i bubbles attached.

Here:

$_{mF}C_i$ is the mathematical combination of m_F and i

Numerical values that are dimensionless are:

Floc (nondimensional) diameter: $F = d_f/d_a$
Concentration of floc diameter F with i attached bubbles: $N_{Fi} = nf_i/n_f$
Concentration of the number of bubbles that do not adhere: $N_a = n_a/n_0$
Maximum number of bubbles attached to the floc of diameter F: $m_f = \alpha_0 F^2$
Dimensionless mixing agitation time: $T = (3/2)\,\pi\beta\,(\varepsilon_0/\mu)^{1/2} n_a d_a^3 \alpha_0 t$
Mixed-diffusion time θ: $d\theta/dT = N_a$ (normalized to account for the reduction in the number of floating bubbles due to the progress of deposition)
Mean number concentration of diameter f flocs, N_{ft}
The mean number concentration of suspended air bubbles, n_a
Viscosity coefficient of water, μ
Factor $\beta = (1/15)^{1/2}$

As a result, the mean number $\bar{i}_F$ of bubbles adhering to the floc of size F is given by Equation 4.86, and the calculation of i_F is easier than the calculation of N_{Fi} using Equation 4.85, and has sufficient practical accuracy. The relationship between the two dimensionless mixing and stirring times θ and T is shown in Equation 4.87.

$$\bar{i}_F = m_F[1-\exp\{-(1+F)^3\theta/m_F\}] \quad —(4\cdot 86)$$

$$\frac{d\theta}{dT} = 1 - N_0 \int m_F[1-\exp\{-(1+F)^3\theta/m_F\}]\, f(F)dF \quad —(4\cdot 87)$$

(4.86–4.87)

Equation 4.86 The average number of bubbles, Equation 4.87 The dimensionless mixing time.

The f(F) distributions of floc size in Equation 4.87 are not as straightforward as discussed by Tambo and Watanabe earlier in the discussion of floc growth functions. Normally, we need to find a numerical solution. The second term on the right-hand side shows the reduction of free-floating bubbles in the liquid system due to the adhesion of bubbles to the floc. Normalized (dimensionless) mixing and agitation time θ is required in any case in the calculation of the adhesion of the bubbles to the floc. Especially in a suspension system with a high concentration, the free-floating bubble concentration in the liquid system rapidly decreases, so it is essential to correct the free-floating bubble concentration with time, and this function corrects the non-dimensional stirring duration T. By contrast, in dilute suspension systems such as water treatment, $\theta \doteqdot T$ is obtained, and the phenomena can be described only by the dimensionless agitation time.

Figures 4.145(a) and 4.145(b) are examples of the bubble deposition process calculated by Equation 4.85.

4.4.2.4 Calculation of floating rate of air bubble-solid coupled floc

The necessary overflow rate of the flotation tank can be calculated by determining the flotation velocity of the flocs with bubbles attached, as follows. By combining the effective density of flocs and the buoyancy of attached air bubbles, which vary depending on the floc density functions of Tambo and Watanabe, the rising velocity W_{af} of flocs with bubbles attached is calculated assuming the normal floating velocity of particles is the Stokes region. Given the motion of the gas in the bubble provided by the Rhbczski-Hadmard equation presented for the sedimentation of emulsion, the equation for the ascending velocity of floc with bubbles attached can be written as shown in Equations 4.88, 4.89, and 4.90.

$$W_{af} = \frac{4g}{3\mu k}\,\frac{i(\rho_w-\rho_a)-a(d_f/1)^{-k_\rho}(d_f/d_a)^3}{i+(d_f/d_a)^3}\,d_{af}^2 \quad —(4\cdot 88)$$

$$d_{af} = (d_f^3 + i d_a^3)^{\frac{1}{2}} \quad —(4\cdot 89)$$

$$k = \frac{16 i d_a^2 + 45 d_f^2}{i d_a^2 + d_f^2} = \frac{16i + 45(d_f/d_a)^2}{i + (d_f/d_a)^2} \quad —(4\cdot 90)$$

(4.88–4.90)

Equation 4.88 - 4.90 The flotation velocity of bubble-attached floc.

where g is the acceleration of gravitational force; ρ_w, ρ_a is density of water and air, a; K_ρ is the coefficient of floc density function; d_{af} is the diameter of bubble-attached floc; i is the number of bubbles attached to floc, d_f; and d_a is the diameter of floc and bubble.

In real operation, the number of bubbles adhering to the floc is in the order of $i \doteqdot 10^0$ to 10^2, the effective density of the floc $\rho_e = \rho_f - \rho_w \doteqdot 10^{-3}$ to 10^{-1} g/cm^3, and the relative size ratio of bubbles to floc df/da $\doteqdot$

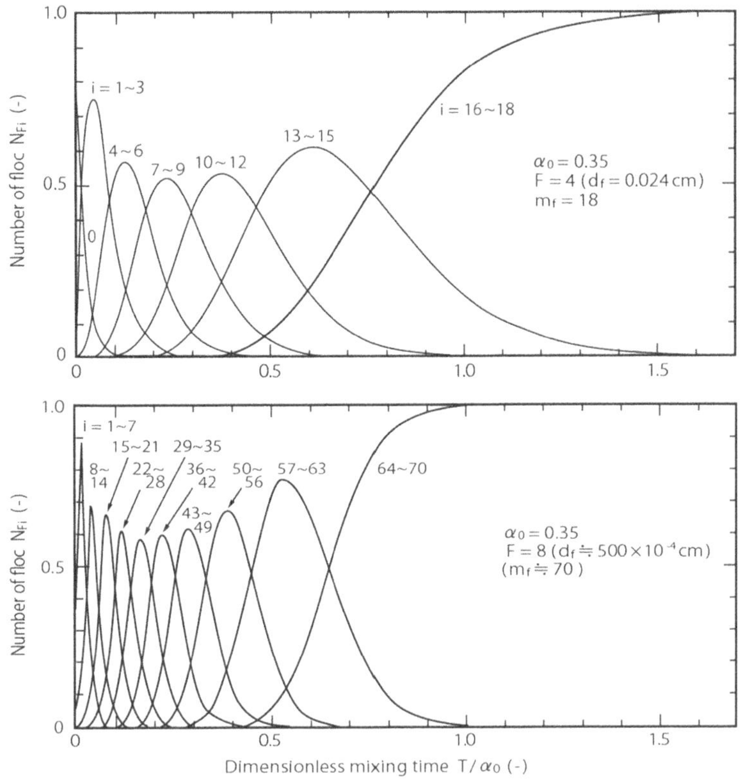

Figure 4.145 Two examples showing relative changes of i-bubble attached to number of floc and dimensionless mixing time.

100 to 10, which means that the floc size and number of bubbles adhering to the floc determine the floating velocity of the bubble adhering floc.

Figure 4.146 shows the simulation results showing how floating velocity is improved by the progression of the dimensionless agitation mixing time T/α from the floc number concentration N_{Fi} (the dimensionless value being the size F floc with i bubbles attached), determined by Equations 4.85 and 4.86.

4.4.2.5 Experimental model validation

To demonstrate the proposed kinetic models, batch flotation tests and small continuous-flow flotation system tests (throughput of 1 ℓ/min) were conducted on kaolinite flocs and color colloid flocs (a dilution of KP black liquor), using a Hokkaido University type flotation tester (see Figure 6.73, Chapter 6) (Tambo *et al.*, 1986).

The effective agitation energy consumption rate estimated from the maximum-growth particle size by way of Tambo and Hozumi (1970b, 1979a) is $\varepsilon_0 = 6 \times 10^{-2}$ erg/cm^{3}sec $= 6 \times 10^{-9}$ W/cm^3 in batch mode and $\varepsilon_0 = 0.6$ erg/cm^{3}sec $= 10^{-8}$ W/cm^3 in continuous-flow mode. The energy consumption rate of the agitating mixing section of the continuous flow experimental apparatus was higher than that of the batch type, and some floc breakage was observed.

The particle size distributions before and after the flotation operation of color flocs are as shown in Figure 4.147, and the amount of generated air bubbles was determined from Henry's law. When the

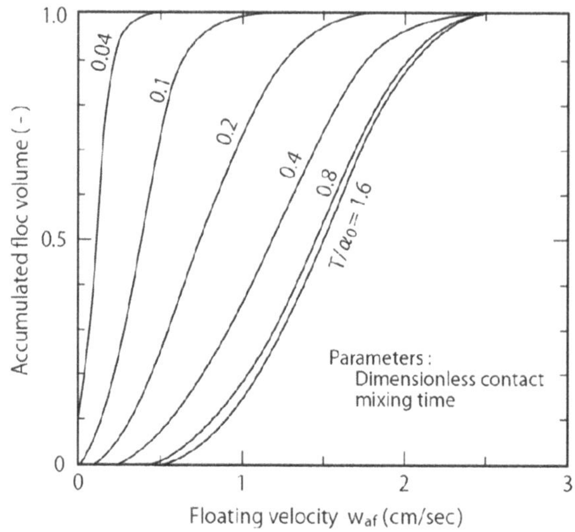

Figure 4.146 Simulation results of the removal rate improvement (accumulated floc volume) according to floating velocity (overflow rate).

circulation ratio was 0.1 and pressure 392 kPa was applied, the generated air bubbles were 7.4×10^4 cm^3. The flotation velocity measurement for the bubble diameter and floc density function was obtained by microscope videography. The factor that has not been clarified in the operation of the kinetic equation is the collision-adhering probability α. In a clay-based system, α_0 was previously estimated to be about 0.35, but the values for color flocs need to be estimated. When we calculated the floating rate of the color floc by varying the values of α_0, we found that the simulation results assuming assumed α_0= 0.4 (shaded part in Figure 4.148) and the actual measurement results (circles in the Figure) were well suited, and that the values were almost the same (slightly larger) as $\alpha_0 = 0.35$ that was assumed when clay aluminum floc floated. From the two Figures in Figure 4.148, it can be observed that if the aluminum content is increased over a wider range, the α_0 value increases.

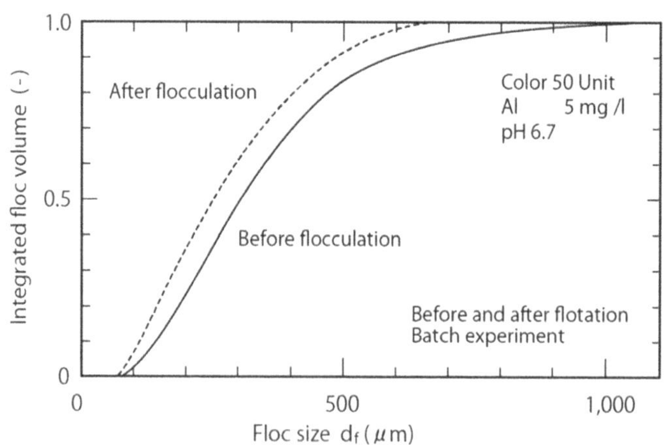

Figure 4.147 The particle size distribution of color flocs (a dilution of KP black liquor).

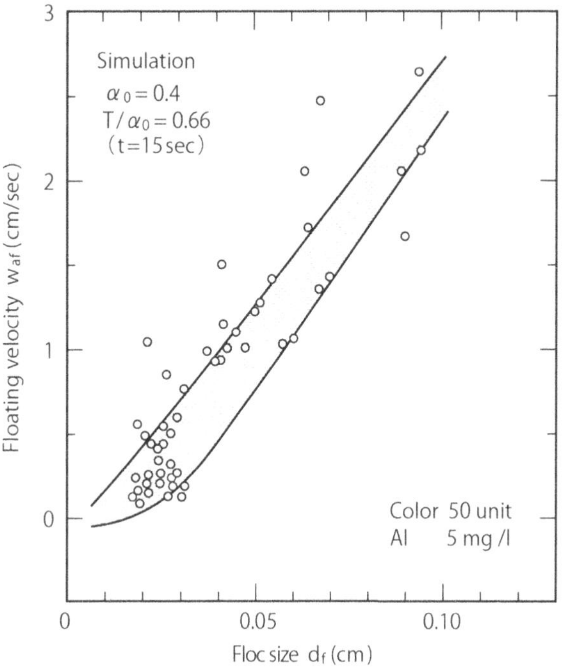

Figure 4.148 Floating velocity and particle size of color flocs with bubbles attached.

Figure 4.149 verifies how the floating rate of clay aluminum flocs proceeds with increasing bubble mixing times using the kinetic modeling and batch floating test results. Experimental results demonstrate the effectiveness of the kinetic models.

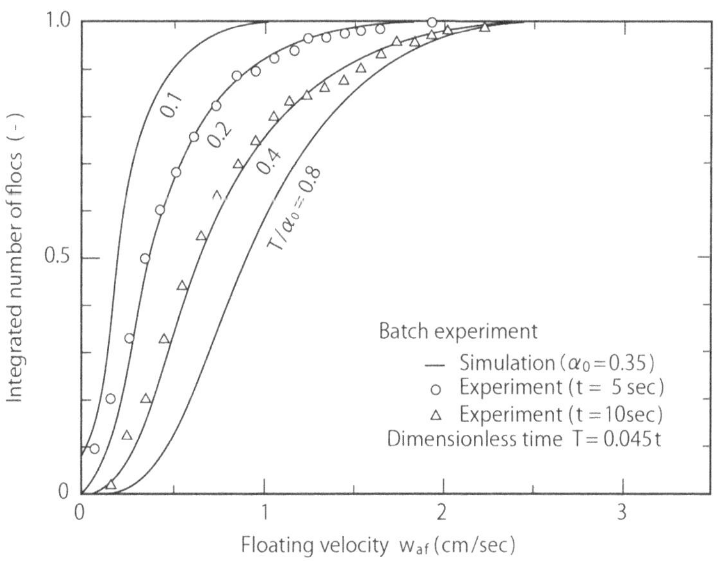

Figure 4.149 Floating velocity distribution of clay aluminium floc.

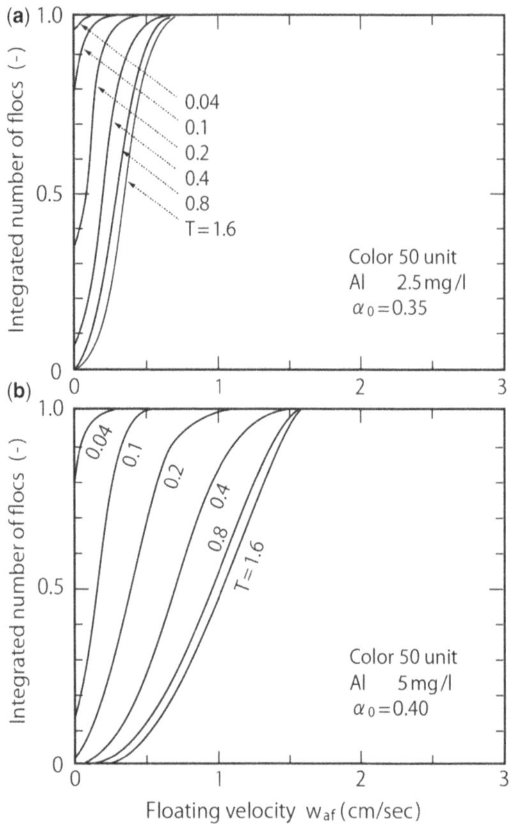

Figure 4.150 Simulation of floating velocity distribution.

Figure 4.150 shows the results of simulations of the process of increasing the floating velocity profile with the transition of the mixing time of color aluminum flocs. The results of simulations show that the floating velocity was significantly increased when the addition rate of aluminum was 5 mg/ℓ, and $\alpha_0 = 0.40$. When the injection rate was 2.5 mg/ℓ, $\alpha_0 = 0.35$. Figure 4.151(a) and (b) show how this result appears in a continuous floating experiment. Although both have almost the same removal rates, it can be seen from Figure 4.150(a) that, when the aluminum content is low at 2.5 mg/l, there is a slight sedimentation floc on the bottom of the floating tank and the floating effluent water color is 14.8 degrees; with a high injection rate of 5 mg/l, the effluent water color is 11.1 degrees and the bubble deposition condition α_0 appears to scarcely meet the flotation requirements at critical state. In addition, with a high injection rate of 5 mg/l and $\alpha_0 = 0.4$, the floating sludge was stable, and the scraper was used to ensure removal and no sedimentation to the bottom of the floating tank was observed. This can also be predicted from the flotation tests shown in Figure 4.144, indicating that the operation can be accurately predicted using the Hokkaido University type Floating Tester.

Figures 4.152(a) and (b) are used to estimate the floating removal rate in a floating tank when the two aluminum additive rates differ, by Equations 4.85 to 4.90. The mixing time (t sec, or dimensionless stirring time T) of bubbles and flocs groups is taken in the horizontal axis to show how the removal

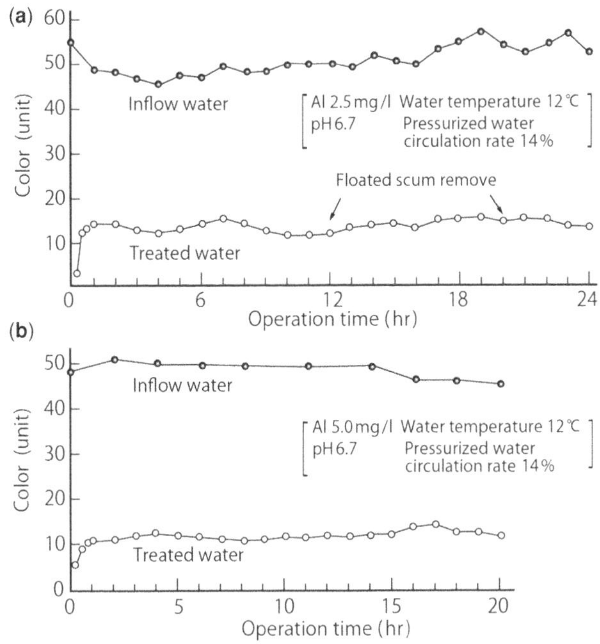

Figure 4.151 Color colloid continuous flotation experiments.

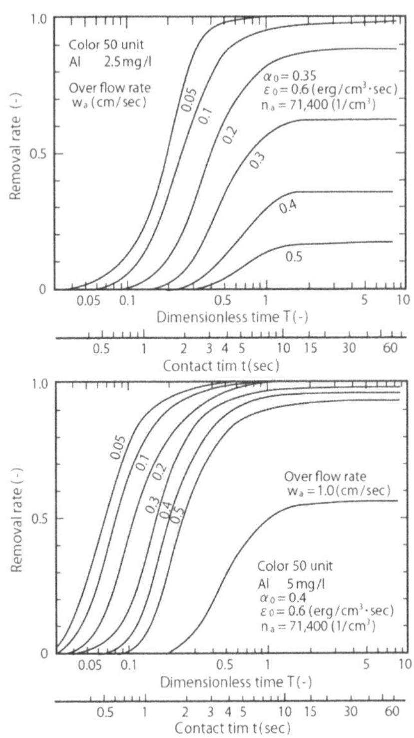

Figure 4.152 Floating bed removal rate simulation.

rate improves for each load rate. It was found that the maximal removal rate was reached at about T = 1.6, and the significance of increasing α_0 from 0.35 (Al: 2.5 mg/l) to 0.40 (Al: 5 mg/l) was clearly expressed.

Based on these findings, the authors believe that their kinetic formula can be accurately used as a basis for the theoretical design of flotation treatment processes, combined with batch flotation experiments.

FURTHER READING

Amirthrajaha A. and Tambo N. (1991). Mixing in water treatment. In: Mixing in Coagulation and Flocculation, A. Amirtharajha, M. M. Clark and R. R. Trussell (eds), (Chapter 1). American Water Works Association, Denver.

Bachelor G. K. (1946). Kolmogoroff's Theory at Locally Isotropic Turbulence. Trinity College, Cambridge.

Black A. P. and Christman R. F. (1963). Characteristics of colored surface water. *Journal American Water Works Association*, **55**(6).

Degramont (1991). "Water Treatment Handbook", Vol.1 (3.1Coagulation-Flocculation), pp. 131–146.

Faust S. D. and Aly O. M. (1998). Chemistry of Water Treatment 2nd edn, (Chapt. 6 Removal of Particulate Matter by Coagulation), pp. 219–270, Lewis Publishers.

Fukushi K. and Tambo N. (1987). Evaluation of Air-Solid ratio by analysis of the microbubble attachment process. (in Japanese) *Journal of Japan Water Works Association*, **630**, 48–51.

Fukushi K., Tambo N. and Kiyozuka M. (1986). Experimental evaluations of the kinetic process of pressure floatation separations (in Japanese). *Journal of Japan Water Works Association*, **607**, 32–41.

Gauden A. M. (1957). "Flotation. 2nd edn McGraw-Hill, Kyouritsu shuppan, Tokyo.

Hasegawa T., Hashimoto K., Onitsuka T. and Tambo N. (1991). Characteristics of metalpolysilicate coagulants. *Water Science and Technology*, **23**, 1713–1722.

Ide T. (ed.). (1990). Water Treatment Engineering. 2nd edn. (1.8, Coagulation Separations) (in Japanese), pp. 42–80, Gihoudo Publishing, Tokyo.

James M. Montgomery Consulting Engineers, Inc. (1985). "Water Treatment Principles & Design (6. Precipitation, Coagulation, Flocculation)" pp. 116–134, John Wiley & Sons, Inc.

Kitahara F. and Watanabe M. (1972). Interfacial Electric Phenomena-Basic, Measurement and Application. (in Japanese) Kyoritsu Publishing Co. Ltd. (Korean translation is also published), Kyouritsu shuppan, Tokyo.

Melbourne J. D. and Zabel T. F. (eds) (1977). Flotation for Water and Waste Treatment. Water research Centre, Medenham, UK.

Metijevic E., Mathai K. G., Otterwill R. H. and Kerker M. (1961). Detection of Metal Ion Hydrolysis by Coagulation III, Aluminum. *Journal of Physical Chemistry*, **65**(826).

Mysels K. J. (1959). Introduction to Colloid Chemistry. Interscience Publishers, Inc, New York.

Sapiro J. (1963). Effect of yellow organic acid on iron and other metal in water. *Journal American Water Works Association*, **55**(6).

Stumm W. and Morgan J. J. (1981). Aquatic Chemistry. (Chapt. 5 Precipitation and Dissolution) pp. 230–322, John Wiley & Sons, Inc.

Sullivan J. H. and Singley J. E. (1968). Reaction of metal ions in dilute aqueous solution: Hydrolysis of aluminum. *Journal American Water Works Association*, 1281–1287.

Sutherland D. N. (1966). Comment on Vold's simulation of floc formation. *Journal of Colloid and Interface Science*, **2**, 300.

Sutherland D. N. (1967). A theoretical model of floc structure. *Journal of Colloid and Interface Science*, **25**, 373.

Tambo N. (1965a). Basic study of floc formation process (I), floc formation in conventional flocculator (in Japanese). *Journal of Japan Water Works Association*, **372**, 10–19.

Tambo N. (1965b). Basic study of floc formation process (II), measurement of floc formation rate, (in Japanese). *Journal of Japan Water Works Association*, **381**, 14–22.

Tambo N. (1965c). Basic study of the coagulation in water treatment. (III) coagulation of clay-based turbidity by aluminium sulfate. *Journal of Japan Water Works Association*, **365**, 27–39.

Tambo N. (1966a). Fundamental studies of floc-forming processes (III), contact floc-forming under forced agitation, (in Japanese). *Journal of Japan Water Works Association*, **382**, 9–21.

Tambo N. (1980a). Water Supply. (New System of Civil Engineering 88). (6–1 Coagulation and Floc Formation), pp. 180–198. Society of Civil Engineers, Edition. Gihoudo Publishing, Tokyo.

Tambo N. (1980b). Water Supply (New System of Civil Engineering 88) (6.2.6 Flotation), pp. 199–202. Civil Engineering Society, Edition. Gihoudo Publishing, Tokyo.

Tambo N. (1991a). Criteria for flocculator design. *Journal of Water Supply Research and Technology – Aqua*, **2**, 97–102.

Tambo N. (1991b). Flocculator-Design index G and GC_0T values. *Journal of Japan Water Works Association*, **685**, 11–18. (in Japanese).

Tambo N. and Abe S. (1969). The behavior of floc blankets in an upstream sedimentation basin (I)-Equilibration of floc blankets suspension, (in Japanese). *Journal of Japan Water Works Association*, No. 415, pp. 7–15.

Tambo N. and Francois R. J. (1991). Mixing, break up and floc characteristics. In: Amirtharajha, A., Clark M.M. & Trussell, R.R (eds), "Mixing in Coagulation and Flocculation" (Chapter 7). American Water Works Association.

Tambo N. and Hozumi H. (1969). The behavior of floc blankets in an upstream sedimentation basin (II)-Effects of operation factors on supernatant water turbidity, (in Japanese). *Journal of Japan Water Works Association*, No. 417 pp. 7–17.

Tambo N. and Kamei T. (1989). Evaluation of extent of humic-substance removal by coagulation. *American Chemical Society, Advances in Chemistry Series*, **219**, 53–471.

Tambo N. and Kamei T. (1998). Coagulation and flocculation on water quality matrix. Water Science and Technology *IAWQ*, **37**(10), 31–41.

Tambo N. and Koichi O. (1985). Water Purification Technologies for Creating Safe Drinking Water, Chapter 2, Coagulation, Floc Formation (in Japanese). pp. 45–64, Gihoudo Publishing, Tokyo.

Tambo N. and Watanabe Y. (1972a). Rational Design of Flocculators-Development of theory by numerical solution (in Japanese). *Journal of Japan Water Works Association*, **449**, 22–37.

Tambo N. and Watanabe Y. (1972b). Rational Design of Flocculators-Floc formation in continuous flocculators with back-mixing, (in Japanese). *Journal of Japan Water Works Association*, **454**, 38–48.

Tambo N. and Watanabe Y. (1972c). Rational Design of flocculators (V)-Functional assessment and design method of Flocculators (in Japanese). *Journal of Japan Water Works Association*, **457**, 49–62.

Tambo N., Hozumi H. and Watanabe Y. (1970a). Rational Design of Flocculators (I)-Theory underlying design (in Japanese) *Journal of Japan Water Works Association*, **431**, 1–8.

Tambo N., Yamada K. and Hozumi H. (1970b). Study on floc strength (in Japanese). *Journal of Japan Water Works Association*, **427**, 4–15.

Tambo N., Hozumi H. and Fukuhara H. (1971). Behavior of floc blanket in upstream sedimentation basin (III)-Mixing Phenomena in sedimentation pond (in Japanese). *Journal of Japan Water Works Association*, **425**, 18–31.

Tambo N., Fukushi K. and Ohota H. (1984). Comparison of dissolved air flotation and sedimentation; Separation by Flotation Tester (in Japanese). *Journal of Japan Water Works Association*, **603**, 17–27.

Tambo N., Igarashi T. and Kiyozuka M. (1985). Electrophoretic study of bubble-attached floc generation. *Journal of Japan Water Works Association*, **604**, 2–6.

Tambo N., Wang X. C. and Matsui Y. (1993). High-speed solid-liquid separations of high-turbidity water by pellet-fluidized bed (in Japanese) *Journal of Japan Water Works Association*, **701**, 34–48.

Verway E. J. E. and Overbeek J. H. J. (1948). Theory of the Stability of Lyophobic Colloid. Elsevier Publishing Co., Amsterdom, New York, London, p. 106.

Vrablik E. R. (1957). "Fundamental principle of dissolved air flotation". 14th Annual Purdue International Conference, p. 744.

Wang X. C. and Tambo N. (1993a). Treatment of high turbidity and color waters by pellet-fluidized bed", (in Japanese). *Journal of Japan Water Works Association*, **700**, 26–37.

Wang X. C. and Tambo N. (1993b). Kinetic models of pellet-fluidized bed separation, (in Japanese). *Journal of Japan Water Works Association*, **706**, 16–27.

Weber W. J. (1972). Physicochemical Processes for Water Quality Control. (2. Coagulation and Flocculation) pp. 61–109, Wiley-InterScience, New York, London, Sidney, Toronto.

REFERENCES

Amirtharajah A. (1981). Initial Mixing. Proc. AWWA Seminar, Coagulation and Filtration back to the Basics, pp. 1–22.

Babitt H. E. and Doland J. J. (1955). Water Supply Engineering. McGraw-Hill Book Co. Inc., New York.

Bachelor G. K. (1958). The Theory of Homogeneous Turbulence. Cambridge University Press, Cambridge.

Ban S., Hatano S. and Kobayashi T. (1968). Fundamental investigation of Polynuclear aluminum chloride. *Journal of Japan Water Works Association*, **404**, 18–29. (in Japanese).

Baylis J. R. (1937). Silicate as aids to coagulation. *Journal American Water Works Association*, **29**, 1355.

Camp T. R. (1953). Flocculation and flocculation basins. *Transactions of the American Society of Civil Engineers,*. **120**, 1–16.

Camp T. R. and Stein P. C. (1943). Velocity gradient and internal work in fluid motion. *Journal Boston Society of Civil Engineers*, **30**, 219–237.

Cockerham P. W. and Himmelblau M. (1974). Stochastic analysis of orthokinetic flocculation. *Journal of Environmental Engineering*, **100**, EE.2, 279–293.

Edzwald J. K., Becker W. C. and Wattier K. L. (1985). Surrogate parameter for monitoring organic matter and THM precursors. *Journal American Water Works Association,* **77**(4), 122–132.

Edzwald J. K., Malley J. P. and Yu C. (1990). A conceptual model for dissolved air flotation in water treatment. *Water Supply*, **8**, 141–150.

Einstein A. (1905). Uber die von der molekularkinetischen Theorie der Warme geforderte Bewegung von in ruhenden Flussigkeiten suspendieten Teilichen. *Annalen der Physik*, **322**(8), 549–560.

Fair G. M. and Gemmell R. S. (1964). A mathematical model of coagulation. *Journal of Colloid Science*, **19**, 360–372.

Fair G. M. and Geyer J. C. (1956). Water and Waste–Water Disposal (Chapter 23, Chemical Coagulation, Precipitation, Ion Exchange, and Stabilization). John Wiley & Sons, Inc., New York. pp. 616–655.

Fukushi K., Tambo N. and Matsui Y. (1995). A kinetic model for dissolved air flotation in water and wastewater treatment. *Water Science and Technology*, **31**(3–4), 37–47.

Goto K. (1958). Estimation of specific surface area of particles in colloidal silica sols from the rate of dissolution. *Bulletin of the Chemical Society of Japan*, **31**(8).

Goto K. (1960). Effect of pH on polymerization of silicic acid. *Journal of Physical and Colloid Chemistry*, **54**(6).

Graveland A. (1980). Water softening by christallization. *Aqua*, **29**(2), 3–4.

Graveland A. *et al.* (1983). Development in softening by means of pellet reactor. *Journal American Water Works Association*, **75**, 619–626.

Heinz J. O. (1959). Turbulence. McGraw-Hill, London.

Hudson H. E., Jr. (1965). Physical aspect of flocculation. *Journal American Water Works Association*, **57**, 885–892.

Iler R. K. (1955). The Colloid Chemistry of Silica and Silicates. Cornell University Press, New York.

Itoh E., Hashizume H. and Tambo N. (1981) "A fundamental Study of hetero coagulation- an electrophoretic study of multi-component solution of natural organic color colloids and phosphates(ortho and tri-phosphate)" (in Japanese) Research report of Hokkaido Public Health Institute, Sapporo.

Kruyt H. R. (ed.) (1960). Colloid Science Vol. I, Chapt. VII, Kinetics of flocculation (by Oveerbeek J.Th.). Elsevier Publishing Co., Amsterdom, New York, London, pp. 279–301.

Kolb M., Botet R. and Jullien R. (1983). Scaling of kinetically growing clusters. *Physical Review Letters*, **51**, 1123.

Lagvankar A. L. and Gemmell S. (1968). A size-density relationship of flocs. *Journal American Water Works Association*, **60**(9), 1040–1046.

Lamer V. K. and Healy T. W. (1963). Adsorption-flocculation reaction of macromolecules at the solid-liquid interface. *Pure and Applied Chemistry*, **13**, 112–132.

Langelier W. F. (1921). Coagulation of water with Alum by prolonged agitation. *Engineering News-Record*, **86**, 924–8.

Levich V. G. (1962). Physicochemical Hydrodynamics (see Chapter I.4 Turbulent flow; "Chapter III Diffusion rates in Turbulent flow; Chapter V Certain Problems of Coagulation of Dispersions involving Liquid and Gas). Prentice-Hall Inc., New York.

Mandelbrot B. B. (1977). Fractals: Form, Chance, and Dimension. W.H. Freeman & Co., San Francisco. 365pp.

Mandelbrot B. B. (1982). The Fractal Geometry of Nature. W.H. Freeman & Co., San Francisco.

Matsui Y. and Tambo N. (1990). On-line measurement of floc diameter for coagulation control. *Japan Water Works Association*, **664**, 12–19. (in Japanese).

Matsushita M. (2006). Revisit of Fractal coagulation. *Journal of the Japanese Society of Physics*, **14**, 842–844 (In Japanese.).

Meakin P. (1983). Formation of fractal clusters and networks by irreversible diffusion-limited aggregation. *Physical Review Letters* , **51**, 1119.

Prker D.S. *et al.* (1971). Physical conditioning of activated sludge flocs. *Journal of Water Pollution Control Federation*, **43**(9), 1817–1833.

Prker D.S. *et al.* (1972). Floc break up in turbulent flocculation process. *Journal of Sanitary Engineering Division*, **98**(1), 79–99.

Riddick T.M. (1961). Zeta Potential and Its Application to Difficult Waters. *Journal American Water Works Association*, **55**(8), 1007–1030.

Rook J. J. (1972). Production of potable water from a highly polluted river. *Water Treatment and Examination*, **21**(Part 3), 259.

Rook J. J. (1974). Formation of haloform during chlorination of natural water. *Water Treatment and Examination*, **23**(Part 2), 234.

Smoluchowski M. (1917). Versuch einer mathematischen Theorie der Koagulationskinetik kolloider Lösungen. *Zeitschrift für Physikalische Chemie*, **92**, 129–168.

Stumm W. and O'Meria C. R. (1968). Stoichiometry of coagulation. *Journal American Water Works Association*, **60**, 514–539.

Swift D. L. and Friedlander S. K. (1964). The coagulation of hydrosols by Brownian motion and laminar shear flow. *Journal of Colloid and Interface Science*, **19**, 621–647.

Tambo N. (1959). Fluid mixing effect in a flocculator (An experimental verification of Camp's GT-value). *Journal of Japan Water Works Association*, (in Japanese), No. 301, 45–501.

Tambo N. (1964a). Basic study of the coagulation in water treatment (I) theoretical results and discussions. *Journal of Japan Water Works Association*, (in Japanese), No. 361, 1–11.

Tambo N. (1964b). "A. Fundamental Investigation of Coagulation and Flocculation (Chapter 2)" Doctor Engineering Thesis, Hokkaido University.

Tambo N. (1964c). Basic study of the coagulation in water treatment. "(II) determination of zeta potential by microscopic electrophoresis". *Journal of Japan Water Works Association*, (in Japanese), 353, 1–23.

Tambo N. (1965d). Basic study of the coagulation in water treatment. (IV) coagulation of natural organic colored water with aluminum sulfate. *Journal of Japan Water Works Association*, No. 367, 43–50.

Tambo N. (1966b). Mechanisms and conditions for generation of activity of silica sol (Mainly on the Baylis silica sol) (in Japanese). *Japan Society of Civil Engineers, Sanitary Engineering Proceedings*, **3**, 46–51.

Tambo N. and Fukushi K. (1986a). "A kinetic study of dissolved air flotation", Proc. World Congress III of Chemical Engineering, Tokyo.

Tambo N. and Fukushi K. (1986b). Pressure flotation separation kinetics. (in Japanese) *Journal of Japan Water Works Association*, No. 606, pp. 22–30.

Tambo N., Yamada K. and Hozumi H. (1970). Strength of flocs. *Journal of Japan Water Works Association*, **427**, 24–35. (In Japanese).

Tambo N. and Hozumi H (1979a). Physical characteristics of flocs—II. Strength of floc. *Water Research*, **13**, 421–427, 1979.

Tambo N. and Hozumi H. (1979b). Physical aspect of flocculation-II -Contact flocculation. *Water Research*, **13**, 441–448.

Tambo N. and Ito H. (1977). Electrophoretic study on the coagulation of natural organic coloured substances (in Japanese). *Journal of Japan Water Works Association*, **508**, 38.

Tambo N. and Kamei T. (1978a). Treatability Evaluation of General Organic Matter: Matrix conception and its application for a regional water and wastewater system. *Water Research*, **12**, 931–950.

Tambo N. and Kamei T. (1978b). Water quality conversion matrix for evaluation of treatability, "[I] Projection and assessment of removal by coagulation of organic components. *Journal of Japan Water Works Association*, **530**, 8–18 (in Japanese).

Tambo N. and Kamei T. (1978c). Water Quality Conversion Matrix for Evaluation of Treatability, "[II] Projection and assessment of organic removal by aerobic biological treatment. *Journal of Japan Water Works Association*, 15–24, No. 531 (in Japanese).

Tambo N. and Kamei T. (1980). Water quality conversion matrix of aerobic biological processes. *Journal Water Pollution Control Federation*, **52**(5), 1019–1028.

Tambo N. and Wang X. C. (1993a). Control of coagulation condition for treatment of high-turbidity water by fluidized pellet bed separation. *Journal of Water Supply Research and Technology – Aqua*, **42**(4), 212–222.

Tambo N. and Wang X. C. (1993b). Application of fluidized pellet bed technique in the treatment of highly colored and turbid water. *Journal of Water Supply Research and Technology – Aqua*, **42**(5), 301–309.

Tambo N. and Watanabe Y. (1967). A study on the density of aluminum floc (in Japanese). *Journal of Japan Water Works Association*, **397**, 2–10.

Tambo N. and Watanabe Y. (1971). Rational Design of Flocculators (II)-Implementation of theory by experimental deployment, (in Japanese). *Journal of Japan Water Works Association*, **441**, 9–21.

Tambo N. and Watanabe Y. (1979d). Physical Characteristics of Flocs-I. Floc density function and Aluminum floc. *Water Research*, **13**, 409–419.

Tambo N. and Watanabe Y. (1979e). Physical aspect of flocculation-I, Fundamental treaties. *Water Research*, **13**, 429–437.

Tambo N. and Watanabe Y. (1984). Physical aspect of flocculation-III –Flocculation process in a continuous flow flocculator with a back-mix flow. *Water Research*, **18**, 695–707.

Tambo N., Watanabe Y. and Shimizu S. (1968). Study on Aluminum Floc Densities (II) (in Japanese), *Journal of Japan Water Works Association*, **410**, 14–17.

Tambo N., Kamei T. and Takahashi M. (1981a). "Behavior and Properties of Metabolic Waste Components from Aerobic Biochemical Processes", "(I) Evaluation of process balance by Gel Chromatography". *Journal of Japan Water Works Association*, **18**(210), 1–10.

Tambo N., Kamei T. and Takahashi M., (1981b). "Behavior and Properties of Metabolic Waste Components from Aerobic Biochemical Processes", "(II) Effects of metabolic waste components on substrate removal rates." *Journal of Journal Japan Sewage Works Association*, **18**(211), 1–9.

Tambo N., Kamei T. and Takahashi M., (1982c). "Behavior and Properties of Metabolic Waste Components from Aerobic Biochemical Processes", "(III) Polysaccharide expressed as biometabolic waste". *Journal of Journal Japan Sewage Works Association*, **19**(212), 1–4.

Tambo N., Fukushi K. and Matsui Y. (1986). Analysis of microbubble adhesion process in the pressure floatation (in Japanese). *Journal of Japan Water Works Association*, **610**, 2–11.

Vold M. J. (1963). Computer simulation of floc formation in a colloidal suspension. *Journal of Colloid Science*, **187**, 684.

von Smoluchowski M. (1906). Zur kinetischen Theorie der Brownschen Molekularbewegung der Suspensionen. *Annalen der Physik*, **326**(14), 756–780.

Weitz D. A. and Oliveria M. (1984). Fractal structures formed by kinetic aggregation of aqueous gold colloids. *Physical Review Letters* **52**, 1433.

Witten T. A. and Sander L. M. (1981). Diffusion-limited aggregation, a kinetic critical phenomenon. *Physical Review Letters*, **47**, 1400.

Chapter 5

Precipitation of soluble inorganic components

5.1 PRECIPITATIVE SEPARATION OF METALS

The process of precipitation is mainly used to remove inorganic components, especially polyvalent metal ions, which are dissolved in water.

The precipitation operations typically used in water treatment are oxidative precipitation of iron and manganese, and alkali treatment of metals, such as in water softening. The precipitation process in a broad sense includes aeration operations to remove volatile odorous substances, hydrogen sulfide and methane dissolved in groundwater and sewage, and other noxious gases generated under anaerobic conditions. These are all operations which involve the solubility equilibrium of the ingredients.

5.1.1 Alkali precipitation of heavy metals

Most dissolved heavy metal ions become insoluble and precipitate as metal hydroxides and carbonates (at a pH corresponding to each metal) when pH is raised from neutral to alkaline by adding an alkaline agent such as Ca $(OH)_2$ or NaOH to the solution. The precipitated (deposited) metal can be removed by solid-liquid separation.

The ratio of soluble and insoluble components of a salt that are precipitated by such an alkaline precipitation process is shown by the chemical equilibrium theory. For example, if Equation 5.1 shows how an n-valent metal ion Me^{n+} forms an insoluble hydroxide $Me(OH)^n$ by the addition of an alkaline agent, the amount of the dissolved Me^{n+} ion is given by Equation 5.2 and given a constant K_{sp} called the solubility product.

Author(s): Norihito Tambo and Koichi Ogasawara
doi: 10.2166/9781789061307_0305

$$Me^{n+} + nOH^- \rightleftarrows Me(OH)_n \quad \text{(Insoluble)} \quad — (5\cdot1)$$
$$[Me^{n+}][OH^-]^n = K_{sp} \quad \text{(Solubility produc)} \quad — (5\cdot2)$$

K_{sp}: Solubility product
$[Me^{n+}],[OH^-]^n$: Molar concentration of metal ion Me^{n+} and hydroxide ion OH^-

(5.1–5.2)

Equation 5.1、5.2 The solubility product.

The dissociation constant of water $K_w = 1 \times 10^{-14} = [H^+]\ [OH^-]$. Substituting the relationship into Equation 5.2 yields Equation 5.3. If pH or pOH is determined, the dissolved metal ion concentration is determined.

$$\frac{[Me^{n+}]}{[H^+]^n} \times 10^{-14} = K_{sp} \quad — (5\cdot3) \tag{5.3}$$

Equation 5.3 The dissolved metal ion concentration.

Aluminium (a light metal) and heavy metals such as lead, zinc, iron and manganese are amphoteric electrolytes. Even if they are precipitated by the addition of an alkaline agent, if the pH is further increased by increasing the amount of alkali added, they become negatively charged complexes and are resolved as shown in Equation 5.4.

$$Me(OH)_n(\text{Insoluble}) + OH^- \rightleftarrows Me(OH)_{n+1}^- \quad — (5\cdot4) \tag{5.4}$$

Equation 5.4 The complex formation.

The solubility equilibrium constant K in this case is expressed by Equation 5.5, and the dissolution rate is determined by the pH.

$$\frac{[Me(OH)_{n+1}^-]}{[OH^-]} = K \quad — (5\cdot5) \tag{5.5}$$

Equation 5.5 The dissolution equilibrium constant.

Tables 5.1 and 5.2 show the solubility products (K_{sp}) and equilibrium constants (K) of some metals that are often used in water treatment (Feitknecht & Schindler, 1963).

In addition to being insoluble as hydroxides by alkaline condensation, metal ions are insoluble as carbonates, such as $FeCO_3$, $MnCO_3$, and $CaCO_3$. This phenomenon is also a geoscientific source of large amounts of marine sediment. The allowable concentration of iron, manganese, etc., in water is set as low as 10^{-6} to 10^{-7} mol/l. However, the concentration of iron, manganese, etc., cannot be reached by precipitation treatment with carbonate compounds, and the iron, manganese, etc., must be in the form of hydroxides. On the other hand, in water softening treatment of hardness components in water, the treatment target is about several tens of mg/l. Therefore, it is unnecessary to treat the hardness components in water at extremely high alkaline levels, such as at pH 13 to 14, which is required to remove them as hydroxides. For this reason, a carbonate ($CaCO_3$) production process is performed to

Table 5.1 The solubility products of metal hydroxides (Ksp).
25°C, Ionic strength (I = 0)

Hydroxides	K_{sp}	Hydroxides	K_{sp}
am. $Al(OH)_3$	6.3×10^{-32}	am. $Fe(OH)_3$	7.9×10^{-40}
$Ca(OH)_3$	3.7×10^{-6}		(I = 3M $NaClO_4$)
β-$Cd(OH)_2$	4.0×10^{-5}	$Mg(OH)_2$	1.3×10^{-11}
$Cr(OH)_3$	1.0×10^{-30}	$Mn(OH)_2$	1.6×10^{-13}
	(I = varid.)	PbO	5.0×10^{-16}*
$Cu(OH)_2$	1.6×10^{-19}	am. $Zn(OH)_2$	$1.0 \sim 2.5 \times 10^{-16}$
$Fe(OH)_2$	7.9×10^{-16}		

*$Me^{n+} + nOH^- \leftrightharpoons Me(OH)^n$ Me: Metal ion $K_{sp} = [Me^{n+}][OH^-]^n$

Table 5.2 The equilibrium constants (K) of metal hydroxide complex ions.
25°C, Ionic strength (I = 0)

Reactions	K (Equilibrium Constant)
$Fe(OH)_2 + OH^- \leftrightharpoons Fe(OH)_3^-$	7.9×10^{-6}
am. $Fe(OH)_3 + OH^- \leftrightharpoons Fe(OH)_4^-$	1.3×10^{-5} (I = 3M $NaClO_4$)
$Cu(OH)_2 + OH^- \leftrightharpoons Cu(OH)_3^-$	2.5×10^{-4}
am. $Zn(OH)_2 + OH^- \leftrightharpoons Zn(OH)_3^-$	$1.3 \sim 3.2 \times 10^{-2}$
β-$Cd(OH)_2 + OH^- \leftrightharpoons Cd(OH)_3^-$	2.0×10^{-5}
am. $Al(OH)_3 + OH^- \leftrightharpoons Al(OH)_4^-$	2.0×10
$PbO + H_2O + OH^- \leftrightharpoons Pb(OH)_3^-$	4.0×10^{-2}
am. $Cr(OH)_3 + OH^- \leftrightharpoons Cr(OH)_4^-$	4.0×10^{-1} (20°C, I = varid.)
$Mn(OH)_2 + OH^- \leftrightharpoons Pb(OH)_3^-$	1.0×10^{-5}

$Me(OH)_n + OH^- \leftrightharpoons Me(OH)_{n+1}^-$ $K = \frac{[Me(OH)_{n+1}^-]}{[OH^-]}$
Me: metal ion

obtain the necessary precipitation of 10^{-5} mol/l at pH10 to 11. The solubility equilibrium in this case is shown in Equations 5.6 and 5.7.

$$CaCO_3\,(\text{Insoluble}) \rightleftharpoons Ca_2^{+} + CO_3^{2-} \quad —(5\cdot6)$$
$$[Ca^{2+}][CO_3^{2-}] = K_{sp} = 4.8 \times 10^{-9} \quad —(5\cdot7) \qquad (5.6–5.7)$$

Equation 5.6, 5.7 The solubility equilibrium of calcium carbonate.

As shown in Figure 5.1, the total carbonic acid concentration in water ($[CO_2] + [HCO_3^-] + [CO_3^{2-}]$) varies greatly depending on the pH. The concentration of CO_3^{2-} approaches a maximum value above pH 11 in Equation 5.6, and the concentration of Ca^{2+} approaches a minimum value in Equation 5.7 when the pH exceeds pH 11. However, these estimates are simply based on the assumption that only the ions are present in the water, and are affected by the coexisting ions in actual processing.

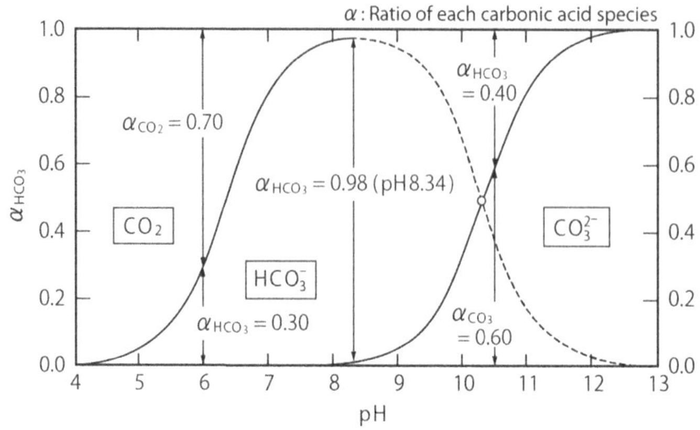

Figure 5.1 Carbonic acid concentration ratios in water, depending on pH.

In summary, the results of calculations by the author Tambo using the Feitknecht and Schindler (1963) equilibrium constants for the insoluble areas of metallic hydroxides and calcium carbonate are shown in Figure 5.2. The diagram simply deals with the equilibrium between Me $(OH)_n$ and Me^{n+}, and does not consider various intermediate species. In a later section, we borrow the Pourbaix diagram (Pourbaix, 1966) to explain how ions of metals are dissolved and precipitated depending on their reduction-oxidation potential (ROP) and pH.

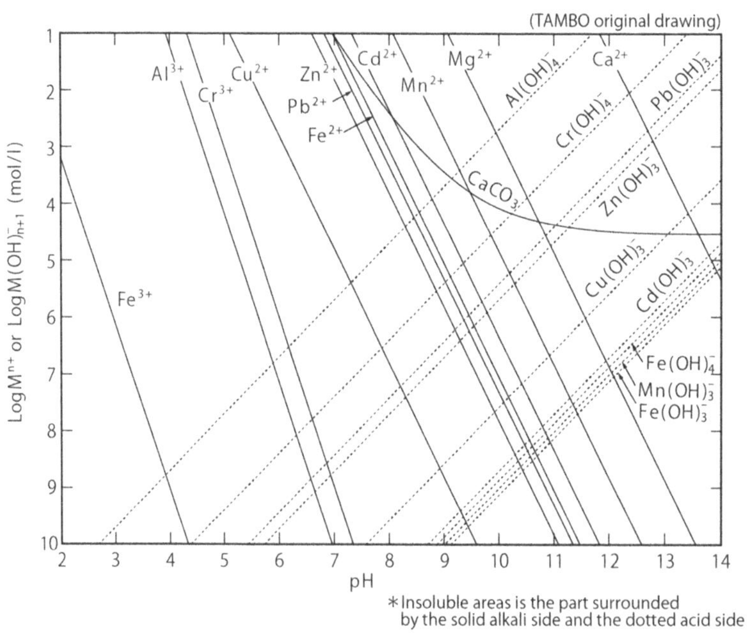

Figure 5.2 The insoluble areas of metallic hydroxides and calcium carbonate state of metal hydroxide.

5.1.2 Hard water softening treatment

Calcium and magnesium in water may cause poor soap foaming and cause scale generation in boilers, which may hinder the use of water. In Japan, hard water, which hinders the use of ordinary water, hardly exists not only in surface water but also in groundwater, and only a small amount of water needs to be disposed of in limestone areas (such as around Mount Ibuki in Shiga Prefecture, and on islands originating from coral reefs such as the Kadena group of wells on the main island of Okinawa). However, hard water is common in Europe and North America, and its control is an indispensable operation for urban and industrial water supply, and it accounts for a large proportion of water treatment.

Divalent metal ions, which exhibit hardness, coagulate soap, and the phenomenon of poor soap foaming is said to be common in hard water areas. Today, hardness is mainly found as a result of Ca^{2+} and Mg^{2+} complexed with EDTA (ethylenediaminetetraacetic acid: a type of chelating agent), and amounts of all divalent ions such as Fe^{2+}, Mn^{2+} and Sr^{2+} are often expressed in equivalent terms of $CaCO_3$ concentrations. For urban water, water is considered hard at 120 mg/l or more, and soft water is considered to be anything less than this level. Water used in ultra-high-pressure boilers in power plants requires hardness-free water, which is purified to the limit by ion exchange to remove dissolved ions.

Total hardness is subdivided into carbonate and non-carbonate hardness:

(1) If the alkalinity (mg/l) is greater than or equal to the total hardness (mg/l), then the hardness is carbonate.
(2) If the alkalinity is less than the total hardness, then carbonate hardness (mg/l) = alkalinity (mg/l).

A significant portion of carbonic acid hardness is precipitated at high temperatures in a boiler, etc., but can be removed by the lime softening method. Carbonic acid hardness is removed by boiling (and so was formerly also called temporary hardness). Components other than carbonate hardness in the total hardness are called non-carbonate hardness (also called permanent hardness because they are not precipitated and removed by boiling).

This section explains water softening treatment by the lime-addition method, which produces water with a hardness of several tens of mg/l, which has been widely used in urban waterworks using hard water in Europe and the United States as raw water.

Carbonic acid hardness reacts as shown in Equations 5.8 and 5.9 when lime Ca $(OH)_2$ is added to raise the pH of the system.

$$Ca^{2+} + 2HCO_3^- + Ca(OH)_2 \rightarrow 2CaCO_3(\text{Insoluble}) + 2H_2O \quad —(5\cdot8)$$
$$Mg^{2+} + 2HCO_3^- + Ca(OH)_2 \rightarrow CaCO_3(\text{Insoluble}) + Mg^{2+} + CO_3^{2-} + 2H_2O \quad —(5\cdot9) \qquad (5.8\text{–}5.9)$$

Equation 5.8, 5.9 The lime softening process.

As shown in Equation 5.9, lime is added to remove residual Mg^{2+} ions and precipitated as shown in Equation 5.10.

Non-carbonate hardness reacts as shown in Equation 5.11. This reaction only changes the hardness of the Mg^{2+} to the Ca^{2+}. To further remove the calcium ion Ca^{2+}, sodium carbonate (soda ash) must be added to proceed with the reaction of Equation 5.12.

$$Mg^{2+} + CO_3^{2-} + Ca(OH)_2 \rightarrow CaCO_3\,(\text{Insoluble}) + Mg(OH)_2\,(\text{Insoluble}) \quad —(5\cdot10)$$
$$Mg^{2+} + SO_4^{2-} + Ca(OH)_2 \rightarrow Ca^{2+} + SO_4^{2-} + Mg(OH)_2\,(\text{Insoluble}) \quad —(5\cdot11) \qquad (5.10\text{–}5.11)$$

Equation 5.10, 5.11 The excess lime process.

$$Ca^{2+} + SO_4^{2-} + Na_2CO_3 \rightarrow CaCO_3\,(\text{Insoluble}) + Na_2SO_4 \quad —(5\cdot12) \qquad (5.12)$$

Equation 5.12 The lime-soda process.

The method using the reactions of Equations 5.8 and 5.9 is called the lime softening method (lime process); the method proceeding to the reactions of Equations 5.10 and 5.11 is called the excess lime process; and the method proceeding to Equation 5.12 is called the lime-soda water softening process.

These reactions are usually carried out at about pH 10.8. However, under conditions where the initial pH of the raw water is low and free carbonic acid is present, as shown in Figure 5.1, in order to remove free carbonic acid prior to the above-mentioned reactions of the water softening process, neutralization procedures such as the following are performed: $CO_2 + Ca\,(OH)_2 \rightarrow CaCO_3$ (insoluble) + H_2O.

These reactions can result in high pH, sedimentation and separated $CaCO_3$ and the water may leave a slight amount of $CaCO_3$ as a suspension, which may clog the filter basin. Therefore, the suspended $CaCO_3$ is resolved by neutralizing the solution with acid in conjunction with pH control (decreasing it) for distribution. In many cases, carbon dioxide is used, and this process is referred to as re-carbonation. The reaction is as shown in Equation 5.13.

$$CaCO_3\,(\text{Insoluble}) + CO_2 + H_2O \rightarrow Ca^{2+} + 2HCO_3^- \quad —(5\cdot13) \qquad (5.13)$$

Equation 5.13 The re-carbonation process.

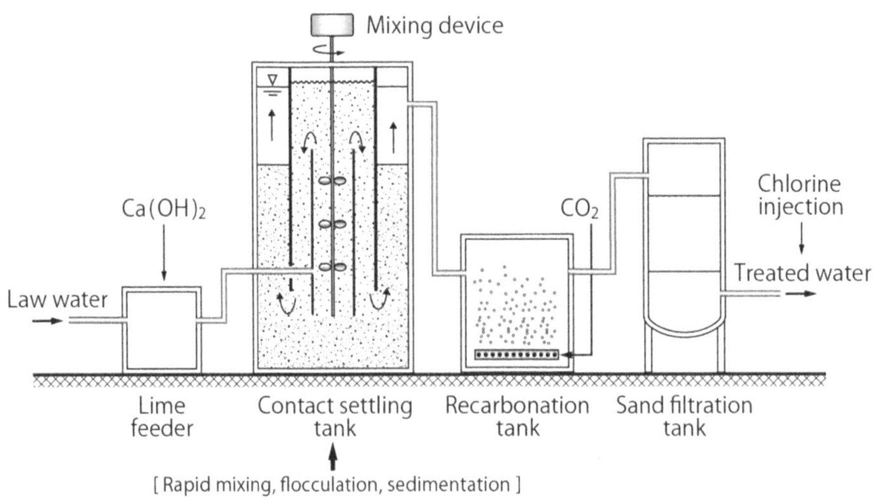

Figure 5.3 A water softening system using the alkaline coagulation method.

The composition of the water softening system using these reactions is similar to that of the rapid sand filtration system: raw water → addition of lime → rapid mixing → flocculation → sedimentation → re-carbonation → sand filtration → chlorine disinfection, and so on, as shown in Figure 5.3.

Both the floc blanket and slurry circulation contact coagulation-sedimentation tanks (see Chapter 6 below) were originally developed for water softening treatment, and then used for usual turbidity treatment. The major difference between the softening treatment and turbidity treatment flow is that the softening treatment increases the pH of the raw water by adding lime, separates the coagulated precipitate into solid and liquid, and then returns the supernatant water to neutral by re-carbonation. This alkaline precipitation method involves the treatment of a large amount of calcium carbonate sludge.

5.2 WATER SOFTENING BY PELLET FLUIDIZED BEDS: METASTABLE STATE TREATMENT

5.2.1 Overview of pellet softening method

In the 1970s, Dr A. Graveland and colleagues at the Amsterdam Water Works, in the Netherlands, proposed a new water softening method using a fluidized bed of pellets, as shown in Figure 5.4, for the alkaline precipitation method described above (Graveland, 1980). In this softening method, sand coated with calcium carbonate is fluidized, and an alkaline agent such as sodium hydroxide is injected into the lower part of the fluidized bed to raise the pH to about 10. Calcium in the water is instantaneously crystallized as calcium carbonate on the surfaces of the fluidized calcium carbonate-coated sand (here after referred to as pellets) and separated from the water. Since the produced pellets are crystalline, they can be separated from the liquid system only by discharging the produced excess pellets from the fluidized bed, and they can be used as calcium carbonate granules without sludge treatment.

In the conventional alkali coagulation precipitation method, the separation rate (overflow rate) was about 2 to 5 cm/min. In the pellet fluidized bed crystallization method, a high separation rate of 100 to 200 cm/min. can easily be obtained, and the removal performance is almost the same as in the conventional precipitation method.

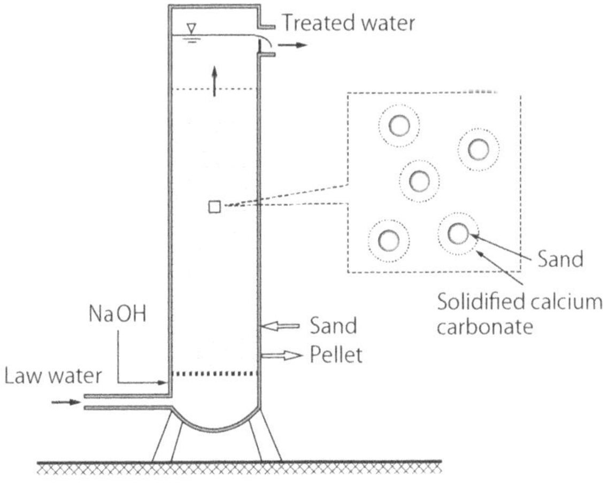

Figure 5.4 The crystallization method using a pellet fluidized bed.

Graveland *et al.* (1983) reported that, when the Langelier saturation index became positive and supersaturated, crystallization progressed as a first order reaction at supersaturated concentrations. However, detailed kinetic mechanisms have not yet been elucidated.

The key to this crystallization process is considered to be the 'metastable state operation' as Tambo has called it, which can be applied to various treatment phases, including pellet agglomeration in high concentration turbidity and color floc formation, as described in Chapter 4.

Taking metal ions as an example, a solution in a supersaturated state precipitates dissolved components easily (for a short period of time). However, in the metastable state around the saturated state, as shown schematically in Figure 4.119, it is necessary to allow a certain aging time before precipitation (coagulation) occurs. However, when a solid component having the same properties as the coagulum is introduced into a solution in a metastable state, the dissolved component of the liquid phase coagulates rapidly on the surface of the introduced solid, and the coagulum formed is often dense, which is an excellent separation characteristic. This process is commonly referred to as crystallization.

The precipitation and growth of minute calcium carbonate particles in the metastable state is thought to depend on the degree of supersaturation. As the degree of supersaturation increases, the limit particle size to which the particles can grow decreases, and crystal nuclei tend to be generated. Conversely, in the presence of large particles, particles can grow at low saturation. In the conventional coagulation treatment method, particles with a large diameter are not present at first, and precipitation does not proceed quickly unless the unstable state has a large degree of supersaturation. In pellet coagulation (crystallization), on the other hand, the limit concentration of precipitation is governed by a group of large-sized pellet particles, and even if they cannot grow themselves, pellet particles coalesce as they crystallize on their surfaces, resulting in a decrease in the calcium concentration of the liquid system. Such metastable zone treatments have special applications as described above for colloidal pellet coagulation, as well as similar procedures for water treatment, such as crystallization dephosphorylation with calcium phosphate pellets.

5.2.2 Removal pattern of the pellet softening operation

The situation under which hardness removal by pellet fluidized bed proceeds was experimentally evaluated using equipment available in the Hokkaido University Laboratory, as shown in Figure 5.5.

Calcium chloride was added to Sapporo City tap water at concentrations ranging from 100 to 600 mg/l as $CaCO_3$, and an equivalent amount of sodium bicarbonate was added as alkalinity to make the test water. The water temperature was adjusted to 10°C. The reaction tower (fluidized bed) was an acrylic resin tube with an inner diameter of 4 cm and a height of 200 cm, and there was a stirring chamber for injecting a pH adjusting agent into the lower part and for rapidly mixing it. The mixture was stirred and mixed with a magnetic stirrer. The reactor was equipped with water sampling holes at intervals of 10 cm to observe changes in water quality over time, and the head loss of the fluidized bed was observed by an open-tube manometer. The treatment pH was set by injecting a predetermined amount and concentration of sodium hydroxide solution into the stirring chamber with a metering pump, and changing the treatment standard pH to a range of 9.5 to 11.0. The flow rate into the reactor was controlled by a needle valve so that the superficial up-flow velocity became 100 m/h, and the flow rate was kept constant even if the head loss of the fluidized bed increased.

Silica sand with a sieve diameter of 0.71 to 0.84 mm was used for the sand constituting the fluidized bed. As the operation continued, calcium carbonate crystallizes on the silica sand surface and the particle

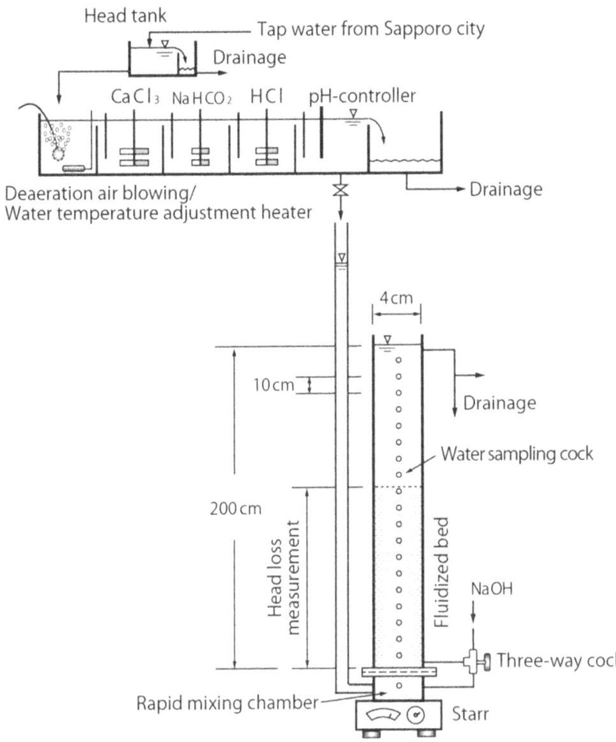

Figure 5.5 Experimental set up used to investigate hardness removal using a pellet fluidized bed.

size increased in the form of pellets. In course of operation, overgrown pellets were pulled from the bottom of the fluidized bed and replenished with fine sand or small chunks of crushed pellets from the top.

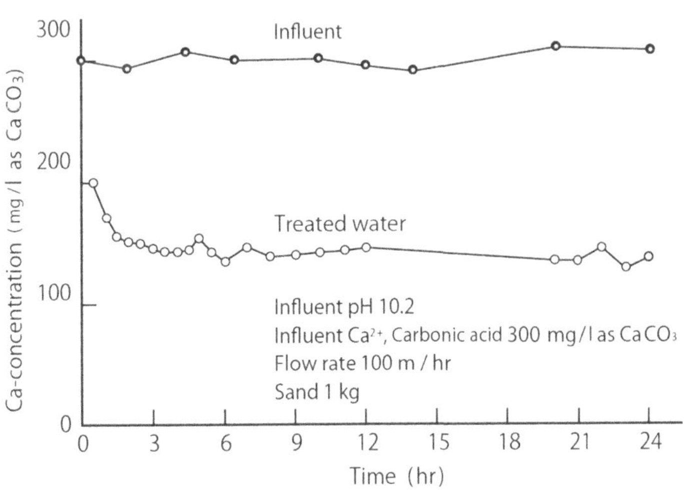

Figure 5.6 Changes in the calcium concentration of the treated water.

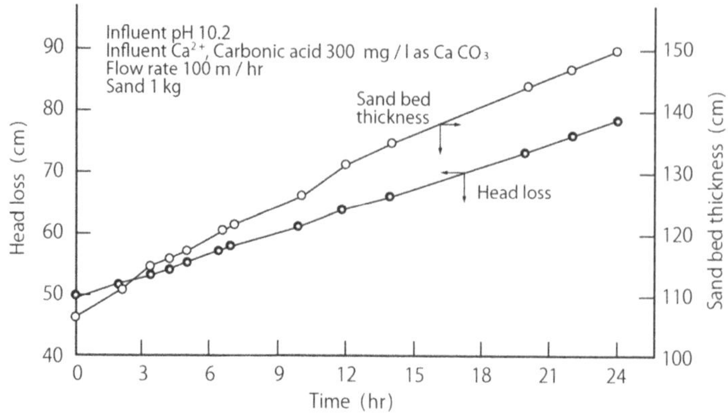

Figure 5.7 Head loss and thickness variations in the fluidized bed.

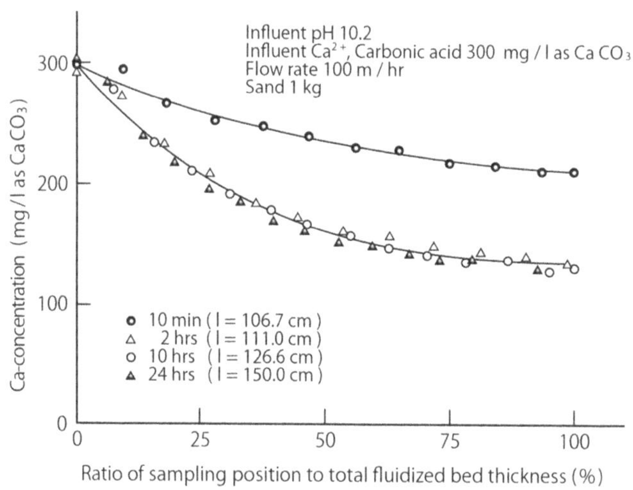

Figure 5.8 Changes in calcium concentration across the thickness of the fluidized bed.

Examples of our experimental results are shown in Figures 5.6, 5.7 and 5.8 for a test water with a calcium concentration of 300 mg/l (as $CaCO_3$), and a void column upflow rate of 100 m/h and a fluidized bed inflow adjusted to pH 10.2.

Figure 5.6 shows the changes in the calcium concentration over time in the fluidized bed and after treatment as a result of starting the operation with 1 kg of silica sand with an initial bed thickness of 106 cm. It can be seen that the hardness of 300 mg/l (as $CaCO_3$) in the raw water was constantly reduced to 150 mg/l (as $CaCO_3$) or less after 2 to 3 hours.

Figure 5.7 shows how the total thickness of the fluidized bed increases over time and also shows how the head loss of the fluidized bed increases. As the calcium pellets grow, the fluidized bed thickness increases almost linearly with time. This shows that the head loss through the flow increases almost linearly with time.

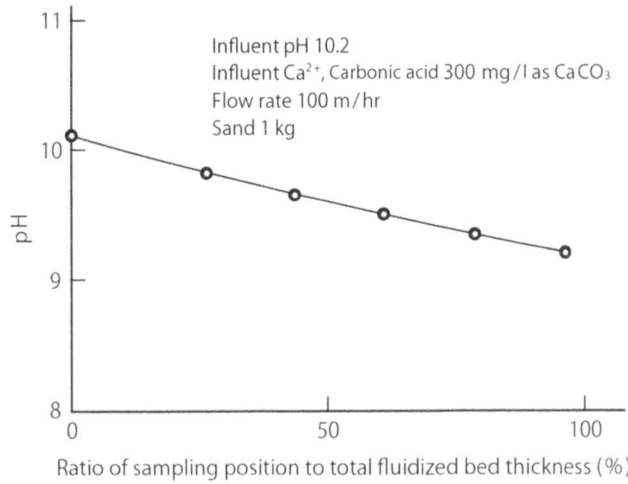

Figure 5.9 Change in pH against fluidized lamina thickness and direction.

Figure 5.8 shows several superimposed plots of the decay of calcium concentration across the thickness of the fluidized bed over time. After 10 minutes of operation, calcium was removed up to about 250 mg/l (as $CaCO_3$). After 2 hours, changes in calcium concentrations across the sand layer thickness showed almost the same results for all positions of the fluidized bed, compared to the total sand layer. This is thought to be due to the fact that the maturation state, which can be represented by a single decay curve, has been reached.

Figure 5.9 shows the change in pH against fluidized layer thickness and direction on the same axis. The initial pH 10.2 at the inflow point decreases moving downstream (the process of the reaction) and drops to a value of pH 9.2 at the outflow end (treated water). Conventional precipitation-type alkaline water softening methods need to add a re-carbonation process, but the pellet bed method has a more neutral outflow than the pH of conventional treatment water, being as low as pH 9.

Table 5.3 summarizes the properties of the calcium pellets produced in this way, as the result of experiments conducted at an influent level of 200 mg/l (as $CaCO_3$) and pH 10 for 154 hours. The table shows the properties of bare sand particles and calcium carbonate-coated pellet particles.

The resulting pellets grew from 0.98 mm to 1.50 mm in diameter, and the density of the calcium carbonate-attached sand was about 2.65, which is almost similar to the density of 2.71 g/cm^3

Table 5.3 Properties of bare sand particles and calcium carbonate-coated pellets.

Materials		Density (g/m^3)	Particle Size (mm)
Sand only	(Upper fluidized bed)	2.63	0.78
	(Lower fluid bed)	2.63	0.98
Calcium sand*	(Upper fluidized bed)	2.63	0.83
	(Lower fluid bed)	2.69	1.50
Calcium carbonate		2.71	—

*Inflowing Calcium concentration = 200 (mg/l as Ca CO_3),
After operating for 154 hrs at pH10.0.

of calcium carbonate. The calcium carbonate-attached sand can be considered to be marble particles, and when left still, it separates instantaneously from water, so sludge treatment is not required.

5.2.3 Equilibrium study of water softening reaction in a fluidized bed

From the equilibrium relationships of alkaline water softening described in Equations 5.6 and 5.7 above, the abundance ratios of the three carbonic acids in water were calculated as shown in Figure 5.1, and the carbonate-CO_3^{2-} ion predominates as pH is increased around pH 10. As a result, the alkaline water softening treatment above can be described as the reaction of calcium ions and carbonate ions to precipitate as solid calcium carbonate.

In order to examine how the precipitation of calcium carbonate solid phase progresses in the fluidized bed, the pH of each height of the fluidized bed is shown in Figure 5.10 on the horizontal axis, and the measurement results of calcium ion concentration and carbonate concentration are shown on the vertical axis. Since the reduction in calcium ions and carbonate concentrations in water is almost one-to-one, we can assume that the reaction proceeds in an equivalent relationship, as in Equations 5.6 and 5.7. As carbonate ions are consumed and the concentration of calcium ions decreases, precipitation of calcium carbonate leads to the dissociation of bicarbonate ions and a decrease in pH in order to maintain the carbonic acid system in equilibrium.

The dotted line (1) in Figure 5.10 represents the locus of change in calcium ion concentration and pH when an alkali added in a predetermined amount reacts and the calcium ion in the liquid system is considered to be calcium carbonate, and represents the concentration of calcium ion that should be present at that pH. Calculations were made using Equation 5.14, which shows the electrical neutral condition of water ions, assuming that all activity coefficients are equal to 1, regardless of the ionic strength of the solvent.

Curve (2) in Figure 5.10 shows the total amount of calcium measured which was present in the liquid system in a soluble state. Therefore, the calcium present between curves (1) and (2) can be considered to be ultrafine calcium carbonate still present in the liquid system in the soluble category, despite the completion of the calcium carbonate formation reaction. Hereafter, we considered this ultrafine particle

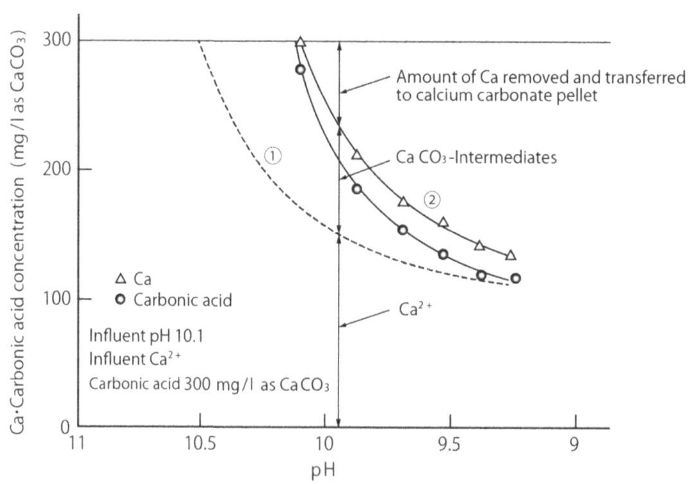

Figure 5.10 Relationship between pH and calcium and carbonic acid concentrations in the fluidized bed.

to be an intermediate product in the pellet formation process, and considered the following mechanisms for the formation of calcium carbonate pellets.

$$[H^+] - \frac{K_w}{[H^+]} + (\alpha_1 + 2\alpha_2)(C_{T_0} - \Delta C) = [Cl^-] - [Na^+] - 2C_0 + 2\Delta C \qquad \text{—}(5\cdot14) \tag{5.14}$$

$[Cl^-]$: Cl^- concentration derived from added HCl and $CaCl_2$ (mol/l)
$[Na^+]$: Na^+ concentration derived from added $NaHCO_3$ and NaOH (mol/l)
C_0 : Initial calcium concentration (Ca^{2+}) (mol/l)
C_{T_0} : Initial total carbonic acid concentration (mol/l)
K_w : Water dissociation constant (mol^2/l^2)
α_1 : Ratio of HCO_3^- to total carbonation (-)
α_2 : Ratio of CO_3^{2-} to total carbonation (-)
$[H^+]$: Hydrogen ion concentration (mol/l)

Equation 5.14 The expression showing electrical neutral condition of water ions.

We believe that the reaction of calcium and carbonate ions which forms calcium carbonate intermediates proceeds instantaneously. However, it takes a certain amount of time for the ultrafine calcium carbonate particles (intermediate products) produced to be transported from the liquid system to the surface of the pellet particles, to become part of the pellet leaving the liquid system, and transfer to the solid system. As a result, a certain number of intermediates (ultrafine calcium carbonate particles in the solubility section between curves (1) and (2)) are present in the liquid system during the transport deposition process. Under these mechanisms, the number of intermediates decreases as the reaction moves to the upper part of the fluidized bed and eventually reaches the calcium concentration and pH calculated in the chemical reaction equilibrium. The concentration of calcium in the upper part of the fluidized bed is mostly unreacted calcium ions, and there are no intermediates.

Assuming the presence of such intermediates (very small calcium carbonate particles in the solubility category), the pellet fluidized bed softening reaction can be described as a two-step reaction, as in Equation 5.15.

$$Ca^{2+} + CO_3^{2-} \rightleftharpoons CaCO_3\,(\text{int.}) \rightleftharpoons CaCO_3\,(\text{solid}) \qquad \text{—}(5\cdot15) \tag{5.15}$$

$CaCO_3$ (int.) : Calcium carbonate intermediate
$CaCO_3$ (solid) : Solid calcium carbonate (pellet)

Equation 5.15 The two-steps reaction in hard water softening.

To examine the dimensions of the calcium carbonate intermediate, water was sampled from the fluidized bed immediately after adjusting the pH to 10.1 by adding NaOH, and the calcium concentration of the filtrate through a membrane filter of 0.1 to 0.8 μm was measured. The results showed they were all the same as raw water, and all the dimensions of the produced intermediates (fine calcium carbonate particles) were found to be ultrafine particles of 0.1 μm or less.

From these findings, we can write the equilibrium relationship between calcium ions, carbonate ions, and calcium carbonate intermediates as shown in Equation 5.16.

$$\frac{[Ca^{2+}][CO_3^{2-}]}{[CaCO_3(int.)]} = K \qquad -(5\cdot 16) \qquad (5.16)$$

Equation 5.16 The equilibrium relationship between intermediates.

Figure 5.10 showed the calcium ion concentration $[Ca^{2+}]$ at each pH, the carbonate ion concentration $[COO_3^{2-}]$, and the calcium carbonate intermediate concentration $[CaCO_3(int)]$ on the basis of actual measurements, and the ratio of the calcium ion concentration to the calcium carbonate intermediate on the horizontal axis, and the ratio of the calcium ion concentration to the calcium carbonate intermediate on the vertical axis. These results were then used to produce Figure 5.11, in which all experimental results were regressed to a straight line passing through the origin, and the slope gave an equilibrium constant K = 1000 (l/mol).

The value of this equilibrium constant is nearly equal to the value known as the equilibrium constant of the previous calcium carbonate formation reaction, indicating that the intermediate formation reaction as shown in Equation 5.15 occurs instantaneously. This suggests that the water softening process by pellet fluidized bed proceeds in two stages: instantaneous equilibration of calcium carbonate (intermediate) formation, followed by transport and attachment of the intermediate to the preformed pellet particles (calcium carbonate solid phase).

Table 5.4 shows an example of a comparison between the results of a pellet softening experiment in which the pH at the time of inflow was changed to around pH 9–11 under different raw water concentration conditions and using the theoretical calculation results based on the above equilibrium theory. As shown in the table, when the pH of the inflow point and the influent calcium concentration are high, the experimental result does not reach the removal rate which is considered to be theoretically reachable. This phenomenon is thought to be caused by the fact that, at high pH and high calcium concentration, the concentration of calcium carbonate produced is not limited to the metastable state, but enters the unstable state at once, precipitates beyond the size of the intermediate, coagulates (random

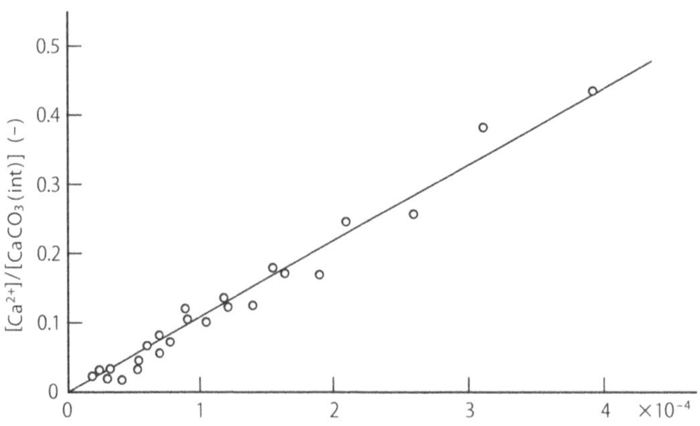

Figure 5.11 Calculation of the equilibrium constant of $CaCO_3$ intermediates.

Table 5.4 Examples of results from a pellet fluidized bed softening experiment.
Inflowing Ca concentration = 300 (mg/l as Ca CO_3)

Inlet pH	9.2	9.6	10.0	10.2	10.5	10.8
Treated water Ca concentration	227	217	164	133	93	116
Treated water Ca theoretical concentration	257	202	145	126	57	8
Supersaturated concentration	80	460	800	580	1090	1270

Inflowing Ca concentration = 400 (mg/l as Ca CO_3)

Inlet pH	9.2	9.6	10.0	10.2	10.5	Two-step dosage	
						No.1	No.2
Treated water Ca concentration	—	287	235	—	200	125	88
Treated water Ca theoretical concentration	—	237	159	—	104	31	—
Supersaturated concentration	—	940	1540	—	1500	2220	890

agglomeration) and cannot adhere to the surface of the pellet, and flows out of the fluidized bed. The subsequent multistage addition of sodium hydroxide (described below) is an effective measure to prevent the deterioration of the removal effect at high pH and high calcium concentration and to perform high-speed treatment.

In the experiments described above, when the product of Ca^{2+} and C concentrations reaches several thousand times the saturated value, colloidal calcium carbonate (fractal agglomerates) precipitates rapidly, which is considered to be the reason for the reduction in the removal rate. Therefore, the amount of sodium hydroxide added is controlled to ensure a lower pH value slightly exceeding the saturation value, and sodium hydroxide (as an alkaline agent) is added again so that the liquid system becomes slightly supersaturated after the calcium carbonate intermediate (ultrafine particles) produced migrates to the pellet surface and the calcium concentration of the liquid system decreases. Theoretically, by repeating this step, the fluidized pellet bed softening method can remove a high concentration of calcium ions at high speed up to the theoretical point of alkaline solubility equilibrium.

The results of the two-step pH-treatment on very hard water with a Ca-concentration of 400 mg/l (as $CaCO_3$) are shown in the lower part of Table 5.4, and in Figure 5.12. This is the result of the operation of the pelletized fluidized bed with single-stage alkaline neutralization for 9 hours, adding the same amount of NaOH in two portions, and injecting the pelletized fluidized bed into the second stage at a position 50 cm from the inflow point. In the single stage system, the elimination was expected to be only up to about 130 mg/l (as $CaCO_3$) by neutralizing pH 10.5, whereas in the second stage, the elimination was eventually achieved to the theoretical level of 90 mg/l (as $CaCO_3$), despite that the first stage is reduced to pH 9.8.

In these experiments, the calcium concentration was rapidly reduced to the required concentration of 100 mg/l (as $CaCO_3$) or less without increasing the alkali injection rate by performing a 2 to 3 step alkali split injection of a high concentration calcium system.

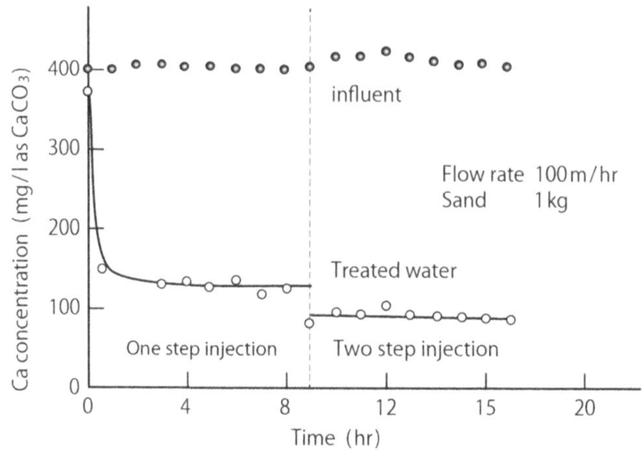

Figure 5.12 Improvement in efficiency of the pellet method using two steps of alkaline agent injection.

5.2.4 Kinetics of the pellet softening process

The process of scavenging calcium ions in the pellet fluidized layer can be considered as a rate-limiting phenomenon in which the formed calcium carbonate intermediate $CaCO_3$(int.) is attached to the pellet surface by the chemical equilibrium established instantaneously, and the total calcium crystallization reaction. Considering the rate-limiting reaction of these intermediates, the kinetic process of the calcium crystallization reaction can be described by Equations 5.17 to 5.19.

In these equations, Q is the calcium carbonate deposition per pellet volume (g/cm^3), ε is the void fraction (-) of the fluidized layer, and a_v is the specific surface area of pellets (cm^2/cm^3), which are given by Equations 5.20, 5.21 and 5.22, respectively. Here, the Richardson exponential equation (in this case, Equation 5.21) is used to calculate the change in porosity (see Equation 6.41 in Chapter 6). The final sedimentation velocity W_t of the single particles in Equation 5.21 is calculated using the Allen equation (see Table 6.2) that can be applied to the transition state from the Newtonian to the Stokes state, based on the measured Reynolds number, because the diameter of the pellets is relatively large.

In the pellet fluidized bed, the particles thicken as time progresses, and the total thickness of the fluidized bed increases. In such a situation, since particles having different initial positions move around each other at a certain position z of the fluidized bed, it is not easy to fix the position of a certain height z and describe the progress of the crystallization reaction of the particle groups having different initial positions which change, one after another. Therefore, we consider how the state of the same particle changes over time using a coordinate axis with the initial particle position as a reference. In this case, as a reference coordinate axis, consider the height ℓ when the initial particle is assumed to be filled with zero porosity. At this time, the relationship between the virtual height ℓ and the actual height z is shown in Equation 5.23. This hypothetical height ℓ can be written as shown in Equation 5.24 as the height from the bottom to the point where particles of the initial particle size d_a are present, with the height ℓm being the height of all initial particles filled with zero porosity.

Material balance equation	$u\frac{\partial C}{\partial z} + \frac{1{,}000}{M} \cdot \frac{\partial(Q(1-\varepsilon))}{\partial t} = \varepsilon D \frac{\partial^2 C}{\partial z^2}$	—(5·17)
Liquid boundary transfer limit reaction rate equation	$\frac{\partial(Q(1-\varepsilon))}{\partial t} = k_f(C_x - K_{st}K_{so})\frac{M}{1{,}000} a_v$	—(5·18)
Head loss in fluidized bed	$\rho wg \frac{dH}{dt} = (\rho_b - \rho_w)g(1-\varepsilon)$	—(5·19)
Amount attachment of calcium carbonate	$Q = \frac{\rho_c(d_3 - d_0^3)}{d^3} = \rho_c\left(1 - \left(\frac{d_0}{d}\right)^3\right)$ $\rho_p = \rho_c(d^3 - d_0^3) + \rho_{p0} d_0^3$	—(5·20)
Porosity of fluidized bed	$\varepsilon = \left(\frac{u}{W_t}\right)^n = \left[\frac{u}{\left(\frac{4(\rho_s - \rho)^2 g^2}{225\rho\mu}\right)^{\frac{1}{3}} d}\right]^n$	—(5·21)
Specific surface area of pellets	$a_v = \frac{6(1-\varepsilon)}{d}$	—(5·22)

(5.17–5.22)

Equation 5.17 - 5.22 The various factors in fluidized bed.

Equation 5.17 - 5.22 Explanation of symbols.

C : Calcium concentration in water (Ca^{2+}+$CaCO_3$(int))(mol/l)
Q : Amount of $CaCO_3$ attached per unit volume of pellets (g/cm³)
ε : Porosity of fluidized bed (-)
D : Longitudinal dispersion coefficient (cm²/sec)
k_f : Liquid boundary mass transfer coefficient (cm²/sec)
K_{st} : Equilibrium constants of calcium carbonate intermediates (mol/l)
K_{so} : Calcium carbonate solubility product (mol²/l²)
a_v : Pellet surface area per unit volume of fluidized bed (cm²/cm³)
d : Pellet size (cm)
d_0 : Pellet initial size (cm)
ρ_w: Water density (g/cm³)
g : Gravitational acceleration (cm/sec²)
H : Head loss (cm)
M : Molecular weight of calcium carbonate (g/mol)
z : Fluidized bed thickness (cm)
t : Operation time (sec)
C : Calcium carbonate intermediate concentration (mol/l)
ρ_p: Pellet density (g/cm³)
ρ_c : Calcium carbonate density (g/cm³)
u : Superficial velocity (cm/sec)
W_t : Terminal settling velocity of single pellet (cm/sec)
$F(d_0)$: Cumulative weight distribution of initial pellet particles (-)
l : Virtual packed bed height (cm)
l_m : Virtual packed bed total thickness (cm)
C_0 : Calcium concentration of influent water (mol/l)
C_T : Total carbonic acid concentration (mol/l)
C_{T0} : Total carbonic acid concentration inflow (mol/l)
α_0 : Rate of H_2CO_3 in total carbonic acid (-)
α_1 : Rate of HCO_3^- in total carbonic acid (-)
α_2 : Rate of CO_3^{2-} in total carbonic acid (-)
α_{00} : α_0 value at the inlet (-)
α_{10} : α_1 value at the inlet (-)
α_{20} : α_2 value at the inlet (-)

$$\left(\frac{dz}{dl}\right) = \left(\frac{d}{d_0}\right)^3 \frac{1}{1-\varepsilon} \quad —(5 \cdot 23)$$

Z : Actual height
l : Virtual height

(5.23)

Equation 5.23 Relationship between virtual height and actual height for fludized bed.

$$l = l_m \int_0^{d_0} \frac{dF(d_0)}{dd_0} \cdot \frac{1}{\rho_{p0}} dd_0 \Big/ \int_0^{\infty} \frac{dF(d_0)}{dd_0} \cdot \frac{1}{\rho_{p0}} dd_0 \quad —(5 \cdot 24)$$

(5.24)

Equation 5.24 The fluidized bed virtual thickness.

The quantitative relationship between calcium ions and calcium carbonate intermediates, which constitute the calcium concentration in water, is necessary in order to obtain changes in the calcium concentration of the liquid system by combining Equations 5.17 and 5.18. The balance of the H^+ ion, OH^-, Ca^{2+}, $CO_3{}^{2-}$, $HCO_3{}^-$, precipitate H_2CO_3, and intermediate $CaCO_3$(int.) involved in the water softening reactions are given by Equations 5.25 and 5.26. At the inflow point, the equilibrium relationship of Equation 5.26 is shown in Equation 5.27.

$$[H^+]_0 - \frac{K_w}{[H^+]_0} + (\alpha_{10} + 2\alpha_{20})(C_{T_0} - C_{x0})$$
$$= [H^+] - \frac{K_w}{[H^+]} + (\alpha_1 + 2\alpha_2)(C_{T_0} - C_0 + C - C_{x0}) \quad —(5 \cdot 25)$$
$$K_{st}(C - C_x)\,\alpha_3(C_{T_0} - C_0 + C - C_x) = C_x \quad —(5 \cdot 26)$$
$$K_{st}(C_0 - C_x)\,\alpha_2(C_{T_0} - C_{x0}) = C_{x0} \quad —(5 \cdot 27)$$

(5.25–5.27)

Equation 5.25 - 5.27 the material balance and equilibrium for water softening reaction.

To operate these equations, the necessary coefficient values are required. When the longitudinal dispersion factor D of Equation 5.17 is measured by the tracer response, D = 1.74 cm^2/sec, and the magnitude of this dispersion term is about 10^{-2} times smaller than that of Equation 5.17, which is obtained from the k_f values described later. Therefore, in this experiment, calculations are performed ignoring the dispersion term. For large diameter implementations, it is necessary to divide the ascending water channels so that the longitudinal dispersion (mixing) does not increase. The porosity of the fluidized bed was obtained by plotting the index n of the Richardson Equation 5.21 from the measured values shown in Figure 5.13, and n = 0.225. This value can be applied to full-scale prototype operations.

Next, we need to evaluate the film mass transfer coefficient k_f as the rate coefficient that governs water softening reactions. The molecular diffusivity D_m of the underlying calcium carbonate intermediate is determined by the Wilke-Chang equation, as shown in Equation 5.28.

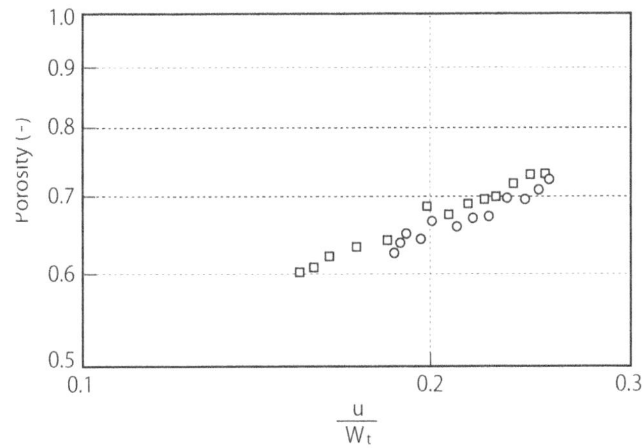

Figure 5.13 Determination of the exponent n in the Richardson equation.

$$D_m = 7.4 \times 10^{-10} \frac{(2.6\,M_w)^{1/2} \cdot T}{\mu V_m^{0.5}} \quad \text{---}(5 \cdot 28)$$

M_w : Molecular weight of solvent (g/mol)
V_m : molal volume of solute at normal boiling point (cm^2/mol)
T : Absolute temperature

(5.28)

Equation 5.28 The Wilke-Chang molecular diffusivity equation.

The relation between the mass transfer coefficient k_f and the molecular diffusivity D_m in a fluidized bed can be determined by mediating the mass transfer j factor (J_d) defined by Equation 5.29. The J_d of the fluidized bed can be obtained from the hydraulic condition of the fluidized bed by using the Yen-Wen Fan equation shown in Equation 5.30.

$$J_d = \frac{k_f}{u}\left(\frac{\nu}{D_m}\right)^{\frac{2}{3}} \quad \text{---}(5 \cdot 29)$$

$$J_d = 1.865(Re(1-\varepsilon))^{-0.48} \quad \text{---}(5 \cdot 30)$$

k_f : Mass transfer coefficient in fluidized bed
D_m : Molecular diffusion coefficient

(5.29–5.30)

Equation 5.29, 5.30 The mass transfer coefficient and molecular diffusivity.

The molecular volume V_m of the calcium carbonate monomer is required to obtain the molecular diffusivity D_m of Equation 5.28. The molecular volume is estimated by dividing the molecular weight of calcium carbonate by the density. Densities were 2.93 g/cm^3 for aragonite, 2.71 g/cm^3 for calcites, and 2.70 g/cm^3 for experimental pellets. From the results, we can estimate the molecular volume $V_m \fallingdotseq 34$ to 37 cm^3/mol.

Substituting these values into Equation 5.26 to calculate the molecular diffusivity D_m yielded values of $D_m \doteqdot (1.26–1.32) \times 10^{-5}$ cm^2/sec. The porosity ε of the fluidized bed and the Reynolds number (Re) are required to determine J_d in Equation 5.30. The porosity ε is given by Equation 5.21. For pellets with diameters of 1 mm, the final sedimentation velocity was $W_t = 14.4$ cm/sec when the fluidized bed was operated at an up flow rate u=100 m/day, $k_f = 0.007$ cm^2/sec.

Considering the increase in the diameter of pellets over time at each position of the fluidized bed, the change in k_f is calculated by Equations 5.29 and 5.30.

Figure 5.14 shows changes of the treatment effect over time, Figure 5.15 shows changes in the fluidized bed thickness and pressure loss over time, and Figure 5.16 shows changes, over time, in the calcium concentration in the fluidized bed for both the experimental (measured) values and values calculated by the kinetic formula. We can see that the kinetic process in our model has been sufficiently formulated for practical use. In the initial maturation stage, a maturation time of about 2 hours is considered necessary.

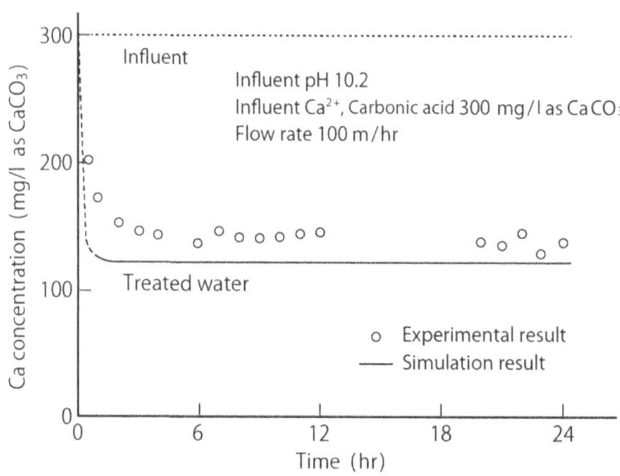

Figure 5.14 Changes of the treatment effect over time.

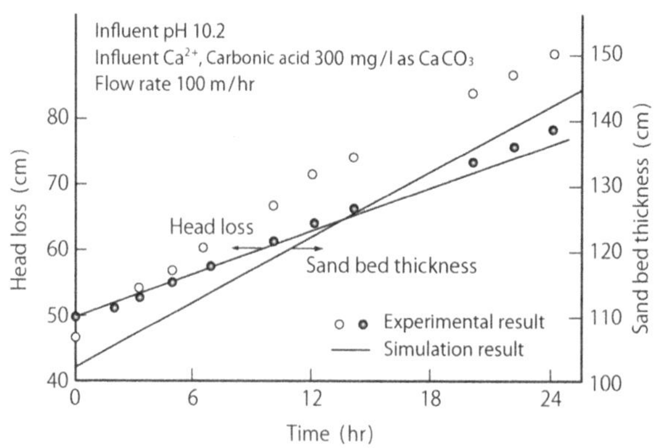

Figure 5.15 Changes in the fluidized bed thickness and pressure (head) loss over time.

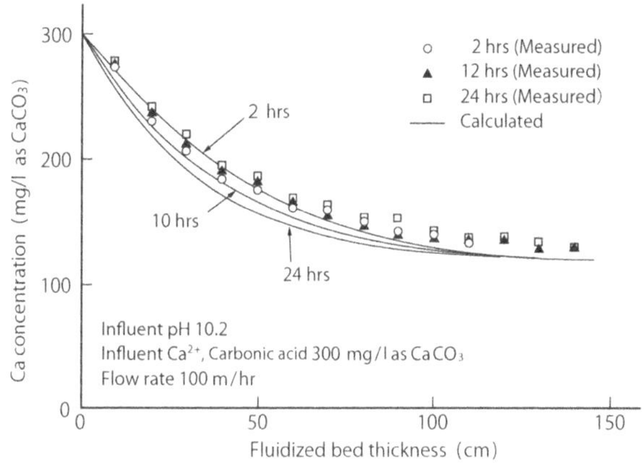

Figure 5.16 Changes, over time, in the calcium concentration in the fluidized bed for both the experimental (measured) values and values calculated by the dynamic formula.

The pellet softening method developed by Graveland at the Amsterdam Water Works (see Photos 1 and 2) was theoretically evaluated by Tambo *et al.* (1988) at Hokkaido University, as described above, and was designed theoretically. In Japan, it was applied to the Kadena group of wells on Okinawa, which originated from coral reefs, and to the well group in the limestone region at the foot of Mount Ibuki, near Kyoto.

Photo 5.1 A prototype pellet water softening system constructed by Dr A. Graveland at the Amsterdam Water Works, the Netherlands (photo by Tambo).

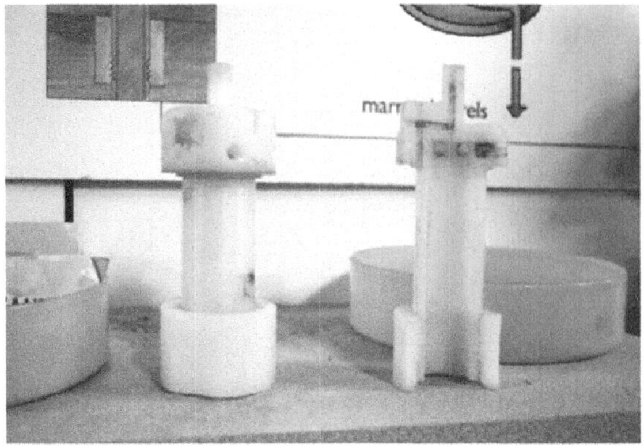

Photo 5.2 A prototype instantaneous mixing nozzle to mix raw water and alkaline agent at the bottom of a fluidized bed, produced by Dr A. Graveland at the Amsterdam Water Works, the Netherlands (photo by Tambo).

5.3 TREATMENT OF IRON, MANGANESE, ETC.

5.3.1 Oxidation of iron and manganese

Iron and manganese in utility water and tap water cause problems mainly because they precipitate as hydroxides and hinder water coloration, laundry, food production, and other applications. Along with hardness, it is a major factor that causes problems when groundwater is used. The taste of iron-containing water is also a problem for beverages and tea water if it exceeds 0.5 mg/l. From the standpoint of water coloration disturbance, it is desirable that the amount of iron is 0.3 mg/l or less and the amount of manganese is 0.05 mg/l or less. Iron, on the other hand, is the major element in hemoglobin, which transports oxygen in the blood, and its physiologically acceptable level may be higher, and it is necessary to obtain approximately 10 mg/day of iron from various foods and beverages for human health.

To accurately determine the presence of iron and manganese, it is easiest to refer to the Pourbaix diagrams of the pH and redox potential (ROP) of the system on both axes (Pourbaix, 1966), and to describe the presence of metals by the relationship between pH and RO potential in the system (see Section 2.1.2 for the definition and measurement of pH and ORP).

As shown in Figure 5.2, both divalent ferrous ions (Fe^{2+}) and manganese ions (Mn^{2+}) are present in a reduced state (in an anaerobic state in the natural system, with negative RO potential). Unless the pH is highly alkaline, such as pH 10 to 11, neither the Fe^{2+} or Mn^{2+} ions will be able to precipitate efficiently, and will remain insoluble as ferric hydroxide or manganese oxide under aerobic conditions. Therefore, if precipitation is attempted without greatly increasing the pH, it is necessary to bring the OR potential into a high range of +200 to +400 MV using the oxidizing agent while maintaining the neutral pH range.

Iron is generally dissolved as a divalent iron ion (Fe^{2+}) under anaerobic conditions of natural water at normal pH (in groundwater, lakes and at marsh bottom, etc.), and is present as a colloidal or suspended (precipitated) trivalent iron hydroxide under aerobic conditions (surface water). Even by using simple aeration by blowing air, oxidation precipitation gradually progresses as the pH rises above 7. In peat areas such as around Hokkaido and in Scotland, iron forms organic complex ions with humic acids and is stably soluble. In groundwater with a high silica content (such as SiO_2 higher than 40 mg/l), oxidation by aeration proceeds as with other cases, but it was reported by Takai (1964) that the produced trivalent iron hydroxide forms colloids associated with silica and dispersions, making it difficult to separate them. Compare this with the coagulation acceleration mechanism of polymeric silica coagulants described in Chapter 4.

The peaty groundwater found in Hokkaido, Japan, which was studied by Tambo and colleagues from the 1960s to the 1980s, falls into three categories:

(1) The first is a soft water which contains very stable colored organics that are almost free of heavy metals such as iron and manganese. The brown colored water that Tambo investigated in the eastern part of the United States in the early 1960s can also be included in this category, and is an exceptional soft water on the US continent.
(2) The second is water which is colorless and transparent when it contains a large amount of iron and has an iron odor, but precipitates as iron hydroxide in a relatively short period of time when it is pumped in a well,or similar, and exposed to the atmosphere.
(3) The third is a water in which metal components such as iron coexist with organic coloring components, and its bonds are considerably strong and stable. It is common in groundwaters in the peaty region of Hokkaido.

The presence of iron in these colored waters is thought to be attributable to the geological structures predominant during peat formation. In general, plants such as reeds, sedges, field horsetails, mosses, and lingonberry are abundant in wetlands and peatlands which are laden with shallow waters. In order to create such a shallow pool of water, a water-impermeable layer such as clay is usually widely and evenly distributed under the shallow flooded water area. Therefore, the main cause of the shallow pooled water on the upper surface of the impermeable layer is usually rainwater or snow-melt, which is a soft water different from the groundwater under the impermeable layer.

Figure 5.17 shows a typical geological cross-section of peatland. The upper part of the clay layer, which constitutes the impermeable layer, is dark brown, whilst the lower part rapidly shifts to a blue clay layer. This is because air/oxygen and water hardly diffuse vertically through the clay layer, and iron in the soil is red as a result of air oxidation on the upper surface of the water-impermeable layer, whereas the lower surface of the clay is anaerobic clay.

In such a case, humus on the impermeable layer forms colored soft water in which humic substances are extracted in surface inundation by mild rainwater, and constitutes the first category of water. Ishikari Peat

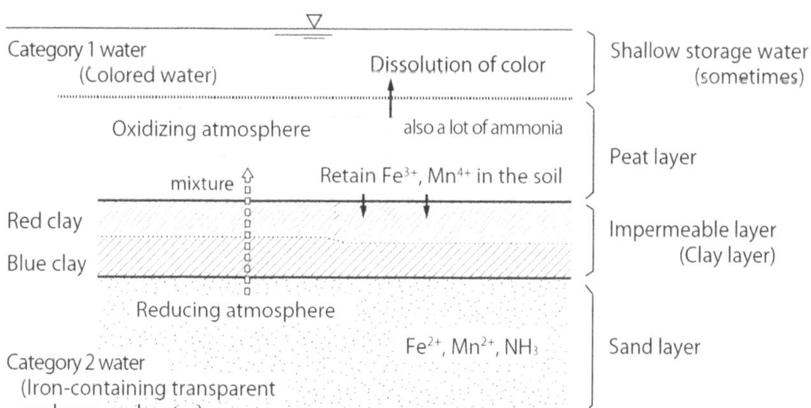

Figure 5.17 A typical geological cross-section of peatlands.

Water has been used for many years by Tambo and others for research into organic colored water and trihalomethane control researches.

Peatland water in the second category is in a reduced (anaerobic) state in which the OR potential is negative in the lower part of the impermeable layer, and constitutes groundwater in which iron, manganese, etc., are dissolved in large quantities in the form of divalent ions. In addition to the general characteristics of the groundwater, which is relatively hard, this groundwater is free of dissolved oxygen. Water in peat areas with little color is often thought of as groundwater. When pumped up in a well and in contact with air, a large amount of iron precipitate is formed, but manganese remains in the water without being oxidized.

The third category of water is colored water containing iron mixed with the water of the first category and the second category which leaks out of the water-impermeable layer in some circumstances. In this case, reducing iron and an organic (humic) coloring component combine to form a colloidal metal-organic complex, which is stably present. It cannot be oxidized and precipitated by simple aeration with air or addition of chlorine. Similarly, potassium permanganate or ozone-induced strong oxidation hardly leads to precipitation. Unfortunately, Tambo's laboratory cannot artificially produce stable organo-color metal complexes (colloids) by adding iron or manganese to colored humic substances. Organo-iron (1.1 mg/l) contained in peaty groundwater in the suburbs of Sapporo city, which has been used for a long time as a test water by Tambo's team, is adsorbed by anion exchange resin (Dowex1) instead of cation exchange resin (Dowex50 W), and is removed by aluminium coagulation by reacting with organic coloring components (negatively charged fumes).

In lakes, marshes and dam lakes that are frozen in winter, the surface of the water is covered with thick ice, and gas exchange with the atmosphere is cut off. In this situation, dissolved oxygen in the reservoir is gradually consumed by organic matter, reduced by inorganic matter in the water or in the bottom mud, and finally, oxygen is lost.

In deep ponds, such as artificial dam lakes, stable winter stratification produces anoxic conditions centered on the lowest water mass at about 4°C. Ice in shallow lakes and marshes demonstrates the full effect of anoxia. In dam lakes with organic bottom mud layers during the summer stratification phase, oxygen-deficit water masses are generated in the lower layers of the lake. Reduced iron Fe^{2+} and manganese Mn^{2+} contained in the soil of the anaerobic lake bottom diffuse into the oxygen-free water masses in the lower layers of the lake. Sometimes we see the generation of hydrogen sulfide. Figure 5.18

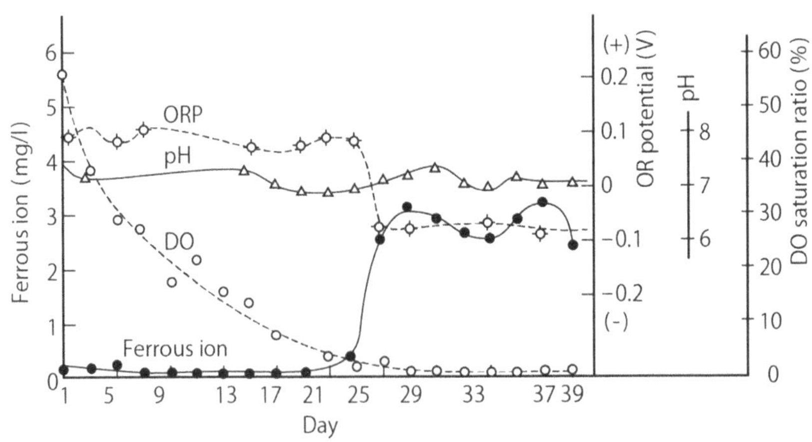

Figure 5.18 Anaerobic bottom mud of the Toyohira River and iron elution.

shows the changes in the composition of the water over time, when the bottom mud of the Toyohira River flowing through the center of Sapporo City is kept at 10°C with a thickness of 1.5 cm under a 15-cm-deep glass container as shown in the figure:

(1) Dissolved oxygen (DO) in water decreases exponentially to nearly zero around day 20;
(2) When the oxidation-reduction potential (ORP) becomes negative, a large amount of iron in the bottom mud is eluted into the water.

When the water loses oxygen and the environment becomes anaerobic, a large amount of Fe^{2+} ions are eluted into the water in a short period of time (about one day). In order to predict when a sediment elution will occur rapidly in an actual reservoir, dissolved oxygen in water decreases exponentially (following the Streeter expression), and the DO concentration of lake water under freezing is measured day by day, and the point (number of days) at which the DO concentration becomes 0 is extrapolated.

In the 1970s, iron elution occurred in a shallow water source lake (2–3 m water depth) in Nemuro City in the eastern part of Hokkaido. If ice formation continued for 150 to 170 days, a high concentration of iron and hydrogen sulfide was suddenly observed in the raw water of the city water supply. How long the freezing continues during the winter is a problem. As a countermeasure, it is necessary to prevent the anaerobic state of the water by constantly breaking the ice to make contact between water and air, or by constantly installing a circulating aeration paddle.

Manganese continues to be a reduced Mn^{2+} ion even in the aerobic range until the OR potential of manganese reaches +200 mV. Air oxidation, such as is the case with iron, and low concentration chlorine oxidation do not insolubilize manganese as trivalent or tetravalent manganese compounds (MnO_2, Mn_2O_3, etc.) or as manganese hydrate. In addition, when the lower layers of lakes and marshes slowly become anaerobic and iron and manganese are eluted into the water, the manganese ions proceed first, followed by the iron ions. It is thought to be ordered according to the time series of ORP lowering. However, among the eluted Fe^{2+} and Mn^{2+} ions, iron is oxidized and precipitated early in the aerobic water mass, whereas manganese continues to exist as a soluble Mn^{2+} in the natural system. In the presence of iron bacteria, both iron and manganese are taken up by the bacteria and precipitated as part of slime-like microbial communities.

Manganese, therefore, is often dissolved as Mn^{2+} ions, and passes through the water treatment plant. However, when chlorinated in treated water, Mn_2O_3 deposited gradually in the water pipe catalyzes the oxidation of Mn^{2+} by chlorination, the Mn^{2+} ion is oxidized and deposited on the inside surface of the water pipe. This in turn catalyzes the insoluble precipitation of manganese in chlorinated water, accelerating it, colors the tap water as a black suspension, and discharges from the tap. At the end of World War II, the US Occupation Force enforced the chlorination of tap water at all Japanese water works. The Japanese water supply, which had not necessarily been chlorinated nationwide before the war, was ordered to contain residual chlorine at a high level of 0.4 ppm, which was the standard on the US battlefield. As a result, the 'black water problem' developed in the water supply nationwide, and became the first water quality problem in Japan in the postwar period.

We will now consider to remove or make divalent iron and manganese ions harmless at normal pH when raw water containing them is used.

Metaphosphate polymerized salts may be added to dissolved metal ions such as Fe^{2+} and Mn^{2+} to perform complex ionization that maintains stable solubility, and sequestration may be performed to prevent damage through the distribution and use of the water. They are also removed by ion exchange with iron and manganese zeolite (described later in the section on ion exchange adsorption; this operation is often used for the treatment of boiler and industrial water).

Table 5.5 Reaction products and their reactivity, depending on oxidants, at normal pH.

Oxidants	Dissolved Oxygen (Aeration) O_2	Chlorine Cl_2	Potassium Permanganate K MnO_4
Fe	$Fe^{2+} \rightarrow FeOOH$ Autocatalytic reaction	$Fe^{2+} \rightarrow$ Amorphous iron hydroxide	$Fe^{2+} \rightarrow$ Amorphous iron hydroxide
Mn	Almost no reaction	$Mn^{2+} \rightarrow MnO_2 \cdot nH_2O$ Autocatalytic reaction (Reaction rate is slow and not practical depending on Cl_2 dosage alone)	$Mn^{2+} \rightarrow$ Amorphous manganese hydroxide

Valences of iron and manganese in water are generally divalent and trivalent (in the case of iron), and divalent and tetravalent (in the case of manganese). The divalent valences are reduced and dissolved, and an oxidized form is precipitated as iron hydroxide or manganese dioxide. Most of the metal ions are removed by an alkali precipitation process raising the pH to alkaline states. If an oxidizing agent is added and the system is operated at a positive OR potential of appropriate magnitude, precipitation can be performed at normal pH and can be removed by sedimentation and filtration.

The precipitation operation, in which the pH of a large amount of raw water is greatly changed in order to remove a small amount of iron and manganese ions in the water treatment process, is not practical, and in actual iron manganese treatment, precipitation treatment is usually performed at a normal pH by the use of an oxidizing agent. Dissolved oxygen, chlorine, and potassium permanganate are commonly used as oxidizing agents, and ozone has also been used recently.

Table 5.5 summarizes the types of oxidizing agents and their reactions. Air for iron and chlorine for manganese are slow to react and impractical in the case of simple addition. However, if precipitated oxides and hydroxides are already present, the autocatalytic reaction leads to rapid oxidation. The following sections describe these oxidation/precipitation reactions used for water treatment.

5.3.2 Treatment of iron at normal pH

5.3.2.1 Precipitation treatment of Fe^{2+} in natural water by oxidizing treatment such as aeration

The air oxidation rate equation in the pH neutral range can be written as Equation 5.31.

$$\frac{d[Fe^{2+}]}{dt} = k\,[Fe^{2+}][O^{2-}][OH^-]^2 \qquad \text{— (5·31)}$$

k : Oxidation rate constant

$k = 8.0(\pm 2.5) \times 10^{13}$ ($min^{-1} atm^{-1} mol^{-2}$) at 20℃

by Stumm·Morgan (1970)

(5.31)

Equation 5.31 The air oxidation rate equation in pH neutral region.

The rate at which iron is oxidized depends heavily on the pH, resulting in a first order reaction of the iron concentration and a second-order reaction with respect to the OH^-. The precipitation reactions proceed rapidly at higher pH, but slowly at neutral pH and at low Fe^{2+} concentrations. When trivalent iron hydroxide deposits are present, the catalytic reaction at the interface of the formed trivalent iron precipitate (considered to be iron oxyhydroxide, FeOOH) proceeds in parallel with aqueous phase and the precipitation of Fe^{2+} is accelerated.

In general, when water in which oxygen is sufficiently dissolved is passed through the filtration layer, iron hydroxide precipitates and is retained near the surface of the filtration layer. If the iron concentration is extremely high, such as several tens of mg/l or more, a large amount of insoluble iron is removed in advance of the sedimentation tank or of the dissolved air flotation method, to prevent rapid blockage of the filtration tank. However, if it is a normal natural water, it can be treated by using only normal sand filtration or a high-capacity filtration tank (two-layer or two-stage filtration tank) equipped with coarse filtration materials, as described in Section 6.2 below.

An example of peaty groundwater from Hokkaido was used as an example of how iron in groundwater cannot be precipitated by simple aeration. Takai (1964) reported an example in which water with 10 degrees of color was removed by simple aeration – even though the author (Tambo) found it hard to precipitate the same colored water. The reason is unknown, and it is unclear why precipitation and non-precipitation occur, because the color of peaty groundwater in Hokkaido is high and the color component is a polymer with low electrical conductivity of around 1500 to 1000 Da, and the colored water that Takai handled is normal hard groundwater. Since the author was not successful in mixing iron ions and peaty groundwater to form soluble stable iron-humic organic complexes, it seems necessary to further study the mechanism of groundwater containing high concentrations of iron and organic (humic) complex iron.

Figure 5.19 summarizes the relationships between iron content and humic concentrations (ultraviolet absorbance E260) in various colored waters in Hokkaido. There is a linear correlation between the two, indicating that iron is protected from air oxidation and remains soluble in proportion to the amount of organic matter exhibited by the ultraviolet absorbance at 260 nm. If the iron concentration exceeds this protection level, soluble iron present in the groundwater in a reduced state is considered to be insoluble after oxidation by exposure to air and precipitation. Therefore, most of the soluble iron present in peaty ground surface water is present as a stable organic complex (iron humate) protected by humic substances at the concentrations shown in Figure 5.19 (Tambo & Tsutsumi, 1982).

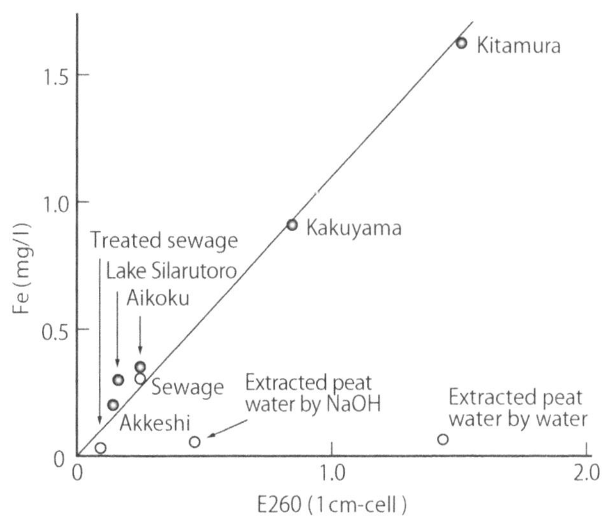

Figure 5.19 The UV absorbance (E 260) and iron content of colored water (color substances extracted by NaOH from peaty soil).

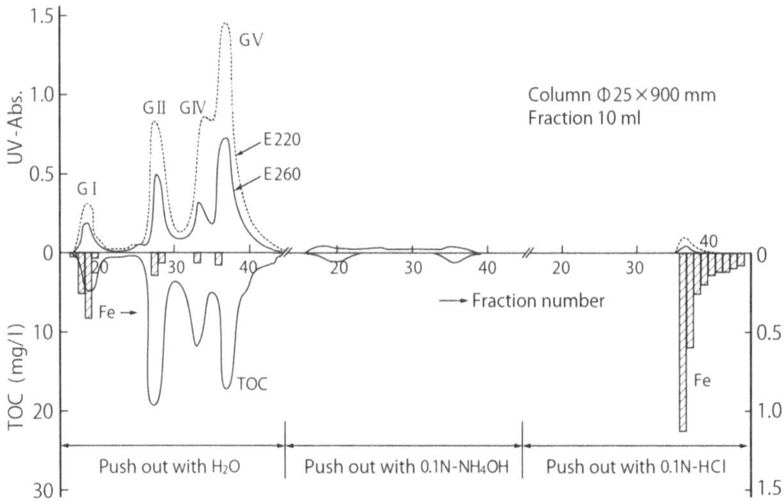

Figure 5.20 A gel chromatogram for Kitamura peat water, using Sephardex G15 gel.

Figure 5.20 shows a gel chromatogram (Sephardex G15) of Kitamura peaty groundwater from the suburbs of Sapporo city. The chromatogram indicates that iron coexists with the polymer organic matter (1500 Da or more) of fractions 16 to 20 which cannot diffuse into the gelled particles of the void volume ($K_d = 0$). The organic coloring components are scarcely present in the second group of the gel-adsorbed component extruded with NaOH or in the third group extruded with HCl, and a large amount of iron appears alone in the last HCl-extruded group. This is probably because the organic iron is almost bound to the humic polymer substance, and when passing through the Sephardex G15 gel column, it remains bound to the humic substance and remains in the polymer image group 1, is adsorbed on the Sephardex G15 gel, is redissolved by hydrochloric acid, leaves the organic complex, and becomes an iron-only group that flows out of the gel column by HCl extrusion.

To confirm this, we can see in Figure 5.21 that only the polymer part fractionated into the first group of water extrusion was subjected to G15 gel chromatography again. As described above, iron is basically bound to the polymer part of humic substances and dissolved in peaty groundwater as a stable humic-iron complex. This indicates that dissolved iron in peaty groundwater is strongly complexed with E260 sensitive organic matter having a large molecular weight, or behaves as one body for ion and polymer adsorption. Figure 5.21 shows that when iron bound to humic substances was adsorbed on a gel column, it redistributed with the Sephardex G15 gel, and the part once adsorbed on the gel was redissolved in the HCl extrudate, and the iron was extruded alone. For peaty groundwater, humic substances are considered to be the main organic substances sensitive to E260.

Iron in peaty ground water is not adsorbed by cation exchange resins, but is adsorbed by anion exchange resins. Therefore, iron in raw water appears to be bound to humic substances and exists as an anionic organic complex.

Figure 5.22 shows the changes in the water quality factors of Kitamura peat groundwater, which has been concentrated by a rotary evaporator at five times normal temperature, when treated with ozone. Despite the significant reduction in UV E260 absorbance, and the absence of any coloration, the reduction due to iron oxidative precipitation and the reduction in TOC are slight. This is probably because even if the humic color disappears, the body structure of the remaining organic matter (colorless) continues to have a protective

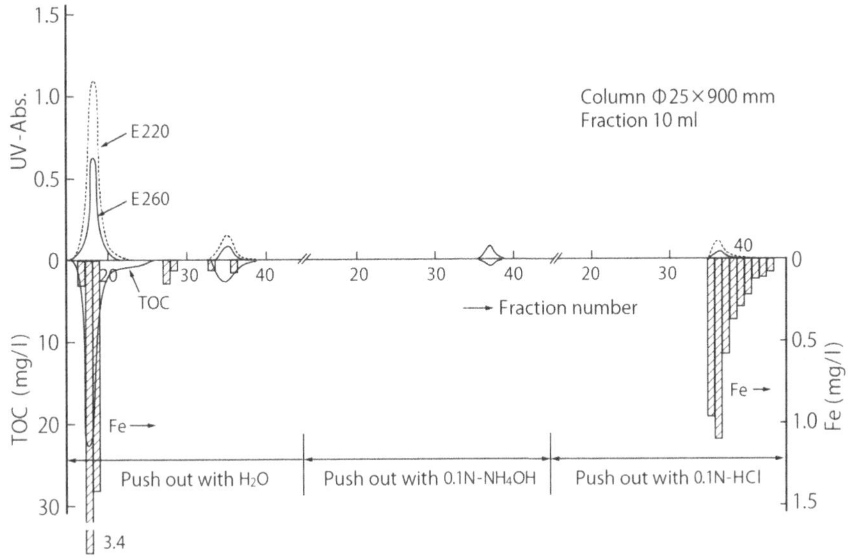

Figure 5.21 A gel chromatogram for Kitamura peat water fraction No.1, using Sephardex G15 gel.

ability to suppress the oxidation of iron. In addition, the relatively low precipitation and removal of iron compared to the relatively large reduction in TOC suggests that the protective capacity of the colorless organics produced from humic substances after ozone oxidation is actually increased.

Figure 5.23 shows a gel chromatogram of the sample after ozone treatment for 60 minutes. It is clear that TOCs, which are reduced in molecular weight and migrate to the 2, 3 (several hundred Da) regions of the chromatogram group and show little ultraviolet E260 absorbance, retain a large amount of iron, demonstrating the hypothesis considered in Figure 5.21.

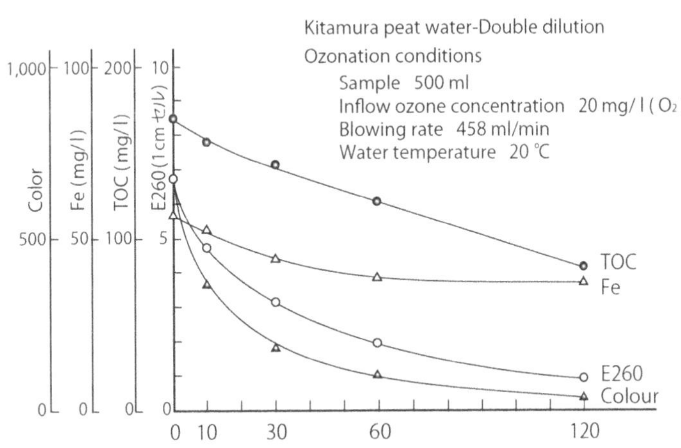

Figure 5.22 Changes in water quality factors due to ozonation.

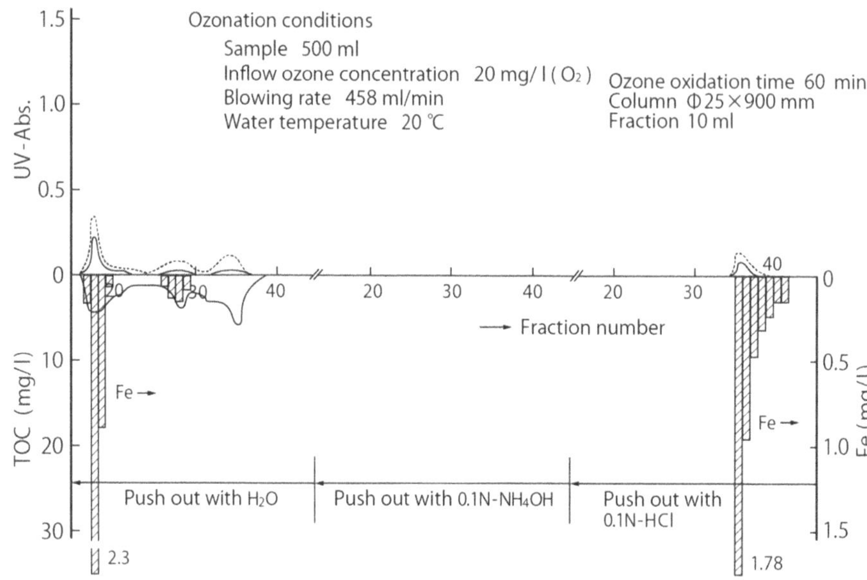

Figure 5.23 A G15 gel chromatogram after ozonation of components of Kitamura peat water fraction No.1.

Figure 5.24 shows the precipitation of TOCs and Fe according to the amounts added when chlorine (Cl_2) and potassium permanganate ($KMnO_4$) are added to Kitamura peaty groundwater. Within 30 minutes of reaction time, there is little change. This is a natural consequence of the fact that iron humate complexes which do not precipitate even when strong oxidizing ozone is added cannot be oxidized and precipitated by Cl_2 or $KMnO_4$. This suggests that humic iron complexes behave similarly to humic substances, as described in Section 3.5.4 on water conversion matrices, and therefore it is reasonable to treat them by coagulation-flocculation-sedimentation, or by a dissolved air flotation method.

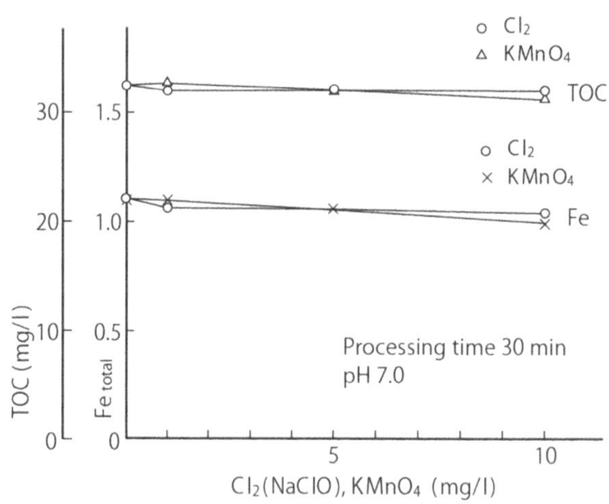

Figure 5.24 The oxidation of Kitamura peat water by Cl_2 (NaClO) and $KMnO_4$.

Takai (1964) reported a case in which colorless water with a high iron (Fe^{2+}) concentration (10 mg/l or more) in the soluble silica (d-SiO_2) range of 4 to 50 mg/l produced opalescent colloidal particles (silica protective colloids) that were not precipitated by aeration and could not be separated by sedimentation, paper filtration, or aluminium coagulation.

5.3.2.2 *Chemical oxidation treatment of iron*

Most of the soluble iron Fe^{2+}, except that which is strongly protected by humic acid complexes, is rapidly precipitated into iron hydroxide by adding chlorine or an oxidizing agent such as potassium permanganate, and is separated and removed. An amount of 0.63 mg Chlorine (Cl_2) or 0.57 mg of potassium permanganate ($KMnO_4$) is required to oxidize 1 mg of Fe^{2+}. However, when potassium permanganate is added, some amount of Mn^{2+} is newly added to the solution, hence, to remove this, 0.94 mg of potassium permanganate per mg of Fe^{2+} is additionally required.

The oxidation/precipitation of iron by chlorine is shown in Equation 5.32, where the presence of chlorine is as HClO. The reaction with potassium permanganate results in Equation 5.33, which requires an additional reaction with manganese, Equation 5.34, to remove the Mn^{2+} formed.

The floc of iron hydroxide, which coagulates by adding chlorine or potassium permanganate to a rapid mixing pond, is extremely fragile, does not grow easily in flocculators, and easily leaks out of the filter layer. Therefore, it is necessary to strengthen the flocs using a polymer flocculant aid, using multi-layer filters, and performing low-speed filtration.

Oxidation of iron by chlorine

$$2Fe^{2+} + HClO + 5H_2O \rightarrow 2Fe(OH)_3 + 5H^+ + Cl^- \quad — (5\cdot32)$$

Oxidation of iron by potassium permanganate

$$5Fe^{2+} + KMnO_4 + 5H_2O \rightarrow 5Fe(OH)_3 + K^+ + Mn^{2+} + H^+ \quad — (5\cdot33)$$

Oxidation of Mn^{2+}

$$3Mn^{2+} + KMnO_4 + 2H_2O \rightarrow 5MnO_2 + 2K^+ + 4H^+ \quad — (5\cdot34)$$

(5.32–5.34)

Equation 5.32 - 5.34 The oxidation·precipitation of iron by chlorine and potassium permanganate.

5.3.2.3 *Iron removal by filtration*

Groundwater treatment to remove the reducing iron Fe^{2+} generally involves spraying raw water in the air, blowing dispersed air into a vessel, or using a cascaded fixed bed to dissolve air, oxidizing the Fe^{2+} ions, and separating the iron by granular bed filtration.

The surfaces of the granular filter media are covered with iron oxyhydroxide (FeOOH) which has been precipitated in advance, and this is used as an auto catalyst, and the Fe^{2+}, which flows in, is adsorbed on the interface, oxidized and removed. The iron hydroxide thus removed by the interfacial heterogeneous reaction (as described in the section on pellet coagulation in Chapter 4) is filtered (accumulated) in a form of metastable zone treatment, and the coagulum formed is not a low-density, swellable, fragile random floc like a coagulum in a homogeneously dispersed system, but rather a high-density precipitate which does not significantly increase the head loss of the filter layer.

Fe^{2+} is oxidized and precipitated and removed by simple aeration. However, because of its relatively slow reaction rate, if the complete reaction is difficult to proceed, a chlorine-containing oxidizing agent is added to perform the same treatment. Microbial de-ironing and de-manganisation using iron bacteria are described below.

5.3.3 Manganese treatment at normal pH

5.3.3.1 Oxidation by potassium permanganate

The manganese ion Mn^{2+} becomes insoluble in water and precipitates as Mn_2O_3 or MnO_2 when it reaches pH 12 or higher. However, oxidizing agents must be added at neutral pHs. The oxidation-precipitation reaction with the addition of potassium permanganate is shown in Equation 5.34. Therefore, theoretically, 1.92 mg of potassium permanganate $KMnO_4$ is required for 1 mg of Mn^{2+}.

The resulting MnO_2 or $Mn\,(OH)_4$ is thought to undergo an oxidative reaction by autocatalytic action. The authors' old experiments confirm that the reaction is completed at about 60–70% of the equivalent amount in the vicinity of pH 7. The addition of potassium permanganate has little effect on the pH of the system.

Figure 5.25 shows the Mn^{2+} concentration of the sedimentation supernatant liquid when potassium permanganate is added to pH 7.0 groundwater containing Mn^{2+}. On the alkaline side, the Mn^{2+} concentration becomes zero with the addition of $KMnO_4$ at an amount 1.94 times the Mn^{2+} concentration, but on the acidic side, it has been shown that the amount added to make the concentration become zero is 1.94 times or more, which is considerably less than the equivalent (stoichiometric) amount. The effective value, which may be less than this theoretical amount, may be attributed to the autocatalytic reaction, and as the pH increases, the reduction width increases linearly, as shown in Figure 5.26. When the pH reaches 9.0, precipitation progresses at about 30% of the theoretical amount (equivalent). This can be said to be an additive effect with alkaline precipitation.

At a later date, the widespread use of atomic absorption spectroscopy made it possible to measure manganese concentrations accurately, enabling more detailed experiments. When an equivalent amount of potassium permanganate equivalent to 1.92 times of manganese Mn^{2+} by weight is added, manganese dioxide is precipitated by reducing potassium permanganate to 100% manganese oxide in 1/8 of equivalent weight, and the Mn^{2+} of water from which precipitated manganese is filtered is measured. The results of the same experiments as those in Figure 5.25 (Matsui *et al.*, 1986a, b) are summarized in Figure 5.27. The line a–b in Figure 5.27 shows the amount of manganese oxidized and precipitated by

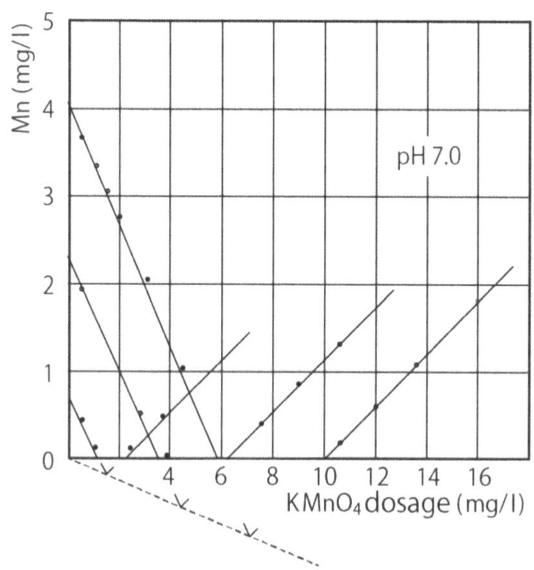

Figure 5.25 $KMnO_4$ dosage and residual Mn^{2+}concentration.

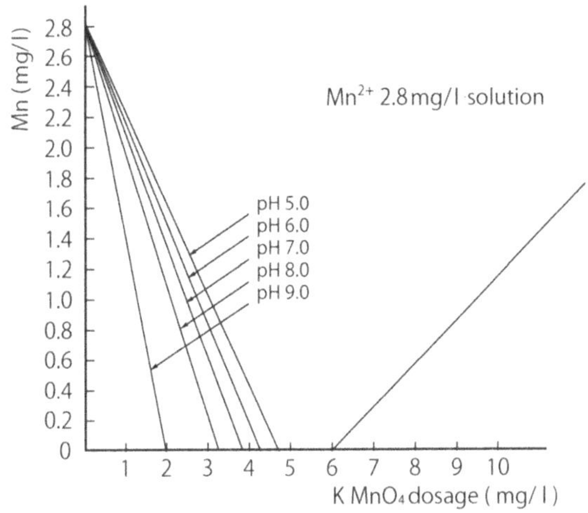

Figure 5.26 The difference in pH between the theoretical and actual Mn dosage.

potassium permanganate, and c–d the amount of manganese remaining in the liquid system as a soluble Mn^{2+} without reacting. The dotted line shows the trajectory of the oxidized precipitated portion, corresponding to the amount of reaction equivalent to the manganese added to the beaker. Therefore, the part of the total amount of manganese (a–d) added to the beakers that corresponds to the remaining amount of manganese (c–b) after subtracting the segment (c–d) corresponding to the unreacted soluble Mn^{2+} actually present in the liquid system is manganese adsorbed on the precipitated manganese particles.

This adsorption portion is shown as the actual addition amount discounted in proportion to the total amount of manganese. Figure 5.25 explains the finding that manganese ions are generally sufficiently

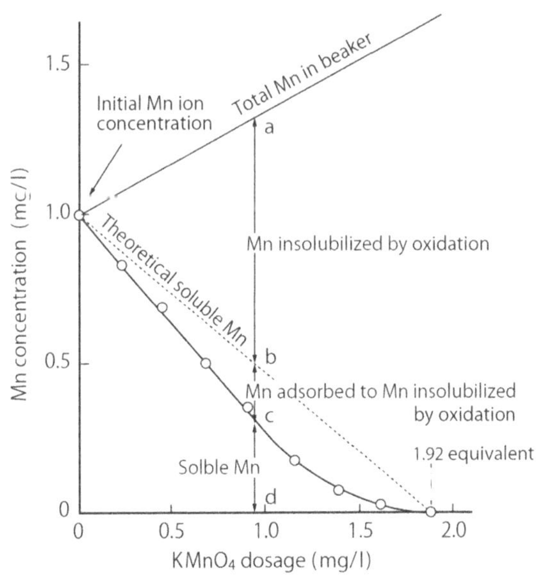

Figure 5.27 $KMnO_4$ dosage and Mn amount in each state.

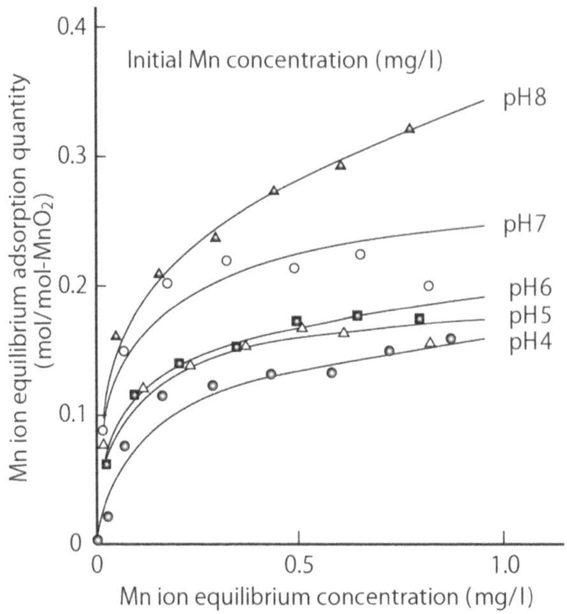

Figure 5.28 Mn ion adsorption isotherm.

removed with a lower potassium permanganate dose to reach the equivalence point, and the degree of removal increases with increasing pH. The discount rate does not increase linearly with concentration, but rather increases with a gradual curve as shown in Figure 5.27.

In order to evaluate the adsorption of Mn^{2+} on the surfaces of precipitated manganese dioxide (hydrate), Figure 5.27 shows the adsorption isotherm of manganese ions per precipitated manganese dioxide [mol/mol-MnO_2] from data under the various conditions of the amount of insoluble manganese (a–b), the amount of manganese adsorbed on the insoluble manganese (b–c), and the amount of manganese remaining in the liquid system (c–d). These isotherms can also be generalized (linearized) with plots like those in Figure 5.29 based on the Langmuir Equation (Equation 5.35). (See also Chapter 7, Section 7.1.2, below).

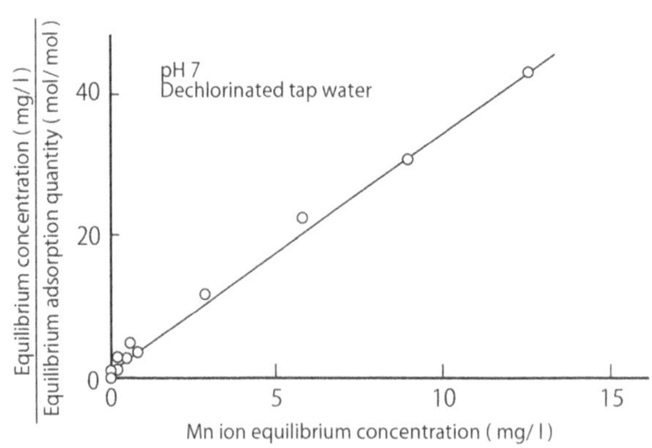

Figure 5.29 A Langmuir plot for the Mn ion adsorption of MnO_2.

As shown in Figure 5.29, in the adsorption phenomenon linearly described by Equation 5.35 of the Langmuir-type adsorption isotherm, under pH 7 conditions, the slope of the straight line is $1/q_{max} = 0.3378$ and the intercept is $1/q_{max}K$, resulting in a maximal adsorption rate per mole of manganese dioxide deposited of $q_{max} = 0.296$ [mol/mol-MnO_2]. Substituting the obtained $q_{max} = 0.296$ [mol/mol-MnO_2] and the manganese concentration in water M (mol/l) into Equation 5.35, the reaction rate constant K = 10 l/mg is obtained for the reaction rate constant $K = k_1$ (adsorption rate constant)/k_2 (desorption rate constant), and the adsorption isotherm is established.

$$\frac{M}{q} = \frac{M}{q_{max}} + \frac{1}{q_{max}K} \quad —\ (5\cdot35) \qquad (5.35)$$

M : Mn concentration in water (mol/l)
q : MnO_2 adsorption amount (mol/mol-MnO_2)
q_{max} : Maximum adsorption amount of MnO_2 (mol/mol-MnO_2)
K : Reaction rate constant
= Adsorption rate constant/Desorption rate constant

Equation 5.35 The Langmuir-type adsorption isotherm.

Figure 5.30 shows the mobility of a suspension of potassium permanganate-oxidized Mn^{2+} (random agglomerates) by electrophoresis (Rank Brothers, Inc., Particle Micro-Electrophoresis Apparatus Mark II). In a wide range from the weakly acidic state to the alkaline state, the mobility in dechlorinated tap water is −1.0 (μm/s)/(V/cm) and in distilled water is −1.8 (μm/s)/(V/cm). Hydrated manganese dioxide is considered to be capable of adsorbing Mn^{2+} ions. The zeta potential (ZP) (mobility) of hydrated manganese dioxide in the suspension system is not affected by the amount of potassium permanganate added.

Figure 5.31 shows how quickly this reaction proceeds. At neutral pHs, the potassium permanganate oxidation reaction with Mn^{2+} can be completed in less than 2 minutes.

Micro flocs of manganese hydroxide produced by potassium permanganate coagulation are fragile fractal particles with a negative ZP of about −7 mV in the neutral pH range, and are precipitated and separated by an aluminium coagulant, or similar. Silica polymers may be added as flocculation aids to give strength to the flocs, or a small amount of clay or perlite cinder may be added to accelerate sedimentation. Similar to color flocs, they may be separated by flotation.

In the early 1960s, potassium permanganate sedimentation and filtration methods were standardized by Sapporo City Water Supply and the Hokkaido University Tambo Group. This method is still useful when combined with (MF/UF) membrane separation.

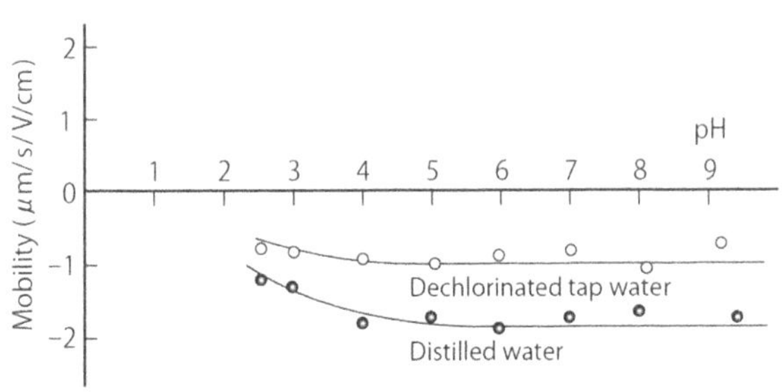

Figure 5.30 Change in pH and mobility of Mn^{2+} suspensions.

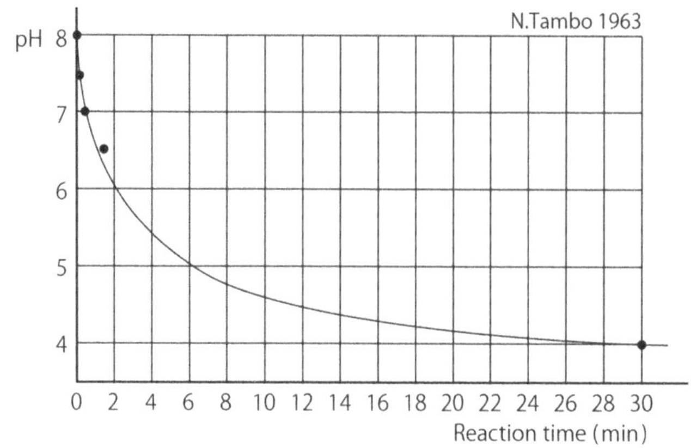

Figure 5.31 The rate change of Mn^{2+} by pH, of the $KMnO_4$ oxidation reaction.

In a water purification plant with a sand filter pond, chlorine-manganese sand filtration (established by Kyoto Water Supply staff members and Professor Nakanishi, as described below) is a prominent method in a rapid filtration system.

5.3.3.2 Catalytic sand filtration method for Mn^{2+} removal using chlorine

Even if chlorine is simply added to water containing Mn^{2+}, the rate of oxidation is extremely slow, and this method alone is not practical. However, when chlorine is added to water containing Mn^{2+} and sand filtration is continued, the surfaces of the filter sand are gradually covered with manganese oxide coating, which does not peel off even in backwash. This fixed oxide catalyzes the oxidation of chlorinated Mn^{2+} and effectively lowers the concentration of manganese in the filtrate. In order to achieve this effect quickly, the filter bed should be immersed in a solution containing about 3% of manganese chloride ($MnCl_2$). Enough potassium permanganate should be added to the solution to form a manganese oxide coating on the surface of the filter sand. When water containing chlorine is passed through the resulting manganese sand layer, the Mn^{2+} in water is adsorbed on the sand surface and then reacts with chlorine in water to form hydrated manganese dioxide, thereby regenerating the adsorption force, by the mechanisms as shown in Equations 5.36 and 5.37. Professor Nakanishi (1967a, b, c) revealed a removal mechanism that repeats this cycle. Takai (1964) also conducted chemical studies but did not elucidate the processing capacity and kinetic processes required for design.

Adsorption process

$$Mn^{2+} + MnO(OH)_2 \leftrightharpoons MnO_2Mn^{+} + 2H^{+} \quad — (5\cdot36)$$

Oxidation process

$$MnO_2MnO^{+} + HOCl + 2H_2O \rightarrow 2MnO_2\cdot H_2O + H^{+} + Cl^{-} \quad — (5\cdot37)$$

(5.36–5.37)

Equation 5.36, 5.37 The reaction of manganese contact oxidation filtration.

Equations 5.36 and 5.37 indicate that the theoretical amount of chlorine required to reactivate (oxidize) manganese sand is 1.29 mg/mg of Mn^{2+}. Therefore, we attempted to establish the basis of the theoretical design by:

(1) examining the process of producing manganese sand, its interfacial physicochemical properties (electric potential), and its processing capacity, and
(2) examining the kinetics of the manganese removal process (Matsui *et al.*, 1986a, b).

5.3.3.2.1 Preparation of experimental manganese sand

First, 100 l of a 0.5% solution of manganese chloride was added to 100 g of filtered sand, and allowed to stand for 1 hour. The solution was then removed and 100 l of a 0.5% solution of potassium permanganate was added, mixed, and allowed to stand for 1 hour to remove the solution. This cycle was repeated 1 to 6 times to produce manganese sand with different immersion times.

Four types of filter sand (silica sand) with typical diameters of 0.65 mm, 0.77 mm, 0.92 mm, and 1.09 mm were used as a base material to confirm the particle size effect, and the manganese sand of the actual sand filter was evaluated using the filter sand of filtration plants from Tokyo, Osaka and Sapporo City Water Works (i.e. from throughout Japan).

5.3.3.2.2 Amount of attached manganese dioxide

Figure 5.32 shows how much manganese dioxide is deposited on the 1 g of sand. The number of immersions and the amount of immersion become linear on both logarithmic plots, and they are more accurate as the number of cycles increases. It can be seen that the manganese coating increases (1 mg per g-sand at a time). For reference, the manganese coating condition of the filter sand in each city is indicated by the asterisk (*) on the graph.

5.3.3.2.3 Surface potential of manganese sand

The manganese sand was packed into a fixed bed column (50 cm depth). Water was passed through the column to make the manganese sand conform to water, and then the fluidization potential generated at the upper and lower ends of the packed bed was measured. The filtration speed was changed in five steps, at pH 7.0 and with the electrical conductivity set to 20 μS/cm, and the potential was measured after a certain period of time had elapsed for each filtration speed. The base potential at zero filter speed was subtracted to obtain the flow potential E corresponding to each flow speed. Using the filtration head

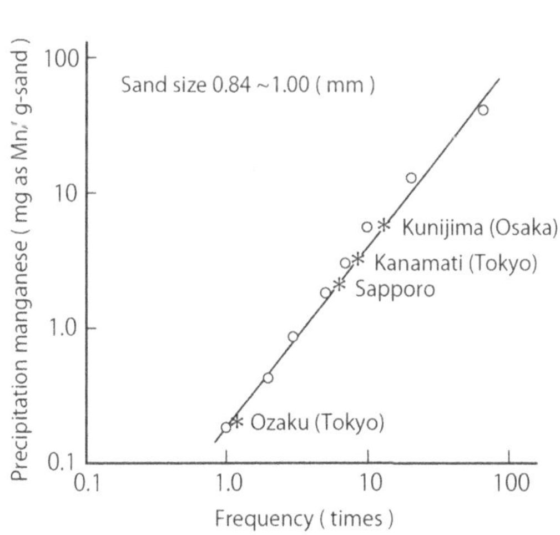

Figure 5.32 Amount of manganese precipitated and the number of emersions.

loss measured simultaneously, the ZP of the manganese sand surface was calculated using the Helmholtz-Smoluchouski equation (Equation 5.38). For the practical unit (cgs) system, Equation 5.39 was used.

$$\zeta = \frac{\eta K E}{\varepsilon P} \quad \text{— (5·38)}$$

$$\zeta = \frac{4\pi\,\eta K E \times 9 \times 105}{80 \times 980 \times H} = 1.443 \times 102 \times \frac{\eta K E}{H} \quad \text{— (5·39)}$$

ζ : Zeta potential (V)
E : Streaming potential (V)
P : Pressure to make liquid flow (Pa)
η : Liquid viscosity (Pa·s)
ε : Liquid dielectric constant ($C^2N^{-1}m^{-2}$)
Value of dielectric capacity of water in vacuum
8~10/36π ($C^2N^{-1}m^{-2}$)×80
K : Liquid conductivity ($V^{-1}Am^{-1}$)

(5.38–5.39)

Equation 5.38, 5.39 The zeta potential of manganese sand.

Figure 5.33 shows that the positive flow potential increases with increasing filtration rate. The values in the table are the results of the ZP around a flow rate of 1.0 cm/sec, calculated from the flow potential E. The surface of the floc seems to have a ZP around −15 mV, which is almost the same as that of the coagulated manganese hydroxide microflow shown in Figure 5.30.

5.3.3.2.4 Adsorption isotherm of manganese sand

The manganese sand was placed on a mesh tray inside a 1ℓ beaker, as shown in Figure 5.34, and a batch adsorption test was performed in which the Mn^{2+} in the solution adsorbed to the 1 g manganese sand with a concentration of 500 ml manganese chloride solution of 0.08 to 20 mg/l. The solution, adjusted

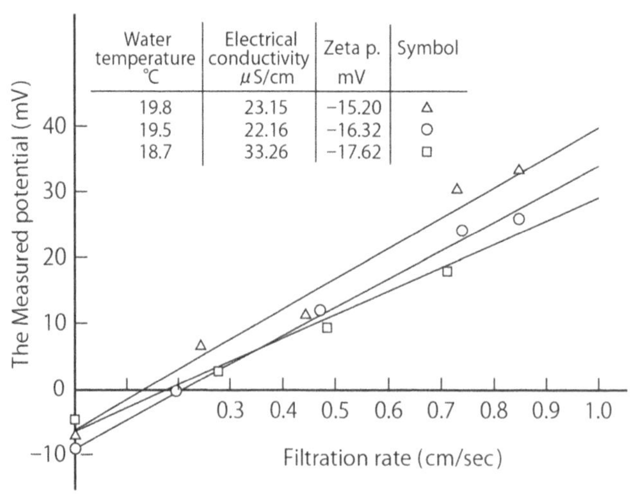

Figure 5.33 Measured ZP and filtration rate.

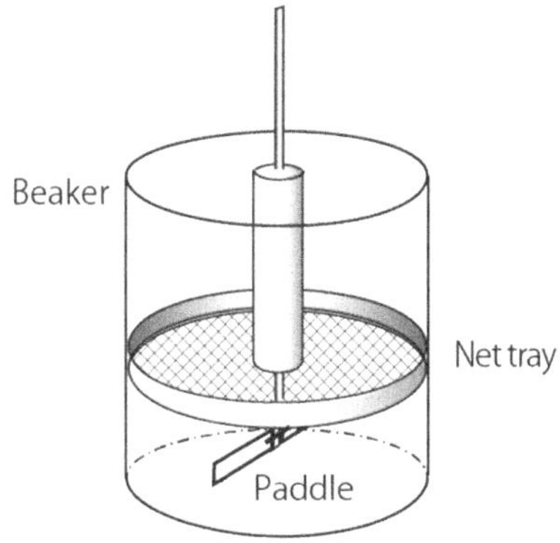

Figure 5.34 Experimental equipment to perform a batch adsorption test.

to pH 7.0, was stirred continuously for 3 to 5 hours, and the decrease in the concentration of Mn^{2+} in the solution over time was measured by atomic absorption spectroscopy. The point at which the concentration of the Mn^{2+} in the liquid system was no longer changed is regarded as the adsorption equilibrium, and the amount of adsorption and the residual concentration are calculated as equilibrium values.

Using a fixed bed adsorption test system as shown in Figure 5.35, a new manganese sand was installed to control the pH at pH 7.0 by flowing a predetermined concentration of manganese chloride solution, and

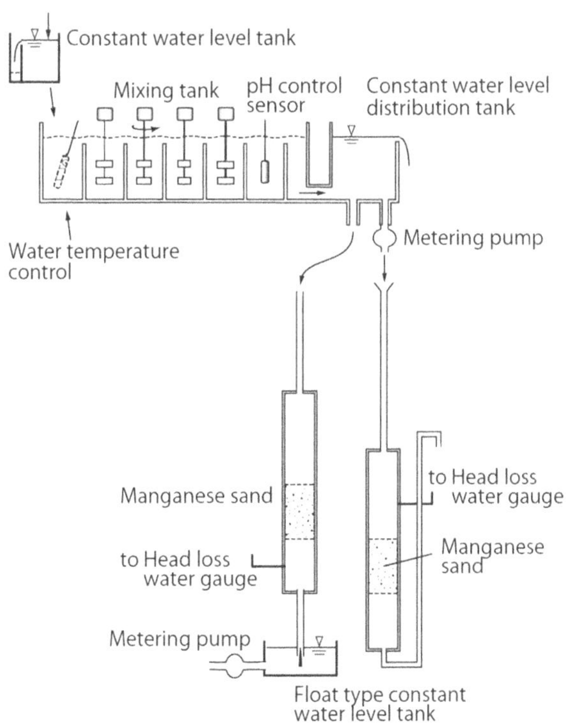

Figure 5.35 A continuous fixed bed adsorption test system.

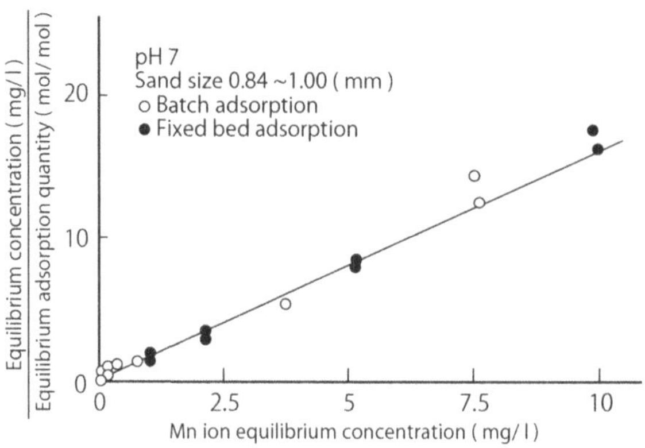

Figure 5.36 A Langmuir plot of Mn ion equilibrium adsorption values for manganese sand. (Solid circles indicate results of fixed bed adsorption; open circles indicate results of batch adsorption).

quantitative filtration was performed at a void flow rate of 200 m/day. The Mn^{2+} concentration of the effluent was measured at regular time intervals, and the test was discontinued when the concentration of the effluent reached 95% of the influent concentration. The integral difference of the concentration of the inflow and outflow water up to this point is the total adsorption amount, and it is considered that, at this point, the adsorption equilibrium was established between the concentration of the influent and the outflow water.

A Langmuir plot of the adsorption equilibrium values obtained from the experiment, in accordance with Equation 5.35, show that the Langmuir-type adsorption reaction is progressing (as shown in Figure 5.36; see also Chapter 7 below).

In the case of manganese sand, the Langmuir adsorption isotherm is established, as in the case of the fine floc of manganese hydrate deposited in the liquid system. The maximal adsorption is estimated to be 0.60 mg/g-sand at pH 7.0 from the slope of the plots. The rate factor $K = k_1/k_2 = 10$ l/mol was the same as for the suspension. Adsorption tests at pH 5.0 on the same sand showed that the maximum adsorption volume was 0.35 mg/g-sand, and the reduction in pH greatly reduced the adsorption capacity of the sand.

The amount of adhered manganese and the amount of saturated manganese adsorption we re plotted for laboratory manganese sand and sand from the three water purification plants previously described. Figure 5.37 shows that the filtration sand in the water purification plants has a considerably higher Mn^{2+} adsorption capacity. The relationship between the amount of attached manganese and the maximum adsorption amount for all manganese sands differing in origin is one-to-one. It is considered that the manganese dioxide (manganese hydroxide) attached to the sand is almost homogeneous and has the same adsorption capacity of the manganese ion (Mn^{2+}) uniformly, including inside. Therefore, the total adsorption of manganese ions in the chlorine-manganese sand process is determined in linear proportion to the amount of attached manganese oxide, which plays a catalytic role.

The Langmuir isotherm results in Figure 5.37 can be summarized as: $q_{max} = 0.17$ (mol/mol-MnO_2). The rate constants for adsorption and desorption K are constant values of 10 l/mol $= 1.82 \times 10^{-7}$(cm^3/mol) for all sands.

The maximum adsorption amount of coagulated floc in the suspension is $q_{max} = 0.296$ (mol/mol-MnO_2) as shown above, whereas the maximum adsorption amount of manganese oxide film adhering to the sand

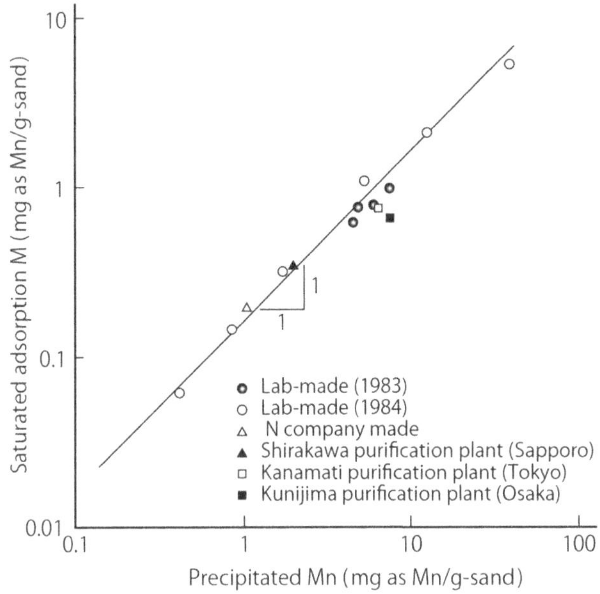

Figure 5.37 Relationship between attached Mn and maximum adsorption Mn.

surface is $q_{max} = 0.17$ (mol/mol-MnO_2), and the saturated adsorption amount is small. This is probably because a significant portion of the adsorption sites are used to adhere to the sand.

The second part of the investigation (investigating the two-step manganese sand chlorine catalytic filtration reaction) also involved several stages.

5.3.3.2.5 Process for removing dissolved manganese by the chlorine manganese sand method

Removal can be thought of as a two-step reaction of (1) ion exchange adsorption of manganese ions to manganese sand and (2) the resulting oxidation of the adsorbed manganese by chlorine. Therefore, in order to more accurately describe the decrease in manganese, we evaluated the adsorption rate and the oxidation rate individually, and attempted to grasp the total removal phenomenon as a whole. The experiment was performed with two sets of equipment and operations shown in Figures 5.34 (batch experiment) and 5.35 (continuous flow experiment).

5.3.3.2.6 Manganese ion adsorption rate

The column of the continuous fixed bed unit was packed with new manganese sand at 2, 5, and 10 cm depth, and the inflow manganese ion concentration was variously changed to 1, 2, 5, and 10 mg/l, whilst the adsorption test was carried out at a superficial flow rate of 200 m/day. As shown in Figure 5.38, when filtration was started ($t = 0$), the manganese ion concentration decreases linearly with sand layer thickness in a semi-logarithmic plot, and the adsorption rate is the first order response of the manganese ion-Mn^{2+} concentration. As a result of examining manganese sand with different amounts of attached manganese, the first order reaction constant for manganese ion adsorption showed a constant value regardless of the amount of attached manganese oxide on manganese sand, as shown in Figure 5.39.

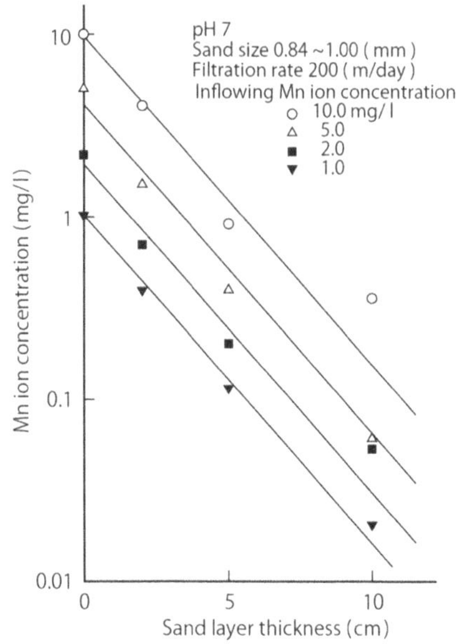

Figure 5.38 The change of Mn ion concentration by sand layer thickness (t = 0).

Despite the fact that the maximum adsorption rate q_{max} is proportional to the amount of deposited manganese oxide, the characteristic that the adsorption rate is independent of the maximum adsorption rate suggests that the reaction rate is dominated by phenomena at the uppermost surface of the manganese sand.

Therefore, we investigated how the first order reaction constant of adsorption varies depending on the filtration rate and the diameter of the filter medium, and the results are shown in Figures 5.40 and 5.41. Since the primary constant of adsorption decreases in proportion to the $-2/3$ power of the filtration rate and the $-5/3$ power of the diameter of the filter medium, it is estimated that the transport at the liquid boundary layer is the rate-limiting step.

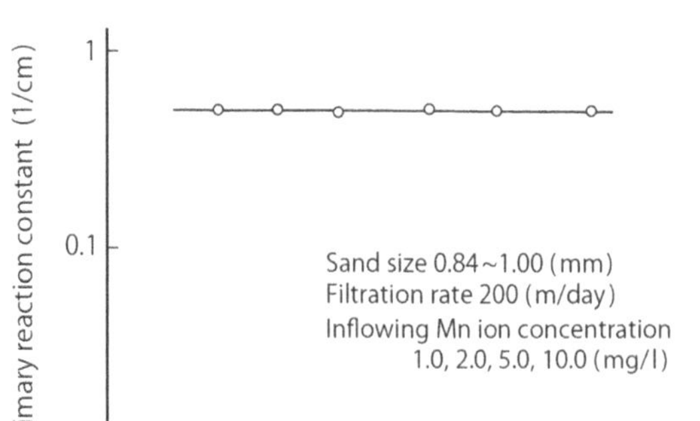

Figure 5.39 Attached Mn and the primary reaction constant of adsorption.

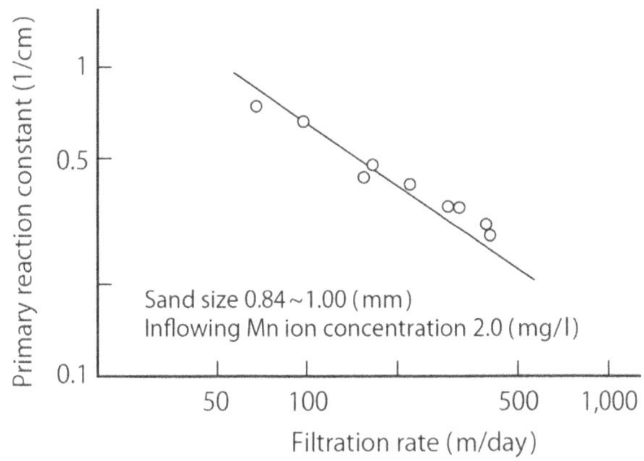

Figure 5.40 The filtration rate and primary reaction constant of adsorption.

Given that the movement of material across the boundary layer controls the progression of the phenomenon, the apparent first order reaction constant can be expressed as Equation 5.40 using the k_f of the boundary layer mass transfer coefficient. Using the reaction mass transfer factor k_f, mass transfer factor J_d, Equation 5.41 can be used to relate the J_d to the Reynolds number Re, and finally the first order kinetic constants can be expressed as Equation 5.43, using the Wilson Equation (Equation 5.42).

From Equation 5.43, the first order kinetic constants (cm^{-1}) related to the void filtration rate and media diameters should be a straight line with a slope of $-2/3$ and $-5/3$ on both the log-log plots. Figures 5.40 and 5.41 prove the relationship by experiment.

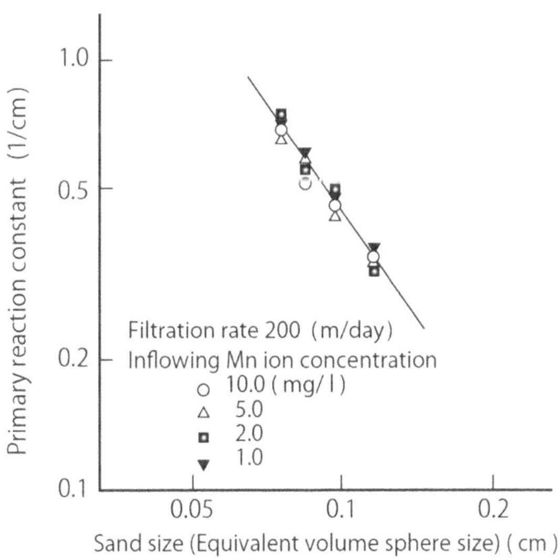

Figure 5.41 The primary reaction constant of adsorption and filter medium size.

$$[\text{Primary reaction constant}] = k_f a_v / v \quad (1/\text{cm}) \qquad (5 \cdot 40)$$

$$k_f = J d_v \left(\frac{D_m}{\nu}\right)^{2/3} \qquad (5 \cdot 41)$$

$$J_d = 1.00\,\varepsilon^{-1} Re^{-2/3} \qquad (5 \cdot 42)$$

$$[\text{Primary reaction constant}] = 6.5\left(\frac{1-\varepsilon}{\varepsilon}\right) D_m^{2/3} v^{-2/3} d^{-5/3} \qquad (5 \cdot 43)$$

k_f : Boundary mass transfer coefficient (cm/sec)
$a_v = 6(1-\varepsilon)/d$: Filter bed specific surface area (1/cm)
v : Superficial filtration velocity (cm/sec)
J_d : Mass transfer factor (-)
D_m : Molecular diffusion coefficient (cm^2/sec)
ε : Porosity of filter bed (-)
Re : Reynolds number (-)

(5.40–5.43)

Equation 5.40 - 5.43 The primary reaction constant.

The molecular diffusivity D_m can be estimated by the Wilke-Chang Equation (Equation 5.44), using the metallic manganese density as the molar volume. Then, the molecular diffusivity $D_m = 3.61 \times 10^{-5}$ cm^2/sec. The primary reaction constant for adsorption is 1.161/cm ($v = 200$ m/day, $d = 0.98$ mm), which is roughly double the experimental value of 0.491/cm. This is probably due to the underestimation of the molar molecular weight by ignoring the hydration of the manganese ion.

$$D_m = \frac{7.4 \times 10^{-10} (2.6 \times M_w)^{1/2} T}{\mu V_w^{0.8}} \qquad (5 \cdot 44)$$

V_m : Molar molecular volume (cm^3/mol)
M_w : Molecular weight of solvent (g/mol)
μ : Solvent viscosity (g/cm·sec)
T : Absolute temperature (T)

(5.44)

Equation 5.44 The molecular diffusivity.

From the above, it can be considered that mass transfer at the liquid boundary film is the process that governs the adsorption rate.

5.3.3.2.7 Rate characteristics of chlorine oxidation

A process in which a reduced part of the surface of manganese sand adsorbed with manganese ions is oxidized by chlorine and activated as manganese dioxide was now considered. The manganese sand was adsorbed and saturated with a manganese ion-Mn^{2+} by flowing manganese chloride solution through a continuous fixed-bed experimental system equipped with new manganese sand. Water obtained by adjusting the concentration and pH of sodium hypochlorite was then passed through the column at a filtration rate of 100 m/day, and the chlorine concentration of the effluent from the column with a packed bed thickness of 2, 5, and 10 cm was measured over time. An example with an inflow chlorine concentration of 1 mg/l and showing the decrease in chlorine concentration due to sand layer thickness at $t = 0$ is linearly displayed in a semi-logarithmic plot in Figure 5.42. Therefore, as in the case of the

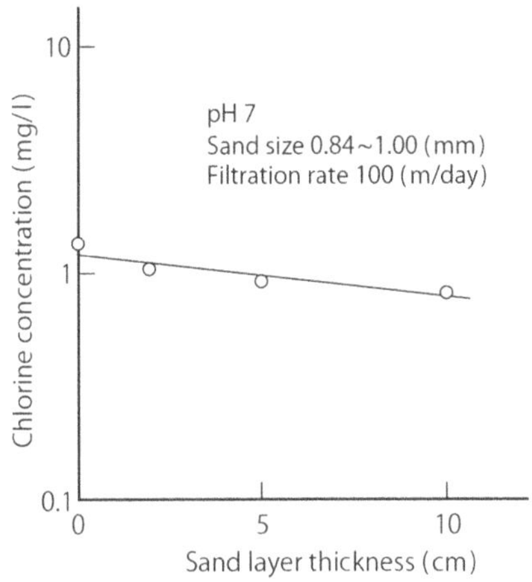

Figure 5.42 The change in chlorine concentration due to sand layer thickness (t = 0).

adsorption reaction, oxidation proceeds in the first order reaction. However, the slope of the straight line is remarkably small compared to the case with adsorption, and the oxidation rate is considered to be considerably smaller compared to the adsorption rate.

Figures 5.43, 5.44 and 5.45 show the results of a study on how the amount of manganese adhering, the filtration rate, and the diameter of the filter medium all affect the oxidation reaction rate. Figure 5.43 suggests that the oxidative regeneration reaction of manganese sand is constant regardless of the amount of attached manganese, as is the case with the adsorption reaction of manganese ions, and that only the surface of the filter medium is involved in the reaction. Figures 5.44 and 5.45 show that the oxidation-regeneration reaction constant is inversely proportional to the filtration rate and the filtrate diameter, and decreases with a slope of -1. This change is considered to be due to the rate-limiting chemical reaction on the surface of the manganese oxide layer attached to the manganese sand, unlike the case of Equation 5.43, which shows the rate-limiting diffusion rate of manganese ion adsorption.

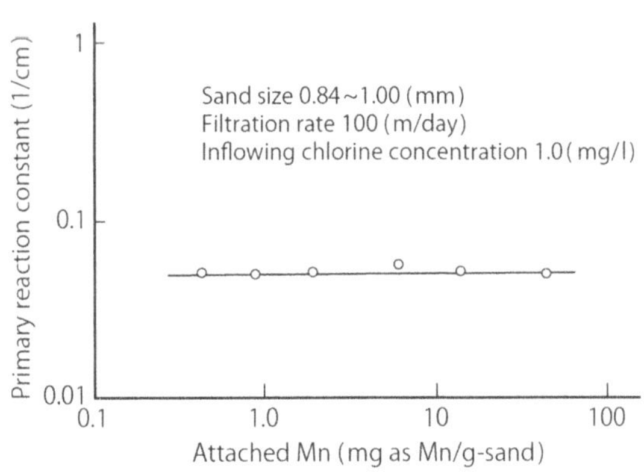

Figure 5.43 Attached Mn and primary reaction constant of oxidation.

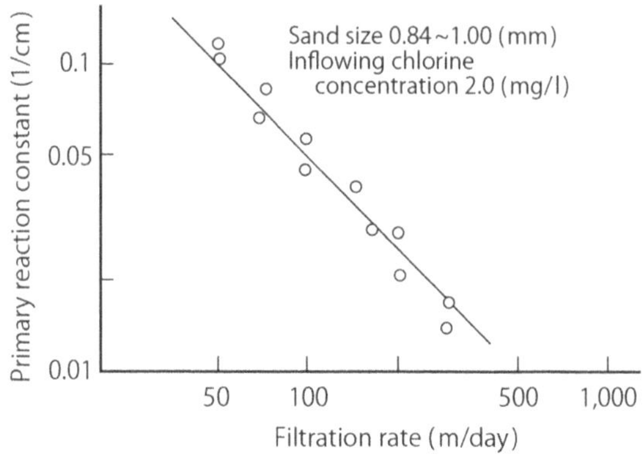

Figure 5.44 The filtration rate and primary reaction constant of oxidation.

In addition, since the primary reaction coefficient of the oxidization (regeneration) reaction of the adsorbed Mn^{2+} ion is lower than the rate coefficient of the adsorption reaction (diffusion-limiting rate of the boundary film) by one order of magnitude or more, the liquid boundary film mass transfer phenomena are not considered to be the rate limiting phenomena discussed.

5.3.3.2.8 Removal of manganese ions and regeneration of manganese sand in a system containing chlorine and manganese ions simultaneously

From the above examination, it became clear that the rate at which Mn^{2+} ions are removed by manganese sand filtration depends on the filtration rate and the diameter of the filter medium, but is not related to the amount of manganese oxide adhering thereto.

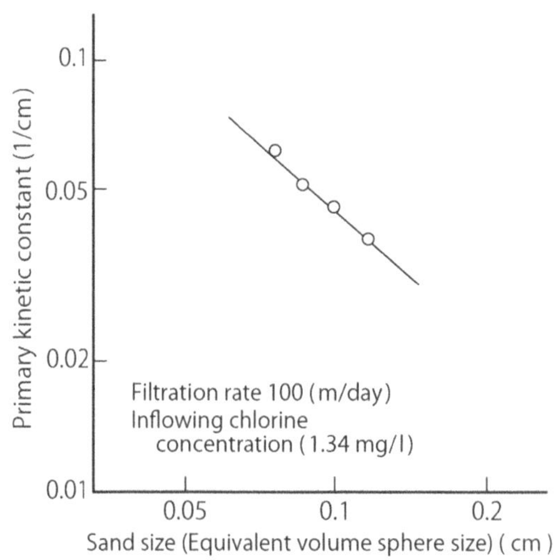

Figure 5.45 The primary reaction constant of oxidation and sand (filter) size.

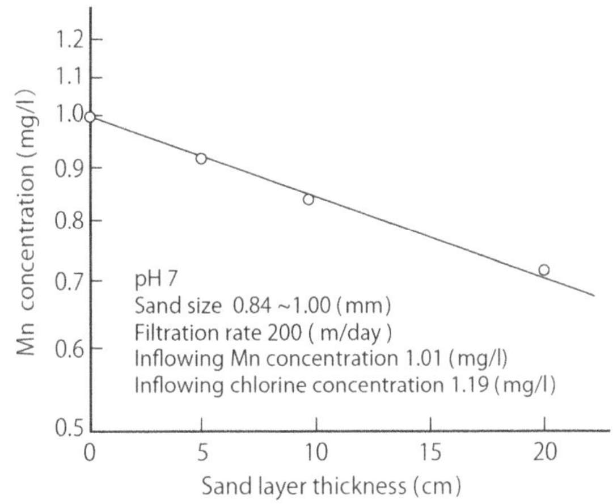

Figure 5.46 Change of Mn concentration by sand layer thickness (t = 0).

Next, we show a mechanism for describing the removal condition by combining both the above methods, and consider a kinetic model. Using the continuous fixed-bed experimental system shown in Figure 5.35, new manganese sand was filled to a thickness of 5, 20, and 20 cm, and water containing manganese ions and chlorine was flowed at a filtration rate of 200 m/day to carry out the experiment. After a certain period of time after the start of water flow, the adsorption rate of manganese ions in the sand layer and the oxidation rate of manganese adsorbed in the manganese granular layer become equal to each other as a result of the chlorine, and the steady state of the removal rate is reached. Plotting the downward changes in the flow of manganese ions in the sand lamina at this steady state, Figure 5.46 shows that the manganese ion Mn^{2+}changes in the first order response. Figure 5.47

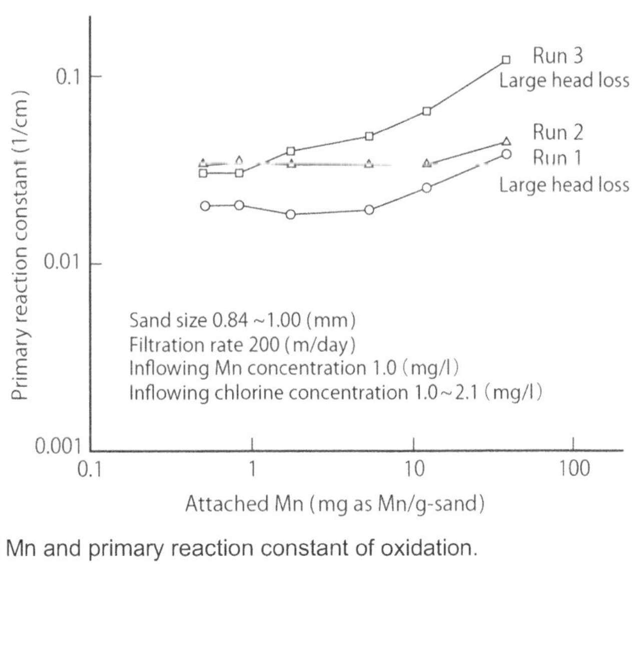

Figure 5.47 Attached Mn and primary reaction constant of oxidation.

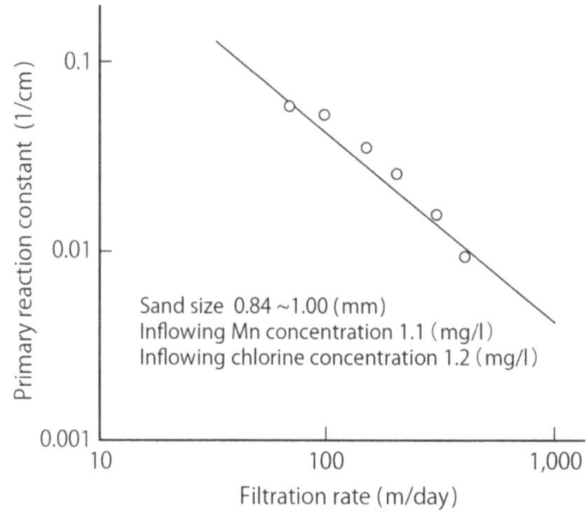

Figure 5.48 Filtration rate and primary reaction constant.

shows the relationship between the first order reaction rate constant of manganese ion removal and the amount of attached manganese. It shows that the removal rate is almost constant over a wide range of the amount of attached manganese, regardless of the amount attached to the sand, which suggests the reaction is going on on the surface of the filter medium, as described above. A slight increase in the reaction rate in areas with high manganese oxide deposition can be understood as the effect of increasing the gap flow rate and the specific surface area of the filter layer due to the thickening of the filter media. Therefore, filtration is stopped and the original value is restored by backwash.

Figure 5.48 shows that the first order reaction constant for removing manganese ions is inversely proportional to the filtration rate. The fact that the catalytic oxidation process, which proceeds to the two stages of adsorption of manganese ions and oxidation by chlorine, proceeds in inverse proportion to the filtration rate means that the regeneration oxidation process is the rate-limiting process of all phenomena.

The kinetics of manganese catalyzed oxidation (the catalytic oxidation rate formula), in which the conditions governing the maximum adsorption rate and reaction rate described above occur at different locations, were formulated as shown in Equations 5.45, 5.46 and 5.47. However, since no clear relationship with pH was found, k_2 was used only as a factor of pH dependency.

In order to use these kinetic equations, it is necessary to obtain values related to the properties of manganese, such as the maximum specific adsorption quantity q_{max} [mol/mol-MnO_2] and the rate constants of adsorption, desorption and oxidation $k'_1, k'_2,$ and k'_3 (which are independent of the operating conditions), and the specifications of sand, such as the quantity β of manganese oxide which is directly involved in the reactions on the sand surfaces.

The experimental results in the previous section show that the maximal amount of manganese ions adsorbed is $q_{max} \doteqdot 0.17$ mol/mol-MnO_2, but it is difficult to measure the amount of manganese oxide β actually involved in the reactions. Since $k'_1, k'_2,$ and k'_3 are used in the form of products in the equation, k_1, k_2, and k_3 corrected by β (such as $k_1 = \beta k_1'$) are replaced by the rate constants of adsorption, desorption, and oxidation, respectively, and evaluated from experimental values.

$$-v\frac{\partial M}{\partial z} = k'_1 \beta a_v M(q_{max} - q) - k'_2 \beta a_v q \quad \text{— (5·45)}$$

$$-v\frac{\partial C}{\partial z} = k'_3 \beta a_v C q \quad \text{— (5·46)}$$

$$\rho(1-\varepsilon)Q\frac{\partial q}{\partial t} = -v\frac{\partial M}{\partial z} + v\frac{\partial C}{\partial z}$$
$$= k'_1 \beta a_v M(q_{max} - q) - k'_2 a_v q - k'_3 \beta a_v C q \quad \text{— (5·47)}$$

q_{max} : Mn-ion ratio maximum adsorption amount of sand attached hydrated manganese dioxide (mg-Mn^{2+}/mg-MnO_2 as Mn)
q : Amount of adsorbed and inactivated Mn-ions (mg-Mn^{2+}/mg-MnO_2 as Mn)
M : Mn concentration (mol/cm^3)
C : Chlorine concentration (mol/cm^3)
ρ : Sand density (g/cm^3)
ε : Porosity of sand bed (-)
Q : Amount of attached Mn per unit mass of sands (mol/g-sand)
t : Duration of filtration (sec)
z : Sand bed direction thickness (cm)
a_v : Sand surface area per unit sand volume (=6(1 − ε)/d, cm^2/cm^3)
d : Equal volume spherical sand particle size (cm)
β : Amount of Mn oxide directly controlling the reaction rate in the vicinity of the Mn sand surfaceMn (mol/cm^3)
k'_1 : Adsorption rate constant (cm^3/sec/mol)
k'_2 : Desorption rate constant (considered to change depending on pH, l/sec)
k'_3 : Oxidation rate constant (cm^3/sec/mol)

(5.45–5.47)

Equation 5.45 - 5.47 The kinetics of manganese catalytic oxidation.

Figures 5.40 and 5.41 show that $k_1 = 1.98 \times 10^{-2}$ (cm^3/sec/mol) (mol/cm^2) is obtained when v = 200 m/day and d = 0.099 cm for the rate constant k_1 of adsorption. Since k_1 is proportional to $v^{-1/3}$ and $d^{-2/3}$, using the above values to determine the constant, the generalized equation for the adsorption rate constant k_1 is given by Equation 5.48.

In the previous section, the adsorption equilibrium of manganese ions is represented by the Langmuir adsorption isotherm, and the ratio K of the adsorption and desorption rate constants that define the equilibrium is, for the most part, $K = k_1/k_2 = 10$ l/mol $= 5.49 \times 10^8$ cm^3/mol (as shown by the desorption rate constant $k_2 = 3.61 \times 10^{-11}$ (1/sec) (mol/cm^2)).

General formula Equation 5.49 is obtained in the same way as Equation 5.48. Also, from Figures 5.44 and 5.45, $k_3 = 1.06 \times 10^{-3}$ (cm^3/sec/mol) (mol/cm^2). This value is considered to be almost constant regardless of the operating conditions.

$$k_1 = 6.91 \times 10^{-3} v^{-1/3} d^{-2/3} \text{ (cm/sec)} \quad \text{— (5·48)}$$

$$k_2 = 1.26 \times 10^{-11} v^{-1/3} d^{-2/3} \text{ (mol/sec·cm}^2\text{)} \quad \text{— (5·49)}$$

v : Superficialf iltration rate (cm/sec)
d : Manganese sand size (cm)

(5.48–5.49)

Equation 5.48, 5.49 The adsorption and desorption rate constant.

The column of the continuous fixed bed experiment system was packed with new manganese sand or manganese sand saturated with adsorbed manganese ions, and a constant concentration of manganese

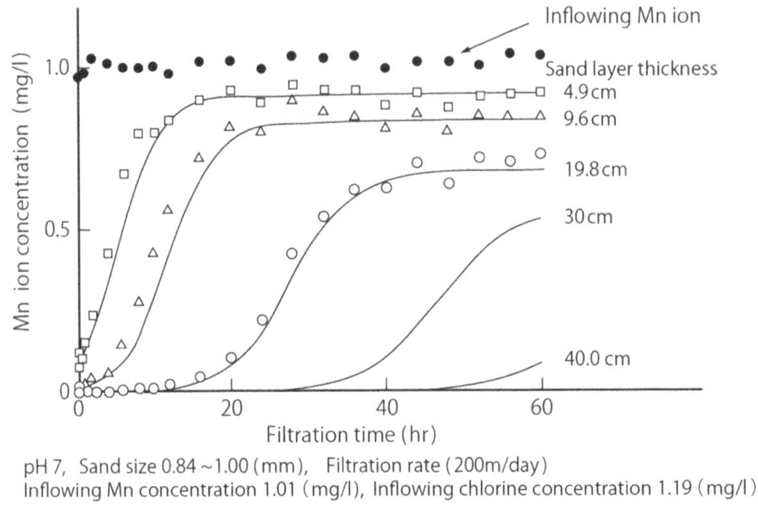

Figure 5.49 Filtration time and outflowing Mn ion concentration.

ions and chlorine was introduced into the column. The condition changes over time in the concentration of manganese ions and chlorine flowing out of the filter bed were measured and the results compared with the kinetic model to examine the model's validity.

Figures 5.49 and 5.50 compare the model calculated manganese ion concentrations and chlorine concentrations in 4.9 cm to 40 cm sand layers of a new manganese sand, with a particle size of 0.84 to 1.00 mm, with the effluent water in each of the actual experimental sand layers of 4.9 to 19.8 cm. The manganese removal rate reached a steady state after a sharp decline in a certain period of time, and the removal rate continued to be maintained after a certain period of time. This experiment bed was much thinner than the actual water purification plant, which has a sand layer thickness of about 60 cm. In the

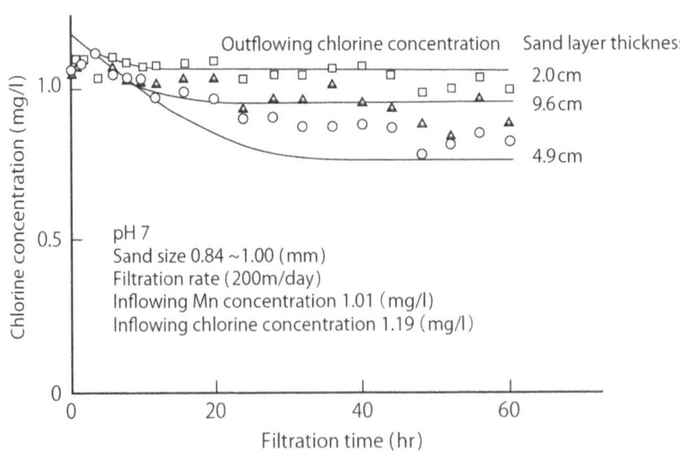

Figure 5.50 Filtration time and outflowing chlorine concentration.

actual pond, the sand layer thickness is so large that a larger removal rate can be obtained, and it slowly reaches a steady state over a long period of time.

In the actual operation of the manganese sand method, the filtration cycle starts from the state where manganese sand adsorbs manganese ions at a certain steady level and is regenerated with chlorine. The equations required to discuss steady state elimination are given by Equations 5.45, 5.46 and 5.47 (above), assuming the conditions of Equations 5.50 and 5.51.

$$\frac{\partial q}{\partial t} = 0 \qquad \text{— (5·50)}$$

$$6(1-\varepsilon)q_{max}\frac{\partial z}{\partial v} = \frac{1}{k_1}\ln\frac{M_0}{M} + \frac{1}{k_3}\ln\frac{C_0}{C} + \frac{k_2}{k_1 k_3 (C_0 - M_0)}\left(\ln\frac{M_0}{M} - \ln\frac{C_0}{C}\right) \qquad \text{— (5·51)}$$

q_{max} : Mn ion ratio maximum adsorption amount of sand attached hydrated manganese dioxide (mg-Mn^{2+}/mg-MnO_2 as Mn)
C_0 : Chlorine inflow concentration (mol/cm^3)
M_0 : Mn Chlorine inflow concentration (mol/cm^3)
C : Chlorine outflow concentration (mol/cm^3)
ε : Sand bed porosity (-)
z : Sand bed direction thickness (cm)
d : Equal volume spherical sand particle size (cm)
k_1 : Adsorption rate constant (cm/sec)
k_2 : Desorption rate constant (mol/cm^2/sec)
k_3 : Oxidation rate constant (cm/sec)

(5.50–5.51)

Equation 5.50, 5.51 The manganese removal equation in steady state.

Given the manganese ion concentration in the raw water, M_0, and the influent chlorine concentration, C_0, in the filtration bed, the layer thickness of manganese sand required to achieve a certain removal rate of Mn^{2+} can be directly calculated by Equation 5.51, using the four kinetic characteristics of q_{max}, k_1, k_2, and k_3 that govern the chlorine-manganese sand method, depending on the operating condition. For maximum specific adsorption of manganese oxide on sand:

- $q_{max} - 0.17$ (mol/mol MnO_2)
- $k_1 = 1.98 \times 10^{-2}$ (cm^3/sec/mol)
- $k_2 = 3.61 \times 10^{-11}$ (1/sec) (mol/cm^2)
- $k_3 = 1.06 \times 10^{-3}$ (cm^3/sec/mol) (mol/cm^2)

Figure 5.51 shows the results of calculation of the filtration operating conditions (sand thickness) necessary to achieve a manganese ion concentration of 0.01 mg/l in treated water using the values obtained in the previous tests.

The vertical axis on the left side of the figure is the dimensionless depth expressed as the ratio of the required bed thickness z to the media diameter d. Naturally, the necessary bed thickness will be increased or decreased depending on the filtration speed. The filtration speed can be corrected on the horizontal axis. The parameters on the right show the molar ratio of added chlorine to the manganese ion in the raw water and the excess added to the reaction equivalents. If chlorine is added in a predetermined proportion to manganese ions, it is possible to determine what kind of bed thickness is required.

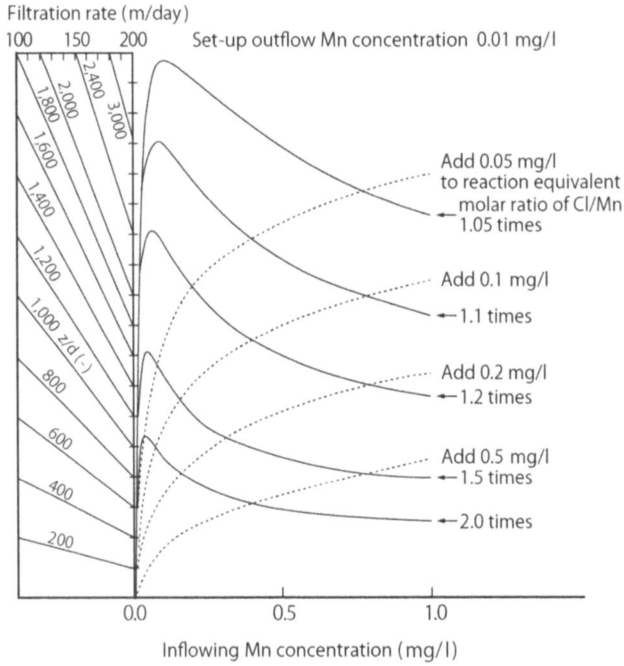

Figure 5.51 The sand layer thickness required for Mn removal by the chlorine-manganese sand method.

Figure 5.51 can be used in a variety of ways, such as to determine the necessary filter bed thickness and Cl_2 to M^{2+} ratios. For example, to attempt to obtain 0.01 mg/l of treated water by treating with a manganese concentration of 0.5 mg/l at a filtration rate v = 150 m/day, a media diameter d = 0.7 mm, and a bed thickness z = 60 cm, from the left side of the figure, z/d = 60/0.7 = 857, and if this value is drawn horizontally to the right side of the figure, it can be seen that adequate removal can be obtained if chlorine is added to an equivalent of about 0.2 mg/l (assuming residual chlorine of about 0.2 mg/l) and chlorine is added to a molar ratio of Cl_2 to Mn^{2+} of about 1.25.

For details on non-routine operation and batch operation, refer to the original papers by Matsui *et al.* (1986a, b).

5.3.4 Treatment of iron, manganese, etc., by microbial filtration

It is known that if a filter bed is operated for a long period of time, soluble iron Fe^{2+} and manganese Mn^{2+} are reduced to some extent by the activity of iron bacteria and the like that live in the filter bed. However, under a system that operates a backwash once every 24 to 72 hours to keep the filter bed clean at a high filtration rate of 120 to 200 m/day for the rapid sand filter, iron bacteria cannot be present in the filter bed in an amount sufficient to work effectively. This suggests that the ability of ordinary rapid sand filters to de-iron and de-manganese by microorganisms cannot be expected.

In recent years, Nagaoka Co., Ltd, Japan have proposed a deep air oxidation system (Figure 5.52) with a filtration rate comparable to rapid sand filtration, which can bio-oxidize, insolubilize, and separate Fe^{2+} and Mn^{2+} in groundwater by aeration alone, and with simultaneous nitrification of ammonia. This system was devised to make widespread use of simple aerated microorganism filtration without injection of chemicals.

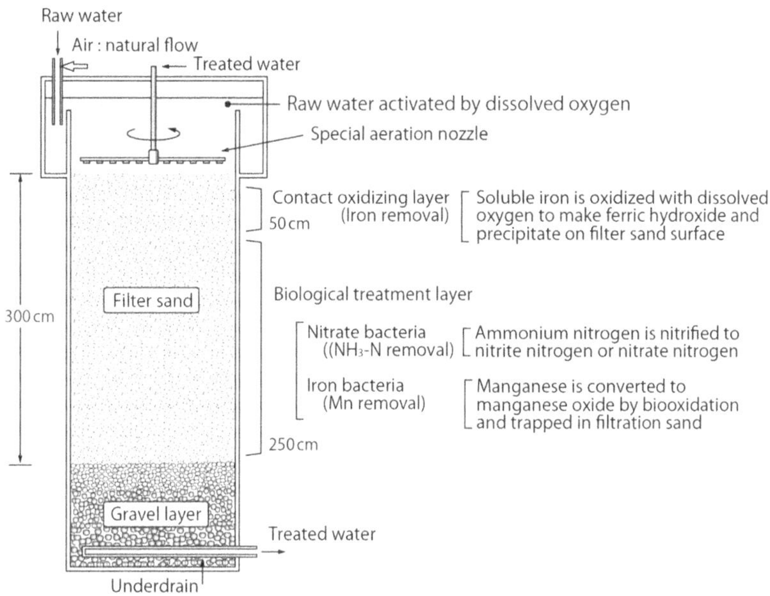

Figure 5.52 A microbial filtration system. (Taken from the Nagaoka Co., Ltd, catalog; see Hayashi *et al.*, 2015).

This is a proposal for a new high-speed biochemical de-iron and de-manganese method, and is a powerful proposal awaiting detailed quantitative evaluation of microbial groups and their functions, and kinetic studies that formulate them and link them to design. Having not studied this process ourselves, we can only refer to the research results and catalogs published by Nagaoka Corporation (see Hayashi *et al.*, 2015).

It seems that various experiments and implementations have been carried out, and there are examples of filtration speeds ranging from 120–400 m/day. Here, we refer to an example in which groundwater was treated under conventional conditions of rapid sand filtration, at 120 m/day. Concentrations of raw water, such as iron, should range from 10 to 20 mg/l for Fe^{2+}, 0.4 to 0.7 mg/l for Mn^{2+}, and 0.28 to 0.76 mg/l for NH^{4+}-N.

The following is a brief introduction to the system, borrowing materials from Nagaoka Corporation (Hayashi *et al.*, 2015). The main features of this system are:

(1) Air is mixed with the inflow groundwater to produce oxygen-saturated raw water.
(2) The water then flows into the upper hydrostatic (pressurized) part of the rapid filtration tank, which has a deep sand layer with a diameter of 0.6 mm and a thickness of about 300 cm.
(3) Water is passed through for a certain period of time to wait for the generation of iron bacteria in the filter layer.
(4) In order to oxidize and precipitates the Fe^{2+}, a series of reactions are developed to form a film on the sand grains of the produced iron oxyhydroxide and to coagulate ferric hydroxide using the same as an autocatalyst. These reactions occurs relatively quickly, as follows:

$$2Fe^{2+} + 1/2O_2 + 4OH^- + H_2O \rightarrow 2FeOOH \cdot H_2O \text{ (iron oxyhydroxide)}$$

$$2Fe^{2+} + 1/2O_2 + 4OH^- + H_2O \rightarrow 2Fe(OH)_3 \text{ (ferric hydroxide: coagulation)}$$

(The retention of the insoluble iron oxide in this form is concentrated in the surface layer of about 50 cm).

(5) Oxidation and precipitation of the Mn^{2+} occur after a relatively long aging (acclimation) period of the filter bed (depending on the conditions, this can be 2–3 months); (the manganese removal area is between 50–250 cm above the top of the tank, deeper than for iron):

$$2Mn^{2+} + O_2 + 4OH^- \rightarrow 2MnO_2 + 2H_2O$$

(6) The oxidation of ammonia also proceeds from NO_2 to NO_3:

$$NH_4^+ + 3/2O_2 \rightarrow NO_2^- + H_2O + 2H^+$$

$$NO_2^- + 1/2O_2 \rightarrow NO_3^-$$

(Removal of ammonia occurs after a similar aging period as that for manganese; Figures 5.53 and 5.54); see Hayashi *et al.* (2015).

In order for these phenomena to proceed steadily, it is necessary to establish operating standards based on detailed kinetic studies such as the degree to which the dissolved oxygen concentration in the filter layer can be maintained. In particular, oxygen-deficient conditions occur easily in raw water with high concentrations of NH_4^+-N. Designers seem to be considering measures for recycling of treated water.

The emphasis on establishing this method appears to be the adoption of two types of cleaning methods:

(1) The first is a group of micro flocs of ferric hydroxide that rapidly accumulate in the interstices of the sand layer in about 40 cm above the top of the sand layer, which is gently swept away by a slow backwash while retaining the attached microorganisms on the surface of the sand grains. At the same time, or a little earlier, the surface is cleaned by water jetting to improve the cleanliness of the sand particle surfaces to some extent. It is important to determine the level of energy consumption at which to maintain the cleanliness of the gap between the filter layers,

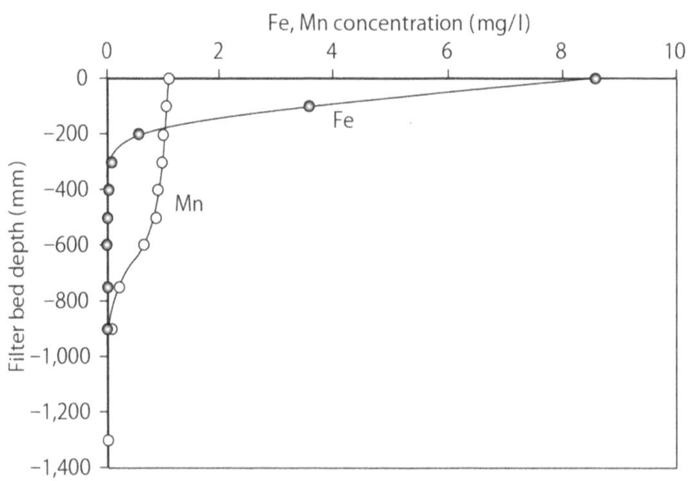

Figure 5.53 Filter bed depth and decreasing Fe^{2+} and Mn^{2+}.

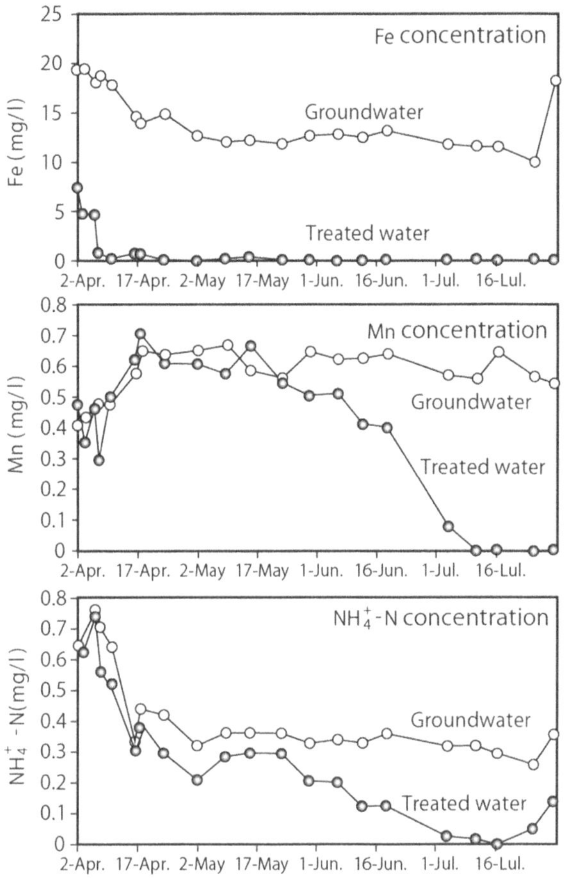

Figure 5.54 Daily variations in Fe^{2+}, Mn^{2+} and NH_4^+-N concentration.

while leaving an adequate attached microbial layer. Partial cleaning is performed two or three times a day.

(2) In the second method, a full-layer backwash is performed at 24 to 48 hour intervals to push out all layers of blocked components. The larger the diameter of the filter, the longer the extrusion of the suspended components by thorough mixing during reflux, and the longer the wash time required; the author imagines that difficulties would exist in making a larger diameter filter pond; see Figure 5.55.

In the previous section, it was mentioned that the precipitation of Fe^{2+} only progressed by simple aeration. However, by taking a sufficient acclimatization period of about 60 days, it seems that not only Fe^{2+} but also Mn^{2+} and NH_4^+ can be treated with a no-chemical injection using microbial communities based on iron bacteria and others, and at a high filtration rate of 100 m/day or more. Is it a Columbus egg?

We expect the inventor to adopt a long acclimatization time and to refine the washing method to keep the biological community. We also expect a variety of new methods to be devised, since it is thought that there is still another approach to the method of standing the eggs of Columbus. Once the

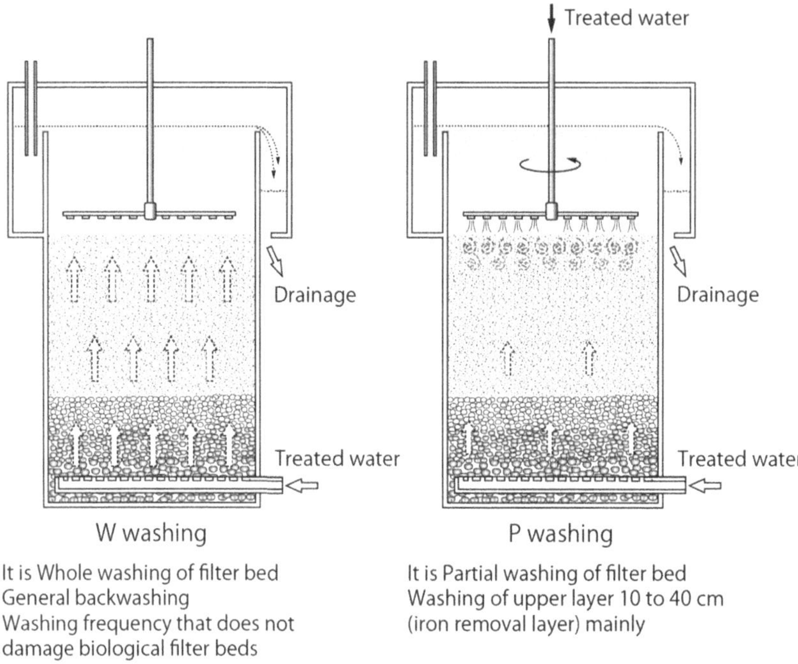

Figure 5.55 Washing a microbial filtration system (modified from the Nagaoka Co., Ltd, catalog; see Hayashi *et al.*, 2015).

identification and function of a clear microorganism population has been confirmed and a reliable kinetic process that can respond quickly to a wide range of raw water conditions has been created, we believe that this method could have a vital position in metal ion processing.

5.3.5 Arsenic precipitation treatment

Arsenic (As) is a nitrogen group element (Group 15 of the Periodic Table) that is often contained in soil. Arsenic (As) is highly toxic to living things and has long been used as a rodenticide. In addition to being a trace essential element necessary for living organisms, it is also a carcinogenic element, as well as exhibiting acute and chronic toxicity.

The pioneering environmental issue concerning arsenic was the reporting, in 1964, of black foot disease as an endemic disease in southern Taiwanese areas, with eczema and coloration on the skin, at the Second International Water Pollution Conference held in San Francisco. During the 1990s, similar reports were made in many areas using arsenic-containing groundwater, worldwide. In Japan, there was an arsenic contamination incident caused by arsenic wastes from the Toroku Mine in Miyazaki Prefecture, Kyushu. Groundwater contamination with arsenic content in well water is a particular problem that has occurred throughout Bangladesh: when groundwater is used to avoid polluted turbid surface water, it can bring a wide range of arsenic-contaminated water from groundwater aquifers. It seems that there are quite a number of such problems in the Indian subcontinent and on mainland China. In addition, as geothermal energy is used in various ways, arsenic-containing water has emerged on the ground surface due to the pumping of high-temperature water deep into the ground. In geothermal power generation wells, the reduction of spent water to the ground is inevitable. In addition, many hot spring waters and mine wastewaters may contain arsenic in high concentrations.

Arsenic in water can occur in a +5, +3, 0, or trivalent state, but is generally present in the trivalent As(III) and pentavalent As(V) states. Arsenic is present in most areas in the form of oxyanions ($H_2AsO_4^-$, $HAsO_4^{2-}$, AsO_4^{3-}, H_2AsO^{3-}, $HAsO_3^{2-}$, and AsO_3^{3-}) and behave as anions.

When arsenic is present in natural water, its presence coexisting with various cations (such as iron, manganese, and sulphur) rather than with a pure H_2O system is a problem. In particular, in systems in which thresholds of low arsenic concentrations (such as 0.001 mg/l)mg/l are of concern, the presence of cations such as iron and manganese is of great significance to the behavior of arsenic, particularly if there is a large ratio, such as Fe/As = 100. In water in which oxygen is sufficiently dissolved, arsenic usually appears as pentavalent arsenic As(V) and, when it is incorporated into an ecosystem, it is present as organic arsenic (methylated arsenic, dimethylated arsenic, trimethylated arsenic, etc.).

Gupta & Chen (1978) showed the presence of As(III) and As(V) as oxyanions changing with pH. Figure 5.56 shows this distribution in more detail.

Pentavalent arsenic As(V) behaves as a strong acid that accounts for the majority of arsenic's chemical species in oxidative environments, and $H_2AsO_4^{2-}$ appears to be the preferred species in the pH range of 7–11.5 in fresh water. Below pH 7, $H_2AsO_4^-$ is the dominant species.

Trivalent arsenic As(III) (arsenous acid) is a weak acid, and exists as the dominant species of trivalent arsenic in a totally undissociated H_3AsO_3 below pH 8. Arsenous acid accounts for a large proportion of the reducing atmosphere of groundwater. As is the case with As(V), it cannot be removed satisfactorily

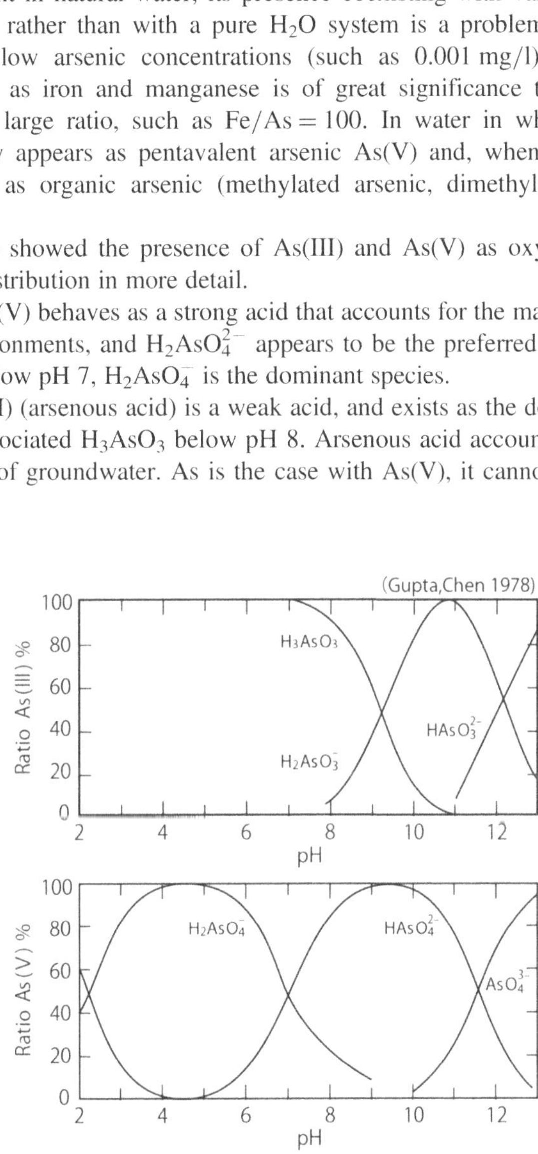

Figure 5.56 the presence of As(III) and As(V) as oxyanions changes with pH.

by precipitation or an activated aluminium oxide (alumina) adsorption method. As a result, groundwater containing arsenic needs to be preliminarily oxidized to As(V) in any treatment system.

The standard value for arsenic in drinking water is 0.01 mg/l in Japan, whilst the USEPA maximum contaminant level (MCL) is 0.010 mg/l. Treatment of arsenic in groundwater as raw water includes the following removal processes: oxidation, precipitation, filtration, granular bed adsorption, anion exchange, and reverse osmosis.

5.3.5.1 Oxidation of As (III) to As(V)

Trivalent arsenic in groundwater needs to first be oxidized to pentavalent arsenic As(V) to ensure that the subsequent treatment processes work effectively. Chlorine, potassium permanganate, ozone and other solid oxidizing agents are appropriately selected for oxidation. Simple aeration, chlorine dioxide, monochloramine and UV cannot effectively oxidize As(III) to As(V).

Oxidation by chlorine is achieved by the addition of chlorine gas or liquid hypochlorite. Chlorine oxidizes As(III) rapidly, within 1 minute. The amount of chlorine (Cl_2) required to oxidize 1 mg of As (III) is 0.95 mg, and pH dependency of oxidation reactions is not observed in the neutral range of pH 6.3–8.3. It is necessary to predict in advance the chlorine demand of any coexisting inorganic ion which reacts more actively with chlorine, and to add the amount needed for the coexisting inorganic ion to the amount required for As(III) oxidation. In the case of iron, 1 mg of Fe^{2+} requires 0.64 mg of chlorine, 1 mg of manganese-Mn^{2+} requires 1.29 mg of chlorine for, and 1 mg of hydrogen sulfide H_2S requires 2.21 mg of chlorine. When chlorine oxidation is used, it is difficult to use a separation method using an organic membrane that is vulnerable to chlorine in any solid-liquid separation process subsequent to the oxidation process.

Oxidation by potassium permanganate can be considered in combination with removal of iron and manganese. Many different waters contain components that require (consume) oxidizing agents in addition to As(III). The potassium permanganate $KMnO_4$ required to oxidize 1 mg of As(III) is 1.06 mg, which is not affected by pH in the neutral pH range. It is necessary to add 0.71 mg of potassium permanganate to iron, 1.44 mg to manganese, and 2.84 mg of potassium permanganate to hydrogen sulfide per mg of each ingredient requiring an oxidizing agent. If these items are included, the above mentioned amount should be added to the amount required for As(III).

Ozone is a strong oxidant, and the reaction proceeds in a very short time; 0.64 mg ozone (O_3) is required to oxidize 1 mg of As (III), which is largely unaffected by pH in the neutral pH range. An additional ozone loading of 0.43 mg for 1 mg iron oxidation, 0.88 mg for manganese, and 1.5 mg for hydrogen sulfide is required. Ozone also reacts with TOC, so it is also necessary to take the amount required for that into account (see Table 5.6).

A manganese sand (MnO_2) filter bed for iron and manganese removal can also effectively oxidize As(III) and filter As(V) if dissolved oxygen is present in water.

Table 5.6 The oxidants required for oxidation of 1 mg of As(III).

Oxidants	**mg**	**Oxidants Demand for 1 mg of Coexistent Ions mg**					
Cl_2	0.95	Fe^{2+}	0.64	Mn^{2+}	1.29	H_2S	2.21
$KMnO_4$	1.06		0.71		1.44		2.84
O_3	0.64		0.43		0.88		1.5

5.3.5.2 Oxidation/filtration (co-precipitation of arsenic with iron)

Many arsenic-containing waters, especially groundwaters, often contain significant amount of iron and manganese. Iron has two valences, Fe(II) and Fe(III), whilst manganese has five: Mn(II), Mn(III), Mn (IV), Mn(V), and Mn(VI). The reduced Fe(II) and Mn(II) forms of iron and manganese are soluble. This precipitation treatment causes coloring of the water, which is a serious obstacle to its use. Therefore, in the Drinking Water Quality Standards (USEPA, 1979), the MCL is defined as 0.3 mg/l for iron and 0.05 mg/l for manganese.

In the process of oxidation, precipitation, flocculation, solid liquid separation process of iron and manganese, arsenic is adsorbed on the surface of the iron coagulum, and both are removed in parallel. This co-removal is thought to be due to two mechanisms: adsorption and co-precipitation. The adsorption is the mechanism by which arsenic adsorbs on the surface of the precipitated Fe (III) particles, and the co-precipitation may be considered to be due to the adsorption of arsenic to Fe(III) and the incorporation of arsenic into the internal structure. This phenomenon is considered to be a phenomenon in which adsorption/desorption/coagulation (precipitation) is mutually entangled, and detailed kinetic study is desired.

The key to effective co-precipitation is the ratio of As(III) to Fe(II) concentrations in groundwater. With reference to USEPA 'State Implementation Guidance for the Arsenic rule (2002)' shows in the neutral range of pH 5.5–8.5, iron concentrations of 0.3 mg/l or more and an iron-to-arsenic concentration ratio $\geqq$ 20:1 can successfully co-precipitate to achieve the MCL of 0.010 mg/l. In another expression, 1 mg/l iron can co-precipitate 0.05 mg/l arsenic. This is a value that can be artificially applied to the treatment of iron salt.

Adjusting the ORP is crucial in removing iron and co-precipitating arsenic. Both Fe(III) and As (V) must be in a fully oxidized state. Iron oxidation can be oxidized by aeration, but it is not effective for the oxidation of arsenic. Therefore, it is necessary to oxidize chemicals with chlorine, potassium permanganate, ozone, etc., and it is necessary to install a reaction tank to secure the time required for the oxidation of arsenic. When chlorine is used as the oxidizing agent, it is recommended that the reaction time be about 20 minutes before the filtration operation.

The manganese precipitation reaction occurs in the same way as iron, but it does not seem to be very effective from the viewpoint of co-precipitation removal with arsenic.

5.3.5.3 Other arsenic treatment

In addition to precipitation and filtration, granular media filtration such as activated alumina and anion exchange resins are used for adsorbtion and ion exchange. The details of adsorption ion exchange are described below (see Chapter 7). As(V) in its oxidized state is effectively removed in the same manner as in the precipitation/coagulation process. For ion exchange, high base Cl anion exchange resins are used. Treatment with an Reverse Osmosis (RO) membrane is also effective. These systems can be operated at water purification plants, at water supply (tap). Depending on the application, these systems can eliminate hazardous pollutants such as arsenic from drinking water and redistribute them to non-drinking water for other uses.

FURTHER READING

Aris R. and Amundson N. R. (1957). Some remarks on longitudinal mixing or diffusion in fixed beds. *AIChE Journal*, **2**(2), 280–282.

Faust S. D. and Aly O. M. (1997). Chemistry of Water Treatment (2nd. Edition). Chapter 8 Removal of hardness (pp. 315–351), Chapter 9 Removal of inorganic contaminants (pp. 355–380). Lewis Publishers. Boca Raton, London, New York, Washington D.C.

Fukunaga A. and Takai Y. (1981). Form of manganese catalyst in contact filtration manganese removal method. *Journal of Japan Water Works Association*, **578**, 42–50. (In Japanese).
Gravel A. and van Dilk J. C., et al. (1976). Water Softening by means of Pellet Reactor: Developments in the Netherland. *Journal AWWA*, **75**, 619–625.
James M. Montgomery Consulting Engineers, Inc. (1985). Water Treatment Principles & Design (15 Inorganics). John Wiley & Sons, Inc, Pasadena, California, U.S.A. pp. 333, 338, 342.
Miki S. and Takagi K. (1964). Chlorine Oxidation Test. *Journal of Japan Water Works Association*, **352**, 29. (In Japanese).
Nagayama M. and Goto K., et al. (1968). On the air oxidation of ferrous ions in water. *Industrial Water*, **120**. (In Japanese).
Ngayama M. and Goto K. (1955). Oxidation reaction of ferrous ions in natural water. *Journal of Industrial Water Association*, **58**, 239. (In Japanese).
Stumm W. and Morgan J. J. (1981). Aquatic Chemistry, John Wiley & Sons, Inc., New York, Chichester, Brisbane, Toronto, pp. 171–229, 230–374.
Tambo N. (1977). Water in Hokkaido (in Japanese). *Journal of Japan Water Works Association*, **515**, 65–75. (In Japanese).
Tambo N. (1980). Water Supply (New System of Civil Engineering 88) (7.3 Coagulation). Gihoudou Publishing, Tokyo (Society of Civil Engineers, Edition). pp. 242–253, (In Japanese).
Tambo N. and Matsumoto T. (1960). Research on demanganation by Potassium permanganate. *Journal of Japanese Industrial Water Association*, **24**, 55–59. (In Japanese).
Tambo N. and Matsumoto T. (1963). Manganese Treatment in Water Supply by Potassium Permanganate. *Journal of Japan Water Works Association*, **342**, 43–56. (In Japanese).
USEPA (2010). Guideline for Arsenic Removal for Small Public Drinking Water System. United States Environment Protection Agency, Ohio.
Wilke R. and Chang P. (1955). Correlation of diffusion coefficients in dilute solutions. *AIChE Journal*, **1**(2), 264–270.
Wilson E. J. and Geankoplis G. C. (1966). Liquid mass transfer at very low Reynolds number in packed bed. *Industrial and Engineering Chemistry Fundamentals*, **5**(1), 9–14.

REFERENCES

Feitknecht W. and Schindler P. (1963). Solubility constants of metal oxides, metal hydroxides salt in aqueous solution. *International Union of Pure and Applied Chemistry*, **6**(2), 130–199.
Graveland G. (1980). Water softening by Crystallization. *Aqua*, **29**(2), 3–4.
Graveland G., et al. (1983). Development in softening by means of pellet reactor. *Journal AWWA*, **75**, 619–626.
Gupta S. K. and Chen K. Y. (1978). Arsenic removal by adsorption. *Journal of Water Pollution and Control Federation*, **50**, 493–566.
Hayashi N., Imao K., Mukai K., Cai H. and Kawaguchi T. (2015). Effective use of groundwater by chemical-free high speed biological water treatment. IWA 2011 Singapore International Water Week, Catalog of Chemiless Groundwater Biofiltration, Nagaoka Inc., Sakai City Japan.
Matsui Y., Tambo N. et al. (1986a). The dynamics of manganese removal by chlorine-manganese sand method (I) Basic properties of manganese sand. *Journal of Japan Water Works Association*, **621**, 15–22 (In Japanese).
Matsui Y., Tambo N. et al. (1986b). Dynamics of manganese removal by chlorine-manganese sand method (II), Kinetic Study of elimination process. *Journal of Japan Water Works Association*, **622**, 31–43 (In Japanese).
Nakanishi H. (1967a). Study of Manganese Removal by Catalytic Oxidation (I). *Journal of Japan Water Works Association*, **388**, 36–47. (In Japanese).
Nakanishi H. (1967b). Study of Manganese Removal by Catalytic Oxidation (II). *Journal of Japan Water Works Association*, **389**, 19–27. (In Japanese).
Nakanishi H. (1967c). Analysis of the exchange adsorption rate of manganese ions in continuous catalytic filtration method. *Journal of Industrial Chemistry*, **70**, 407–410. (In Japanese).

Pourbaix M. (1966). Chapters I–III: Introduction, Establishment of diagram, Use of potential-pH equilibrium diagrams. In: Atlas of Electrochemical Equilibria in Aqueous Solution (Translated from French by James A. Franklin), Pergamon Press, Oxford, London, New York, Toronto, Paris, Frankfurt.

Takai Y. (1964). Air and Chlorine oxidation in Iron breakdown treatment. *Journal of Japan Water Works Association*, **356**, 12. (In Japanese).

Tambo N. and Tsutsumi Y. (1982). Peaty region humic organic iron and ozone. *Journal of Japan Water Works Association*, **568**, 50–55. (In Japanese).

Tambo N., Matsui Y., Makiya K. and Sugimoto T. (1988). Dynamic Study of Water Softening by Pellet Fluidized Beds: Removal of Calcium by Treatment of Metastable state. *Journal of Japan Water Works Association*, **646**, 10–20. (In Japanese).

USEPA (1979). Drinking Water Quality Standards. United States Environment Protection Agency, Ohio.

Chapter 6

Solid–liquid separation processes

Water treatment is a process that attempts to preserve the value of water by transferring impurities into a solid or gas state. By transferring to a solid state, components that spread through water and can cause damage at concentrations as low as 10 to 100 mg/l in a liquid state are converted into high-concentration components that occupy a small volume in a solid state, and in many cases, they can be disposed of locally and intensively in the ground. When transferred into the air, in a gas state, such components are diffused into the atmosphere, and are dispersed over a wide area for disposal. From the standpoint of waste control, it is desirable to solidify impurities to minimize their presence. In most environmental controls concerning water treatment, the primary option is to transfer the load to a solid state. The basic process for this purpose is 'solid–liquid separation'.

6.1 SEDIMENTATION AND GRAVITATIONAL SEPARATION

6.1.1 Sedimentation

Sedimentation broadly refers to the phenomena in which a suspension, in which solid particles are suspended, is separated into a supernatant and a thickened suspension. Sedimentation phenomena in the field of environmental engineering have a wide range of applications, but here we discuss the gravitational sedimentation of suspensions (suspended matter), which have a higher specific gravity than water. An operation in which a suspension is placed in a container and particles are allowed to settle by gravitational force to separate solids and liquids is collectively referred to as a sedimentation process, or simply as sedimentation. Among sedimentation treatments, 'clarification' aims to obtain clear supernatant water with a low solid concentration of raw water (by overflow), and 'thickening sedimentation' (or simply 'thickening') aims to obtain a highly concentrated solid (sludge) from raw water, in a process in which thickened solids are discharged from the bottom of a container (as underflow) in the form of a slurry (Table 6.1).

Author(s): Norihito Tambo and Koichi Ogasawara
doi: 10.2166/9781789061307_0367

Table 6.1 Sedimentation treatments by gravity.

Solid Concentration of Raw Water	Classification of Treatment	Purpose
Low	Clarification	Clear supernatant water production
High	Thickening	Thickening solids production

6.1.1.1 Classification of sedimentation

When sedimentation and/or thickening etc. is classified according to the settling pattern (regime) of the particles, it can be roughly classified into four types (as shown in Figure 6.1).

(1) When there are low concentrations of suspended particles, and the streamlines of particles settling side by side do not interfere with each other and there is free settling of single separated particles (i.e. they do not mutually interfere with each other as in coagulation or/and flocculation), such settling is referred to as 'single particle free settling' (or discrete free settling);

(2) Even in free sedimentation, where the streamline interferences around particles settling side by side at low concentrations are negligible, differences in particle size (sedimentation rate) can cause large particles to collide, causing both particles to coagulate/agglomerate and grow into larger particles, resulting in faster sedimentation. The sedimentation of a dilute system with collisional agglomeration is called 'coagulation–flocculation free settling'.

(3) As the concentration of the suspension increases, the sedimentation rate of all particles becomes slower than that of a single particle due to the streamline interference of surrounding particles.

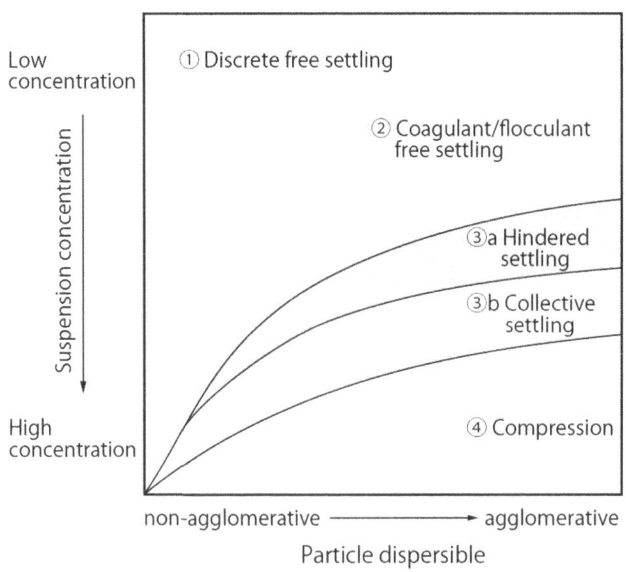

Figure 6.1 Classification of sedimentation according to settling regime.

The sedimentation of these regions is called 'hindered settling'. At higher concentrations and higher interferences, the particles can no longer settle at their intrinsic sedimentation rates, making it difficult to change their position relative to each other, and the groups of particles settle together. When individual particles can move, this is 'hindered settling'; when the interference becomes greater, it is referred to as 'mass sedimentation' (or collective settling).

(4) As the particles settle to very high concentrations at the bottom of the settling tank, the individual particles contact each other and the weight of the upper particles causes the underlying particles to undergo compression deformation and continue to settle and thicken while draining the interstitial water upwardly. This type of sedimentation is called 'compression sedimentation' (or, simply, compression).

The transition to the sedimentation mode in (2), (3) and (4) above progresses at lower concentrations with higher coagulative particles. Coagulation is a very crucial property of the water treatment process, as discussed in Chapter 4.

In actual water treatment, discrete free settling appears in sedimentation basins in water supply and sewerage etc., such as in the plane sedimentation basin of a slow sand filtration system or in the primary settling tank in activated sludge treatment. Coagulation–flocculation free settling appears in horizontal flow type chemical (coagulation) sedimentation basins etc., in municipal water and industrial water supply. Hindered and collective settling is found in blanket type upward-flow sedimentation basins in municipal, industrial water supply, and in final sedimentation basins in activated sludge wastewater treatment plants. Hindered settling and compression mainly occurs in the thickening tanks and sludge discharge areas of sedimentation basins.

The modes of sedimentation described above are based on the idea of sedimentation in hydrostatic water (quiescent settling). In actual sequential operation, the sedimentation pattern and the water flow pattern of the sedimentation pond are synthesized to design a system. Therefore, the water flow of the pond needs to be further examined. A sedimentation pond can be considered as a type of open channel, though the flow velocity is very small, and the ratio of width to length and the ratio of depth to length are remarkably small compared to an ordinary open channel. Therefore, the deviation of short-circuit flow etc. caused by density flow and blown (wind) flow etc. largely influences the flow conditions, and difference from the theoretical flow patterns becomes a major problem. In the latter part of this chapter, we describe the hydrology of ponds, after the sedimentation regime, and consider the design and process of sedimentation basins, by synthesizing both.

6.1.1.2 Single particle settling formula

As a basic equation of the sedimentation phenomenon, the 'free sedimentation equation of a spherical single particle' is used, and it is derived under the following conditions:

(1) Particles are spherical in shape with no compression and no deformation;
(2) The fluids in which the sedimentation takes place are stationary and non-compressed, and the sedimentation of the particles is not affected by the vessel walls;
(3) The acceleration force on the particles is given by a uniform gravitational field;
(4) The particles are free to move and do not affect the simultaneous absence or presence of other particles.

If these conditions are met, the settling velocity depends only on the nature of the single particle and the fluid in which it is suspended. Under these conditions, the force F to move the particles suspended in a fluid is expressed as the resultant forces shown in Equation 6.1.

$$F = F_g - F_b - R \quad \text{—(6·1)} \tag{6.1}$$

F : Resultant force trying to move particles
F_g : Gravity acting on particles
F_b : Buoyant force acting on particles
R : Resistance to particle motion (drag force)

Equation 6.1 The force acting on particles.

Now, considering spherical particles, (F_g – F_b) is represented by Equation 6.2.

$$F_g - F_b = \frac{\pi d^3}{6} \cdot g(\rho_s - \rho) \quad \text{—(6·2)} \tag{6.2}$$

d : Particle diameter (cm)
g : Gravitational acceleration (cm^3/sec^2)
ρ_s : Particle density (g/cm^3)
ρ : Water density (g/cm^3)

Equation 6.2 [$F_g - F_b$].

Resistance R is considered to be a function of four independent variables: the water viscosity coefficient μ, density ρ, particle diameter d, and settling velocity w, and can be expressed as Equation 6.3 using an exponential model.

$$R = K(d^p w^q \rho^r \mu^s) \quad \text{—(6·3)} \tag{6.3}$$

K : Constant
w : Particle setling velocity
μ : Water viscosity

Equation 6.3 Resistance R.

Using Buckingham's π theorem (see Langhaar, 1951), we can propose a function of the resistive force, as in Equation 6.4, by performing a dimensional analysis using the equation enumerating the dimensions of each element in Equation 6.3 to determine the power number from p to s in Equation 6.4.

Buckingham's π theorem

$$[mlt^{-2}] = K[l^p(lt^{-1})^q(ml^{-3})^r(ml^{-1}t^{-1})^s] \quad \text{—(6·4)} \tag{6.4}$$

m : Mass
t : Time
l : Length

Equation 6.4 The dimensional analysis of resistance force R.

In order for the dimensions of both sides of Equation 6.4 to be equal, the sum of the dimensions of each element (m, t and l) must be equal on both sides, which leads to Equation 6.5.

Since the dimensions of both sides of Eq.4.4 must be equal,

$$\left.\begin{array}{ll}\text{Mass m} & : 1 = r + s \\ \text{Time t} & : -2 = -q - s \\ \text{Length l} & : 1 = p + q - 3r - s\end{array}\right] \Rightarrow \left[\begin{array}{l} r = 1 - s \\ q = 2 - s \\ p = 2 - s \end{array}\right.$$

Substitute r, p, q into Eq.4 and 3

$$R = K(d^2 w^2 \rho)\cdot(dw\rho)^{-s} \mu^s = K(d^2 w^2 \rho)\cdot\left(\frac{\mu}{dw\rho}\right)^s$$

$$= K(d^2 w^2 \rho)\cdot \Phi(\mathrm{Re}) \quad \text{—}(6\cdot 5)$$

$$\left[\begin{array}{ll}\mathrm{Re} = dw/\nu & : \text{Reynolds number} \\ \nu = \mu/\rho & : \text{Kinematic viscosity}\end{array}\right]$$

According to Newtonian flow
"Resistance of the solid in the fluid is proportional to square of velocity"

$$R = C_D A \rho w^2/2$$

$$\left[\begin{array}{ll}C_D & : \text{Coefficient of resistance } (= \Phi(\mathrm{Re})) \\ A & : \text{Cross-sectional area perpendicular to direction of motion } (= 2Kd^2)\end{array}\right]$$

(6.5)

Equation 6.5 The equation of resistance foece R.

In addition, the commonly used expression for Newtonian flow, in which the resistance of solids in fluids is generated in proportion to the square of the velocity, and the proportionality constant C_D defined as the coefficient of resistance, yield Equation 6.6.

Upon adapting to spherical particles R is

$$R = \frac{C_D A \rho w^2}{2} = C_D \frac{\pi}{4} \frac{d^2 \rho w^2}{2}$$

$$= C_D \frac{\pi d^2 \rho w^2}{8} \quad \text{—}(6\cdot 6)$$

$A = \pi d^2/4$: Particle cross-section area

(6.6)

Equation 6.6 The drag force equation to spherical particles.

Substituting Equations 6.2 and 6.5 into Equation 6.1 gives the sedimentation equation, Equation 6.7.

Substituting Eq.4·2 and Eq.4·5 into Eq.4·1,
from Newton's motion equation $F = m\alpha$

$$F = \frac{\pi d^3}{6} \rho_s \frac{dw}{dt}$$

$$= \frac{\pi d^3}{6} g(\rho_s - \rho) - C_D \frac{\pi d^2 \rho w^2}{8} \quad \text{—}(6\cdot 7)$$

(6.7)

Equation 6.7 The sedimentation equation.

Therefore, sedimentation acceleration is given by Equation 6.8.

$$\frac{dw}{dt} = \frac{\rho_s - \rho}{\rho_s} g - \frac{3}{4} \frac{C_D \rho w^2}{d \rho_s} \quad \text{—(6·8)} \tag{6.8}$$

Equation 6.8 The sedimentation acceleration.

The settling velocity of the particles gradually increases over time until reaching a practically constant rate. This speed asymptotically approaches the final (terminal sedimentation) speed $dw/dt = 0$ (terminal velocity), and is given as the sedimentation velocity equation shown in Equation 6.9.

$$w = \left(\frac{4}{3} \frac{g}{C_D} \frac{(\rho_s - \rho)}{\rho} d \right)^{\frac{1}{2}} \quad \text{—(6·9)} \tag{6.9}$$

Equation 6.9 The sedimentation velocity equation.

The coefficient of resistance C_D in Equation 6.9 is a function of the Reynolds (Re) number, and it is known that the resistivity factor C_D takes the form of a characteristic function in different regions of the Reynolds number.

6.1.1.3 Determination of the coefficient of resistance C_D and various settling velocities

The values of the coefficient of resistance $C_D = \Phi(\text{Re}) = \Phi(wd/\nu)$ are calculated by Equation 6.9 by comparing the experimental measurements of the final velocities w of various particles and relationships with Reynolds numbers, and the coefficient of resistance C_D are plotted as shown in Figure 6.2. As can

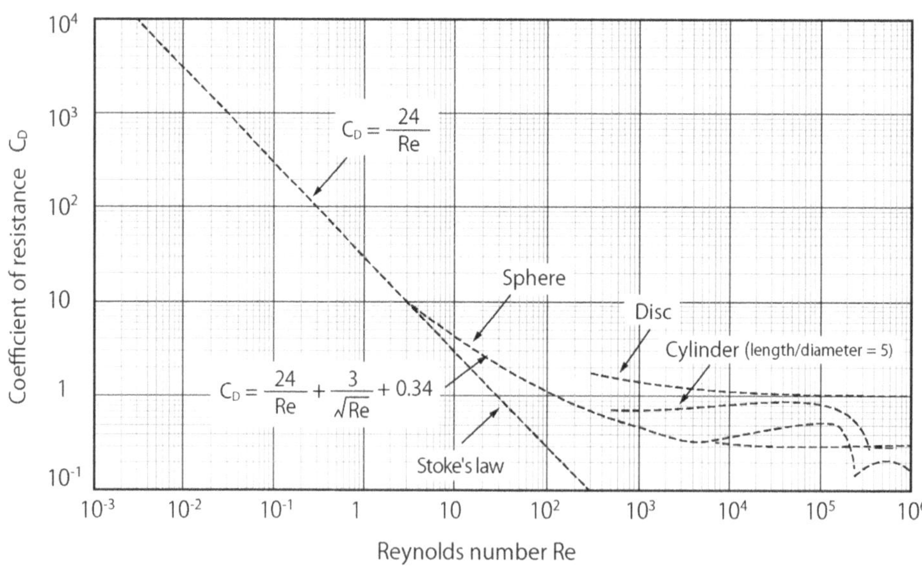

Figure 6.2 Relationships between Reynolds numbers and coefficient of resistance (Rouse, 1951).

Table 6.2 Equations for settling velocity.

Re Number Range	Friction Factor C_D	Drag Force R	Settling Velocity Equation	Name of the Equation
$10^{-4} \sim 2.5 \times 10^5$	$\Phi(Re) = \frac{k}{Re^n}$	$\frac{\pi d^2}{8} \rho w \Phi(Re)$	$\left[\frac{4}{3}\frac{g}{C_D}\frac{\rho_s - \rho}{\rho}d\right]^{\frac{1}{2}}$	General equation
Approximately <10°	$\frac{24}{Re}(n = 1)$	$3\pi\mu wd$	$\frac{g}{18}\frac{\rho_s - \rho}{\mu}d^2$	Stokes' equation
Approximately 10°~10^3	$\frac{12.65}{Re^{0.5}}(n = 0.5)$	$\frac{12.65}{8}\pi\sqrt{\mu\rho}(wd)^{1.5}$	$0.223\left[\frac{(\rho_s - \rho)^2 g^2}{\mu\rho}\right]^{\frac{1}{3}}$	Allen's equation
Approximately $10^3 \sim 2.5 \times 10^5$	0.4 (n = 0)	$0.05\pi\rho(wd)^2$	$1.82\left[\frac{\rho_s - \rho}{\rho}dg\right]^{\frac{1}{2}}$	Newton's equation

be seen, in the region $Re \ll 10^0$ where the Reynolds number is small, the C_D decreases in proportion to the -1 power of Re. However, when the Reynolds number increases to the region of $10^0 < Re < 10^3$, the slope of the C_D decays. When the Reynolds number reaches the region of $10^3 < Re < 10^5$, the C_D becomes almost constant. In the case of a sphere, $C_D = 0.4$.

If these relationships are expressed as the relationships between Reynolds numbers and the coefficient of resistance C_D corresponding to the respective characteristic range, the values in the first and second columns of Table 6.2 are substituted into Equation 6.6, and the resistance (drag) force R of each respective range is obtained as in the third column. The resistance force R in the region where the Reynolds number is smaller than 10^0 is proportional to the first power of the settling velocity w, and in the region where Re is larger than 10^3, the resistance is proportional to the second power of the settling velocity w. For the former, the experimental results agree with those theoretically derived by Stokes under the condition that 'resistance is governed only by viscosity and inertial force can be ignored', referred to as Stokes' law of resistance.

The flow around the settling spheres in this range is perfect laminar flow, without streamline separation and without wake. In contrast, particles with a relatively large particle size and a high Reynolds number that settle at a high settling rate are completely turbulent at the boundary layer with vortices behind the spheres (Figure 6.3). In this range, the resistance due to the pressure differential across the sphere due to vortex

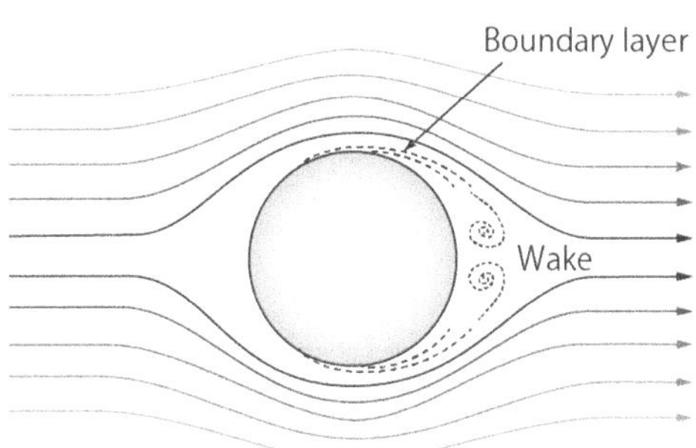

Figure 6.3 Flowlines around a settling particle.

generation is superior to the resistance due to viscosity, and as Newton thought, the resistance is proportional to the square of the velocity, indicating the resistance law of the inertial range. This is called Newton's law of resistance. In the middle of this range, the delamination of the boundary layer is incomplete, and the law of resistance – which cannot be ignored in both viscous resistance and eddy resistance – is considered to be between 1 and 2 powers of the sedimentation rate, and various equations have been proposed. Table 6.2 shows an example of an experimental formula produced by Allen (Allen's equation), which is determined to be proportional to the 1.5th power of the settling velocity.

When trying to calculate the settling velocity of a particle using a variety of settling velocity equations, the settling velocity w is calculated using the formula that seems to be appropriate. To do this, it is necessary to check the Re number and confirm that there are phenomena in the application range for that formula. If the resulting Re number is outside the application of the formula, then the appropriate formula for that Re number must be re-chosen and calculations repeated until the appropriate condition is reached. Since particle size and particle density in the sedimentation process are typically in the range of 50 μm to 1 mm and 1.005 to 2.65 g/cm^3, respectively, it is convenient to know in advance in which area the Re number is located, and the appropriate settling velocity formula can be selected immediately, and the schematic settling velocity can be obtained as shown in Figure 6.4.

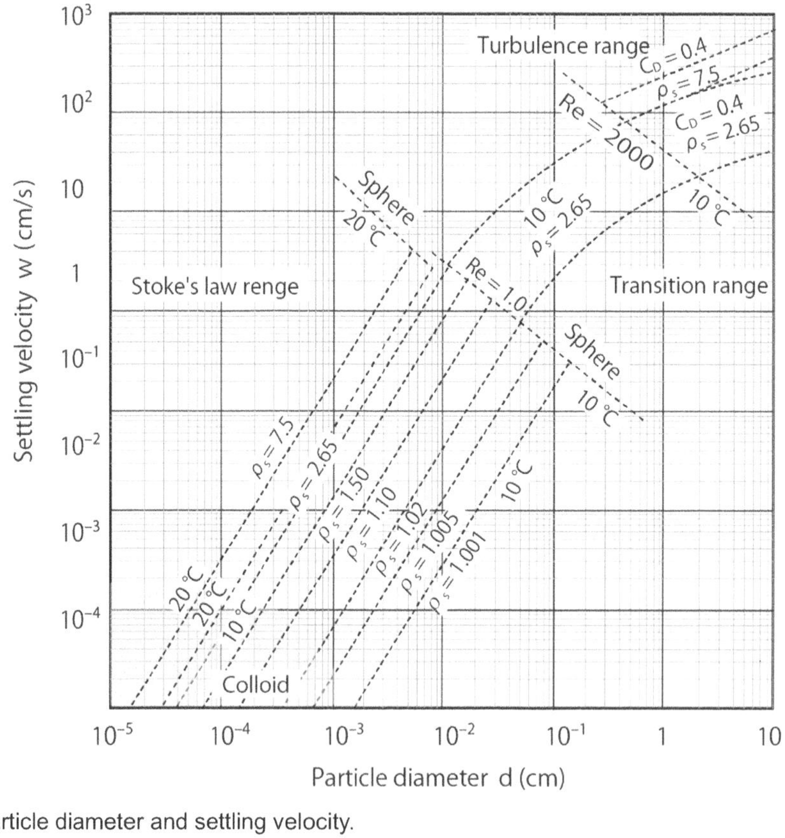

Figure 6.4 Particle diameter and settling velocity.

6.1.1.4 Correction for particle shapes other than spheres

The shape of particles other than spheres is defined as a distortion from a sphere, using the shape factor (coefficient of sphericity) ψ as an index (Equation 6.10).

$$\varphi = \frac{d_a}{d_e} \cdot \frac{1}{n} \qquad (6 \cdot 10) \tag{6.10}$$

φ : Shape factor
d_a : Representative diameter of nonspherical particle
d_e : Diameter of sphere having same volume as particle (Equivalent sphere diameter)
n : Particle specific surface area / Equivalent sphere specific surface area

Equation 6.10 The shape factor.

Since it is difficult to easily determine the values of items in Equation 6.10, we can also estimate the coefficient of sphericity ψ by referring to Table 6.3 and estimating the coefficient of resistance C_D in the sedimentation of aspherical particles by using Figure 6.5 produced from the Reynolds' number, where the equivalent spherical diameter d_e is taken as the length dimension.

6.1.1.5 Sedimentation separation of free settling particles

Consider the fundamental law of settling particle clusters in areas such as those exhibiting single-particle free sedimentation (see Figure 6.1). Actual sedimentation sites contain a mixture of particles ranging

Table 6.3 Particle shape and shape factors.

Particle Shapes		Shape Factor	d_e/d_a
Sphere		1.00	1.00
Octahedron		0.847	0.965
Cube		0.806	1.24
Prisms	a × a × 2a	0.767	1.564
	a × 2a × 2a	0.761	0.985
	a × 2a × 3a	0.725	1.127
Cylinders	h = 2r	0.874	1.135
	h = 3r	0.860	1.31
	h = 10r	0.691	1.96
	h = 20r	0.580	2.592
Disks	h = 1.33r	0.858	1.00
	h = r	0.827	0.909
	h = r/3	0.594	0.630
	h = r/10	0.323	0.422
	h = r/15	0.254	0.368

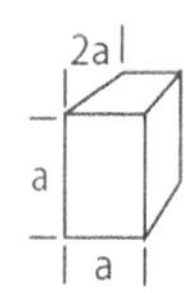

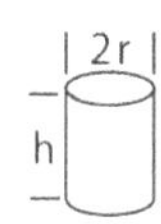

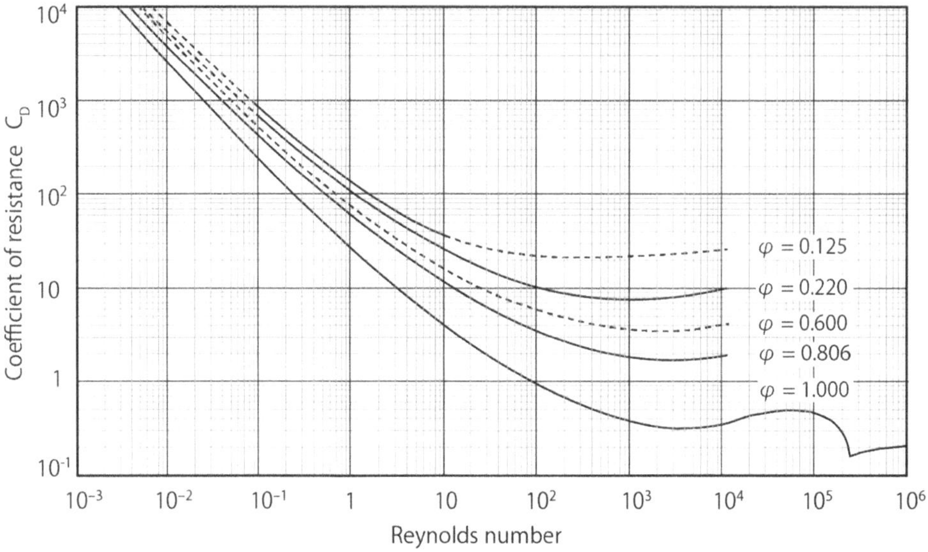

Figure 6.5 Coefficient of resistance versus Reynolds number for particles of different sphericities (Rouse, 1951).

from those in the Stokes range to those in the Newtonian range. In the design of a sedimentation separating process, we define the settling velocity distribution function f(w), which shows the basic particle group characteristics, and consider the settling velocity distribution of the particles as shown in Figure 6.6, where f(w)dw is the mass of a group of particles with a settling velocity between w and w + dw.

If we assume that particles with a sedimentation velocity distribution function f(w) are uniformly dispersed in a hydrostatic sedimentation basin with a depth h as shown in Figure 6.7, and sedimentation

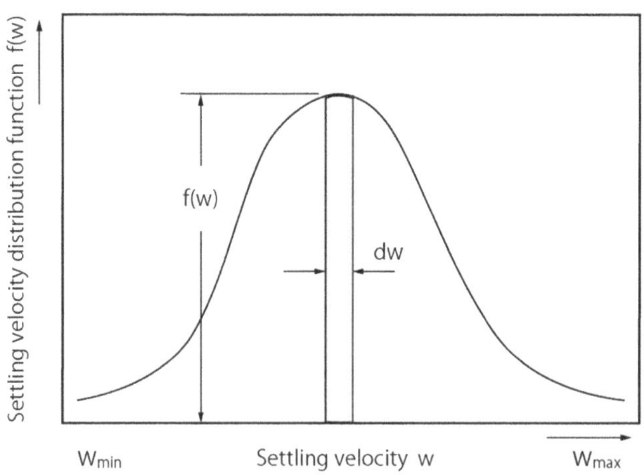

Figure 6.6 Settling velocity distribution f(w).

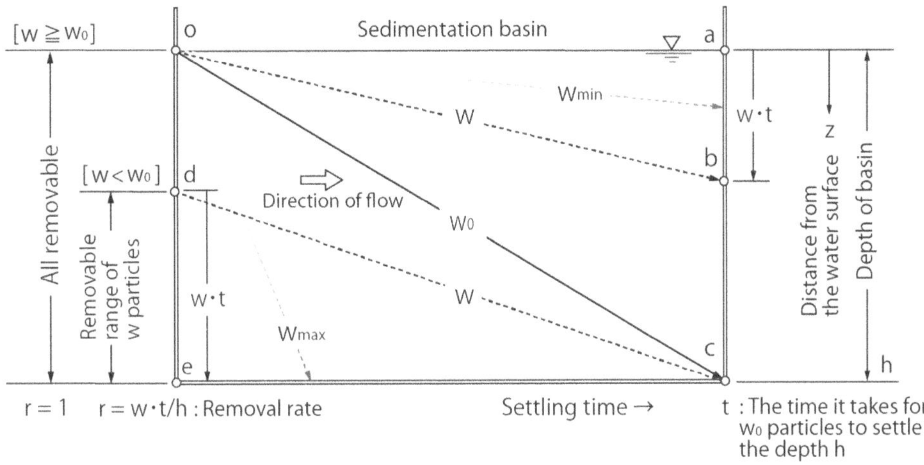

Figure 6.7 The behavior of particles in a sedimentation basin.

starts at time t = 0, when the settling period t elapses, the quantity q(t) of settled particles deposited at the bottom of the tank at depth h is given by Equation 6.11.

$$q(t) = \int_{W_0}^{W_{max}} f(x)\,dw + \int_{W_{min}}^{W_0} \frac{wt}{h} f(w)\,dw \quad \text{—(6·11)} \tag{6.11}$$

q(t) : Removal amount up to settling period t
w : Particle settling velocity
h : Settling distance of particles (depth of the basin)
W_{max} : Maximum settling velocity of particles in suspension
W_{min} : Minimum settling velocity of particles in suspension
W_0 : Settling velocity of particle that reachs depth h at time t

Equation 6.11 The removal quantity up to settling period t.

As can be seen from Figure 6.7 and Equation 6.11, the settling velocity of water (in a sedimentation basin) of depth h at time t is w_0, and any particles with a higher settling velocity will settle out as the first term on the right side of Equation 6.11 indicates. For lower settling velocity particles with $w < w_0$, we can see from Figure 6.7 that by drawing a line dc parallel to ob, only ab/h (= de/h) of the total volume is removed at the point where the suspension below the water depth d is removed. From the congruence of the triangles shown in the figure, $ab/h = w/w_0$, and the removal rate of particles with a settling velocity slower than w_0 is $r = w/w_0 = w_t/h$.

6.1.1.6 Measurement of sedimentation velocity distribution of particle groups

The design of a sedimentation unit begins by knowing the settling velocity profile f(w) of the suspended particles which are to be separated from the suspension. There are a variety of experimental methods, but sedimentation is the most direct method in terms of its purpose and nature. Pipetting is one of the most commonly used methods for water treatment sedimentation equipment design. This method is a highly practical sedimentation assay that can be applied to isolated particles, and also to coagulated particles for both low-concentration and high-concentration systems.

The suspension is dispersed uniformly in a cylindrical precipitation vessel, as shown in Figure 6.8, used for injection and settling under static conditions. Samples are taken in small quantities from multiple water sampling holes as shown in the drawing at various interval times, and the concentrations are measured. Samples will not contain particles that have a velocity greater than the velocity required to sink from the surface to the sampling hole in the sampling time. Therefore, the maximum settling velocity of the particles contained in the sample is given by Equation 6.12.

$$w_t = \frac{Z_n}{t} \quad \text{—(6·12)} \qquad (6.12)$$

w_t : The maximum settling velocity of particles in sampling water at t
t : Water sampling time after settling started
Z_n : Sampling depth

Equation 6.12 The Maximum settling velocity of particles.

Since the concentration profile of the initial suspension is uniform for each ingredient, the concentration of particles with a sedimentation rate lower than w_t in each sample will be equal to the initial concentration. Therefore, the cumulative settling velocity distribution curve shown in Figure 6.9 can be plotted, as shown in Table 6.4.

The integral of the second term on the right side of Equation 6.11 is the shadow area below the x_0 of the cumulative settling velocity profile as shown in Figure 6.9.

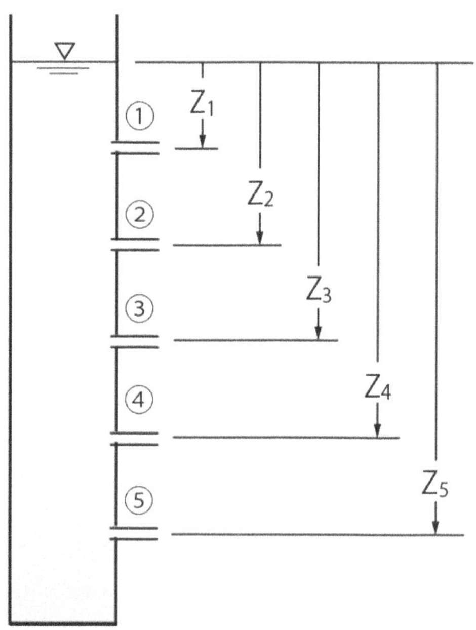

Figure 6.8 A settling analysis cylinder.

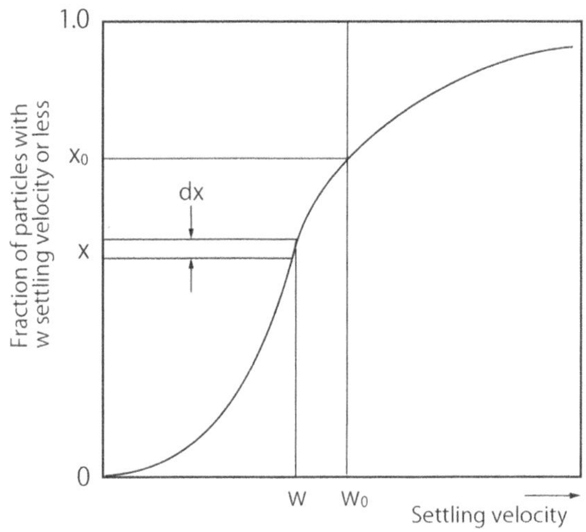

Figure 6.9 A cumulative settling velocity distribution curve.

Table 6.4 The cumulative settling velocity analysis curve and removal rate.

Assuming that the concentration is c_1 at time t_1, c_2 at t_2 and c_n at t_n,
The concentration between $w_{t,n} = Z_n/t_n$ and $w_{t,n-1} = Z_n/t_{n-1}$ is $(c_n - c_{n-1})$ C

Therefore,
The ratio Δx of the portion having the settling velocity of $w = (w_{t,n} + w_{t,n-1})/2$ is
$\Delta x = f(w)dw = (c_n - c_{n-1})/c_0$ c_0: Initial concentration
This makes it possible to draw a distribution curve,
The over-all removal rate R_0 at the water depth Z_n is calculated from Equation 6.11.

$$R_0 = (1 - x_0) + \frac{1}{w_0}\int_0^{x_0} w\,dx$$

X_0: Presence ratio of settling velocity w_0 particles (100% removed)

6.1.1.7 *Coagulated sedimentation: agglomeration phenomena in the sedimentation process*

If there is a difference in sedimentation velocity between the different suspended particles, the fast settling particles collide with the slow settling particles during the sedimentation process, and there will be coagulation between the particles, the particle size of the suspended particles increases as the sedimentation process progresses, and the sedimentation speed increases. Flocs of metallic hydroxides and microbial flocs (discussed in later chapters), which are widely handled in water treatment, are typical groups of particles in which agglomerates grow during such sedimentation. These particles are referred to here as coagulation (or flocculent) particles, and coagulation sedimentation (or flocculent settling) is a process in which suspensions change particle size and sedimentation rate increases as the sedimentation process progresses.

In coagulation suspensions, even if sedimentation is started from uniform particle size distributions throughout the water depth, as the sedimentation process progresses, large diameter particles with a higher (faster) settling rate collide with small diameter particles with a lower (slower) settling rate and agglomerate to reduce the number of particles, resulting in large diameter and high velocity settling particles, and the composition changes from time to time. Therefore, for flocculent suspensions, it is impossible to deal with problems based on steady settling velocity distributions, such as separated particles, and it is necessary to devise different handling methods.

Equation 6.13 presents a classic qualitative equation proposed by T. R. Camp for the impingement frequency in the unit volume due to velocity differences between settling particles.

$$N = \frac{\pi}{4} n'n'' (d' + d'')^2 (w' - w'') \quad \text{—(6·13)} \qquad (6.13)$$

N : Impingement frequency between spheres of d'and d" in unit time, unit liquid volume
d'.d": Diameter of each sphere
n'.n": Each number of spheres in unit liquid volume
w'.w": Settling velocity of ech sphere

Equation 6.13 The impingement frequency equation of T.R.Camp

As shown in the equation, it is qualitatively understood that agglomeration during sedimentation progresses more intensely as the number of particle groups increases, the particle size increases, and the size differences between the particles increase. However, since the combination of particle sizes in actual suspension is infinite and changes from time to time, it is difficult to derive a practical mathematical expression from this qualitative expression, and enormous numerical calculations seem almost meaningless considering the cost-effectiveness.

6.1.1.8 Experimental assessment of coagulation (flocculent) settling

Figure 6.10 shows the sedimentation condition of the iso-concentration interface of the flocculent particle group, and the figures within it show the removal rates of the initial concentration. For example, the first (left) line is an iso-removal rate line at which 30% of the original concentration is removed. If

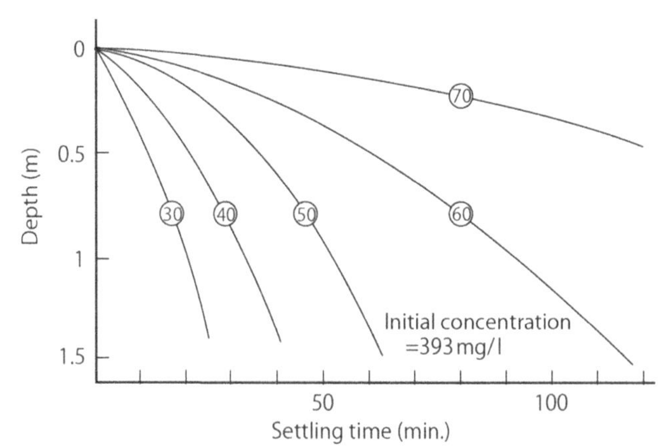

Figure 6.10 The iso-concentration lines of flocculent particle groups.

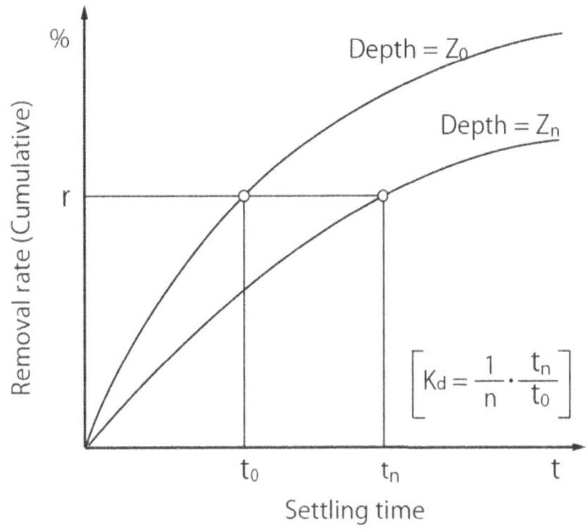

Figure 6.11 Flocculent particle settling analysis and correction factor.

the particles are not flocculent, the isolines are tangential to these curves at a depth of $Z = 0$ and at time $t = 0$, and the removal rate after a certain time is easily given. However, in the coagulation (flocculent) sedimentation group, the iso-removal rate line curves downward over time, and curves steeper as the water depth and suspension concentration increases (which is strongly affected by coagulation).

We next consider how the settling velocity of coagulation particles changes with depth. To do so, we assume that the water depth of the sedimentation tube to be used as a reference is Z_0, and the cumulative percentage curve obtained from this water depth is used as a reference curve; we can estimate the cumulative sedimentation curve at any depth by adding some correction to this curve. The coagulation sedimentation depth correction factor 'K_d' is defined as the time required to obtain a certain removal percentage or cumulative percentage (r%) when the suspension is settled and separated in a hydrostatic sedimentation column, where the water depth is increased n times in the case of the separated particle group ($K_d \times n$), but the water depth is increased in the case of the coagulation particle group (here, $K_d < 1$). For the same suspension, curves with differing depth Z_0 and removal percentage (r%) at $Z_n = nZ_0$ are obtained, as shown in Figure 6.11. In order to obtain the same removal percentage r%, the depth Z_0 requires time t_0 and Z_n requires time t_n. From the definitions, the K_d values are given by Equation 6.14.

$$K_d(Z_0/Z_n.r) = \frac{t_n}{t_0} \cdot \frac{Z_0}{Z_n} = \frac{1}{n} \cdot \frac{t_n}{t_0} \qquad \text{—(6·14)}$$

Z_0 : Reference measurement depth (cm)
Z_n : n times the depth of Z_0 (cm)
r : Assumed removal percentage(Cumulative) (%)
t_0 : Time taken to obtain removal rate r % at water depth Z_0 (min)
t_n : Time taken to obtain removal rate r % at water depth Z_n (min)
$K_d(Z_0/Z_n.r)$: Correction coefficient for multiplying r % by the observation value of water depth Z_0 to obtain the required settling time in Z_n

(6.14)

Equation 6.14 The water depth correction coefficient "K_d" for flocculent settling.

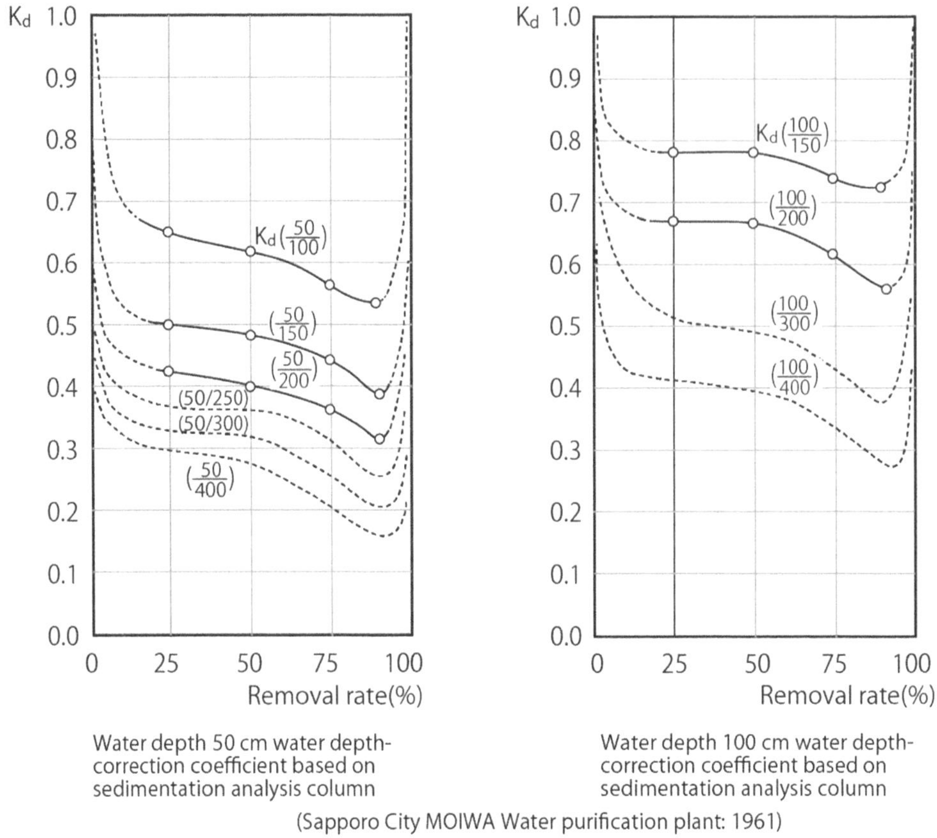

Figure 6.12 Examples of the water depth correction coefficient K_d of a chemical sedimentation basin.

The K_d values differ depending on the removal percentage of particles of the same composition, and theoretical analysis is not possible. In fact, it is necessary to collect a large number of observations in ponds to obtain the values required for practical use. Figure 6.12 shows values for the coagulation sedimentation depth compensation factor K_d from actual coagulation sedimentation ponds obtained by a sedimentation test using a 200 cm long undisturbed sampling tube using sedimentation basin influent water (flocculator outlet water) at a raw water turbidity of 15 to 20 mg/l from the Sapporo City Water Supply Purification Station as a sample. Theoretically, when $K_d \doteqdot 1$ it indicates near full removal (100%) and a very low removal rate (near 0%). However, for relatively high removal rates (around 95%), the time-accumulated curves for different water depths practically converge near a single line, and therefore, empirically, the time-accumulated curves may be considered to have a $K_d \doteqdot 1/n$.

The author deduced from the graphs in Figure 6.12 that, in the sedimentation of coagulation particles in clarified sedimentation basins of rapid sand filtration system, deeper ponds can be used to obtain more supernatant water in the same sedimentation time. The effects of this phenomenon on the actual process are described in the sedimentation basin design section below. This study was conducted in the early 1960s as a graduation study by K. Koyama (Kawakita) and S. Hiraga under the guidance of Tambo in the Department of Sanitary Engineering, Hokkaido University (Tambo, 1962).

6.1.1.9 *Hindered settling and zone (stratified) settling*

As the concentration of suspended particles increases, the streamlines around the individual particles that settle out interfere with each other. When this happens, the upward movement of water between the particles is not negligible, and the sedimentation rate of the particles is lower than in the case of single-particle free sedimentation. This phenomenon is called hindered settling. Similar phenomena, known as wall effects, occur in cylinders of small cross-section as relatively large particles settle. It does not matter if the Reynolds number of the particle sedimentation is large or when the settling particle size is 1% or less of the cylinder diameter.

In typical water treatment processes, isolated particles are not very likely to have suspensions as high as 1% of the cylinder diameter that cause interfering sedimentation (although there are extreme examples, e.g. Yellow River water), and this sedimentation is actually more often seen in flocs from chemical coagulation and flocs from microbial treatment. In such a group of flocculant particles, if the solid content exceeds about 500 mg/l, the particles interfere strongly with each other, and it becomes difficult for the fast (large) sediment particles to overtake the slow (small) particles and settle, and sedimentation progresses as one whole. This phenomenon is called zone settling. As will be explained later in the section on sedimentation units, this phenomenon is observed in upflow sedimentation basins with a floc blanket and in the final sedimentation basin of the activated sludge process.

6.1.1.10 *Sedimentation patterns of highly concentrated particles (consolidation)*

Next we consider sedimentation with a highly concentrated flocculant suspension uniformly dispersed in cylinders. Particles with a high sedimentation rate settle first, and a layer with a high concentration rate is formed in the lower part of the cylinder and collectively settles. When a high initial concentration suspension (slurry) is put into cylinders and settled, there is no sedimentation as individual particles at the onset of sedimentation. Instead, there is a collective sedimentation state in which a clear interface exists between the supernatant liquid and the settling suspension, and a stratified sedimentation (zone settling) state in which the separating interface subside with time. A batch sedimentation curve showing the interfacial sedimentation of such stratified sedimentation is shown in Figure 6.13.

The concentration distribution of suspensions exhibiting such batch sedimentation curves is classified into four levels, shown schematically in Figure 6.14, and transitions from rapid sedimentation to consolidation

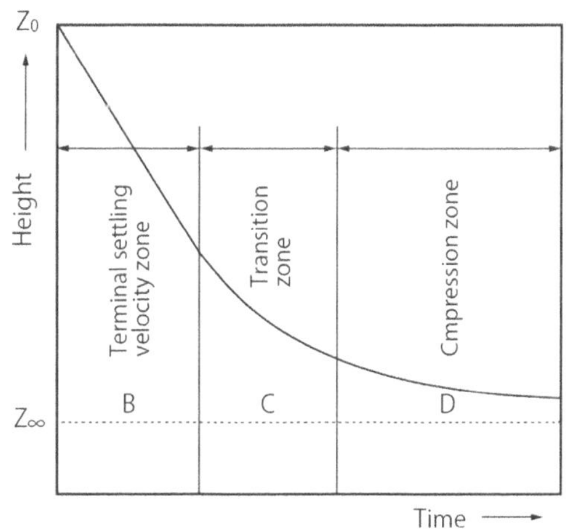

Figure 6.13 The batch sedimentation curve.

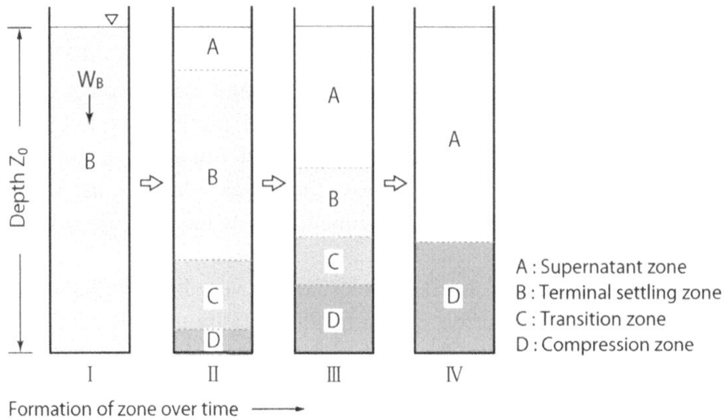

Figure 6.14 Creation of zones for four levels of concentration.

sedimentation with the passage of settling time. Initially, the suspension is suspended at a uniform concentration B over the entire depth Z_0 of the cylinders, and sedimentation proceeds at the group's constant settling rate W_B, creating a clear interface with supernatant A and initiating interfacial settling at a constant rate. The interface between the A and B layers is clearer the more uniform the size of the particles (single particle sedimentation rate). The rate of sedimentation of the suspensions in layer B is uniform and is present in equal concentration because of the collective behavior of the group of particles. The sedimentation of this area is called mass sedimentation (collective settling). In contrast to the formation of the finer part A in the uppermost layer, the C and D layers thicken more than the B layer appear to in the lower part over time. The C layer is a transition region (transition zone) in which the settling velocity decreases due to the increase of the concentration of the suspensions and, as the concentration further increases, the settled particles become supported directly by the lower deposited particles, and the D layer develops, in which the sedimentation proceeds very slowly by eliminating the pore water of the settled particles due to the self-weight of the particles. This region is called the compression sedimentation region (compression zone). Together, this is collectively referred to as stratified sedimentation (zone settling).

6.1.1.11 Theories expressing the state of zone settling

The relation between slurry density and settling velocity is a phenomenon that changes complicatedly depending on the size, density, and coagulation of the suspended particles, as well as their distributions within the group of particles. It is considered that a theoretical expression is almost impossible to obtain, and there is little practical use for it. Therefore, as in the case of the coagulation (flocculant) sedimentation described in the preceding section, it is common practice to estimate the batch sedimentation curve at the necessary water depth using a batch sedimentation curve obtained at a certain water depth and concentration for the suspensions considered. Three examples of estimates are shown below.

The first of these examples is the Work-Kohler equation, which is a rule of thumb for describing the correlation of sedimentation curves for different water depths for a given suspension of the same concentration. The relationship between sedimentation curves for the same suspension of the same concentration with different initial liquid level Z_0 is such that $0Z_0/0Z_0' = 0Z_A/0Z_A' = 0A/0A' = 0B/0B' = 0C/0C = \cdots$ as shown in Figure 6.15. This relationship holds throughout constant rate settling zone as well as during the reducing rate sedimentation zone. It is useful for estimating actual sedimentation behavior of prototype ponds from actual measurement results in small settling test tubes.

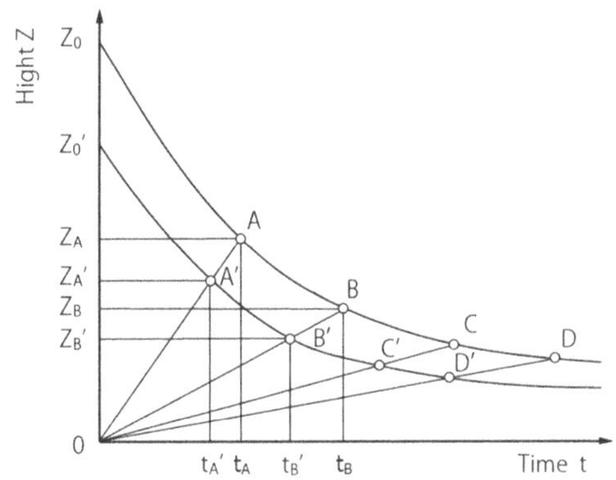

$$\left[\frac{0Z_0}{0Z_0'} = \frac{0Z_A}{0Z_A'} = \frac{0A}{0A'} = \frac{0B}{0B'} = \frac{0C}{0C'} = \cdots\right]$$

Figure 6.15 The use of the Work-Kohler equation for settling curves.

The second method of estimation involves Kynch's assumption. Kynch (1952) omitted the mechanisms of batch sedimentation from the assumption that the settling velocity of a layer of homogeneous slurries with zone sedimentation is a function only of the local concentration of the part considered, and for the same suspensions proposed estimating zone sedimentation curves for various other concentrations from the results of settling tests with certain concentrations.

Consider a sedimentation tube as shown in Figure 6.16. At height Z from the bottom of a layer in the sedimentation tube, the settling velocity is w, the concentration is c, and the mass balance at time dt in

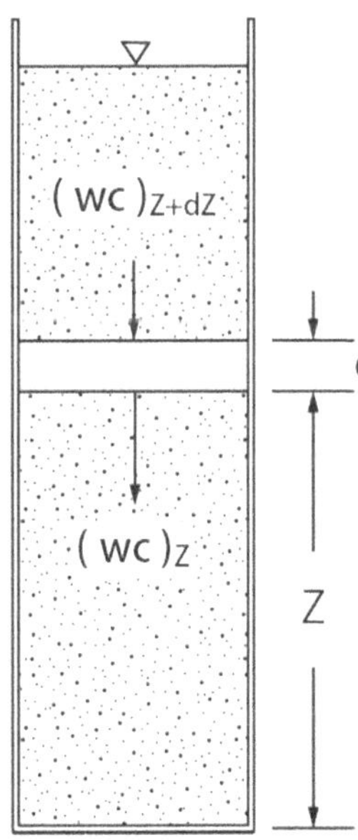

Figure 6.16 A test cylinder for an interface settling test.

the horizontal layer of thickness dZ at height Z is considered. The change in concentration in the layer is expressed as Equation 6.15. The coefficient W represents the sedimentation characteristics of Equation 6.15, and thus represents the depth and time of a suspension developed on the basis of the mass balance equation, is also a function of the concentration c, and has the same dimension as the settling velocity.

The mass balance equation at time dt in the horizontal layer of thickness dZ at height Z is

$$\frac{dc}{dt}dZ = (wc)_{Z+dZ} - (wc)_Z = \frac{\partial wc}{\partial Z}dZ$$

In the case of the same suspension, w can be assumed to be a function of concentration only

$$\frac{\partial wc}{\partial Z} = \frac{\partial wc}{dc} \cdot \frac{dc}{\partial Z}$$

By expressing in partial differential form how concentration changes with depth and time

$$\frac{\partial c}{\partial t} + \mathbf{W}\frac{\partial c}{\partial Z} = 0 \qquad \mathbf{W} = -\frac{dwc}{dc} = -w - c\frac{dw}{dc} \qquad \text{—}(6 \cdot 15)$$

Z : Height from the bottom of one layer during settling
w : Settling velocity of the layer
c : Concentration of the layer

(6.15)

Equation 6.15 The change in concentration in the layer.

Now consider how a layer of concentration c moves in the Z direction with time t. This is expressed by drawing an iso-concentration line as shown in Figure 6.17 on a two-dimensional plane with t in the horizontal axis and Z in the vertical axis. Since the concentration c is a function of the height Z and the

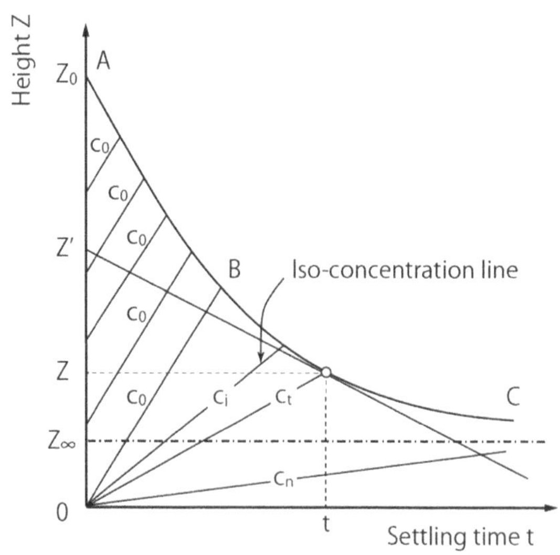

Figure 6.17 The rising pattern iso-concentration surface of zone settling slurry.

time t, the equation for the iso-concentration layer is Equation 6.16. Substituting this relationship in Equation 6.15 gives the relationship of Equation 6.17, indicating that the layer of each concentration C_i increases linearly with time.

$$dc = \frac{\partial c}{\partial Z} dZ + \frac{\partial c}{\partial t} dt = 0 \quad —(6\cdot16) \tag{6.16}$$

Equation 6.16 The equation for iso-concentration layer.

Substitute Eq. 4.15 for Eq. 4.16

$$W = \left(\frac{dZ}{dt}\right)_{c=\text{constant}} \quad —(6\cdot17) \tag{6.17}$$

Equation 6.17 The Equation of constant settling velocity zone.

The iso-line of the initial concentration c_0 rises at a rate equal to the initial interfacial settling velocity w_0 of the settling test. As the sedimentation progresses, the position of the level of the bed from the bottom of the tank, where the suspensions have settled to the bottom of the vessel and become more concentrated as $c > c_0$, rises at a rate of $Z = W_t$ reaching concentration c at the interface between clarified water and settling slurry at time t.

The total amount q of suspended matter that has settled down through the surface at depth Z, which is the concentration c, by time t, has a relative rate of settling $(W + w)$, and after time t, all particles in the slurry interface settle to clear water. Therefore, it is assumed that all the particles in the c_0Z_0 passed through this interface and the amount is described as in Equation 6.18.

Assuming that all initial particles have settled through the depth Z plane, the total amount of passed particles was

$$q = c_0 Z_0 = c_t(W+w)A \quad —(6\cdot18) \tag{6.18}$$

q : The total amount of particles passed through the interface
A : Horizontal section of the settling cylinder

Equation 6.18 The total amount of suspended solids passing through a interface

Substituting and arranging the relationship $Z = W_t$ as described above, the concentration c of the considered iso-concentration curve is given by Equation 6.19, which gives the sedimentation rate of the layer with the tangential gradient (Equation 6.20) for the concentration c at time t of the batch sedimentation curve, so that the relationship $cZ' = c_0Z_0$ is derived, where Z' is the height of the suspension when all slurries have the same concentration, as shown in Figure 6.17. In summary, the tangential slope of the batch sedimentation curve at sedimentation time t gives the value of the corresponding concentration c and the settling velocity w of the suspension. By arbitrarily selecting the time t, it is possible to determine the relation between the concentration c and the sedimentation rate w_c of suspensions for various concentrations $(c > c_0)$ which settle collectively from a single batch sedimentation curve.

$$c = \frac{c_0 Z_0}{Wt + Z} \quad —(6 \cdot 19) \tag{6.19}$$

Equation 6.19 The Concentration of iso-concentration line.

Tangential gradient of batch settling curve at time t

$$w = \frac{dZ}{dt} = \frac{Z' - Z}{t} \quad —(6 \cdot 20)$$

Tangential gradient gives the settling velocity of the concentration c layer

$$cZ' = c_0 Z_0 \tag{6.20}$$

Equation 6.20 The tangential gradient of batch settling curve.

The third method of estimation is Roberts' empirical expression for the progress of the compression dewatering process, which can be described by Equation 6.21.

Roberts empirical equation

$$-\frac{dZ}{dt} = K(Z - Z_\infty) \quad —(6 \cdot 21)$$

Expression of progressing process of compression sttling

$$\frac{Z - Z_\infty}{Z_c - Z_\infty} = e^{-K(t - t_c)} \tag{6.21}$$

Z_∞ : Layer thickness at completion of compression sedimentation
K : coefficient
t_c : Compression settling start time
Z_c : Slurry layer thickness at start of compression sedimentation

Equation 6.21 The empirical expression of Roberts in the compression dewatering process.

6.1.1.12 Basic theory of horizontal flow sedimentation basin removal: overflow rate

The typical equipment used for sequential sedimentation of particles is a horizontal flow sedimentation basin (an ideal settling tank) and this section discusses the theories behind them.

An idealized horizontal flow sedimentation basin is composed of four parts, as shown in Figure 6.18:

(1) the inlet zone: an inflow area in which all suspended matter is assumed to be uniformly distributed across the depth of the pond;
(2) the settling zone: the area where valid sedimentation is performed;
(3) the outlet zone: where clarified water is collected and flows out; and
(4) the sludge zone: where deposits settle in the lowest part of the sedimentation basin, and where sludge discharge (underflow) is performed.

In such an idealized sedimentation basin: the flow in the settling zone is uniform in all parts with a horizontal flow velocity V and complete extrusion flow (piston flow); the suspended particle

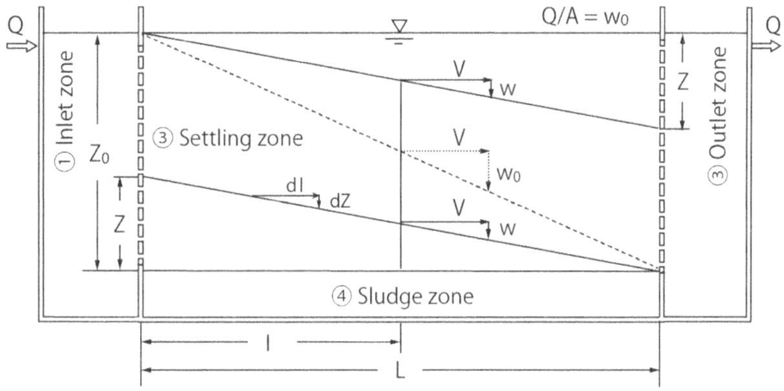

Figure 6.18 An ideal horizontal flow sedimentation basin.

concentrations of a respective size are uniform throughout the entire water depth from the inlet zone to the settling zone; and there is no resuspension of the particles once settled in the sludge zone.

The inflow of single free settling particles in such an idealized basin forms the basis of the 'surface load rate (overflow rate) theory' when considering removal from an ideal horizontal flow continuous sedimentation basin.

The trajectory of a particle as it passes through the settling zone is represented by the vector of the horizontal flow velocity V and settling velocity w. Particles with a settling velocity w inflow from the pond surface and exit the outflow zone at a depth Z, while all particles with a sedimentation velocity w which inflow below a depth Z from the surface of the sedimentation basin are removed completely from the sedimentation zone to the sludge zone. Now, as shown in Figure 6.18, if the sedimentation velocity of particles that inflow from the top of the sedimentation basin (water surface) and reach the bottom of the pond at the outflow end of the sedimentation basin is w_0, particles for which $w \geqq w_0$ are 100% removed, and the sedimentation rate w_0 of the smallest sedimentation rate particles, which can be removed 100% in this pond, is called the 'surface load or overflow rate' and is determined by Equation 6.22. The removal rate r of particles with a settling velocity w smaller than w_0 is w/w_0 (the settling velocity divided by the overflow rate) and is called the Hazen number (Hazen, 1904). If the settling curves shown in Figure 6.9 are obtained by the batch settling test, the removal rate of an idealized horizontal flow sedimentation basin with a surface load rate w_0 can be calculated as r_0 in Equations 6.22 and 6.23.

Table 6.5 Conditions of an ideal horizontal flow sedimentation basin.

Flow Condition etc.	Conditions in the Basin
Flow of settling zone	Horizontal flow velocity is uniform over all cross sections Complete piston flow
Inlet suspension concentration	Each size particles uniformly distributed in the cross section
Behavior of settling particles	All particles settling downward No re-floating from the settling zone

The detention time t_0 of the sedimentation basin is

$t_0 = L/V$ L : Length of the basin
V : Flow velocity of the basin

The velocity w_0 of the particle falling through the full depth of the settling zone is

$$w_0 = \frac{Z_0}{t_0} = \frac{Z_0}{L/V}$$ Z_0 : Settling zone depth

Sedimentation basin flow rate Q is

$Q = V \cdot B \cdot Z_0$ B : Width of the basin

Therefore, the 100% removable minimum settling velocity w_0 is

$$w_0 = \frac{Q}{L \cdot B} = \frac{Q}{A} \quad \text{[Overflow Rate]} \qquad —(6 \cdot 22)$$

The removal ratio r of particles smaller than w_0 is

$$r = \frac{w}{w_0} = w\frac{A}{Q} \quad \text{[Hazen Number]}$$

From the settling curves in Figure 6.9, the removal ratio r_0 of the basin with the overflow rate w_0 is

$$r_0 = (1 - x_0) + \frac{A}{Q}\int_0^{x_0} w\,dx \qquad —(6 \cdot 23)$$

x_0 : Fraction of particles larger than the settling velocity w_0

(6.22)–(6.23)

Equation 6.22, 6.23 The overflow rate and removal ratio.

As is clear from Equations 6.22 and 6.23, the smallest particles can theoretically be removed regardless of the depth of the pond, and the value $w_0 = Q/A$ is obtained by dividing the treatment flow rate Q of the settling zone by the surface area A of the pond. Based on this value, the total removal rate is calculated from the obtained settling velocity distributions. The numerical value defined in Equation 6.22 is called the surface loading (or the overflow rate) of the sedimentation basin, and is the first theoretical index for calculating sedimentation removal. Therefore, in a horizontal flow sedimentation basin, specifications such as the retention time t of the pond and the depth Z_0 of the pond are not related to the removal rate, in theory. This relationship is illustrated by the examples in Figures 6.19 and 6.20.

The depth of the basin shown in Figure 6.19 is half of the prototype pond, $Z_0/2$ (Figure 6.18). In this case, the horizontal flow velocity is doubled to 2 V, and the retention time is halved to $t_0/2$. However, the settling distance to reach the bottom is also halved to $Z_0/2$. As a result, the removal rate of the original pond and the basin with half the water depth is calculated as a pond with the same surface load rate and the equivalent removal rate, even though the position of the particles with sedimentation rate w_0 bottoming on the pond remains unchanged. In an idealized flow, the removal rate can be calculated only by the surface load rate, regardless of the water depth/horizontal flow velocity/retention time. However, in an actual pond, the water depth/horizontal flow velocity/retention time are significant design factors, as described later, due to the turbulence of the water flow and excavation of sediment.

When an intermediate basin floor is horizontally inserted at the point where water depth $Z = Z_0/2$ (i.e. half of the prototype pond), as shown in Figure 6.20, the distance required for settling will be half that of $Z_0/2$, although the horizontal flow velocity V will not change. Therefore, the sedimentation rate (surface

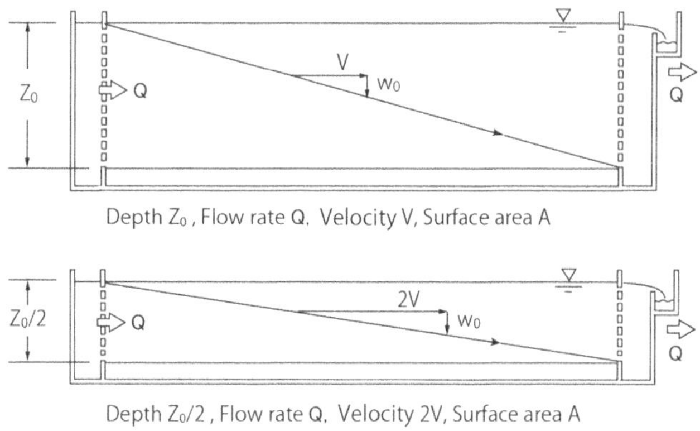

Figure 6.19 Changes in removal pattern due to halving the depth in a sedimentation basin.

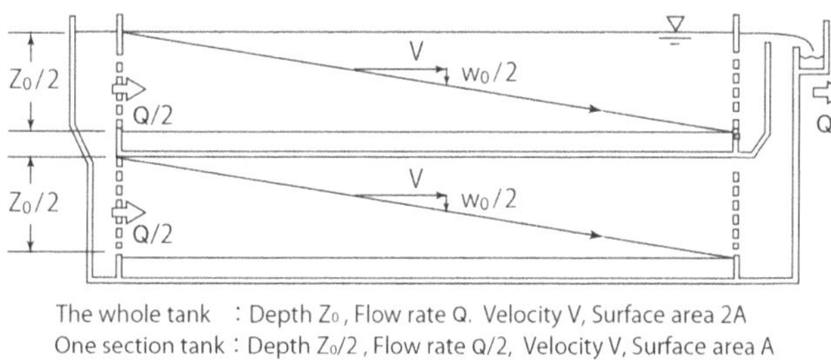

Figure 6.20 The removal pattern of a two-layer sedimentation basin.

load rate) of particles that reach the bottom of the pond at the end of the sedimentation basin will be halved ($w_0/2$). The trajectory of the 100% removal particles is doubled as shown by the dotted line. This means that the floor area of the operating sedimentation basin is doubled to 2A. The surface load ratio $w = Q/2A = w_0/2$ of a two-floor tank, and the 100% sedimentation rate can be changed to include particles with half the settling velocity of particles in the original pond. By constructing such a multi-storey tank sedimentation basin (multilayer settling tank), it is possible to reduce the surface load rate (overflow rate), and even very small particles can be targeted for removal. This idea leads to proposals for various devices such as inclined plates or inclined tube sedimentation basins, also referred to as multi-level sedimentation basins.

Actual design of sedimentation sites needs to cope with various realistic hydrological phenomena such as sedimentation behavior of particles, turbulence of water flow, deviation of water flow, and excavation of sediment. The following is a description of some of the basic actions that have been taken so far.

6.1.1.13 Setting the minimum water depth

Removal has been discussed on the assumption that, once settled, particles do not re-suspend in an idealized sedimentation pond. However, in an actual sedimentation basin, if the pond is made too shallow and the lateral flow velocity exceeds a certain limit (whilst the surface load rate is unchanged), the flow of the settling zone and the frictions on the sludge zone boundary increase beyond a certain limit and the turbulent flow causes scouring phenomena such as the migration and re-suspension of sediment particles. Therefore, in actual sedimentation basins, water depth is required to keep the lateral flow velocity below the limit of scouring.

Theories about determining the critical flow velocity for a sedimentation pond and open channel scouring include the idea proposed by T. R. Camp (1946) based on the critical sweeping force theory of Shield used in river engineering, and the idea of 'rolling up' due to turbulent fluctuations, developed by A.C. Ingersoll and colleagues (Ingersoll *et al.*, 1955) in the sedimentation laboratory of the California Institute of Technology. Since coagulated particles in water treatment sedimentation basins, such as chemical and biochemical flocs, are very light and very small in diameter, the mechanisms for them being agitated by turbulent fluctuations in the water flow seem to be reasonable from observations of the phenomena. Ingersoll *et al.* (1955) proposed Equation 6.24 as a condition under which re-washing does not occur, if the particle sedimentation velocity w is sufficiently larger than the turbulence velocity v* near the bottom, quoting the finding that the shear velocity $v^* = (\tau_0/\rho)^{1/2}$ (where τ_0 is the sedimentation basin channel bottom shear force and ρ is the water density) of the flow was approximately equal to the turbulent velocity v′ on the bottom surface that was observed by Laufer (1950) in a wind tunnel experiment with Reynolds number (Re = 12,300 to 61,600), approximately equal to sedimentation basins.

$$\frac{w}{\sqrt{\frac{\tau_0}{\rho}}} = \frac{w}{\left(\sqrt{\frac{f}{8}}\right)\cdot V} > 1.2 \sim 2.0 \qquad \text{—(6·24)} \tag{6.24}$$

w : Settling velosity of particles
f : Friction coefficient of basin bottom
V : Horizontal flow velocity of basin

Equation 6.24 The conditional equation for puffing velocity.

Assuming that the frictional coefficient of a basin bottom is f ≑ 0.025, which is roughly equal to the roughness of a concrete surface, the limiting horizontal flow velocity permitted for a pond from the standpoint of preventing re-suspension is around V < 9 W to 15 W. Accordingly, the necessary minimum water depth is determined by dividing the process flow Q by the critical horizontal flow velocity V_c, as in Equation 6.25.

$$Z_{min} = \frac{Q}{V_c \cdot B} \qquad \text{—(6·25)} \tag{6.25}$$

Z_{min} : Minimum water depth not to scour
V_c : Critical horizontal flow velocity
B : Sedimentation basin depth

Equation 6.25 The minimum depth for non-re-scouring.

6.1.1.14 Design of a coagulated (flocculant) particle sedimentation basin

During water treatment, coagulated (flocculant) particles are often subjected to sedimentation. In such cases, the dimensions of a sedimentation pond cannot be determined simply by the surface load (overflow) rate, and correction is required to take into account for the fact that the settling particles grow (agglomerate) as the water depth of the pond increases. In such a system, it is difficult to theoretically consider only the surface load rate. As described in the 'Experimental assessment of coagulation (flocculant) settling' section above, a batch sedimentation test of the suspensions concerned is performed, and the settling curve of the actual pond in the water depth Z_n is estimated using the obtained time-cumulative curve at the water depth Z_0 as shown in Figure 6.11, and the retention time t_n required for the actual pond is estimated from the time t_0 of the settling test required to obtain the predetermined removal rate r_0.

Now, let's think of an example for a chemical (coagulated) sedimentation basin in a water supply system. Let's consider the design of a 2 m deep sedimentation basin with a 90% removal rate for a given coagulated floc population. If the sedimentation analysis was carried out in a settling analysis tube with a depth of 50 cm and the time t_0 required to obtain a 90% removal rate was 120 minutes, the sedimentation time t_n required for an ideal flow sedimentation pond with a depth of 200 cm would be as follows: K_d (50/200, 90%) ≑ 0.32 (obtained from Figure 6.12) and the retention time t_n required for a sedimentation pond to process four times the flow rate is $t_n = 0.32 \times 4 \times 120 = 154$ minutes. In the case of separate particle groups, to obtain the same removal rate, if the water depth increased 4 times, required retention times must be increased by a factor of 4, while in the case of coagulated particle groups, the retention time must be increased by a factor of $154/120 = 1.28$. In other words, the same effect is obtainable by increasing the overflow rate by a factor of $1/K_d = 1.00/0.32 ≑ 3.13$, by increasing the depth of the pond.

Figure 6.21 shows how the retention time requirement and the surface load factor (overflow rate) vary with depth when the target is 90% removal of the same group of particles. As can be seen from the figure, the removal in the chemical sedimentation basin of a water purification plant increases by a factor of $1/K_d$ as the water depth increases, taking into account the collisional growth in the settling process of the coagulated particles, and the surface load rate at which the same removal rate can be obtained (90% in this figure) increases. According to actual measured examples, it seems that the processible quantity can be increased almost without changing the retention time by increasing the water depth within the range of the practical target removal rate which is usually considered. In a sedimentation pond, which is used to handle coagulated particles such as with city water supply, retention time is conventionally adopted as the design criteria and a standard retention time of 4 hours was used widely in the early 20th Century.

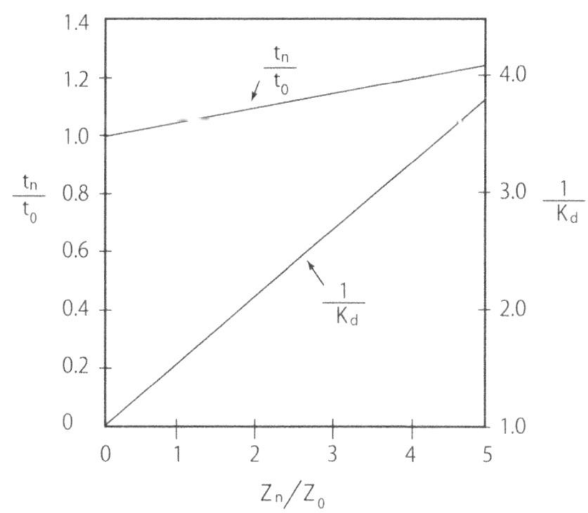

Figure 6.21 The use of water depth and detention time in removing flocculant particles.

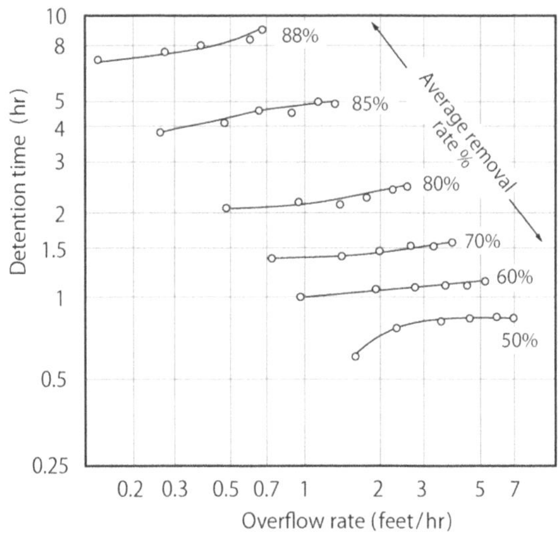

Figure 6.22 The removal rate of 'chalk' suspension, overflow rate and retention time.

Figure 6.22 shows the relation between overflow rate (surface load rate) and the retention time required to obtain a constant removal rate in a horizontal flow sedimentation basin by changing the depth to about 2 m in the settling analysis of a suspension of 408 mg/l of chalk powder (whitewash). It can be seen that the removal rate of such high concentrations of coagulation particles is largely determined by retention time, and is not based on the overflow rate (surface load rate) theory. McLaughlin conducted similar experiments on municipal sewage water and discussed suspensions with concentrations of 500 mg/l and mean settling velocity of around 1 cm/min; and suggested that removal is often dominated by retention time.

As mentioned above, when discussing the settling of coagulated particles, it is not possible to design a sedimentation basin using only the surface load rate for a single particle, and retention time must also be considered. Of course, the compositional changes of floc groups during the sedimentation process differ depending on the settling velocity distributions, concentrations, coagulation of particles, depth of ponds, etc. Specifically, a settling test should be conducted and the numerical value obtained must be used for any advanced design.

6.1.1.15 Turbulence and flow deflection in a pond

Actual sedimentation basin water flow differs from idealized flow conditions in that there is always an uneven distribution of the kinetic energies of the inflow water, as well as differences in the densities of the inflow water and the pond water, and resistances of the pond walls and the pond component, resulting in turbulence. Attempts to avoid the influence of such disturbances leads to procedures such as setting the pond depth (limit horizontal flow rate) in consideration of the scouring phenomenon, as described above.

Consider the disturbance of a sedimentation pond in terms of Reynolds numbers. If we take a chemical sedimentation pond with a water depth Z_0 of 400 cm, a mean horizontal flow velocity $V = 30$ cm/min $=$ 0.5 cm/s, water temperature of about 20°C, and kinematic viscosity $\nu = 0.01$ cm^2/s, the Reynolds number is $Re = Z_0 V/\nu = 400 \times 0.5/0.01 = 20{,}000$, which is much higher than the laminar flow limit $Re = 500$ of

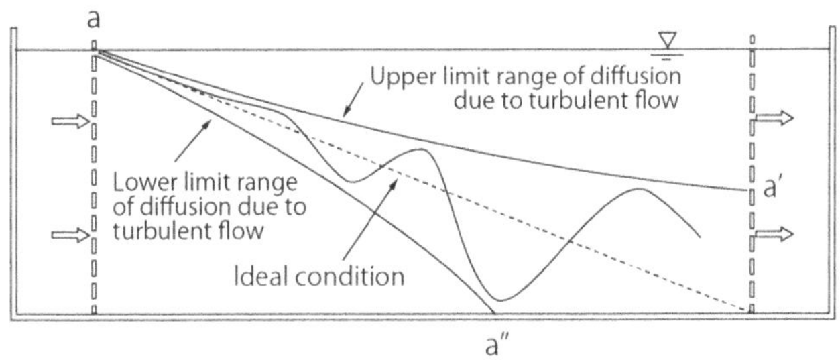

Figure 6.23 Diffusion settling of particles as a result of turbulent flow.

the open channel, and the water flow is completely turbulent even at low speed. It has been very difficult to theoretically address this type of sedimentation pond flow in low speed turbulent regions, and it has been approached by limit phenomena or by replacing it with a hypothetical pattern model.

Figure 6.23 schematically illustrates the settling condition when turbulent diffusions are added to the flow of an idealized sedimentation basin. Particles with a settling velocity w_0 (overflow rate Q/A) inflow from the sedimentation pond surface and reach the pond bottom just at the outflow end (in an idealized sedimentation pond), but under turbulent conditions particles are dispersed by turbulent diffusion due to random movement of water molecules and settle in the region in the figure bounded by the lines a-a′ and a-a″. As a result, some particles deposit faster than they do without disturbance, but the removal rate decreases as a whole because some particles are unable to deposit and go into the outlet zone.

Based on the W. D. Dobbins' solution (Dobbins, 1944), the reduction of removal rate by a turbulence applied to parallel flow, T. R. Camp (1946) drew a diagram, shown in Figure 6.24. This is a case in

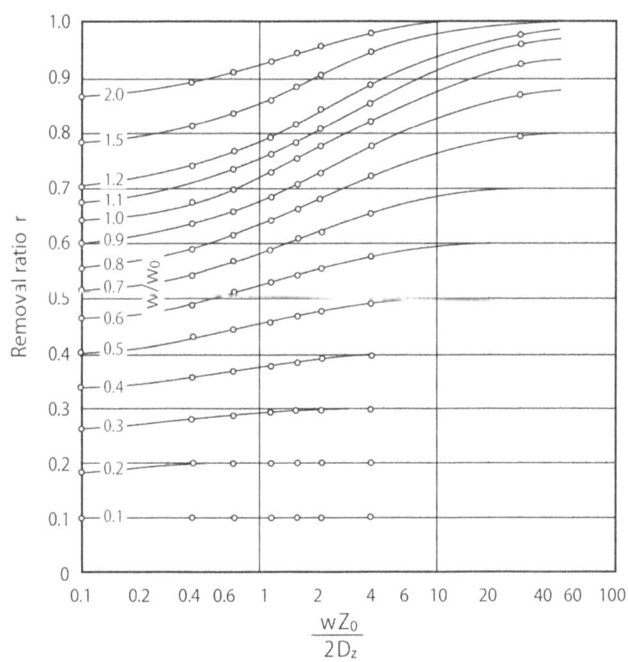

Figure 6.24 The reduction in removal efficiency due to turbulent dispersion (Dobbins, 1944; Camp, 1946).

which there is no re-suspension, and the reduction of a sedimentation basin (removal rate decrease) causes only turbulent dispersion. The calculation is carried out using two dimensionless numbers w/w_0 (a Hazen number) and $wZ_0/2D_z$. In wZ_0 and $2D_z$, assuming that the flow velocity is parabolic in the open channel, D_z becomes a constant throughout the entire water depth and, as a practical expression, the horizontal axis is calculated by $wZ_0/2D_z = (120 \text{ to } 140)\ (Z_0/L)\ (w/w_0)$, and the removal rate – which decreases by turbulent flow diffusion when the particles having the considered settling velocity w pass through the pond of length L and depth Z_0 – can be estimated by using the Hazen number w/w_0 as a parameter. D_z is the vertical turbulence diffusivity.

In a real sedimentation basin, the uneven distribution of kinetic energies due to inflow, differences in densities of inflow water, and the presence of wall-friction etc., cause short-circuit currents and stagnant areas in the pond, resulting in a reduction in mean retention time. As a comprehensive representation of these phenomena, the tracer method shown in Figure 6.25 is used. This concept uses the ratio of the theoretical retention time T (retention time of the plug flow) to the actual flow time to pass through the pond, as an index of the degree of short-circuit flow with volume efficiencies from the C curve showing changes in the concentration distribution during the flow process of tracers. The definitions and qualitative significance of these indicators are as follows: a tracer injected into the inflow end of the sedimentation basin flows out at the theoretical retention time given by $T = L/V$ in the case of an ideal sedimentation pond; however, in the case of an actual basin, the tracer disperses due to mixing in the pond, and flows out from the basin exhibiting a time concentration profile like a C-curve with mean detention time $t = \bar{t}$ (the center of gravity of the C-curve). No matter how ideal it is, as long as the water is a viscous fluid, it has a velocity gradient. Therefore, it may be difficult to achieve volume efficiencies of 85% in normal chemical sedimentation and sewage sedimentation ponds, even under the best conditions.

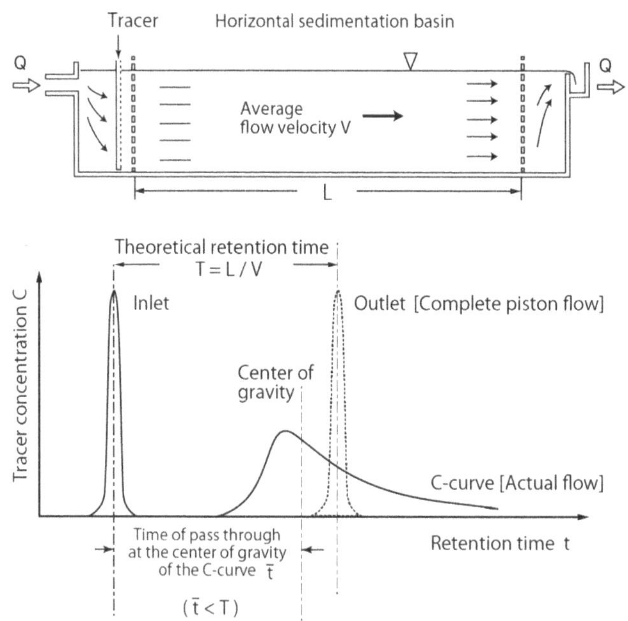

Figure 6.25 A concentration distribution C-curve, using the tracer method applied to a rectangular horizontal sedimentation basin.

There are two typical phenomena that may adversely affect the process: (1) a reduction in the overflow rate due to short-circuit flow, and (2) an inhibition of sedimentation due to an increase in turbulence due to the occurrence of short-circuit high-speed flow and a resuspension of the sediment.

We now consider deflection in the horizontal and vertical cross sections. Flow deviations in a cross section are mainly caused by non-uniform kinetic energy distributions during inflow, and in ponds with large width/length ratios, they are strongly affected by wind blows. As shown schematically in Figure 6.26, deflection occurs either (a) in a manner in which the high-speed portion and the low-speed portion are in parallel, or (b) it increases until a backflow portion is created. In (a), it is sufficient to consider the removal rate as being from parallel sedimentation basins having different overflow rates, which leads to a decrease in the removal rate of particles having a sedimentation rate w ranging from 'Q_2/b_2L' to 'Q_1/b_1L'. Therefore, as Q_1 and Q_2 allocations differ significantly (i.e., the deflection becomes stronger), the overall removal rate decreases significantly. In (b), when the low speed part becomes extreme, reverse flow occurs and circulating flow is created, the closed vortex part becomes a dead area, and the overflow rate of the whole pond becomes $w_0 = Q/BL$ to $w_0' = Q/b_1L$, and the minimum settling velocity w_0' that can be removal increases, the turbulence also increases due to the increase of the horizontal flow velocity, and the removal capacity of the pond is markedly impaired.

Difficulties with longitudinal short-circuit flow are caused by the difference in water temperature between the water in the sedimentation pond and the inflow water, and/or the difference in density due to the clarity of the settled sewage and inflow activated sludge suspension. This is a fundamental issue that cannot be avoided, particularly in horizontal flow sedimentation ponds where high turbidity influents are treated, and it is difficult to theoretically deal with dynamic states in which sedimentation separations are progressing constantly. However, we believe that various experimental observations can be used to approximate the state of density flow. Energy difference between inflow water and clear water in a pond is converted to density flow velocity, and a simple qualitative formula can be obtained by assuming the type of density flow shown in Figure 6.27.

As an example, Nakagawa Tokyo Water Works proposed the experimental Equation 6.26. In ponds with perforated baffle inlet walls, the flow velocity of dense flow can be estimated by considering the values $\alpha = 0.5$ and $k = 1.5$ to 2.0, and the relation $\Delta h = 0.5Z_0$.

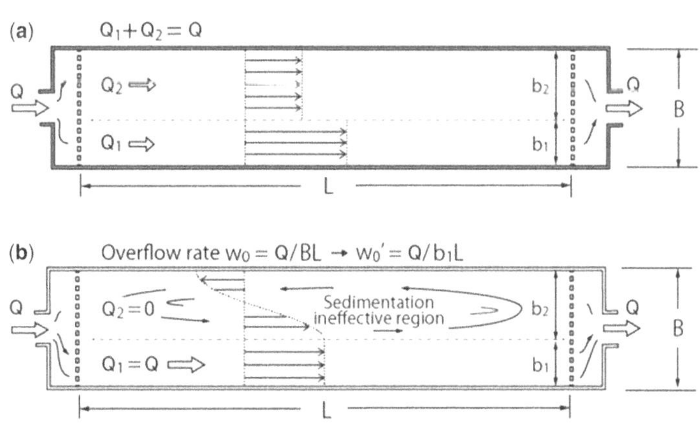

Figure 6.26 Horizontal deflection flow. Deflection occurs either (**a**) in a manner in which the high-speed portion and the low-speed portion are in parallel, or (**b**) it increases until a backflow portion is created.

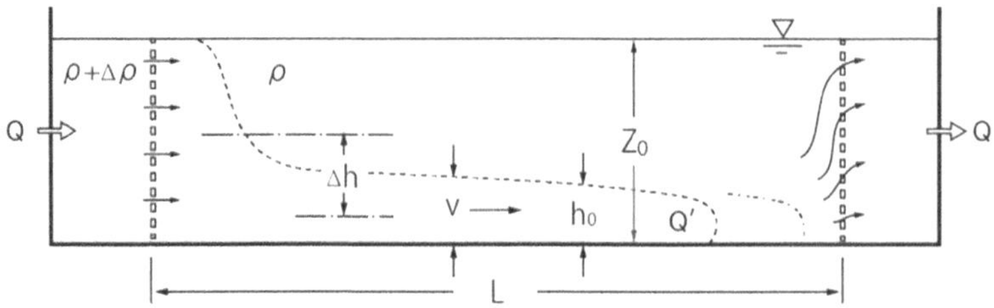

Q : Flow rate Q′ : Density flow rate v : Density flow velocity
ρ : Water density Δρ : Density difference between influent and basin water
Δh : Average sinking height of influent water Z_0 : Basin depth

Figure 6.27 Mechanisms for the occurrence of density flow.

$$v = \sqrt{2\alpha\left(\frac{\Delta\rho}{\rho}\right)\left(\frac{g\Delta h}{k}\right)} \quad \text{—(4·26)} \tag{6.26}$$

v : Density flow velocity (cm/s)
α : The rate at which the position energy of influent water is converted to kinetic energy
Δρ : Density difference between Influent water and basin water (g/cm³)
ρ : Water density (g/cm³)
g : Gravitational acceleration (980cm/s²)
Δh : Average sinking height of influent water (cm)
k : Q′/Q (Q : Flow rate Q′ : Density flow rate)

Equation 6.26 The Nakagawa's Density flow equation.

Tambo proposed an experimental formula (as shown in Equation 6.27), which gives the volume efficiency $\bar{t}/T$ of density flow, after numerous experiments with sedimentation ponds of dimension $Z_0/L = 0.106$ to 0.058 with perforated baffle inlets, Froude number $Fr = V/\sqrt{gL} = 10^{-3}$, $d\rho/\rho = (1.061 \text{ to } 22.6) \times 10^{-5}$

$$\frac{\bar{t}}{T} = 0.02\left(\frac{L}{Z_0}\right)\left\{\log\left(\frac{\Delta\rho}{\rho}\right) - 3.0\right\} + 0.1 \quad \text{—(6·27)} \tag{6.27}$$

$\bar{t}$: Actual detention time of density flow
T : Theoretical detention time
L : Basin length Z_0 : Basin depth ρ : Water density
Δρ : Actual detention time of density flow

[Applicable scope range]

$Z_0/L = 0.106 \sim 0.058$
Froude number $Fr = v/\sqrt{gL} \fallingdotseq 10^{3}$
$\Delta\rho/\rho = (1.061 \sim 22.6) \times 10^{5}$ g : Gravitational acceleration
v : Flow rate

Equation 6.27 The Tambo's experimental formula for the volume efficiency.

Since the overflow rate does not change due to the occurrence of density short-circuit flow occurring in the vertical cross section, the sedimentation efficiency decreases due to the increase in turbulence from the increase in the horizontal flow velocity in the short-circuit flow, and the surface coagulation efficiency decreases due to the decrease in the growth opportunities associated with the decrease in the actual settling distances for the coagulation particles, explained above in Section 6.1.1.14. In addition, the occurrence of a bottom layer short-circuit flow increases the occurrence of bottom sediment scour (see Table 6.6).

Table 6.6 The influence of occurrence of bottom layer short-circuit flow.

Decrease in settling efficiency	Increase turbulence of horizontal flow velocity increase in short-circuit current Reduced growth opportunities of floccurated particles by decreasing settling distance Scouring and scattering of sludge layer
Overflow rate	No change

An idealized flow sedimentation basin is considered to be the perfect extrusion (plug) flow for a pond water stream, and plug flow models exhibit the highest removal rate. On the other hand, complete mixing models assume the worst water flow conditions in which inflow water mixes and diffuses instantaneously throughout a pond (up and downstream, top and bottom layers). Now, assuming a basin with severe turbulence due to wind, density flow, inflow drift, etc., and considering the worst-case water flow in which a pond is completely mixed, the concentration of water in a pond and the concentration of water flowing out will be equal to each other. Assuming a sedimentation removal of a pond/tank with an overflow (surface load) rate $w_0 = Q/A$ under these worst-case conditions of thorough mixing, the removal rate of the pond/tank is shown in Equation 6.28.

$$r = 1 - \left\{\frac{1}{1 + w/(Q/A)}\right\} \quad \text{—(6·28)} \tag{6.28}$$

r : Removal ratio
w : Particle settling velocity
Q : Flow rate
A : Surface area of tank

Equation 6.28 The removal ratio of complete mixing tank.

When n such tanks/ponds are arranged in series with a semi-complete mixing flow, it is referred to as a mixing model in which the extrusion flows are mixed thoroughly in each of the compartments. If n such thoroughly mixed ponds are arranged in series (n ponds continuous thorough mixing models), the residual rate of the first pond is given by Equation 6.29. In addition, if connected in series up to the nth pond, the removal rate r(n) of the n-continuous complete mixed sedimentation ponds in series is given by Equation 6.30.

Residual rate of the first pond from Equation 6.28 is

$$(1 - r)_1 = \left\{\frac{1}{1 + w/(Q/A)}\right\} \quad \text{—(6·29)} \tag{6.29}$$

Equation 6.29 The residual rate in the first complete mixed tank.

The removal ratio when ponds are connected in series up to the n th is

$$r(n) = 1 - \left\{\frac{1}{[1 + (1/n)w/(Q/A)]^n}\right\} \quad \text{—(6·30)} \tag{6.30}$$

Equation 6.30 The removal ratio of continuous complete mixing model sedimentation pond.

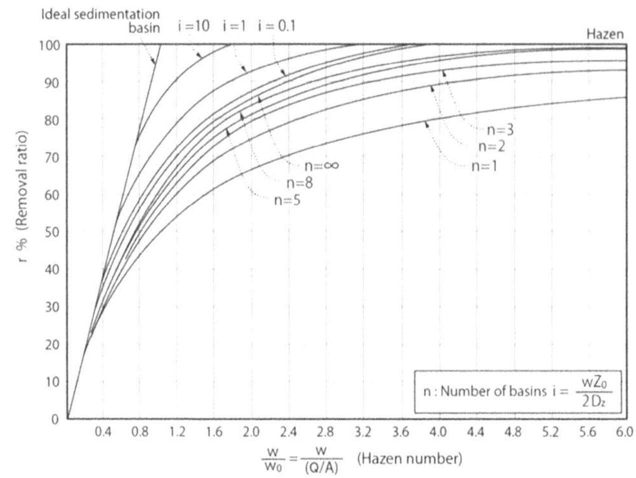

Figure 6.28 Hazen numbers and removal ratios under various water flow conditions.

An infinite number of ponds/tanks in series is composed of $n = \infty$. This means a pond that mixes completely only in the vertical (depth) direction without mixing at all in the (plug) flow direction. Ponds with $1 < n < \infty$ indicates they remain mixed vertically, and Fair and Geyer (1956) rated basins as $n = 1$ (bad), $n = 2$ (slightly bad), $n = 4$ (good), $n = 8$ (excellent), and $n = \infty$ (top rank). The fact that $n = \infty$ corresponds to the fact that the dimensionless number $wZ_0/2D_Z$, (which shows the magnitude of the disturbance in Figure 6.23) is nearly zero. In addition, the condition under which the dimensionless number $wZ_0/2D_Z$ is set to infinity indicates an idealized flow sedimentation basin. Figure 6.28 shows these relationships.

6.1.1.16 Types of horizontal flow sedimentation basins and substructures

There are two typical types of horizontal flow sedimentation basins: parallel flow rectangular sedimentation basins and radiant flow circular sedimentation basins.

An example of a parallel flow rectangular sedimentation basin is shown in Figure 6.29, and they are typically constructed in width-to-length ratios of 1:4 to 1:5. The greater the ratio of length to pond width, the more stable the horizontal flow conditions. Some sedimentation ponds with slow sand filtration systems are as long as 100 m. However, a length of 30–40 m is common in chemical (coagulation) and biological floc sedimentation basins. The width of the pond is usually around 10 m, which is often influenced by the usable width of the sludge scraper. The depth of the pond is usually 3 to 4 m, taking into consideration planned component depths, such as the flocculator and aeration agitator. Shallower and longer ponds are less susceptible to density flow. However, shallower ponds have to be considered to have negative effects, such as increased horizontal flow velocity even at the same surface load rate, resulting in increased turbulence and scouring, as well as a reduced positive flocculant effect due to reduced settling depth with respect to the coagulated particles.

Figure 6.30 shows an example of a radial flow circular sedimentation pond; such ponds are usually as large as 30 m in diameter. Inflow is from the center of the pond, and water flows radially to the periphery, flowing out from the circumference. In a circular pond (and also in a square pond with center inflow), the flow becomes unstable because the streamline interval gradually spreads to the downstream

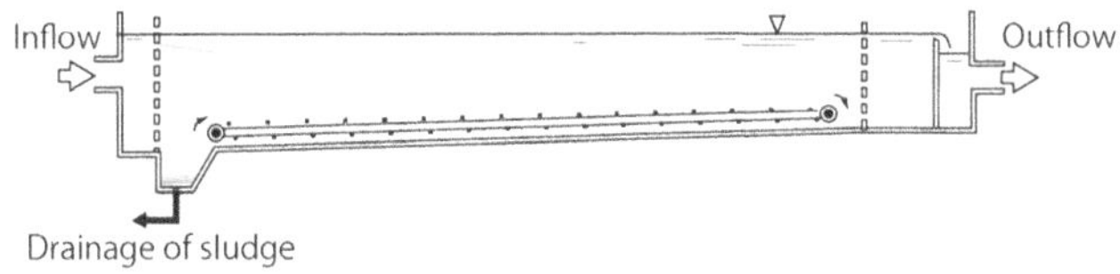

Figure 6.29 An example of a parallel flow rectangular sedimentation basin.

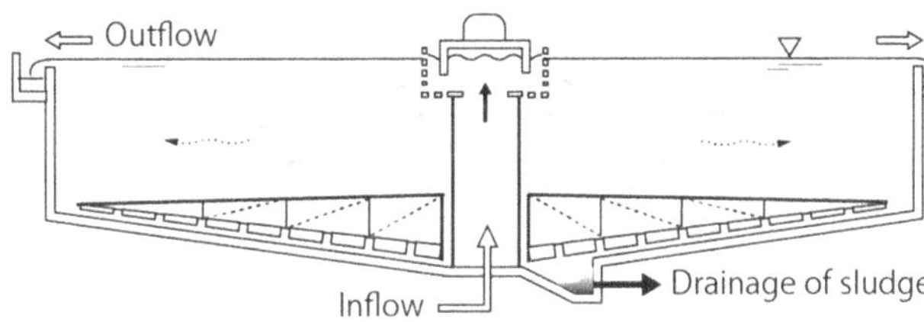

Figure 6.30 An example of a radial flow circular sedimentation basin.

side (divergent flow). In addition, if there are no flow guide walls opposing the horizontal direction deflection, the current will be deflected by the blowing action of winds. In addition, a circular pond generally uses a rotating sludge scraper, as shown in Figure 6.30. Therefore, application to low concentration clarified sedimentation is not recommended because a rotating flow is constantly generated and the flow conditions become unstable. Circular ponds are widely used in treatment systems that generate large amounts of sludge, such as sewage and industrial wastewater treatment, with the benefit of easy scraping and discharging of sludge in high-concentration thickening sedimentation ponds. In addition, thickening ponds with higher suspensions are supposed to have an upward flow, which requires a different approach than horizontal flow sedimentation ponds (see the description of an upflow sedimentation pond below). Circular pond diameters are often limited by the length of the sludge scraper arm.

In order to maintain high sedimentation efficiencies (even closer to an idealized sedimentation pond), a number of measures are taken with regard to the inflow and outflow. To be highly efficient, the inflow structure (inlet) must meet the following requirements: (1) the flow rate and suspension load should be uniformly distributed to a plurality of sedimentation ponds; (2) the water and the suspension load in the inflow to each pond should be uniformly distributed across the whole cross section of the pond, so as to make the water flowing down the sedimentation pond as close as possible to the parallel velocity flow; and (3) turbulence in the pond caused by the non-uniform velocity gradients of the inflow should be minimized, and any turbulence that occurs in the pipes and conduits before the inflow port to the sedimentation pond should be introduced as little as possible and as weakly as possible, so that sedimentation phenomena are not reduced by turbulent diffusion.

In order to distribute load uniformly between two or more sedimentation ponds, which is the first requirement, the flow is divided (as shown in Figure 6.31a) as proposed by Fair and Geyer (1956), and the flow of water reaches each inflow port (A, B, C and D) with the same head loss (equal distance), or the head loss is reduced as much as possible when the inflow (lead) channel is transported from D towards A (as shown in Figure 6.31b), and the head loss of each inflow port (A, B, C, and D) is designed so as to show as uniform head loss as possible (Table 6.7).

Assuming that the velocity of the head of the inflow channel (between D and A) is negligible, the head loss of the nth inflow opening (an inflow opening or porous plate) is given by Equation 6.31a so that the flow

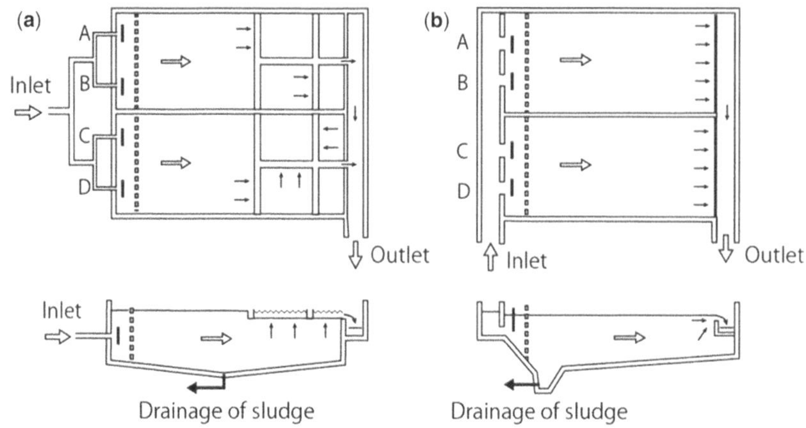

Figure 6.31 The equalization of inlet and outlet flow at each opening.

Table 6.7 Inlet structure requirements.

Flow load	(i) Even distribution of load to each basin
Flow status	(ii) Piston flow in whole cross section (iii) Reduce the turbulence in the basin

q_n at the n-th inflow port remains a factor of m ($m < 1$) of the first inflow q_1. The inflow head from the first inflow port to the nth inflow port is $h_n = h_1 - h_0 = m^2 h_1$ (Equation 6.31b), and the relation between the head loss at the first and nth inflow ports is given by Equation 6.31c, using the ratio of the head loss h_1 at the inflow opening to the head loss of inflow channel h_0 (from the first opening port to the n-th port) to give the ratio of the uneven inflow from the first port to the n-th port.

The head loss h_n of the nth inlet is

$$h_n = kq_n^2 = k(mq_1)^2 = km^2h_1 \quad (6\cdot31a)$$

Assuming that the head loss between the first and the nth port is h_0

$$h_n = h_1 - h_0 = m^2 h_1 \quad (6\cdot31b)$$

$$h_1 = \frac{h_0}{1 - m^2} \quad (6\cdot31)$$

q_1 : Flow rate of the first port
q_n : Flow rate of the nth port
m : Value of inlet flow nonuniformity difference ($m<1$)

(6.31)

Equation 6.31 The inflow orifice and channel head loss.

The second requirement – the structure of the inflow port to bring the flow conditions of individual basins closer to the plug (extrusion) flow – can be considered in various ways. However, the theoretical approach is extremely difficult. Here, we will try various model experiments using the Froude number as a similarity factor, and refer to the tracer test results etc. to select an appropriate form. Figure 6.32 shows an

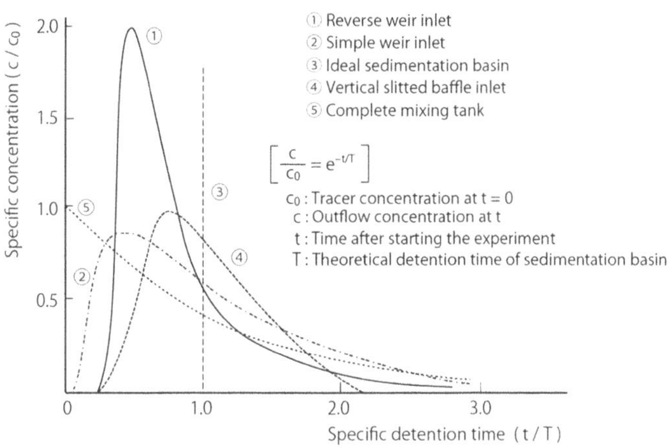

Figure 6.32 Examples of time-tracer density response curves (C-curve).

experimental example of a 'time-tracer density response curve (C-curve)' in which a variety of inflow structures at present used in sedimentation basins with the same surface load rate were examined. Tracers were injected at the inflow port and measured at the outflow end. The inflow of the simple overflow weir shown at (2) on the graph shows flow conditions that are not very different from those of the complete mixing pond, indicating a poor inflow pattern. The inverted weir inlet (a reversed weir) indicated at (1) has narrow peaks, and is close to the extrusion flow with less mixing, indicating the best flow among the structures considered.

A third goal is to keep the degree of turbulence in a basin as low as possible. The turbulence caused by inflow is as a result of the mixing between the inflow jet and the in-pond hydrostatic water. Using perforated baffle walls, the inflow jets immediately interfere with the neighboring jets, mix with each other, and lose their jet characteristics, i.e. their directional turbulence, and turn into a state of isotropic turbulence. The attenuation of turbulence fluctuation velocity under these conditions is shown in Equation 6.32.

$$\frac{V}{\sqrt{v'^2}} = \frac{5X}{SA} + B \qquad \text{—(6·32)}$$

V : Average flow velocity of pond
$\sqrt{v'^2}$: Turbulent fluctuation velocity
S : Center spacing of holes in the baffle wall
X : Distance from the baffle wall
A,B : Coefficient determined by wall opening ratio and hole shape

(6.32)

Equation 6.32 The attenuation of turbulence fluctuation velocity under isotropic turbulence.

It can be seen from this equation that (1) turbulence fluctuation velocity decreases rapidly in the vicinity of perforated baffle walls and decreases with distance, and (2) the turbulence intensity decreases with decreasing baffle wall hole interval and the attenuation rate increases. This is a fundamental finding for inlet baffle wall structure design, and it shows the disadvantage of inlet openings in which a few thick tubes are installed, and also shows semi-quantitatively the advantage of making holes as small as possible and distributed uniformly, as long as the wall opening ratio is constant.

Rather than having a small number of openings at the outlet, overflow weirs can draw water uniformly across the entire width of a pond, thereby greatly improving the flow pattern of the pond. When an outflow weir is installed across the entire surface of the outflow end of a rectangular sedimentation basin, the flow condition is as is shown in the flow diagram in the top part of Figure 6.33.

Consider two types of particles, (a) and (b), which inflow to the top of these basins. Particle (a) with a settling velocity w_a reaches the pond bottom at point a in the idealized sedimentation basin. Considering the effect of the outlet as shown in the streamline diagram, the settling lines traveling at the combined vector of w_a and flow velocity v is affected by the upward convergence of the streamline in the outflow section, and gradually reaches the pond bottom at distance L (the total length of the pond) to the point a' at the pond end, which is the limiting point at which particle (a) can gradually settle to the pond bottom. When this pond is considered to be an idealized sedimentation pond, particles reaching the pond bottom at length L of the sedimentation zone ($w_0 = Q/A$, equivalent to the overflow rate) will flow out if the upward divergence rate to the outflow zone is greater than the settling velocity. So we need to know the magnitude of the sedimentation rate w_a for 100% removal in this pond. Figure 6.33 can be considered as an idealized sedimentation pond (overflow rate $w_a = vZ_0/L_a$) with a small surface area, with an ineffective part $\Delta A = (L - L_a) \times B$ (B is the pond width). (It should be noted that the size of the ineffective part of an actual sedimentation basin varies greatly depending on the settling velocity w).

Regarding the size of such ineffective portions of sedimentation ponds, the US National Research Council (NRC Subcommittee on Sewage Treatment, 1948) proposed an experimental formula, shown in Equation 6.33, for the overflow weir outflow regions of a rectangular pond.

$$\Delta A = B(L - La) = K \cdot \frac{q^2}{w^2} \quad \text{—} (6 \cdot 33) \tag{6.33}$$

ΔA : Surface area of ineffective part
B : Width of basin
L : Settling zone length of basin
La : Distance required for settling of w particles in ideal sedimentary basin
q : Weir load (flow rate per unit width of weir)
w : Settling velocity of the particle considered
K : Constants (0.35 to 0.55 depending on the shape of basin)

Equation 6.33 The surface area by the ineffective part of horizontal sedimentation basin.

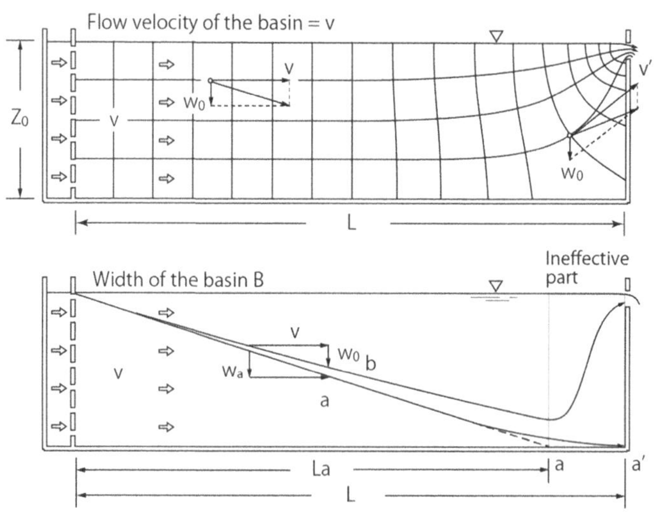

Figure 6.33 Flow diagram of the outflow end of an idealized sedimentation pond.

As can be seen from this equation, the smaller the weir load, the smaller the outlet area, the smaller the impact on the reduction of the settling removal rate. Therefore, many attempts have been made to reduce weir load by installing a composite overflow weir at the end of the sedimentation basin, as shown in Figure 6.31(a). However, even in such cases, the streamlines toward adjacent weirs interfere with each other and, therefore, even if many weirs are provided close to each other, it is not possible to greatly improve the sedimentation effects. This type of weir is widely used when high concentration suspensions collectively settle out from the inflow and spread over the whole surface of a pond due to the effects of densities, as in the case of the final sedimentation pond in the activated sludge system in a sewage treatment plant, and zone settling sedimentation is performed. This is the outflow weir structure when the first half of a pond is considered to be the forming section (by density current) and the second half is considered to be the (vertical) settling section. In a sedimentation pond at a water purification plant, the structure shown in Figure 6.31(b) is normally taken and no excessive demands are made for reducing the weir load.

Since the weir load is smaller in a circular sedimentation pond with radial flow than in a rectangular sedimentation pond with parallel flow, and the destabilization due to wind which occurs in diffuse radial flow can be compensated for, circular ponds are widely used for zone sedimentation and thickening ponds where the controlling effect of weir load is large when taken together with the ease of sludge discharge.

In addition to the inflow/outflow structures described above, various basin structures have been devised to bring pond water streams as close to plug flow as possible and to minimize turbulence. The effects of these structures can be explained using three indices: the overflow rate of the pond (w_0), the Reynolds number (Re), and the Froude number (Fr) (Table 6.8).

Reducing the overflow rate reduces the minimum particle size (sedimentation rate) that can be removed in a pond, and reducing the Re number suppresses turbulence. In addition, the Fr number indicates the magnitude of the density current and the stability of the pond. The larger the Fr number, the higher the volume efficiency t/T (where t is the actual retention time, and T is the theoretical retention time), and it is recommended that the Fr number should be 10^{-5} or more (Fisherström, 1955). Since an increase in the Fr number is generally accompanied by an increase in the Re number, the harmony of the combination of two parameters is important. Some typical examples of flow improvement follow below.

The vertical flow guide wall (training wall) shown in Figure 6.34 aims to obtain a horizontal parallel flow by preventing horizontal deflection due to inflow inhomogeneity. The aim is to increase the width–length ratio of the pond, increase the Fr number, and decrease the Re number in order to improve the flow situation. Flow deflection can be prevented in many cases by installing a training wall in the first half of the pond.

Table 6.8 Evaluation indexes for sedimentation flow.

Overflow rate	$w_0 = \frac{Q}{A}$	The smaller value, the smaller particle diameter that can be removed	Basic index of removal ratio evaluation
Reynolds number	$Re = \frac{vR}{\upsilon}$	The smaller value, the less disturbance in basin	Index of Laminar/Turbulent flow
Froude number	$Fr = \frac{v^2}{gR}$	The larger value, the higher capacity efficiency	Index of Subcritical/Supercritical flow

Q: Flow rate; A: Pond surface area; v: Pond avarage flow velocity; R: pond hydraulic radius; *v*: Kinematic viscosity; g: Gravitational acceleration.

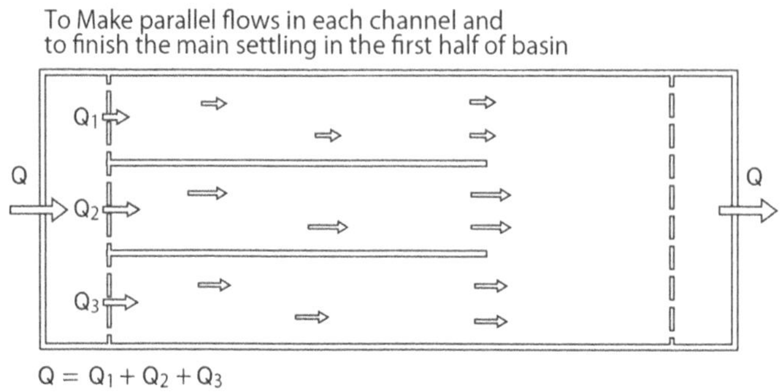

Figure 6.34 A sedimentation basin with a vertical training wall.

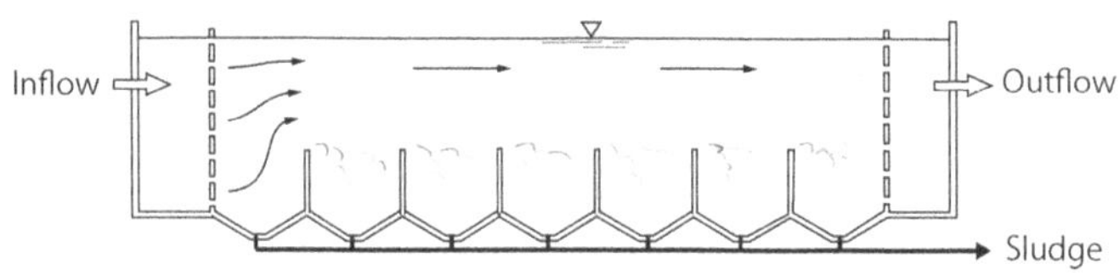

Figure 6.35 A sedimentation basin with baffle plates at right angles to flow.

As shown in Figure 6.35, a large number of baffle plates, placed at right angles to the flow, can be installed at the bottom of a sedimentation pond to enhance sedimentation effectiveness. Even with the nstallation of baffles, the overflow rate remains unchanged. Even if turbulence increases due to the increase in the horizontal flow velocity inside the pond, the maintenance of actual retention time, and the avoidance of sediment scouring and resuspension of sediments is expected by suppressing high-speed short-circuit density flow running on the pond bottom. Experimental results, as shown in Figure 6.36, demonstrate the effectiveness of bottom baffles. Sludge discharge hoppers can be used between the baffles. The height of the baffles should be minimized to ensure sufficient free-flow sewerage depth.

As shown in Figure 6.37, a porous wall is created at right angles to the flow in the middle of a sedimentation basin, and the pond is divided into a series of smaller ponds (e.g., the Higashimurayama Water Purification Plant in Tokyo). Basically, this method may be considered to be a pond in which the

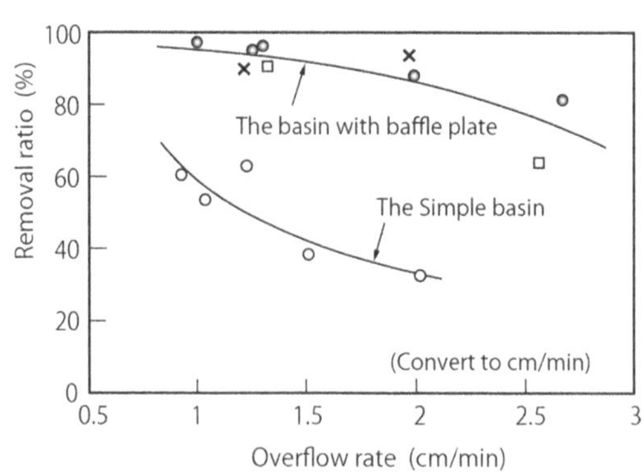

Figure 6.36 Hyden's example of the effectiveness of bottom baffles.

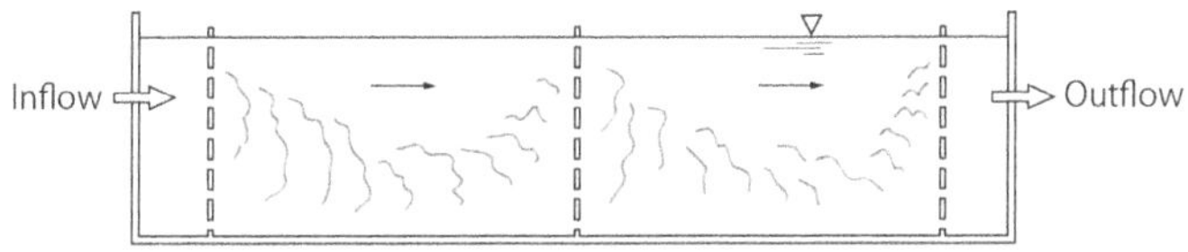

Figure 6.37 An intermediate perforated baffles sedimentation basin.

mixing model shown in Equation 6.28 is arranged in series, and it is considered that deflection of short-circuit flow occurs in the first pond and flow is introduced to the next pond by weakened deflection. At the intermediate percolated walls, an up-and-down mixing flow is always generated, but even in this situation, sedimentation is dropped in the first half of the pond and a suspension system with a small density difference leads to the next stage, so that the generation of a strong density flow can be remarkably reduced.

By inserting n intermediate beds into a parallel-flow horizontal sedimentation basin, as shown in the two-layer sedimentation pond in Figure 6.20 above, the overflow rate of a pond can be reduced to 1/(n + 1) and greatly improve the removal of small particles of low settling velocity in the same volume of pond. Figure 6.38(a) shows how increasing the number n of beds improves the removal efficiency (overflow rate) of particles. This shows the change in removal rate with an increasing number of layers. Figure 6.38(b) shows the relative size of the smallest particles with 100% removal on each layer as the constant particle density. A slant plate (tube) is also represented (described below).

As can be seen from Figure 6.38, settling efficiencies improve markedly at 0 to 1 intermediate bed. A multi-storey sedimentation pond is shallower and longer in depth than a simple pond, while maintaining the same horizontal flow velocity. This results in a pond with a small Re number and a large Fr number, which greatly improves removal efficiency. However, the removal of sludge and keeping the equal (even) distribution of flow rates to each layer is difficult, and structures to remove sedimented sludge and to equalize the inflow need to be devised. Two-storey sedimentation basins are widely used.

In order to make sludge discharge easy and to realize a multi-storey, hierarchical sedimentation basin, a sedimentation basin with a slanted flow wall, as an intermediate type, was conceived, as shown in

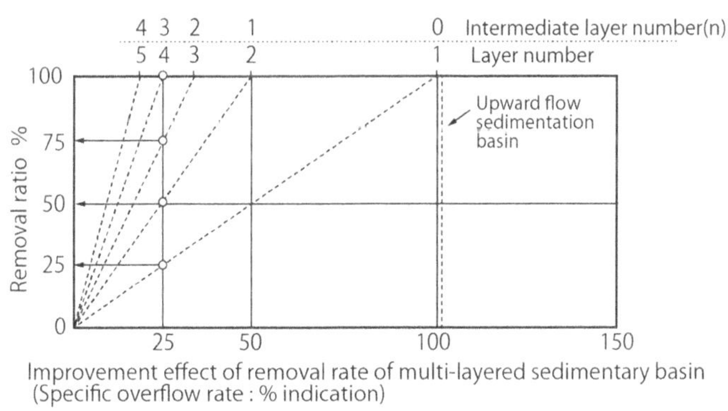

Overflow rate (Settling velocity) 25% example	Story number	5	4	3	2	1
	Removal ratio (%)	100	100	75	50	25

Figure 6.38 Ideal sedimentation basin removal rates for (**a**) a multi-layer horizontal flow and (**b**) an ideal sedimentation basin.

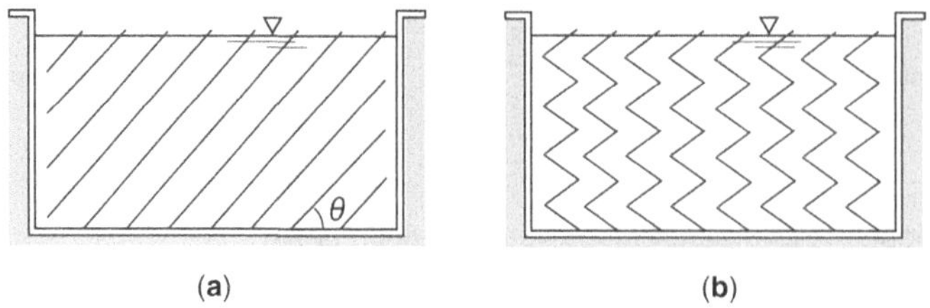

Figure 6.39 A sedimentation basin with slanted flow walls.

Figure 6.39. The overflow rate can be considered as an oblique intermediate layer 'n cos θ' (θ is the inclination angle of the n intermediate layer), and a pond in which θ is about 30 to 60 degrees can be produced. Such a basin could be referred to as a multi-level slanted plate sedimentation basin. Since the practical range of this pond is limited when the incline plate is made as shown in Figure 6.39a, the mid-layer is often folded into a squared shape as shown in Figure 6.39b. In the 1960s, Yanase (1964a, b) introduced a folding inclined multilayer system into conventional horizontal flow sedimentation ponds in Chousi City, Japan. Since then, it has been widely used all over Japan because of its high effectiveness, and has been described in the design guideline of Japanese Water Supply facilities as a 'slanted plate sedimentation pond'.

In recent years, for clarified sedimentation, a tilted tube sedimentation basin or a tilted plate sedimentation pond are often used, in which flowing water is divided into small portions (channels) so that the flow becomes laminar.

In a horizontal laminar flow sedimentation basin, commonly referred to as a slanted plate sedimentation pond in Japan, as shown in Figure 6.40, the flow of horizontal water and the flow of sludge falling to the

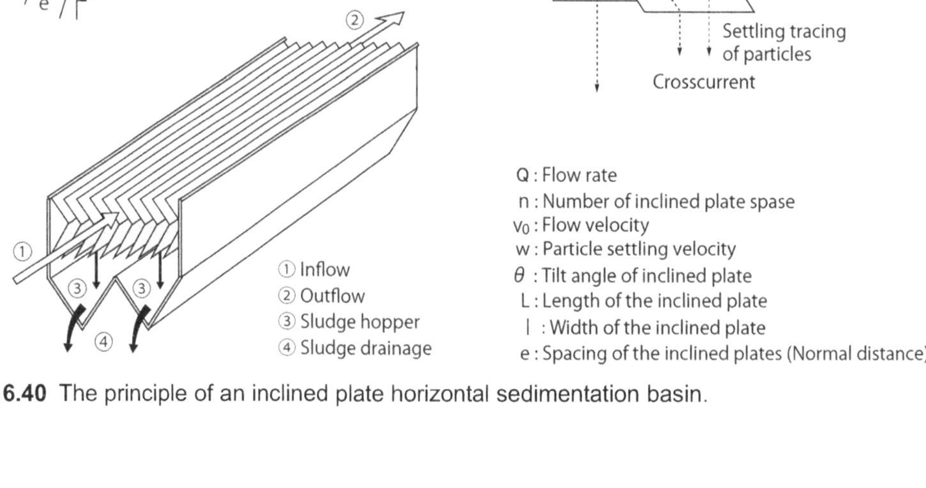

Figure 6.40 The principle of an inclined plate horizontal sedimentation basin.

pond bottom when sedimentation sludge slides on the upper surface of the slanted plate are perpendicular to each other. In this case, the overflow rate of the incline plate, which is the reference value when the suspension is settled and separated, is as in Equation 6.34.

In order to be a laminar sedimentation basin, the Re number in the channel must be $Re = V_0R/\nu \doteqdot V_0\, e/\nu < 2300$ (pipeline) and 5000 (parallel plate channel). In the case of inclined plates, a channel of $R = 2e$ is considered to be the characteristic length (cm), where R is the depth of the channel between the incline plates; e is the interval between the slant plates (cm); V_0 is the horizontal flow velocity of the sedimentation pond (cm/s); and ν the kinematic viscosity of water (20°C, 0.0010 cm^2/s). Since the limit Re number for water channels to be laminar between parallel plates is considered to be 5000, if the horizontal flow velocity V_0 is considered to be about 0.2 cm/s as a common sense value, the condition $Re = 0.2e/0.0010 < 5000$ must be satisfied in order to achieve laminar flow conditions, and the interval e of the inclined plates must be 25 cm or less.

$$W_0 = \frac{Q}{n \cdot L \cdot l \cdot \cos\theta} \quad \text{—— (6·34)}$$

Q : Flow rate
n : Number of inclined plate sections
L : Length of inclined plate
l : Width of inclined plate
θ : Tilt angle of inclined plate

(6.34)

Equation 6.34 The overflow rate of inclined plate one section.

In a horizontal flow inclined plate sedimentation basin, a density short circuit flow is easily generated. Therefore, measures such as the installation of inlet and intermediate perforated walls, as shown in Figure 6.41, and short-circuit cutout baffle walls in the bottom sludge sedimentation zone are required. These horizontal flow systems are vulnerable to dense flow, so they are not used for sewage and wastewater treatment with high concentrations of suspensions, but are used exclusively for clarifying water with low sedimentation concentration. In addition, when the high concentration of suspended matter and settled sludge that has settled on the upper surface of an inclined plate are moved downward due to the difference in density, the clarified water is moved upward as a countercurrent, and a swirling flow is generated between the inclined plates. The authors experimentally reported that the generation of cross-sectional circulating flow between inclined plates near to hydrostatic water makes it difficult to achieve a theoretical overflow rate, which is a weak point of horizontal flow inclined plate sedimentation basins.

A counter-current type of laminar sedimentation basin that separates under ascending flow through an upwardly inclined plate or an inclined tube is probably the most stable and highly efficient sedimentation

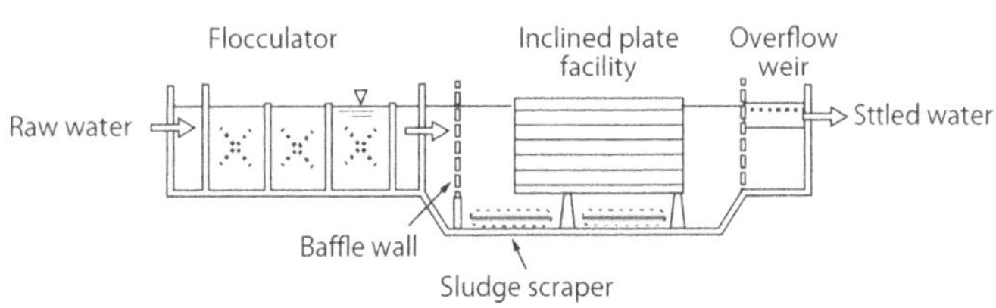

Figure 6.41 The basic system of an inclined plate horizontal sedimentation basin.

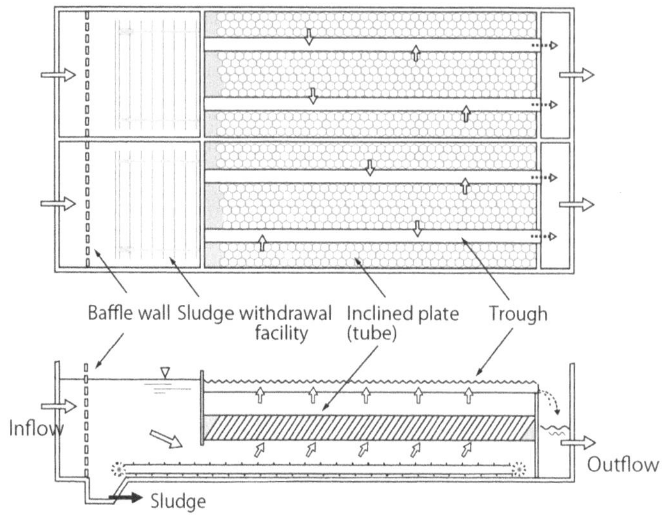

Figure 6.42 An inclined plate (tube) sedimentation basin with sludge withdrawal facility.

device among the various types of current gravitational clarification sedimentation ponds. A variety of pond structures are conceivable, but if we consider the development of a large conventional horizontal flow rectangular sedimentation basin, it would take the form shown in Figure 6.42. Even if a dense stream of inflow water flows into the lower part of a sedimentation basin, except for the limited first half of the inflow zone, the stream comes into inflow with upstream sedimentation separation blocks which turn the flow up at right angles along almost all the tank length. Upwardly inclined plates or inclined tubes are mounted on the whole surface of the second half of the pond, and water travels upward through the inner channels of the incline plates or the slanted tubes with a small cross section so that separation can proceed in a laminar flow condition. The suspension is brought into sedimentation with the lower channel of the slanted plates/tubes, and the settled sludge becomes a counter current that slides down the surface of the inner channel against the water flow by gravitational force. The coagulated sediment sludge from the slant plates/pipes slides down the inner channel, becomes a dense agglomerate due to the synergistic effects of coagulation and pelletization, settled down through the bottom channel in a short time, and accumulates on the pond bottom.

The slant plates/tubes for laminar flow separation consist of parallel slant plates of 8 to 10 cm intervals inclined at an angle $\theta = 60$ degrees, or slant tubes of about 100 cm in length, in square sections. The slant plates/tubes are formed into blocks as shown in Figure 6.43, and are arranged/installed on the support racks of the sedimentation basins.

The overflow rate w_0 of the separator band of an upstream sloping tube sedimentation pond is given by Equation 6.35, referring to Figure 6.44.

Generally, the cross-sectional size e of an inclined tube is about 5 to 10 cm, the angle of inclination ά is about 60 degrees, and the height of the inclined tube blocks is about 100 to 120 cm. The water flow in an inclined pipe is not an extruded flow, but a flow velocity distribution as laminar flow is formed temporarily from the inflow end of the lower part, and the flow velocity distribution is not uniform in the pipe. In addition, there is an interface disturbance that occurs between the sediment sludge sliding down the lower surface of an inclined pipe and the countercurrent flow of water. In actual design, the overflow

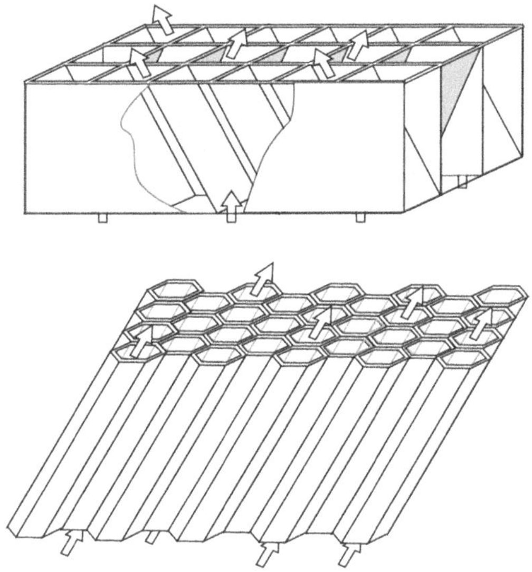

Figure 6.43 An example of an upflow inclined plate (tube) facility.

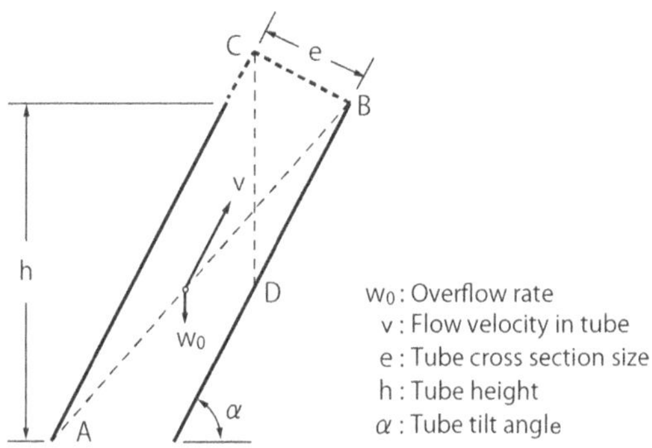

Figure 6.44 An upflow inclined tube.

rate must be considered by correcting the deviation from the ideal condition. A safety factor of 0.7 to 0.8 is multiplied, and the overflow rate is discounted to obtain the required horizontal cross-sectional total block-laying area (the sedimentation zone).

$$w_0 = \frac{Q}{A} \cdot \frac{e}{h\cos\alpha + e\cos^2\alpha} \quad \text{— (6·35)}$$

w_0 : Overflow rate
Q : Flow velocity in tube
A : Horizontal area of inclined tube
e : Tube cross section size
h : Tube height
α : Tube tilt angle

(6.35)

Equation 6.35 The overflow rate of one section in upflow inclined tube facility.

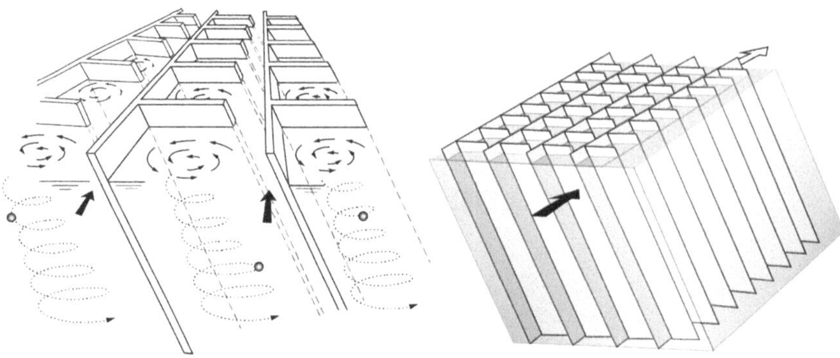

Figure 6.45 A horizontal flow finned channel separator.

In the early 1980s, K. Hashimoto *et al.* (1982a, b, c, 1983) at Suido-Kikou Co., Tokyo, proposed a prototype for a new slanted plate sedimentation pond. The theoretical analysis and rational design procedures for the dynamic separation was performed through joint research with Tambo's Laboratory at the Hokkaido University. This was the world's first sedimentation device with a separation rate which exceeded the overflow rate theory. Hashimoto was made Doctor of Engineering at Hokkaido University as a result of this study.

This separation process is based on a cross-flow incline plate sedimentation pond. The typical construction of this sedimentation unit with horizontal flow fins (finned channel separators) is shown in Figure 6.45, in which a large number of fins are erected on the bottom surface of the horizontal flow slanted plate at regular intervals and at right angles to the flow direction. Figure 6.46 shows a schematic representation of the flow of a finned channel that cuts a portion of the flow in a plane perpendicular to the incline plate surface. Upflow finned channel separators are also constructed.

Figure 6.47 shows an example of how the finned channel separator differs from the underlying tilted plate sedimentation basin. Cross-flow finned incline plates clearly improve the rate of separation, and there is an addition of kinetic separation mechanisms that cannot be explained by surface load (overflow rate) theories.

Now, let us consider the mechanism by which much higher-rate separation can be achieved, far exceeding sedimentation based on overflow rate theories under laminar flow. It can be assumed that, as shown in Figure 6.46, the flow after a laminar wake passes through fins and the circulating flow induced between fins may both play a major role in improving sedimentation efficiencies. For this reason, the flow structure in the fin-unit compartment of the separation channel, which is considered to drive new dynamic separation, is presented schematically in Figure 6.48, with reference to the results of multiple

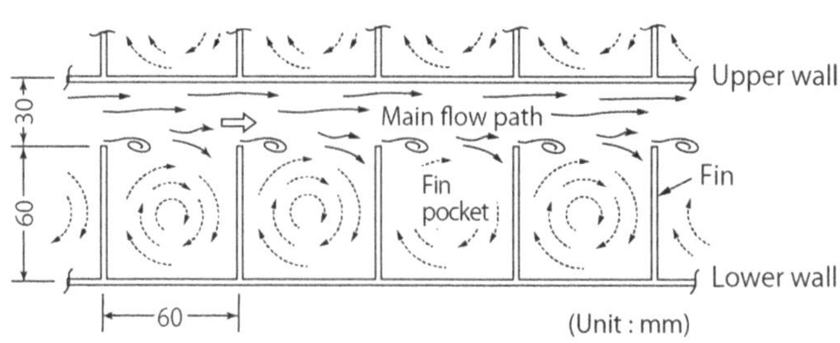

Figure 6.46 A longitudinal section of a horizontal flow finned channel separator.

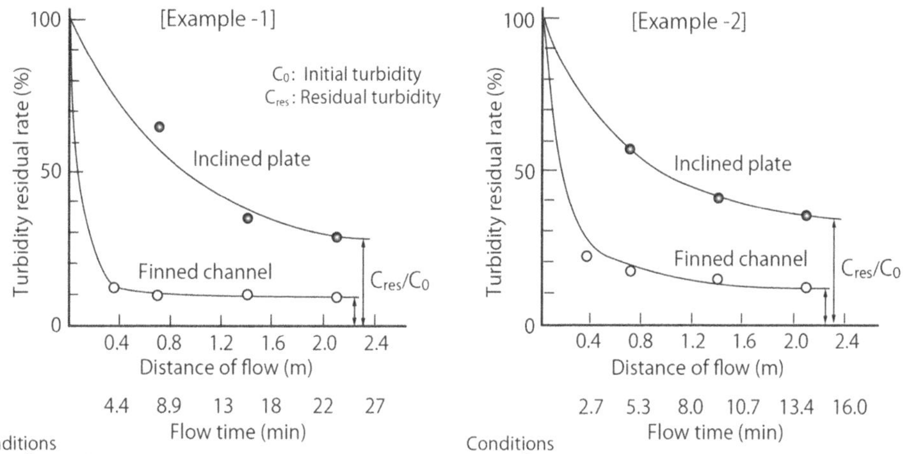

Example -1: Conditions — Raw water

Temperature	6.5 ℃	Treated water amount	150 m^3/d
pH	7.0	Superficial horizontal velocity	0.09 m/min
Alkalinity	18 mg/l	Kaolin dosage	- mg/l
Turbidity	13 度	PACl dosage	20 mg/l

Example -2: Conditions — Raw water

Temperature	9.0 ℃	Treated water amount	250 m^3/d
pH	6.9	Superficial horizontal velocity	0.15 m/min
Alkalinity	14 mg/l	Kaolin dosage	50 mg/l
Turbidity	4.4	PACl dosage	20 mg/l

Figure 6.47 Comparative experimental results for an inclined plate and finned channel separator.

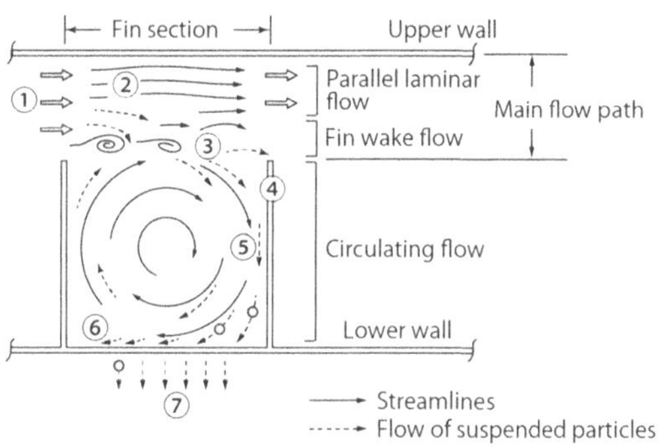

Figure 6.48 Schematic diagram of suspended particle separation in a fin section.

measurements of local flow velocity distribution using a microflow velocimeter and the results of flow visualization experiments using a hydrogen-bubble method.

Based on this schematic flow structure, we consider that the process of separation proceeds in seven stages:

(1) The suspended particles coming from the preceding fin section inflow to the fin section of interest and are divided by a section inflow to a parallel laminar flow area (Poiseuille flow) near the upper wall surface of the main flow path and to a section separated from the main flow path, and then directly entrained to the fin wake section.

(2) The suspended particles which inflow into the parallel laminar flow region settle down under the effect of gravitational force, and some of the suspended particles reach the fin wake vortex train. Separation in this part is evaluated in Hazen numbers (w/w_0) according to overflow rate theory.

(3) Suspended particles which were entrained in the fin wake region of the preceding section and gravitationally settled from the laminar flow region are now entrained in the rotating eddy flow between fins, and transported to the vicinity of the upstream surface of the downstream side of the fin.
(4) Suspended particles transported to the vicinity of the downstream fin go around the section between the upstream and the downstream fin (fin pocket) on a vortex that periodically turns around and up and down. By this action, some particles go to the next fin section and some remain in the fin pocket, transported by circulating flow.
(5) The suspended particles remaining in the fin pocket rotate on the flow driven by momentum from the main flow and move to the vicinity of the bottom wall surface by the gently circulating Rankin vortex laminar flow.
(6) The suspended particles that are transported on a slow velocity circulating (swirling) flow (having a maximal flow rate of about 20% of the parallel laminar flow rate of the main channel) in the vicinity of the bottom wall surface are under a condition that can be considered to be a laminar flow sedimentation basin, having a smaller overflow rate of about one fifth of the main flow portion, and is rapidly deposited by gravitational sedimentation and removed from the water system.
(7) The deposited particles (sludge) slide down to the bottom of the inclined fin pocket and quickly migrate to the sludge pool.

Figure 6.49 illustrates the process of separating high concentrations of non-coagulative silt particles by depicting the states of concentration reduction and particle size distribution. By adding a kinetic separation process to a laminar flow sedimentation pond (an incline plate sedimentation pond), separation exceeding the overflow rate theory can be stably obtained. See the original paper for flow visualization experiments, removal process modeling, and design theories by Hashimoto *et al.* (1982a, b, c, 1983).

The mechanism of a finned inclined plate sedimentation device has been explained in some detail. This separation mechanism is similar to removal by granular bed filtration, and sedimentation and filtration using these two systems could be considered as a series of sequential operations. We are considering this approach to build theories on the ideas of a series of continuously operating solid–liquid separation processes, transforming the conventional idea of design/operation as a two-step process in which sedimentation and filtration are clearly separated from each other.

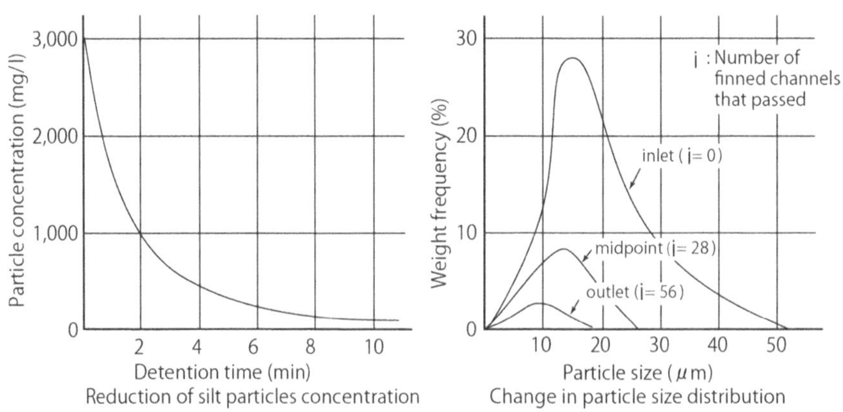

Figure 6.49 An example of separating concrete aggregate using a horizontal flow finned channel sedimentation separator.

Both Section 6.2 below (which looks at filtration) and also Section 4.2 above (which addressed flocculation theory) discuss the existence of similar separation mechanisms, as well as load allocation to sedimentation and filtration through floc formation (flocculation).

Sludge deposits on the bottom of sedimentation basins differ depending on the settling velocity distributions of the suspended particles and on the flow conditions of the basins. The sludge deposit profile S(x) in a downstream direction x in a horizontally flowing idealized sedimentation pond is given by Equation 6.36, where the cumulative suspension time profile q(t) at a given depth is given by Equation 6.11.

$$S(x) = \frac{dq(x)}{dx} = \frac{1}{v} \cdot \frac{dq(t)}{t} \quad \text{---}(6 \cdot 36) \tag{6.36}$$

S(x) : Deposit-sludge profile per time
dq(x) : Cumulative curve of suspended solids
dq(t) : Time cumulative curve of suspended solids
v : Horizontal flow velocity of the basin
x : Distance of flow (=vt)
t : Settling time

Equation 6.36 The deposit-sludge profile in horizontal flow sedimentation basin.

In a radial flow circular sedimentation basin, Equation 6.37 applies.

$$S(x) = \frac{dq(r)}{dr} = \frac{r}{K} \cdot \frac{dq(t)}{t} \quad \text{---}(6 \cdot 37) \tag{6.37}$$

dq(r) : Cumulative curve of suspended solids
r : Q/πht
Q : Treated water amount
h : The basin depth
K : hπ
t : Settling time

In general, $q(t) \fallingdotseq A(1-e^{-Bt})$ ---(6·37a)
A, B : Constant

Equation 6.37 The deposit-sludge profile in radial flow circular sedimentation basin.

In general, q(t) has a shape that can be approximated by Equation 6.37a, so the amount of the deposit S(x) is large in the first half of the pond and small in the second half, as shown in Figure 6.50.

The depth of the deposition zone (the sludge zone) and the sludge withdrawal time interval are determined by deciding how long the down flow deposition should continue. On the other hand, septic-tank sedimentary sludge, such as sewage, must discharge as quickly as possible. For high-concentration systems with deposition volumes exceeding 0.1% it is also necessary to consider sequential sludge discharge methods such as mechanical scrapers. In ponds without a continuous sludge scraper, use of the pond is stopped when the depth of sludge reaches a certain level, the pond water is removed, and the sedimented sludge is collected by natural flow down into a sludge discharge well provided in one corner of the pond, or by spraying water. Examples of various mechanical sludge scraping systems are shown in Figures 6.29, 6.30, 6.31, 6.35, 6.41, and 6.42 (above). The speed of a scraper should be 20 cm/min or less to prevent sediment from re-suspending. The sludge collected in the sludge discharge wells (sludge sump) by the scraper or water stream is discharged out of the pond by gravitational force or a pump, and treated by dewatering, drying, etc.

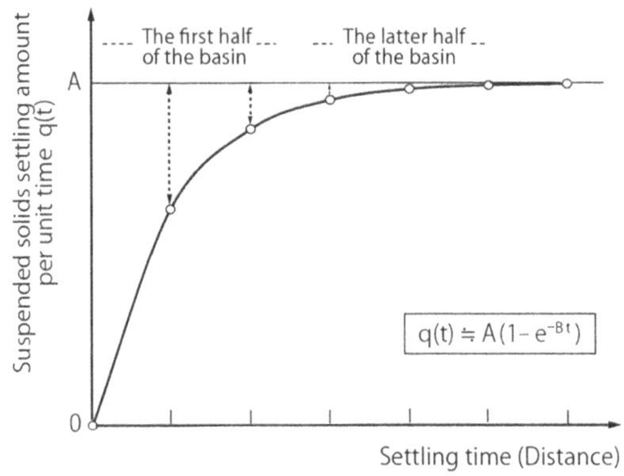

Figure 6.50 The transformation of sediments for settling.

The bottom of the sludge settling zone is scraped and flushed as described above. For ease, a gradient of about 8% is made in a circular pond bottom and about 1% in the longitudinal direction of rectangular ponds. Sludge discharge wells (sludge sumps and hoppers) are built with vertical-to-horizontal cross-sectional distances ranging from 1.2:1 to 2:1 so that accumulated sludge collects at the bottom quickly under flow by its own weight.

6.1.1.17 Characteristics of an upflow clarifier

In the basic theory of horizontal flow sedimentation basins, 'overflow rate = surface load ratio' explains that the total removal rate r_0 of a horizontal flow sedimentation pond is calculated by adding 100% of particles having a settling velocity larger than the overflow rate w_0 of the pond, and some particles having a settling velocity smaller than w_0 are removed with the removal ratio w/w_0 (Hazen number). In an upflow sedimentation basin, however, lower settling velocity particles below the overflow rate cannot be removed, and the removal rate r_0 is calculated as in Equation 6.38 and is always lower than that of a horizontal flow sedimentation basin with the same overflow rate.

Removal rate of the upflow basin

$$r_0 = 1 - x_0 \qquad \text{---}(6 \cdot 38)$$

Removal rate of the horizontal flow basin

$$r_0 = (1 - x_0) + \frac{A}{Q}\int_0^{x_0} w\,dx \qquad (6.38)$$

r_0 : Removal ratio
x_0 : Rate of particles larger than w_0
w_0 : Overflow rate (=A/Q)
A : Surface area of the basin
Q : Flow rate

Equation 6.38 The removal rate of upflow and horizontal flow sedimentation basin.

As can be seen from Equation 6.38, an upflow sedimentation basin cannot remove particles with settling velocities less than the overflow rate, so the shape of the pond becomes too large if the same removal target is set for a group of particles with a wide settling velocity distribution. Therefore, the actual use of upflow sedimentation ponds is primarily for removal of particles having nearly uniform settling velocities. As will be described later, the water in these ponds is used in a variety of ways, taking advantage of their excellent hydrological properties.

Typical examples of separation of particles with a nearly uniform distribution in water treatment include:

(1) high-rate coagulation sedimentation ponds (either floc blanket or circulating slurry types, as shown in Figures 6.51 and 6.52), in which particles with a nearly uniform particle size distribution inflow the separation zone of sedimentation basins after contact flocculation; and
(2) in the final sedimentation tanks of activated sludge systems (Figure 6.53), in which particles are suspended in a high concentration and are instantaneously settled on the pond bottom by a dense flow immediately after inflow, even in a horizontal flow sedimentation basin, and then the particles are not allowed to move vertically due to interfering particles; the separation proceeds as zone settling of the whole particle group of sewerage (Figure 6.53). In many cases, particles are coagulated prior to separation.

Compared to conventional horizontal flow sedimentation ponds, upflow sedimentation ponds have some beneficial hydrological characteristics. In a horizontal flow sedimentation basin, there is almost no way to

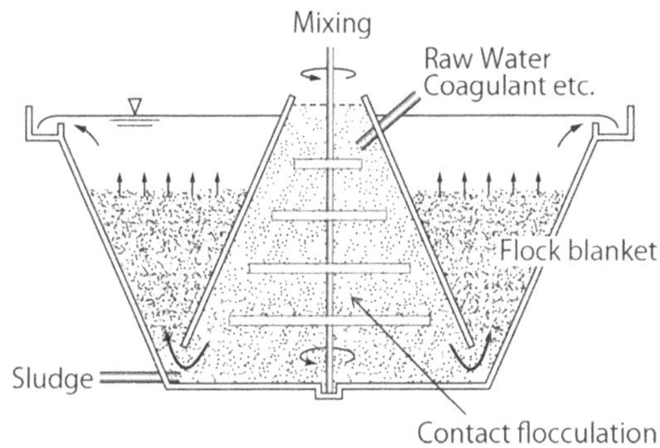

Figure 6.51 A floc blanket sedimentation basin.

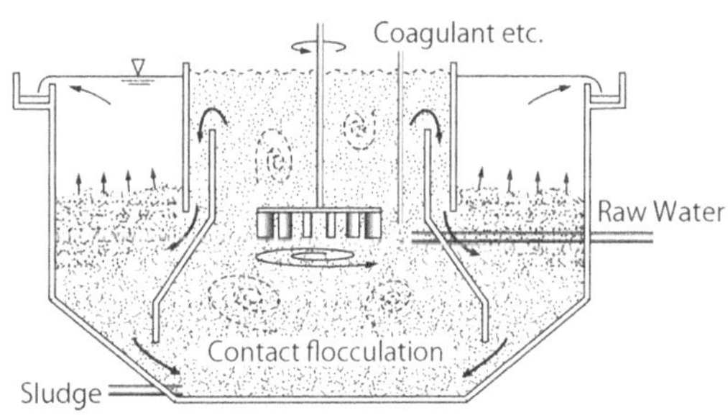

Figure 6.52 A circulating slurry sedimentation basin.

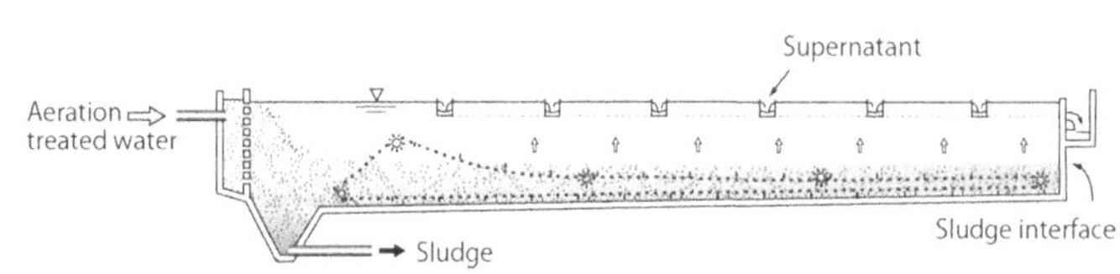

Figure 6.53 The final sedimentation basin in a conventional activated sludge treatment process.

arrest the unavoidable density shortcut flow that is the worst-case factor preventing access to idealized pond flow conditions. By contrast, in the final sedimentation tank in an activated sludge system, because of the high solid concentration (10^3 mg/l) of inflow water, it is possible to utilize zone settling of high concentration suspensions, in which the suspension immediately forms a high-density zone, and water is separated upward from the lower high-concentration layer.

Sometimes (as we have seen above) a perforated baffle wall or a conducting (training) wall is installed in a pond to suppress water flow bias and turbulence, as in the case of horizontal flow. Sludge then accumulates on the bottom of the pond by its own weight against the flow of water. In addition, flow turbulence can be reduced by dividing the ascending flow channel into smaller sections to reduce Re numbers, thereby improving sedimentation efficiency.

However, upflow sedimentation does have some disadvantages. Since the depth of a pond cannot be made too large due to the structure and the amount of earth work, it is difficult to obtain sufficient height for the transition section required for rectifying turbulent inflow eddies. And because of the susceptibility of suspended particle groups to local convection and flotation due to thermal convection, it may be difficult to make large devices with large horizontal cross-sectional areas.

Upflow sedimentation basins should have overflow troughs around the entire circumference of the basin to minimize weir loading of outflow and minimize suspension carryover. In addition, non-uniformity in the approach flow velocity between overflow weirs can cause easy short-circuits in a pond and drastically reduce treatment efficiencies, requiring careful design and daily control of the overflow water level.

The sludge deposits against the water stream at a location lower than the inflow port or in the dead area of the pond, and under flow from the pond bottom occurs after a predetermined thickening. However, it is usually difficult to increase the volume of the sedimentation area in upflow sedimentation ponds, and thickening time is insufficient. Therefore, sludge density at the time of sludge discharge cannot be increased, resulting in large flow rate losses due to sludge discharge. This limits the maximum processable concentration of high-concentration suspensions.

6.1.1.18 Floc blanket (fluidized bed) sedimentation basins

In order to obtain good clarified water in an upflow sedimentation basin (particularly with coagulated particles), various methods have been devised in which inflow particles are trapped in the suspended particle group of the fluidized bed and separated when the inflow suspended water is passed through a fluidized bed in which high concentration particles are suspended. The structure of the equipment used for inflow suspension trapping in the fluidized bed is described in detail in the coagulation section above (See Section 4.2.7), whilst the basic hydraulic behavior of a suspended fluidized bed is described here. Usually, such a bed of suspended particles (a fluidized bed) is composed of a group of coagulated particles, which is referred to as a floc blanket. In addition, in pellet coagulation and pellet crystallization processing of suspensions, described above, a fluidized bed of fine particles forms the basic equipment.

Separating mechanisms are described in detail in Sections 4.1 and 4.3, and the functions of floc blankets in upflow sedimentation ponds are outlined in order to understand the hydrological characteristics of fluidized beds.

Since a fluidized bed is composed of particles with a uniform settling velocity with high concentration, smaller particles are grown to a uniform diameter in the blanket. Particles or micro flocs in the inflow as dilute suspensions are trapped by the fluidized bed particles in the form of fluidized bed filtration, and separated from the inflow water. The blanket also has a function to absorb disturbances from the inflow port and decelerate inflow turbulence.

We now consider the hydraulics of floc blankets (fluidized beds). Consider placing a group of particles (a suspension) in a vertically mounted tube, as shown in Figure 6.54, and inflowing water from the bottom of the tube to allow water to overflow from the top.

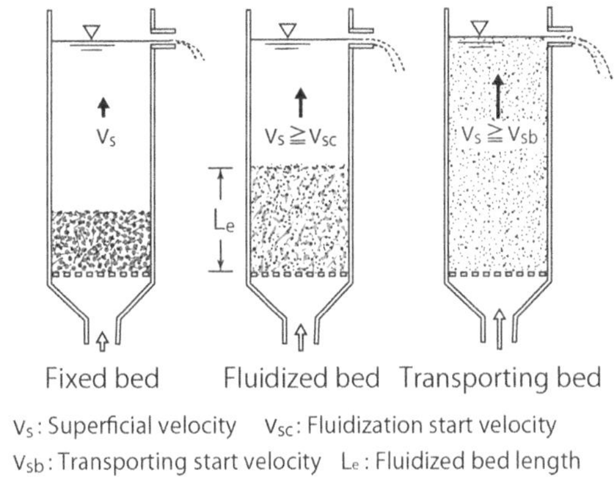

Figure 6.54 The three states of fixed bed, fluidized bed and transporting bed.

When the upflow velocity v is low, particles are still settled at the bottom, and water rises between the particles. To proceed with the treatment under these conditions, referred to as the fixed bed process, the pressure loss Δh as water passes through the bed increases in proportion to the superficial velocity of the upflow (the value obtained by dividing the passing flow rate Q by the cross-sectional area A of the cylinder $V_s = Q/A$, the value obtained by dividing the superficial velocity of the cylinder by the mean upflow velocity of the cylinder when there are no particles, and the value obtained by the overflow rate w_0 of the sedimentation pond).

When the flow rate reaches a certain level, the effective weight (weight in water) of the particles (as a group) exceeds the frictional force of the water (pressure losses through the fixed bed), exceeding the point of equilibrium and then the particles (group) begin to flow out from the vessel (entering the transport phase).

Such a limit point is referred to as a fluidization point (or point of fluidization), and the ascending flow velocity V_{sc} at this time is referred to as the flow initiation velocity. A fluidized bed is a suspension of particles that is lifted by a stream of water and becomes a transport phase.

At the point of fluidization, the weight of particles in the water and the lift force (loss of head) due to the upward flow are balanced, and Equation 6.39 is established.

$$\Delta p A = gLA(1-\varepsilon_{sc})(\rho_s-\rho_0) \quad \text{or}$$

$$h_f = \frac{g}{g_c} \cdot L(1-\varepsilon_{sc}) \cdot \frac{\rho_s-\rho_0}{\rho_0} \qquad \text{—(6·39)} \tag{6.39}$$

Δp : Pressure loss of water passing through particle bed (g/cm·sec^2)
g : Gravitational acceleration (980g/cm·sec^2)
L : Thickness of the particle bed at static state (cm)
A : Cross-sectional area of tube (cm^2)
ρ_s : Particle density (g/cm^3)
ρ_0 : Water density (g/cm^3)
h_f : Head loss when water passes through particle bed (cm)
g_c : Newton's Law conversion factor (980·cm/g force sec^2)
ε_{sc} : Porosity at fluidization start (-)
Approximately equal to the porosity of fixed bed

Equation 6.39 The weight of particles in water and the lift force due to upward flow at the starting point of fluidization.

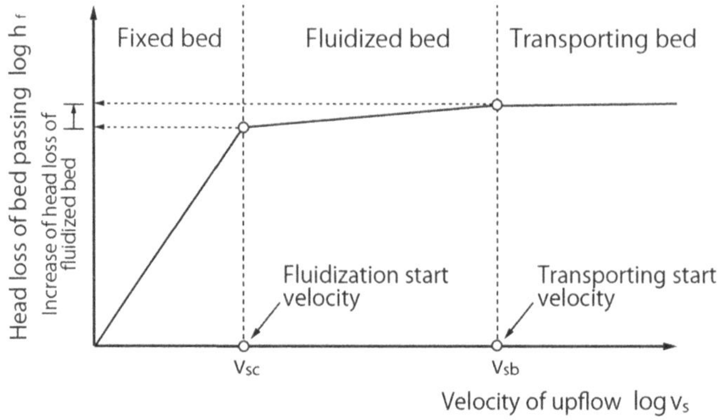

Figure 6.55 The change of head loss for fixed, fluidized and transporting beds.

When the ascending flow velocity V_s is increased beyond the fluidization point, the particles are completely suspended to form a fluidized bed. As V_S increases, the thickness L_e of the fluidized bed increases. However, as shown schematically in Figure 6.55, the characteristic features of the fluidized bed can be used for many processes.

The relationship between the fluidized bed thickness L_e with porosity ε_e and the fixed layer thickness L with porosity ε_{sc} at the starting point of fluidization is given by Equation 6.40.

$$LA(1-\varepsilon_{sc})\rho_s = L_e A(1-\varepsilon_e)\rho_s \quad \text{or}$$

$$L_e = \frac{L(1-\varepsilon_{sc})}{1-\varepsilon_e} \qquad \text{---}(6 \cdot 40)$$

L : thickness of the stationary particle bed (Fixed bed thickness) (cm)
L_e : Fluidized bed thickness (cm)
A : Tube cross-sectional area (cm^2)
ε_{sc} : Porosity at fluidization start (-)
ε_e : Porosity at fluidized bed thickness L_e (-)

(6.40)

Equation 6.40 The fluidized bed thickness and the porosity.

The magnitude of ε_e in Equation 6.40 is expressed by the relation between the final sedimentation velocity w_t and the upflow velocity v_s (= sedimentation velocity w of a group) of the single particles constituting the group. Experimental examples for groups of non-coagulation particulate particles are shown in Figure 6.56. In the Re number range of normal operation, the empirical expression shown in Equation 6.41 is obtained.

$$\varepsilon_e = \left(\frac{v_s}{w_t}\right)^n \qquad \text{---}(6 \cdot 41)$$

ε_e : Fluidized bed porosity
v_s : Upflow velocity
w_t : Terminal settling velocity of single particle

(6.41)

Equation 6.41 The fluidized bed porosity, particle settling velocity and upflow velocity.

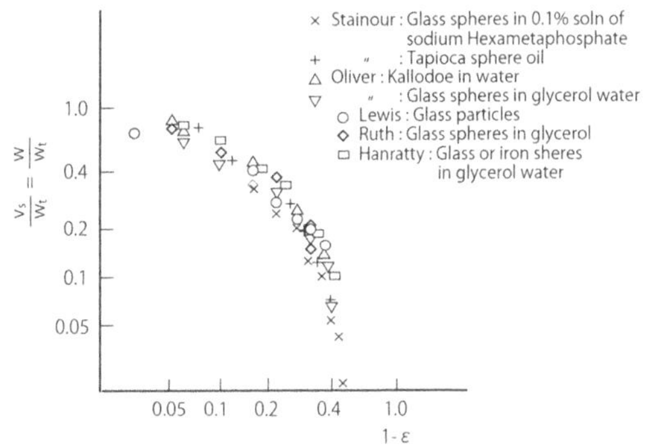

Figure 6.56 Fluidized bed porosity of non-coagulation particles and upflow velocity.

The magnitude of the exponent n in this empirical expression varies depending on the Re number of settling particles. The exponent n has n ≑ 0.18 at an Re < 0.1 and increases with an increasing Re number. In the area of $1 < Re < 500$, $n = Re^{0.1}/4.45$.

The characteristics of a fluidized bed as described above are given for discrete (non-coagulation) particle groups and can be applied directly to processes such as backflow washing of rapid sand filtration (Section 6.2) and pellet coagulation, crystallization and granulation (Section 4.2). The same qualitative behavior can be expected for common coagulation particles (e.g., coagulated floc sedimentation and activated sludge flocs).

Figure 6.57 shows the results of measurements of several particles in which the apparent floc-volume ratio of activated sludge flocs and chemical coagulation flocs at the time of sedimentation for 30 minutes is 10–40% in actual upflow sedimentation ponds. The results are plotted by plotting the logarithm of the floc-volume ratio on the horizontal axis and the logarithm of the sedimentation velocity of the particles on the vertical axis. From this diagram, we can roughly estimate the bed (zone) settling velocity v_s (the upflow velocity of the fluidized bed) by the linear representation on the log-log plot, as shown in Equation 6.42, using a representation of floc-volume-ratio C_v* – which has been widely used empirically in water-treatment – such as a 30-minute settling volume.

The constants in Equation 6.42 are considered to be determined by the settling rate of the single particles that make up the floc group. However, since actual measurement is difficult, the settling ratio based on the

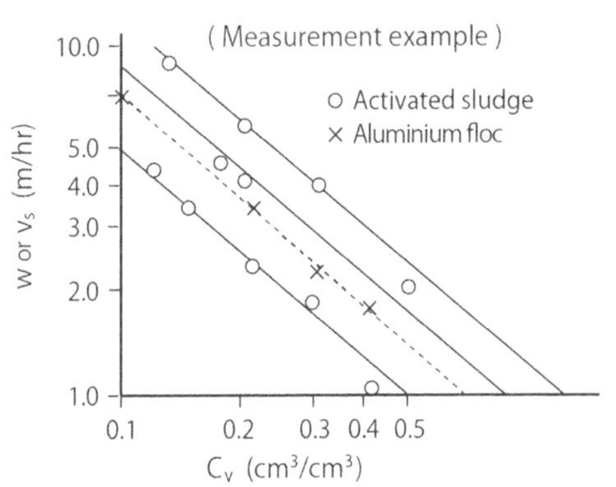

Figure 6.57 Floc volume ratio and zone settling velocity.

sedimentation volume ratio (e.g., $C_v^* = 0.1$) for a given period of time is estimated. For convenience, a C_v^* of sedimentation for 5 minutes or 30 minutes is often used to indicate the floc volume ratio C_v^* for comparisons of floc groups of differing concentrations.

$$\log v_s \fallingdotseq -\log C_v^* + \text{Const.}$$

or

$$v_s C_v^* \fallingdotseq \text{Const.} \quad -(6 \cdot 42) \qquad (6.42)$$

C_v^* : Floc volume ratio

Equation 6.42 The floc volume and zone settling velocity.

However, the ratio Z/Z_0 (the ratio of the interface height Z to the initial height Z_0) at any time (e.g., 5 minutes, 30 minutes, etc.) of a group of particles with differing slopes and shapes of sedimentation curves as the typical volume fraction of the group of flocs is not quantitatively (theoretically) significant, and therefore a logical route for obtaining the actual volume fraction needs to be examined.

The point at which zone settling (interfacial sedimentation) curves move from the constant settling zone to the consolidation settling zone can be thought of as the point at which the contact of all the floc particles is approximately completed (consolidation starting point). At that point, consolidation has already progressed at the bottom. Therefore, using the interfacial sedimentation curve as shown in Figure 6.58, the intersection C of the sedimentation curve and the bisector of the intersection angle at the intersection C′ between the extension of the interfacial sedimentation line of the constant settling rate region and the asymptote of the compaction sedimentation curve, is defined as the consolidation starting point. When the interfacial concentration reaches this consolidation starting point, C is the concentration at which the group of particles completes the contact, and if all the particles are present at that concentration, the volume is given by the point Z_c' at which the tangent at point C cuts the Y-axis.

The theoretical porosity ε of a uniform -diameter sphere under these conditions is between 26% and 48%, depending on the arrangement (see Section 6.2.1 below). However, it is assumed that irregular flocs will have values close to these values, but we need to collect some examples of actual measurements.

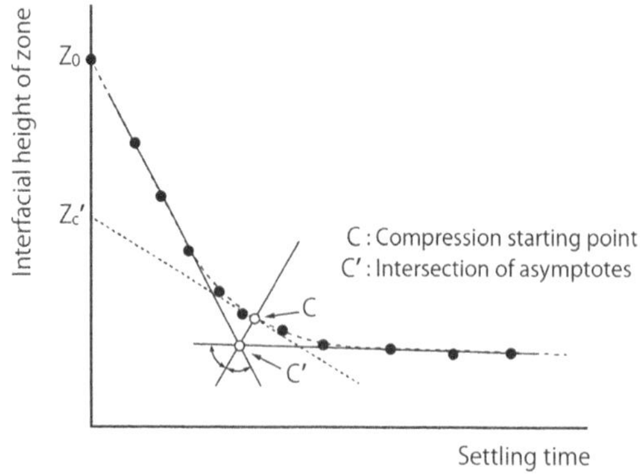

Figure 6.58 The zone settling curve and the starting point of the volume of compression settling.

Therefore, we first determine the interfacial sedimentation curves for a large number of floc groups with different concentrations, and then determine the settling velocity $W = V_s$ in the constant settling region and the hypothetical zone height Z_c' for the concentration C at the compression starting point of the curve described above. Assuming an arbitrary porosity ε_0 at the time of Z_c', we can calculate the porosity ε at any given density using Equation 6.40 based on it, and carry out a log-log plot; the relation with a corresponding sedimentation rate yields a straight line that generally satisfies Equation 6.41. In the floc particles group, the slope of this straight line is about 0.2 as described in Equation 6.40, so if we take this value and calculate ε_c, we can calculate the floc volume ratio C_v as in Equation 6.43 as the product with thickness Z_c assuming a floc group with an equal volume at the beginning of consolidation. This factor was derived by the authors from the results of a number of experiments with a coagulated floc particle group.

$$C_v = 1 - \varepsilon_c \fallingdotseq (0.65 \sim 0.55) Z_c'/Z_0 \qquad \text{—(6·43)} \tag{6.43}$$

C_v : Floc volume concentration (-)
ε_c : Concentration and porosity of zone at the compression start point (-)
Z_c' : Hypothetic isoconcentration height indicated by floc consentration at the compression starting point (cm)
Z_0 : Initial height of suspension (cm)

Equation 6.43 The floc volume concentration and porosity.

When the upflow velocity of a fluidized bed is further increased and reaches close to the final sedimentation velocity of single particles, the fluidized bed (floc blanket) breaks down, and the dispersed particles are caused to flow out from the separation tank with a much higher flow rate. In actual operation, the appearance of the fluidized bed interface being shown in Figure 6.54 becomes unclear as the upward flow velocity increases because the fluidized bed is not composed of uniformly distributed particles. In a fluidized bed of coagulation flocs, the uniformity of the particle size composition of the constituent particles makes the interface unclear when approaching this condition.

To consider flocculation in a fluidized bed, a floc blanket up-flow sedimentation basin is used to introduce raw water, in which coagulated fine flocs are suspended, into a fluidized bed, and then collision-agglomeration occurs to remove the particles that make up the bed. Whereas the filtration process in the sand bed described in Section 6.2 below is a collision one in a fixed bed, the separation process in a floc blanket is carried out in a fluidized bed. The removal mechanism in the fluidized bed is basically collisions between large and small particles with differences in settling rate (as described in Equation 6.13, which was described earlier in Section 6.1.1.7 on Coagulation Sedimentation) and is a phenomenon that holds for a group of particles with coagulation.

For simplicity, Equation 6.13 is expressed in terms of the Stokes sedimentation region for two groups of large flocs of diameter D and small flocs of diameter d which inflow from the bottom of a fluidized bed (floc blanket), and Equation 6.44 is obtained to evaluate the reduction rate dn/dt of the number of small floc particles. Generally, since $d \ll D$, Equation 6.44 can be expressed as Equation 6.45, and if we describe how the number of small flocs decreases at distance x from the bottom of the floc blanket, we can obtain Equation 6.46 by replacing $V_s t = x$ (time by distance).

$$\frac{dn}{dt} = -C\frac{\pi g}{72} \cdot \frac{\rho_s - 1}{\nu}(D+d)^3(D-d)Nn \quad \text{---}(6 \cdot 44)$$

Because, $d \ll D$,

$$\frac{dn}{dt} \fallingdotseq -C\frac{\pi g}{72} \cdot \frac{\rho_s - 1}{\nu} D^4 Nn = -C\frac{\pi}{4} V_s D^2 Nn \quad \text{---}(6 \cdot 45)$$

When time is replaced with distance and $V_s t = x$,

$$\frac{dn}{dx} = -C\frac{\pi}{4} D^2 Nn = -C\frac{3}{2} \cdot \frac{C_v}{D} n \quad \text{---}(6 \cdot 46)$$

t : Time since microparticles flowed into the floc blanket (sec)
x : Distance from the bottom of the floc blanket (cm)
n : Number of micro particles (floc) in unit volume water ($1/cm^3$)
N : Number of large particles (floc) constituting the floc blanket ($1/cm^3$)
D : Diameter of large particle (floc) constituting the floc blanket (cm)
d : Diameter of inflowing micro particle (floc) (cm)
C_V : Volume rate of suspended particles (floc) in the floc blanket (In the case of flocculent particles, volume rate is floc volume) (-)
ρ_s : Suspension density (g/cm^3)
V_s : Upward flow velocity of the fluidized bed (equal to settling velocity w of particles constituting fluidized bed) (cm/sec)
C : Collision-agglomeration coefficient of D and d particles (floc) in the floc blanket (-) (According to the actual measurement example, it is considered to be about 10^{-1} in floc blanket operation)

(6.44)–(6.46)

Equation 6.44 - 6.46 The reduction rate of number of small floc particles.

Generally, there is a proportional relationship between the concentration of the number of fine flocs, n, and the corresponding turbidity, T, so that the concentration of the number of fine flocs at the inflow of the fluidized bed n_0, and the turbidity T_0, passes through the blanket layer, x cm and remains in raw suspension (not adsorbed), as shown in Equation 6.47. According to actual measured examples, the values in the right-hand brackets of Equation 6.47 are about 10^{-2} (1/sec). Therefore, if a floc blanket with a transit duration of about 6 to 7 min is made, nearly complete adsorption (collision agglomeration) of turbidity can be achieved.

$$\frac{n}{n_0} = \frac{T}{T_0} = \exp\left(-C\frac{3}{2} \cdot \frac{C_V}{D} x\right) \quad \text{---}(6 \cdot 47)$$

n_0 : Number concentration of fine floc flowing into the fluidized bed (floc blanket)
T_0 : Turbidity of microparticles flowing into the fluidized bed
x : Moving distance in the fluidized bed
n : Number concentration of non agglomeration fine flocs in x
T : Turbidity of non agglomeration fine flocs in x

(6.47)

Equation 6.47 The Change in the concentration of fine flocs in the floc blanket.

Unlike a fluidized bed in which catalytic reactions are carried out, in a floc blanket upflow sedimentation basin fine flocs are constantly adsorbed onto large floc particle groups constituting the fluidized bed to increase the bed's concentration. Therefore, in order to stably maintain the fluidized bed, surplus suspended particles must be removed from the system. In water treatment equipment, changes in raw water or wastewater concentrations are large and sometimes rapid, but it is necessary to automatically perform this operation in response to these changes.

There are various methods for such an operation, with the aim of keeping the position of the upper interface of the floc blanket nearly constant, at a process flow rate (upflow rate) which is generally also kept constant:

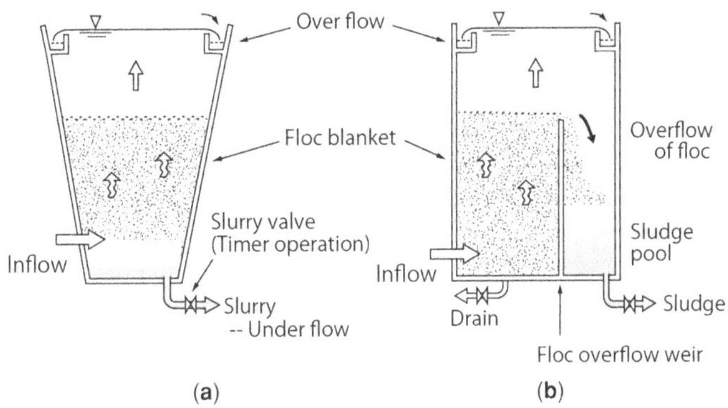

Figure 6.59 Examples of floc blanket surface regulation.

(1) as shown in Figure 6.59(a), a sludge discharge valve is provided at the bottom of the sedimentation pond, and excess sludge is discharged (as under flow) from the bottom of the pond in a timely manner by setting a timer to keep the floc blanket nearly steady.

(2) as shown in Figure 6.59(b), a floc blanket overflow weir is provided to regulate the surface (interface) of the fluidized bed, and the steady state of the floc blanket is automatically maintained. In addition, the floc group overflowed from the weir is discharged intermittently from the sludge pool after sludge thickening.

Whether sludge is withdrawn from the bottom of the fluidized bed or from the sludge pool, a system space is required to ensure adequate sludge thickening. Otherwise, the ratio of drainage water (under flow) to treatment capacity due to the extraction of sludge with insufficient thickening and low concentrations may be increased, thereby impairing water treatment efficiencies (recovery rate). However, we need to be aware that inadequate clearance of sludge thickening can still be a weakness of upflow sedimentation ponds and can often occur during the treatment of high-concentration suspensions. We consider sludge thickening in the next section.

In addition, when a fluidized bed is composed of a floc of clay, with a high settling rate, and a small density colored floc, and is composed of non-uniform particle settling rate groups, the fluidized bed is formed in two layers. Focusing on the interface between a fluidized bed of height settling rate particles in the lower part and light flocs formed in the upper part, the fluidized bed composed of heavy flocs will continue to accumulate in the lower part and operation will not be continued if control is carried out by the method shown in Figure 6.59 (b). It is difficult to operate fluidized beds on a steady basis if they are composed of multiple-component systems that differ greatly in settling rates, and it is prudent not to employ such a system. A small volume of heavy particles can be handled by operating a sludge blow off from the bottom, but this is not a judicious process choice. An up-flow floc blanket sedimentation pond, called a 'pulsator', has faced extreme difficulties in the treatment of high-color, high-turbidity two-component systems in Africa and Southeast Asia, as experienced by the author (Tambo). A sludge discharge unit was added to a water purification plant in Japan, Yokohama Water Works, and operation was performed with difficulty.

6.1.1.19 Thickening ponds (thickeners)

In water treatment, sludge that settles to the bottom of a pond after treated water is discharged (as outflow) to obtain as high a density as possible is itself discharged (as underflow), and then finally disposed of through thickening and dewatering processes.

Because of the lower concentrations of suspensions in a typical clarifying sedimentation pond, the primary determinant of pond size is the overflow rate, which is defined under the conditions of static sedimentation. In horizontal flow sedimentation ponds that handle dilute suspensions, the cross-sectional area of the bottom required to allow thickening to proceed is usually ensured by an adequate pond size, which is itself determined by the overflow rate. However, when dealing with high concentration suspensions consisting of particles with a high settling rate, the thickening sedimentation rate of the particles at the bottom of the pond is the limiting factor that determines the lateral cross-sectional area of the pond. If the thickening velocity is less than the cumulative velocity of the particles that settle rapidly, the thickness of the sludge deposit zone will increase over time while the underflow of the thickening suspension from the bottom is kept constant, eventually causing the concentrated suspension to overflow and the sedimentation operation to fail. In such cases, the volume drawn from the bottom will be increased and the slurry will be discharged with the slurry concentration remaining low.

The design theories of processes in which the sludge discharge from the bottom of such concentrated suspensions is carried out at as high a concentration as possible have been investigated historically in the mining field. In the field of water treatment, the treatment of high concentration suspensions and discharge of concentrated thickened sludge is required as part of the design of the sludge settling zone of a sedimentation pond in an activated sludge system, and as part of the thickening pond process used to conduct sludge withdrawn from various sedimentation operations to the dewatering operation. Such thickening operations are often aimed at solids with weight concentrations of greater than 10 to 20%.

A typical thickening sedimentation basin or pond has a structural composition as shown in Figure 4.60. Suspensions (slurry) is supplied at the center of the pond, and the supernatant flows out from the water surface into the overflow troughs. As described in Section 6.1.1.10 above, the inflow slurry settles vertically in the manner shown in Figures 6.13 and 6.14, which show the movement of the settling interface in the batch sedimentation operation and the changes in the internal concentrations of slurries during the same settling process, respectively. In continuous-type thickeners, a certain amount of settled sludge is drawn off every hour, and a certain amount of supernatant water flows out from the upper part of the tank. As shown in Figure 6.14, the concentration distribution of the slurry layer – which is similar to the concentration distribution of the slurry layer observed in batch-type sedimentation – is constant, in accordance with the conditions of operation.

The thickening of a clarifying sedimentation pond (such as the final sedimentation tank used for activated sludge; see Figure 6.53) which handles high concentration suspensions also behaves similarly, except that there are portions where discrete particles settle on top of the rapid settling layer (layer B in Figure 6.14).

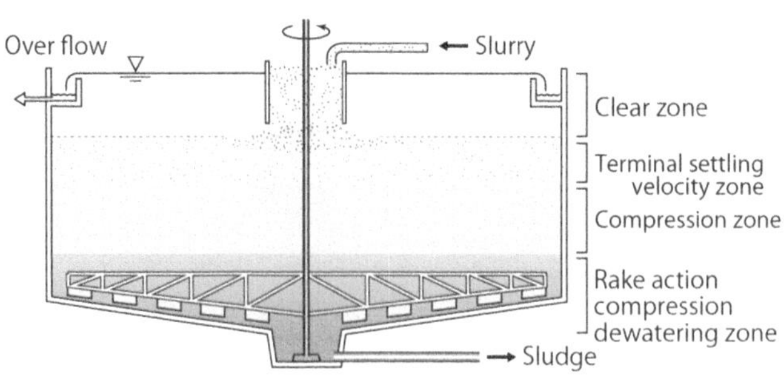

Figure 6.60 The basic structure of a thickener.

Sludges deposited on the pond bottom are collected in the vicinity of the sludge discharge port by a sludge scraper (rake) and discharged out of the pond by hydrostatic pressure or pump aspiration.

In a clarification sedimentation pond, the necessary horizontal area can be calculated by the overflow rate (cm/sec, m/day) in both horizontal and upward flow types. In a thickening sedimentation pond, the required horizontal cross-sectional area is calculated by calculating the maximum horizontal cross-sectional area required to process a given mass of solids (suspensions) per unit time, i.e., the maximum solid handling capacity per unit horizontal cross-sectional area of the thickening pond, measured in kg solid/m^2, using the following procedures.

Material balance of suspended solids in inflow and outflow is

$$Q_f C_f = Q_u C_u$$

or

$$Q_u = \frac{Q_f C_f}{C_u} \quad \text{---}(6 \cdot 48)$$

Q_f, Q_u : Flow rate of feed slurry and drainage sludge
C_f, C_u : Suspended solids concentration of feed slurry and drainage sludge

Material balance for water is

$$Q_f(\rho_f - C_f) = Q_0 \rho + Q_u(\rho_u - C_u) \quad \text{---}(6 \cdot 49)$$

ρ : Water density
ρ_f : Feed slurry density
ρ_u : Drainage of sludge density

From Eq.4·48 and Eq.4·49

$$Q_0 = \frac{Q_f C_f[(\rho_f/C_f) - (\rho_f/C_u)]}{\rho} \quad \text{---}(6 \cdot 50)$$

(6.48)–(6.50)

Equation 6.48 - 6.50 The material balance for water of a thickener.

The mass balance of the inflow and outflow solids in the schematic thickening sedimentation pond shown in Figure 6.61 is shown in Equation 6.48. The material balance for water is shown in Equation 6.49. Equation 6.48 is used to eliminate the Q_u from Equation 6.49 and rearrange it as Equation 6.50.

Generally speaking, for a group of coagulated particles treated in water treatment, we can approximate that the $\rho_f \doteqdot \rho_u \doteqdot \rho \doteqdot 1.0$. If we substitute this relation into Equation 6.50 and divide the two sides of the equation by the horizontal cross-sectional area A (m^2) of the pond, the overflow rate of the pond is Q_o/A. In order for suspended particles of the considered concentration to settle against this upflow rate, the sedimentation rate w $> Q_o/A$ should be satisfied for the particle group. Equation 6.50 is a formula concerning the upper surface of

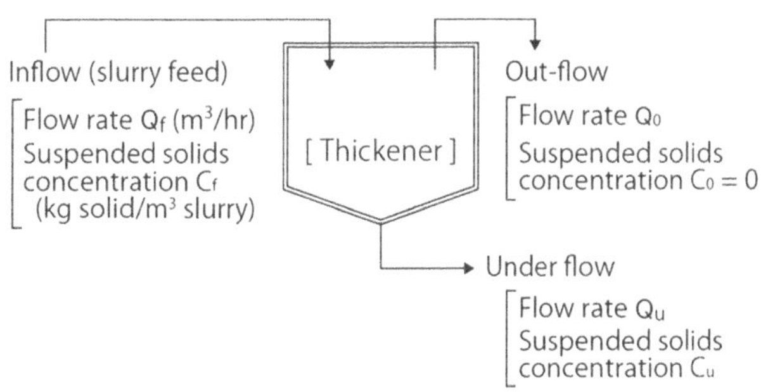

Figure 6.61 The material balance diagram of a thickener.

the slurry. However, for C_i – the layer of the concentration in which thickening is proceeding – Equation 6.51 can be used as a similar material balance equation by considering the hypothetical mud feed rate (slurry flow rate) Q_i at concentration C_i and the solids feed rate is equal to $Q_f C_f$. The sedimentation rate w_i is determined for each of the layers of C_i. Also, since $Q_iC_i = Q_fC_f$, the required thickening horizontal cross-sectional area A_i for the C_i slurry-layer is given by Equation 6.52.

$$w_i = \frac{Q_iC_i}{A}\left(\frac{1}{C_i} - \frac{1}{C_u}\right) \quad \text{—(6·51)}$$

w_i : Settling velocity of suspended particles in the zone with concentration Ci
Q_i : Flow rate in the zone with concentration Ci
A : Horizontal cross section area of the thickener

Since the required concentration horizontal sectional area Ai in the slurry zone of the concentration Ci is $Q_iC_i = Q_fC_f$.

$$A_i \geq \frac{Q_fC_f}{w_i}\left(\frac{1}{C_i} - \frac{1}{C_u}\right) \quad \text{—(6·52)}$$

(6.51)–(6.52)

Equation 6.51, 6.52 The required concentrating horizontal cross-sectional area in slurry layer.

In actual design, the C_i and w_i of a given suspension are obtained by settling analysis, and the values of C_u, C, and Q_f are used to calculate the required A_i for various C_i. The largest required horizontal cross-sectional area value ($A_{i.max}$.) among the obtained values is defined as the horizontal cross-sectional area of a thickening sedimentation basin.

To determine the depth of the thickener, to obtain the thickness of the discrete particle settling layer and the supernatant water layer so that the turbulence caused by the raw slurry feed does not affect the thickening layer:

(1) The feed slurry should not reach the overflowing supernatant water, and should absorb the change in the thickness of the thickening layer due to the variation in the amount of feed slurry that is unavoidable;
(2) The thickening layer should have a margin of height within a certain range, and the sludge discharge (underflow) should have a volume that ensures enough compaction time to reach concentration C_u. In this way, design and operation are performed to prevent low concentration sludge discharge from the bottom, even if the amount of sludge is increased.

A reasonable design method taking all these factors into account has not yet been established, but to meet requirements 1 and 2 above, take 30 to 100 cm for the upflow part of the supernatant water, 30 to 60 cm for discrete particle settling part, and 30 to 60 cm for the lowest sludge collection part (thickening is accelerated by the rake action during mechanical sludge collection), and add to these values the thickness of the compression (compaction) bed required to reach the sludge discharge density C_u calculated below to obtain a total water depth.

Assuming that the required volume of the compaction part is V, and the mean density of the slurry in the compression layer is ρ_m, the total slurry weight in the compression layer is $V\rho_m$, and the compression settling time (retention time in the compaction layer) required for the sludge discharge to reach a predetermined concentration is T, the flow rate in the compression layer is Equation 6.53, and the total amount of slurry is calculated as the sum of the amount of solids (suspensions) and the amount of water as Equation 6.54.

Assuming that the cross-sectional area of the compression layer is A and the thickness is H, V = AH. If this relation is substituted into Equation 6.54, the required thickness H of the compression layer can be obtained by Equation 6.55.

Amount of water in the compression layer $= \left(V - \frac{Q_f C_f T}{\rho_s}\right)\rho$ —(6·53)

Total amount of slurry $V\rho_m$ (sum of total solids and total water) is

$$V\rho_m = Q_f C_f T + \left(V - \frac{Q_f C_f T}{\rho_s}\right)\rho \quad \text{—(6·54)}$$

The length H of the compression layer is

$$H = \frac{V}{A} = \frac{(\rho_s - \rho)Q_f C_f T}{\rho_s(\rho_m - \rho)A} \quad \text{—(6·55)}$$

V : Required volume of the compression layer
ρ_m : Slurry average density in the compression layer
T : Compression settling time (detention time of compression layer) required for the drainage sludge to reach a predetermined concentration
$Q_f C_f T$: Amount of solids in the compression layer
ρ_s : Suspended solids density of the compression layer
ρ : Water density
A : Horizontal cross section area of the compression layer
V : Amount of solids in the compression layer

(6.53)–(6.55)

Equation 6.53 - 6.55 The required thickness of a compression layer.

Now, determining the time T required for compression and the mean slurry density ρ_m is a matter of reaching the following target. In general, batch sedimentation tests using the same slurry are performed, and the results are obtained according to the following procedure.

(1) The zone sedimentation in the compression domain during batch sedimentation analyses can be described as in Equation 6.21 proposed by Roberts, and is amended and expressed in logarithmic terms in Equation 6.56. Let T be the time from the start of the compression dewatering process until the target sludge discharge concentration C_u is reached, and ρ_m is approximately the mean density of sludge discharge concentration C_u and sludge concentration C_c at the start of compression sedimentation.

$$\mathrm{Log}\left(\frac{Z - Z_\infty}{Z_c - Z_\infty}\right) = -K(t - t_c) \quad \text{—(6·56)}$$

Z : Interface height at the settling start time t (Slurry zone length)
Z_c : Interfacial height at the compression starting time
Z_∞ : Interfacial height at the end of compression settling
t_c : Compression settling start time
K : Coefficient indicated by common logarithm

(6.56)

Equation 6.56 The Roberts empirical equation of the compression dewatering process.

(2) The slurry batch zone sedimentation curves are plotted semi-logarithmically as shown in Figure 6.62. The interfacial height of the starting point of compression sedimentation, Z_c, at time t_c at the onset of compression sedimentation, with interface height at the end of compression sedimentation Z_∞, and logarithmic compression coefficient K are all estimated to determine the interfacial height Z at time t after the onset of sedimentation. In this case, the interface height Z_∞ value at the time of the final thickening is determined by trial and error since the sedimentation curves obtained by the semi-logarithmic plot in Figure 6.62 are calculated using various estimated values Z_∞, and the relation between the vertical axis value $(Z - Z_\infty)/(Z_0 - Z_\infty)$ and the sedimentation time t is considered to be a compression area (Roberts equation holds), and the semi-logarithmic plot becomes a straight line.

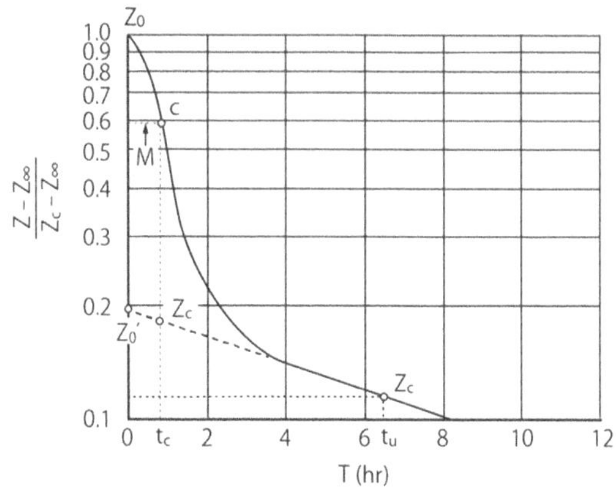

Figure 6.62 Estimation of the required depth of a thickener.

(3) The results are used to estimate the height, Z_c, of the bed at which compression sedimentation begins and its onset, t_c, by empirical methods, as shown by Roberts in Figure 6.62. That is, the straight line of the compression sedimentation section is extended to the vertical axis, and the intersection point is defined as Z_0'. The mean value M of Z_0' and the initial water depth Z_0 of the sedimentation test is obtained, and the time indicated by the point c at which a line parallel to the horizontal axis passing through M intersects with the sedimentation curve is defined as the compression sedimentation onset time t_c, and $t_u - t_c = T$ is the necessary retention time.

(4) Since the particle density ρ_s, the initial weight concentration C_f, the initial sedimentation height Z_0, the compression starting point height Z, and the sludge discharge concentration C_u are all obtained, the mean density ρ_m of the slurry in the compression bed can be obtained by the procedures shown in Equations 6.57 to 6.60.

(5) Substituting these values into Equation 6.55 gives the required compression depth H.

(6) To this value, add 120 to 220 cm for the height of the supernatant, the discrete sedimentation, and sludge collection section, as described above, and the required water depth of the thickener is determined.

Slurry density ρ_f at the time of slurry feed ($t = 0, C = C_f$) is

$$\rho_f = \left(\frac{C_f}{\rho_s} + (1 - C_f)\right)^{-1} \quad \text{—(6·57)}$$

Slurry density ρ_c of compression point ($t - t_c, Z = Z_c$) is

$$\rho_c = \frac{\rho_f Z_0 - (Z_0 - Z_c)}{Z_c} \quad \text{—(6·58)}$$

Slurry density ρ_u of drainage sludge ($t = t_u, C = C_u$) is

$$\rho_u = \left(\frac{C_u}{\rho_s} + (1 - C_u)\right)^{-1} \quad \text{—(6·59)}$$

Therefore,
Slurry mean density ρ_m of the compression layer is

$$\rho_m = \frac{\rho_c + \rho_u}{2} \quad \text{—(6·60)}$$

(6.57)–(6.60)

Equation 6.57 - 6.60 Mean slurry density in the compression layer.

As in the case of clarified sedimentation ponds, the design of the inflow structures of a thickener should be devised so that the inflow vortex and inflow deflection are quickly attenuated to keep the separating interface and thickening zone calm. In particular, if a concentrated solution is supplied, very often there is a danger that it may run as a density flow over the separating interface and short-circuit the stock solution to the outlet. Therefore, in the design of various thickening ponds, the greatest consideration must be given to ensure measures against this density flow.

Structures like upflow sedimentation ponds need to be designed with particularly small weir loading because the overflow surface and separating interface are short. Referring to Anderson's measurements of activated sludge in sedimentation ponds, it can be seen that a reduction in weir loading linearly improves the water quality of the overflow, as shown in Figure 6.63.

Thickening ponds are equipped with a mechanical scraper to constantly scrape off a large amount of sediment and discharge it from the sludge discharge pit. Estimation of the power required to operate a scraper is difficult to calculate theoretically because of many relevant factors. In many cases, it is based on empirical rules.

For round ponds, there is a semi-theoretical equation (Equation 6.61) for the sludge collection power P proposed by Chelminski for heavier sandy sludge. The calculation formula is explained by referring to Figure 6.64.

The required sludge collection power P* is calculated by $P^* = P/E$ obtained by dividing the theoretical sludge collection power P (Equation 6.61) by the sludge collection efficiency factor E, as in Equation 6.62. The frictional coefficient f_2 between the deposited component and the smooth steel plate (blade) in the equation is obtained by measuring the angle φ at which the deposited component is placed on the steel plate

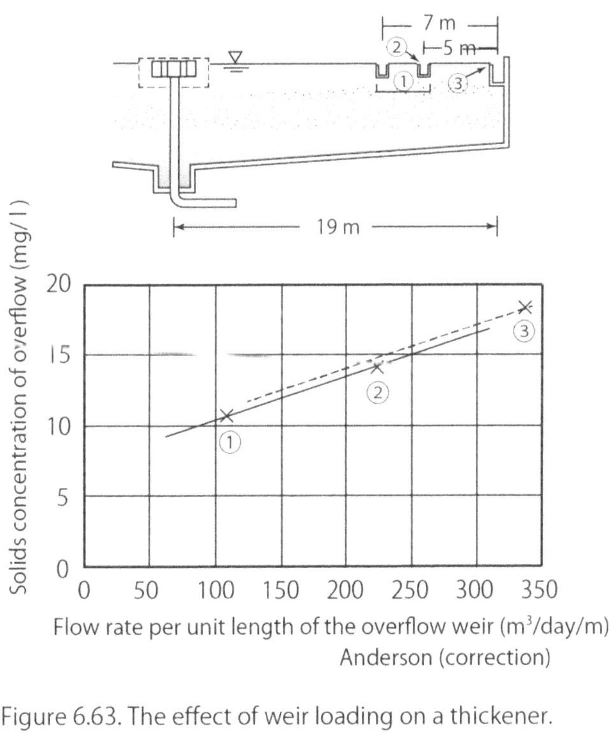

Figure 6.63 The effect of weir loading on a thickener.

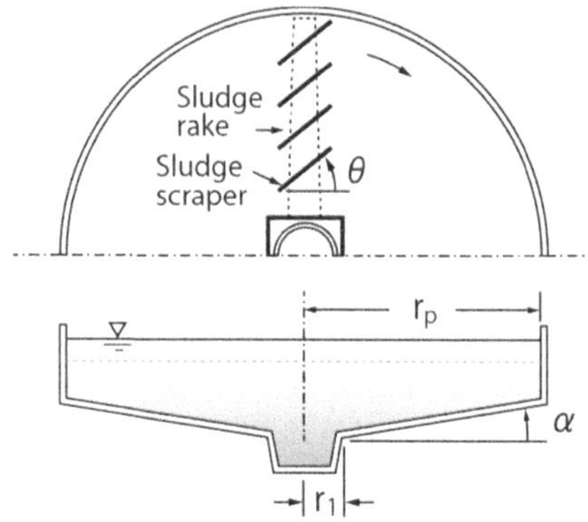

Figure 6.64. A sludge collecting mechanism for a circular thickener.

Figure 6.64 A sludge collecting mechanism for a circular thickener.

and the steel plate is tilted until the deposited material starts to move, and measuring the angle φ at that point.

$$P = \frac{T}{3}\,\frac{(f_2\cos\alpha - \sin\alpha)(2r_p^3 + r_1^3 - 3r_1r_p^2)}{r_p^2 - r_1^2} \qquad \text{—(6·61)}$$

T : Amount of solids settled to the bottom of thickener in unit time ($Q_f\,C_f$ kg/m²)
α : Angle which the scraper makes with the horizontal plane (Bottom gradient)
f_2 : Frictional coefficient between deposited sludge and the scraper
r_p : Radius of thickener
r_1 : Outer radius of sludge hopper

Sludge collection efficiency factor E

$$P^* = \frac{P}{E} \qquad \text{—(6·62)}$$

P : Theoretical sludge collection power
P* : Required sludge collection power

(6.61)–(6.62)

Equation 6.61 The theoretical sludge collection power and the required collection power.

$$E = \frac{f_2\cot\alpha - 1}{\{A\sin(\theta+\phi)(\cot\theta + \sin(\theta+\phi)/[\cos(\theta+\phi) + 1/A])\}} \qquad \text{—(6·62)}$$

A : Characteristic value determined by Equation 6. 63
ϕ : It is expressed as $\tan\phi = f_2$ by the resultant force given by the sludge scraper and the angle formed by Its vertical component.

$$A = -\cos(\theta+\phi) + (f_2\cot^2\alpha - \sin^2(\theta+\phi)) \qquad \text{—(6·63)}$$

(6.62)–(6.63)

Equation 6.62, 6.63 The sludge collection efficiency factor.

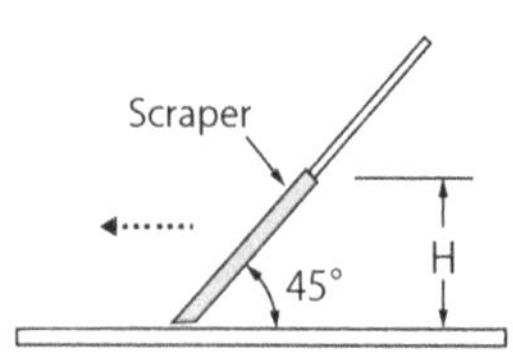

Figure 6.65 A scraper type of sludge collector for horizontal linear movement.

The sludge collection power P of a scraper type clarifier, such as that shown in Figure 6.65, is obtained by linear upstream and downstream movement in a rectangular thickening pond, as shown in Equation 6.64.

The total required power of a sludge scraping system is determined by adding the power consumed by the scraper drive system to the power used for sludge collection, estimated as above. The scraper should scrape at a rate of 0.5–0.6 m per minute or less to avoid re-floating of sludge once settled. In addition, the sludge collector always takes a large amount of load. If excessive load occurs due to unexpected situations such as excessive thickening or foreign matter contamination, it may immediately cause the sludge collector to break down and lead to problems such as burning out the drive motors. For this reason, precautions must be taken such as installing an automatic emergency stop unit and a raising unit for the rake in the event of excessive load.

$$P = f_3 B \rho_s \left(\frac{H}{2\tan\beta} + \frac{H}{2} \right) \quad -(6\cdot64) \qquad (6.64)$$

f_3 : Friction factor between the sludge and thickener bottom
B : Scraper length
ρ_s : Sludge density
H : Scraper height
β : Repose angle of sludge in the scraper operation area

Equation 6.64 The sludge colletion power P of the horizontal linear scraper.

6.1.2 Flotation

6.1.2.1 Overview of various flotation treatment

Flotation units are widely used to separate solid and liquid particles from the liquid phase (water). If the density of solid particles or different types of liquid particles suspended in water is smaller than that of water, 'upward sedimentation' (flotation) will occur, in a manner opposite to that of ordinary sedimentation. The buoyancy inherent to the suspended particles will initiate upward movement and they will collect on the surface of the suspending liquid (water), where it will be collected and removed from the system. For suspended matter that has a small difference in density from the suspending medium (water) and are difficult to separate by simple flotation or sedimentation, microbubbles such as air are often attached to the suspending matters to increase buoyancy (a difference in density), and the separation speed is increased to cause rapid flotation and separation.

Air is commonly used to generate such bubbles in the suspension. Bubbles introduced into the suspension may adhere to the suspended particles or may become trapped inside floc particles, as discussed in Section 6.1.2. The density of the formed mixture of suspensions and air bubbles is much lower than the density of water, and floats on the water surface with its increased buoyancy. In order to carry out such an operation, the size, density and concentration of suspended particles to be separated, and adhesiveness between bubbles and the surfaces of particles are process factors, as well as the process to generate an appropriate size and adequate number of bubbles in the water.

If the particles suspended in water are sufficiently less dense than water, the same process as described for gravitational sedimentation is performed. Sludge produced by the suspended matter and any matter being floated is thickened as a scum (floated sludge) on the water surface and discharged out of the system.

In water treatment, it is rare that such a difference in density between suspensions and water exists naturally. In most cases, air bubbles are attached to suspensions having a specific gravity close to 1, which only have a small difference in density from water, and the suspension is separated by deliberately increasing the difference in density.

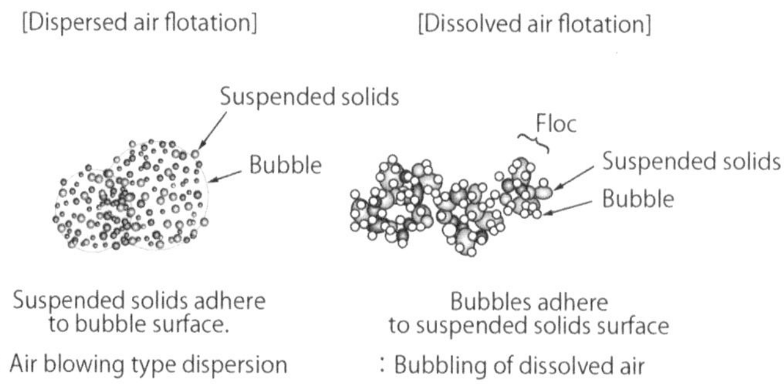

Figure 6.66 The two main concepts of flotation: dispersed and dissolved air flotation.

Broadly speaking, the flotation process used for water treatment can be classified into two methods: dispersed air flotation and dissolved air flotation, as shown in Figure 6.66.

Dispersed air flotation is a method in which a required amount of air is fed into a suspension through sparger rings or porous gas diffusers that perform strong agitation. Suspended matter is adsorbed on the surface of the dispersed and suspended air bubbles, and the air bubbles with adsorbed particles are separated by flotation from the water body. This process has long been used as a flotation method in the mining and metallurgical fields to classify relatively concentrated slurries of ore powders. The bubbles used in such a process are approximately 1 mm in diameter, and the suspended particles to be removed (of several microns) adhere to the surfaces of the foam/bubble groups (froth) thus formed, float and separate and, after separating, the foam is destroyed, thereby recovering highly concentrated thickenings.

The 'dissolved air flotation' method is a process in which air dissolved under pressure in water is brought into a supersaturated state by reducing the pressure of the system; supersaturated air is then precipitated as a large number of fine bubbles (microbubbles) at the solid interface of the suspended particles, and individual particles are floated and separated. Since the precipitated microbubbles generally have a size of 10^1 micrometers and are generated from the outset on the surfaces of the suspended particles to be separated when the suspension is depressurized, there is no need for strong agitation to mix and stir to bring the bubbles and the suspensions into collision, as is necessary in dispersed air flotation. Therefore, dissolved air flotation is considered to be a process capable of floating flocs with fragile fractal structures such as chemical coagulation flocs and aerobic microbial treatment (activated sludge) flocs (described in Section 4.4 above) without destroying them. It is widely used in the water treatment field, in particular in the separation process for the pretreatment of seawater desalination which has become increasingly necessary recently, or in the final separation of aerobic microbial treatment using seawater. The dissolved air flotation method is frequently used because the density of seawater is higher than that of fresh water and it is therefore difficult to separate suspensions by simple gravitational sedimentation.

6.1.2.2 Binding of suspended particles and air bubbles

There are two possible mechanisms for the binding of suspended particles and air bubbles:

(1) The first mode of bonding is the main mechanism of action of the dispersed air flotation method, in terms of adherence of suspended particles to air bubbles. In many years of study of foam separation methods used for flotation, numerous investigations were made in the mining and metallurgical engineering fields on the elucidation of the mechanism of adhesion and application theory.

Attachment is caused by intermolecular attraction (interfacial tension) acting on the interface between the solid and liquid phases. By considering the equilibrium of interfacial tension between the three interfaces of suspending medium (water), air bubbles, and suspended particles, the degree of difficulty of attachment can be discussed.

(2) The second form of bonding occurs primarily in the flotation of coagulated particles, where the generated microbubbles adhere to floc structures during the flocculation process (see Chapter 4) or form in the interstices of the structural (fractal) floc particles. In this case, bubble-particle bonding can be thought of as the main factor by trapping bubbles in the particle structure gap. When the suspensions form flocs, it is generally required that the zeta potential (ZP) of the original particles is lowered to -10 mv or less (see Chapter 4), but the interface potential of the microbubbles has the same sign and a high value, which may cause strong mutual repulsion. The ZP of microbubbles in water is about -150 mV in Japanese soft tap water, in our measurements. However, under turbulent flow conditions, no collision hinderance has been observed between air bubbles with -150 mV ZP and a critical coagulation potential of floc of about -10 mV, because the bubble size is as large as 50 µm, which is much greater than the critical coagulation size. (See Coagulation, in Chapter 4.)

The basic issues which need to be considered for these adhesion phenomena are explained in order below.

Wetting (contact angle). Figure 6.67 shows the presence of air bubbles on the surface of solid particles in water. In the diagram, the largest angle θ formed by the gas–liquid interface sandwiching water is called the contact angle at the point of contact of the three solid–liquid-vapor phases. Equations 6.65 and 6.66 are obtained because the interfacial tension (surface tension) between the three phases at the contact point is balanced with the attached bubbles.

The relationship in Equation 6.65 is called a 'cosine scalar', whilst the relationship in Equation 6.66 is called a 'sine scalar'.

If the contact angle $\theta = 0$, the water completely covers the surface of the solid, and the contact between the air and the solid is cut off. This is the state in which the solid is completely wetted by water. On the other hand, if the contact angle is 180 degrees, complete wetting of the air and solids will result in the contact of liquids and solids being cut off. However, solids with contact angles greater than 110 degrees between the gas and water phases are not known. In this way, the contact angle θ is a measure of the wetting of a solid

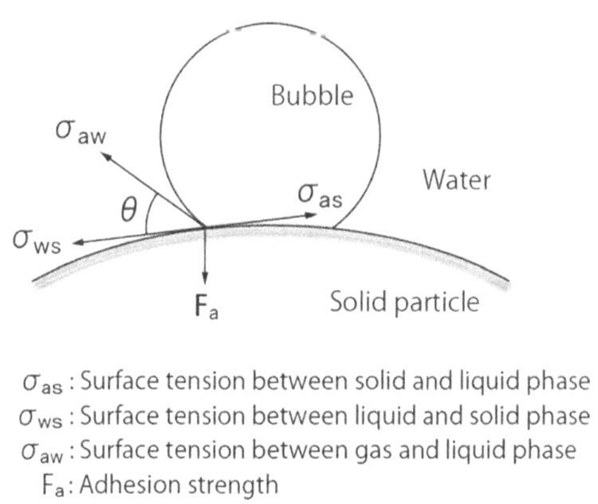

Figure 6.67 The contact angle (θ).

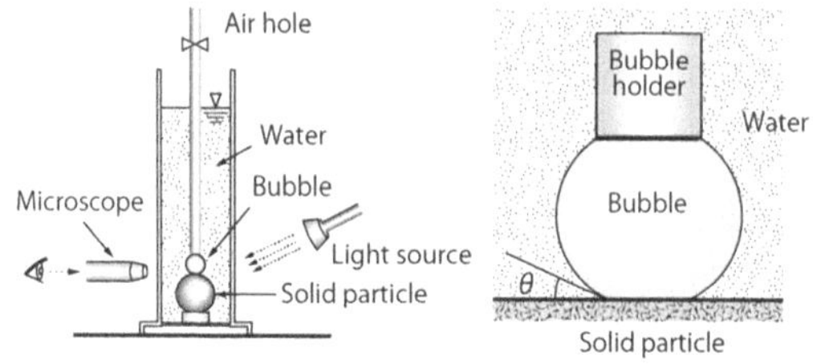

Figure 6.68 Measuring a contact angle.

with a liquid. If the contact angle θ is large, the solid is difficult to wet with water, and if it is small, it is easy to wet. At $\theta = 90$ degrees, solids and liquids do not work (react).

$$\sigma_{as} = \sigma_{ws} + \sigma_{aw} \cos\theta \quad \text{—(6·65)}$$

$$\sigma_{aw} \sin\theta = F_a \quad \text{—(6·66)}$$

σ_{as} : Surface tension between gas and solid phase (dyn/cm)
σ_{ws} : Surface tension between liquid (water) and solid phase (dyn/cm)
σ_{aw} : Surface tension between gas and liquid (water) phase (dyn/cm)
F_a : Adhesion strength (dyn/cm)

(6.65)–(6.66)

Equation 6.65, 6.66 The surface tension for three-phase between gas,liquid and solid phase.

A profile method is widely used as a method for measuring the contact angle, and typical examples are the captive bubble method and the Mack method:

(1) The captive bubble method uses a microscope to measure the contact angle. A glass rod is gently lowered into water to hold bubbles in the recess at the tip, as shown in Figure 6.68, when a solid surface and a bubble contact are placed on the bottom. Various measures are taken, such as magnification projection and photographic measurements.
(2) The Mack method is used when the captive bubble method is difficult to apply due to very small bubbles. The size of attached bubbles are measured using a projected or photographed image, as are the height h of attached bubbles and the diameter b of the attached portions of spherical segments, and $\theta = 2\tan^{-1}(h/b)$ is calculated, as shown in Figure 6.69.

Attachment (adhesion) phenomenon. In a system in which bubbles and particles are present separately, the sum of the surface free energies (interfacial energy) after deposition must be less than in the separated

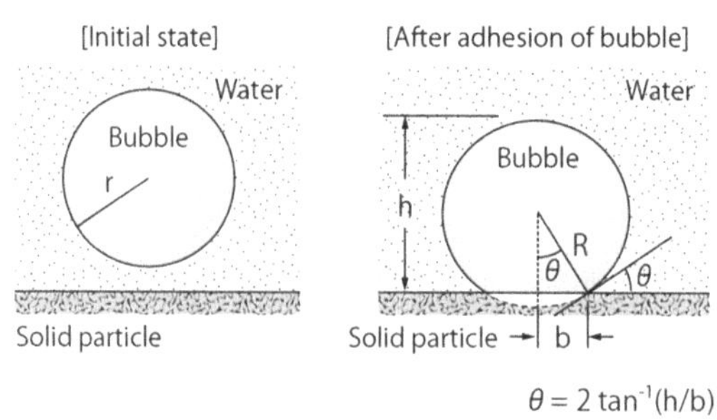

Figure 6.69 The attachment of microbubbles to large particles.

state in order for the bubbles to adhere to the particles. This change can be calculated from the three interfacial energies of solid–liquid, gas–liquid, and gas–solid, and from the interfacial area between the three phases. By introducing a measurable contact angle θ, the unmeasurable terms of solid–liquid and gas–solid interfacial energies can be eliminated, and calculations can be made from three known terms: the surface energy of fluids (surface tension), and the change in the interfacial area of the three phases and between the contact angles. The change in energy (work) of the system during deposition indicates a different result for two different forms of operation: when the bubble size is small relative to the suspended particle size (i.e., dissolved air flotation) and when the bubble size is large relative to the suspended particle size (i.e., dispersed air flotation or foam isolation).

When the bubble diameter is smaller than the diameter of the suspended particles, suppose that bubbles with radius r adhere to a solid–state surface of area A with a large radius of curvature, as shown in Figure 6.69, and the height of the fan is R and the contact angle is θ. The decrease in free energy caused by such an attachment phenomenon is the total interfacial energy before attachment minus the interfacial energy after attachment. Here, $b = R \sin \theta$, height $h = R(1 + \cos \theta)$, and the surface area is $4\pi R^2(1+\cos\theta)/2$. The change in free energies before and after deposition (work done) ΔW is expressed by Equation 6.67.

$$\Delta W = (AE_{sw} + 4\pi r^2 E_{aw}) - [(A - \pi b^2)E_{sw} + \pi b^2 E_{as} + 4\pi r^2 E_{aw}(1+\cos\theta)/2] \quad \text{—}(6\cdot67)$$

E_{sw} : Solid-liquid Interfacial energy per unit volume (erg/cm³)
E_{as} : Gas-solid Interfacial energy per unit volume (erg/cm³)
E_{aw} : Gas-liquid Interfacial energy per unit volume (erg/cm³)

(6.67)

Equation 6.67 The change in free energies before/after deposition.

Further, if Equation 6.68 is derived and the equation is transformed in accordance with the procedures noted in Equations 6.69–6.71, the 'work done' can be calculated by Equation 6.72 or Equation 6.73 by finally measuring the actually measurable gas–liquid contact angle.

① From the relationship of surface tension
T = interfacial energy/surface area (= erg/ cm² = dyne/ cm), Equation 6 · 65 is

$$E_{sw} - E_{as} = E_{aw}\cos\theta$$

② Substituting this relationship and $b = R\sin\theta$ into Equation 6 · 67,
the "work done ΔW" can be expressed by Equation 4·68 using θ and E_{aw} value that can be measured.

$$\Lambda W = \pi R^2 \sin^2\theta \cos\theta E_{aw} + 2\pi r^2(1-\cos\theta)E_{aw} \quad \text{—}(6\cdot68)$$

③ In addition, when put to $R = fr$ and $\sin^2\theta = 1 - \cos^2\theta$.

$$\Delta W = 4\pi r^2 E_{aw}[1 - \frac{f^2}{4}(2 + 3\cos\theta - \cos^3\theta)] \quad \text{—}(6\cdot69)$$

④ If the volume of bubbles does not change before and after adhesion,
Since the height $h = R(1+\cos\theta)$ and the volume of segment of a phere of radius $V = \pi h^2(3R - h)/3$,

$$\frac{4}{3}\pi r^3 = \frac{1}{3}\pi R^2(1+\cos\theta)^2[3R - R(1+\cos\theta)] \quad \text{—}(6\cdot70)$$

⑤ Substituting $R = fr$,

$$f^3 = \frac{4}{(2+3\cos\theta - \cos^2\theta)} \quad \text{—}(6\cdot71)$$

⑥ Therefore, Equation 6 · 69 becomes as follows,

$$\Delta W = 4\pi r^2 E_{aw}(1-F) \quad \text{—}(6\cdot72)$$

$$F = \frac{1}{f} = \left[\frac{(2+3\cos\theta - \cos^3\theta)}{4}\right]^{\frac{1}{3}}$$

(6.68)–(6.72)

Equation 6.68 - 6.72 The change in interfacial energy before/after Work done.

Table 6.9 Contact angles and the (1-F) value.

θ	1–F
1	0.000 000 006
2	0.000 000 08
5	0.000 003 6
10	0.000 057
15	0.000 29
30	0.004 4
45	0.020
60	0.055
90	0.206
120	0.461
180	1.000

θ: degree.

The values of $(1 - F)$ are shown in Table 6.9, and are positive for all contact angles between $\theta = 0$ and 180 degrees, thus indicating absorptivity at all times. As is clear from the table, ΔW decreases rapidly with decreasing contact angle.

The water surface tension σ_{aw} ($erg/cm^2 = dyn/cm$) is shown in Table 6.10, and σ_{aw} is equal to the surface energy E_{aw} per area.

Using these values, Table 6.11 shows an example of how the work of deposition changes with changes in the bubble system and contact angle using Equation 6.72. In the case of microbubbles, adhesion is difficult to achieve in areas where the work amount changes by about 10 KT or less. In the upper left column of Table 6.11, these difficult attachment conditions emerge in the areas of the small bubble diameter and the small contact angle.

When air bubbles are deposited on the surfaces of particles by the dissolved air flotation method, the particles must be hydrophobic to a certain degree, or more (i.e. with a large contact angle). However, the inclusion of the deposited air bubbles differs depending on the agglomeration condition of the particle group, and the structure of the coagulated particle agglomerates (flocs) is significant. (Refer to Section 4.4 in Chapter 4 to understand coagulation processes and floc structure formation).

When the bubble diameter is greater than the attached particles, the bubble diameter can be considered not to change, and the air–water interface and the solid–water interface are reduced by the area S of adhesion, and the solid–air interface is increased by S. Therefore, the difference between the surface energies before and after deposition, i.e. the amount of work of deposition ΔW, is the sum of the changes in the three interfaces, and is given by Equation 6.73 using the relation $E_{sw} - E_{as} = -E_{aw}\cos\theta$ described above.

Table 6.10 Water surface tension (dyn/cm).

Temperature °C	0	5	10	15	20	25	30
Surface tension σ_{aw}	75.6	74.9	74.2	73.5	72.8	72.0	71.2

Table 6.11 The work of adhesion due to changes in the bubble diameter and gas–liquid contact angle.

(10 KT erg)

Contact Angle (θ)	Bubble Diameter (μ)			
	1	**10**	**100**	**1000**
0.5	3×10^{-3}	0.3	30	3×10^{3}
1	3.4×10^{-2}	3.4	3.4×10^{2}	3.4×10^{4}
2	0.44	44	4.4×10^{3}	4.4×10^{5}
5	2.1	2.1×10^{3}	2.1×10^{5}	2.1×10^{7}
10	3.2×10^{2}	3.2×10^{4}	3.2×10^{6}	3.2×10^{8}
15	6×10^{2}	6×10^{4}	6×10^{6}	6×10^{8}
30	1×10^{4}	1×10^{6}	1×10^{8}	1×10^{10}
90	1.2×10^{6}	1.2×10^{8}	1.2×10^{10}	1.2×10^{12}

K: Boltzman constant 1.381×10^{-16} erg/°C.
T: Absolute temperature.

$$\Delta W = S(-E_{as} + E_{aw} + E_{sw}) = SE_{aw}(1 - \cos\theta) \quad \text{---}(6 \cdot 73) \tag{6.73}$$

S: contact area
E_{sw}: Solid-liquid Interfacial energy per unit volume (erg/cm³)
E_{as}: Gas-solid Interfacial energy per unit volume (erg/cm³)
E_{aw}: Gas-liquid Interfacial energy per unit volume (erg/cm³)

Equation 6.73 The amount of work of deposition.

Equation 6.73 shows that increasing the diameter and contact angle of the deposited particles results in an increase in the deposition energies. Table 6.12 shows the approximate work (in orders of magnitude) for

Table 6.12 Approximate work when suspended particles adhere to large bubbles.

Per adhesion of one suspended particle (erg)

Contact Angle (θ)	Particle Diameter (μ)			
	1	**10**	**100**	**1000**
1	10^{-13}	10^{-11}	10^{-9}	10^{-7}
2	10^{-12}	10^{-10}	10^{-8}	10^{-6}
5	10^{-12}	10^{-10}	10^{-8}	10^{-6}
15	10^{-11}	10^{-9}	10^{-7}	10^{-5}
30	10^{-10}	10^{-8}	10^{-6}	10^{-4}
60	10^{-9}	10^{-7}	10^{-5}	10^{-3}
90	10^{-9}	10^{-7}	10^{-5}	10^{-3}

large bubbles with a single suspended particle of differing size, assuming a cubic surface area of 1/10 of the diameter shown in the table with contact area S.

From the equation for the adhesion strength of the sine scalar in Equation 6.66, the adhesion force of the bubble of size l is as shown in Equation 6.74.

$$F = l\sigma_{aw}\sin\theta \quad \text{(dyne)} \quad \text{—}(6\cdot 74)$$

l : Bubble diameter
σ_{aw} : Interfacial tension between gas and liquid phase (water) (dyn/cm)

(6.74)

Equation 6.74 The adhesion force of bubbles.

6.1.2.3 *Flotation rate of solid and gas agglomerated particles*

If solid/gas bubbles are bonded together and a particle floats, the particle will not float if the buoyancy is insufficient even if the bubble and the particle adhere; equally, if the adhesion is insufficient, even if the buoyancy is sufficient, the particle and the gas bubble will separate in the course of the process and the treatment cannot be continued.

The force F_f acting upward on the bubbles and the force F_{sa} acting upward on the solid–gas mass are represented by Equation 6.75 and Equation 6.76, respectively.

Force acting upward on bubbles

$$F_f = V_b g(\rho_l - \rho_g) \quad \text{—}(6\cdot 75)$$

Force acting upward on solid particle mass

$$F_{sa} = \frac{g\,[(V_b - V_s)\rho_l - (V_b\rho_g + V_s\rho_s)]}{V_b\rho_g + V_s\rho_s} \quad \text{—}(6\cdot 76)$$

V_b, V_s : Each volume of bubbles and solid particles
ρ_l, ρ_g, ρ_s : Each density of liquid, solid, and gas
g : Gravitational acceleration

(6.75)–(6.76)

Equation 6.75, 6.76 The upward force acting on bubbles and solid particles.

By letting F be the strength of the bubble and particle adhesion, we can determine whether or not to float according to the inequality shown in Table 6.13. Generally, F is calculated as shown in Equation 6.74, but in the case of coagulated agglomerates (flocs), it is assumed that the upward force is larger than F due to the trapping of complex floc surfaces.

When both adhesion and flotation conditions are met, the solid particles can be floated and separated from the liquid phase. The flotation speed in this case can be calculated using the same equation as in the case of sedimentation, and the ascending speed w_f is shown in Equation 6.77.

Table 6.13 The floating possibilities of flocculated particles.

① In the case of Floating: $V_g\rho_g + V_s\rho_s < (V_g + V_s)\rho_l$ and $F > F_f$
② In the case of Settling: $V_g\rho_g + V_s\rho_s > (V_g + V_s)\rho_l$ and $F > F_f$
③ In the case of Gas and solid particles separating : $F < F_f$

$$W_f = \left(\frac{4}{3}\frac{g}{C_D}\frac{(\rho_{sa}-\rho_l)}{\rho_l}d\right)^{\frac{1}{2}} \quad \text{—}(6\cdot77)$$

W_f : Floating velocity of solid-gas flocculated particles (cm/sec)
d : Diameter of solid-gas flocculated particles (cm)
ρ_l : Liquid density (g/cm³)
ρ_{sa} : Solid-gas flocculated particles density (g/cm³)
C_D : Drag coefficient

(6.77)

Equation 6.77 The floating velocity equation of solid-gas bonded particles.

Combinations of air bubbles and solid particles that float may cause air bubbles to adhere (by adsorption or trapping) to the surfaces of solid or floc particles, or may cause air bubbles to be entrapped in the interstitial spaces of floc particles, as described above. In the latter, the density of the coagulated floc/bubble mix is hardly affected by the size of the solid particles. If the ascending speed is in the Stokes range, the floating speed is increased by n^2 when the solid particle size is increased n times. On the other hand, assuming that the amount of bubbles deposited per unit surface area of solid particles is constant regardless of the solid particle size, the rate of change in the rate of increase in the Stokes region is calculated by Equation 6.78 when the particle size increases n times. The density $\rho_{sa}[n]$ at n times the particle size is given by Equation 6.79, where $\rho_\ell \fallingdotseq 1$ and $\rho_a \fallingdotseq 0$, and the density of solids is ρ_s.

$$\frac{W_f[n]}{W_f[1]} = \frac{\rho_l-\rho_{sa}[n]}{\rho_l-\rho_{sa}[1]}\left(\frac{nd}{d}\right)^2 = \left(\frac{\rho_l-\rho_{sa}[n]}{\rho_l-\rho_{sa}[1]}\right)n^2 \quad \text{—}(6\cdot78)$$

$W_f[1]$, $W_f[n]$: Each floating velocity of solid particles and bubbles with diameters of 1 time and n times
$\rho_{sa}[1]$, $\rho_{sa}[n]$: Each density of solid particles and bubbles with diameters of 1 time and n times

The density $\rho_{sa}[n]$ when the particle diameter becomes n times bigger.
If $\rho_l \fallingdotseq 1, \rho_a \fallingdotseq 0$ and the solid particle density ρ_s is

$$\rho_{sa}[1] \fallingdotseq \frac{\rho_s V_s}{V_a+V_s} \text{、} \rho_{sa}[n] = \frac{\rho_s V_s n^3}{V_a n^2 + V_s n^3} \quad \text{—}(6\cdot79)$$

(6.78)–(6.79)

Equation 6.78, 6.79 The floating velocity change rate and density when the particle diameter becomes n times bigger

However, in the case of coagulated floc particles (random agglomerates), the void fraction increases with increasing floc diameter, and the structural density of floc particles decreases, as described in Chapter 4. Therefore, the density must be calculated by taking into account the floc density functions of 'Tambo and Watanabe', described above in Section 4.2.3 as Equation 4.38.

6.1.2.4 Chemicals for flotation treatment

In this section, we briefly explain the chemicals used for flotation treatment according to the minimum required functions. For efficient flotation, chemicals needed for coagulation or floc formation (coagulants and flocculants), foaming agents (frothers), scavengers (collectors), activators, and inhibitors (depressants) are all used as required. Activators and inhibitors are sometimes collectively referred to as accelerators (promotors). Together, all of these chemicals are generically referred to as flotation agents:

(1) Coagulants are added to a suspension to neutralize the surface charge of the particles, reduce the repulsion potential between suspended particles, and play a cross-linking function to strengthen

the agglomeration bond. Aluminium salts, ferrous salts, and polymeric coagulants, which are used for sedimentation and depth filtration, are also used for flotation. By forming a strong floc of about 100 μm with an appropriate size for flotation, air bubbles can be contained in the floc structure or in the irregular outer periphery to enable stable separation.

(2) Foaming agents are the main additive used in the foam separation (froth flotation) method. They are used to reduce the interfacial tension of the gas–liquid interface in order to promote the formation of stable foam on the water surface. The chemicals used are compounds in which one side of the molecules has no polarity and the other end has a strong polarity. Generally, a linear organic material is used. Typical examples include amyl alcohol $C_5H_{11}OH$ and cresol $CH_3C_6H_4OH$. They are widely used for froth flotation in metal mines.

(3) Collectors (scavengers) are chemicals that adsorb to the surface of the suspensions, making the surface difficult to wet (which increases the contact angle), and promoting stable adherence of bubbles. Therefore, they are required to react with hydrophobic compounds or substances that are hydrophobic per se, or suspensions that do not have hydrophobic groups to modify hydrophobic surfaces. Common scavengers are surface-active substances composed of hydrophobic and active groups which are adsorbed on the surface of suspensions by the active groups, with the hydrophobic groups facing outward. As hydrophobic groups, alkyl groups and the like are used, and as active groups, those having high affinity with the surface of the suspensions are selected.

(4) Accelerators or activators are used to further enhance the effectiveness of scavengers, and inhibitors are used to remove the effectiveness of scavengers for certain suspensions to enable specifically focused scavenging. They are widely used for refining metals, but are not often used for water treatment.

6.1.2.5 Dissolved air flotation

The most widely used process for water treatment is dissolved air flotation. This method for depositing micro air bubbles into a suspension is roughly classified into vacuum flotation and pressure flotation methods.

In vacuum flotation, the treating water (inflow raw water and/or raw wastewater) is first aerated at normal pressure, air is dissolved into water to saturate it (this process is not necessary if the raw water is already saturated with air), and then it is introduced into a covered vacuum tank. Since the tank is decompressed to near vacuum, dissolved air in the water becomes supersaturated, fine bubbles are precipitated and adhere to suspended particles, which become solid–gas-adhering particles and float to the surface of the water. However, the amount of air that can be used in this process is only the amount that can be dissolved at atmospheric pressure (1 atm). Also, it is difficult to create a large vacuum separation tank, and it is not easy to extract the air–solid particle mass that has floated and separated to the outside of the tank. Therefore, the practical value of the air–solid particle mass is low. On the other hand, since it is separated in a vacuum tank, it has suitable features for the treatment of harmful wastewater that emits odors and toxic gas components to the open atmosphere.

In pressure flotation, portions of treated sewage or raw water, or all of the raw water, are introduced into a pressure water tank and pressurized air at about 2 to 5 atmospheres is introduced to dissolve air in the water. After air is absorbed to near saturation under pressure, the pressurized air absorbing solution is drawn out of the tank, the pressure is reduced instantaneously to normal pressure, and the generated microbubbles are attached to the suspensions, which are separated by flotation under atmospheric pressure. With this method, a large amount of air can be obtained by increasing the pressure of the air dissolving tank. In addition, in order to carry out the separation at normal (atmospheric) pressure, unlike with the vacuum flotation method, a large and expensive pressure reduction tank is not required, and there are many advantages, including that it is easy to remove the flotation-separated scum (flotation sludge), and the pressure flotation method is used in most water treatment apparatus.

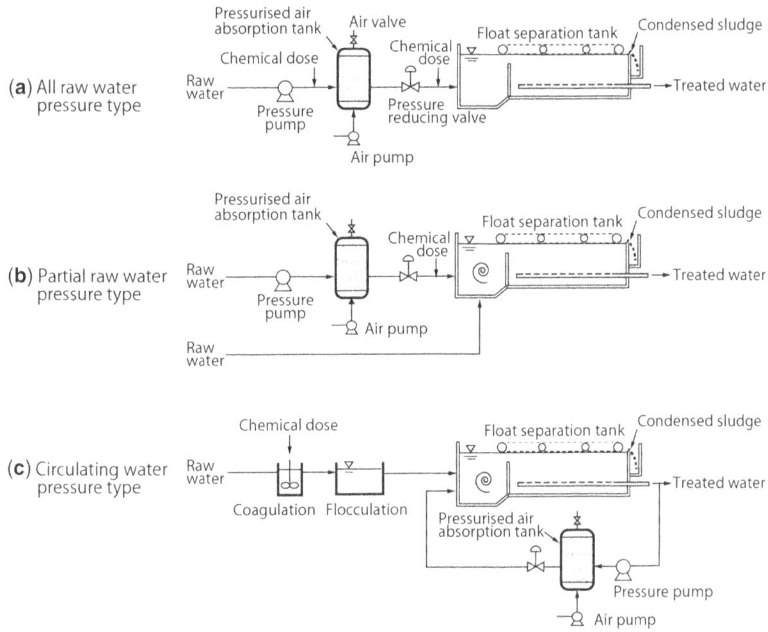

Figure 6.70 The three main methods of pressure dissolved air flotation.

Pressure flotation treatment systems can be divided into process configurations such as full raw water pressurization, partial raw water pressurization, and circulation water pressurization, depending on the process at which the air is pressurised-dissolved and added to the raw water. Figure 6.70 shows a conceptual diagram of the three typical methods.

The total raw water pressure method is a method of introducing the entire amount of raw water into a pressure tank to dissolve air. This method has the disadvantage of increasing the volume of the air absorbing tank. In order to overcome this drawback, in the partial raw water pressurization method only a part of the raw water is diverted and introduced into a pressurizing tank. Air dissolution is carried out under pressurizing conditions capable of obtaining the necessary amount of dissolved air. After that, large amounts of microbubbles are generated by depressurizing to atmospheric pressure, and then the water is mixed with the remaining unpressurized raw water, and air bubbles are attached to the entire suspension (flocs) to perform flotation. This system, which generally pressurizes about half of the raw water, offers the advantages of a smaller design for the pressurized air-dissolving tank. The direct raw water pressurization method is not considered to be applicable when flocs are handled by pump pressurization of raw water by which fragile flocs can be broken.

In order to avoid these drawbacks of the raw water pressure method, the circulation water pressure method was considered. In this method, a part of the treated water that does not contain suspensions is subjected to pressurized air absorption. Once the pressure is reduced and the microbubbles are precipitated, it is mixed immediately before the inflow of the flotation tank, and microbubbles are attached and adhere to the flocs. According to this method, the raw water containing flocs does not pass through the pressurizing pump and the depressurizing valve, and the removal rate is not lowered by the breakage of coagulated particles (flocs). However, in the circulation method, the treatment flow rate of the flotation tank is increased by the combination of circulation water and raw water to increase the flow rate, which increases the size of the tank and increases the required power. Generally, the circulation flow rate is set to 20–30% or less of the original flow rate, the air dissolution pressure is increased, and the system is designed to suppress the need for an increase in the size of the tank. This method is the most widely used method for achieving high overall treatment efficiency.

6.1.2.6 Dissolution of air and precipitation of air bubbles

The main components of air are nitrogen and oxygen, which are poorly soluble in water. The solubility of air components increases in proportion to gas pressure, according to Henry's law. Air as a whole is the sum of the behavior of the principal components, and the relationship between air pressure p[atm] and solubility x also holds to Henry's law, as shown in Equation 6.80, and the theoretical saturation solubility can be calculated using Henry's law constant for air, shown in Table 6.14.

$$p = Hx \qquad \text{—}(6 \cdot 80) \tag{6.80}$$

p : Partial pressure of gas phase of dissolved gas (atm)
H : Henry's law constant (atm/fraction)
x : Solubility (mol fraction of solute gas in liquid phase)

Equation 6.80 The Henry's Law.

Figure 6.71 shows the theoretical results of these procedures for calculating the saturation solubility of air. When air is pressurized and dissolved and released to atmospheric pressure, the amount released varies depending on the pressure, temperature, and types and concentrations of dissolved substances in the system. If a large amount of soluble substances are present, the amount of air released will be less than the pure water.

Table 6.14 Henry's law constant for air.

Temperature (°C)	**0**	**10**	**20**	**30**	**40**	**50**	**60**	**70**	**80**	**90**
$H \times 10^{-4}$	4.32	5.49	6.64	7.71	8.70	9.46	10.1	10.5	10.7	10.8

$H \times 10^4$ (atm/mol fraction)
Reference: Average molecular weight of air 28.97 Molecular weight of water 18.02

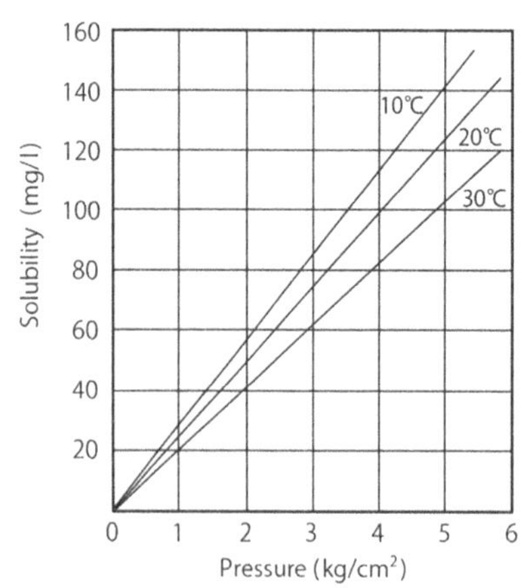

Figure 6.71 Theoretical air solubility.

In our laboratory, a study in the 1960s in which dissolved and released air into treated sewage showed no differences from theoretical levels until BOD 40 ppm, a slight decrease with increasing solids concentrations, and a clear decrease above 1000 ppm total solids. We believe that the circulation water pressure method can be expected to emit almost as much air as the theoretical amount of air.

As a method of introducing air and pressure-dissolving it:

(1) introduce air into the suction side of the pump, mix it with pump impeller, and introduce it into the dissolution tank;
(2) pull air into the line by ejectors (using a venturi nozzle, etc.) and insert it into the dissolution tank;
(3) the water is blown into the tank from the top of the tank using a showerhead, and the water is rapidly dissolved in the pressurized air in the tank as the shower water falls;
(4) alternatively, a process of continuously introducing pressurized air bubbles from a diffuser plate or nozzles installed at the bottom of a pressurized dissolution tank filled with water to dissolve air in the water in the tank is also conceivable.

Figure 6.72 lists examples of various pressurized air dissolution tanks.

When introducing air into the suction side of a centrifugal pump, it appears that the amount of air introduced needs to be 4–8% or less of the amount of water at normal pressure in order to avoid air failure (air binding). According to this method, 60–80% of the saturated dissolved amount is dissolved in water, and this amount is about 35–45% of the added amount of air. In the static equipment shown in Figure 6.72, the air dissolution efficiency stops at 50%. In a dynamic pressurization and dissolution tank, efficiency can be increased to 90% or more. However, theoretical design of air dissolution by ejectors is difficult, and dissolution efficiencies do not appear to be any higher than 5–10%. Inflow of water from the top of a tank filled with pressurized air in a shower-like manner increases the danger of clogging of nozzles and makes application difficult for wastewater treatment. The bubble column, in which pressurized air is introduced from the bottom and air ascends, is expected to be highly efficient as a

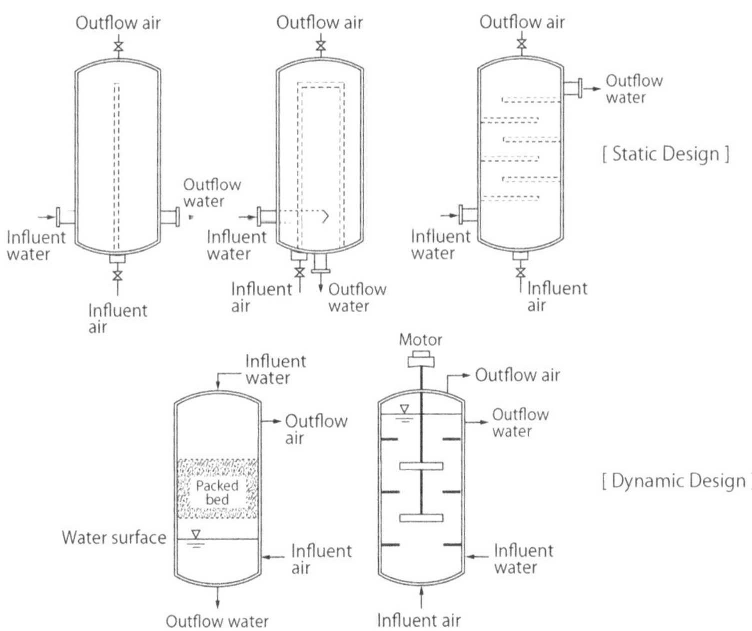

Figure 6.72 Various pressurized air dissolution tanks.

dynamically pressurized dissolution tank. It is expected that taking measures against aerator clogging is one of the most effective methods. Mechanical agitation can be used in combination to achieve a dissolution efficiency of 90% or more.

The dissolution efficiencies of air described above are governed by the retention time of the pressurized air dissolution tank. However, in practice, many air dissolution tanks only have a retention time of about 1–2 minutes, and at most about 3–4 minutes.

After the pressurized air has been absorbed, the pressure of the water is reduced to atmospheric pressure through a vacuum valve. A large number of microbubbles are precipitated and adhere to the solid particles, and the water floats them as solid–gas-bound particles.

In the 1980s, we fabricated a flotation tester (the Hokkaido University model) as shown in Figure 6.73, and conducted basic research on the properties of microbubbles generated by depressurization with a pressurized air nozzle, solid particle generation, and separation. This study was developed into the basic study of bubble floc coagulation separations during flotation which was described in Section 4.4 in Chapter 4.

Figure 6.74 shows the size distributions of microbubbles generated by reducing the pressure to atmospheric pressure through the needle valve of a flotation tester using stereo-microscopy. This system can use pressurized air and water to 10 kg/cm^2, but Figure 6.74 shows an example of the precipitation of a 4 kg/cm^2 pressurized air bubble. The size distribution was approximated by probability distribution, the average diameter was approximately 55.5 μm, and the average volume diameter was 62.8 μm. The distribution was hardly affected by the opening degree of the needle valve. In addition, the results of measuring the bubble size when the microbubbles formed under the same conditions adhere to the coagulation flocs and begin to float as solid gas masses are shown in Figure 6.75. The particle size distribution of the attached air bubbles is almost the same as in Figure 6.74. The generated micro-floc does not grow over time and has a stable distribution as it is generated, and it binds to the suspended matter (floc, etc.) and adds buoyancy.

The basic characteristics of the pressurized air flotation method are that the microbubbles generated by the vacuum nozzles do not form large bubbles even if they collide with each other. The microbubbles can

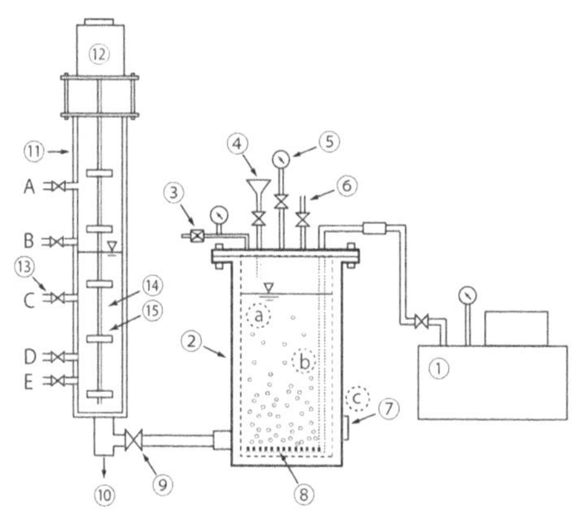

① Compressor
② Pressure tank (Capacity 10l)
③ Pressure regulating valve (Needle valve)
④ Water inlet
⑤ Pressure gauge
⑥ Exhaust valve
⑦ Observation window (a~c)
⑧ Diffuser pipe
⑨ Pressure reducing valve
⑩ Drain
⑪ Flotation cell (70×70×500 mm)
⑫ Variable-speed motor
⑬ Exhaust port (A~E)
⑭ Sample water (1.5 or 2l)
⑮ Impeller

Figure 6.73 The Hokkaido University model flotation tester.

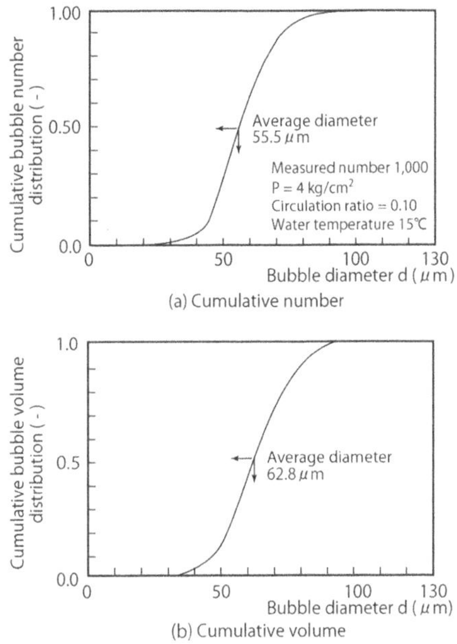

Figure 6.74 Size distribution of microbubbles.

adhere and coalesce to the suspensions and form a stable solid–gas combination by transport. Figure 6.76 shows the ZP distributions of microbubbles generated in Sapporo City water as measured by the ascending flow potential method. The ZP distributions have large negative surface charges in the neutral-pH range of about −150 mV (which is much higher than the critical coagulation potential level (−10 mV) defined by

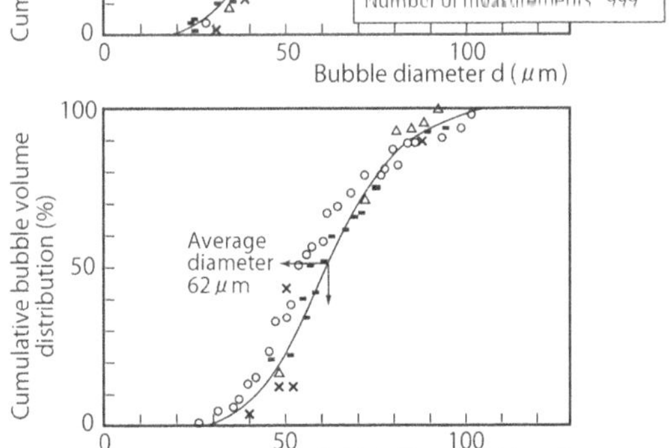

Figure 6.75 Particle size distribution of attached air bubbles. The ALT ratio indicated in the top graph refers to the added coagulant (aluminium) concentration/(dimensionless) suspension concentration.

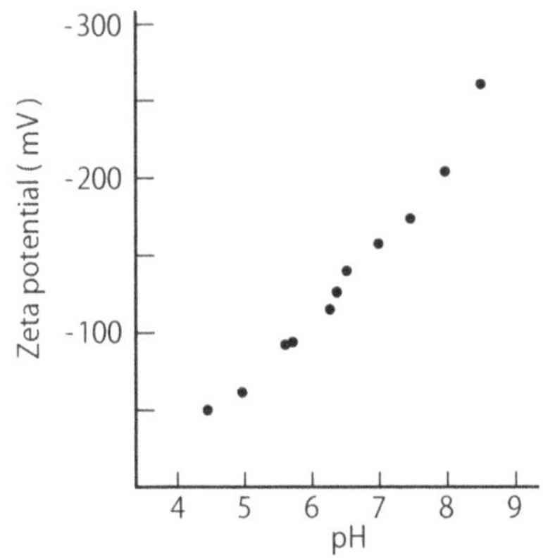

Figure 6.76 The zeta potential (ZP) of microbubbles.

Tambo; see Chapter 4, Figure 4.8) and collisions and agglomeration cannot be formed due to the high repulsion of the surfaces to each other. The microbubbles suspend stably in water and attach onto the particle surfaces without changing the size distributions, as a result of the very high ZP.

6.1.2.7 Process requirements for the dissolution air flotation method

The main process factors for flotation treatment are the solids concentrations of the raw (waste) water, the pressurized air pressure, the circulation flow ratio, and the overflow rate (retention time) of the flotation tank. These optimum values differ depending on the type of raw (waste) water, the type of equipment used, etc., and cannot be generally determined. It has been reported that the air dissolution pressure normally used is in the range of 2–5 kg/cm^2 and the dissolution amount is approximately 0.5% (5000 ppm) of raw (waste) water, and that flotation treatment can obtain both good sludge thickening and treated water quality as a high concentration treatment process (Vrablik, 1957). The amount of air needed to be introduced is closely related to the ratio of air to suspended matter in the raw (waste) water, and thus the gas–solid ratio (air to solid ratio) is often used as an important index of operation.

This section introduces the application of activated sludge to the terminal flotation treatment by Eckenfelder. An assessment of the process and design of this theoretical treatment was described precisely, with the authors' own work, in the coagulation and flocculation section in Chapter 4.

The gas to solid ratio is the ratio A/S of air, A, discharged from the pressurized air dissolving tank divided by the suspended mass S in the treated water, and is defined as in Equation 6.81.

$$\frac{A}{S} = \frac{CS_a(fp-1)R}{S_aQ} \quad \text{—(6·81)} \tag{6.81}$$

A : Amount of outflow air from the absorption tank (mg/l)
S : Amount of suspended solids in treated water (mg/l)
C : Coefficient (-)
S_a : Saturated dissolved air at atmospheric pressure (mg/l/atm)
f : Ratio of [Actual dissolved air/Saturated dissolved air]
p : Pressure (atm)
Q : Flow rate of treated water (l/sec)
R : Flow rate of Circulating (Pressure) water (l/sec)

Equation 6.81 The air to solid ratio.

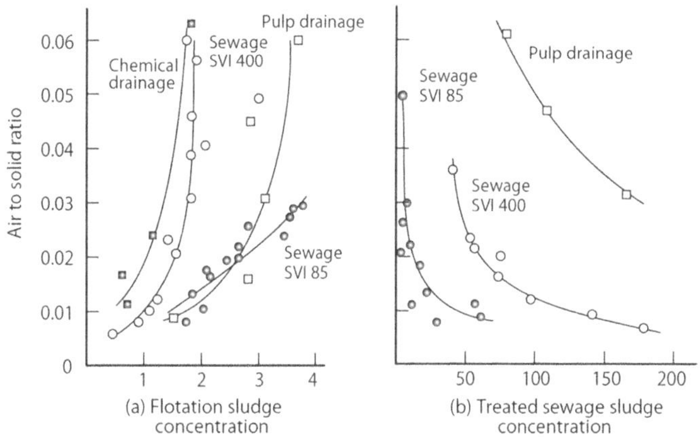

Figure 6.77 The gas-solid ratio in concentrations of (a) flotation sludge; and (b) treated sewage sludge (Eckenfelder & O'Connor, 1961).

Figure 6.77 shows examples of gas–solid ratios in concentrations of (a) flotation sludge, and (b) treated sewage sludge. Both concentrations were obtained by flotation treatment and vary in the same manner in terms of the gas–solid ratio, but the treatment results differ depending on the concentration of the raw wastewater. However, when looking at these two figures, it appears that in the area of $A/S < 0.02$, the concentration of suspensions in the treated water rapidly deteriorates due to the decrease in the amount of air, and in the area of $A/S > 0.03$, the increase in the amount of air hardly improves the concentration of floating sludge nor the concentration of treated sewage. Therefore, the process of flotation treatment of activated sludge under such circumstances should be operated with an A/S ratio of about 0.03 to 0.04.

6.2 FILTRATION

Filtration, in which a suspension is flowed through a porous bed to separate a solid and a liquid, is widely used in engineering. The filtration process can be roughly divided into two types:

(1) Depth filtration (inner filtration) – a process in which a suspension of particulate matter such as clay particles is passed through thick layers of filter media and the suspension is trapped and removed in the voids between the filter media. The size of the suspended particles trapped is much smaller than the void size of the filter media. Inner filtration or depth filtration is named after the mechanism of removal. Suspension removal with such a process is limited to dilute systems (0.1% or less) and is suitable for clarification filtration, most often being used in water treatment.

(2) Cake filtration (surface filtration) – a process in which a muddy layer (a sludge mat) called a cake (or filter cake) is formed on the surfaces of porous walls that itself becomes a filter medium, and the suspension is passed through the cake layer to separate and clarify the suspension, with more cake being formed at the same time. Because the suspension is removed on the surface of the filter medium, it is referred to as surface filtration. This process is often used to treat suspensions (e.g., slurries) with higher volume concentrations (e.g., as much as 1% or more) as is the case with sludge dewatering.

Filtration processes are classified into gravitational force filtration, vacuum filtration, pressure filtration, and centrifugal force filtration, according to the propulsive (driving) force employed. From this point of view, the filtration processes used for water treatment can be classified as shown in Table 6.15.

In addition, there is a large new group of membrane filtration technologies. Membrane filtration is a process carried out by a functional separation membrane, which was first introduced at the start of the 21st Century and is now widely used. This filtration method has been developed for use over a wide range of applications ranging from suspended to dissolved matter with molecular weights ranging from 10^5 Da to 10^0 Da, using micro-filtration (MF), ultra-filtration (UF), nano-filtration (NF) and reverse osmosis (RO), with molecular weight as a main index, and is now widely used in the separation process of water treatment.

6.2.1 Depth filtration (inner filtration)

Historically, sand filters (both rapid and slow sand filtration systems) have been widely used in water supply systems, as part of granular bed (inner) filtration. Recently, however, they have also been widely used for the recycling and advanced treatment of treated sewage.

The activated carbon adsorption tower and ion-exchange tower described in Section 7.1 are processes which have the same hydrological characteristics as a standard fixed bed granular system. Therefore, the hydraulics of the fixed bed are explained in this section, and we define the characteristics of the group of granular materials used, and consider the flow in a fixed bed and as well as the flow in a fluidized bed.

6.2.1.1 Granular filtering materials

The most common particulate matter (granular materials) in a filter bed for water treatment is quartz-sand, but anthracite, garnet, etc., are also used as components of a multilayer filter bed. Others materials include ion-exchange resin and granular activated carbon which is used for adsorptive treatment.

Particle size. To define the size of particles used, the following diameters are used for non-spherical particles, depending on the purpose of the measurement:

(1) one-dimensional diameter: by projecting particles onto a two-dimensional plane and measuring a large number of particles with a diameter measured in a given direction, a statistically useful average value can be obtained.

(2) mean diameter: there are two mean diameters, using the major and minor axes, with a third axis added to indicate depth size.

Table 6.15 The classification of filtration processes used in water treatment.

	Depth Filtration (Inner Filtration)		Cake Filtration (Surface Filtration)			
Driving Force	**Gravity**	**Pressure**	**Pressure**		**Vacuum Negative Pressure**	**Centrifugal Force**
Device example	Rapid sand filtration Slow sand filtration	Tank filtration	Diatomaceous earth filtration	Filter press	Rotary drum filtration	Centrifugal dewater
Application	Separating clear water			Dewatering and thickening of sludge		

(3) equivalent diameter : this is the sphere diameter linked by a certain physical law. For example, the diameter of a sphere with a sedimentation velocity equal to that of the particles being considered may be defined as the equivalent settling sphere diameter, as described in the Section 6.1.1 on sedimentation, above.

Usually, the particles being handled do not consist of particles of the same size, but a mixture of various sizes. For such particle groups, the size distributions described in the next section are used to strictly describe them. However, the following definitions of mean sizes are often used as a typical expression to show the characteristics of the particle groups.

Assuming that the number of particles of particle diameter d_i is n_i, typical examples of various mean diameters are shown below:

(1) The length mean diameter (arithmetic mean diameter) d_a is given by Equation 6.82.

$$d_a = \frac{\Sigma (n_i d_i)}{\Sigma n_i} \quad \text{—(6·82)} \tag{6.82}$$

d_a : Arithmetic mean diameter
n_i : Number of particles of diameter d_i

Equation 6.82 The length mean diameter.

(2) Area mean diameter (mean surface diameter) d_s is an indicator for adsorption, etc. and is represented by Equation 6.83.

$$d_s = \sqrt{\frac{\Sigma (n_i d_i)^2}{\Sigma n_i}} \quad \text{—(6·83)} \tag{6.83}$$

d_s : Area mean diameter
n_i : Number of particles of diameter d_i

Equation 6.83 The area mean diameter.

(3) The mean volume or mass diameter d_v is given by Equation 6.84.

$$d_v = \sqrt[3]{\frac{\Sigma (n_i d_i)^3}{\Sigma n_i}} \quad \text{—(6·84)} \tag{6.84}$$

d_v : Mean volume (mass) diameter
n_i : Number of particles of diameter d_i

Equation 6.84 The volume mean diameter.

In addition, the following simple mean diameters are used for the particle size distributions described in the next section (these values differ depending on whether the distribution is represented by a number distribution or a weight distribution):

(1) Largest frequency diameter (mode diameter);
(2) Medium diameter (50% diameter);
(3) Characteristic diameter depending on the objectives in practical use : 10% diameter (effective size).

Particle size distribution and practical particle size indicator. Granular media are usually not only of constant size but also of fairly wide size distribution. As shown in Figure 6.78a, the particle size distribution is obtained by plotting the particle size d_p in the horizontal axis and the abundance ratio for a particle group $f(d_p)\%$ in the vertical axis. The frequency distribution function is the abundance ratio $f(d_p)$ of particles with a diameter d_p. The cumulative frequency distribution q (non-excess probability) is shown in Figure 6.78b, which shows the proportion of particles with a diameter smaller than d_p; q can be expressed as Equation 6.85.

$$q = 100\int_{d_{min}}^{d_p} f(d_p)\,dd_p \quad (\%) \qquad \text{—(6·85)}$$

q : Ratio of particles smaller than diameter dp to total particles (Non-excess probability)
$f(d_p)$: Frequency distribution function

(6.85)

Equation 6.85 The cumulative frequency distribution of particles smaller than diameter d_p.

The ratio R of particles larger than diameter d_p to the total particles is shown in Equation 6.86. Here, if $f(d_p)$ is defined, whether the number distribution or the weight (mass) distribution is used depends on the purpose.

$$R = 100 - q = 100\int_{d_p}^{d_{max}} f(d_p)\,dd_p \quad (\%) \qquad \text{—(6·86)}$$

R : Ratio of particles larger than diameter d_p to total particles (Excess probability)
$f(d_p)$: Frequency distribution function

(6.86)

Equation 6.86 The cumulative frequency distribution of particles larger than diameter d_p.

The cumulative frequency distribution of naturally occurring particle clusters is generally an S-shaped curve on a graph. If we can create a graph that can display such a distribution as a straight line, we can extrapolate or interpolate the measured value to estimate the unmeasured part. In extreme cases, only a

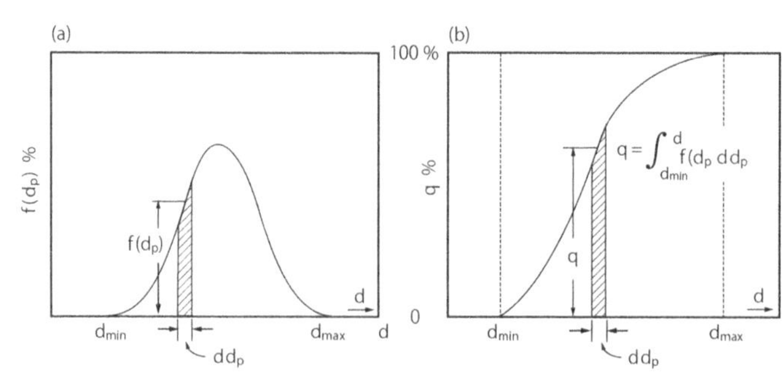

Figure 6.78 Particle size distribution.

minimum of two points can be determined and the distribution assumed. For this purpose, we use the US logarithmic random distribution (logarithmic probability distribution) and the Rosin-Rammler (R-R) distribution, which is the German standard. For coarse particles such as sand, the log probability distribution is often used for linearization, and for particle distributions of 100 μm or less, the R-R distribution is often used.

If the particle size distribution (based on weight) follows a log probability distribution, plotting the log of the particle size d_p in the horizontal axis and the cumulative weight percentage q in the vertical axis with a normal probability scale yields a straight line, as shown in Figure 6.79.

The frequency distribution function $f(d_p)$ and the non-excess probability q% are given by Equations 6.87 and 6.88, respectively.

Frequency distribution function $f(d_p)$

$$f(d_p)\,\% = \frac{100}{\sqrt{2\pi}\,\log\sigma_g}\exp\left[-\frac{(\log d_p - \log d_m)^2}{2(\log\sigma_g)^2}\right]d(\log d_p) \quad \text{—(6·87)}$$

Non-excess probability q %

$$q\,\% = 100\int_0^{d_p}\frac{1}{\sqrt{2\pi}\,\log\sigma_g}\exp\left[-\frac{(\log d_p - \log d_m)^2}{2(\log\sigma_g)^2}\right]d(\log d_p) \quad \text{—(6·88)}$$

σ_g: Geometric standard deviation
d_m: Geometric mean diameter

(6.87)–(6.88)

Equation 6.87 The frequency distribution function., Equation 6.88 The non-excess probability.

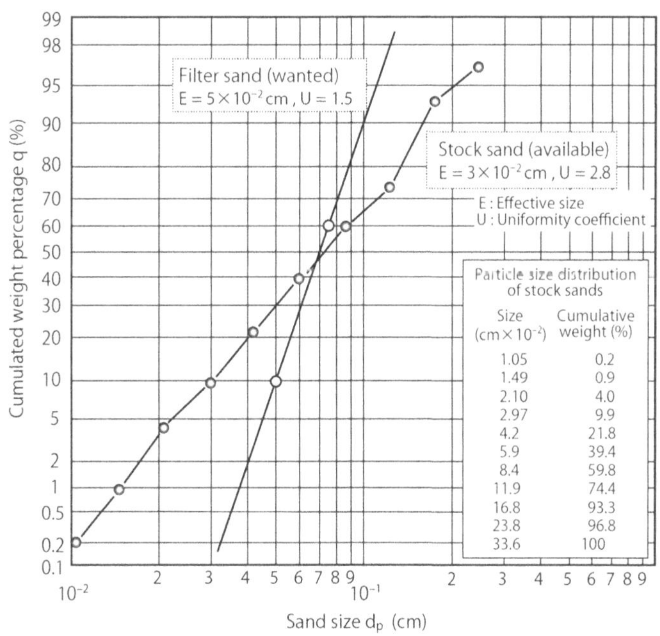

Figure 6.79 A linear plot of particle size distribution.

Assuming that d_m is the median diameter of the weight corresponding to $q = 50\%$, σ_g is given by Equation 6.89 or Equation 6.90, as well as by the standard deviations of normal distributions.

$$\log \sigma_g = \log d_m - \log d_{g=15.9\%} = \log\left(\frac{d_m}{d_{g=15.9\%}}\right) \quad \text{—(6·89)}$$

or

$$\log \sigma_g = \log d_{g=84.1\%} - \log d_m = \log\left(\frac{d_{g=84.1\%}}{d_m}\right) \quad \text{—(6·90)} \qquad (6.89)\text{–}(6.90)$$

Equation 6.89, 6.90 The standard deviation and weight median diameter.

Therefore, the standard deviation σ of the particle size distribution is given by Equation 6.91.

$$\sigma = \frac{q = 50\% \text{ particle size } p_{50}}{q = 15.9\% \text{ particle size } p_{15.9}} = \frac{q = 84.1\% \text{ particle size } p_{84.1}}{q = 50\% \text{ particle size } p_{50}} \quad \text{—(6·91)} \qquad (6.91)$$

Equation 6.91 The standard deviation of particle size distribution.

In the fields of water treatment engineering and soil engineering, the logarithmic probability distribution representation method – in which the effective diameter is defined as the particle size p_{10} of $q = 10\%$, proposed by Hazen, and the ratio of the diameter of $q = 60\%$ to $q = 10\%$ $[p_{60}/p_{10}]$ is defined as the uniformity coefficient – has been conventionally used since the beginning of the previous century. This was proposed by Hazen as a hydraulically significant diameter, based on the empirical rule that the water head loss of a flow passing through a fixed bed containing a mixture of large and small particles is governed by a diameter of 10% (effective diameter p_{10}); p_{10} (cumulative weight 10% diameter) is nearly equivalent to medium in number distributions.

Using the 10% diameter p_{10} to replace the 50% diameter d_m, p_{60}/p_{10} would give a standard deviation $\sigma_g = p_{84.1}/p_{50}$, but if the distribution is normally distributed, it would be the same using either notation. However, according to the Hazen approach, it is possible to define an area that contains 50% of the total volume of the center of the distribution from p_{10} to p_{60}, and it is also possible that the actual distribution can minimize errors in natural systems that do not exhibit a strict lognormal distribution. Equation 6.92 (effective diameter) and 6.93 (uniformity coefficient) show this correlation.

$$\text{Effective size} \quad E = p_{10} = \frac{d_m}{\sigma_g^{1.282}} \quad \text{—(6·92)}$$

$$\text{Uniformity coefficient} \quad U = \frac{p_{60}}{p_{10}} = \sigma_g^{1.535} \quad \text{—(6·93)} \qquad (6.92)\text{–}(6.93)$$

Equation 6.92, 6.93 The effective size and uniformity coefficient.

Particle size analysis. There are various methods for determining the particle size distribution of particulate matter, such as sieving, microscopy, and settling analysis. For a relatively large group of particulate matter with a size of tens of microns or more, the standard sieving method, which can be measured most quickly and easy as an engineering tool, is commonly used. Detailed measurements of particulate sedimentation are described in Section 6.1.

The principle of the sieving method is to determine the percentage of particles that is stopped by two sieves, using the weight percentage of the total particle group that remains on the next (smaller) sieve as $f(d_i)$%, and the geometric mean diameter d_i of the two sieves as the representative diameter d for $f(d_i)$. There are various standards for the standard sieves used for measurements, but typical standards include the International Standard (ISO) R20 series (used by Germany, France, and Canada), and the R40/3 series (used by the United States, the United Kingdom, Australia, South Africa, India, and other countries), as well as the Japanese JIS standard sieve (Z8801-1982) in which the openings are made up of a series of ratios, with the ratio being $\sqrt[4]{2}$.

Size adjustment. When natural filter media are used, such as sand, it is common that the particulate matter in its natural state is too fine or non-uniform with respect to the required particle size. Therefore, if a filter medium with an effective diameter E and an uniformity coefficient U (or a mean diameter d_m and a standard deviation σ_g) is to be obtained, the raw material is sieved to obtain a particle size distribution cumulative curve as shown in Figure 6.79, and the particle size distribution curve is adjusted to the filter medium with the required particle size by the following procedure:

(1) since the amount of particles between p_{60} and p_{10} accounts for 50% of the total amount of particles with a specified particle size distribution, the percentage F of usable particles in the raw material is given by Equation 6.94, where q_{60} and q_{10} are the non-excess probabilities of the raw materials corresponding to the specified 60% and 10% diameters, respectively.

(2) since 10% of the available raw material fraction F is less than or equal to the specified filter size p_{10}, the amount S of raw material that is too fine to be used is given by Equation 6.95, and the diameter d_s corresponding to $q = S$ of the particle size accumulation line of the raw material is the lower limit of the diameter of the raw material particles that can be used.

(3) Since 40% of the quantity F of raw materials available has a size of p60 or more, the ratio of raw materials that are too coarse to be used is given by Equation 6.96. Therefore, a diameter dG corresponding to $q = 100 - G = q10 + 1.8$ (between q60–q10) of the grain size addition line of the raw material is used as the maximum particle size.

Percentage of usable particles in stock sand : F

$$F = 2(q_{60} - q_{10})\ \% \qquad \text{—(6·94)}$$

Percentage of sand too fine to use out of stock sands : S
The size corresponding to q = S is the smallest particle size used.

$$S = q_{10} - 0.1F$$
$$= q_{10} - 0.2(q_{60} - q_{10})\ \% \qquad \text{—(6·95)}$$

Amount of unusable coarse size particles in stock sand : G

$$G = 100 - F - S$$
$$= 100 - [q_{10} + 1.8(q_{60} - q_{10})]\ \% \qquad \text{—(6·96)}$$

The size corresponding to the following equation of particle size cumulative line of stock sands is maximum size used.

$$q = 100 - G$$
$$= q_{10} + 1.8(q_{60} - q_{10})\ \%$$

(6.94)–(6.96)

Equation 6.94 - 6.96 The particle size adjustment of filtration sands.

Based on these calculations, sieving particles smaller than d_s and larger than d_G yields a population of particles between S% and (100 − G)% of the raw material, and a filter media with particle size distributions of p_{10} and p_{60}/p_{10}, respectively.

Particle shape. Shape factor is an indicator of the difference of a particle's shape from that of a sphere. There are various definitions of shape difference coefficients, but commonly used is sphericity, φ, which is defined as the surface area of a sphere having the same volume as the particle considered divided by the actual surface area of the particle considered. The sphericity of a sphere is $\varphi = 1$, and the other particles always have $\varphi < 1$. This is defined in Equation 6.10 (above) and is illustrated in Table 6.3.

The flow resistance of a particulate (fixed) bed is proportional to the surface area of the particles in the laminar flow area, as is the chemical reaction, mass and heat transfer. Therefore, the surface area of all particles in a particle layer of a unit volume cm^2/cm^3 is referred to as the specific surface area 'a', and the specific surface area a of a particle layer of a particle size d, with sphericity φ, and a porosity ε_s is expressed by Equation 6.97. If the volume V_p and surface area A_p of a single particle are known, the sphericity φ is defined by Equation 6.98.

Assuming that the typical diameter of the particles to be considered is d, the area shape coefficient is α, and the volume shape coefficient is β, the surface area can be expressed as $A_p = \alpha d^2$ and the volume $V_p = \beta d^3$. By substituting these into Equation 6.98, the relation between the sphericity φ, the area shape coefficient α, and the volume shape coefficient β becomes Equation 6.99.

Specific surface area $$a = \frac{6(1-\varepsilon_s)}{\varphi d} \quad \text{—(6·97)}$$

a : Specific surface area (cm^2/cm^3)
d : Particle size
φ: Sphericity
ε_s: Porosity

Sphericity $$\varphi = 4.87\left(\frac{V_p^{2/3}}{A_p}\right) \quad \text{—(6·98)}$$

A_p : Surface area of one particle
V_p : Surface volume of one particle

Assuming, d: Representative diameter, α: Area shape factor, β : Volume shape factor.

$$A_p = \alpha d^2$$
$$V_p = \beta d^3$$

Substituting these equation into Equation 6·98,

$$\varphi = 4.87\left(\frac{\beta^{2/3}}{\alpha}\right) \quad \text{—(6·99)}$$

(6.97)–(6.99)

Equation 6.97 - 6.99 The particle shapes.

Table 6.16 shows the approximate values of the shape coefficients α and β and the sphericity φ for sand grains.

Table 6.16 Shape coefficients and sphericity of sand grains.

Description	β	α/β	φ
Rich in surface	0.64	6.9	0.81
Sharp	0.77	6.2	0.85
Worn	0.86	5.7	0.89
Rounded	0.91	5.5	0.91
Spherical	0.52	6.0	1.00

α: Area shape factor, β: Volume shape factor, φ: Sphericity

Table 6.17 The mean shape factor (sphericity) φ_s of various sands used in filtration.

Material	Nature of Grain	φ_s
Sand: average		0.75
Flint sand	jagged	0.65
Flint sand	jagged flakes	0.43
Ottawa sand	nearly spherical	0.95
Sand	angular	0.70~0.75
Sand	rounded	0.83
Coal		0.55~0.70
Pulverized coal		0.73
Natural coal dust	up to 3/8 Inch	0.65
Flue dust	fused, aggregates	0.55
Flue dust	fused, spherical	0.89
Berl saddles		0.30
Mica	flakes	0.28
Fusain fibres		0.38
	spheres	1.00

Table 6.17 shows the mean shape factor (sphericity) φ_s of various sands used for filtration in water treatment plants, which were determined by the permeation method using the pressure losses in fixed beds.

Porosity of the granular layer (fixed bed). When particles are filled into a space, the percentage of voids to the total apparent volume is referred to as the porosity and is generally expressed as ε. Figure 6.80 shows an arrangement of uniform-diameter spherical particles illustrating various porosity values ranging between (1) the coarsest filling of a cubic array with a maximum porosity of 0.476 and (4) the closest filling with a diamond (rhombohedral) array with a minimum porosity of 0.2595. A mixture of large and small particle sizes will cause small particles to enter between the large particles, resulting in a lower porosity. Porosity also increases as the shape of the particles deviates from being spherical.

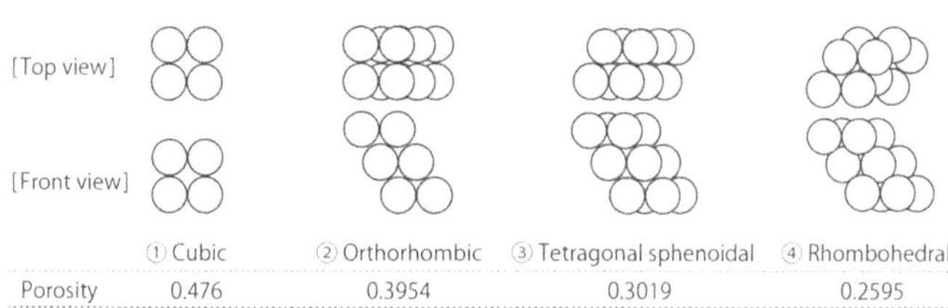

Figure 6.80 Various arrangements of particles (spheres) and their porosities.

6.2.1.2 *Flow through a fixed bed*

A granular bed is roughly divided into a fixed bed and a fluidized bed. In Section 6.1.1.18 above, we described fluidized bed hydraulics for floc blankets (see Figures 6.54, 6.55, and 6.56, and Equations 6.39 and 6.40). In filtration, in a gravitational field, if the flow direction is in the downward flow, the particle arrangement of the filter layer remains fixed regardless of the increase in the flow velocity. In the case of an upward flow, as shown in Equations 6.39 and 6.40, the particle array remains fixed in the fixed bed until a certain flow velocity is reached. When the upward flow velocity exceeds that value, the particle layer cannot keep its position fixed by its own weight against the water flow, and the layer starts to expand due to the viscous frictional force of the water flow. We have previously explained that such a state is referred to as a fluidized bed.

In this section, we will first describe the flow resistance caused by water flowing through a fixed bed (here, it is expressed as head loss). Various equations have been proposed to express the head loss of a granular bed since Darcy in the mid-19th Century, but the Carman-Kozeny equation (derived by Carman using the Kozeny equation as a starting point) is the most widely used equation (Kozeny, 1927) (Carman, 1937).

Kozeny considered a model for replacing the hydraulic head of frictional losses as fluids pass through a granular bed with parallel small channels, where the surface area of the granular bed is equivalent to the total surface area of the packed particles, and the total volume of the water channel is equivalent to the total void space of those packed particles. Such small water channels are only hypothetical, and it would not be very meaningful to discuss their geometric structure. However, the length of a waterway is also discussed by various people, including by Carman (1937).

The head loss h_r due to flow through a fixed layer is represented by the Darcy–Weisbach equation (Equation 6.100) as the loss of a channel whose channel length is equal to the fixed bed layer (filtration layer) thickness L with a hydraulic radius R.

$$h_f = f \frac{L}{4R} \frac{\bar{V}^2}{2g} \quad \text{—(6·100)} \tag{6.100}$$

h_f : Head loss (cm)
f : Friction factor (-)
L : Fixed bed thickness (cm)
R : Hydraulic radius (cm)
V : Flow average velocity (cm/sec)
g : Gravitational acceleration (cm/sec^2)

Equation 6.100 The Darcy - Weisbach Equation.

The depth of the virtual channel is determined as follows. The total volume of the virtual channel is considered to be equal to the total void space of the fixed bed (filter bed), and is expressed by Equation 6.101. The total wetted surface area of a virtual channel is given by Equation 6.102, where A is the surface area of a single particle, and the depth R of the virtual channel can be defined as the total volume of the virtual channel divided by the wetted surface area: therefore, it is given by Equation 6.103.

The value of V_p/A_p can be estimated from the diameter of the particles in Equation 6.104 if the individual particles are considered spherical, whilst if the individual particles are non-spherical Equation 6.105 is used with mean shape factor (sphericity) φ_s.

$$\text{Total channel volume} = \frac{\varepsilon}{1-\varepsilon} NV_p \quad \text{—}(6 \cdot 101)$$

ε : Fixed bed porosity (-)
N : Total number of particles in the fixed bed
V_p : Volume of one particle in the fixed bed (cm^3)

$$\text{Total wetted surface area of channel} = NA_p \quad \text{—}(6 \cdot 102)$$

A_p : Surface area of one particle in the fixed bed (cm^2)

$$\text{Hydraulic radius (R)} = \frac{\varepsilon}{1-\varepsilon}\frac{NV_p}{NA_p} = \frac{\varepsilon}{1-\varepsilon}\frac{V_p}{A_p} \quad \text{—}(6 \cdot 103)$$

$$\text{Spherical: } \frac{V_p}{A_p} = \frac{\pi d^3/6}{\pi d^2} = \frac{d}{6} \quad \text{—}(6 \cdot 104)$$

$$\text{Non-spherical: } \frac{V_p}{A_p} = \varphi_s \frac{d}{6} \quad \text{—}(6 \cdot 105)$$

φ_s : Average shape factot

(6.101)–(6.105)

Equation 6.101- 6.105 The hydraulic radius of virtual channel.

Equation 6.100 is rewritten as Equation 6.106, since the mean flow velocity $\bar{V}$ of the virtual channel is calculated by dividing the superficial velocity V_s by the fixed bed porosity ε following the Dupute assumptions, such as $\bar{V} = V_s/\varepsilon$.

$$h_f = f'\left(\frac{L}{\varphi_s d}\right)\left(\frac{\varepsilon}{1-\varepsilon}\right)\frac{\bar{V}_s^2}{2g} \quad \text{—}(6 \cdot 106) \tag{6.106}$$

Equation 6.106 The head loss of virtual channel.

Since the frictional coefficient of a virtual conduit is expressed as a function of the Reynolds number Re, as in general hydraulic phenomena, the Reynolds number of a virtual conduit is defined by Equation 6.107 from the mean flow velocity $\bar{V} = V_s/\varepsilon$ and the hydraulic radius $R = \varepsilon/\ (1 - \varepsilon)\ \varphi_s d/6$.

$$R_e = \frac{R\bar{V}}{\nu} = \frac{\bar{V}_s \varphi_s d}{\nu(1-\varepsilon)} \quad \text{—}(6 \cdot 107)$$

Average velocity $\bar{V} = \bar{V}_s/\varepsilon$
Hydraulic radius $R = \varepsilon(1-\varepsilon)\varphi_s d/6$
ν : Kinematic viscosity

(6.107)

Equation 6.107 The Reynolds number of virtual channel.

Under laminar flow conditions, the coefficient of friction f′ is inversely proportional to the Reynolds number and is written as Equation 6.108.

$$f' = \frac{k_1}{R_e} \quad \text{—}(6 \cdot 108) \tag{6.108}$$

k_1 : Constant

Equation 6.108 The friction coefficient under laminar flow conditions.

Therefore, the fluid head loss flowing through a filter bed under laminar flow conditions is given in Equation 6.109 and is widely used as the Carman-Kozeny equation (named after Josef Kozeny and Philip C. Carman, who described the phenomena around the Reynolds number region). Here, we propose values such as 180 for Carman and 150 for Ergun as the constant k_1.

$$h_f = k_1 \frac{L}{(\varphi_s d)^2} \frac{(1-\varepsilon)^2}{\varepsilon^3} \frac{\nu \bar{V}_s}{g} \quad -(6 \cdot 109) \qquad (6.109)$$

Equation 6.109 The head loss under laminar flow conditions.

It has been widely accepted in turbulence research that the coefficient of friction remains constant in a completely turbulent flow regime. However, perfect turbulence conditions hardly ever occur in a fixed bed in an actual water treatment process, but it may be encountered in a process such as gas exchange. The fluid resistance head loss of a turbulent fixed bed is given by Equation 6.110, where the constant k_2 is the coefficient of fluid resistance loss of a turbulent fixed bed.

$$h_f = k_2 \frac{L}{\varphi_s d} \frac{1-\varepsilon}{\varepsilon^3} \frac{\nu \bar{V}_s^2}{g} \quad -(6 \cdot 110) \qquad (6.110)$$

Equation 6.110 The head loss under turbulent flow conditions.

As an equation for expressing the resistivity coefficients for a wide range of Reynolds numbers from the laminar region to the turbulent region, taking the experimental results of Burke & Plummer with $k_2 \doteqdot 1.75$ in the turbulent region, and synthesizing the straight line of the Carman-Kozeny equation for the laminar region (as shown in Figure 6.81) gives Ergun's equation, Equation 6.111.

$$\begin{aligned} f' &= \frac{150}{R_e} + 1.75 \\ &= \frac{150(1-\varepsilon)\nu}{\bar{V}_s \varphi_s d} + 1.75 \quad -(6 \cdot 111) \end{aligned} \qquad (6.111)$$

Equation 6.111 The Ergun equation.

The above equations are derived on the assumption that the particle size of the granular (fixed) layer is uniform. In order to deal with the flow of a mixed layer, in which large and small particle sizes are mixed, it is necessary to introduce appropriate values for $\varphi_s d$ representing diameter. Assuming the mean (representative) particle size $d_{ave.}$ of all layers from Equations 6.105 and 6.112 is obtained, when the mean particle size $d_{ave.}$ of Equation 6.112 is replaced with Equation 6.106, the head loss of the fixed layer (filter bed) is represented by Equation 6.113. The mean $(A'/V')_{ave.}$ for all fixed layers is given by Equation 6.114, where $(A'/V')_{ave.}$ is determined by sieving, d_i is the geometric mean diameter of adjacent sieves, and w_i is the weight of particles remaining between sieves.

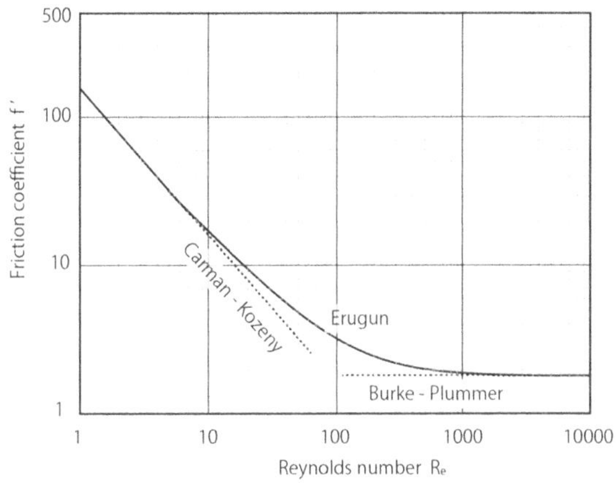

Figure 6.81 Reynolds number and friction coefficients for a fixed bed.

$$d_{ave} = \frac{6}{\varphi_s}\frac{NV_p}{NA_p} = \frac{6}{\varphi_s}\frac{V'}{A'} \quad —(6 \cdot 112)$$

V' : Total particle volume in the granular bed
A' : Total particle surface area in the granular bed

$$h_f = f' \frac{L}{6}\left(\frac{V'}{A'}\right)_{ave} \frac{(1-\varepsilon)}{\varepsilon^3}\frac{\bar{V}_s^2}{2g} \quad —(6 \cdot 113)$$

$$\left(\frac{V'}{A'}\right)_{ave} = \frac{6}{\varphi_s}\sum_{i=1}^{i}\frac{xi}{di} \quad —(6 \cdot 114)$$

xi : Amount of particle size di
i : Series number of weight distribution

(6.112)–(6.114)

Equation 6.112 - 6.114 The head loss of heterogeneous particle layers.

6.2.1.3 Fluidized bed flow

The flow of a fluidized bed was explained above in Section 6.1.1.18; see Figures 6.54 and 6.55, and Equations 6.39 and 6.40. This section explains the process used to wash a particulate bed that is constantly operated as a fixed bed, such as a sand filter or an ion-exchange resin adsorption tower. Washing is a process that introduces clean water from the bottom of the granular media, causes the bed to float and swell to fluidize, washes out impurities trapped between particles and elutes any impurities, restoring the function of the granular media. It is commonly referred to as backwashing.

Various equations have been proposed for the upward flow velocity V_{sc} at which fluidization begins. Yagi and Shirai (1952) proposed an experimental equation in which fluidization starts as Equation 6.115 for the range of Reynolds number from 0.5 to 200 for the particles that comprise the granular layer, which is commonly used in water treatment.

$$V_{sc} = 0.012\,(w\,\varepsilon_{sc})^{1.5} \quad —(6 \cdot 115)$$

V_{sc} : Upflow velocity at the point of fluidization
ε_{sc} : Porosity at the point of fluidization
w : Terminal settling velocity of particles

(6.115)

Equation 6.115 The empirical upflow velocity equation at the fluidization start.

The relation between the thickness L of the fixed layer and the porosity Ɛ, and the thickness L_e of the fluidized expanded granular layer and its porosity ε_e is expressed by Equation 6.116, as in Equation 6.40, because the particle volume before and after expansion is unchanged.

$$L_e = L \frac{1-\varepsilon}{1-\varepsilon_e} \quad \text{—(6·116)} \tag{6.116}$$

L : Fixed bed length
L_e : Length of the fluidized expanded bed
ε_e: Porosity of the expanded bed

Equation 6.116 The fluidized expanded layer thickness and porosity.

To determine the expansion of the granular layer, it is necessary to know the relation between the porosity e_ε and the upflow velocity V_s, and Equation 6.41 (described above and reposted here) is used to determine the expansion of the granular layer.

$$\varepsilon_e = \left(\frac{V_s}{W_t}\right)^n \quad \text{—(6·41)} \qquad \text{(6.41; reposted)}$$

ε_e: Fluidized bed porosity
V_s : Upflow velocity
W_t : Terminal settling velocity of single particle

Equation 6.41 The fluidized bed porosity, particle settling velocity and upflow velocity.

where n is a constant determined by the Reynolds number $Re = w_t d/\nu$ for the terminal sedimentation velocity w_t of a particle of particle size d. The values of n, which are determined by the Reynolds number, shown by Richardson and Zaki (1954) are shown in Table 6.18.

6.2.1.4 Overview of suspension removal mechanisms in granular bed filtration

In granular bed filtration, fine particles suspended in water (impurities) are trapped in the interstices of other particles (filter media) with a much larger particle size which make up the filter bed. Figure 6.82 is a schematic representation of the size of the filter media and suspended (removed) particles involved in filtration.

As a process in which suspended (fine) particles of up to about 10^0 μm flow into a filter medium space of the order of 10^2 μm for removal, two steps need to be considered: the transportation of particulate matter to the surfaces of the filter media; and the attachment of those transported particulates onto the surfaces of the filter media particles.

Table 6.18 The n values determined by the Reynolds number. (Richardson & Zaki, 1954).

Re Range	1/n Value
$Re < 0.2$	4.65
$0.2 < Re < 1$	$4.36\ Re^{-0.03}$
$1 < Re < 500$	$4.45\ Re^{-0.1}$
$500 < Re < 7000$	2.39

Richardson & Zaki.

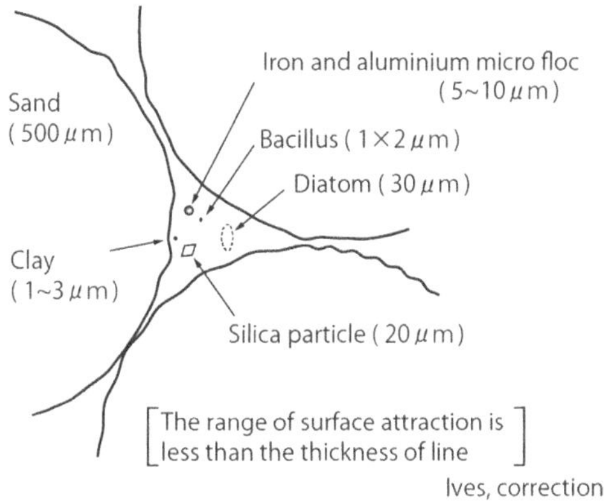

Figure 6.82 The size of filter media and suspended solids.

Table 6.19 lists the factors involved in the transport and attachment stages, which are considered to be the main processes of suspension removal in a filter bed. The transport process is physically and hydraulically controlled by phenomena related to material transport. By contrast, the process of attachment is largely governed by chemical factors and may involve physicochemical/surface chemical factors.

6.2.1.5 *Filter bed transport process*

The transport process in a filter bed proceeds by one of the various factors shown in Table 6.19, or by a combination of these factors. The effects of these various factors are intricately intertwined, and clear quantitative analysis is still inadequate. Therefore, we have tried to partly quantify and understand the roles of the factors, as far as possible. Generally, the flow in a filtration bed is approximately laminar flow to the weakly transitional region, and the transport process of suspended matter by turbulence may be negligible. Below, we consider each factor.

Sieving (straining). This is an excellent function when the filter medium is finer than the suspended particles, and is a basic function of cake filtration and membrane filtration (which will be described further below). In granular bed filtration, straining occurs when suspended particles pass through the interstitial spaces of wedge-shaped filter media particles – but only when the suspension passes between fine particles that have already closed the filter media gaps or where the gaps between these accumulated fine particles and neighboring filter media particles are smaller than the size of the suspended particles.

Table 6.19 Main factors in the transport and attachment processes.

Transport [Physical, Hydraulic]	Attachment [Physicochemical, Surface Chemical]	
Straining	Mechanical holding	
Brownian movement	Surface flocculation	
Streamline contact	adhesion	Chemical adhesion
Inertia collision	coagulation	Chemical bonding, Crosslinking etc.
Gravitational settling	flocculation	Physical adsorption
		Coulomb force, Van der Waals forces etc.
	Biological action	

Inner filtration (depth filtration) which is also used in water treatment is considered to account for less of the filtration effect. However, when a suspension passes through a fine granular layer such as a slow sand filter at a low speed, microorganisms in the water are first separated and trapped on the surface of the filter layer, and then grow to form a biofilm surface on the sand layer. As a result, the main mechanism of separation is the trapping (straining) by a surface filtration film, which is similar to cake filtration (described later). The biological filtration layers formed on the surfaces of these granular layers are called Schmutzdecke (the German term is commonly used) and have a pronounced sieving effect on suspended substances in the interstitial spaces.

Slow sand filters take the form of granular bed filtration, but should be understood as a process that is quite distinct from a rapid sand filter system which exerts a separating mechanism close to cake filtration and an aerobic microbial treatment mechanism, and which will be described further below.

Brownian motion. Brownian motion is a phenomenon in which the impact of the thermal motion of water molecules is transmitted to suspended particles to induce irregular motion, and colloidal particles with a diameter of about 1 to 2 μm or less are observed (as described above). According to Einstein, the diffusivity D_{br} due to Brownian motion is shown in Equation 6.117. The diffusion coefficient at room temperature of 20°C ($T = 293$ K and $\mu = 0.010$ g/cm/sec), can be calculated by Equation 6.118. In addition, the time t required for suspended particles to move by the distance L due to Brownian motion can be calculated by Equation 6.119.

Einstein's Brownian diffusivity: D_{br}

$$D_{br} = \frac{KT}{3\pi\mu d_p} \qquad \text{—}(6\cdot117)$$

K: Boltzmann's constant (1.38×10^{-16} erg/K)
T: Absolute temperature (K)
μ: Water viscosity (g/cm·sec)
d_p: Suspended particle diameter (cm)

D_{br} at 20℃ (T=293K, μ=0.01 g/cm/sec)

$$D_{br}(20℃) = \frac{4.3\times10^{-13}}{d_p} \ (\text{cm}^2/\text{sec}) \qquad \text{—}(6\cdot118)$$

Time required for suspended particles to move distance L: t

$$t = \frac{L^2}{D_{br}} \qquad \text{—}(6\cdot119)$$

L: Moving distance of the particle
t: Moving time of the particle

(6.117)–(6.119)

Equation 6.117 - 6.119 The Brownian movement equation.

Generally, the coarse suspensions of clays etc. present in natural water and flocs produced by coagulation treatment have diameters of several micrometers or more, and therefore Brownian motion is not considered as the main factor of transport. However, in the case of small (colloidal size) particles of 1 to 2 μm or less (and with a filter bed of pore size of about 0.4 to 0.1 mm), Equation 6.119 indicates that all suspended particles can reach the surface of the filter media in a short time, such as 1 second or less. It is considered that Brownian motion serves as an excellent transport mechanism for colloidal particles and microorganisms on the surface of a slow sand filter without using a coagulant. As a result, a Schmutzdecke is formed on the limited thin surface layer of the sand bed at the beginning of the filtration, and after that, the surface filtration (similar to cake filtration) separates other particles.

Inertial collision. As the suspension passes through the tortuous channels between the filter media particles, the density of the water and the density of the suspension differ, causing the suspended particles to travel in a straight direction without completely following the bending of the water stream and so collide with the filter media surfaces for removal. The denser the suspended particles and the larger the size of the particles, the more prominent the transport effect is. This filtration mechanism is the primary mechanism for the removal of airborne dust in air, where the difference in density between suspensions and fluids is large. The difference in density between water and suspensions is generally not so large, so that the percentage effectiveness is more limited than it is in air.

Gravitational settling. Gravity-induced sedimentation, like inertial collision, is negligible for small suspensions that do not differ significantly from water densities. When the first filter layer is composed of a large filter medium of 1 to 5 mm, as in a dual media filter using two stages of a coarse filter medium and fine filter medium in series (described below), and the removal of floc particles of order 10^1 μm by direct filtration without a sedimentation basin is attempted, the individual filter media gaps act like an idealized sedimentation pond with in series laminar flow, and sedimentation is an excellent mechanism of removal.

In normal filtration, the overflow rate of a preceding sedimentation basin is usually lower than that of filters. However, let us consider a sedimentation pond in which only the Hazen number w/w_0 (w being the settling rate of particulate suspensions, and w_0 being the overflow rate of filter medium voids) is removed by sedimentation in the first filtration layer. Even the rate of removal in that one layer is much smaller than expected; the total filter layer thickness L is composed in series of the multiple layers of the filter medium diameter d arranged in series with $k = (L/d)$ stages. The process can be determined to be an arrangement of sedimentation ponds in series and in which removal efficiencies are not very high. It is easier to understand when we think of a series of fully mixed sedimentation ponds consisting of multiple, inefficient tanks.

Professor Ken Ives of University College London developed an upflow filtration experiment in which we can observe that flocs adhere to the upper surface of sand particles and emphasize the effect of sedimentation in a real operation (Ives, 1970).

Streamline contact by water flow. In normal rapid filtration, in which the diameter of the suspended particles is large enough to ignore Brownian motion (d »1 μm), and in which sedimentation and inertial collisions may not affect the predominantly large diameter particles, it is likely that the primary transport mechanisms are contact deposition by water flow.

The main removal of floc in a filter layer occurs when suspended floc approaches the surfaces of sand grains or a floc particle layer which has already adhered to the surface of sand grains by being caught/pulled in the water stream, and contact (streamline constriction/branching/meandering contact) occurs. Using the results of a rod filter (two-dimensional model), as shown in Figure 6.83, P.C. Stein (1940) discussed how the flow line between sand grains creates a sort of sand-grain stenosis and most frequently an attachment of suspended particles to the filter material. The physicochemical theory proposed by Prof. D. Mints (1966) also seems to assume contact removal mechanisms that are less dependent on the diameter of the suspensions. In addition, the author's own experimental results suggest that the main mechanism of suspension transport is contact by water flow, in which the removal rate hardly decreases even if the inflow carried only micro flocs in the filtration. From the practical results of the micro-floc method, it is considered that the main mechanism of suspension removal (transport step) would be water flow (stream line contact).

Professor Ebie (1975, 1976a, 1976b, 1977), at the Kitami Institute of Technology, Japan, has studied in detail the contributions of streamline contact and sedimentation in a number of experiments using a similar rod filter unit to that used by Stein. At the Sanitary Engineering Symposium of the Japan Society of Civil

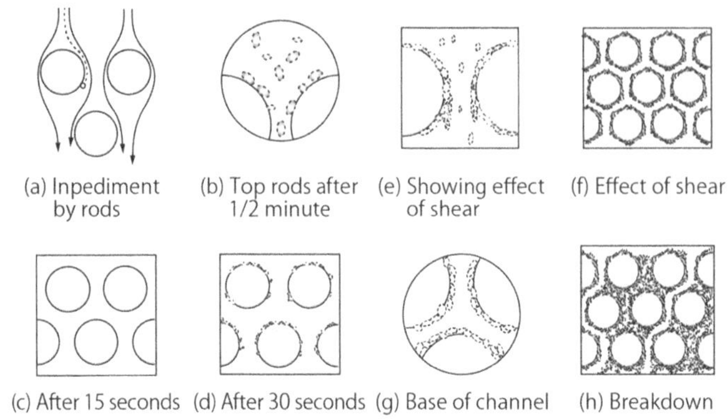

Figure 6.83 Stein's schematic explanation of particle transport and attachment by sand grain in a rod filter. It would be a natural procedure in a model rod filter.

Engineers in the 1970s, Professor Sumitomo of Kyoto University stated that turbulence in the flow of water through the interstices between filter media particles was likely to be the main mechanism of transport. Since the flow of a low Reynolds number filter layer cannot be a turbulent flow, the author considers that the transport phenomena are caused by water flow fluctuation (immediately decaying) and does not proceed to turbulent flow. This may be a very difficult research objective to theorize in a transition fluctuation flow regime.

If the contact caused by water flow is the subject of the transport action, the removal rate should change inversely proportional to the d^2 of the filter medium, since the size of the filter medium surface area is the direct removal controlling factor. In addition, the ratio of the diameter of suspended contact particles (the floc diameter) to the particle diameter of the filter medium is generally low, so that the diameter of suspended matter may not have a significant effect on the removal rate.

6.2.1.6 Attachment step of filter bed removal

If the suspended particles reach the solid–liquid interface of the filter media (the surface of the filter media or the interface of previously attached suspended particle layers) by any of the transport mechanisms described in the preceding section, the suspended particles are fixed to the surface of the filter media and removal is performed if the properties of the solid–liquid interface of the filter media and the surface of the suspended particles both satisfy attachment conditions. If the interface conditions are not adequate for attachment, the suspended particles are returned to the suspended particle system.

The actions that govern the deposition process are determined by factors such as those in column 2 of Table 6.19. The most powerful of these four attachment mechanisms are considered to be adhesion and coagulation, which are considered to be due to similar driving forces that are difficult to separate. Consider it to be a two-stage phenomenum: (1) the adsorption of coagulation particles on the surface of the filter medium; and (2) the adhesion of coagulation particles which have newly arrived in the inflow and have adhered to the surface of previously adhered coagulation particles on the filter grains.

Mechanical holding. There are also a small number of suspensions that are mechanically retained by hydrodynamic and gravitational forces in the constrictions and recesses between the filter media particles. In one experiment in which clay particles of a few microns were suspended at a concentration as low as several mg/l, and were passed through a sand layer so that the attachment mechanism (described in Section 6.1.7 below) is almost negligible, turbidity was removed by about 20–30%. This may be thought of primarily as a result of the mechanical retention of suspended particles in the filter bed. Although the rate of removal by this mechanism in a rapid sand filter is not high, in the Shumutzdecke in slow filtration, a mechanical deterrent similar to that of cake filtration may play a significant role in conjunction with the biological action.

Surface flocculation (adhesion/coagulation). The suspended particulate matter that has been transport to the surface of filter media particles by various transport mechanisms may adhere to them by the same mechanism as described above for the agglomeration of colloidal particles (see Figures 4.13, 4.14, 4.15, etc. in Chapter 4) or may further adhere to the previously adhered suspended particulate, and the process of coagulation (adhesion) may proceed on the surface of the filter media. Similar to the coagulation between suspended particles (discussed in detail in Chapter 4), if an electric double layer is formed between the surface of the filter media particles and the interface between the suspended particles, or between the interface of the suspended particles attached to the filter media particles and the suspended particles that have been subjected to subsequent transport, and the interfacial electrochemical condition (potential balance between attraction and repulsion) and the appropriate cross-linking condition (chemical bonding) are satisfied, attachment coagulation of the suspended matter to the filter media surface is established. According to Ives and Gregory (1966) the distances at which the filter media particles and suspended particles begin to interact are expected to be of the order of 2000 Å for water with low ionic strength, such as distilled water, and 100 Å for natural water.

If the filter media particles (generally negatively charged) and suspended particles (coagulated particles are positively charged) are oppositely charged to each other, there is no energetic barrier to adherence from the outset, but the transport of coagulation suspended particles to the surface of the filter media particles covered by coagulated suspended particles leads to further coagulation of particles with the same charge. If we continue to removal a large number of suspended particles, we will experience the same charge neutralization process as the coagulation/flocculation of suspended systems described in Chapter 4.

When iron removal filtration proceeds by passing iron hydroxide particles (positively charged) through a filter layer of sand particles (negatively charged), the first layer of iron hydroxide adheres to the sand without any problem of repulsion potential. However, subsequent iron hydroxide particles having the same positive charge approach the surface of the filter medium covered with an iron hydroxide adhesive layer, and bonding is prevented by the mutual repulsion potential of the positive charges. Therefore, filtration cannot be continued unless the positive repulsion potential is lowered to the level where iron hydroxide particles can coagulate each other. This condition is almost the same as the potential condition that iron flocs form in a suspension system. If the surface of the filter media and suspended particles have the same charge sign, an adhesion condition for lowering the repulsion energy potential barrier between the filter media particles and suspended particles and lowering the repulsion potential of both suspended particles is required. Figure 6.84 shows the measurements of energy barriers on the surfaces of sand particles that absorb iron-floc from Ives and Gregory (1966). A detailed description of the author's theory is presented in Chapter 4 above.

Filter particles such as sand and anthracite show a ZP in the range of −10 to −20 mv in normal water. Crushed coal with a new interface, crushed sand, or quartz-sand cleaned with acid show a high negative ZP of −35 to −60 mv.

As with coagulation flocculation between suspended particles, maintaining effective adherence against external forces requires the presence of cross-linking substances capable of exerting chemical

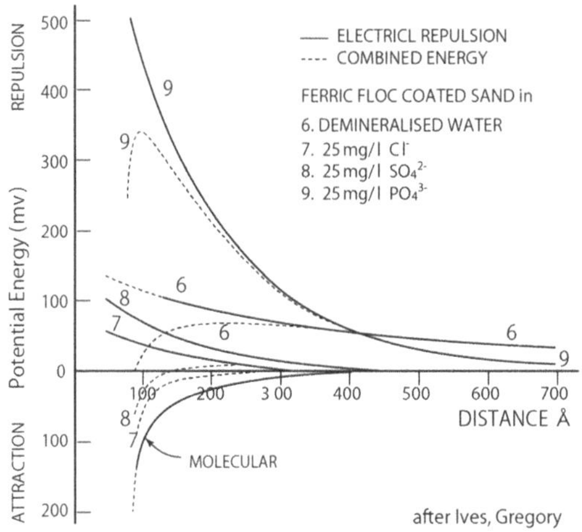

Figure 6.84 The surface energy barriers of ferric floc coated sand particles.

bonding forces, such as hydrogen-bonding, as well as van der Waals bonding forces. Therefore, the presence of a polymeric agent or a polymerized metal hydroxide complex compound that co-adsorbs to the filter media particles and suspended particles is necessary to enhance the retention capacity of the filter media surface.

As the amount of retention on the surface of the filter media increases, the water channel between the filter media becomes narrower, and the attached suspensions are subjected to greater shear forces. For this reason, retention can be increased only to the point where the binding force of the suspended coagulation mass attached to the filter medium surface and the shearing force by the above-mentioned water flow are just balanced with each other. In a flocculator, for the formation of large flocs exhibiting high settling velocities in sedimentation ponds, the agitation process is performed under optimal turbulent agitation to grow floc particles against turbulent shear forces. However, since the formation of a floc particle layer on the surface of the filter medium can proceed to a certain thickness against the shear forces in the laminar region, the amount of co-adsorption (polymer cross-linking substances) required to provide a binding force to suspended particles is smaller in filtration than in the sedimentation process. Since coagulants are mainly added to reduce the repulsion potential, aluminium sulfate of 10^1 mg/l is required to form a floc for sedimentation when the raw water turbidity is of the order of 10^1 TU, but in the case of direct filtration, a separation process can be carried out of the order of 10^0 mg/l, and that separation process can be carried out with just a small amount of coagulant using the interface potential as the main index of operation.

Biological action. In slow sand filtration, when a gelatinous cell film layer is formed on the surfaces of the granular filter media, attachment mechanisms are developed mainly for adsorption to biological cell groups, as in the case of bio flocs formed by activated sludge and biofilter beds. Because the gaps between the produced organisms are extremely narrow, the retention function of the biofilm formed on the surface of the filter bed is excellent, and biodegradation, uptake of adsorbed organic matter into cells, and bio-oxidation of inorganic matter, etc., all proceed at the same time.

The biological effects of these extensive attachment/anabolic/catabolic mechanisms can also occur in limited spaces in a Schmutzdecke on the filter as thin as 1 cm in thickness, making the total quantitative capacity small and progressively more responsive to low concentrations of contaminants. On the other hand, upgrading systems by making use of the multifaceted capabilities of different finishing treatments can also prove to be very useful.

6.2.1.7 Kinetics of suspension retention

The basic equation for the patterns of removal and retention of suspensions in filter beds has been discussed by many researchers of the world, starting from the first-order equation of suspensions concentrations published by Iwasaki, working at the Tokyo Water Works, published in the journal of Society of Civil Engineers as early as 1937 (Iwasaki, 1937). Equation 6.120 and 6.121 are basic equations that describe retention patterns in the filtrate layer of suspended matter, and which are today commonly used and referred to as filtration equations.

$$\frac{\partial C}{\partial Z} = -\lambda C \quad \text{[Rate equation]} \quad \text{—}(6\cdot120)$$

$$\frac{\partial C}{\partial Z} + \frac{1-\varepsilon}{v}\cdot\frac{\partial q}{\partial t} = 0 \quad \text{[Continuity equation]} \quad \text{—}(6\cdot121)$$

$C = C(Z, t)$: Concentration of suspended solids (mg/cm³)
Z : Distance into the filter from the bed surface (cm, m)
t : Time (sec, day)
λ : Rejection rate (1/cm)
-- change due to the deposition of suspended solids
$q = q(Z, t)$: Deposit weight per unit filter volume (mg/cm³)
v : Filtration velocity (cm/sec, m/day)
ε : Sand bed porosity (-)

(6.120)–(6.121)

Equation 6.120, 6.121 The basic equation for filtration.

Here, the reaction rate coefficient (rejection rate) λ [1/cm] is not a constant value, as will be described later, but a function that changes as the retention of the suspensions progresses even if the conditions of the suspensions and the type of the filter media are kept constant.

Equation 6.120 is a rate equation in which the removal of suspensions in a filter layer progresses in linear proportion to the concentrations of suspensions in that filter layer. Its rate of progress is evaluated by the reaction rate coefficient λ. This first order kinetic factor λ of granular layer filtration is called the rejection rate, and the progress of the filtration is expressed by considering various rejection rate functions. Equation 6.121 is a continuous equation that shows a decrease in the concentration of suspensions in water in a layer from the sand surface to a depth z that is equal to the mass of suspensions retained in the sand bed for a certain period of time.

When Equation 6.120 is solved for the initial rejection rate $\lambda = \lambda_0$ of initial $t = 0$ of filtration and with concentration $C = C_0$ of inflow to the bed at $z = z_0$, the concentration profile of suspensions in the filter bed at the beginning of filtration is given by Equation 6.122. The initial rejection rate λ_0 is a value that varies depending on the size of the filter medium, the filtration rate, and the nature of the suspension. As a result of these combinations, the rate of rejection can be changed by a variety of removal mechanisms, which will be discussed later. In relation to filtration velocity v and filter medium diameter D, the initial rejection rate λ_0 is generally found to be similar to Equation 6.123.

As filtration progresses, the rejection rate λ varies from the initial rejection rate λ_0 to the amount q of the suspended matter in the filter bed, which changes as shown in the generalized form, as in Equation 6.124.

$$\frac{C}{C_0} = \exp(-\lambda_0 Z) \qquad \text{—(6·122)}$$

$$\lambda_0 \propto \frac{1}{vd^2} \qquad \text{—(6·123)}$$

$$\lambda = F(\lambda_0, q) \qquad \text{—(6·124)} \qquad (6.122)\text{–}(6.124)$$

C : Suspended solids concentration distribution
C_0 : Inflow concentration
λ_0 : Initial rejection rate
v : Filtration rate
d : Sand diameter
Z : Distance into the filter from the bed surface (cm, m)

Equation 6.122 - 6.124 The suspended solids concentration and rejection rate in filter bed.

Although various functional forms have been proposed to express the rejection rate λ in Equation 6.124, we would like here to introduce a series of Ives's proposals (Ives, 1969, 1970, 1975) as a good classical study that adequately expresses the mechanism of various removal.

The retention effect of a filter layer on a suspension is governed almost independently of the retention mechanism of the filter layer from the surface area to the interstitial area onto which the suspension is retained. Ives considered the relation between the trappable surface area in the filter bed and the interstitial flow velocity due to the progress of filtration to be as follows:

(1) in the early stages of filtration, the filter media (particles) are spherical and the retained suspended particles uniformly cover their surfaces;
(2) as the quantity of suspended particles retained increases, the gaps between the filter medium particles are filled, the water passage through which the remaining suspension passes becomes cylindrical, and the inner diameter of the cylindrical water passage decreases as the retention progresses;
(3) in this way, the gap flow rate increases continuously, reaching a threshold flow rate at which water flow shearing forces increase and eventually retention cannot occur.

Considering the progress of this filtration, Ives proposed a generic equation, as shown in Equation 6.125, to take into account the three effects of the change of the geometric shape of the suspended matter deposition on filter surfaces, pore water channels, and pore flow rate.

$$\frac{\lambda}{\lambda_0} = \left(1 + \frac{\alpha}{\varepsilon} q^*\right)^y \left(1 - \frac{q^*}{\varepsilon}\right)^z \left(1 - \frac{q^*}{q_u^*}\right)^x \qquad \text{—(6·125)} \qquad (6.125)$$

α : Constant depending on the packing of grain (Void ratio)
ε : Initial (clean) filter bed porosity
q^* : Specific deposit volume per unit filter volume
q_u^* : Ultimate specific deposit which causes the velocity to rise to its limiting,inhibiting velocity (Maximum detained rate)
x, y, w : Emperical exponents

Equation 6.125 The rejection rate function (Ives).

The first term of Equation 6.125 is the change in the surface area of the spherical filter material, the second term is the deformation of the cylindrical channel, and the third term is the change in the pore flow velocity. Each term is formulated by the following procedures:

(1) Change in the surface area of the spherical filter medium. Assuming the volume V_0 of the filter media particles and initial porosity Ɛ, the number of filter media particles per unit volume is $(1 - \varepsilon)/V_0$, and therefore the suspended particle mass per filter media particle is $q^* \times V_0/(1 - \varepsilon)$. Thus, the volume of each particle covered by the suspended matter is Equation 6.126. Since the gap ratio is $\varepsilon/(1-\varepsilon)=\alpha$, the volume of one filter medium containing the retaining suspending is $V = V_0(1 + \alpha q^*\varepsilon)$. Accordingly, the surface area ratio of the filter particle area at the beginning of the unit filtration layer S_0 to the particle area at the given retention time S are given in Equation 6.127.

Number of filter media particles per unit filter bed volume $= (1 - \varepsilon)/V_0$

Deposited solids volume per a filter media particle $= q^* V_0(1 - \varepsilon)$

Volume of filter media particles that deposited the suspended solids V

$$V = V_0 + \frac{q^* V_0}{1-\varepsilon} = V_0\left(1 + \frac{q^*}{1-\varepsilon}\right) \quad \text{—(6·126)}$$

Void ratio $\alpha = \varepsilon/(1 - \varepsilon)$

When, $1/(1 - \varepsilon) = [\varepsilon/(1 - \varepsilon)]1/\varepsilon = \alpha/\varepsilon$,

$$V = V_0\left(1 + \frac{\alpha q^*}{\varepsilon}\right)$$

Ratio of the specific surface area after filtration to the initial filter specific surface area : S/S_0

$$\frac{S}{S_0} = \left(\frac{V}{V_0}\right)^{\frac{2}{3}} = \left(1 + \frac{\alpha q^*}{\varepsilon}\right)^{\frac{2}{3}} \quad \text{—(6·127)}$$

V_0 : Volume of a filter media particle

(6.126)–(6.127)

Equation 6.126, 6.127 The change of the filter media particles surface area ratio.

(2) Change in specific surface area in the filter bed when considering the cylindrical water channel set forth in the second term of the Equation 6.128. If there are N small water channels per unit horizontal cross-sectional area, the length per unit depth is l, and the inside diameter is r, the specific surface area of the channel when it is coated is as shown in Equation 6.128.

Initial porosity $\varepsilon = \pi r^2 N l$

Initial specific surface area $S_0 = 2\pi r N l$

Amount of suspended solids in channel $q^* = \pi r^2 N l - \pi(r - \theta)^2 N l$

Deposit layer thickness in channel $\theta = r[1 - (1 - q^*/\varepsilon)^{1/2}]$

Specific surface area after deposited $S = 2\pi(r - \theta)Nl$

$= S_0 - 2\pi\theta N l$

$= S_0 - 2\pi r N l + 2\pi r N l(1 - q^*/\varepsilon)^{1/2}$

N : Number of water channels per unit cross area of filter

l : Length per the unit depth of filter

r : Inner diameter of channel

Therefore,

$$S = S_0\left(1 - \frac{q^*}{\varepsilon}\right)^{\frac{1}{2}} \quad \text{—(6·128)}$$

(6.128)

Equation 6.128 The specific surface area of Coated channels.

In practice, the two models are not independent but interrelated, with the first term predominating at the beginning of the filtration and the second term predominating at the end of the period. In addition, since the particle size of the filter medium is irregular, both indices cannot be definite values, as described above. Therefore, as shown in Equation 6.129, the indices are expressed using the experimentally determined indices y and z.

$$S = S_0\left(1+\frac{\alpha}{\varepsilon}\right)^y\left(1-\frac{q^*}{\varepsilon}\right)^z \quad \text{—}(6\cdot129) \qquad (6.129)$$

Equation 6.129 The change of actual specific surface area.

(3) Change in the flow velocity in the gap, shown in the third term. If the superficial velocity v and the initial gap flow velocity $v_i = v/\varepsilon$, the minimum gap flow velocity in the filter bed when the maximum retention q* is reached is $vc = v/\ (\varepsilon - qu^*)$. Generally, since the rejection factor λ is known to be inversely proportional to the flow rate, we can assume an exponential function represented by the difference between the reverse of the initial pore flow velocity and the inverse of the marginal pore flow velocity, as in Equation 6.130. Furthermore, when $q^* = 0$ the filter rejection rate $\lambda = \lambda_{0,}$ and the $\lambda_0 \sim (qu^*/v) \times$ is given by Equation 6.131 for the change in the rate of rejection with increasing deposition.

Thus, by multiplying the above three terms, we obtain a generalized expression for the Ives Equation shown in Equation 6.125.

$$\lambda \backsim \left(\frac{1}{v_i}-\frac{1}{v_c}\right) \backsim \left(\frac{\varepsilon-q^*}{v}-\frac{\varepsilon-q_u^*}{v}\right) \backsim \left(\frac{q_u^*-q^*}{v}\right)^x \quad \text{—}(6\cdot130)$$

v : Filtration rate
v_i : Initial channel flow velocity (v/ε)
v_c : Channel critical velocity $(v/(\varepsilon - q_u^*))$
q_u^* : Maximum deposit of filter bed

Since $\lambda = \lambda_0$ when $q^* = 0$

$$\lambda_0 \backsim (q_u^*/v)^x$$

Therefore, The change in the rejection rate due to increase the deposits are

$$\frac{\lambda}{\lambda_0} \backsim \left(\frac{q_u^*-q^*}{q_u^*}\right)^x \backsim \left(1-\frac{q^*}{q_u^*}\right)^x \quad \text{—}(6\cdot131)$$

(6.130)–(6.131)

Equation 6.130, 6.131 The change of rejection rate with increasing deposits.

Various formulas have been proposed by various researchers depending on experimental conditions (e. g., filter media particle size, filtration rate, type of suspensions, etc.). These equations are summarized in the Ives generalized equation, as shown in Table 6.20.

To understand the state of suspensions retained in filtration media, we would now like to consider some actual examples. Figure 6.85 is a schematic diagram showing the ratio C/C_0 of the residual suspensions concentration C to the inflow suspensions concentration C_0 at each sand bed position Z when coagulated particles (flocs, concentration C_0) are brought into inflow with the rapid filtration of a water supply system.

The slope of the curves $\delta(C/C_0)/\delta(Z)$ at the depth z is considered to be the rejection rate λ at a time point (Z, t). As this diagram shows, the removal rate λ in a normal rapid filtration process increases once at the beginning of filtration and decreases steadily after reaching a certain retention level, with a peak λ_{max}. The second term in the Ives equation above represents the increase in the rate of removal as retention progresses, and the third term represents the decrease. The overall result is a relation between the amount retained q* and the rate of rejection λ, as shown in Figure 6.86.

These patterns of rejection (removal) are universal, but not necessarily all stages appear within a practical filtration duration (48 to 96 hr), depending on the composition of the filter bed, the magnitude of the

Table 6.20 Examples of proposed rejection rate equations.

Ives : $x = y = z = 1$,

$$\lambda = \lambda_0 + a_1 q^* - \frac{a_2 q^{*2}}{\varepsilon - q^*}$$

Mintz, Maroudas : $y = z = 0$, $x = 1$,

$$\lambda = \lambda_0 \left(1 - \frac{q^*}{q_u{}^*}\right)$$ Coarse filter media High speed filtration

Iwasaki : $z = 0$, $y = 1$,

$$\lambda = \lambda_0 (1 + a_1 q^*)$$ Fine filter media

a_1, a_2 : constant Low speed filtration

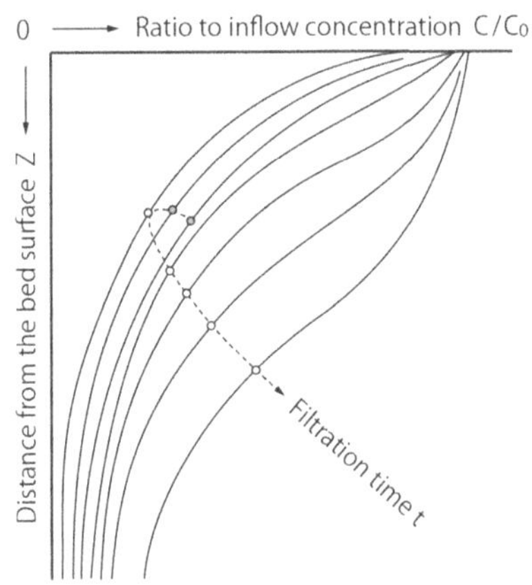

Figure 6.85 Changes in the residual suspended solids concentration ratio in a filter bed.

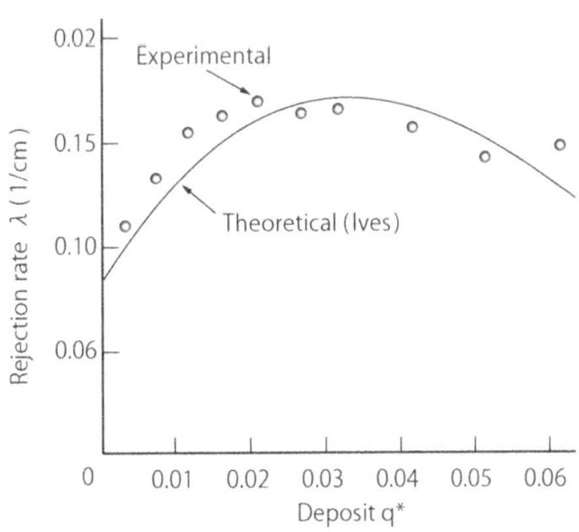

Figure 6.86 Relationship between deposits and rejection rate.

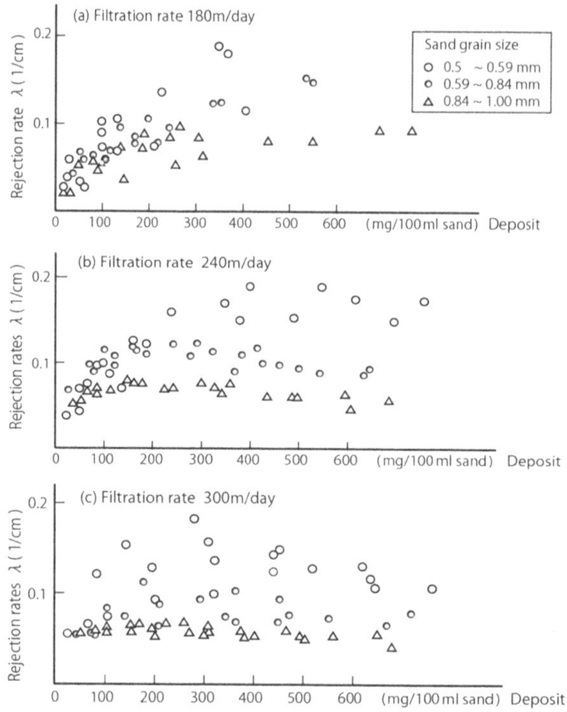

Figure 6.87 Relationship between the rejection rate (deposit) and filtration rate of (**a**) 180 m/day (**b**) 240 m/day; and (**c**) 300 m/day.

filtration rate, the type of suspensions, etc. Figures 6.87a–c show how the rejection rate λ changes with increasing filter layer retention q* when a micro floc group of coagulated kaolin suspension with aluminium salt is subjected to direct filtration and the particle size and filtration rate of the filter layer are changed in three steps. The Iwasaki model in Table 6.20 is for low filtration velocity (standard rapid filtration) composed of small-sized filter media. For high-speed filtration using large-particle-size media, a few patterns of the Ives type are seen, but the simplest patterns, such as $\lambda = \lambda_0$, exist for a considerable amount of time. This means that in high-speed filtration, the process can be operated at a constant rejection rate, and the vertical distribution of the filter layer can be treated as an exponential distribution. Therefore, it is easy to estimate the removal rate of the sand layer. Therefore, the change in the Mintz rejection rate can be considered to be a good representation of fast (high-speed) filtration. However, it is recommended to increase the thickness of the filter layer in an actual high-speed filtration to avoid the first short-term spillage of turbidity from the filter layer.

For filters with high rejection rates, as shown in Figure 6.88a, removal of suspended matter proceeds mainly near the surface of the filter layer, reaching the retention limit S_{max} in the vicinity of the surface in a short period of time, and then the main area of retention gradually shifts to the lower area. In general, however, under these circumstances, there is a significant increase in hydraulic head loss near the surface, resulting in a local negative pressure in the filter bed, which can lead to the escape of retained turbidity, or a rapid increase in total head loss, which can lead to an inability to continue filtration (see filtration head loss in Section 6.2.8 below). For this reason, if the rejection rate is high, the deep filter layer is not necessary. Conversely, at a lower rejection rate, the rejection rate itself varies little, and the retention profile is exponential over a considerable period of time, as shown in Figure 6.88(c), and even though the retention capacity remains

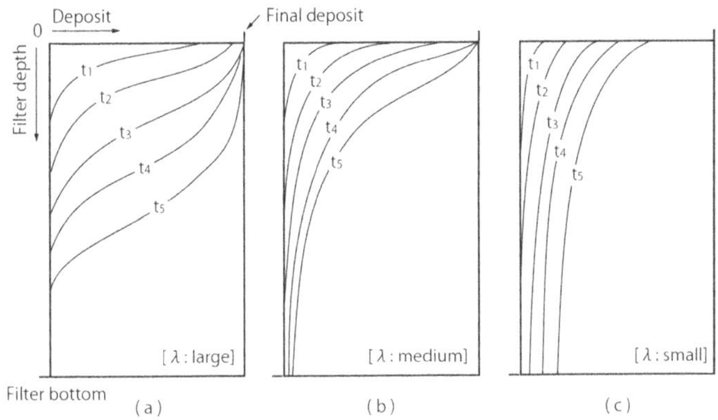

Figure 6.88 Relationship between rejection rates and deposit profiles.

near the surface of the filter layer, there is significant leakage of suspended matter from the bottom of the filter bed, resulting in loss of filtration function. Therefore, the most favorable process condition for a single filter bed is that the turbidity begins to leak only when the head loss reaches the threshold.

6.2.1.8 Filtration head loss by retained suspensions

If the water passage between the sand particles becomes blocked by the retention of suspended matter in the filter media gaps, the pressure difference (head loss) required to obtain the predetermined flow rate increases, and filtration is finally disabled. Therefore, we need to know how the head loss of the filter bed progresses as filtration progresses. Figure 6.89 shows the hydrostatic pressure in the filter layer gradually decreasing as filtration progresses. When a filtration pond is filled with water, a hydrostatic pressure distribution as shown in the straight line (1) is obtained, and at the beginning of filtration, a hydrostatic pressure distribution corresponding to the resistances of the respective granular layers occurs, as shown in the curved line (2). As the retention progresses near the surface of the sand layer, the hydraulic head loss at the upper part of the filter layer becomes larger than that at other parts, and the hydrostatic pressure distribution is shown in curves (3) or (4). Around curve (4), the hydrostatic head available between the inlet and outlet of the filtration pond, which is required to obtain a predetermined treatment capacity, is almost used up, and the filtration has to be discontinued from the point of maintaining water flow capacity (see Figure 6.89). In addition, even if the pressure differential across the filtration pond is somehow sufficient to maintain water flow capacity, as shown in curve (5), in parts of

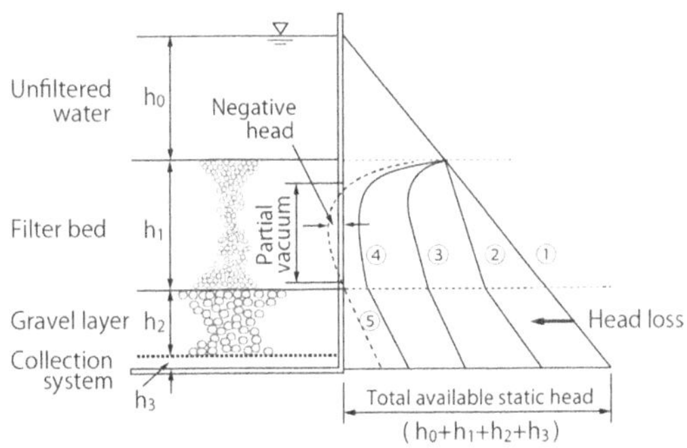

Figure 6.89 Head loss patterns in a filter.

the sand layer where clogging is notable and where the hydrostatic pressure generated is locally lower than the atmospheric pressure, dissolved air in the water will precipitate (see Section 6.1.2 above), causing air bubbles to close the gaps between the filter beds, or causing the adherent suspensions near the filter bed surface to flow off due to air bubbles rising or by deformation, thereby causing the filtration water quality to be seriously impaired. Such local negative pressure generation phenomena tend to occur when the rejection rate λ is large and retention concentrates on the filter bed surface zone or the process is carried out with too small an overlying water head (depth) h_0. Generally, a sand water depth h_0 of 1.0 to 1.5 m or more is taken as the necessary hydrostatic head so as not to cause any hindrance.

The Carman-Kozeny equation (Equation 6.109) is used to calculate the head loss of the filtration layer because the filtration process usually proceeds in laminar flow. However, as filtration progresses and suspensions accumulate in the filter media particle gaps, calculations must be made taking into account the retention profile of the suspension accumulation q* in the filter media particle gaps and any temporal changes. For this reason, many researchers have attempted to correct the Carman-Kozeny porosity and particle size terms by making them functions of the amount of suspension deposited in the voids. Assuming that the increase in loss of water head in the filtration layer is a function of the volume of suspended matter accumulated in the filter pores, the empirical formula for expressing the pattern of increased loss of water head has two versions: Equations 6.132 and 6.133

$$\frac{\partial H}{\partial Z} = \left(\frac{\partial H}{\partial Z}\right)_0 + F(q^*) \qquad \text{—}(6\cdot 132)$$

$$\frac{\partial H}{\partial Z} = \left(\frac{\partial H}{\partial Z}\right)_0 \times F(q^*) \qquad \text{—}(6\cdot 133)$$

H: Filter bed head loss (cm)
Z: Thickness of the filter bed considered (cm)
$(\partial H/\partial Z)_0$: Head loss gradient of the clean filter (initial) obtained by Carman-Kozeny equation
$\partial H/\partial Z$: Head loss gradient
q*: Deposited solids volume ratio

(6.132)–(6.133)

Equation 6.132, 6.133 The empirical equations of increased head loss.

A typical form of Equation 6.132 is Equation 6.134, following Ives, where $F(q^*) = Kq^*$ (and where K is a constant). A head loss equation (a function of q*, z, and t) for the whole of a filter layer with a uniform particle size composition has been numerically solved and Equation 6.135 has been proposed.

$$\frac{\partial H}{\partial Z} = \left(\frac{\partial H}{\partial Z}\right)_0 + Kq^* \qquad \text{—}(6\cdot 134)$$

K: Constant

$$H = H_0 + \frac{KvC_0t}{1-\varepsilon} \qquad \text{—}(6\cdot 135)$$

H: Total head loss (cm)
H_0: Initial total head loss (cm)
v: Filtrationrate (cm/sec)
C_0: INflow concentration (vol/vol)
t: Filtrationtime (sec)

(6.134)–(6.135)

Equation 6.134, 6.35 The head loss equation [I] by Ives.

The proposed Mints, Stein, Camp, etc. equations belong to the same form as Equation 6.133, but can be summarized by the head loss equation presented in Equation 6.136.

$$\frac{\partial H}{\partial Z} = \left(\frac{\partial H}{\partial Z}\right)_0 \left[\frac{1}{1-(Kq^*)^n} + F(q^*)\right] \quad —(6 \cdot 136) \tag{6.136}$$

n : Exponent obtained in experiment

Equation 6.136 The head loss equation [II].

With the exception of the filter layer, which is very close to the surface layer, it is reasonable to consider that within a typical filtration time, the increase in the head loss of a filter layer proceeds almost linearly. In addition, the total hydraulic head loss also increases almost linearly, and in particular, in high-speed coarse filter medium filtration with a low rejection rate, the linearity of the head loss increase is not lost until critical head loss is reached.

6.2.1.9 Composition of the filter layer

Single particle size filter beds. Filtration progression is discontinued when (1) suspensions leaking (breaking through) into the filtered water reach a certain limit or (2) when the filtration head loss reaches a certain limit. For filtration at normal speeds, either factors (1) or (2) determine the shutdown of the filtration process. In rare cases, during filtration at low filter speeds, fine filter media, and large suspensions, etc., the retentions concentrate on the surface layer, and a partial vacuum occurs in the filter bed, causing the retention matter (flocs, etc.) to decay and leak out, making it impossible to continue with filtration.

Figure 6.90a shows the linear growth of the total head loss of a filtration bed over time during filtration. Figure 6.90b shows changes in the filtrate water turbidity. At the beginning of filtration there is a small increase in the suspended matter concentration of the filtrate water (breakthrough), and the filter maturation (coating with suspended material) of the filter media ensures that the filtration water quality remains almost constant at the lowest level. After a certain period, however, a rapid increase in breakthrough is observed.

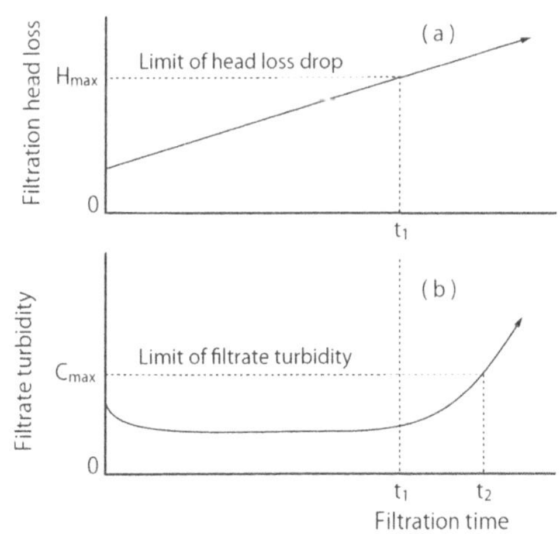

Figure 6.90 Filtration head loss and breakthrough.

The time to reach the maximum allowable head H_{max} is t_1, and the time to reach the maximum allowable filtrate water turbidity (break through) C_{max} is t_2. The optimal filtration condition for the above two process (design) elements is $t_1 = t_2$. However, in actual operation, if the limiting head is chosen as the design and operation parameter so that the filtration is aborted shortly before breakthrough reaches the allowable limit, the aborting time of the filtration can be confirmed without compromising the intrinsic water quality. Therefore, a condition such as $t_1 \leqq t_2$ is assumed in design.

If $t_1 > t_2$, (1) the filter media should be roughened, (2) the thickness of the filter bed should be reduced, (3) the coagulant adjuvant content should be reduced to reduce the adherence of the suspensions (as detailed in Coagulation, Chapter 5), and (4) an attempt should be made to increase the filtration rate to approach $t_1 = t_2$. When $t_1 < t_2$, the opposite operation is used as shown in Figure 6.91.

It is very difficult to describe theoretically how to determine filtration velocity, the thickness of the filter bed, or the particle size of the filter medium, etc., according to the nature of the suspensions to be removed. However, in Sections 4.2.3, 4.2.4 and 4.2.5 we have summarized and explained the results of our half-century studies on the modification of suspensions, such as coagulation and flocculation, and would now like to offer an approach to filtration theory. In general terms, it is a practical method to examine the correspondence between external and operation conditions that vary, depending on the pilot test rig used and the actual design followed. Even in this case, it is only possible to carry out an appropriate experiment based on a theoretical understanding.

It is not always difficult to determine the most efficient method to be used for removal, depending on the nature of the suspension (floc, etc.), the velocity of the filtration, the thickness of the filter bed, and the diameter of the filter medium, etc., but it is not easy to accumulate the data to be applied to the actual design under a wide range of conditions considering varying external conditions. Therefore, as a route for systematically accumulating data, we consider how to deal with the relationship between the filtration rate (velocity), the thickness of the filter layer, and the diameter of the filter medium for a given suspension.

A schematic diagram for obtaining the condition $t_1 = t_2$ under which the limit time t_2 for maintaining the predetermined removal rate C/C_0 and the time t_1 for reaching the limit head are equal by changing the filter bed thickness L (when other conditions are equal) is shown in Figure 6.92.

Next, we can see the relationship shown in Figure 6.93, which schematically shows how this relationship occurs when the filtration rate changes between v_1 and v_n. Figure 6.94 shows a schematic diagram of the condition in which the diameter of the filter medium changes from d_1 to d_n. If a designer draws these diagrams for each target ingredient of the filtration, they can use these procedures to properly define the

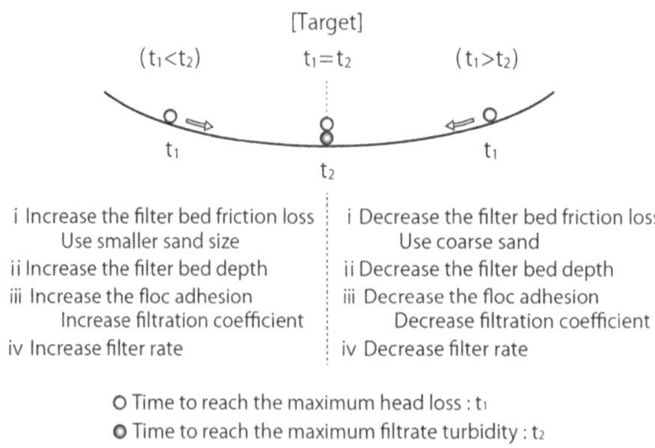

Figure 6.91 Time to reach maximum head loss and maximum filtrate turbidity.

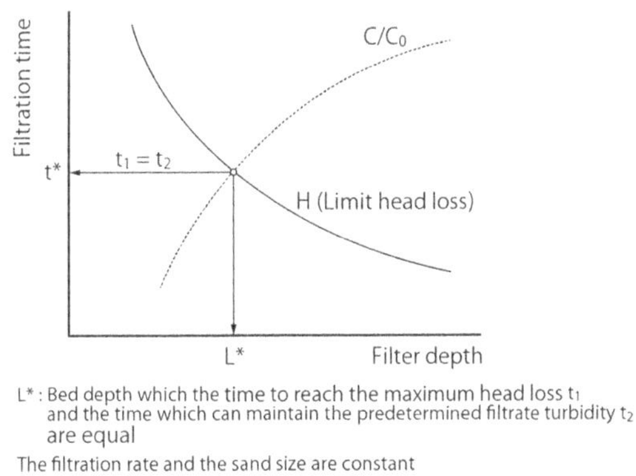

Figure 6.92 Relationship between the limit removal rate and the limit head loss.

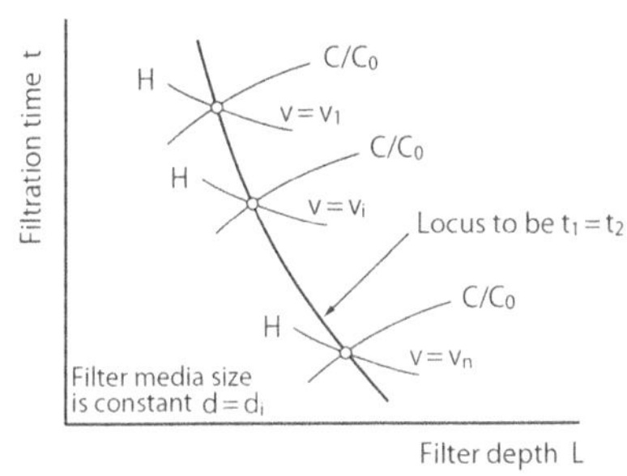

Figure 6.93 The filter depth and filter medium size at which $t_1 = t_2$.

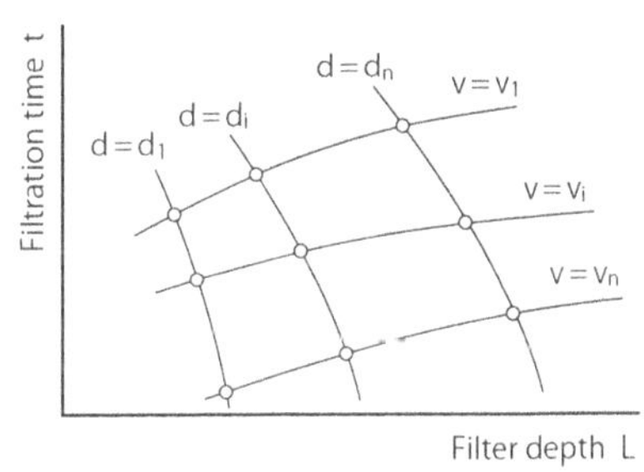

Figure 6.94 The filtration rate and filter media size at which $t_1 = t_2$.

composition of the filter bed. Indicating the physical properties of suspensions (flocs, etc.) is itself a difficult issue and is discussed in detail in the Coagulation section of Chapter 4.

The optimal conditions for filter design are obtained by adding dimensions of the floc physical properties and finding the conditions under the smallest build/operation costs among various filtration processes in which $t_1 = t_2$.

The authors have not yet ever seen a completely logical design path for all filtration ponds, and in many cases, design and operation are based on long-term experience and rules of thumb, but clearly that is something that could still be improved.

The common sense values for a rapid sand filter for water supply systems are: filtration rate (speed): 120 to 180 m/day; sand layer thickness: 60 to 70 cm; effective diameter (10% non-excess diameter of particle size distribution) of the filter media particles composing the sand layer: 0.6 mm; uniformity coefficient of filter media particle size distribution (60% non-excess diameter d_{60} to 10% non-excess diameter d_{10} ratio d_{60}/d_{10}) when removing turbidity flocs in a normal rapid filtration pond: about 1.5 to 1.3; and operation is usually cut off ($t_2 > t_1$). Under high-speed filtration conditions such as filter media diameter: 0.8 to 1.0 mm and filtration speed: 250 m, the condition $t_1 = t_2$ is reached. A slow sand filter is a superficial filtration with a very high λ, so t_1 is always much shorter than t_2, and the filtration is always terminated only by head loss.

Multi-layer filter beds. Even if a single-layer filter is designed under the optimal conditions presented in the preceding paragraph, the retention saturation level q_{max} is reached near the surface of the filter layer at the final phase of filtration (refer back to Figure 6.88b) and, at the same time, leakage of suspended matter occurs from the bottom of the filter at the maximum permissible level. Accordingly, even with optimal operation, as shown in Figure 6.95a, the used capacity of the middle and lower part of the filter layer is limited to a level far from the saturation point even when the filtration is discontinued, and the working capacity in the run is not effective (or satisfactory). Therefore, a multi-layer filter as shown in Figure 6.95b and Figure 6.95c can be constructed, consisting of an upper layer (coarse grain filter media) with a small rejection rate λ, whose main objective is to increase the total amount of retention in the filtration layer, and a lower layer (fine grain filter media) with a large rejection rate λ, whose main objective is to prevent turbidity leakage.

Generally, in rapid sand filtration, a 'backflow washing' is performed in which water flows upward from the bottom of the filter bed to fluidize the granular bed and wash out any retained suspensions. The granular layer fluidized by this washing process is regenerated by redistributing the fine particles on the top and coarse particle filter medium on the bottom of the granular layer. Therefore, the sedimentation rate of the coarse upper layer filter media must be smaller than the sedimentation rate of the lower-layer fine particle filter media in order to stably configure the multilayered bed after washing. For this purpose, it is necessary to select a high-density fine filter medium, which has a density much higher than that of the coarse filter medium and a high sedimentation rate, even if it has a small diameter. Typical practical applications are a three layer filter of anthracite with a specific gravity of 1.4 from the upper layer, quartz sand with a specific gravity of 2.6 for the middle layer, and garnet with a specific gravity of 3.4 to 3.7 on the bottom, or for two layers of anthracite and sand.

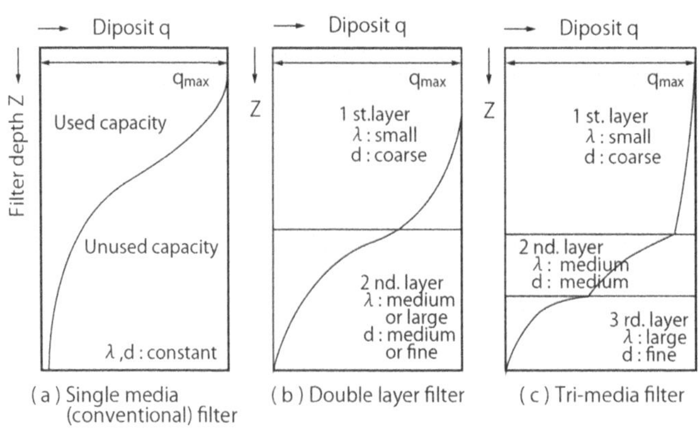

Figure 6.95 A deposit profile of different sand sizes, by layer.

High volume capacity filtration. The concept of high volume capacity filtration that the authors have been trying to develop is as follows:

(1) Normal filters, whether single or multi-layered, usually have a depth of about 1.0 to 1.5 m of water above the filter bed for a pressurized gravitational driving force. On the other hand, solid–liquid separation by filtration is performed only in a 20 to 30 cm area of the filtration bed, and three-quarters of the filtration bed volume is not an active space for suspensions (flocs) removal. Therefore, by devising some effective structures for removal in the pressurizing space above the sand surface, it becomes possible to drastically increase the retention amount (capacity).

(2) The effective water depth (head h_0) above the sand bed must be kept constant, as a key removal mechanism to maintain the water flow rate even if the filtration head loss of the sand layer increases as retention progresses. Therefore, when considering removal in this pressurized space, the design must be such that the increase of the pressure head can be almost ignored even as the retention progresses.

(3) In order to fit a pressurized space in, in alignment with the underlying filtration layer and to allow the performance of the separating function, the process to be considered needs to have a surface load (overflow) rate Q/A (where Q is the treatment flow rate m^3 per day, and A is the horizontal cross-sectional area of filter m^2) very close to being equal to the filtration rate of the sand layer body, and a quantitatively large retention capacity can be expected.

(4) Since the removal equipment loaded in the pressurized space is intended for the removal of a large amount of suspensions, the separated suspensions are required to be easily discharged to the outside of the filter, in a renewable way and with a flow rate and energy consumption lower than the backflow washing of the filter bed.

Practical examples of such filtration ponds are shown in Figure 6.96. If setting up a slanted plate or slanted tube into the sand water depth is attempted, to obtain an effective removal with a sufficiently low overflow rate corresponding to the filtration rate, multiple steps (30 to 40 steps) will be needed if the installation surface is set at 60 degrees horizontally, with gaps between plates as small as 2 to 3 centimeters. Therefore, by using a finned channel separator (as described above), it is possible to operate at an overflow rate of about 150 to 200 m/day with a normal inclined plate interval of about 5–10 cm (see Figure 6.47 for an example). Settled suspensions can be deposited in an inverted washing trough (discussed below) and drained out of the filter either simultaneously with or independently of backwash by lowering the water level of the filter to the overflow.

In addition, if a dissolved air flotation process is introduced into the pressurized space above the filter bed itself, the separation rate of the sand layer is balanced with the separation rate of the flotation overflow rate of

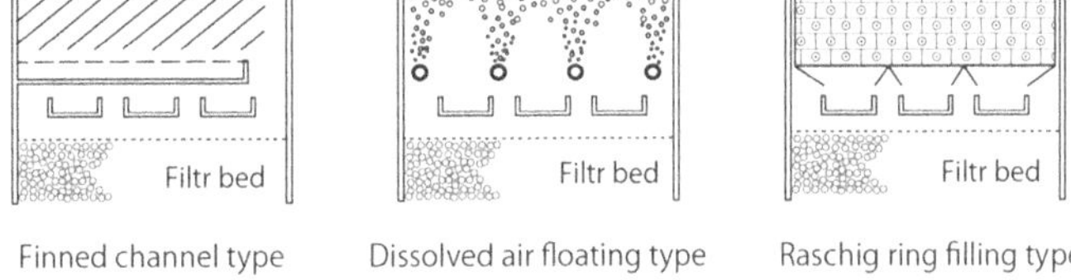

Figure 6.96 Solid–liquid separation using the pressurized space above the filter bed.

about 10 to 15 cm/min (144 to 216 m/day). Operation can be performed only when turbidity is high; but the process can be performed only with the lower sand filtration layers have low turbidity.

A super coarse filter medium with a large filling porosity of several millimeters to several centimeters, such as a Raschig ring (a porcelain/plastic cylindrical or saddle-type filler that is commonly used in a fixed-bed reactor process), can be filled into the pressurized space to about 1 meter. The porosity of the sand filter layer is only about 40 to 45%, but it reaches 70 to 80% in the case of Raschig rings, suggesting that the retention of a large amount of suspensions can proceed with a small increase in head loss.

As a removal mechanism for deep filtration, various methods such as sedimentation, obstruction (collision) and sieving can be considered, and it is considered that a Raschig ring layer is a process which emphasizes the sedimentation method. Micro-sedimentation ponds which have an extremely low removal rate and an overflow rate of 10–20 cm/min can be connected in series (of several tens or more) and, as a result, a higher removal rate can be obtained as a whole. Since finned channel separators can obtain a removal rate (velocity) exceeding overflow rate theory by adding a contact separating mechanism to swirl flow to the sedimentation mechanism, there are proposals to use these high volume capacity filtration ponds as a conventional two-part treatment process, such as sedimentation and filtration – which can actually be regarded as one continuous process.

The authors now propose a two-layer filtration pond equipped with a Raschig ring, as shown in Figure 6.96 and Figure 6.97, which has a high retention capacity to cope with high concentrations of raw water, beyond the usual constraints of a traditional granular-bed filter intended for clarifying waters that have been subjected to sedimentation or to treat low turbidity raw water.

The results of this process are presented to demonstrate its effectiveness. Raschig ring media was used to fill 3-, 6- and 9-mm diameter cylindrical Raschig rings and a diameter-to-length ratio of 1:2 with a Raschig ring bed thickness of 90 centimeters so that the void fraction is approximately 70%. The lower sand layer was covered with relatively coarse quartz sand with a diameter of 0.84 to 1.00 mm and a thickness of 50 cm. A wastewater trough was installed between the Raschig ring bed and the sand floor, and the suspension retained in the Raschig ring spaces was discharged by gravity by draining the water into the trough. Suspensions trapped in the sand layers were removed by conventional backflow washing through the same trough after the Raschig ring drain.

Figure 6.98 shows the effectiveness of a Raschig ring bed in the first stage for a high-speed filtration, with a filtration speed of 200 m/day, for high turbidity (kaolin) suspensions, such as 80 TU, indicating that the available removal can be continued for as long as 6 to 10 hours with only a few centimeters or less of water head loss. Figure 6.99 shows the successive changes in turbidity between the inflow of the Raschig ring bed and the turbidity of the out flow in the lower sand layer when filtration is continued, washing the Raschig

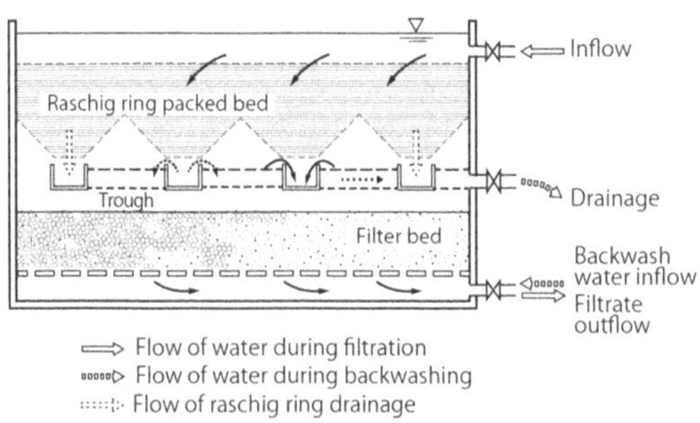

Figure 6.97 High retention capacity filtration with Raschig rings.

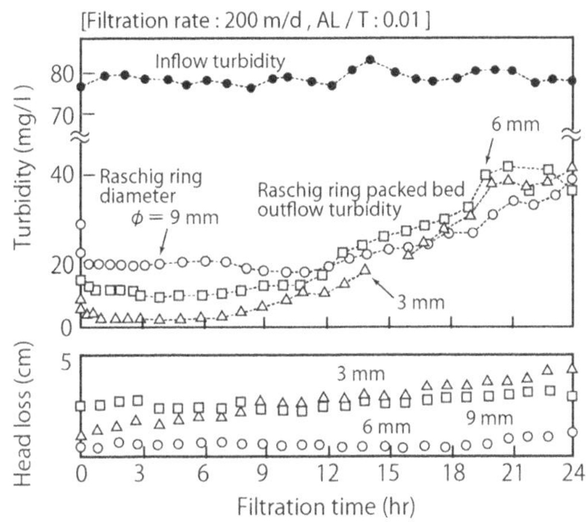

Figure 6.98 Turbidity removal and head loss in a Raschig ring packed bed. (Note: The AL/T ratio in this and other figures is defined by Tambo and colleagues in terms of coagulant injection rates and is described in detail in Section 4.2.3, on the Aluminium/Turbidity Concentration' ratio, which is today used worldwide).

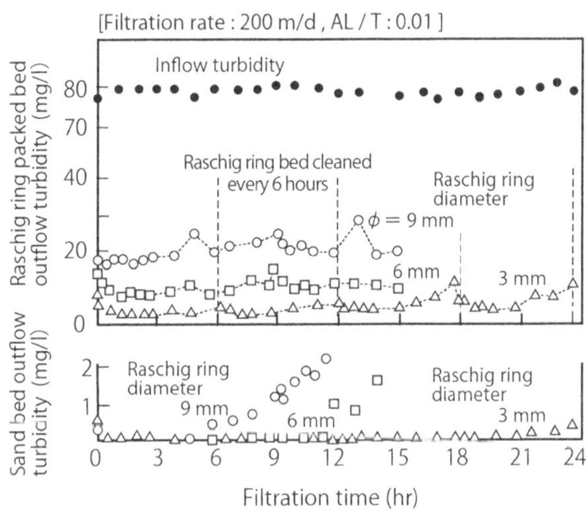

Figure 6.99 Removal of turbidity in a raschig ring packed bed and sand bed.

ring bed every 6 hours. We have been able to effectively treat highly turbid water as high as 80 TU for one day and night. High-concentration coagulated Kraft pulp wastewater (50 degree color unit) can also be effectively treated.

The removal process which governs the rejection rate λ in the Raschig ring filter bed was monitored by video, and flow was shown to be laminar, and suppression was carried out entirely on the upper surfaces (of both the inner and outer surfaces) of the Raschig ring, and it was presumed that sedimentation was the dominant mechanism among the various filtration retention mechanisms.

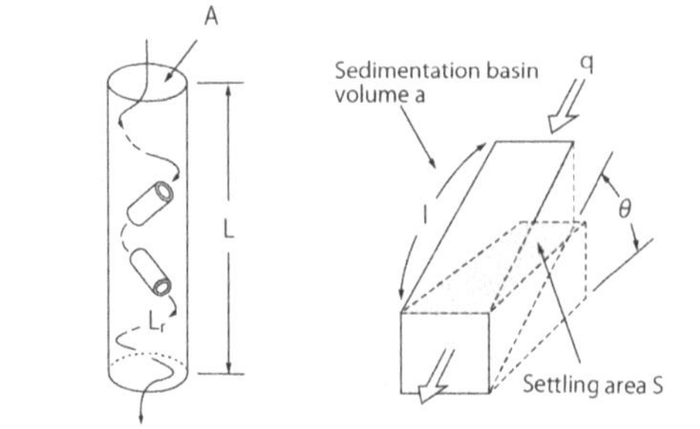

Figure 6.100 Basic concept of a Raschig ring packed bed.

Therefore, as shown in Figure 6.100, the authors considered a ladder structure of virtual microscopic sedimentation chambers, and considered a large number of microscopic sedimentation chambers with a relatively large overflow rate (and hence a small removal rate for an individual Raschig ring) which were arranged in series, resulting in the achievement of a valid removal. The fluids were assumed to mix thoroughly between these series of hypothetical micro-sedimentation chambers (Raschig rings), and the suspension inflow to subsequent Raschig rings was assumed to be in uniform concentration. It was also assumed that the flow direction was the shaft direction of the Raschig ring and that removal proceeded by gravitational sedimentation (overflow rate dominated).

As the suspension flowed down the L-depth of the Raschig ring bed, it passed through the m Raschig rings in shaft directions, in series. In this case, the average sedimentation area S of a micro-sedimentation chamber is as shown in Table 6.21, with the number of series passes m of the micro-sedimentation chambers passing down the L_r and L-depth ladder flooring passing through the reservoir.

The mean volume occupied by a virtual micro-sedimentation chamber in the filter bed is shown in Equation 6.137.

Table 6.21 Virtual Raschig ring packed sedimentation basin specifications.

Total flow distance of m Raschig rings: L_r

$$L_r = L \sin \bar{\theta}$$

Number of serial passes of Raschig ring: m

$$m = \frac{\beta L_r}{l} = \frac{\beta L \sin \bar{\theta}}{l}$$

Average settling area of
the imaginary micro sedimentation basins: S

$$S = (d_0 + d_1) l \cos \bar{\theta}$$

L : Raschig ring packed bed depth
$\bar{\theta}$: Average settling area of virtual sedimentation ponds
β : Coefficient related to packed condition of Raschig rings (dimensionless)
l : Virtual micro sedimentation pond length
d_0, d_1 : Outer and inner diameter of Raschig ring respectively

$$a = \frac{\varepsilon A L}{A L(1-\varepsilon)/[(\pi/4)(d_0^2 - d_i^2)l} = \frac{\pi}{4}\frac{\varepsilon}{1-\varepsilon}(d_0^2 - d_i^2)l \quad \text{—(6·137)}$$

A : Filter bed area

(6.137)

Equation 6.137 The mean volume occupied by virtual micro sedimentation chamber in filter bed.

The flow rate q passing through the virtual micro-sedimentation pond is given by Equation 6.138.

$$q = \left(\frac{v}{\varepsilon}\right)\left(\frac{L_r}{L}\right)\left(\frac{a}{l}\right) = \frac{\pi v (d_0^2 - d_i^2)}{4(1-\varepsilon)\sin\theta} \quad \text{—(6·138)}$$

(6.138)

Equation 6.138 The flow rate passing through virtual micro sedimentation pond.

Therefore, the overflow rate of a virtual micro-sedimentation chamber, $w_0 = q/S$, is similar to Equation 6.139.

$$w_0 = \left(\frac{q}{S}\right) = \frac{\pi v (d_0 - d_1)}{2l(1-\varepsilon)\sin\bar{\theta}} \quad \text{—(6·139)}$$

(6.139)

Equation 6.139 The overflow rate of virtual micro sedimentation pond.

From these equations, the total removal rate R for the inflow of a group of particles with a sedimentation rate w on a Raschig rings bed of depth L is given by Equation 6.140. Since we think that the removal rate $r = w/w_0$ of each micro-sedimentation chamber is small, we have omitted the higher-order term of the expansion equation and calculated the initial blocking rate λ_0 as shown in Equation 6.141.

$$R = 1 - (1-r)^m = 1 - \left(1 - \frac{w}{w_0}\right)^m = 1 - \exp(-\lambda_0 L) \quad \text{—(6·140)}$$

r : Removal rate of imaginary micro sedimentation basin (w/w_0)
w : Settling velocity of the particle
λ_0 : Initial rejection rate

(6.140)

Equation 6.140 The total removal rate of raschig ring packed filter bed.

$$\lambda_0 = \left(\frac{m}{L}\right)\ln(1-r) = \frac{m}{L}\left(r + \frac{r^2}{2!} + \frac{r^3}{3!} + \cdots\cdots\right) \fallingdotseq \frac{m}{L} r = \alpha\beta\frac{4(1-\varepsilon)\cos\bar{\theta}}{\pi(d_0 - d_i)}\frac{w}{v} \quad \text{—(6·141)}$$

α : Coefficient related to adhesion between micro sedimentation basin and the suspended solids (dimensionless)
β : Coefficient related to the packed condition of raschig rings (dimensionless)

(6.141)

Equation 6.141 The initial rejection rate of raschig ring packed filter bed.

As shown in Equations 6.140 and 6.141, the initial rejection rate of the filter shows that the suspended matter concentration decreases exponentially as the suspensions flow down the Raschig ring bed. In either Figure 6.101 (turbidity removal) or Figure 6.102 (color removal), it is recognized that the initial rejection rate λ_0 takes a constant value according to the filtration condition; λ_0 can be determined from the gradient of the plot, and the characteristics of the filtrate can be estimated from various filtration conditions. From the results of various experiments, the relationship $\lambda_0 \propto v^{-1}$ is almost demonstrated.

As removal in the filter bed progresses, deposits accumulate in and out of the Raschig ring, reducing the porosity of the filter bed and increasing the actual flow rate, thereby reducing filtration efficiencies, and we will now quantitatively evaluate the evolution of such phenomena in the Raschig ring filter bed. As shown in Figure 6.103, when we consider a filter bed layer with a small thickness of ΔL, and consider the balance of

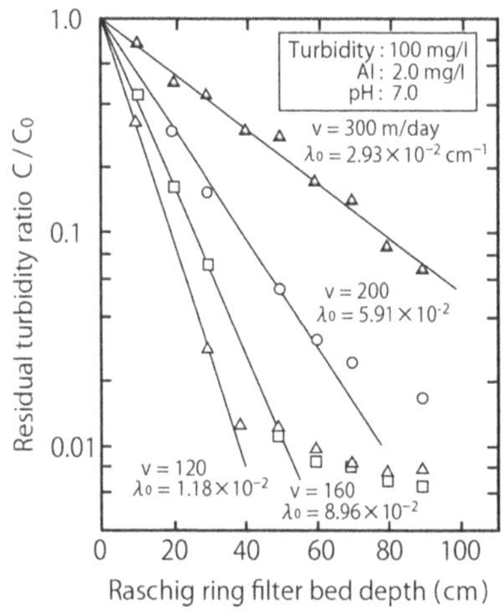

Figure 6.101 Decreased turbidity in relation to filter bed depth.

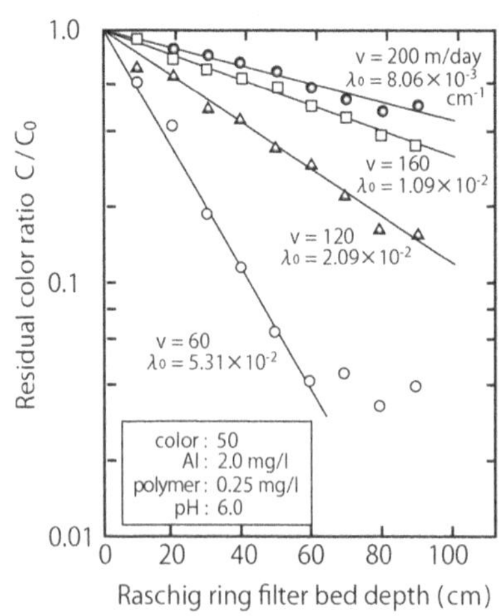

Figure 6.102 Decrease in color in relation to filter bed depth.

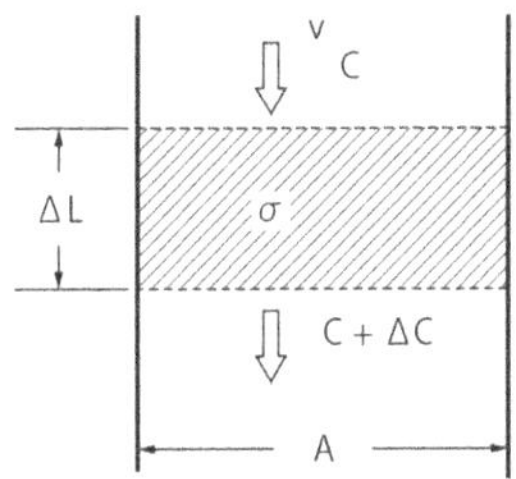

Figure 6.103 Material balance of a micro filter layer.

the suspended matter that is retained by inflow and spillage into this microfilter layer during time Δt, we have Equation 6.142.

$$\frac{\partial C}{\partial L} = -\lambda_0 C + \frac{Z(\sigma)}{v} \quad \text{—}(6 \cdot 142) \tag{6.142}$$

σ : Suspended solids retained in unit volume of filter bed (mg/cm³)
$Z(\sigma)$: Scouring function (mg/cm³·min)

Equation 6.142 The material balance equation of filter bed.

If deposition and scouring are mutually independent phenomena, Equation 6.142 can be used. However, in discussion with Mintz, Ives made inferences that 'even if the water is passed through at the same speed without inflow suspending in the absence of filtration, peeling does not occur' and 'peeling and deterrence are closely related interconnected phenomena,'(Ives, 1969, 1975) and an expression such as Equation 6.142 (Mintz phenomenon) dos not show any reality here. Therefore, when $C_0 = 0$ without inflow suspensions, we can consistently express the change in the rate of filtration as the retention progresses in Equation 6.143 by modifying Equation 6.142, considering the condition that $\delta C/\delta L = 0$.

$$\frac{\partial C}{\partial L} = -\lambda_0 \left(1 - \frac{Z(\sigma)/C}{v\lambda_0}\right) C \quad \text{—}(6 \cdot 143) \tag{6.143}$$

Equation 6.143 The modified material balance equation of filter bed.

Examining the form $Z(\sigma)$ of the scouring function included in Equation 6.143 (from the experimental results using Equation 6.144), it was found that a linear relationship $Z(\sigma)/C = k\sigma$ exists between the deposition amount σ and the modified scouring function $Z(\sigma)/C$, as shown in Figure 6.104, where k is the experimental constant and is immediately derived from the slope in Figure 6.104.

$$Z(\sigma) = [C_{out} - C_{in} \exp(-\lambda \Delta L)] \frac{v}{\Delta L} \quad \text{—}(6 \cdot 144) \tag{6.144}$$

C_{out}, C_{in} : Outflow/inflow suspended solid concentration from ΔL of bed depth (mg/l)

Equation 6.144 The scouring function.

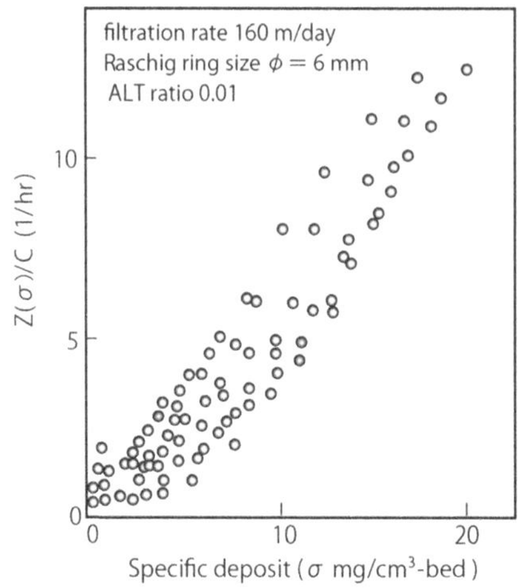

Figure 6.104 Retained flocs and scouring equation.

Therefore, substituting the linear relation $Z(\sigma)/C = k\sigma$ into Equation 6.143 rewrites the equation describing the course of removal in the Raschig ring filter bed as Equation 6.145.

$$\frac{\partial C}{\partial L} = -\lambda_0\left(1 - \frac{k\sigma}{v\lambda_0}\right)C \qquad \text{—(5·145)} \qquad (6.145)$$

Equation 5.145 The removal equation in raschig ring filter bed.

As the filtration time increases, the total amount of trapped material σ in the filter bed increases, the retention capacity gradually decreases, and finally, the filter bed loses its removal capacity when the maximal retention capacity σ_u is reached. That is, when $\sigma = \sigma_u$ is reached $\delta C/\delta L = 0$, and when this condition is applied to Equation 6.145, the maximal retention potential σ_u is given by Equation 6.146.

$$\sigma_u = \frac{v\lambda_0}{k} \qquad \text{—(6·146)} \qquad (6.146)$$

Equation 6.146 The maximum detention capacity.

If we know the gradient k in Figure 6.104, we can estimate the magnitude of σ_u. The maximal retention capacity σ_u of coagulated kaolin floc is 30 to 25 mg/cm^3 at a high rate filtration of 100 to 300 m/day, which is higher than the normal 10 mg/cm^3 of a rapid sand filter. Substituting Equation 6.146 into Equation 6.145, the equation for Raschig ring filtration described above results in the filtration equations previously proposed by Mintz.

6.2.1.10 Washing the filter with water

Even with good retention of the suspensions, the filtration function cannot continue endlessly without proper regeneration washing of the filter bed. Historically, a variety of regeneration washing schemes have been proposed, and during the creation of sand filters, a variety of washing schemes were proposed that are no longer in use today.

The suspensions retained in the filter media can be divided into those that are merely retained in the gaps between the particles of the filter media (narrow channels) and those that are attached to the surfaces of the filter media. Figure 6.105 is a schematic representation of the suspensions retained in the filter bed, showing both conditions. We have devised a characteristic method of washing to respond appropriately to each case, and we begin by considering the washing of suspended components that are simply trapped in the gaps between particles.

Most recent rapid filtration systems use an upward back washing flow (using filtrate water) introduced from the lower part of the filter bed to swell and fluidize the filter bed, and wash out the suspended matter retained in the filter media gap. Retained suspensions are smaller in diameter, less dense, and have less sedimentation velocity than filter media particles, and are carried away easily by an upward flow through the filter media. Therefore, the magnitude of the washing velocity (flow velocity) seems to have little effect on the separating function in the normal flow velocity range (expansion coefficient). Our study suggests that the most effective termination of backflow washing with minimal flow rate can be achieved by setting the upward flow rate to fluidize a bed of sand with a thickness of about 50 to 60 centimeters so that the rate of expansion is about 120 to 135%, followed by backflow washing for about 5 to 7 minutes.

Filter media gaps with a bed expansion of about 120 to 135% may be considered the smallest channel size required for the retained suspension to escape the filter media gaps without difficulty, meaning that there is an adequate constant fluidized bed expansion rate with varying ascending flow rates, almost independent of the size and density of the filter media particles. In an actual rapid sand filter with a large horizontal cross-sectional area (several meters square) relative to the thickness of the filter bed (several tens of centimeters), the water flow in the pond at the time of reverse washing for what is considered to be a thorough (complete) mixing is governed by the Reynolds number, which is characterized by the horizontal cross-sectional length of the filtration pond. Accordingly, under commonly used operating conditions, a backflow washing duration of 5 to 7 minutes which rinses the fluidized sand bed in a state of complete mixing (a turbulent state) proceeds at an upward flow rate of 60 cm/min. Under these conditions, in our experiments, the vacant column rose by about 3 to 4 m in about 5 to 7 minutes of washing time, resulting in rinsing approximately 5 to 7 times with complete mixing. The washing wastewater concentration drops exponentially over time, and the cleaning water concentration at washing discontinuation falls below $1/e^{(5 \text{ to } 7)}$ of the initial suspension concentration.

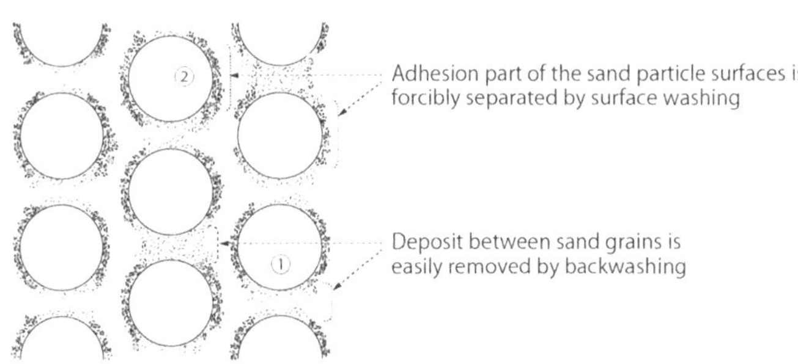

Figure 6.105 Schematic representation of flocs detained in a sand layer.

If the thickness of the sand layer is L and the porosity is ε, and the thickness of the sand layer expanded by the back washing is L_e, the necessary back washing flow velocity v_b is determined by Equation 6.147.

$$\text{Expansion rate} = \frac{L_e}{L} = \frac{1-\varepsilon}{1-\left(\frac{v_b}{w_t}\right)^n} \quad \text{---}(6\cdot147) \tag{6.147}$$

L : Filter depth (cm)
L_e : Sand layer length expanded by backwashing (cm)
ε : Filter bed porosity
w_t : Terminal settling velocity of particle size d (cm/min)
v_b : Backwash superficial velocity (cm/min)
n : Constant determined by Reynolds number Re
$R_e = w_t d / \nu$ (ν : Kinematic viscosity coefficient)

The n value by Richardson

$R_e < 0.2$	$1/n = 4.65$
$0.2 < R_e < 1.0$	$1/n = 4.36\, R_e^{-0.03}$
$1 < R_e < 500$	$1/n = 4.45\, R_e^{-0.1}$

Equation 6.147 The filter bed expansion rate and backwashing rate.

When the interstitial suspension is suspended once again in the backflow of the washing water and discharged after complete mixing (Figure 6.106), it begins to rinse after a highly uniform mixing system with a high concentration is dispersed in the entire sand layer, and the exponentially decreasing cleaning water turbidity continues until the approved backwash level is reached. Therefore, the overflow surface level at the time of backwash should be lowered so that the water surface comes to be about 40 to 50 centimeters from the expanded sand surface, and the volume to be mixed should be as small as possible for the washing operation. To this end, we devised troughs to be used for an inverse washing system, as shown in Figure 6.107 and Figure 6.108, and the above-mentioned action is shown in Figure 6.109.

The wastewater trough is arranged as shown in Figure 6.110 so as to maintain an upward flow velocity as uniformly as possible at the washing of the backflow, as is the case with the effluent baffles of a sedimentation pond. Since the settling velocity of the filter media particles differs drastically from that of the suspended particles in the detached backwash wastewater, the top of the overflow of the trough may be located at the lowest point where the filter media particles do not flow out during the backflow washing. Therefore, the backwash water level is lower than the filtration water level and is defined by the upper surface of the troughs. This eliminates wasted water, as shown in Figure 6.109, and eliminates

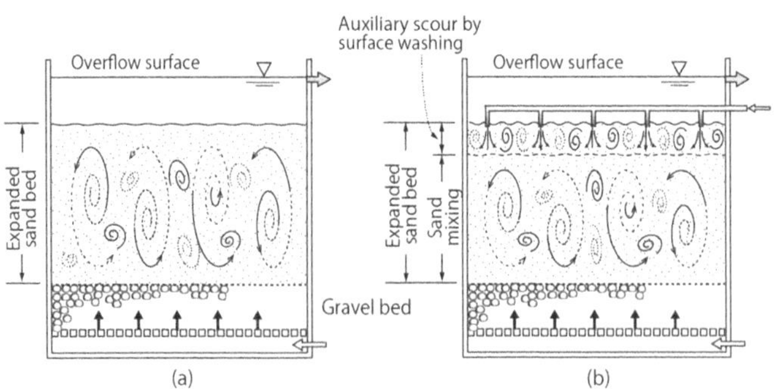

Figure 6.106 Schematic of complete mixing of the sand layer during backwashing.

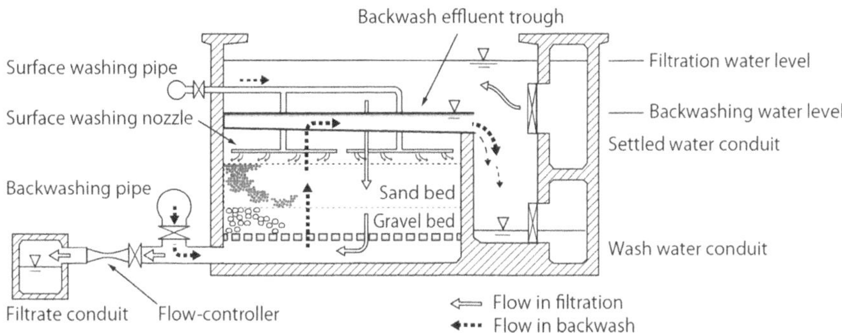

Figure 6.107 The water flow in a filter pond.

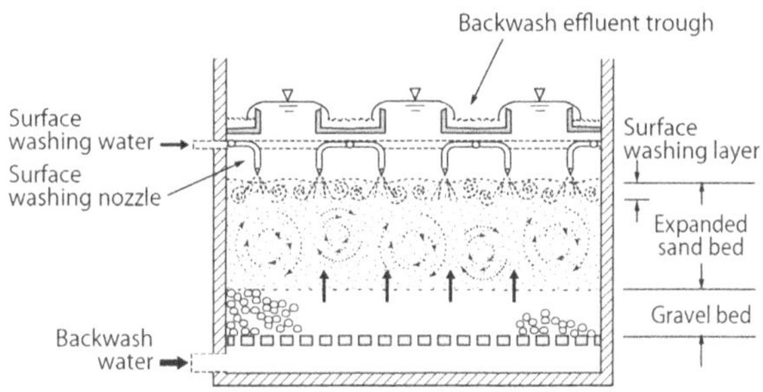

Figure 6.108 The backwashing of a filter pond.

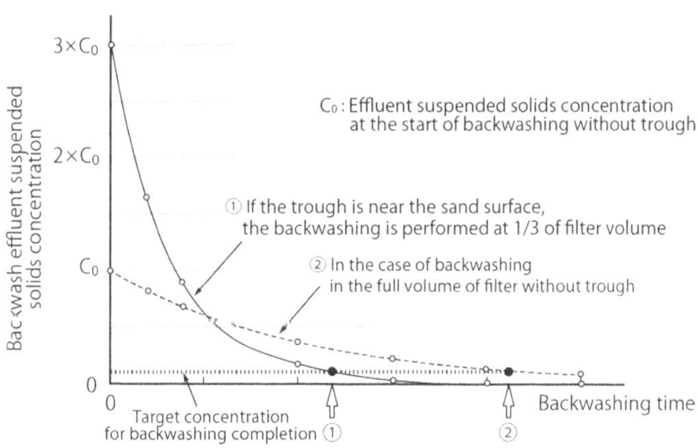

Figure 6.109 Trough effluent concentration and backwashing time.

only wastewater from the upper part of the expansion bed and water from the lower part of the troughs during backwash when thoroughly mixed. The troughs are hydrologically open channels with a lateral inflow, and were first developed using the T.R. Camp formula, and various other calculation formulas and calculation diagrams have since been proposed. As shown in Figure 6.110, when the wastewater of the troughs comes into inflow with the wastewater pond, the water level of the wastewater pond can be lowered to maintain the free overflow condition. If the wastewater drainage capacity is low, the water level in the troughs may be affected by the back water, which may cause the wastewater capacity to decrease, which is not preferable.

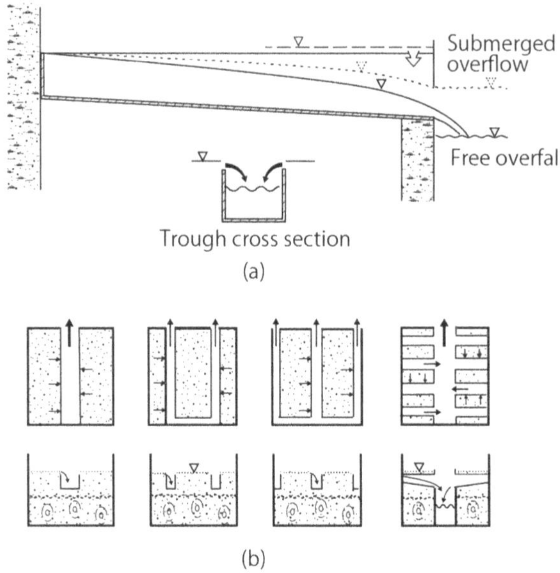

Figure 6.110 Trough arrangements and flow.

With a high rate of filtration, large-diameter filter media with a particle size of 1 mm or more are required to prevent excessive backwash flow, and anthracite, which is much less dense than sand, can be used. In addition, we have devised a lighter-weight plastic filter material that can achieve an expansion rate of 30 to 50% without increasing the backwash flow rate, and a filter material that has been granulated with fibers.

The expansion of the sand layer is a function of the ratio of the backwash flow rate to the terminal sedimentation rate w_t of the filter particles (V_b/w_t), as shown in Equation 6.147. The sedimentation rate w_t of the filter particles varies inversely with the viscosity factor μ of water according to the Stokes equation (as described in Section 6.1). The viscosity of water is 1/1.6 in relation to water of 30°C/10°C and therefore, the required expansion from 130 to 150% of the backwashed sand bed cannot be obtained in the hot summer season or in tropical water purification plants unless the backwash rate is increased by more than 1.6 times (Figure 6.111).

Table 6.22 illustrates the differences in water temperature and sand bed expansion rate for the same backwash velocity, and the index n of Equation 6.147 is calculated as 0.22.

In many urban areas in Southeast Asia, rapid sand filters in tropical water treatment plants designed using techniques (such as design guidelines) established in northern developed countries such as Japan and the EU in low temperature areas have resulted in a shortage of backwash flow velocity in most cases, which has resulted in inadequate sand expansion. The higher the water temperature, the lower the viscosity, but the density change with temperature is limited. Because the performance of pumps, which are designed based on inertial forces (water densities), vary little with water temperature, tropical regions must compensate for the reduced viscosity by ensuring a higher backflow washing flow rate, which requires a larger pump. In the past, there were many water treatment plants in countries near the equator that did not reflect this understanding of principles in practical use.

Removal of retained suspensions attached to filter media surfaces. For the removal of interstitial suspensions, a surface washing operation using high-intensity turbulent washing (jet washing) is widely used in conjunction with back washing, which generates a high energy dissipation rate that is 10 times

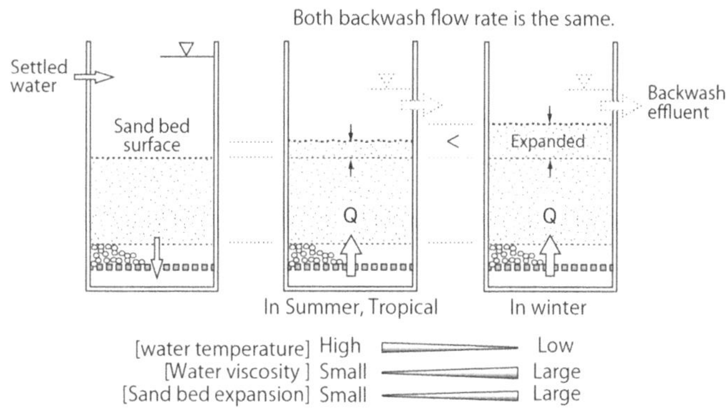

Figure 6.111 Water temperature and sand bed expansion rate.

Table 6.22 Examples of sand bed expansion rates.

Water Temperature °C	Viscosity c Pa.s	Density g/cm^3	Settling Velocity cm/s	Expansion Rate %
10	1.310	1.000	17.2	23
30	0.801	0.996	28.1	11
30/10°C	0.611	0.996	1.63	0.48

Water temperature°C,30°C Sand: Density 2.65 g/cm^3 Size 0.05 cm
Sand bed porosity: 0.4 Backwash superficial velocity: 0.5 m/min (0.83 cm/sec)

higher than the back washing on a sand layer with a particularly large retained quantity on a 10 to 30 cm surface (as shown in Figure 6.106b above).

The first technique devised to remove the attached suspensions from the surface of filter media and take them back out into the backwash water stream and wash them out of the filtration pond is a process in which the filter media collide with each other under turbulent flow conditions. It has long been thought that a coefficient of expansion of as much as 150% would be a suitable value for the collision between the filter media.

When turbulence is weak, which is considered to be in an area where the particle size is smaller than the microscale λ_0 of the turbulence vortex (Equation 6.148), it is known that the number of collisions of particles under turbulent conditions is proportional to the square root of the energy dissipation rate in the unit volume of fluid in units time.

$$\lambda_0 = \left(\frac{\nu^3 \rho}{\varepsilon}\right)^{\frac{1}{4}} \quad \text{—(6·148)} \tag{6.148}$$

ν : Water kinematic viscosity (cm^2/sec)
ρ : Water density (g/cm^3)
ε : Energy dissipation rate of turbulence (erg/cm^3 sec)

Equation 6.148 The microscale of turbulent vortex.

The $\acute{\varepsilon}_b$, which is the energy dissipation rate lost per unit time per unit volume of sand bed fluidized during the back washing, is estimated from the head loss of water in the sand bed during the back wash.

Approximating Equation 6.39 (as described earlier in this chapter), the head loss h_r at the time of expansion of the sand layer is expressed by Equation 6.149. Therefore, the energy dissipation rate $\acute{\varepsilon}_b$ in the expanded sand layer of unit volume per unit time is Equation 6.150.

Approximate expression of Eq.6·39

$$h_f \fallingdotseq \frac{L(1-\varepsilon)(\rho_s-\rho)}{\rho} \quad —(6\cdot149) \tag{6.149}$$

h_f : Sand bed head loss during backwashing (cm)
L : Sand bed length (cm)
ρ_s : Filter media density (g/cm³)
ε : Porosity of the expanded sand bed (-)

Equation 6.149 The head loss during expanded sand bed.

Energy per expanded sand bed unit volume and per unit time (Energy dissipation rate)

$$\varepsilon_b \fallingdotseq m g h_f \quad —(6\cdot150) \tag{6.150}$$

m : Water mass (g)
g : Gravitational acceleration (cm/sec²)

Equation 6.150 The energy dissipation in expanded sand bed.

Actually, since it is a suspension, the values of m and $\acute{\varepsilon}_b$ with the suspension are slightly larger than those of water alone. Assuming a sand density of 2.65 g/cm^3, a backwash flow rate of 1 cm/sec = 60 cm/min, and $\varepsilon = 50\%$, the energy dissipation rate of the expanded sand bed is 8×10^2 $erg/cm^3/s$. This is the intensity of the slow agitation levels of conventional jar testers used in the coagulation test, which does not appear to be high enough to completely remove the strongly adhered suspensions on the media surfaces. On the other hand, as shown in Figure 6.108, the typical process of a modern rapid sand filter is to spray washing water jets onto the surface of the filter bed prior to the backflow washing or for a short period of time in the early stages of the backflow washing to wash the surface of the filter bed.

The energy dissipation rate of turbulent agitation is calculated to compare it with the backflow washing intensity by calculating the intensity at which the surface washing peels off the deposits on the surface of the filter media. The calculations proceed with examples of fixed-surface washing devices in the Japanese Water Works Design guideline. Assuming that the spray water pressure is 10 to 20 m hydrostatic head h, the spray flow rate Q is 0.3 $cm^3/sec{\cdot}cm^2$, the spray water flow penetrates into the expanded sand layer to about 20 cm, and the total jet energy is consumed in the 20 cm surface layer, the energy dissipation rate of the jet in the expansion sand layer of unit volume $\acute{\varepsilon}_s \fallingdotseq mgh \fallingdotseq 3 \times 10^4$ $erg/cm^3 sec$. Therefore, the agitation energy intensity (dissipation rate) provided by the superficial jet washing is more than 30 times that provided by backflow washing and the highest agitation intensity level of a jar tester, and is considered to be the main contributor to removing attached flocs.

If backflow washing is not performed sufficiently, suspended matter collects at the center of the circulation vortex with a size of the order of the depth of the expansion sand bed, or is drawn into the water flow at the corners of the filtration pond. The suspended matter agglomerates at the core of the rotating vortex and grows into large particles that cannot be taken out of the system at a flow rate of the order of the backwashing flow. This is called a mudball, and it occurs in insufficient filtration equipment and is often seen in filtration ponds in Southeast Asia and Africa.

The introduction of surface jet washing is a crucial measure to prevent mudballs from being produced for a number of reasons. First, surface washing can stop mudball formation because the dirtiest surface filter media layer is cleaned by strong local washing. Second, a small-scale high vortex strength layer (of a thickness of 20 to 30 cm) is formed at the top of the filter layer during washing. A large-scale circulating vortex caused by a backwashing current with a small intensity cannot enter the high energy dissipation rate layer at the top. As a result, a large amount of suspended matter retained at the top of the sand layer cannot be moved into the large-scale mixing eddies in the backwash water but move straight to the wastewater trough. This ensures that suspensions detached from the surface washing will not become entrained in the lower circulating vortex, and that the lower expansion sand layer will be rinsed at low levels of complete mixing without introducing large amounts of trapped material from the upper part. (In a similar way, to improve the effectiveness of air conditioning, department stores are equipped with air curtains at the front doors to create high-strength micro eddy boundaries, and the cooling/warming of the air due to the loose convective vortices inside is blocked from leaking into the outside air).

Blowing air bubbles from the bottom during reverse washing has long been used to help backflow washing. The authors are not aware of any quantitative mechanical analysis having been carried out, but it is considered that large (of the order of millimeters) air bubbles can improve the cleanliness of a sand bed as a result of the combined action of the deformation of the air bubbles as they pass through the fluidized sand bed at high speed, the impact/shear force due to ruptures in the surface layer, and the rectifying action of the vertical upward flow of the bubbles to suppress the rotation of the circulating water flow.

6.2.1.11 Operation of gravitational filtration

The flow of gravitational filters during operation are summarized in Figures 6.107 and Figure 6.108. The hydraulic systems used to operate a filtration pond are roughly classified into filters that are composed of (1) a lower water collector for collecting clear water and (2) a washing device.

In rapid sand filtration, a large amount of clean water is pushed from the lower collecting system to fluidize the bed during washing, so the lower collecting unit also acts as a washing unit. In addition, since the flow rate in backflow washing is 7 to 8 times the filtration flow rate, if the water supply condition in the backflow washing is satisfied, the water collection condition is also satisfied hydraulically. The main objective of the design of these units is to move the specified flow rate uniformly into and out of the filter bed.

Under drain (filter bottom) collectors. Historically, porous pipes or slabs, Wheeler's filter bottoms, or perforated blocks and other systems have been used as filter bottom water collectors. Products using materials such as concrete, metal, and plastic have been proposed and adopted. Except for porous slabs composed of sintered porous plates that can completely equalize the inflow and outflow, multipoint dispersion type inflows and outflows are common in filter bottom water collectors. In many cases, different size gravels of about 2 to 50 mm are laid on the lower part of the sand layer to about 20 to 50 cm depth so that the grain size increases downward, and the cleaning backwash water jets flowing out from multiple points at the time of washing are divided and made uniform by the gravel layer before it reaches the sand layer.

Porous tube type water collectors. The simplest type of lower water harvester is a porous pipe in which pore branches (small holes of about 10 mm in diameter, directed downward at about 45 degrees, and with an opening ratio of about 0.2 to 0.3% of the cross-sectional area of the filtration pond) arranged as shown in Figure 6.112 are laid, and backwash cleaning water is fed from the main pipe to collect filtrate water. Of these components, the porous tube type is the simplest system, but the backwash flow rate distribution of the filter bed may be locally uneven due to changes in the flow rate reduction of the branch pipes caused by water jetting from the small holes in the middle of the pipes and differences in the jetting volume (flow rate) from the respective holes caused by changes in the head loss between the branch pipes and

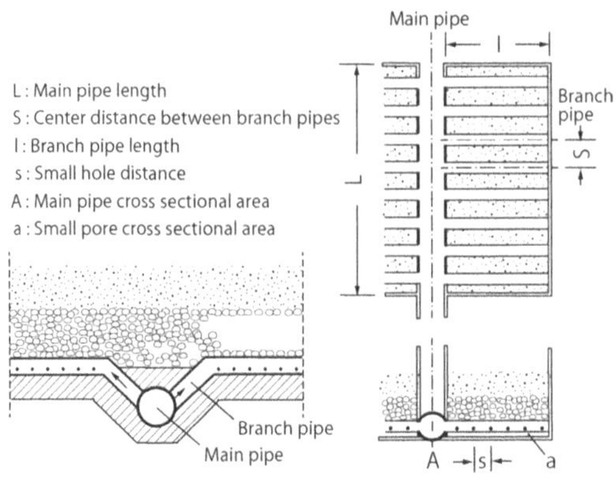

Figure 6.112 A lower perforated water collection pipe device.

the main pipes. To prevent this, the diameters of the main and branch pipes should be as large as possible, and the smallest size of holes should be considered so that water flow losses and kinetic energies in the pipes are practically negligible compared with water flow resistances when passing through the holes. In the case of a rapid sand filtration site in a normal water purification plant, it is common practice that the area of the hole is about 0.15 to 0.5% of the area of the filter bed, the interval between holes in the branch pipe is about 10 cm, the diameter of the holes is about 6 to 15 mm, and the interval of the branch pipes is about 2 to 4 times the interval of the hole.

Professor Sueishi (Kyoto University, Japan) (Sueishi, 1940) considered the water flow of a porous pipe collector to be a uniform outflow from the horizontal side which solves problems by considering the frictional water head loss of the kinetic energy head, hydrostatic pressure head, frictional loss water head, and inlet hole (nozzle) in the pipe (ignoring frictional resistance of the main pipe and under conditions where the branch pipe length is about 60 times the inside diameter of the branch pipe). Figure 6.113 shows the results of calculations of washing uniformity obtained under various conditions after obtaining the relationships shown in Equation 6.151. Using these diagrams to determine the size of a filtration pond, and defining the main and branch arrangements as size L (main pipe) and size l (branch pipe), a design conditions with a washing uniformity through nozzles E can be obtained close to 1 (uniform).

Sueishi's equation

$$\gamma\left(\frac{Ll}{A}\right)=\frac{(1+E)\sqrt{3E^2-1}}{2E\sqrt{2(K+3)}} \quad \text{—(6·151)} \tag{6.151}$$

$\gamma = cf/sS$: Ratio of total effective area of outflow holes to the filtration area

c : Pore outflow coefficient (0.6~0.8)

f : Total area of outflow holes per meter of branch pipes (m^2)

s : Center distance between branch pipes (m)

L : Main pipe length (m)

l: Branch pipe length (m)

A : Main pipe cross sectional area (m^2)
1/2 value when the branch pipes are arranged symmetrically on both sides of the main pipe

E : Uniformity of jet distribution
Flow ratio between maximum and minimum outflow holes

$K = 2gHA^2/U^2L^2l^2$: Specific head loss

H : Water head loss in the water collection devices during washing (m)

U : Washing velocity (m/sec)

Equation 6.151 The uniformity for backwashing.

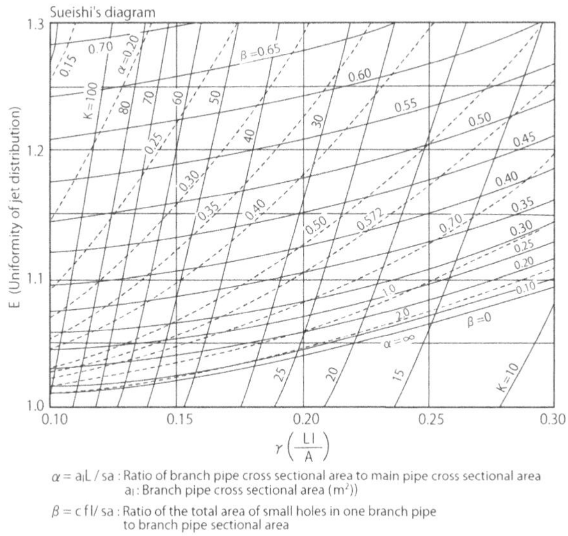

Figure 6.113 The degree of uniformity achieved by backwashing with a perforated water collection pipe device (Sueishi, 1940, 1957).

Porous slabs. Various materials and technologies have been used to form porous slabs or floor plates in lower water collectors, and have been widely used in the United States. As shown in Figure 6.114, the bottom of a filter floor is made into a water chamber separated by a porous plate, and by injecting pressurised backwash water into this large water chamber, the backflow cleaning water can be raised from the slab holes to the sand layer uniformly with respect to the entire horizontal cross-section. Although clogging and corner damage were often an issue due to the use of a fired porous plate in prototype systems, improved sintered materials and the development of new materials have resulted in the production of plates with high strength and accurate pore distributions, and the number of applications has gradually increased. With porous plates, the gravel layer does not have to be placed under the sand layer, so that the filter layer can be made shallower.

Wheeler's bottom. Wheeler's filter bottom was developed in the early 1900s in the early stages of the rapid sand filtration process. It was the first washing device with an integral pressure chamber which, like the perforated plate type, allows uniform pressure to be applied to the outflow holes. As shown in Figure 6.115, two types of small and large porcelain spheres are mounted in a quadrangular pyramid to disperse the rising water from the lower part and feed it into the overlying gravel layer to create a uniform flow of water to uniformly expand the sand layer.

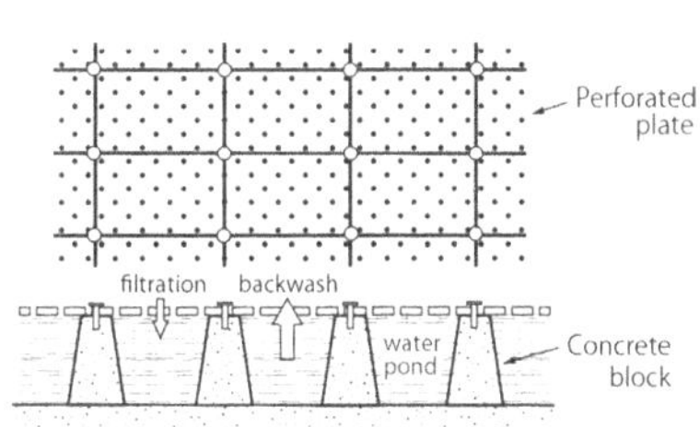

Figure 6.114 Porous slabs being used on a filter bed floor.

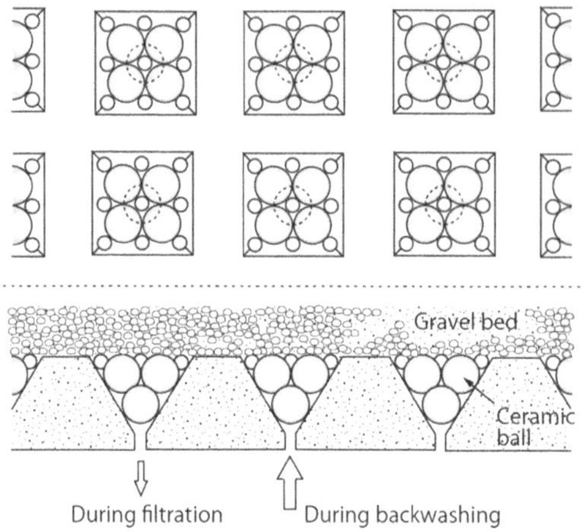

Figure 6.115 Wheeler's filter bottom.

As mentioned above, a filter bottom collector literally serves to collect water, but the flow rate of the down-flow of clean water (filtrate) is only 10% or more of the upflow in back-flow washing, so all hydraulic design conditions are performed by the back-flow washing time flow rate. For low-speed granular bed filtration, such as a slow sand filter or intermittent wastewater filtration, described later, the washing of the filter bed is usually done by scraping the surface layer, and the lower collector is used only to collect filtrate water. Generally, since the filtration rate is extremely low, the water head loss in the water collection system is also small, and it is possible to arrange a perforated pipe in the lower part of the gravel layer relatively loosely, and the design can be thought of as being equivalent to an infiltration gallery collecting groundwater.

Back wash pump and elevated water tank. The head loss required to expand and fluidize a filter bed to perform backflow washing is approximately 70 to 100 cm at most. However, the pump pressure required for backflow washing and the water level of the overhead water tank must be determined in consideration of the head loss of the water collector, the intermediate piping, and the valves etc. Normally, the water head should be set to 2 to 5 m with the upper surface of the washing troughs as the reference, and installing an elevated tank at 10 m or more is excessive.

If the number of filter ponds in a water purification plant is as small as around 10 ponds, it is uneconomical to have a pump and electric facilities to send a large amount of flow rate only at the time of washing. Therefore, an elevated water tank is installed to store water to ensure the flow rate required for a single washing. The required washing water volume is pumped up slowly over an extended time and then stored, and a large amount of reverse cleaning water is then sent to the filtration pond by gravitational force during washing in a short time. By contrast, water purification plants with a large number of filters, such as those with 20 to 30 ponds, are not equipped with overhead water tanks because of the high frequency of back wash required, and washing is carried out by direct delivery using a back wash pump.

In addition, back wash systems that do not require overhead water tanks or back wash pumps have been developed and put into practical use. This type of system combines multiple (8 to 10) deep filters with a large water depth above the filter surface, such as is the case in naturally balanced filtration ponds (trade name: Green leaf filters). These systems use the water head of several connected filters' clean water conduits to send a required clean water volume for filtrate to be back washed (Figure 6.116).

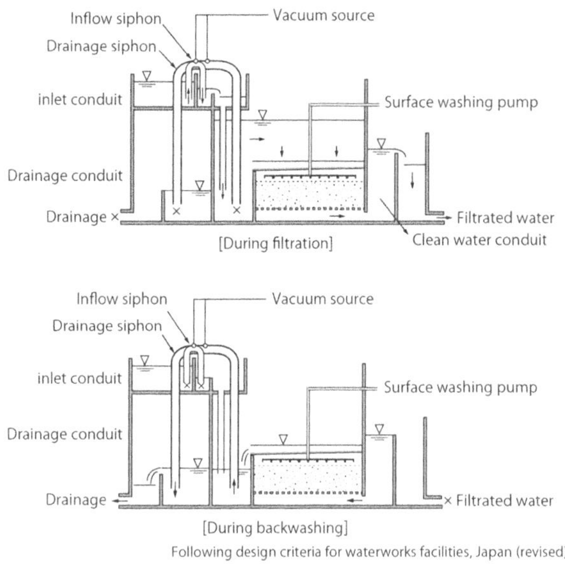

Figure 6.116 Filtration and backwashing of a naturally balanced filter.

Control of filtration flow rate. The operation of filtration is controlled in some way to keep the process flow rate constant or within a certain range of variation. Typically:

(1) In a damped filtration system, after the opening of the outflow valve is initially set to an appropriate flow rate it is not adjusted, and the drive hydrostatic pressure is decreased by the increasing head loss due to clogging of the filter bed, and the treatment flow rate gradually decreases.
(2) A flow meter and a regulating valve are installed at the outlet of the filtration pond, and the regulating valve is opened by the amount of increased hydraulic head lost in the sand bed to reduce the flow-out resistance of the valve, and ensure a constant flow filtration to keep the filtration volume constant.
(3) In a constant flow rate filtration system in which the outlet is installed higher than the upper surface of the filter floor, the inflow flow rate is controlled to be constant, and the process is carried out at a variable sand surface water depth that naturally balances with the outflow (treatment) amount.

In whichever filtration system, as shown in Figure 6.107, the downstream side outlet is used as the free outflow end, and is used as a cascade-type outflow that cuts off the water surface of the downstream filtrate reservoir (clean water storage) so that the influence of the downstream storage level does not affect the filtration rate control.

6.2.1.12 Slow sand filtration and rapid sand filtration

Slow sand filtration was first used on 14 January 1829, with a sand filter of area of 1 acre (about 4,000 m^2) at the Chelsea Water Company in London, under the direction of 30-year-old chief engineer James Simpson. This filter became the first English Slow Sand Filter which became the basic water treatment technique for water supply systems worldwide for more than a century. This was a basic technology behind the rise of the British Empire. A microbial filtration layer (a Schmutzdecke) of a few cm on the surface of a filtration pond behaves as the key mechanism of the system which is a sort of natural aquo-ecosystem.

By the end of the 19th Century, the development of the east coast of North America had become remarkable, and the speed of filtration had been increased to meet the growing demand for water in large cities. In the US, attempts were made to commercialize a method of quickly starting a slow-speed filtration by capturing a gel-like floc obtained by hydrolyzing an aluminium salt or an iron salt in water on the surface of a filter bed, and quickly creating an artificial aluminium hydroxide or iron hydroxide Schmutzdecke. However, they failed. Subsequently, iron salt and aluminium salt were continuously added to the raw water, and as a result of the continuing operation (now called coagulation), the rapid filtration method, which was expected to be effective for retention to the deep part of a filter bed, had been developed.

From the last 10 years of the 19th Century to the beginning of the 20th Century, numerous proposals were made in cities in the eastern United States, creating a prototype of today's rapid sand filtration process. However, the basic process behind it is coagulation itself, and it was only in the 1950s after World War II that the process of flocculation of coagulated fine particles was added to form larger agglomerations (flocs) to improve the efficiency of the preceding sedimentation ponds. Many parallel studies were carried out on this and many people were involved, and a variety of ideas were gathered. We can only imagine the people of The Louisville Experimental Station (1895–97) and the chief chemist and bacteriologist George W. Fuller, at their center, who represents one of the leading creative people behind modern rapid filtration systems. Chemical coagulation was a major advancement in increasing the subsequent sand filter speeds to over 100 m/day and in allowing the treatment of water containing humic substances of higher color that could not be accommodated by slow sand filtration. On the other hand, 20th Century water treatment systems did not integrate a slow sand filtration method that could cope with trace organic components such as unusual odors and tastes, or trace inorganic metals such as iron/manganese. The slow sand filtration method has a low capacity but a wide variety of compatibility, and rapid filtration and slow filtration were developed rigidly as two separate larger systems.

In the latter half of the 20th Century, we finally began constructing advanced treatment systems using rapid and slow filtration, or microbial treatment and physical chemical treatment in series. The water quality conversion matrix created by the author (see Chapter 3) proposes a basic approach to integrating these matrices.

Table 6.23 shows the basic features and differences between slow and rapid sand filtration, the two standard big filtration methods.

6.2.2 Cake filtration (surface filtration)

The main final products of the water treatment process are clarified supernatant water and sludge (slurry) separated from the liquid system. In Section 6.2.1 granular bed filtration used primarily for clarification filtration was described, while in this section (Section 6.2.2) cake filtration, which handles the separation of high-concentration ingredients, is described. Cake filtration is sometimes referred to as surface filtration.

6.2.2.1 Solid thickening from a water treatment process

The thickened solids generated from the water treatment process are:

(1) relatively coarse ($d > \mu m$) and dense ($\rho > 2.6\ g/cm^3$) clays suspended in river water, etc. and swollen color floc groups in which humic substances are treated with coagulation whilst the density hardly changes in water;

Table 6.23 Features of both slow and rapid sand filtration.

Items	Slow Sand Filtration	Rapid Sand Filtration
• Filtration rate	• 4 ~ 8 m/day	• 120 ~ (300) m/day
• Filter area	• about 2000 m^2	• 100 m^2 or less
• Filter bed (conventional)	Gravel bed : about 30 cm Sand bed : Operate until the sand bed length is 50 to 60 cm	45 cm or less 60 ~ 70 cm
• Sand size	Effective size : 0.3 ~ 0.35 mm Uniformity coefficient: about 2.5	0.45 mm or more 1.5 or less
• Filtration mechanism	• Biological action at the top layer	• Deep filtration
• Under drain system	• Slip tile laterals laid in coarse stone and discharging into a tile or concrete main drain	① Perforated pipe ② Diffused plate bottom ③ others, also washing water distribution
• Available head loss	• Initial : about 6 cm • Final : 1.1 ~ 1.5 m	• Initial : about 30 cm • Final : 2 ~ 3 m
• Length filter run	• 20 ~ 60 days	• 12 ~ 72 hours
• Penetration depth of suspended solids into the filter bed	• Filter surface only (Filter coefficient λ : large)	• To the depth of filter bed (medium or large)
• Method of cleaning	• Scraping off dirty surface layer (Small parts may be washed by Special equipment)	• Fluidizing the bed by backwashing water • Mechanical washing in combination with surface scour of top layer
• Amount of wash water used	• 0.2 ~ 0.6% of filtered water	• 1 ~ 5%
• Pre-treatment	• Plane sedemention or not	• Coagulation, Flocculation, and Sedimentation sometimes Coagulation only, Prechlorination
• Supplementary treatment	• Chlorination	• Chlorination
• Cost of construction	• Lower (depend highly upon land cost)	• Higher
• Cost of operation	• Lower	• Higher

(2) inorganic calcifications and other components insolubilized by water softening treatment (see Chapter 5, Precipitation Treatment);

(3) coarse organic matter removed in sewage/wastewater sedimentation basins and swollen microorganism flocs with densities close to unity removed in biological treatment sedimentation ponds.

These thickened components most often contain large amounts of water in the structure (floc) and do not easily release the water on their own. Flocs usually have a fractal structure, and upon thickening they continue to retain water in the structure while plastically deforming the honeycomb structure (see Coagulation and Floc Formation in Chapter 4) Therefore, in order to bring these sludges to final disposal, the process of reducing the water content by dewatering treatment is a crucial process in the final stages of water treatment systems.

In addition to dewatering, sludge treatment and disposal systems involve intermediate processing such as thickening, drying and incineration, and final disposal such as landfill etc.

In terms of dewatering, the two most problematic factors are the water content and the amount of sludge generated. However, the viscosity of sludge for transport and the heat content or water content for incineration are also important. The weight W_s of solids in sludge, the sludge moisture content p and the sludge solids concentration c for the water-containing weight W_w are defined in Equations 6.152 and 6.153.

$$p = \frac{100\,W_w}{W_w + W_s} \quad —(6 \cdot 152)$$

$$c = 100 - p = \frac{100\,W_s}{W_w + W_s} \quad —(6 \cdot 153)$$

W_s : Solids concentration in sludge
W_w : Moisture content
p : Sludge moisture content
c : Sludge solids concentration

(6.152)–(6.153)

Equation 6.152, 6.153 The sludge moisture content and sludge solids concentration.

When the specific gravity of a solid is γ_s and the specific gravity of water is γ_w, the specific gravity γ of a sludge containing water is as shown in Equation 6.154. In addition, the total volume V of sludge is the sum of the volume of water and solids, and is expressed by Equation 6.155.

$$\gamma = \frac{p + (100 - p)}{p/\gamma_w + (100 - p)/\gamma_s} \quad —(6 \cdot 154)$$

$$V = \frac{W_s}{\gamma_s S_w} + \frac{W_w}{\gamma_w S_w} \quad —(6 \cdot 155)$$

γ : Specific gravity of sludge containing water
γ_s: Specific gravity of solids
γ_w: Specific gravity of water
V : Total volume of sludge
S_w : Unit weight of water

(6.154)–(6.155)

Equation 6.154, 6.155 The specific gravity of sludge containing water and total volume of sludge.

Therefore, if the water content of sludge having water content p_0% in sediment V_0 is removed to make water content P%, the volume change ratio V/V_0 of the sludge is expressed by Equation 6.156.

$$\frac{V}{V_0} = \frac{100\,\gamma_w + p\,(\gamma_s - \gamma_w)\,(100 - p_0)}{100\,\gamma_w + p_0\,(\gamma_s - \gamma_w)\,(100 - p)} \quad —(6 \cdot 156)$$

V : Sludge volume when moisture content reaches p %
V_0 : Sludge volume when moisture content reaches p_0 %

(6.156)

Equation 6.156 The volume change ratio of sludge by dewatering.

By contrast, typical solids content (weight)% of sludge is as shown in Table 6.24.

Table 6.24 The typical solids content of sludge.

(weight %)

Processes	State of the Sludge	Weight of Solid Content %
Flocculatlon/sedimentation of water purification	Drained sludge	0.1
	Thickened sludge	1 ~ 2
		2.5 ~ 5
Plane sedimentation of sewage treatment	Drained sludge	
	Thickened sludge	8 ~ 10
	Digested sludge	10 ~ 15
Flocculation/sedimentation of sewage treatment	Drained sludge	2 ~ 5
	Digested sludge	10
Trickling filtration	Drained sludge (peeling)	5 ~ 10
	Thickened sludge	7 ~ 10
	Mixed with primary tank sludge	3 ~ 9
	Digested sludge	10
Activated sludge	Drained sludge	0.5 ~ 1
	Thickened sludge	2.5 ~ 3
	Digested sludge	2 ~ 3
	Drained sludge of primary tank	4 ~ 5
	Thickened sludge of primary tank	5 ~ 10
	Digested sludge of primary tank	6 ~ 8

6.2.2.2 Cake (dewatering) filter

Cake filtration is driven by pressure differences between the front and back of the porous material (filter media), and the process proceeds in such a way that solid particles in the suspension are trapped on the surfaces of the filter media, and the particles fixed by the trapping are separated into a mass (a cake) and the filtrate permeates through it. Cake filtration is carried out with the following process cycle:

(1) Cake formation of suspended particles on filter media surfaces;
(2) Dewatering of fluid in the trapped particles;
(3) Release (cake discharge) of the filtered slag (or cake) that has been caught and adhered above a certain level, for regeneration of the filter media.

Various types of filtration systems have been put into practical use depending on the type of sludge used, and cake filtration equipment for sludge dewatering includes sludge dryer beds, continuous-cylinder vacuum filtration machines, belt dewatering equipment, and filter presses (with or without sludge chamber squeezing mechanisms). Various types of equipment are used depending on the type of sludge and pre-treatment used. Generally, pretreatment is performed to form agglomerates (coagulated clumps) by adding organic polymers and inorganic coagulant to improve filtration efficiencies (see the Coagulation process in Chapter 4 for more information.)

Sludge drying beds use a 1000 m^2 basin with a sand layer of 1 to 2 mm in particle size and 15 cm thickness placed on top of a gravel layer of 20 to 30 cm, as shown in Figure 6.117, to inflow sludge to a thickness of 20 cm for dewatering. Initially, it begins as an intermittent sand filter process, and gradually shifts to cake filtration in which dewatering sludge is generated, and the moisture content decreases. A batch process of days to months is performed, and the dehydrated sludge is finally mechanically scraped

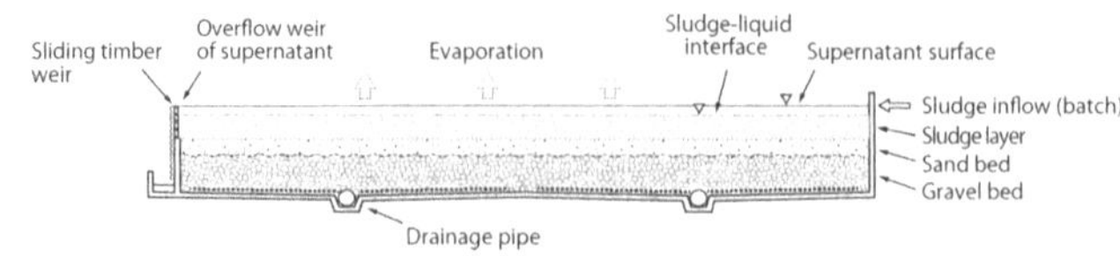

Figure 6.117 A sludge sun drying bed.

off. Over time, water evaporation from surfaces also contributes significantly to the reduction of the final water content.

Continuous cylindrical vacuum filters are one of the most widely used pieces of sludge dewatering equipment, of which the Dollar-Oliver system is popular; see Figure 6.118.

Their operation is as follows. A drum with a relatively large surface area of less than 100 m^2 is placed in a semi-dipped tank filled with the sludge to be dewatered. The area marked (a) in Figure 6.118 is in contact with the sludge in the sludge basin. The drum is composed of a parallel assembly of a number of separate vacuum aspiration sections, the drum surfaces being covered by a single filter cloth belt. The drum rotates with the filter cloth belt at a constant rotational speed (around 10 rpm). A vacuum pump is connected to the individual vacuum suction portions of the drum (degrees of vacuum >500 mm of mercury) through the state separators. When the vacuum suction portions are in the area (a) as shown, the sludge is sucked and trapped on a filter cloth and grown to cake layers as thick as 3 cm.

When the drums rotate and the filter cloth passes through section (b) shown in the drawing, sludge is not supplied, and only the water in the cake layer is sucked in and the water content of the cake layer decreases. In section (c) of the drawing, the filter cloth leaves the drum, and the compressed dewatering cake formed to a thickness of about 1 cm is finally peeled off from the filter cloth by a peeling knife (scraper knife) and discharged to the cake transport line.

The exposed surface of the filter cloth after stripping the cake is washing with water jets, and the surface returns to the cake-forming section (a). Depending on the type of sludge, sludge from water purification and sewage treatment plants seems to have a throughput of about 20 to 30 kgDS (dry solids)/m^2 h and a dewatering cake contains about 25 to 30% DS.

Filter press machines have become widely used in recent years. This is because it has become necessary to make dewatering cakes with a much lower moisture content in response to the increase in the necessity of dewatering, such as because of the shortage of disposal sites, for heat recovery of organic sludge, and the reuse of sludge. In addition, the popularity of automatic control will also contribute to the advances of the process design, which previously had lower throughput rates on a batch basis compared o today's developed high-efficiency automatic systems. Large dewatering plates of 1.5 m × 1.5 m or 2 m × 2 m square are arranged in multiple stages of 100–150 plates, and horizontal high-pressure multiple filter

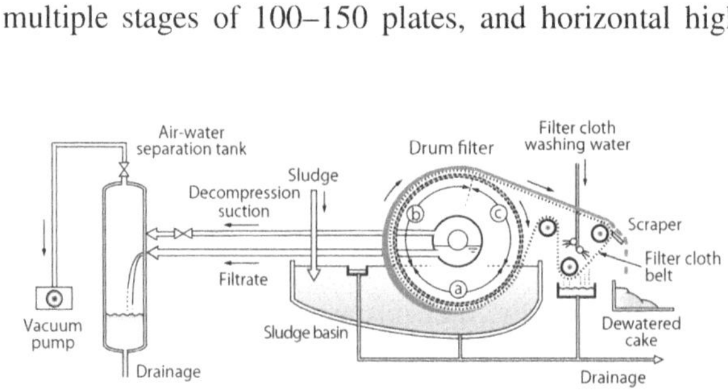

Figure 6.118 An example of a continuous cylindrical vacuum filter.

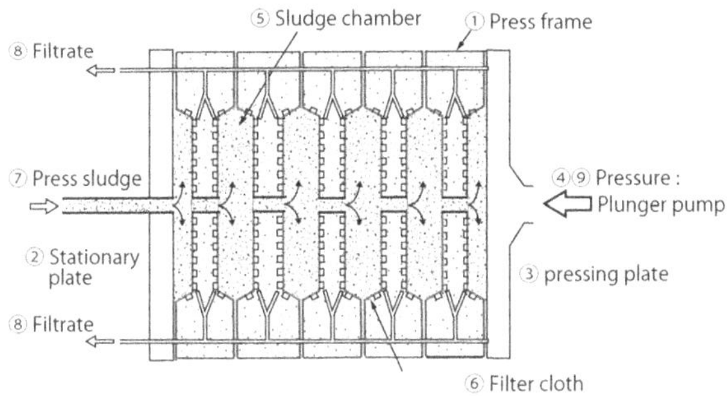

Figure 6.119 An example of a filter press.

presses with a total filtration area of 1000 m^2 and 15 atmospheric pressure have become widely used in wastewater treatment systems. Multiple processing speeds of about 2 to 4 $kgDS/m^2$ h have also been used to achieve adequate throughput, and dewatering cakes exceeding 30 to 50% DS have easily been obtained. In many cases, the operation is carried out at 10 atm or less.

The components of a prototype filter press are illustrated in Figure 6.119. In this filtration system, there are multiple rectangular vertical and parallel press frames(1); plates (2) are fixed at one end (a fixed panel); the other end of the plate (3) is pushed by a pressurizing plunger (4) with an electric/pneumatic drive. The pressure of sludge fed into the sludge chamber (5) between the press frames is received while under pressure. The water is squeezed out of the sludge by passing through the eyes (about 10 to 100 μm) of a filter cloth (6) which is inserted into the surface of each press frame. The filter cloth is stretched on the front surface of the uneven fine water channel engraved on the surface of the press frame for each frame. Sludge (7) is dispensed from the center feed shaft of the filter press into the respective filter chambers for a short period of time, and replenishment of sludge (by the volume reduced by dewatering) is also continuously injected from the center feed shaft. Water squeezed out of the compression sludge (about 30 mm thick, sometimes 40 to 50 mm thick) flows through the filter cloth into the fine channel, collects, and flows out of the system as filtrate water (8). When the compression of the sludge can no longer progress (in many cases for 2 to 4 hours or even 6 to 8 hours depending on the type of sludge), the injection of the raw sludge is stopped. The pressurizing plunger is retracted (9), and the parallel filter plates are sequentially opened from the end. The dewatered cake formed between the filter plates is gravitationally dropped and recovered as a dewatering cake. After that, the filter cloth surfaces and the filter chamber are subjected to water washing to finish one cycle.

Recently, in order to further reduce the final moisture content, a compression type cake filtration has been used to add a highly compressing process to the cake product by applying additional pressure from the back surface of the filter frame (filter cloth) at the end of the press.

6.2.2.3 Cake filtration theories

Hydraulics of caking filtration. The filtration water flow in cakes can be considered as a special case of the fixed bed flow described in the previous section. Consider Darcy-Weisbach flow (Equation 6.100) as a generic formula. Since the size of the cake forming particles is very small, the passages through which the filtration water passes are extremely narrow, and the flow is laminar. Therefore, the filtration head loss h_f or the pressure loss Δp when passing through the cake-bed after the expansion from Equation 6.101 to Equation 6.108 in the previous section is shown by the Carman-Kozeny equation, shown in Equation 6.109 or Equation 6.157.

$$h_f = k_1 \frac{L}{(\varphi_s d)^2} \frac{(1-\varepsilon)^2}{\varepsilon^3} \frac{\nu \bar{V}_s^2}{g_c} \quad \text{—(6·109)}$$

$$\Delta p = k_1 \left[\frac{(1-\varepsilon)^2}{g_c \varphi_s^2 d^2 \varepsilon^3} \right] \mu L \bar{V}_s = k_1 \left[\frac{(1-\varepsilon)^2}{36 g_c (V_p/S_p)^2 \varepsilon^3} \right] \mu L \bar{V}_s \quad \text{—(6·157)}$$

h_f: Filtration head loss (m)
Δp: Pressure loss (Pa)
k_1: Constant
L: Cake thickness (m)
φ_s: Shape factor
d: Particle syze (m)
ε: Cake porosity
g_c: gravitational conversion factor (9.8m/sec²)
V_p: Single particle volume (m³)
S_p: Single particle surface ares (m²)
μ: Viscosity (Pa·s)
$\bar{V}_s^2$: Superficial velocity (m/sec)

(6.109), re-supplied (6.157)

Equation 6.157 The Carman - Kozeny equation.

The parentheses in Equation 6.157 can be considered as characteristic values determined by the nature of the cake itself, and are constants for the same cake.

The apparent flow velocity $\bar{V}_s$ through the cake surface at a given time during filtration is given by Equation 6.158, and if Equation 6.157 is substituted, the filtrate quantity dV/dt flowing out of the cake per unit time is given by Equation 6.159. Here, K (water permeability coefficient m^2) is a constant indicating the properties of the cake. It is a numerical value indicating the flow rate at which water with unit viscosity passes through the cake with unit area and unit thickness in unit time under unit pressure differences. As shown in Equation 6.157, the value is governed by the porosity ε of the cake, the particle size d, and the shape coefficient φ_s.

$$\bar{V}_s = \frac{1}{A} \frac{dV}{dt} \quad \text{—(6·158)}$$

$$\frac{dV}{dt} = \frac{K A \Delta p}{\mu L} \quad \text{—(6·159)}$$

$\bar{V}_s$: Apparent flow velocity through cake surface (m/sec)
V: Filtrate quantity (m³)
A: Cake surface area
K: Permeability (m²)
t: Operation time (sec)
L: Cake thickness(m)

(6.158)–(6.159)

Equation 6.158, 6.159 The filtrate quantity per unit time.

Generally speaking, when discussing cake filtration, it is easier to describe the characteristics of such cake in the form of resistance rather than permeability, as will be described later, to deal with the condition of synthesis with the resistance of the filter medium (filter cloth or filtration bed). Therefore, as the reciprocal of permeability, the cake resistivity R[1/m^2] is defined as in Equation 6.160, and Equation 6.159 is rewritten as in Equation 6.161.

$$R = \frac{1}{K} \quad \text{—(6·160)}$$

$$\frac{dV}{dt} = \frac{1}{R} \frac{A \Delta p}{\mu L} \quad \text{—(6·161)}$$

R: Cake specific resistance (1/m²)

(6.160)–(6.161)

Equation 6.160 The cake resistance, Equation 6.161 The filtration rate.

In the actual cake filtration process, there are two types of resistance in series: the resistance exhibited by the cake and the resistance of the filter medium (such as filter cloth) supporting the cake. Therefore, in practical use, Equation 6.161 is used after deformation as Equation 6.162.

$$\frac{dV}{dt} = \frac{A\Delta p}{\mu(RL + R_f)} \quad —(6 \cdot 162)$$

R_f : Filter media (filter cloth) specific resistiance (1/m)

(6.162)

Equation 6.162 The practical filtration rate equation.

Since the cake thickness L varies with the duration of filtration, the cake thickness L can be obtained from the volume ratio r_v [dimensionless] of solids present in the unit filtrate as shown in Equation 6.163, and as a result Equation 6.162 can be rewritten as shown in Equation 6.164.

The cakes are generally compressive, and it is practically difficult to treat the volume ratio r_v [dimensionless] as a constant over the entire time of the filtration. The cakes deformation Equation 6.164 as Equation 6.165 using the dry solids weight resulting from the volume of liquid per unit volume as an unchanged amount. The resistivity R can then be replaced by a new dimensional resistivity, such as the mean resistivity R' [sec^2/kg].

$$L = \frac{r_v V}{A} \quad —(6 \cdot 163)$$

$$\frac{dV}{dt} = \frac{A^2 \Delta p}{\mu(r_v VR + AR_f)} \quad —(6 \cdot 164)$$

$$\frac{dV}{dt} = \frac{A^2 \Delta p}{\mu(wVR' + AR_f)} \quad —(6 \cdot 165)$$

r_v : Volume ratio of solids in unit filtrate (-)
w : Unit volume dry cake weight (kg g_c/m^3)
g_c : gravitational conversion factor (9.8m/sec^2)
R' : Average specific resistance (sec^2/kg)

(6.163)–(6.165)

Equation 6.163 - 6.165

Cake formation and removed filtrate. When cake filtration (sludge dewatering) is performed continuously, cake formation, cake dewatering and cake stripping (cake discharge) are performed continuously, as described above. During this time, we need to know how much cake is formed and how much water is removed from the suspension in the process.

As an example, we consider a classic continuous-cylinder vacuum filtration machine (vacuum drum filter). In this type of sequential process, constant pressure filtration is performed to keep the pressure differential constant during cake formation. As shown in Figure 6.118, the flow rate V/A transmitted through the cake per unit filtration area at time t in the cake forming stage (a) is given by Equations 6.165, 6.166, 6.167, and 6.168. Equation 6.168 is the amount of filtrate that passes through the unit cake cross-sectional area during time t. The amount of cake that forms during this time is calculated as in Equation 6.169 by multiplying the passing flow rate by the solid weight w contained per unit filtrate.

By taking the time t of Equation 6.169, with the total time t of the cake production process, we can know the total amount of solids that are caked by one sludge treatment cycle. At the same time, the total amount of dewatering water is obtained.

Resistivity (Specific resistance). As can be seen from Equation 6.167, the mean resistivity R' has a direct effect on cake formation volume and the formation rate, and is a major factor in controlling cake filtration unit efficiencies. Dividing both sides of Equation 6.167 by the volume of filtrate V permeated between times t gives Equation 6.170 or Equation 6.171.

$$\int_0^t dt = \int_0^V \left(\frac{\mu w V R'}{A^2 \Delta p} + \frac{\mu R_f}{A \Delta p} \right) dV \quad \text{—(6·166)}$$

$$t = \frac{\mu w R'}{2A^2 \Delta p} V^2 + \frac{\mu R_f}{A \Delta p} V \quad \text{—(6·167)}$$

$$\frac{V}{A} = \left(\frac{2 \Delta p t}{\mu w R'} - \frac{2 R_f V}{w R' A} \right)^{\frac{1}{2}} \quad \text{—(6·168)}$$

$$\frac{Vw}{A} = \left(\frac{2 \Delta p w t}{\mu R'} - \frac{2 w R_f}{R' A} \right)^{\frac{1}{2}} \quad \text{—(6·169)} \qquad (6.166)\text{–}(6.169)$$

Equation 6.166 - 6.169

As can be seen from Equation 6.171, for an isobaric (constant pressure) filtration, a linear relation can be obtained by plotting the elapsed time t at the cake-forming stage and the filtration flow rate V passed through the cake between the time t at the cake-forming stage, with t/V as the vertical axis and V as the horizontal axis (Figure 6.121). Knowing the gradient b shows the mean resistivity R', a major process factor for cake filtration, from the relation of Equation 6.171. In actual design, we use this relation to determine the mean resistivity of each sludge using the Buchner test, leaf test, pressure test, etc. and design the filtration machine. In advance, we test and evaluate improvements in sludge filtration by various pretreatments.

Most sludge treated in the water treatment field has compressive properties, and the specific resistivity (size of cake voids) of cake differs when the pressure of the filter changes. Therefore, we need to know the relation between filtration drive pressure Δp and the mean resistivity R'. Experimentally, it is known that there is an exponential relationship between filtration pressure Δp and the mean resistivity R', as shown in Equation 6.172. From a straight line obtained by plotting the relationship between R' and Δp logarithmically, the resistivity R' for a given Δp can be estimated by interpolation and extrapolation. The slope of this straight line is the compression factor (coefficient of compressibility) in Equation 6.172, and the value of R' when $\Delta p = 1$ is C.

$$\frac{t}{V} = \frac{\mu w R'}{2A^2 \Delta p} V + \frac{\mu R_f}{A \Delta p} \quad \text{—(6·170)}$$

or

$$\frac{t}{V} = bV + a \quad \text{—(6·171)} \qquad (6.170)\text{–}(6.171)$$

$$b = \frac{\mu w R'}{2A^2 \Delta p}, \quad a = \frac{\mu R_f}{A \Delta p}$$

Equation 6.170, 6.171

$$R' = C\,\Delta p^{s} \qquad (6.172)$$

C : Cake constant
s : Coefficient of compressibility

(6.172)

Equation 6.172 The coefficient of compressibility.

Experimental test apparatus for cake filtration. The dewatering efficiencies of sludge differ greatly depending on the nature of the raw sludge to be treated and the presence or absence of pretreatment and methods for sludge dewatering processes (described below), and adequate information on dewatering is required prior to design of a system. In order to improve the dewatering, a small quantity of standby samples must be tested repeatedly for a number of conditions.

Various tests are currently in use for vacuum filtration:

(1) the Büchner funnel test;
(2) The leaf test
(3) the standard test method for pressurized filtration, which is not well established but various methods have been proposed to cope with filtration equipment actually being used.
(4) in addition, the Capillary Suction Time (CST)-test is often used as a simple technique to evaluate the dewatering of the smallest samples.

Büchner funnel test. This test method is used to estimate the average resistivity, which is an important indicator for determining the amount and rate of cake production during the cake formation process. The test equipment is constructed as shown in Figure 6.120.

The test procedure is as follows:

(1) Install water-moistened filter paper on the Büchner funnel;
(2) Adjust the air valve so that the degree of vacuum reaches the specified level and aspirate;
(3) Injection the test sludge into the Büchner funnel to a thickness of 2 to 3 cm;
(4) Measurement and recording of the filtrate quantity V corresponding to the elapsed time t are started immediately;
(5) Measurements are continued until the cake in the Büchner funnel cracks and vacuum breakage occurs.

In this process, the initial concentration and the final moisture content (concentration) of sludge are measured. The relationship between t/V and V is summarized by Equation 6.171, as illustrated in Figure 6.121, and the average resistivity R′ is obtained. Experiments are repeated with varying degrees

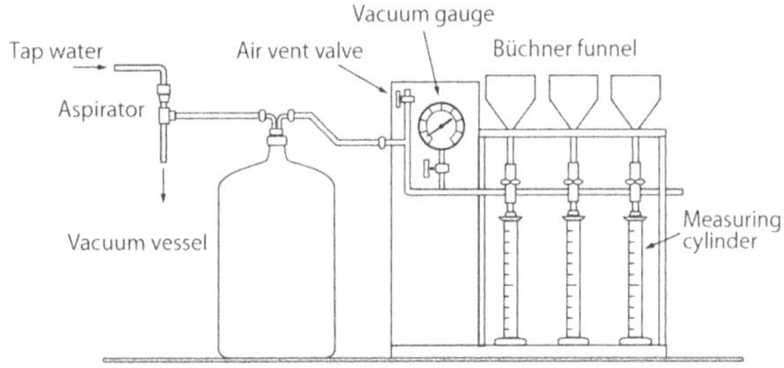

Figure 6.120 Büchner funnel test equipment.

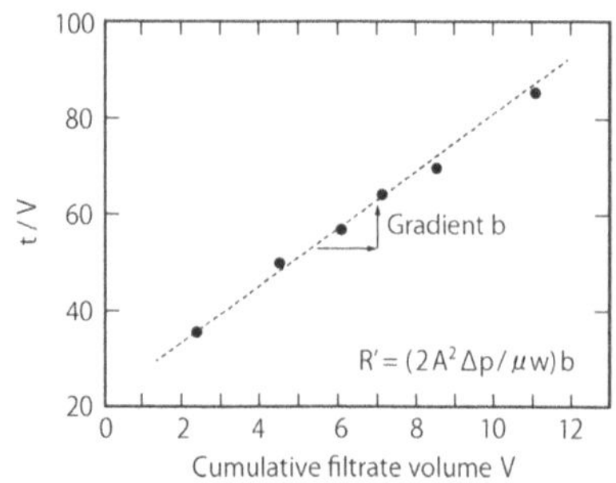

Figure 6.121 An example of results from a Büchner funnel cake filtration test.

of vacuum and the cake compression factor s, and cake constant C, are calculated according to Equation 6.172.

Leaf test. A leaf test is a test method that simulates all cycles of vacuum filtration. The above-mentioned Büchner test and pressure test (below) are tests that apply to the cake-forming part of the entire filtration cycle (cake formation, dewatering, and peeling), and do not represent the entire cycle of the vacuum dewatering process. Therefore, the leaf test shown in Figure 6.122 was devised to simulate all cycles in a batch test.

In accordance with the filtration cycles:

(1) A test leaf (a disk-shaped suction unit with a filter medium stretched on its frame) is immersed in the slurry container, and sludge is sucked onto the surface of the filter medium at a predetermined vacuum pressure;
(2) After a predetermined cake formation time, the test leaf is moved up out of the suction air;
(3) During this period, the relation between the cake-forming time t and the filtrate volume V is measured using filtrate cylinders;
(4) When the caking termination time t_c has been reached, aspiration is continued for the dewatering process time while keeping the test leaf facing upward;
(5) The cake thickness and moisture content are measured at the time of discontinuation of suction. The initial concentration is measured in advance;
(6) By repeating these measurements, the resistivity can be determined in the same manner as in the Büchner test;

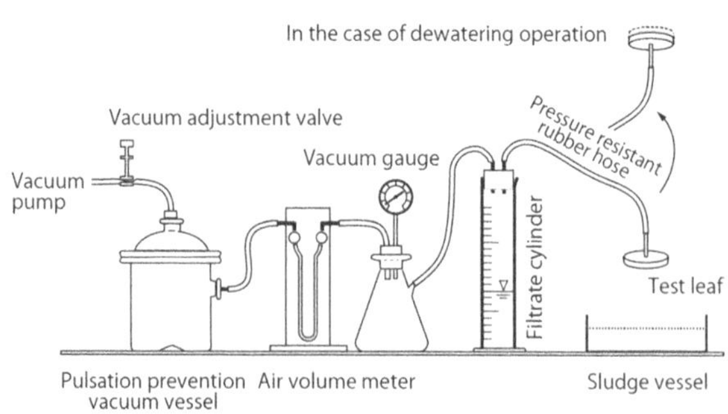

Figure 6.122 An example of equipment used in a leaf test.

(7) In addition, operation materials including final moisture content and disposal cake content can be obtained by arbitrarily combining cake forming and process times of dewatering;
(8) This is a test method devised in an age when continuous-drum vacuum filtration machines are frequently used.

Pressure filtration test. In contrast to the Büchner test which is performed by vacuum aspiration, the pressurized filtration test differs in that sludge is injected into sealed pressurized cylinders and the filtration is advanced by pressurizing air with compressors, or similar. Figure 6.123 shows an example of a system that the author's laboratory used in the past. A variety of commercially available products have been manufactured in accordance with the same principles and with different pressurization methods.

Capillary Suction Time (CST) Test. In order to evaluate the dewatering of sludge in a semi-quantitative manner by using the capillary force of water diffusing and moving through filter paper as a drive force, this simple dewatering test, as shown in Figure 6.124, has been widely used since the 1980s.

The test procedure is as follows.

(1) Thick filter paper (Whatman No.17) is used as the filter medium and is placed on the lower base block. A sludge inserting cylinder (held by a cylinder support block and stabilized) is placed on the filter paper, and a predetermined volume of sludge is injected into the cylinder with an inner diameter of 10 mm or 18 mm;
(2) Electrode 1 is placed on a concentric circle near the outer edge of the sludge cylinder, and Electrode 2 is placed on a concentric circle at predetermined distances from the cylinder;
(3) The dewatering from the inserted sludge gradually spreads while wetting the filter paper in concentric circles by the capillary suction of the filter paper;
(4) The change in the conductivity of the filter paper is measured until the wetting circle of the filter paper passes through Electrode 1 and reaches Electrode 2.

The diameter of the concentric circle of Electrode 1 is 35 mm, the diameter of Electrode 2 is 45 mm, and the duration of the permeate diffusing between the two concentric circles is generally about 5 to 10 seconds. The effect of the filter paper on the filtration evaluation is small, and the rate of capillary flow spread in this device is thought to be largely due to sludge filtration resistivity. Therefore, when the CST value is low, the caking filtration has good dewatering, and by correlating the CST value with the mean resistivity (calibrating) beforehand, the dewatering can be quickly evaluated semi-quantitatively.

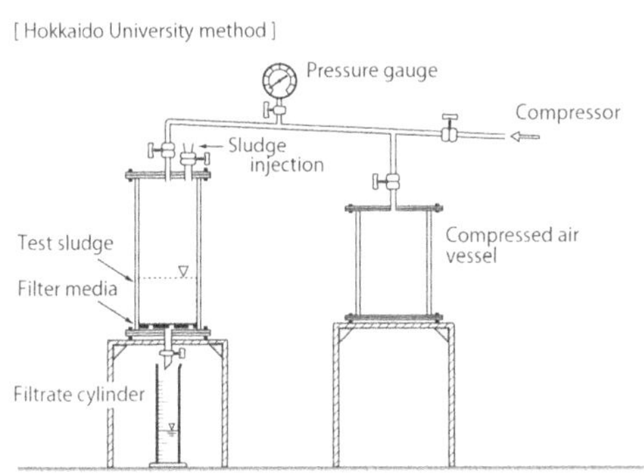

Figure 6.123 Test equipment for pressurized cake filtration used in Tambo's laboratory at Hokkaido University.

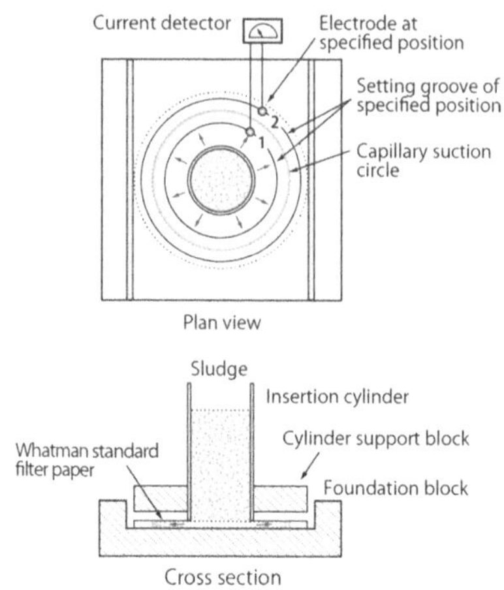

Figure 6.124 Capillary Suction Time (CST) Test equipment.

6.2.2.4 Filtration pretreatment (sludge conditioning)

Since the dewatering of the cake filtration is governed by the particles that make up the cake, the size and water retention of the cake constituent particles immediately affect the resistivity and final moisture content. Therefore, except in the dewatering of coarse, less water-retentive slurries, it is common to perform sludge conditioning of the slurries in some way prior to filtration to coarsen the particles, and add treatments that reduce water retentivity. The basic mechanisms are discussed in Chapter 4 and here we limit discussion to an outline of caking filtration pretreatment.

Preparation of inorganic sludge is a process widely used to increase the permeation rate by adding mineral and polymeric coagulants to the sludge and coagulated agglomerates of the sludge particles to increase the diameter of the water permeable channels of cakes. The basic mechanism for the coagulation and agglomeration of slurry particles is the colloidal agglomeration process described in Chapters 4 and 5. This mechanism is explained by considering the functions of surface charge neutralization, adsorbed water removal (to change hydrophilic colloids into a hydrophobic state), and the formation of interparticle cross-links. However, sludge is extremely concentrated, has a large chance of contact between particles, and the mechanisms of floc breakdown are attributed to frictional drag, unlike the turbulent breakdown in floc-forming ponds (flocculators) discussed in Chapter 4. Floc density (structural density, including water) is a critical factor, as water retention problems in floc structures ultimately affect the final moisture content of cakes.

In coagulated sedimentation sludge such as silt with metal hydroxides, if the sludge is crystalline and has a particle size as large as 10 μm, the water content in the agglomerate can be reduced by dissolving the metal hydroxides (coagulant) in the sludge by acid and washing away. However, the water permeability channel is reduced and the resistivity increases. In cases such as fine clays, the water path between the cakes is too small from the beginning and the resistivity is large, so operations such as adding lime with low water retention and coagulating it again are performed. The reasons cannot be clearly explained, but the addition of lime to a supersaturated slurry appears to give good results. However, as much as 15 to 20% of lime per weight of solids in slurries needs to be added, resulting in the disadvantage of increasing sludge volume. In contrast, polymeric coagulants, which have a high degree of polymerization, can produce cakes with a low moisture content higher than that of lime under low resistivity operating conditions by adding a specific weight of solids of several parts per million, eliminating the need for a dissolution process by acid.

In biological treatment processes in particular, sludge from activated sludge treatment is generated in large quantities. In addition, the resistivity of sludge in the original condition is too high, making any dewatering process by caking filtration difficult. Therefore, the dewatering is improved by thorough pretreatment, leading to dewatering processes such as caking filtration and centrifugation.

Three conventional conditioning pretreatments are possible:

(1) Anaerobic digestion; the primary objective of this process is to decompose and stabilize the undecomposed organic matter present in the filtrate but this process significantly improves the dewatering (final moisture content and specific resistivity) of cake filtration.

(2) Chemical coagulation; this method should be considered to be almost the same as that of a cake made of inorganic fine particles. Generally, ferric chloride $(FeCl)_3$, ferric sulfate $(Fe_2(SO_4)_3)$, aluminium sulfate $(Al_2(SO_4)_3\ 18H_2)$, and lime $Ca\ (OH)_2$ are added, followed by coagulation and dewatering processes. Polymer coagulants can also be added in quantities on the order of parts per million (ppm) to achieve effective coagulation centering on cationic and weak cationic properties. Figure 6.125 shows an example of a change in the resistivity of the dewatering cakes produced when ferric chloride $(FeCl_3)$ is added to the sludge extracted from the final sedimentation tank using the activated sludge method. Coagulation is performed with $FeCl_3$ and $Ca\ (OH)_2$ alone or in combination.

(3) Water washing (elutriation); one feature of biological treatment sludge is that the sludge has a very high pH buffering capacity (typically in alkalinity), and prior to a coagulant, such as $FeCl_3$, reaching an optimal coagulation condition (coagulation pH in a weakly acidic area), a large amount of iron chloride is consumed as a weak acid for neutralizing the alkalinity, resulting in an excessive amount of iron added, resulting in a waste of coagulant. In addition, excessive iron hydroxide is generated, which increases the resistivity of the sludge and the compression of the sludge. Therefore, the sludge is washed out with water in advance, as described below, to wash away the adhering components such as protein particles that increase the alkalinity of the sludge.

In this process, clean water is added to sludges that have been stirred and suspended, single-stage or multi-stage washing is performed, and the suspension is subjected to sedimentation thickening and fed to

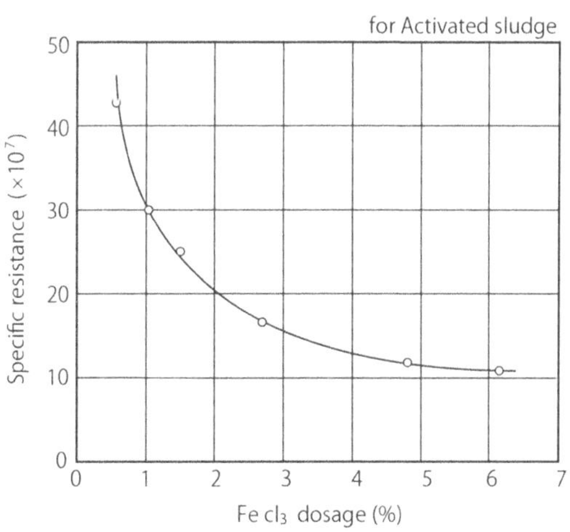

Figure 6.125 The resistivity of dewatering cakes to chemical coagulation.

the dewatering system. At this time, a large amount of backwash wastewater is generated and must be recycled to the sewage treatment processes without being discharged to the outside.

FURTHER READING

Awaya Y. and Kusuda T. (1969). Density flow and dead water in a rectangular sedimentation basin. *Journal of Japan Society of Civil Engineers*, **168**, 21. (In Japanese).

Baker M. N. (1949). The Quest for Pure Water. American Water Works Association, Denver, CO. pp. 64–124; 179–247.

Cadle R. D. (1955). Particle Size Determination. Interscience Publishers, Inc., New York & London.

Camp T. R. (1953). Flocculation and Flocculation basins. *Proceedings of the American Society of Civil Engineers*, **79**(273), 1–18.

Degrement. (1991). Water Treatment Handbook, Vol. 1, pp. 159–177. Lavoisier Publishing, Paris.

Fukushi K. and Tambo N. (1985). Separating high-concentration activated sludge by pressure flotation. *Journal of Japan Society of Sewerage Works*, **17**(196), 31–39. (In Japanese).

Gaudin A. M. (1957). Flotation, 2nd edn. McGraw-Hill, New York, p. 152.

Goda K. (1951). Sedimentation basin purification effects. *Journal of Society of Civil Engineers*, (6), 39–43. (In Japanese).

Gregory J. (1984). Solid–Liquid Separation. Ellis Horwood Limited, Chichester.

Ide T. (ed.) (1990). Water Treatment Engineering, 2nd edn. Gihoudo Publishing, Tokyo, pp. 83–99, pp. 101–152. (In Japanese).

Iwasaki T. (1935). Calculating the filtration rejection rate.(in Japanese) *Journal of Japanese Society of Civil Engineers*, **24**(8).

James M. Montgomery Consulting Engineers, Inc. (1985). Water Treatment Principles & Design. John Wiley & Sons, Inc. pp. 149–173.

Japan Society of Civil Engineers (1999). Handbook of Hydraulics (4.1). Maruzen Co. Ltd., Tokyo, pp. 386–388. (In Japanese).

Johnstone R. E. and Thering W. M. (1957). Pilot Plants, Model, and Scale-up Method in Chemical Engineering. McGraw-Hill Book Co. Inc.

Kitahara F. and Watanabe K. (1972). Interfacial Electric Phenomena. Kyoritsu Publishing, Tokyo, p. 212. (In Japanese).

Kuratsuka Y. (1953). Water Purification Engineering. Chapters 9 and 10; pp. 281–407. Iwanami Book Store. (In Japanese).

Levich V. G. (1962). Physicochemical Hydrodynamics. Prentice-Hall Inc, pp. 20–36. pp. 207–221.

Maroudas A. and Eisenklam P. (1965). Clarification of Suspensions: a study of particle deposition in granular media. *Chemical Engineering Science*, **20**, 867–873.

McGauhey P. H. (1956). Theory of sedimentation. *Journal of the American Water Works Association*, **48**(4), 437–448.

Melbourne J. D. and Zabel T. F. (eds). (1976). Flotation for Water and Waste Treatment. The Water Research Centre, Medmenham, UK.

Montgomery, James M, Consulting Engineers, Inc. (1985). Water Treatment Principles & Design. **7**. Gravity Separation. John Wiley & Sons, Inc, pp. 135–148.

Nagai M. and Hirga K. (1965). Washing effectiveness of rapid sand filters. *Journal of Japan Water Works Association*, (364), 39.

Nakagawa Y. (1962). Theoretical Study on the sedimentation effects of rectangular sedimentation basin (Part 1). *Journal of Japan Water Works Association*, (331), 15–23. (In Japanese).

Orr C., Jr. and Dallavalle J. M. (1960). Fine Particle Measurement. The Macmillan Company.

Rich L. G. (1961). Unit Operations of Sanitary Engineering Chapt. 4: Sedimentation (pp. 81–116). John Wiley & Sons, Inc.

Rich L. G. (1961). Unit Operations of Sanitary Engineering. John Wiley & Sons, Inc.

Shirato M. (ed.) (1980). Chemical Engineering: Fundamentals of Mechanical Operation. Maruzen Book Co. Ltd, Tokyo, pp. 103–111, pp. 112–138, pp. 138–146, pp. 163–186. (In Japanese).

Shirato M. (ed.) (1984). Filtration Technology: Advancement in Chemical Engineering 18 (4. Press and Squeez). Japan Society of Chemical Engineers, Maki Bookstore, pp. 61–83. (In Japanese).
Tambo N. (1964). A. Fundamental Investigation of Coagulation and Flocculation. Chapt.2. Doctor of Engineering Thesis, Hokkaido University.
Tambo N. (1969). Trends in rapid filtration progress. *Water and Wastewater*, **11**(5), 359–367. (In Japanese).
Tambo N. (1980a). Water Supply. Civil Engineering Series. Japan Society of Civil Engineers. Gihoudo Publishing, Tokyo, vol. **88**. (In Japanese).
Tambo N. (1980b). Water Supply. Civil Engineering Series, Japan Society of Civil Engineers. Gihoudo Publishing, Tokyo, vol. **88**, pp. 203–219, pp. 303–311. (In Japanese).
Tambo N. (1984). Clarified filtration. In: Advances in Chemical Engineering 18: Filtration Technology, Japan Society of Chemical Engineers (eds), Maki book store, Tokyo. (In Japanese).
Tambo N. and Matsui Y. (1983). High Capacity Depth Filter. Proc. PACHEC'83 (Bangkok), Vol.1, p.112.
Tambo N. and Shoji M. (1972). Removal characteristics of inclined multi-plate sedimentation basin. *Journal of Japan Water Works Association*, (459), 6–26. (In Japanese).
Tambo N., Hozumi J. and Abe S. (1969). Behavior of floc blankets in upstream sedimentation basin (I, II). *Journal of Japan Water Works Association*, (416), 7, No.417 p.7. (In Japanese).
Tambo N., Fukushi K. and Ohta H. (1981a). Comparison of dissolved air flotation and sedimentation separation by flotation tester. *Journal of Japan Water Works Association*, (603), 17–27. (In Japanese).
Tambo N., Igarashi T. and Kiyozuka M. (1981b). Electrophoretic Study of bubble-attached Floc. *Journal of Japan Water Works Association*, (604), 2–6. (In Japanese).
Tambo N., Kobayashi M. and Matsui Y. (1982). Study of the heigh volume capacity filter: with Raschig Ring on top layer. (I). Proposal of the concept. *Journal of Japan Water Works Association*, (571), 37.(In Japanese)
Tambo N., Kobayashi M, Matsui Y. and Okamoto Y. (1984). Study of the heigh volume capacity filter: with Raschig Ring on top layer. (II). Assessment of the mechanism. *Journal of Japan Water Works Association*, (598), 16.
Tambo N., Hashimoto K. and others (1986). Performance of Finned Channel Separator. 10th Conference of the International Association on Water Pollution Research in Toronto.
Tambo N., Kobayashi M., Matsui Y. and Ozawa G. (1990). Study of the heigh volume capacity filter: with Raschig Ring on top layer. (III). Operation Condition. *Journal of Japan Water Works Association*, (634), 2.v.
Weber W. J. (1972). *Physicochemical Processes for Water Quality Control*. Chapt.3. Wiley-Interscience, pp. 111–138, pp. 139–198.

REFERENCES

Camp T. R. (1946). Sedimentation and the design of sSttling tanks. *Transactions of the American Society of Civil Engineers.*, 895–981.
Carman P. C. (1937). Fluid flow through Granular beds. *Transactions of the Institution of Chemical Engineers*, **39**, 150.
Dobbins W. E. (1944). Effect of Turbulence on Sedimentation. *Transactions of the American Society of Civil Engineers*, **109**, 626.
Ebie K. (1975). Behavior of detained substances in rapid filtration layers I–IV. *Journal of Japan Water Works Association*, (493), 25.
Ebie K. (1976a). Behavior of detained substances in rapid filtration layers I–IV. *Journal of Japan Water Works Association*, (498), 15.
Ebie K. (1976b). Behavior of detained substances in rapid filtration layers I–IV. *Journal of Japan Water Works Association*, (507), 20.
Ebie K. (1977). Behavior of detained substances in rapid filtration layers I–IV. *Journal of Japan Water Works Association*, (508), 18.
Eckenfelder W. W., Jr. and O'Conner D. J. (1961). Biological Waste Treatment. pp.152–179.
Fair G M. and Geyer J. C. (1956). *Water and Waste-water disposal*. Chapt.22: Sedimentation and Flotation; pp. 584–615; Chapt.24: Filtration (pp. 656–702). John Wiley & Sons, Inc.

Fisherström C. N. H. (1955). Sedimentation in rectangular basins. *Proceedings of the American Society of Civil Engineers*, **81**(687).

Hashimoto K., Haiksegawa Otsuka, Sugawara and Tambo N. (1982a). Find channel separators (I). *Journal of Japan Water Works Association*, **577**, 22 (In Japanese).

Hashimoto K., Hasegawa Otsuka, Sugawara and Tambo N. (1982b) Find channel separators (II). *Journal of Japan Water Works Association*, **578**, 30 (In Japanese).

Hashimoto K., Hasegawa Otsuka, Sugawara and Tambo N. (1982c). Find channel separators (III). *Journal of Japan Water Works Association*, **579**, 29 (In Japanese).

Hashimoto K., Hasegawa Otsuka, Sugawara and Tambo N. (1983). Find channel separators (IV). *Journal of Japan Water Works Association*, **591** (In Japanese).

Hazen A. (1904). On sedimentation. *Transactions of the American Society of Civil Engineers*, **53**, 53.

Ingersoll A. C., McKee J. E. and Brooks N. H. (1955). Fundamental concept of rectangular settling tanks. *Proceedings of the American Society of Civil Engineers*, **81**(590), 1–27.

Ives K. J. (1969). Theory of Filtration. Proc. Special Subject No.7, International Water Supply Congress, Munich.

Ives K. J. (1970). Rapid filtration (review paper). *Water Research*, **4**, 201–223.

Ives K. J. (1975). The Scientific Basis of Filtration. Noordhoff International Pub, pp. 203–224.

Ives K. J. and Gregory J. (1966). Surface forces in filtration. *Proceedings of the Society for Water Treatment and Examination*, **15**, 93.

Iwasaki T. (1937). Some note on sand filtration. *Journal of the American Water Works Association*, **29**(10).

Kozeny J. (1927). Über Kapillare Leitung des Wassers im Boden (Concerning capillary conduction of water in soil). Sitzb Akad.Wiss.Wein., Mat.Naturw.Kasse, 136(2a), pp. 271–306. (Vienna Academy of Sciences, Meeting Reports (Dept of Maths and Science). (In German).

Kynch G. J. (1952). A theory of sedimentation. *Transactions of the Faraday Society*, **48**, 166–176.

Langhaar H. L. (1951). Dimensional Analysis and Theory of Models. John Wiley & Sons, Inc.

Laufer J. (1950). Some recent measurement in a two- dimensional turbulent channel. *Journal of Aeronautical Sciences*, **17**, 277.

Mints D. M. (1966). Modern Theory of Filtration.Proc. Special Subject No.10, International Water Supply Congress, Barcelona. IWSA, London.

NRC Subcommittee on Sewage Treatment (1948). Sewage treatment at military installations: summary and conclusions. *Sewage Works Journal*, **20**(1), 52–95.

Richardson J. F. and Zaki W. N. (1954). Sedimentation and fluidisation : Part I. *Transactions of the American Institute of Chemical Engineers*, **32**, 35–53.

Rouse H. (ed.) (1951). Engineering Hydraulics. John Wiley & Sons, Inc, pp. 1–115.

Stein P. C. (1940). A Study of the Theory of Rapid Filtration of Water Through Sand. PhD Thesis, MIT.

Sueishi T. (1940). Hydraulic design of porous pipe water collection system. (In Japnese). Japan Society of Civil Engineers, Paper No.63, 36.

Sueishi T. (1957). Flow and pressure distributions of inverse cleaning water by porous pipe water collection system. *Journal of Japan Water Works Association*, (271), 19. (In Japanese).

Tambo N. (1962). Study on coagulative settling. *Journal of Japan Water Works Association*, (335), 31–38. (In Japanese).

Vrablik E. R. (1957). Fundamental Principle of Dissolved Air Flotation. Proc.14th Purdue Industrial Conference, p. 199.

Yagi S. and Shirai T. (1952). Start of fluidization. *Kagaku-Kikai (Chemical engineering machine)*, **16**, 2–6. (In Japanese).

Yanase I. (1964a). Reform (upgrading) of conventional horizontal sedimentation basin by introduction of a folding inclined multilayer separation system into conventional horizontal flow sedimentation basins (I). *Journal of Japan Water Works Association*, (361), 42. (In Japanese).

Yanase I. (1964b). Reform (upgrading) of conventional horizontal sedimentation basin by introduction of a folding inclined multilayer separation system into conventional horizontal flow sedimentation basins (2). *Journal of Japan Water Works Association*, (363), 44. (In Japanese).

Chapter 7

Adsorption and ion exchange

7.1 ADSORPTION

7.1.1 Adsorption phenomenon

Adsorption and ion exchange in water treatment are processes in which substances in a target fluid are removed by mass transfer between solids and fluids (liquids and gases). There are many similarities between the two, but their actions and mechanisms differ considerably.

Among the phenomena generally treated as adsorption, chemical adsorption (described below) is often indistinguishable from ion exchange, and the number of both natural and artificial adsorbents with ion exchange capacity is not small. As described in Chapter 6, the precipitation treatment of manganese and iron also causes a phenomenon in which an intermediate insolubilized by hydrolysis adsorbs soluble ions, and the process proceeds to the next oxidative coagulation process.

The collection of solutes and gas molecules from the liquid and gas phases onto solid surfaces is called adsorption, and absorption occurs when solutes and gas molecules are incorporated into a solid and liquid body through the surfaces. When both occur at the same time – or when it is uncertain whether one of them is greater or not – it is called sorption. When adsorbed substances return to the liquid or vapor phase it is called desorption (Table 7.1) Desorption is thought to be in the category of the sorption process group.

Table 7.1 Adsorption phenomena.

Adsorption	A phenomenon in which solutes and gas molecules are collected on the solid surface.
Absorption	A phenomenon in which substances are incorporated into solid and liquid.
Sorption	A uncertain phenomenon of adsorption or absorption.
Desorption	A phenomenon that the adsorbed substances returns to a liquid phase/gas phase

Author(s): Norihito Tambo and Koichi Ogasawara
doi: 10.2166/9781789061307_0517

7.1.1.1 Heat of adsorption

The adsorption reaction is always an exothermic reaction, and the solute before adsorption is aggregated and fixed to a narrow boundary surface from a non-equilibrium state in which it is free to move in a liquid system, and the entropy shifts from a large state to a small state. In this case, the phenomenon of adsorption progresses and the free energy of the entire system reaches a certain stable level, toward the lower side. If the amount of decrease in free energy is ΔF, the amount of decrease in kinetic energy is ΔE, and the amount of change in entropy is ΔS, the relationship (of Equation 7.1) is established under the absolute temperature T:

$$\Delta F = \Delta E - T\Delta S \quad \text{—(7·1)} \tag{7.1}$$

ΔF: Amount of decrease in free energy
ΔE: Amount of decrease in kinetic energy
ΔS: Amount of change in entropy
T: Absolute temperature

Equation 7.1 The free energy and kinetic energy for adsorption.

In water treatment, most such phenomena are thought to occur under constant pressure. Therefore, the amount of change in free energy under constant pressure (Gibbs free energy) ΔG is expressed by Equation 7.2, where ΔH is the amount of change in enthalpy.

$$\Delta G = \Delta H - T\Delta S \quad \text{—(7·2)} \tag{7.2}$$

ΔG: Amount of change in free energy under constant pressure
ΔH: Amount of change in enthalpy
ΔS: Amount of change in entropy
T: Absolute temperature

Equation 7.2 The amount of change in free energy under constant pressure.

For adsorption to occur, ΔF or ΔG is always negative, entropy ΔS is also decreased, $\Delta S < 0$, so ΔH or ΔE must always be negative, and heat is always released (exothermic reaction). Such heat generation is called heat of adsorption, and the amount varies depending on the type of adsorption. The adsorption heat always takes a positive value, so the adsorption amount increases as the temperature decreases.

7.1.1.2 Physical sorption and chemical sorption

Depending on the mechanisms by which adsorption occurs, adsorption phenomena are divided into two types: physical and chemical adsorption.

Physical adsorption is usually caused by a weak intermolecular force such as Van der Waal's force, and adsorption occurs almost regardless of the type of solutes, resulting in a very low calorific value. In contrast, chemical adsorption may or may not occur due to solutes, but the amount of heat generated during adsorption increases, and the adsorption phenomenon often progresses even at high temperatures. It has activated energies and is also called activated adsorption.

The main adsorption phenomenon used for water treatment is physical adsorption, but chemical properties are also at work at the same time. In Section 3.5, on the water quality matrix, it was stated that E260 sensitive total organic carbon (TOC) (i.e. organic substances with unsaturated bonds) exhibiting ultraviolet absorption adsorbs on activated carbon, while sugars and alcohols of saturated carbohydrates which do not exhibit ultraviolet absorption (E260 insensitive TOC) do not adsorb at all.

7.1.1.3 Adsorption equilibrium and adsorption rate

Two of the most important designing factors for water treatment using adsorption phenomena are the adsorption equilibrium of the extent to which substances (solutes) to be removed in water are adsorbed under considered states, and the adsorption rate at which removal proceeds to reach equilibrium. The equilibrium adsorption quantity q of a solute is determined for the same adsorbent by the concentration C at the adsorption equilibrium of the adsorbate (solute) and the absolute temperature T at that time. Therefore, this relationship is expressed as Equation 7.3, and is evaluated by calculating the relationship between q and C under constant temperature conditions.

$$q = f(C, T) \qquad \text{—}(7 \cdot 3)$$

q : Equilibrium adsorption quantity of solutes
C : Equilibrium adsorption concentration of solutes
T : Absolute temperature

(7.3)

Equation 7.3 The adsorption isotherm equation.

This relation is called the adsorption isotherm and it is the basis for indexing the progress of the adsorption equilibrium. The following section describes in detail how to evaluate the adsorption rate.

7.1.1.4 Adsorbent

The general properties of the materials which are used as adsorbents (see Figure 3.17 in Chapter 3) are:

(1) Solid particle of high porosity and remarkable internal surface area; and
(2) Large adsorbent capacity solid.

Activated carbon is typically used for water treatment. Activated carbon can be made from most organic materials. Adsorbents with various adsorption and operating characteristics are produced by different types of raw materials, activation (imparting and recovering adsorption properties) and granulation methods. Therefore, it is important to evaluate which adsorbent should be selected according to the purpose of the treatment.

Activated carbon can be made from charcoal, coal, animal bones, coconut shells, and other raw materials. Gas-activated carbon activates adsorption by steam at high temperatures around 900°C, and chemical-activated carbon carbonizes woody raw materials by immersing them in chemicals such as zinc chloride and sulfuric acid. Granular charcoal is characterized in terms of its geometric dimensions, and can be roughly classified into granular carbon and powdered carbon.

As shown in Figure 3.17, macropores of about 10^{-5} to 10^{-7} m in diameter lead from the surface of the activated carbon, and micropores of 10^{-8} to 10^{-10} m in diameter form further branches derived from the macropore walls, forming a network of capillaries. The inner surface area of activated carbon is governed by the inner surface of the micropores, and the adsorbate diffuses into the macropores from the outside and then diffuses into the micropores to adsorb them. Therefore, the size of the adsorbate, the inner diameter of the micropores, and the size of its inner surface area have a major influence on the amount of adsorption equilibrium and the adsorption rate. The micropores have a volume of approximately 0.3 to 0.6 ml/g activated carbon, but have a specific surface area of 700 to 1400 m^2/g

Table 7.2 Inside activated carbon.

	Diameter m	Volume ml/g-Carbon	Specific Surface Area m^2/g-Carbon
Macropores	$10^{-5} \sim 10^{-7}$	$0.3 \sim 0.6$	$1 \sim 200$
Micropores	$10^{-8} \sim 10^{-10}$	//	$700 \sim 1400$

carbon. On the other hand, macropores have almost the same volume, but the specific surface area is only about 1 to 200 m^2/g carbon, as shown in Table 7.2.

When activated carbon is used for water treatment, whether to use powdered or granulated activated carbon is the point of discussion. The reason for this is that the adsorption process in water treatment often handles components (organic matter, etc.) with a slow adsorption rate, and the adsorption rate can be increased if the particle size of the activated carbon is smaller, corresponding to a limited residence time. However, it is not easy to collect and regenerate powdered activated carbon after use, and it must be disposed of and, therefore, using powdered activated carbon at all times is not acceptable. Adsorption treatment with granular activated carbon has been carried out using a fixed-bed or fluidized-bed method on the premise of regeneration. In order to use granular activated carbon in fixed and fluidized beds, it is necessary that the shape and particle size distribution be as uniform as possible, that the hydraulic characteristics be good, and that the particles have a high hardness with less damage or wear and tear caused by regeneration and cleaning processes.

Recently, MF and UF membranes have been placed after powdered activated carbon adsorption to enable strict solid-liquid separation. Therefore, activated carbon can be atomized to increase the adsorption rate. Technology development for uniformly pulverizing coal, energy consumption and the possibility of recycling, and other processes using pulverized coal are potentially technological innovations in water treatment systems.

7.1.1.5 Adsorption characteristics (absorbability)

The absorbability depends on two factors: how much of the solute is adsorbed on the activated carbon (equilibrium) and at what rate the adsorption proceeds (rate). Figure 7.1 (Eguchi, 1966) schematically shows the factors that govern adsorption in the liquid phase. Organic ingredients targeted for water treatment are natural humic substances, organic substances and detergents in sewage water, pesticides, and low-concentration off-flavor and odor ingredients.

Very small amounts of odors and tastes are caused by the generation of large amounts of algae and actinomycetes in natural waters. Geosmin and 2-methylisoborneol (2-MIB) have been identified as causative agents of fungal odors and soil odors. These substances are highly absorbable components that can be removed from water systems relatively quickly by activated carbon adsorption. Substances such as phenol, an artificial odorant, and cyclohexylamine, which are generated by chlorination and emit a very unpleasant odor even in very small amounts of several μg/l, also exhibit good adsorption properties.

Other organochlorine compounds include pesticides such as BHC, DDT, and endrin, which together with organophosphorus-based pesticides are major risk ingredients for drinking. These ingredients are often present at concentrations as low as μg/l, but they are highly adsorbing and can be removed almost completely by passing through an activated carbon bed.

As described in Chapter 3, organic substances derived from sewage and natural organic coloring components can be expected to be adsorbed and removed by unsaturated compounds and other

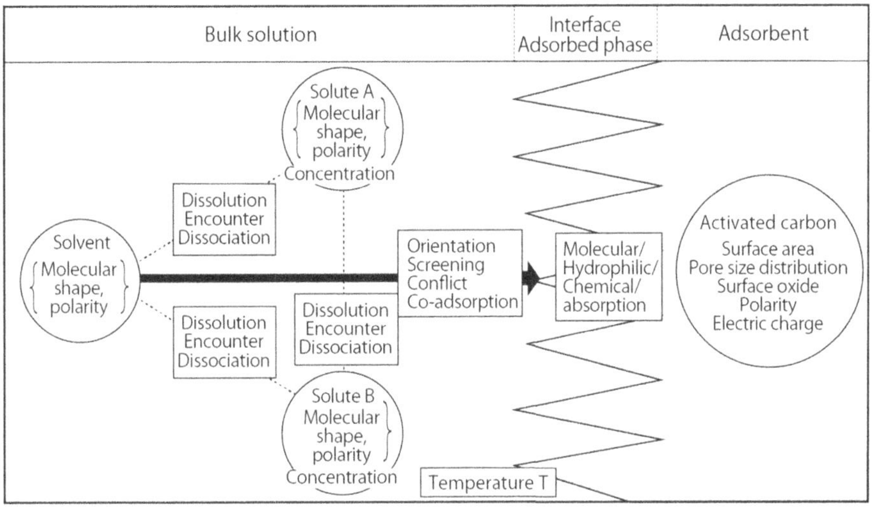

Figure 7.1 The factors that govern adsorption in the liquid phase (Eguchi, 1966).

compounds that show an ultraviolet extinction E260 nm sensitivity, having 1500 Da or less sizes. On the other hand, low molecular weight compounds such as sugars and alcohols, which have linear structures that do not absorb ultraviolet E260 nm (i.e. are ultraviolet E260 insensitive), do not adsorb at all. Higher molecular weight components exceeding 1500 Da cannot diffuse into the micropores and are not subject to adsorption treatment. Therefore, they must be removed in advance by coagulation treatment. Therefore, as described in Section 3.5.4 on water quality conversion matrices, organic matter of 1500 Da or more must be coagulated, and organic matter (such as sugars) that is ultraviolet E260 insensitive, having molecular mass of 1500 Da or less, must be removed by aerobic biological treatment and then activated carbon treatment must be added. Through this process, most of the water can be regenerated.

Trace heavy metals in water are also removed to a considerable extent by passing through the activated carbon layer. The adsorption of these heavy metals is thought to be removed in the form of hydroxides, which are often aquo complexes, and the adsorption properties are largely governed by the pH. The adsorption is scarcely removed in the state where simple hydrated ions are present, and in the insolubilized state, the molecular size becomes too large to diffuse into the pores, and the adsorption seems to proceed in the metastable state.

Halogen compounds such as chlorine and iodine and ozone are almost completely removed by passing them through an activated carbon filter bed. With these oxidizing agents, it is thought that the activated carbon is oxidized and becomes chlorine ions or iodine ions, thereby losing its activity. Boron cannot be removed.

7.1.2 Theory of adsorption

7.1.2.1 Adsorption equilibrium

When designing a water treatment system using the adsorption phenomenon, two items should be addressed:

(1) Evaluation of the extent to which adsorption proceeds with sufficient time; and
(2) Kinetic evaluation of how fast adsorption proceeds (to a certain level in a certain time).

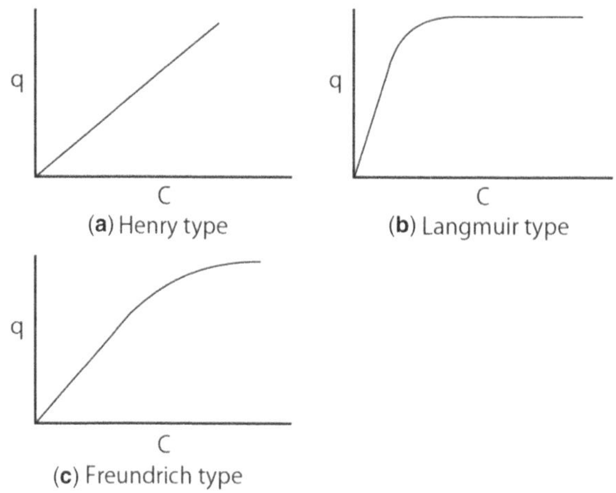

Figure 7.2 Widely used adsorption isotherms for liquid systems.

In this section, we first introduce how to express the adsorption equilibrium in the form of the adsorption isotherm shown in Equation 7.3.

The most widely used adsorption isotherms for liquid systems are the Henry, Langmuir, and Freundlich types, as shown in Figure 7.2.

(a) Henry adsorption isotherm
If the equilibrium adsorbed quantity is q and the concentration of the liquid system when the adsorption equilibrium is reached is C, then the adsorption isothermal equation is given by Equation 7.4. It is called Henry adsorption because it is a function of Henry's solubility law (see Chapter 2, Table 2.3, etc.), in which the gas solubility in liquid is proportional to the partial pressure. The adsorption amount per surface area is expressed as a low adsorption amount range of about 10% or less of the maximum adsorption amount q_{max}, and can be regarded as an approximate form corresponding to the lower part of Langmuir adsorption.

$$q = aC \qquad \text{—(7·4)}$$

q : Equilibrium adsorption quantity
C : Concentration of liquid system in adsorption equilibrium

(7.4)

Equation 7.4 The Henry adsorption isotherm equation.

(b) Langmuir adsorption isotherm
One of the most commonly used isotherms, the Langmuir adsorption isotherm can be written as Equation 7.5. If this equation is transformed into Equation 7.6 and the measured values (experimental values) for C/q and C are summarized, and the relationship in Equations 7.5 and 7.6

hold, then the plot of Equation 7.6 should be a straight line. This is called a Langmuir plot.

$$q = \frac{Kq_{max}C}{1+KC} \quad —(7\cdot5)$$

$$\frac{C}{q} = \frac{1}{Kq_{max}} + \frac{C}{q_{max}} = A + BC \quad —(7\cdot6)$$

q : adsorption quantity
q_{max} : Saturation quantity of adsorption
K : Constant

(7.5–7.6)

Equation 7.5-7.6 The Langmuir adsorption isotherm equation.

Originally, the Langmuir adsorption isotherm equation was an experimental formula but, as will be described later, it has a theoretical background in single molecule adsorption which is important background for understanding – along with the Brunauer-Emett-Teller (BET) adsorption isotherm – how the phenomenon can be extended to multi-molecule adsorption.

If the adsorptive component is a two-component system, the adsorption equilibrium is expressed by the adsorption isotherm shown in Equation 7.7.

$$\left.\begin{aligned} q_a &= \frac{q_{amax}K_aC_a}{1+C_aK_a+C_bK_b} \\ q_b &= \frac{q_{bmax}K_bC_b}{1+C_aK_a+C_bK_b} \end{aligned}\right] \quad —(7\cdot7)$$

q_a ,q_b : Adsorption quantity of a and b components
q_{amax} ,q_{bmax} : Maximum adsorption quantity in a and b components alone
C_a ,C_b : Adsorption equilibrium concentration in a and b components alone
K_a ,K_b : Constants for a and b components alone

(7.7)

Equation 7.7 The Langmuir adsorption isotherm of two-component system.

Various isotherms (including the Langmuir, Brunauer-Emett-Teller, Franker, Jura-Harkins, Gibbs, etc.) have been proposed as liquid and gas-based adsorption isotherms, and it is known that Langmuir adsorption theories are relatively well suited for the adsorption of organic components (such as sewage and pesticides) to describe adsorption phenomena at the solid-liquid interface applied to water treatment. This section explains the theoretical background.

There are N_s adsorption sites (sites where adsorption is possible) on the inner surfaces of an adsorbent, and the adsorption equilibrium is considered to be established in the dynamic equilibrium between the adsorbed and desorbed molecules. In such an equilibrium, N_a of the N_s adsorption sites are occupied by solute components. In this case, the number of solute molecules that collide with an adsorption site that is free per unit time is shown in Equation 7.8. By contrast, the number of solutes desorbed per unit of time is proportional to the N_a of solutes (adsorbed) per unit of time. Therefore, Equation 7.9 shows the number of solutes desorbed per unit of time and, therefore, since the sum of the two equations is zero for equilibrium, the N_a of the number of adsorption sites where solutes are ultimately adsorbed is given by Equation 7.10, which is then transformed into Equation 7.11.

$$\frac{dn}{dt} = KC(N_s - N_a) \quad \text{—(7·8)}$$

$$\frac{dn}{dt} = K'N_a \quad \text{—(7·9)}$$

$$KC(N_s - N_a) = K'N_a \quad \text{—(7·10)}$$

$$N_a = \frac{N_s(K/K')C}{1 + (K/K')C} \quad \text{—(7·11)}$$

(7.8–7.11)

n : Solute molecules numbers colliding with vacant adsorption sites in unit time

N_s : Adsorption sites numbers on inner surfaces of adsorbent (Adsorbable sites)

N_a : Sites numbers adsorbed

K,K′ : Constant

Equation 7.8 - 7.11 The single molecule adsorption model.

Here, N_s is the same as the maximal adsorption q_{max}, and if $K/K' = a$, the Langmuir Equation (Equation 7.5) is obtained. This approach is called the single molecule adsorption model, because one molecule is adsorbed to one adsorption site, and the thickness of the adsorption layer is one molecule.

7.1.2.2 Adsorption rate

The phenomenon in which solutes (adsorbates) move from the fluid and are adsorbed on the inner surface of the porous adsorbent can be divided into four stages:

(1) Diffusion of the adsorbate to reach the outer surface of the adsorbent;
(2) Diffusion of that adsorbate to diffuse through the fluid boundary layer on the outer surface to reach the internal voids of the porous adsorbent;
(3) Intragranular diffusion that diffuses through the internal voids of the adsorbent and reaches the adsorption point (site) on the internal surface; and
(4) Adsorption reaction on the inner surface.

This is shown schematically in Figure 7.3.

The diffusion rate of the adsorbate in solution is high, not only in the case where the adsorbent is suspended under turbulent flow conditions but also in the case of a fixed or fluidized bed, etc., so it is considered that the boundary layer diffusion process (2) governs the transfer rate of the solute on the outer surface of the adsorbent. Among the processes inside the adsorbent, the adsorption reaction (4) can be considered to reach adsorption equilibrium instantaneously at the position of the adsorption locus. Therefore, the intragranular diffusion process in which solutes are diffused and transported through the

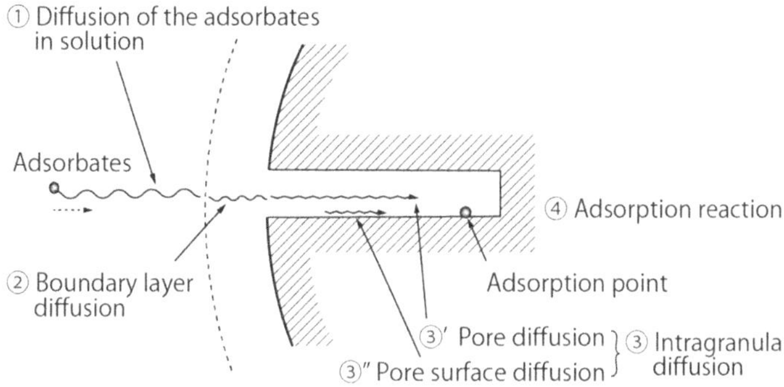

Figure 7.3 The four stages of adsorption.

internal pores of the adsorbent (3) can be considered to control the rate and, therefore, the rate-limiting step that governs the adsorption process is either the liquid boundary film diffusion (2) or the intragranular diffusion (3). There are two types of intragranular diffusion: diffusion through the pores by using the concentration gradient as a driving force and diffusion through the pores by the adsorption gradient as a driving force.

The rate equation that governs the above two steps can be expressed as follows, considering the concentration distribution and adsorption amount distribution, as shown schematically in Figure 7.4.

C_f is the solute (adsorbate) concentration in the body of the solution, C_i is the solute concentration at the solid-liquid interface, q_i is the adsorption amount of the solid surface in the adsorption equilibrium with C_i (assuming that the adsorption reaction reaches equilibrium instantaneously), C_s is the solute concentration in the adsorbent pore, and q_s is the adsorption amount of the pore surface in the adsorption equilibrium with C_s.

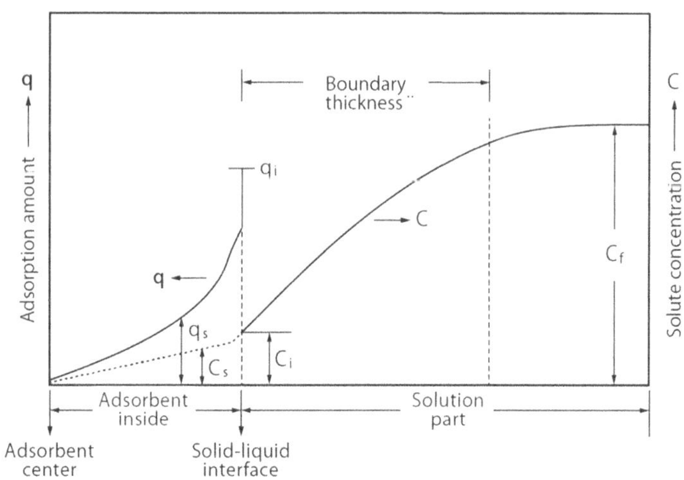

Figure 7.4 Changes in concentration and adsorption quantity through liquid systems and internal pores.

The mass transfer rate dq/dt (g/sec) per unit surface area of adsorbent in a fluid-boundary layer adjoining the surface of a solid is shown in Equation 7.12.

$$\frac{dq}{dt} = k_f (C_f - C_i) \qquad \text{—}(7{\cdot}12) \tag{7.12}$$

k_f: Mass transfer coefficient (cm/sec)
Assuming that the thickness of boundary layer is δ,
$k_f = D/\delta$
C: Solute concentration (g/cm^3)

Equation 7.12 The mass transfer rate per unit surface area of adsorbent.

Diffusion transport and adsorption in the pores are represented in Figure 7.5. The adsorption rate of the adsorbate at point r = r in the adsorbent particle is formulated by the following assumptions and procedures, which refers to Figure 7.5 as a transfer phenomenon due to the molecular diffusion of solutes inside pores:

(1) Adsorbate is diffused and transported in the pores by molecular motion. In this case, the diffusion rate, denoted by the diffusion coefficient, is both local and independent of the adsorption amount at that point.
(2) Adsorption occurs at the pore wall surface, and in this case, the adsorption amount instantaneously reaches equilibrium with the adsorbate concentration of the pore inner surface. This equilibrium relationship is determined by the adsorption isotherm.

Based on this process, consider the mass balance between pore cross-sections I and II in Figure 7.5. The solutes F_I and F_{II} entering and exiting through sections I and II are as shown in Equation 7.13, and the reduced solutes (adsorption matter) in sections I and II are as shown in Equation 7.14. In addition, the change in solute concentration ΔC_p and the amount Δq_p adsorbed between the sections I and II, are

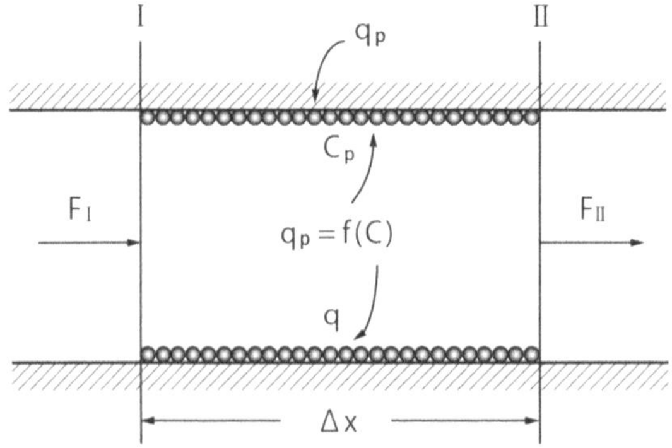

Figure 7.5 The material balance of solute concentration and adsorption quantity in a pore, showing solutes F_I and F_{II} entering and exiting through sections I and II.

expressed as shown in Equation 7.15.

$$F_I = AD\left(\frac{\partial C}{\partial x}\right)_I, \quad F_{II} = AD\left(\frac{\partial C}{\partial x}\right)_{II} \qquad \text{—(7·13)}$$

$$F_I - F_{II} = AD\left\{\left(\frac{\partial C}{\partial x}\right)_I - \left(\frac{\partial C}{\partial x}\right)_{II}\right\} \qquad \text{—(7·14)}$$

$$F_I - F_{II} = \frac{(\Delta q_c + \Delta C) \times A \times \Delta x}{\Delta t} \qquad \text{—(7·15)} \qquad (7.13\text{–}7.15)$$

A : Pore cross section
D : Solute diffusion coefficient
C : Solute concentration

Equation 7.13 - 7.15 The adsorption rate of adsorbates.

Equations 7.14 and 7.15 yield Equation 7.16, which, when differentiated, yields the basic equation of Equation 7.17. Since the adsorbent is mostly granular, it is more practical to use a spherical coordinate system, so a coordinate transformation on the spherical coordinate system yields a generally usable basic equation, given as Equation 7.18.

$$\frac{\Delta q_c}{\Delta t} + \frac{\Delta C}{\Delta t} = D\left\{\left(\frac{\partial C}{\partial x}\right)_I - \left(\frac{\partial C}{\partial x}\right)_{II}\right\}\Big/\Delta x \qquad \text{—(7·16)}$$

Basic equation

$$\frac{\partial C}{\partial t} = D\frac{\partial^2 C}{\partial x^2} - \frac{\partial q_p}{\partial t} \qquad \text{—(7·17)} \qquad (7.16\text{–}7.18)$$

Generally usable basic equation

$$\frac{\partial C}{\partial t} = \frac{D}{r^2} \cdot \frac{\partial}{\partial r}\left(r^2 \frac{\partial C}{\partial r}\right) - \frac{\partial q_p}{\partial t} \qquad \text{—(7·18)}$$

Equation 7.16 - 7.18 The adsorption rate basic equation of adsorbate.

If we solve Equations 7.17 and 7.18 under the given initial and boundary conditions, we can calculate the change in adsorption amount over time. To know the adsorption rate of an adsorbent, we can calculate the diffusion coefficient D for the solute of interest by giving boundary conditions, so we need to determine the intrapore diffusion coefficient first when we discuss the adsorption rate.

The q_p of adsorption in Equations 7.17 and 7.18 can be expressed as a function of the C_p using the appropriate adsorption isotherm, so all can be used as an equation for C. If the equation is solved under appropriate boundary conditions, a general analytic solution cannot be obtained, and the solution is usually obtained by numerical calculation.

As an example, the boundary conditions are defined and a solution is given for the progress of adsorption by a granular adsorbent in a batch adsorption process (described later). Suppose that Equation 7.19 is given as the boundary condition and Equation 7.20 is given as the initial condition.

Boundary condition

$$V(C_0 - C) = \varepsilon \int_0^a 4\pi (C + q_p) r^2 dr \quad \text{—(7·19)}$$

Initial condition

$$C = 0, \text{ at } 0 \ll r \ll a, \ t = 0 \quad \text{—(7·20)}$$

V : Reactor volume
C_0 : Solute initial concentration
C : Solute concentration at certain adsorption quantity
ε : Porosity of adsorbent
a : Adsorbent radius
r : Distance from the center of the adsorbent

(7.19–7.20)

Equation 7.19-7.20 The example of granular adsorbent adsorption by batch adsorption operation.

Under these conditions, dimensionless numbers such as $C^* = C/C_0$, $S = q_p/C_0$, $\rho = r/a$, and $\tau = Dt/a^2$ are introduced, and Equations 7.18, 7.19, and 7.20 are transformed into Equations 7.21, 7.22, and 7.23, and are generally solved.

$$\frac{\partial C^*}{\partial \tau} = \frac{1}{\rho^2} \cdot \frac{\partial}{\partial \rho}\left(\rho^2 \frac{\partial C^*}{\partial \rho}\right) - \frac{\partial S}{\partial \tau} \quad \text{—(7·21)}$$

$$V(1 - C^*) = 2\pi \varepsilon a^3 \int_0^a (C^* + S)\rho^2 d\rho \quad \text{at } \rho = 1 \quad \text{—(7·22)}$$

$$C^* = 0, \text{ at } 0 \ll \rho \ll 1, \ \tau = 0 \quad \text{—(7·23)}$$

$C^* = C/C_0$, $S = q_p/C_0$, $\rho = r/a$, $\tau = Dt/a^2$
(Dimensionless number)

(7.21–23)

Equation 7.21-7.23 The non-dimensionalization of granular adsorbent adsorption equation.

Using the results of experiments conducted with adsorbents of any size, we learn that the time required to reach the same adsorption rate C/C_0 progresses inversely proportionally to the square of the diameter of the dimensionless number $\tau = Dt/a^2$.

If the external adsorbent mass concentration does not change at $C = C_0$ and the adsorption isothermal equation is Henry type ($q = kC$), an analytical solution showing the adsorption amount of the granular adsorbent at time t is obtained, as in Equations 7.24 and 7.25.

$$\frac{q}{q^*} = 1 - \frac{6}{\pi^2}\sum_{n=1}^{\infty}\frac{1}{n^2}\exp(-n^2\alpha t) \quad \text{—(7·24)}$$

$$\alpha = D\pi^2/\rho_p k a^2 \quad \text{—(7·25)}$$

q : Adsorption quantity (g)
q* : Equilibrium adsorption quantity with adsorption component amount in liquid (g)
Maximum adsorption quantity at C_0 : q_{max} at $C=C_0$
ρ_p: Apparent density of adsorbent (g/cm³)
k : Constants of the Henry-type adsorption isotherm equation
a : Adsorbent radius (cm)

(7.24–7.25)

Equation 7.24-7.25 The analysis of adsorption quantity of particle adsorbent after time t.

Except in the initial stage of adsorption, the second and subsequent terms in Equation 7.24 can be omitted, and can therefore be represented by a simple equation such as Equation 7.26. Differentiating and transforming Equation 7.26 yields Equation 7.27, which shows that the adsorption rate progresses in proportion to the difference between adsorption saturation and adsorption.

Accordingly, we can use the rate equation shown in Equation 7.27 for the continuous operation of a low adsorption volume that does not reach adsorption saturation by experimentally determining $\bar{K}$.

$$\frac{q}{q^*} = 1 - \frac{6}{\pi^2}\exp(-\alpha t) \quad \text{—(7·26)}$$

Differentiating

$$\frac{dq}{dt} = \bar{K}(q^* - q) \quad \text{—(7·27)}$$

(7.26–7.27)

Equation 7.26-7.27 The continuous operating rate equation for low adsorption volume.

In general, the adsorption rate of a granular adsorbent proceeds at the very early stage of adsorption by the liquid boundary film diffusion rate control, but as the adsorption amount increases and adsorption to internal pores becomes dominant, the adsorption rate shifts to the intrapore diffusion rate control. In the latter case, the adsorption rate is proportional to the difference ($q_{max} - q$) between the adsorption saturation and the adsorption (Equation 7.27) in the same manner as in Equation 7.12, and the adsorption rate can be expressed as if there is a solid-state boundary film, for the sake of convenience. For this reason, we hypothetically define the mass transfer coefficient k_s (g/cm³/sec) for the virtual solid boundary layer, and consider the mass transfer rate equation of Equation 7.28. The hypothesis of a virtual solid boundary layer is useful when dealing with the adsorption reaction of a packed bed. The relationship between the k_s of the mass transfer coefficient and the internal diffusion coefficient D of the virtual solid boundary layer is written as in Equation 7.29, within the bounds of the Henry-type adsorption isotherm.

$$\rho_b \frac{dq}{dt} = k_f a_p (C - C_i) = k_s a_p (q_i - q) \quad —(7\cdot28)$$

$$k_s a_p = D\pi^2 \rho_b / \rho_p k a \quad —(7\cdot29)$$

ρ_b : Bulk density (g/cm^3)
ρ_p : Apparent density of adsorbent (g/cm^3)
k_s : Mass transfer coefficient ($g/cm^3/sec$)
a : Adsorbent radius (cm)
D : Internal diffusion coefficient (cm^2/sec)
a_p : Particle surface area in packed bed unit volume (1/cm)
k : Coefficients of Henry's adsorption isotherm equation

(7.28–7.29)

Equation 7.28-7.29 The mass transfer rate equation of virtual solid boundary layer.

If we transform Equation 7.28 using Henry's adsorption isotherm equation $q^* = kC$ (where q^* is the saturated adsorption amount at external concentration C), we obtain Equation 7.30 as follows.

$$\rho_b \frac{dq}{dt} = \frac{k_f a_p}{k}(q^* - q_i) = k_s a_p (q_i - q)$$

$$= \frac{q^* - q_i}{\frac{k}{k_f a_p}} = \frac{q_i - q}{\frac{1}{k_s a_p}} = \frac{q^* - q}{\frac{k}{k_f a_p} + \frac{1}{k_s a_p}}$$

$$= k_s a_p (q^* - q) = k_f a_p (C - C^*) \quad —(7\cdot30)$$

q^* : Adsorption equilibrium quantity at externa concentration C according to Henry's law
C^* : Equilibrium concentration at adsorption quantity q
k_s, k_f : Overall transfer coefficient each expressed as solid or liquid boundary layer

$$\frac{1}{K_s} = \frac{k}{k_f} + \frac{1}{k_s}, \quad \frac{1}{K_f} = \frac{1}{k_f} + \frac{1}{kk_s} \quad —(7\cdot31)$$

If $k_f \gg kk_s$, then $K_f = kk_s$: Internal diffusion control
If $k_f \ll kk_s$, then $K_f = k_f$: Boundary layer diffusion control

(7.30–7.31)

Equation 7.30-7.31 The mass transfer rate equation using Henry's adsorption isotherm equation.

7.1.3 Absorber operating theory

There are three main adsorption methods: batch adsorption, fixed bed adsorption, and fluidized bed adsorption.

7.1.3.1 Batch adsorption

In this method, the adsorbent and the solution are mixed and stirred to adsorb the solute, and then the adsorbent suspension is treated by sedimentation or filtration to separate the adsorbent (as a solid) from the liquid system.

The process followed is dependent on the particular method used to contact the adsorbent with the solution, i.e. whether it was a single adsorption process; a multiple co-current adsorption process, or a multiple counter-current adsorption process.

In the case of single adsorption process (in which contact adsorption is performed only once), assuming that the volume of the solution is V (cm^3), the initial concentration of solute is C_0 (g/cm^3 or mol/cm^3), and the amount of adsorbent is m (g), and that the solute concentration is C (g/cm^3 or mol/cm^3), the adsorption amount is q (g/g or mol/g), and $q = 0$ at $T = 0$ when the adsorption process is performed at time t. Equation 7.32 can be obtained from the mass balance.

Equation 7.32 shows the relationship between q and C until the adsorption equilibrium is reached. This expression indicates the progress of the adsorption operation and is called the expression of the operating line. Figure 7.6 shows the relationship between the adsorption isotherm, which shows the equilibrium relationship between q and C, and the adsorption operating line for single adsorption. Point A shows the maximum adsorption under these operating conditions.

$$mq = V(C_0 - C) \quad \text{—(7·32)} \tag{7.32}$$

m : Adsorbent quantity (g)
q : Solute adsorption quantity after t (g/g or mol/g)
V : Liquid volume (cm^3)
C_0 : Initial solute concentration (g/cm^3 or mol/cm^3)
C : Solute concentration after t (g/cm^3 or mol/cm^3)

Equation 7.32 The expression of operation line for adsorption.

In the case of multiple co-current adsorption process, the liquid is contacted with fresh adsorbent many times. Figure 7.7 shows a three-stage batch adsorption operaion, whilst Figure 7.8 shows the relationship between the adsorption isotherm and the adsorption operating line (the relationship between the equilibrium adsorption q_i and C_i in each of the three stages).

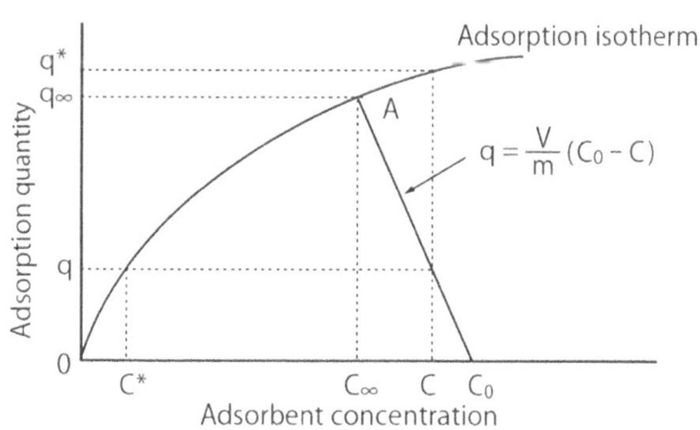

Figure 7.6 The adsorption isotherm and operating line for single adsorption process.

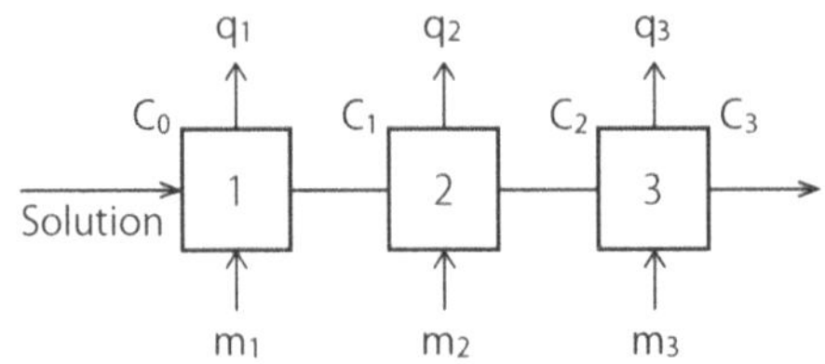

Figure 7.7 A multiple co-current adsorption process flow diagram.

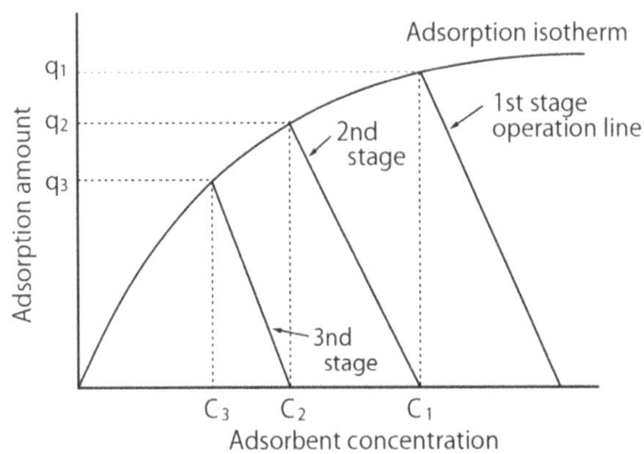

Figure 7.8 The multiple co-current adsorption isotherm and operating lines.

In the equation for the operating line in stage i, $m_i q_i = V\,(C_{i-1} - C_i)$. Therefore, the slope of the operating line for the i-th stage is $-V/m_i$, and the amount of adsorbent for each stage is determined so that the total amount of adsorbent required to obtain the desired final concentrations $\sum m_i$ is minimized in actual design.

In multiple countercurrent adsorption processes, the adsorbent suspension and the adsorbate solution are flowed from opposite directions and come into contact with each other multiple times. The flow diagram for a multiple countercurrent adsorption operation is shown in Figure 7.9, and the adsorption operating line obtained from the material balance of each stage is shown in Equation 7.33. Figure 7.10 shows the relationship between the adsorption isotherm and the adsorption operating line.

All of these values indicate the number of stages when the process is performed until the adsorption equilibrium is reached, and cannot be applied when the adsorption rate is slow. It is also conceivable to use a pseudo-isotherm when a certain adsorption time is specified as a conventional method.

$$m(q_i - q_{i+1}) = V(C_0 - C_i) \quad \text{—(7·33)} \qquad (7.33)$$

Equation 7.33 The adsorption operation line of countercurrent multiple adsorption operation.

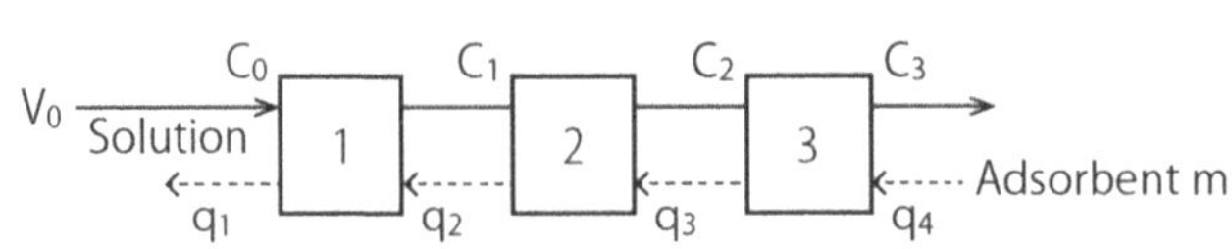

Figure 7.9 A multiple countercurrent adsorption process flow diagram.

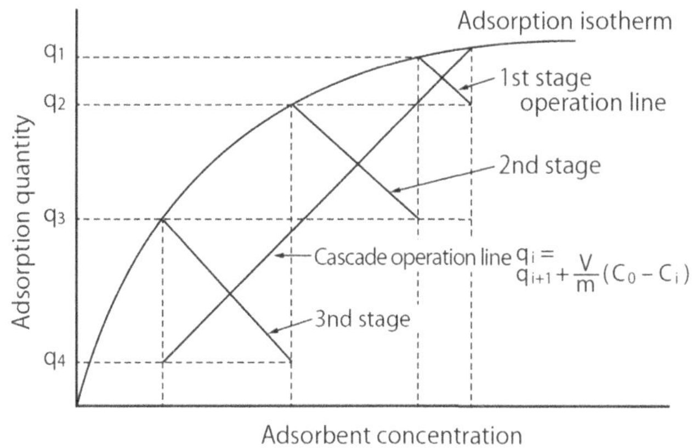

Figure 7.10 The adsorption isotherm of a multiple countercurrent adsorption process, with operating lines and cascade processes for each stage.

7.1.3.2 Fixed bed adsorption

This is an process method in which a solution containing an adsorbate is flowed through a fixed bed filled with an adsorbent, and adsorption proceeds from the upper part of the fixed bed. In fixed bed adsorption, adsorption is actually carried out in a part of the fixed bed, some of which has already reached adsorption saturation, or some of which has not yet been carried out at all. Figure 7.11 shows this relation. The adsorption process is ongoing in the area where adsorption is currently being carried out, and the steadily established length z_1-z_2 of the adsorption descends at a constant rate in the fixed bed without changing the density profile. Finally, when the adsorption zone reaches the outlet of the fixed bed, the remaining adsorbate begins to leak. This point is referred to as the breakthrough point and is shown in Figure 7.12, which also shows the changes in effluent concentrations over time. The change curve of the outflow concentration is called the breakthrough curve.

Since the shape of the adsorption zone (concentration distribution) passes through the fixed bed almost unchanged, the shape of the breakthrough curve is not dependent on the length of the adsorption zone, and

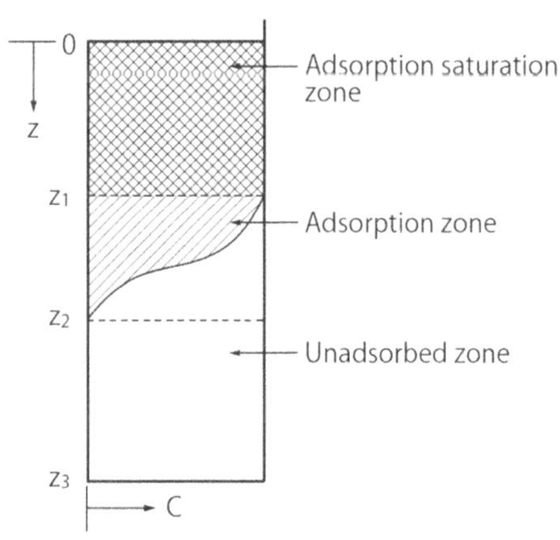

Figure 7.11 The adsorption zone of fixed bed adsorption.

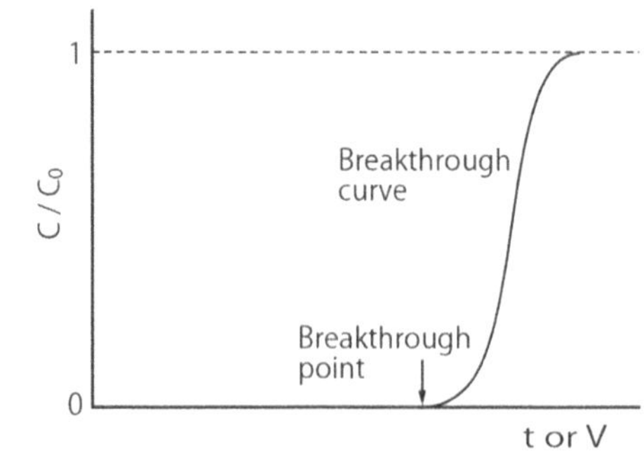

Figure 7.12 The fixed bed adsorption breakthrough curve.

the time t until the breakthrough point is reached, or the amount of extrusion V reaching the breakthrough point, increases in proportion to the length of the fixed bed. The shape of the breakthrough curve shows the concentration distribution of the adsorption zone, and the sharp rising edge of the breakthrough curve in Figure 7.12 shows that the width of the adsorption zone is narrow (the adsorption rate is fast and the distribution width is narrow). Therefore, the structure of the adsorption zone can be estimated from the shape of the adsorption breakthrough curve.

Figure 7.13 shows the results of an experiment in which a substance with a high adsorption rate, such as phenol, is treated at a void velocity of 10 cm/min. during which the usual fixed bed adsorption process is performed. This shows the generation of a steady-state adsorption zone with a width of about 10 cm, the movement of the adsorption zone to the downstream side over time, and the breakthrough curve at each end of the fixed layer over time.

Since the adsorption zone of a substance with a high adsorption rate moves toward the fixed bed terminal at a constant velocity with little change in the concentration distribution, a model in which the adsorption

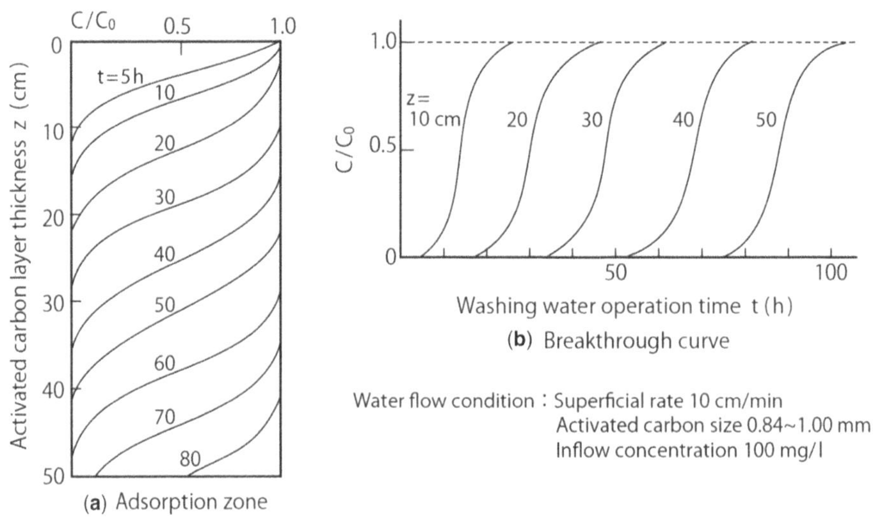

Figure 7.13 The activated carbon adsorption of a phenol solution.

zone rises at a constant velocity and the adsorption zone is stationary instead of moving downward at a constant velocity will be considered for simplification of analysis.

In this model, as shown in Figure 7.14, the values of the concentration of the adsorbent C (g/cm^3), the adsorbent adsorption rate q(g/g), the flow rate Q^* (cm^3/sec) of the fluid, and the virtual flow rate of the adsorption band Q' (g/sec) are all 0, and the adsorbent reaches an equilibrium adsorption rate q_0 at the undiluted solution concentration C at the outlet. Consider a cross-section $B(Cq_B)$ with a concentration close to 0 in the middle of the fixed bed and a cross-section E (C_E, q_E) with a concentration close to the stock solution, and let the distance between the two cross-sections be Z_a.

$$Q^* C_0 = Q' q_0 \qquad \text{—(7·34)}$$

$$Q^* (C_0 - C) = Q'(q_0 - q) \qquad \text{—(7·35)} \qquad (7.34\text{–}7.35)$$

Equation 7.34-7.35 The material balance equation of virtual adsorption zone.

Equation 7.34 is obtained from the material balance of the entire solid layer, and the material balance between any cross-section and the adsorption band inlet is obtained. Furthermore, the amount adsorbed to the micro-adsorption interval dZ in units of time can be expressed as Equation 7.36 using the aggregate mass transfer factor K_f.

From Equations 7.34 and 7.35, the operating line equation of Equation 7.37 is obtained, and the process diagram can be drawn in combination with the adsorption isotherm, which shows the

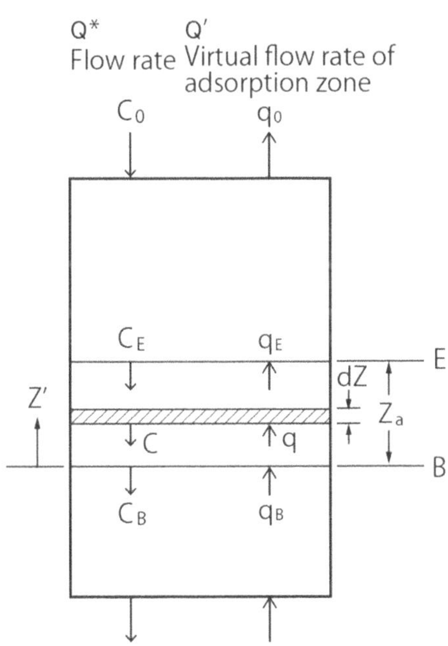

Figure 7.14 The material balance of a virtual adsorption zone.

relationship between C and q, as shown in Figure 7.15.

$$
\begin{aligned}
Q^* dC &= Q' dq \\
&= K_f a_p (C - C^*) S dZ \\
&= K_s a_p (q^* - q) S dZ \qquad \text{—(7·36)}
\end{aligned}
\tag{7.36}
$$

S : Cross section of packed bed (cm^2)

a_p : Packed particle surface area per unit packed bed volume (cm^2/cm^3)

Equation 7.36 The quantity adsorbed in unit time to micro-adsorption interval dZ.

$$
\frac{C}{C_0} = \frac{q}{q_0} \qquad \text{—(7·37)}
\tag{7.37}
$$

Equation 7.37 The operation line equation of fixed bed adsorption.

The amount of total solute adsorbed by the adsorption zone is calculated as in Equation 7.38 by integrating Equation 7.36 from the C_B to the C_E concentration range.

If the density is C at a distance Z′ from the cross-section B, the relation between Z′ and C is expressed by Equation 7.39. This equation also shows the distribution of the concentration C of the adsorbate in the adsorption zone.

In an actual fixed bed, sections B and E move downstream at a constant velocity through the fixed bed. However, because the rate of adsorption Q'/S is negligibly small with respect to the rate of fluid movement u (superficial velocity, cm/sec), the effect of adsorbent movement on Q can be neglected, and Equation 7.38 shows that for a fixed bed, $Q^*/S \doteqdot Q/S = u$ (where Q is the flow rate through the fixed bed), then

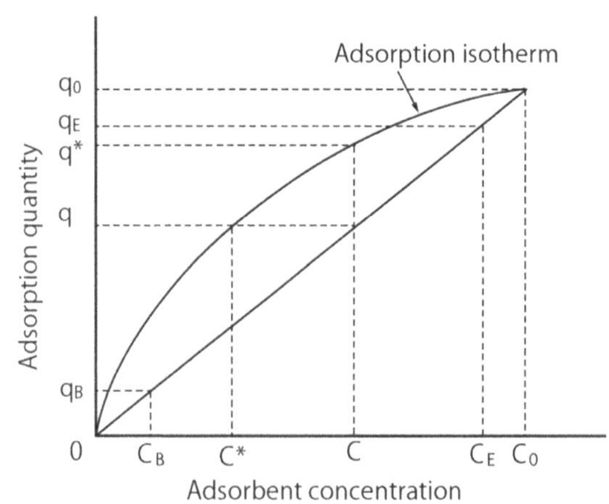

Figure 7.15 The adsorption isotherm and operating line of a fixed bed adsorption process.

Equation 7.40 is obtained.

$$\left.\begin{aligned} \int_{C_B}^{C_E} \frac{dC}{C - C^*} &= \frac{K_f a_p}{Q/S} Z_a \\ \int_{C_B}^{C_E} \frac{dq}{q^* - q} &= \frac{K_S a_p}{Q'/S} \end{aligned}\right\} \text{—(7·38)}$$

$$\int_{C_B}^{C} \frac{dC}{C - C^*} = \frac{K_f a_p}{Q/S} Z' \quad \text{—(7·39)} \qquad (7.38–7.40)$$

$$\left.\begin{aligned} \int_{C_B}^{C_E} \frac{dC}{C - C^*} &= \frac{K_f a_p}{u} Z_a \\ \int_{C_B}^{C} \frac{dC}{C - C^*} &= \frac{K_f a_p}{u} Z' \end{aligned}\right\} \text{—(7·40)}$$

Equation 7.38 – 7.40 The quantity of total solute adsorbed by adsorption zone.

The moving velocity of the adsorption zone is equal to the moving velocity v of the adsorbent, as described above, and is shown in Equation 7.41. Since the adsorption strip moves at the speed shown in Equation 4.41, the time variation of the concentration of any cross-section of the fixed layer is as shown in Equation 7.42.

$$v = \frac{Q'}{S\rho_b} = \frac{Q^*}{S\rho_b k} = \frac{u'}{\rho_b k} \fallingdotseq \frac{u}{\rho_b k} \quad \text{—(7·41)} \qquad (7.41)$$

k : Henry's adsorption coefficient ($k = q_0/C_0$)
u' : Liquid velocity in the model of Fig.7·14
(In the case of the fixed bed $u' \fallingdotseq u$)

Equation 7.41 The moving velocity of adsorption zone.

$$\int_{C_B}^{C} \frac{dC}{C - C^*} = \frac{K_f a_p}{u} Z' = \frac{K_f a_p C_0}{\rho_b} \frac{Z'}{v} = \frac{K_f a_p C_0}{\rho_b q_0}(t - t_B) \quad \text{—(7·42)} \qquad (7.42)$$

Equation 7.42 The time variation of the concentration of any cross section of fixed layer.

Since the adsorbed mass distribution in the fixed bed is symmetrical in the upstream and downstream directions centering on the point of $C/C_0 = q/q_0 = 1/2$, the total length of the fixed bed is Z (cm), and the time t_{50} (sec) at which the outlet concentration becomes $C/C_0 = 1/2$ is calculated by substituting $C = C_0/2$ and $t = Z/v$ in Equation 7.42. The time t_B at which the outlet concentration of the fixed bed reaches C_B is as shown in Equation 7.43, and the time until the considered concentration C_B

breaks down can be calculated.

$$t_B = \frac{\rho_b q_0}{u C_0} Z \left(1 - \frac{u}{K_f a_p Z} \int_{C_B}^{\frac{C_0}{2}} \frac{dC}{C - C^*}\right) \quad \text{—(7·43)} \qquad (7.43)$$

Equation 7.43 The time until the considered concentration C_B breakthrough.

The above equations are integration equations including the integration of $dC/(C - C^*)$. Since adsorption isotherms are complex (except for the Henry type), it is difficult to obtain an analytical solution. Usually, using the operating line and the adsorption isotherm as shown in Figure 7.16, the solution can be obtained by the graphical integration of $1/(C - C^*)$ and C_B-C-q_0 in two-dimensional planes on two axes.

However, in the fixed bed adsorption treatment of components such as organic substances with slower adsorption rates (close to the rate-limiting of intragranular diffusion), including natural organic coloring components breakthrough starts from the lower part of the fixed bed before the typical adsorption zone is completed, and the S-shaped breakthrough curve as described above is not shown. Most of the natural organic components in water treatment have a typical breakthrough curve.

Figure 7.17 shows the concentration distribution in the adsorption layer of the surfactant dodecyl benzene sulfonate (DBS) and its breakthrough curve, which is quite different from that of phenol with a high adsorption rate (Figure 7.13). Before a steady-state adsorption band is formed, breakthrough begins, and the adsorption band cannot be assumed from the breakthrough curve.

Theoretical analysis is difficult because the shape of the breakthrough is largely influenced by two factors: the longitudinal dispersion coefficient and the adsorption rate in the fixed bed. Experimental considerations are essential prior to design. In addition, unlike adsorbates in which the adsorption zone is

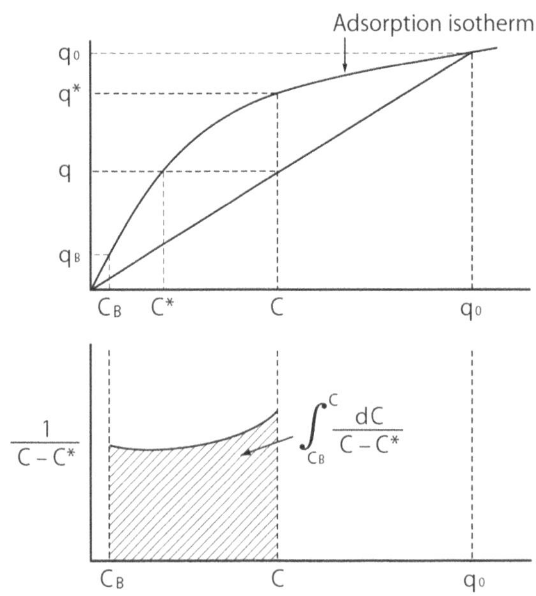

Figure 7.16 Solution of Equation 7.43 by graphical integration.

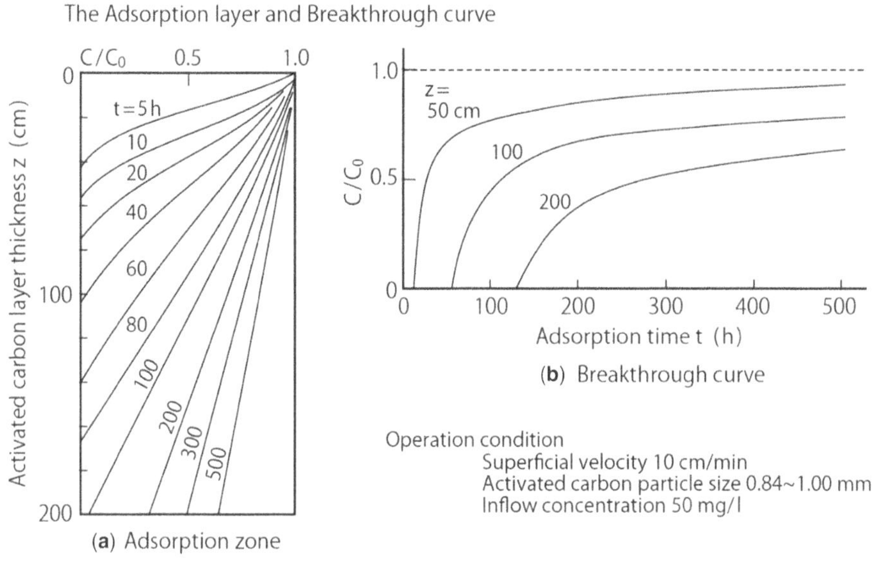

Figure 7.17 Concentration distribution in adsorption layer of DBS solution.

constructed quickly, adsorbates with a slow adsorption rate, in which the long adsorption zone extends more and more loosely must be broken through at an early stage and an adsorption packed column with a deep layer thickness Z is needed to have efficient results. If the adsorption rate is as high as that of phenol, a layer thickness of 100 cm or less is sufficient, but a deep activated carbon filter layer of 400 cm or more is used worldwide to remove components such as humic substances, which are trihalomethane precursors.

7.1.3.3 Fluidized bed adsorption

Fixed bed adsorption is a process in which raw water is passed through an activated carbon packed bed by gravity or pressure in a downward flow. It is widely used because of the advantages of being able to use all types of activated carbon and the ease of the system structure and process management. However, retention of inorganic suspensions in the filter layer and growth of microorganisms can easily lead to blockage of the filter layer, requiring periodic surface cleaning and backwash. In addition, it is necessary to use activated carbon with a relatively large particle size, in consideration of an increase in water flow resistance.

The fluidized bed adsorption process, on the other hand, requires the use of wear-resistant activated carbon and precise flow control to maintain a steady state of flow. However, it has the advantage of using small-particle-size activated carbon with a high adsorption rate, without causing filter bed blockage. Also, there is a mixture of adsorbents in the longitudinal direction which are not in the fixed layer, and the flow rate of the adsorbate increases as a result.

When applying this method to water treatment processes, it is necessary to decide which process to choose, based on sound criteria. As in the case of the fixed bed, we investigated the characteristics of two adsorbates – phenol, with a fast adsorption rate, and DBS, with a slow adsorption rate – by referring to the results of model tests and numerical experiments.

For the inner layer mixing condition, calculations were performed for the full-layer mixing and n-stage mixing model using the application of Equations 7.21–7.23 (above).

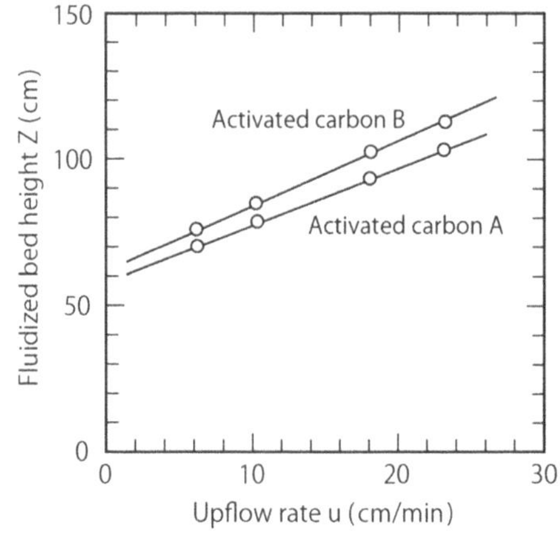

Figure 7.18 Fluidized bed height and upflow rate.

Estimates were made of the longitudinal dispersion factor D_b in the fixed bed, and the Peclet number on a particle size basis as a dimensionless number, indicating the relative magnitude of dispersion $P_{ep} = d_p u/\varepsilon_b D_b$. In a fixed bed, the range is 0.1 to 1.0, but in a fluidized bed, the value is about 0.01 to 0.03, which is two orders of magnitude smaller, and the influence of fluid mixing becomes large. As shown in Figure 7.18, an activated carbon bed with a height of about 60 cm in the fixed bed was fluidized at various upward flow rates (expansion coefficient and fluidized bed thickness), and the test was conducted using adsorbates with two different characteristics: phenol (with a high adsorption rate: 100 mg/l) and DBS (with a low adsorption rate: 50 mg/l).

Figure 7.19 shows the situation in which phenol breaks through with elapsed adsorption time t, and the solid line shows the calculation results when the activated carbon layers are completely mixed. Figure 7.20 shows the breakthrough curve as a function of the treated water volume on the horizontal axis, and shows a group of experimental results that almost coincide around the calculated values expressed in the activated carbon complete mixing model. This was considered to indicate that the carbon in the fluidized bed was

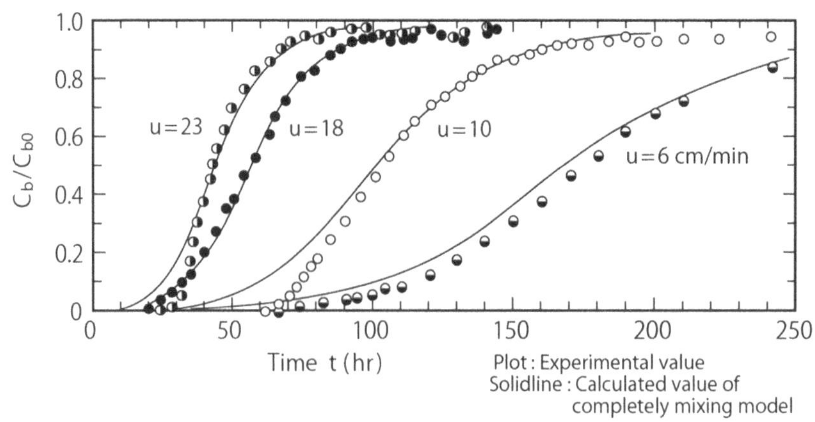

Figure 7.19 The breakthrough curve of the fluidized bed adsorption of phenol.

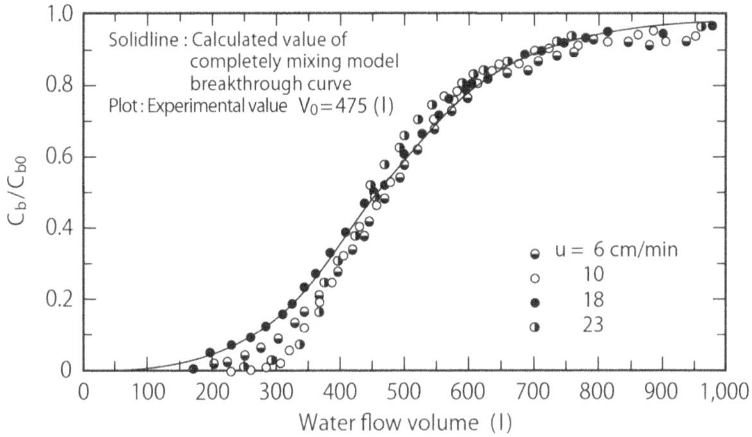

Figure 7.20 The breakthrough curve, as a function of the treated water volume, of the fluidized bed adsorption of phenol.

almost completely mixed, and the solute passed upward through the activated carbon in the fluidized bed. Figure 7.21 compares the results of calculation with experimental values assuming a fluidized bed with n complete mixing sections of activated carbon particles in series. For low ascending flow rates, mixing conditions were expected to be intermediate between the two mixing beds, rather than complete mixing. However, even if the activated carbon particle mixing was not complete, it can be seen that it was close to complete mixing in most parts of the entire layer.

Figure 7.22 shows the experimental results of fluidized bed adsorption of DBS together with the results of calculations assuming the complete mixing state of the activated carbon particles, which agree well with each other. The results of various calculations showed that the values assumed in the fixed bed adsorption study, such as the initial value $D_0 = 3 \times 10^{-11}$ cm^2/sec and the final value $D_f = 4 \times 10^{-12}$ cm^2/sec, of the

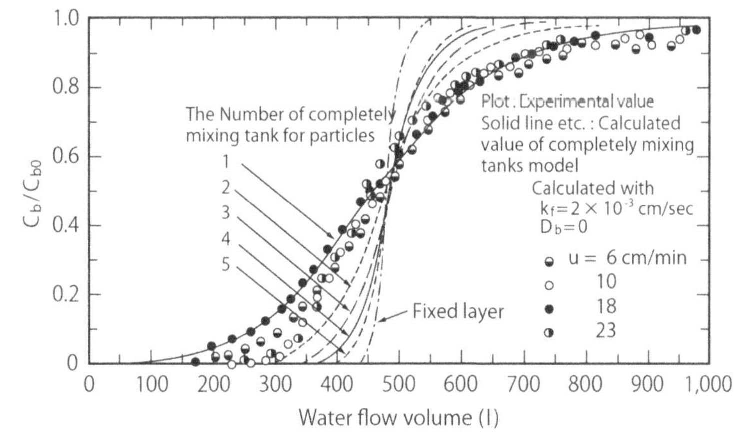

Figure 7.21 The breakthrough curve of the fluidized bed adsorption of phenol, showing the results of both calculations and experimental values.

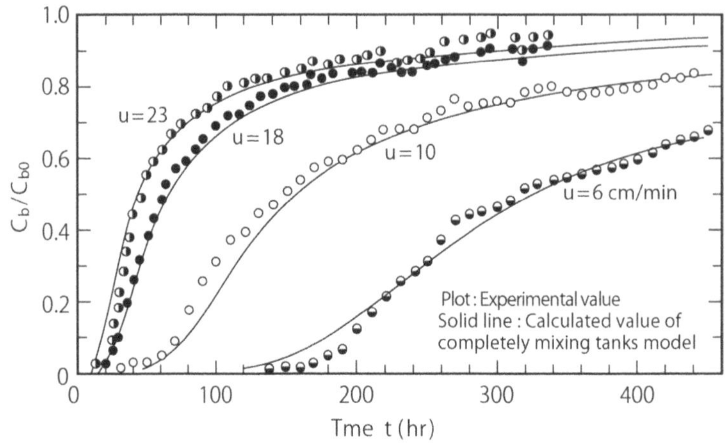

Figure 7.22 The breakthrough curve of the fluidized bed adsorption of DBS showing the results of both calculations and experimental values.

intragranular surface diffusion coefficient of DBS, were similarly established in the fluidized bed adsorption study.

For activated carbon adsorption, as shown in Figure 7.23, the adsorption process is stopped at C_{Bt} when the outlet concentration C_{bZ} reaches a certain set concentration C_B, and the activated carbon is replaced. The amount Q_B adsorbed on the activated carbon by time t_B is smaller than the equilibrium adsorption amount Q_0, and the degree of reduction varies greatly depending on the adsorption rate and operating condition. Therefore, the ratio of the adsorption amount Q_B reached by actual operations to the equilibrium adsorption amount Q_0 with respect to the C_0 is the activated carbon utilization efficiency: $\eta = Q_B/Q_0 \fallingdotseq t_B/t_0 = V_b/V_0$.

The efficiency of the fixed bed adsorption operate and the fluidized bed adsorption process were compared with each other. Figure 7.24 shows the results of calculating the efficiency of activated carbon utilization when the breakthrough concentration is set to $C_B/C_{b0} = 0.1$. Phenol exhibits a high efficiency

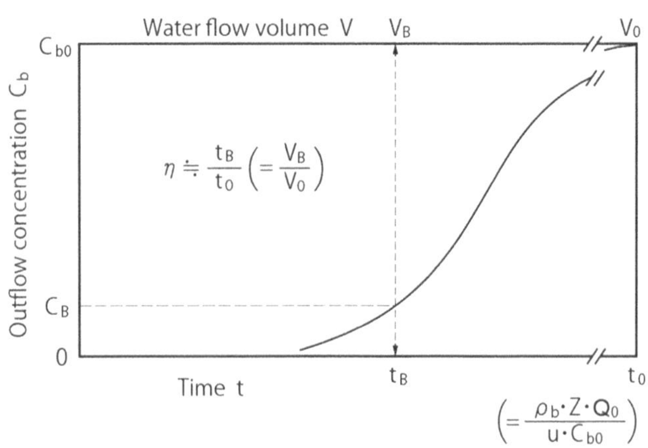

Figure 7.23 The breakthrough curve and activated carbon utilization efficiency.

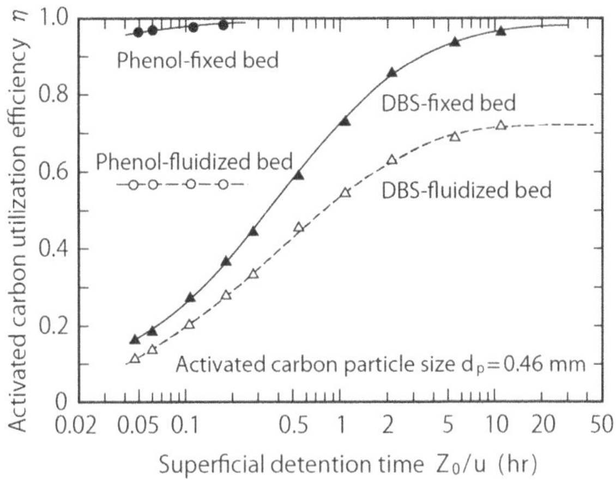

Figure 7.24 Comparison of activated carbon utilization efficiency of a fixed bed and a fluidized bed.

of more than 95% for fixed bed adsorption, but drops to 56% for fluidized bed adsorption. In order to improve the efficiency, it is necessary to reduce the particle size d_p of activated carbon and to increase the retention time Z/u in the bed. While the fluidized bed is operating stably, the operation that satisfies these two conditions cannot be easily achieved, but the process of powdered activated carbon with a functional separation membrane (described below in Section 7.1.5.3) can be performed. This is an activated carbon adsorption process that combines advanced separation methods.

Comparing the fixed bed with the fluidized bed in the case of DBS, which has a slow adsorption rate, the efficiency reduction of the fluidized bed is not as large as that for phenol (which is as much as 40%), but remains small (at 5–10%). However, even if the reaction time is extremely long, the efficiency of the fluidized bed is not more than 72%.

7.1.3.4 Fixed bed adsorption process with repeated backwashing

So far, we have discussed the characteristics of continuous activated carbon adsorption in a fixed bed and in a fluidized bed, and seen that adsorption efficiency of a fixed bed is generally higher than that of a fluidized bed; and, in order to improve efficiency by increasing the throughput, the use of activated carbon with a small particle size in a fluidized bed enables the same measures as in a fixed bed.

Fixed layers are often preferred because of the ease of reactivation of activated carbon and flow control during operation. Fixed layers are often used in biological carbon processes (described below) in order to stably retain attached microorganisms. In such cases, the actual process of fixed bed adsorption requires regular backwashing to recover the filter bed due to an increase in head loss due to the clogging caused by the retention of suspended inflow substances in the upper layer, and the generation of microbial communities in the bed. Therefore, in order to evaluate the actual process of the fixed bed, it is necessary to analyze and evaluate the fixed bed process in consideration of the repeated mixing of the activated carbon filter layer by repetitive backwashing.

Below, we examine the progress of adsorption in repeated washing and mixing of activated carbon filter beds using phenol (fast adsorption rate) and DBS (slow adsorption rate), which have significantly different rates of intragranular diffusion, combining experiments and numerical calculations. For more information on theory and experimentation, refer to Yuasa and Tambo (1959).

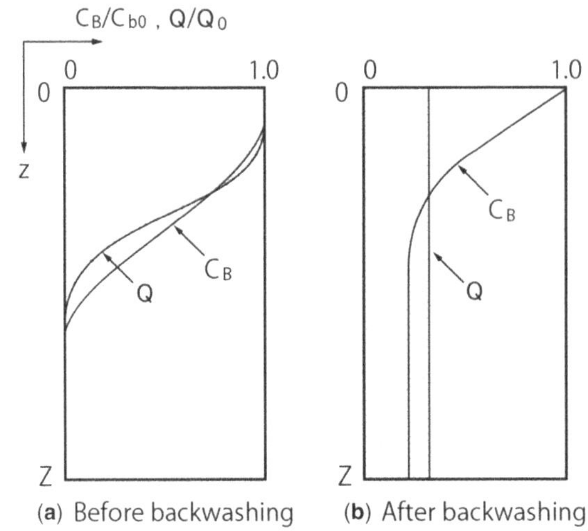

Figure 7.25 The solute concentration distribution and adsorption quantity distribution in fixed bed, before and after backwashing.

In an adsorption-fixed bed prior to backwashing, there are distributions of adsorption quantity Q and external solution concentration C_B, as shown in Figure 7.25(a). The adsorption amount distribution q(r) inside the activated carbon particles has various distributions, as shown in Figure 7.26, depending on the position of the fixed bed. Different activated carbons with different adsorption histories are almost completely mixed by backwash (i.e., in the fluidized bed). The adsorption distribution equation is shown in Equation 7.44 for a mixture of heterogeneous activated carbon (from different sources) after backwashing, as a hypothetical type of activated carbon, as shown in Figure 7.27. Also, the total adsorption amount Q^0 of the virtual activated carbon particles is shown as the mean of the adsorption amount Q(Z) immediately before backwashing, in Equation 7.45. The virtual solution concentration C_s^0 on the particle surface immediately after backwashing can be considered to be in equilibrium with the virtual adsorption amount q_s^0 on the particle surface, and thus their relationship can be established in Equation 7.46. Accordingly, the adsorption rate immediately after the backwash can be expressed

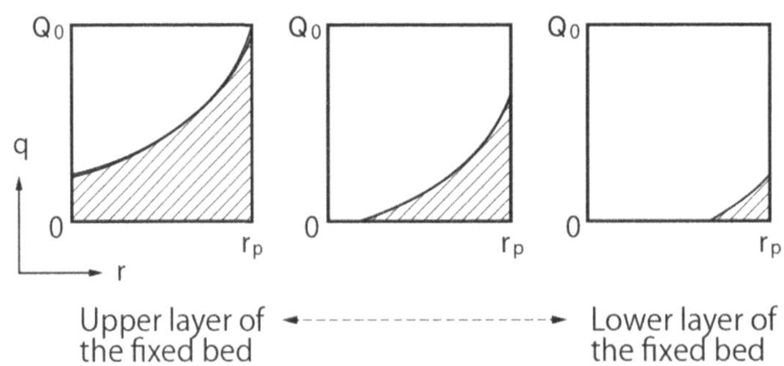

Figure 7.26 The adsorption quantity distribution inside activated carbon particles with different adsorption histories in a fixed bed, just before backwashing.

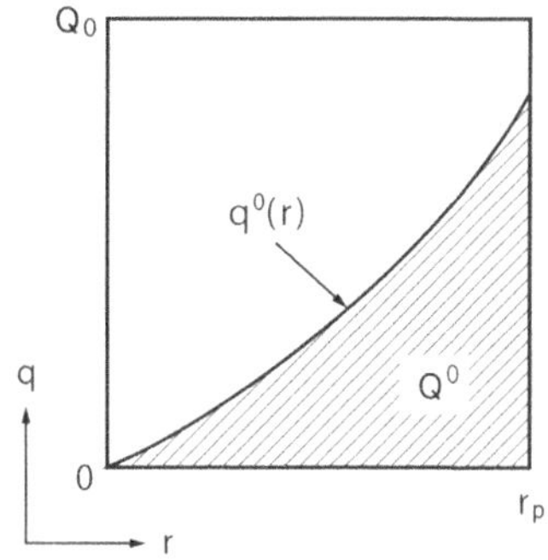

Figure 7.27 The hypothetical adsorption quantity distribution inside activated carbon particles, immediately after backwashing.

as in Equation 7.47.

Adsorption distribution of virtual activated carbon

$$q^0(r) = \frac{1}{Z}\int_0^Z q(r,z)\,dz \qquad 0 \leqq r \leqq r_p \qquad \text{—(7·44)}$$

Total adsorption quantity of virtual activated carbon

$$Q^0 = \frac{3}{rp^3}\int_0^{r_p} q^0(r)\,r^2\,dr = \frac{1}{Z}\int_0^Z Q(z)\,dz \qquad \text{—(7·45)}$$

(7.44–7.45)

Equation 7.44-7.45 The adsorption distribution equation and the total adsorption quantity of different career activated carbon.

Adsorption isotherm

$$C_s^0 = F(q_s^0) \qquad \text{here, } q_s^0 = q^0(r_p) \qquad \text{—(7·46)}$$

Adsorption rate immediately after backwashing

$$\rho_b \frac{\partial Q^{\upsilon}}{\partial t} = k_f a_v (C_b - C_s^0) \qquad \text{—(7·47)}$$

(7.46–7.47)

Equation 7.46–7.47 The Adsorption isotherm and adsorption rate of the virtual activated carbon

Figure 7.25(b) shows the distribution of the adsorbent concentration C_b and adsorption amount Q immediately after resuming water flow after backwashing, and the external solution concentration C_B asymptotes to the virtual particle surface concentration C_s^0 as it approaches the fixed layer outlet.

The changes over time in the distribution of the amount of adsorption in the activated carbon particles after the resumption of water flow show various patterns, as shown in Figure 7.28, depending on the position in the fixed bed.

Within the adsorption zone ($C_s < AC_b \ll C_{b0}$: where C_{b0} is the influent concentration), the concentration C_b increases as the adsorption progresses, and the intragranular adsorption distribution changes over time, as

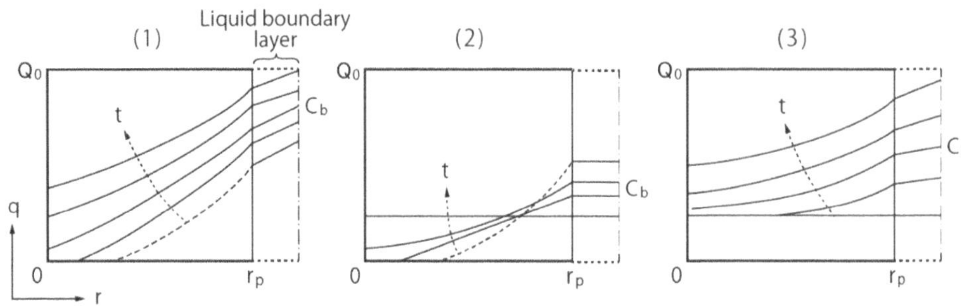

Figure 7.28 Changes over time in the distribution of the quantity of adsorption in activated carbon particles in a fixed bed after backwashing.

shown in Figure 7.28(1). In the area below the adsorption zone ($C_s \doteqdot C_b$), the adsorbate is hardly taken up. However, as the intragranular diffusion progresses, the intragranular adsorption distribution is flattened as shown in Figure 7.28(2). When the inside distribution reaches a flat state, the fixed bed outlet concentration C_{bZ} changes temporarily at a constant value. However, when the tip of the adsorption zone reaches the outlet of the fixed bed, the effluent density C_{bZ} increases, and the intragranular adsorption volume distribution changes as shown in Figure 7.28(3). In this case, changes in the concentration distribution of the solution in the fixed layer over time are shown in Figure 7.29.

Figure 7.30 shows the general shape of the breakthrough curve for fixed bed adsorption after backwashing. The durations of the sections a through b, b through c, and c through d in the figure are greatly influenced by the load of solute inflow into the fixed layer, the fixed layer length, the diffusion

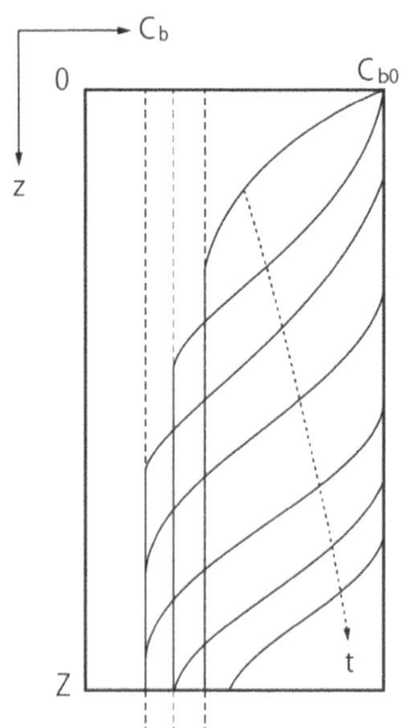

Figure 7.29 Changes over time of solution concentration distribution in a fixed bed after backwashing.

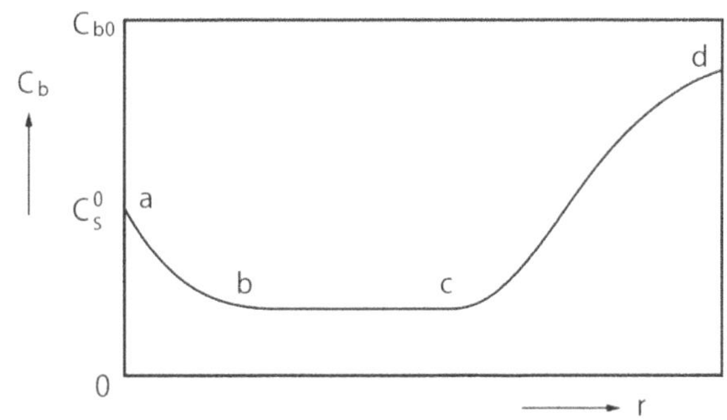

Figure 7.30 The general shape of the breakthrough curve for fixed bed adsorption after backwashing.

rate in the granules, and the backwash cycle. As a result, a broken curve is drawn with various repetitions, as shown in Figure 7.31.

The following presents the results of a model test using phenol and DBS, with high and very low adsorption rates, respectively.

Figure 7.32 shows the results of phenol-adsorption experiments, which show the flow in $C_{b0} = 100$ mg/l activated carbon particle size $d_p = 0.42$ to 0.50 mm, and fixed bed (non-cleaning) and fixed bed backwash operations using columns with a bed length of 10 to 60 cm. The solid line shows numerical simulation results, and a calculation method for accurately expressing the experiment was established. Figure 7.33 shows how the breakthrough curve changes depending on the frequency of backwash during phenol-fixed bed adsorption. As the frequency of backwash increases (repeating for a short period of time), the breakthrough curve approaches the breakthrough curve of the fluidized bed indefinitely. It is important to note that the breakthrough curve is quite different from that of a fixed layer without backwashing. Fluidized bed adsorption has the advantage of reducing the particle size of activated carbon, increasing the adsorption rate, and increasing the throughput without increasing the head loss. This means that the fluidized bed adsorption process can be as effective as the fixed bed adsorption process in which back washing is repeated.

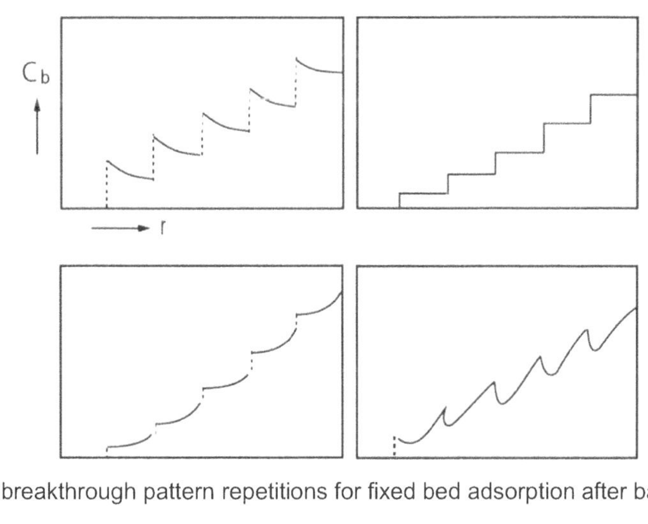

Figure 7.31 Various breakthrough pattern repetitions for fixed bed adsorption after backwashing.

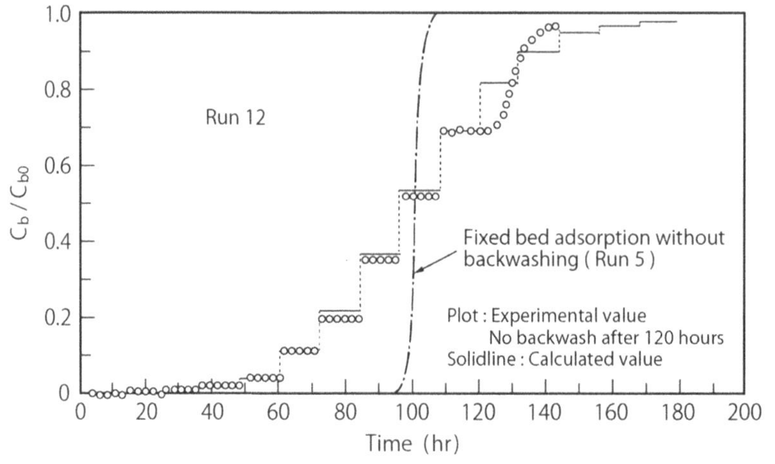

Figure 7.32 The phenol breakthrough curve of fixed bed backwashing repetition.

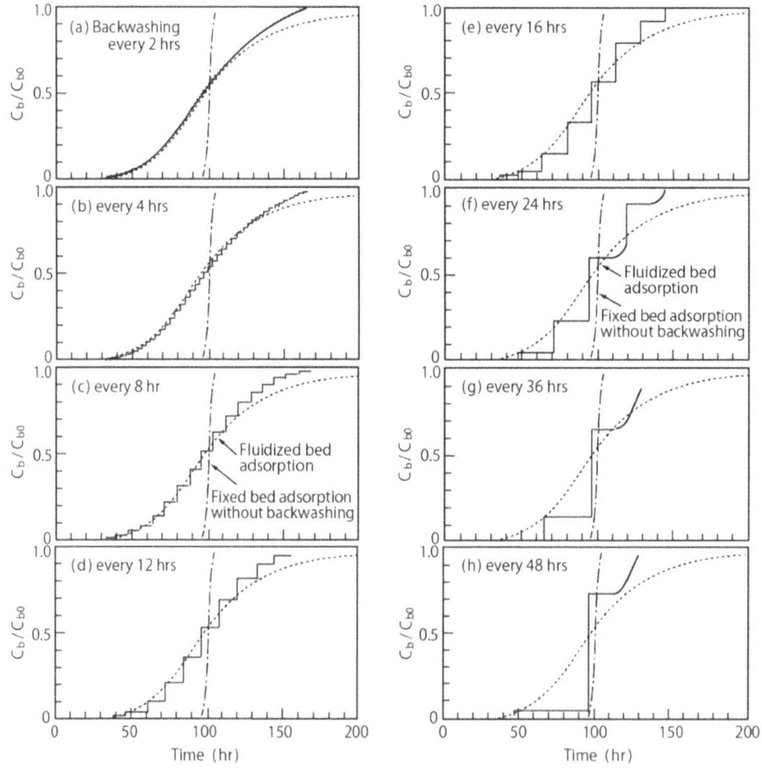

Figure 7.33 Breakthrough curve changes depending on backwashing frequency of fixed bed adsorption of phenol (results of calculations).

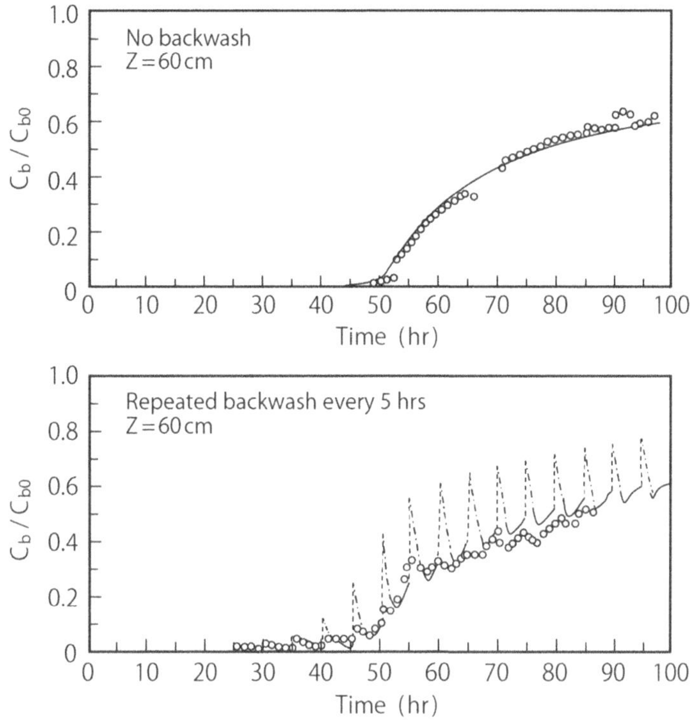

Figure 7.34 Fixed bed adsorption breakthrough curve for DBS (for no backwash and backwash every 5 hours).

Next, the case of DBS, which has a slow adsorption rate, was considered, in which the inlet concentration C = 50 mg/l, adsorbent bed thickness Z = 60 cm, and superficial flow rate u = 10 cm/min. Figure 7.34 shows the results of experiments and calculation simulations with no backwash and repeated backwash every 5 hours. The results of the calculations show that the high concentration peaks immediately after backwashing are exaggerated, but the experimental values do not appear to fluctuate significantly.

Figure 7.35 shows the results of similar simulations for fixed bed (non-backwash), fixed bed (backwash), and fluidized bed adsorption. For DBS with slow adsorption rates, the breakthrough curves are almost

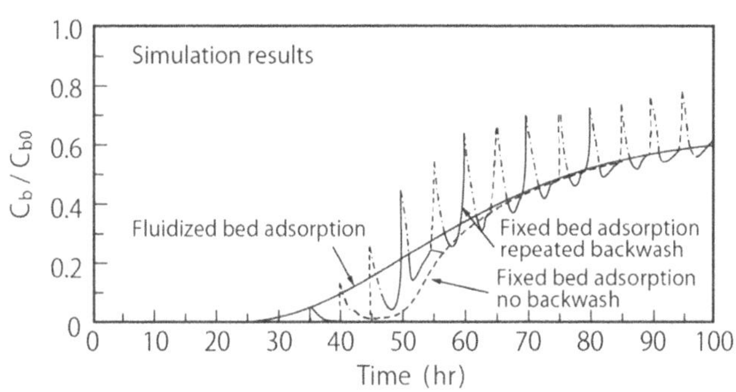

Figure 7.35 Comparison of breakthrough curves for fixed bed and fluidized bed adsorption of DBS.

similar to those for fluidized bed adsorption. In detail, the breakthrough of fixed bed adsorption, which repeats backwashing, is intermediate between fluidized bed adsorption and adsorption of a fixed bed without washing. By increasing the frequency of backwashing, the adsorption efficiency approaches that of a fluidized bed.

7.1.4 Regeneration of activated carbon

7.1.4.1 Recycling of used activated carbon

When powdered activated carbon is used, the used activated carbon is generally disposed of as sludge. On the other hand, in the case of fixed bed adsorption or fluidized bed adsorption process using granular carbon, it is usual to recycle the activated carbon, except in small-scale plants. In addition, even in the case of small-scale plants, the used activated carbon can be collected and sent to regeneration plants for economic reuse.

Activated carbon can be reused without solid waste by regeneration, and the process can be made sustainable by environmental measures such as exhaust gas treatment at the time of regeneration. For reasonable system and operating costs for regeneration equipment, activated carbon consumption is required to be several tons per day at a minimum.

The adsorption process with such a regeneration cycle causes a loss of about 10–40% in the transport process and regeneration process every time regeneration is performed, and therefore it is always necessary to replenish with fresh activated carbon. The main causes of this loss are losses by pulverization and heat treatment, which mechanically crushes the activated carbon during regeneration. The regenerated activated carbon recovers almost 100% adsorption capacity when the adsorbate is mainly organic matter. However, if the adsorbate contains a large amount of inorganic matter or the regeneration process is inappropriate, recovery is only about 50%, making it difficult for the regenerated carbon to operate continuously.

7.1.4.2 Recycling process

Activated carbon is produced using a variety – in fact, almost any type – of carbon-based materials. It is found in a variety of organic matter such as wood, bone, palm nuts, petroleum, coal, blood, peat, and lignin wastewater. Carbon-based raw materials are selected and carbonized at about 600°C or less in an air-free (anaerobic) state, and activated at 800°C to 900°C by the action of an appropriate oxidizing agent (water vapor + air, carbon dioxide gas, etc.). In addition, dehydrated salts (zinc oxide, calcium chloride, magnesium chloride, etc.) are added, and 1 to 6 times the amount of water to the carbonaceous material is added to form a wet state, and carbonization is performed at 400 to 900°C.

As described above, the method of reactivation is not uniform because the methods used and raw materials of activated carbon vary, and the substances adsorbed are different. Commonly used reactivation (regeneration) methods are a combination of four elements: washing with solvents; washing with acid or alkaline agents; using steam; and applying high temperatures.

In the case of adsorbates soluble in solvents, acids, and alkalis, regeneration proceeds by elution, and in the case of volatility, regeneration proceeds by passage of hot water vapor. However, in many water treatment processes, many types of organic substances are treated as adsorbates. Therefore, in these cases, it is necessary to use heat treatment.

For thermal regeneration, the following steps are performed: drying of the activated carbon after use at temperatures of 100°C; an adsorbate broth, at 800°C or less; oxidation of the carbon component of the distributed adsorbate, at 800°C or more. An example of a multi-stage countercurrent regeneration furnace used for this purpose is shown in Figure 7.36.

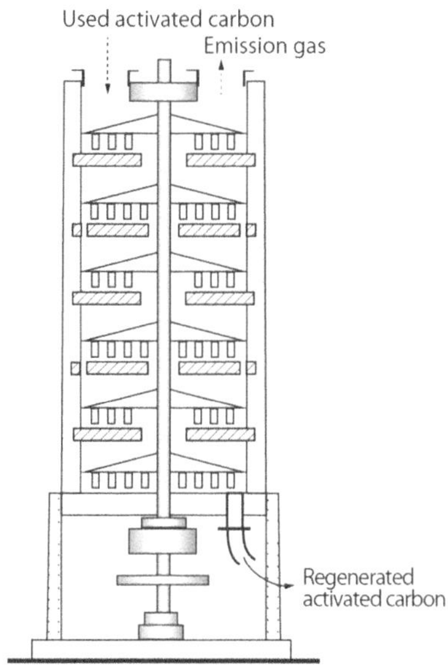

Figure 7.36 An example of a multi-stage countercurrent regeneration furnace.

7.1.5 Humic substances (NOM) and activated carbon adsorption

In Chapter 3 in Section 3.5, on the water quality conversion matrix, it was stated that the main components of the water in the metabolic pathways of natural water and urban water above colloidal size were solids-liquids separated by coagulation + solid liquid separation processes, and the remaining components of the water were colloidal natural humic substances and/or humics generated by microbial treatment. The main treatment method for low molecular weight components is activated carbon adsorption (see Section 3.5). Of course, the main components requiring activated carbon treatment are phenol, DBS, pesticides, insecticides, and other chemical substances mentioned in Section 3.5.

Since Rachel Carson raised issues in her book, *Silent Spring* (1962), the USEPA has investigated the adsorption properties of hundreds of trace organic chemicals and built a vast database of adsorption isotherms and adsorption kinetics. However, trace chemical substances such as pesticides usually coexist under natural conditions in which humic substances are always present as dominant components, and it is not possible to discuss the activated carbon adsorption of trace chemical substances without considering their interactions. We now introduce our work on humic adsorption and competitive adsorption of humics and other trace chemicals that are not addressed in detail in general chemical engineering processes.

7.1.5.1 Adsorption of activated carbon with humic substances and waste components of microbial metabolism

These natural organic components are precursors to organo-chlorine disinfection by-products such as trihalomethanes. Here, we describe the adsorption of humic substances on activated carbon, because humic substances are considered to be the second major organic component in natural and urban water metabolism systems.

In the 1970s, Dr Rook of the Rotterdam Water Supply discovered the existence of chloroform in the river Rhine water, and found that chlorination of tap water resulted in higher concentrations of chlorine by-products such as chloroform and other trihalomethanes (as described in Chapter 2). In this section, we are spreading the use of advanced water purification processes that use activated carbon adsorption as the main process. In Japan, along with the problem of eliminating offensive odors and unusual tastes, the introduction of advanced treatment of tap water using activated carbon has been promoted. The principal quantitative components of biologically persistent organic matter contained in natural water and in discharged water that has undergone biodegradation at sewage treatment plants are humic substances, which are precursors that generate total organic chlorine compounds (TOXs) and trihalomethanes (THMs) by adding chlorine. This is an organic carbon component (as shown in Figure 3.30, in Chapter 3) that is sensitive to ultraviolet absorbance (E260), characterized as TOC ($mg/l/E260_{(1cm\text{-}cell)} = 40$ to 50), and exhibits an infrared absorption spectrum (as shown in Figure 3.35). On the other hand, pesticides and other petrochemicals discharged into the environment account for the vast majority of environmental water (as shown in Figure 3.31). These substances are the most problematic in terms of environmental hygiene safety due to their accumulation in humans and ecosystems. The gel chromatography (Figure 3.30) from Tambo and Kamei (1978a,b,c) for evaluating water quality matrices (Figures 3.32 and 3.33) shows a component adsorbed on Sephadex G15 gel and appearing in the ammonia-extruded Group 5, which is evaluated as an individual component by gas chromatography and mass spectrometry (GC-MS).

In the municipal sewage treatment process, polysaccharides emerge as intermediates in the metabolism of substrates generated during the biodegradation process that coexist with some 50 or so of the above-described humic $TOC/E260_{(1cm\text{-}cell)}$. In the sewage Sephadex G15 gel chromatogram illustrated in Figure 3.30, the substance is composed mainly of arabinose, which is E260 insensitive and appears in Group 1. It is known that the substance is produced in greater quantities when the substrate is nearly completely ingested (see Figure 2.31, Chapter 2) for the behavior of aerobic biodegradable organic materials). Arabinose is a component related to the autoxidation process of microorganisms and is considered to be a general component produced in the metabolic process of microorganisms. It is an ingredient that is not mass-compatible with activated carbon adsorption and has a low flocculation removal efficiency because it is hydrophilic. It is considered to be a TOC ingredient that is often leaked at about 1 mg/l by advanced treatment or recycling of sewage through flocculation and activated carbon adsorption. Therefore, we now examine the adsorption reaction under conditions in which humic substances and polysaccharides coexist.

The final sedimentation tank sludge from an activated sludge sewage treatment plant was collected in a 20-liter polyethylene tank, and long-term aeration was continued for 16 days. As shown in Figure 7.37, the $E260_{(1cm\text{-}cell)}$ value increased rapidly with elapsed time, but the TOC/E260 ratio remained relatively high, around 200. As a final waste component of metabolism in the liquid system, E260 sensitive organic matter (humic substances) increased considerably, resulting in a TOC/E260 ratio of only about 200. The E260 sensitive TOC (polymeric humic substances) present in Group 1 was removed by coagulation, using a 2-fold diluted sample on the 16th day at pH5. 5 and an aluminium injection rate of 40 mg/l. After the coagulation treatment, the water (raw water for adsorption test) had a composition as shown by the gel chromatogram in Figure 7.38, and can be classified as E260 insensitive high-molecular components (polysaccharide substances) in Group 1, and a large E260 sensitive TOC in Group 2.

Simultaneous parallel activated carbon adsorption tests were conducted using 100 conical flasks of 300 ml with test water at 20°C for 100 to 500 hours in a batch-type process, using a constant-temperature shaking incubator. After 33 days of adsorption, the G15 chromatogram of the liquid system reached the equilibrium adsorption states, as shown in Figure 7.39. What is shown to be present is polymer Group 1 E260 insensitive matter such as highly hydrophilic polysaccharides that cannot be sufficiently removed by coagulation treatment. Peaty groundwater contains almost no such ingredients, which are difficult to remove during advanced sewage treatment and recycling. The small residual macromolecular components removed by

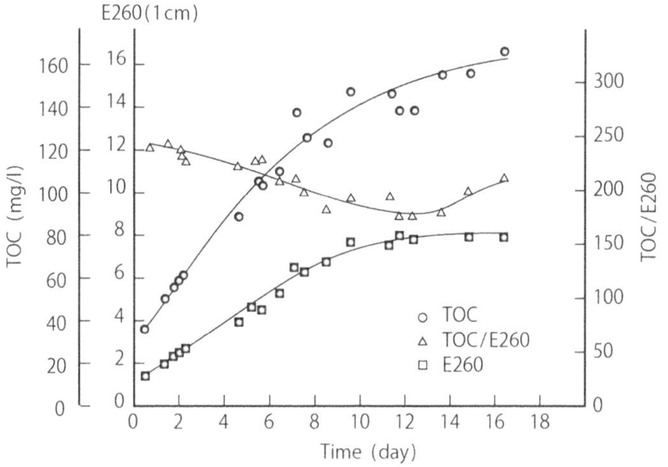

Figure 7.37 Changes in TOC, E260 and TOC/E260 under aerobic degradation of the final sedimentation tank sludge.

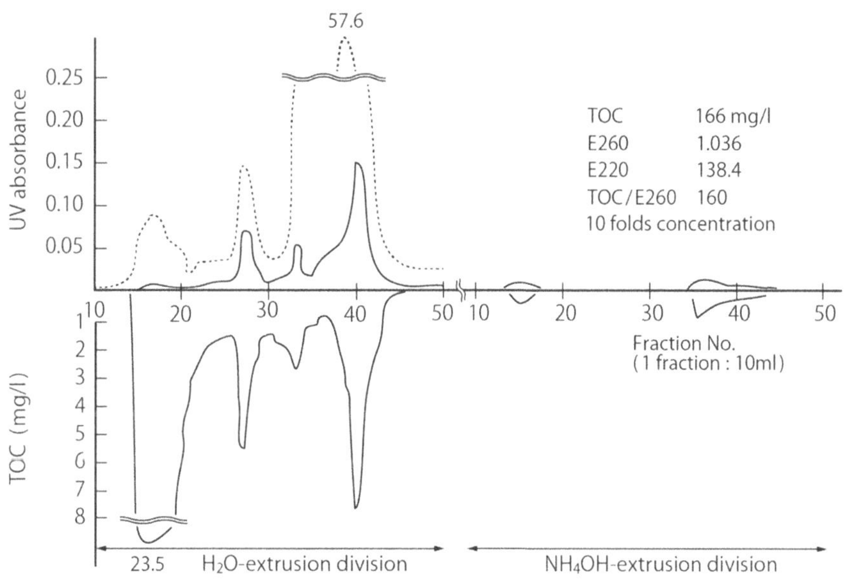

Figure 7.38 A G15 gel chromatogram of raw water from an adsorption test, after coagulation, from an aerobic degradation solution from final sedimentation tank sludge.

coagulation in Group 1 also suggest that they were already removed by adsorption to a certain extent, as explained by Traube's law.

Figures 7.40 and 7.41 show the results of the adsorption isotherms of activated sludge treated sewage water (as shown in Figure 7.37), and the Langmuir constant K in Equation 7.5 shows similar values of 8.70 and 10.9 $E260_{(5cm\text{-}cell)}$/g activated carbon values, respectively. However, the maximal adsorption

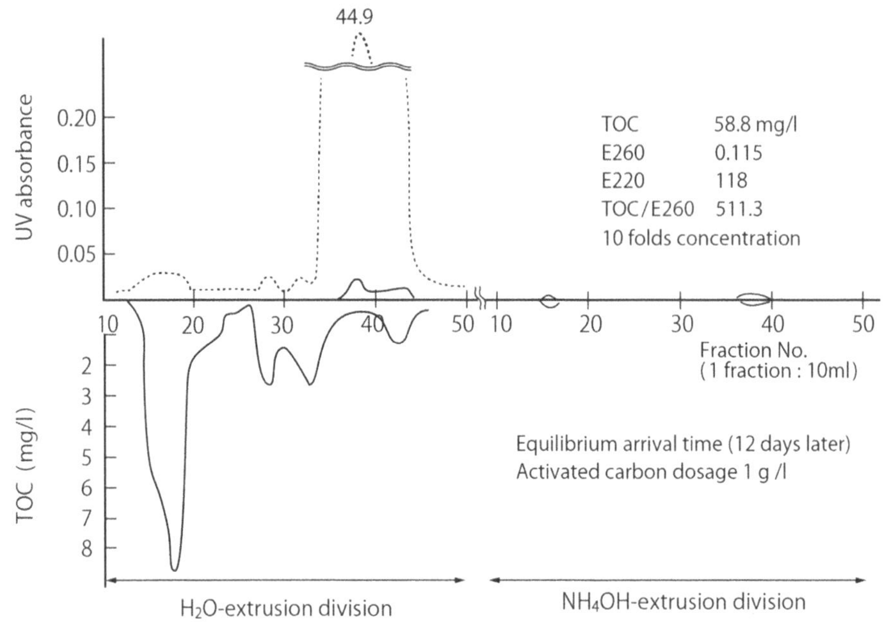

Figure 7.39 A G15 gel chromatogram, after coagulation and adsorption, of a final sedimentation tank sewage sludge aerobic degradation solution.

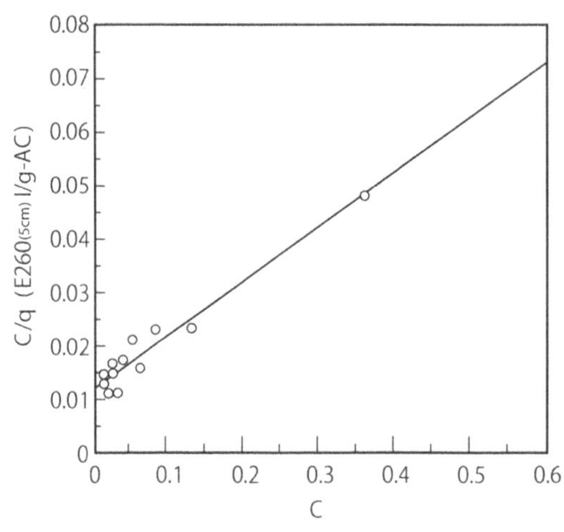

Figure 7.40 A Langmuir plot of Ishikari peatland water.

q_{max} differs greatly between peaty humics and sewage sludge aerobic decomposition solutions, with values of 9.75 and 6.20 $E260_{(5cm\text{-}cell)}/g$ activated carbon, respectively, indicating that adsorption saturations of humics is nearly twice that of the effluent investigated. Competition with E260 insensitive organics results in a larger (50%) reduction in saturation absorptivity than found for humic substances.

Next, as a process that governs the adsorption rate, it is necessary to know the boundary layer film diffusion and intragranular diffusion rate of humic substances. Of these, the intragranular diffusion rate is

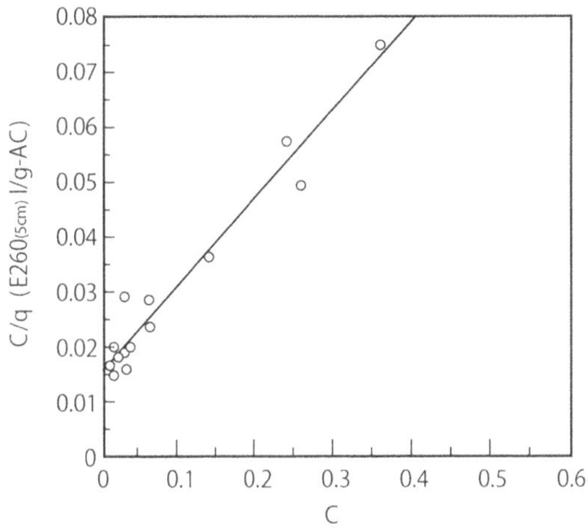

Figure 7.41 A Langmuir plot of sewage sludge aerobic degradation liquid.

determined by the adsorbate and the activated carbon species, and the boundary layer film diffusion rate is mainly governed by the operating conditions rather than the adsorbate type.

In this section, we calculate the intragranular diffusion coefficient D_p (pore diffusion coefficient) or D_s (pore surface diffusion coefficient) of humic substances, using the Yuasa-Tambo method (see Yuasa & Tambo, 1983a, 1983b, 1983c; Yuasa *et al.*, 1983) using either the pore diffusion model (F-P model) or pore surface diffusion model (F-S model). Both models are presented below; see also Figures 7.3 and 7.5 (earlier in this chapter) which show intragranular diffusion conditions.

Note: The symbols for Equations 7.48–7.56 are as follows: r: Distance from particle center (cm), t: adsorption elapsed time (sec), c: f(r,t) (g/cm^3-pore) at pore void free solution concentration, q: f(r,t) (g/g activated carbon) at pore wall adsorption amount, ρ_s: apparent adsorbent density (g/cm^3), Dp: intragranular pore diffusivity (cm^2/sec), δr: radial micro distance inside activated carbon (cm), a: particle radius (cm), K_f: film mass transfer coefficient (cm/sec), M: activated carbon addition concentration (g/cm^3), C_b: external solution concentration (g/cm^3), C_{bi}: initial solution concentration (g/cm^3), Q: total adsorption amount (g/g activated carbon).

(1) Pore-diffusion model (F-P model)

Material balance inside particles

$$\rho_s \frac{\partial q}{\partial t} = \frac{D_p}{r^2} \frac{\partial}{\partial r}\left(r^2 \frac{\partial c}{\partial r}\right) \qquad r \neq 0 \qquad \text{—(7·48)}$$

$$\rho_s = \left(\frac{\partial q}{\partial t}\right)_{r=0} = 6 \frac{D_p}{\delta r}\left(\frac{\partial c}{\partial r}\right)_{r=\delta r/2} \qquad \text{—(7·49)}$$

Material balance of particle surfaces (r=a)

$$\rho_s \left(\frac{\partial q}{\partial t}\right)_{r=a} = \frac{2D_p}{(\delta r)^2}\left\{c_{r=a-\delta r} - c_{r=a} + \frac{K_f \delta r}{D_p}\left(1 + \frac{\delta r}{a}\right)(C_b - c_{r=a})\right\} \qquad \text{—(7·50)}$$

(7.48–7.50)

Equation 7.48 - 7.50 The liquid-boundary layer-pore diffusion model.

(2) Surface diffusion model (F-S model)

Material balance inside particles

$$\frac{\partial q}{\partial t} = \frac{D_s}{r^2}\frac{\partial}{\partial r}\left(r^2\frac{\partial q}{\partial r}\right) \qquad r \neq 0 \qquad \text{—(7·51)}$$

$$\left(\frac{\partial q}{\partial t}\right)_{r=0} = 6\frac{D_s}{\delta r}\left(\frac{\partial q}{\partial r}\right)_{r=\delta r/2} \qquad \text{—(7·52)}$$

Material balance of particle surfaces (r=a)

$$\left(\frac{\partial q}{\partial t}\right)_{r=a} = \frac{2D_s}{(\delta r)^2}\left\{q_{r=a-\delta_r} - q_{r=a} + \frac{K_f\,\delta r}{\rho_s D_s}\left(1+\frac{\delta r}{a}\right)(C_b - c_{r=a})\right\} \qquad \text{—(7·53)}$$

(7.51–7.53)

Equation 7.51 - 7.53 The liquid boundary layer - pore surface diffusion model.

(3) Material balance and initial conditions in a batch-type adsorption process

Initial condition

$$\left.\begin{array}{l} c=0 \quad q=0 \quad \text{at} \quad 0 \leqq r \leqq a \\ C_b = C_{bi} \end{array}\right\} t=0 \qquad \text{—(7·54)}$$

Material balance of liquid system and
Inside activated carbon particles in batch operation

$$M\frac{dQ}{dt} = -\frac{dC_b}{dt} \qquad \text{—(7·55)}$$

$$= \frac{3M}{\rho_s a}K_f(C_b - c_{r=a})$$

Integrate Eq.7·55

$$\frac{3M}{a^3}\int_0^a q\cdot r^2 dr = C_{bi} - c_b \qquad \text{—(7·56)}$$

Equation 7.54 - 7.56 The material balance and Initial conditions of batch adsorption operation.

(7.54–7.56)

If intragranular diffusion is rate-limiting, then $C_{r=a} = C_b$. In order to estimate the diffusion coefficient, etc. by comparing with the experimental values, the above-mentioned format is changed to a balance equation by using the following dimensionless variables (Equations 7.57–7.62).

$$R = ra \quad \text{—(7·57)}$$

$$T_p = \left(\frac{D_p t}{a^2}\right)\left(\frac{C_{bi}}{Q_0 \rho_s}\right) \quad \text{—(7·58)}$$

$$T_s = D_s t a^2 \quad \text{—(7·59)}$$

$$T_f = T_p \gamma_p = T_s \gamma_s \quad \text{—(7·60)}$$

$$\gamma_p = K_f \frac{a}{D_p} \quad \text{—(7·61)}$$

$$\gamma_s = \left(K_f \frac{a}{D_s}\right)\left(\frac{C_{bi}}{Q_0 \rho_s}\right) \quad \text{—(7·62)}$$

Q : The Adsorption quantity corresponding to external solution concentration (initial concentration) C_{bi} (g/g-carbon)

Equation 7.57 – 7.62 The non-dimensionalization of index.

(Equation 7.57) (Equation 7.58) (Equation 7.59) (Equation 7.60) (Equation 7.61) (Equation 7.62)

(7.57–7.62)

(1) Dimensionless F-P models (Equations 7.63–7.65)

$$\rho_s \frac{\partial q}{\partial T_p} = \frac{\partial^2 c}{\partial R^2} + \frac{2}{R}\frac{\partial c}{\partial R} \quad R \neq 0 \quad \text{—(7·63)}$$

$$\rho_s \left(\frac{\partial q}{\partial T_p}\right)_{R=0} = \frac{6}{\partial R}\left(\frac{\partial c}{\partial R}\right)_{R=\delta R/2} \quad \text{—(7·64)}$$

$$\rho_s \left(\frac{\partial q}{\partial T_p}\right)_{R=1} = \frac{2}{(\partial R)^2}\{c_{R=1-\delta R} - c_{R=1} + \gamma_p\, \delta R(1+\delta R)(C_b - c_{R=1})\} \quad \text{—(7·65)}$$

(7.63–7.65)

Equation 7.63 - 7.65 The dimensionless liquid boundary layer - pore diffusion model.

(2) Dimensionless F-S model (Equations 7.66–7.68)

$$\frac{\partial q}{\partial T_s} = \frac{\partial^2 q}{\partial R^2} + \frac{2}{R}\frac{\partial q}{\partial R} \quad R \neq 0 \quad \text{—(7·66)}$$

$$\left(\frac{\partial q}{\partial T_p}\right)_{R=0} = \frac{6}{\partial R}\left(\frac{\partial q}{\partial R}\right)_{R=\delta R/2} \quad \text{—(7·67)}$$

$$\left(\frac{\partial q}{\partial T_s}\right)_{R=1} = \frac{2}{(\partial R)^2}\{q_{R=1-\delta R} - q_{R=1} + \gamma_s \left(\frac{\delta R}{\rho_s}\right)(1+\delta R)(C_b - c_{R=1})\} \quad \text{—(7·68)}$$

Equation 7.66 - 7.68 The dimensionless liquid boundary layer - pore surface diffusion model.

(7.66–7.68)

(3) Material Balance and Initial Conditions for Dimensionless Batch Operations (Equations 7.69 and 7.70)

Initial conditions

$$\left.\begin{array}{l} c=0 \quad q=0 \quad \text{at } 0 \leqq R \leqq a \\ C_b = C_{bi} \end{array}\right\} T_p = T_s = 0 \qquad \text{—}(7\cdot 69)$$

Material balance between liquid system and
Inside activated carbon particles in batch operation

$$3M\int_0^1 q\cdot R^2 dR = C_{bi} - c_b \qquad \text{—}(7\cdot 70) \qquad (7.69\text{–}7.70)$$

Equation 7.69-7.70 The material balance and initial conditions of dimensionless batch adsorption operation.

Using the above equations, we can calculate the intragranular pore diffusion coefficient D_s, the intragranular pore surface diffusion coefficient D_p, and the liquid boundary film mass transfer coefficient K_f as follows.

Using γ_p or γ_s as a parameter, calculate the process of decreasing the external solution concentration C_b with the dimensionless time T_p or T_s using the above dimensionless equation (drawing C_bT curves). Plot the time-change curve (C_b-t curve) of the external solution concentration C_b, which was obtained experimentally. At this time, the relationship between T and t, which gives the same external solution concentration, is plotted on logarithmic paper for each parameter γ. Also at this time, if the assumed value of γ is appropriate, the relation (log T–log t) should satisfy Equations 7.71–7.73, and the slope should be 1. Therefore, we calculate γ_p or γ_s such that the slope of the plot of the relation between log T and log t is 1, and calculate D_p, D_s, and K_f using Equations 7.71–7.73.

$$\log T_p = \log\left(\frac{D_p}{a^2}\,\frac{C_{bi}}{Q_0\rho_s}\right) + \log t \qquad \text{—}(7\cdot 71)$$

$$\log T_s = \log\left(\frac{D_s}{a^2}\right) + \log t \qquad \text{—}(7\cdot 72) \qquad (7.71\text{–}7.73)$$

$$\log T_f = \log\left(\frac{K_f}{a}\,\frac{C_{bi}}{Q_0\rho_s}\right) + \log t \qquad \text{—}(7\cdot 73)$$

Equation 7.71 - 7.73 The diffusion and liquid boundary layer mass transfercoefficient calculation equation.

For the experiments, Filtrasorb 400, an activated carbon manufactured by Cargon Corporation of the United States with the physical properties shown in Table 7.3, was used. The results of the batch adsorption test of raw water with the composition shown in Table 7.4 are shown in Figures 7.42 and 7.43 for peaty water (humic quality) and activated sludge aerobic treated water (Figure 7.39).

For both the F-S and F-P models, examples of how the slope of the (log T–log t) correlations changes with changes in γ_p or γ_s are shown in Figure 7.44 (F-S model of coagulated peaty groundwater) and Figure 7.45 (F-P model of coagulated bio metabolic wastewater). Calculations were made using the F-P and F-S models for all experimental cases. In both cases, the slope of the (log T–log t) correlation remains near 1 when the values of γ_p or γ_s are large, indicating that the organic component (natural

Table 7.3 Physical properties of Filtrasorb 400 activated carbon.

Size	0.35~0.42 mm
Average size	0.385 mm
Real density ρ_t	1.897 g/cm^3
Wet density ρ'	1.310 g/cm^3
Porosity ε	0.654
Bulk density ρ_s	0.657 g/cm^3

Table 7.4 Composition of the batch adsorption experiment water.

Item	Peatland Water Coagulation Treated Water	Biological Metabolism Waste Component Coagulation Treated Water
TOC (mg/l)	5.5	25.2
E260 (5 cm)	0.545	0.560
TOC/E260	50.5	223
pH	6.18	7.0
EC (μS/cm)	401.0	3974.5

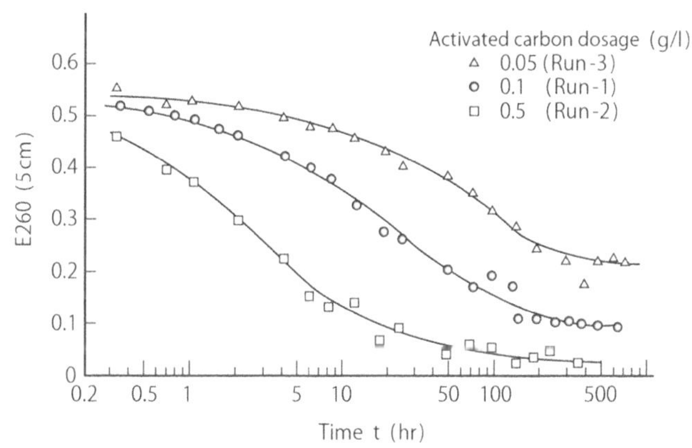

Figure 7.42 Results of batch adsorption experiments of peatland water after coagulation.

organic matter) of interest is an intragranular diffusion-limited type component. Both the F-P model and the F-S model worked effectively, and the superiority and inferiority of these models could not be judged from this experiment with a series of five tests.

For the obtained peaty groundwater coagulation water and coagulated aerobic oxidized activated sludge wastewater, phenols, and DBS, the average values of K_f and D_s obtained by the above-mentioned series of five batch adsorption tests are summarized as shown in Table 7.5, and the results for batch adsorption were obtained by numerical calculations. In this case, the formulas shown in Table 7.6 were used as the optimal

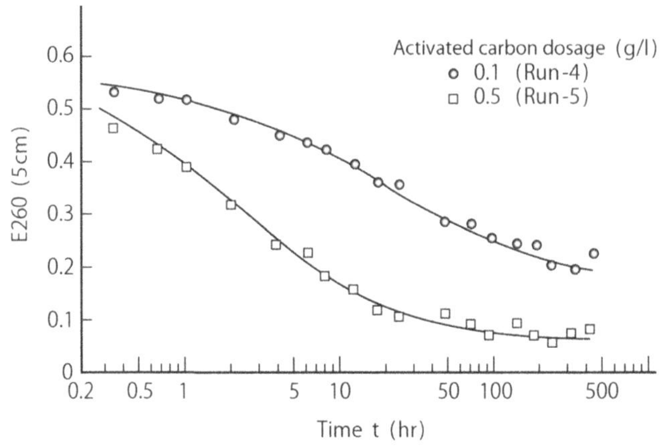

Figure 7.43 Results of batch adsorption experiments of biologically metabolised liquid waste (aerobic activated sludge water).

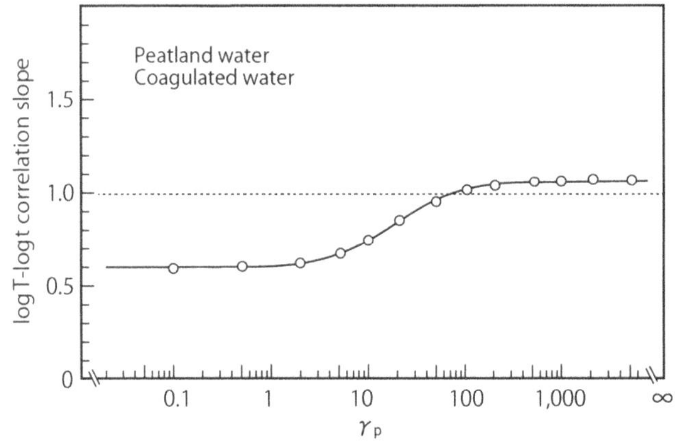

Figure 7.44 The log T–log t correlation slope in the F-S model and relationship with γ_p.

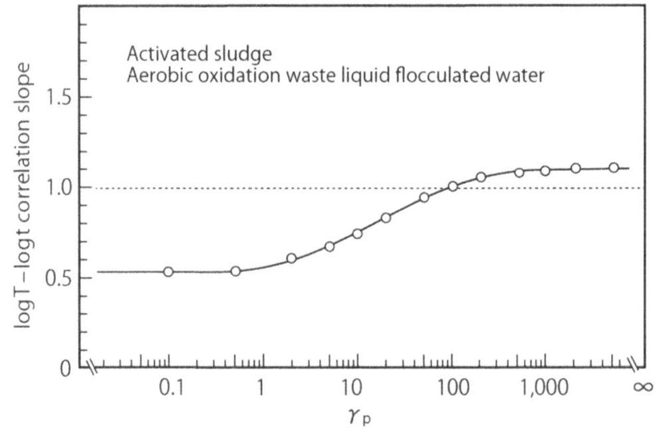

Figure 7.45 The log T–log t correlation slope in the F-P model and relationship with γ_p.

Table 7.5 Average values obtained for initial concentration, K_f and D_s from a series of five batch adsorption tests.

Item	Initial Concentration	K_f (cm/sec)	D_s (cm^2/sec)
Phenol	6.46 (mg/l)	5.6×10^{-3}	1.4×10^{-5}
Peatland water	0.55 (E260 – 5 cm)	2.0×10^{-3}	5.6×10^{-11}
Biological metabolism waste component	0.55 (E260 – 5 cm)		7.5×10^{-11}
DBS	7.98 (mg/l)		8.6×10^{-12}

Table 7.6 Adsorption isotherms used in a series of five batch adsorption tests.

Item	Adsorption Isotherm
Phenol	$q = 50 \times C^{0.34}$
Peatland water	$q = \dfrac{6.20 \times 10.9 \times C}{1 + 10.9 \times C}$
Biological metabolism waste component	$q = \dfrac{9.79 \times 8.70 \times C}{1 + 8.70 \times C}$
DBS	$q = \dfrac{272 \times 0.565 \times C}{1 + 0.565 \times C}$

adsorption isotherm. The operating conditions were as follows: initial TOC concentration was 4.95 mg/l, activated carbon content was 0.1 g/l and activated carbon diameter was 0.385 mm.

The results of these calculations, as illustrated in Figures 7.46 and 7.47, are well reproduced by experimental values, confirming the appropriateness of the evaluation logic and process discussed.

The above-mentioned natural color components were converted to TOCs with a TOC/$E260_{(1cm\text{-}cell)}$ ratio of 45. In tests 1–3 (Figure 7.42), the E260 insensitive TOC is 0, and the 5.5 mg/l TOC is all E260 sensitive, whereas in tests 4–5 (Figure 7.43), the E260 absorbance is almost the same as in Runs 1–3, but the TOC is as

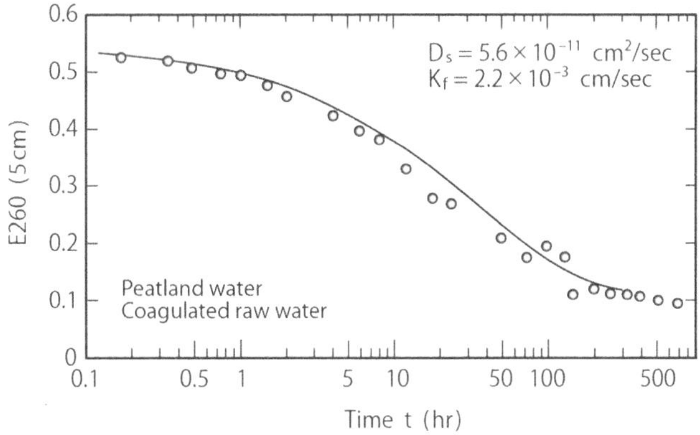

Figure 7.46 Calculated results and experimental values for the F-S model.

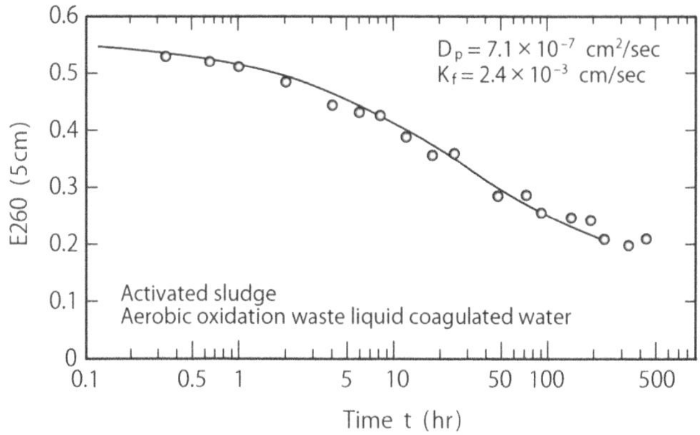

Figure 7.47 Calculated results and experimental values for the F-P model.

high as 25.2 mg/l (Table 7.4). the adsorption equilibrium achievement rate $E = (C_0 - C)/(C_0 - C_\infty)$ on the vertical axis, under the same conditions as in Figure 7.48.

From Figure 7.48, it can be seen that humic substances are considered to be components that adsorb at almost the same rate as DBS, but at a rate as slow as about 1/100 of that of phenol. The adsorption isotherm shows that the adsorption equilibrium decreases to about 50–60% when E260 insensitive TOC are present at about 50%. In addition, it can be inferred from this figure that the intragranular diffusion rate is not significantly affected. Numerically speaking, considering Filtrasorb 400 activated carbon as a reference, the pore surface diffusion coefficient of humic substances is $D_s \doteqdot (6 \text{ to } 7) \times 10^{-11}$ (cm^2/sec), the pore void diffusion coefficient is $D_p \doteqdot 6.6 \times 10^{-7}$ (cm^2/sec), and $D_p \doteqdot 6.4 \times 10^{-7}$ (cm^2/sec) for the surfactant DBS, which is about the same as that of Pittsburg activated carbon, and the adsorption rate is nearly two orders of magnitude smaller than that of phenol, for which $D_p \doteqdot 9.0 \times 10^{-6}$.

7.1.5.2 Activated carbon adsorption characteristics of trace hazardous organic components in coexistence with humic substances

One of the characteristics of adsorption processes in water treatment, especially water purification, is the removal of artificial harmful trace organic components, such as pesticides, in the presence of humic

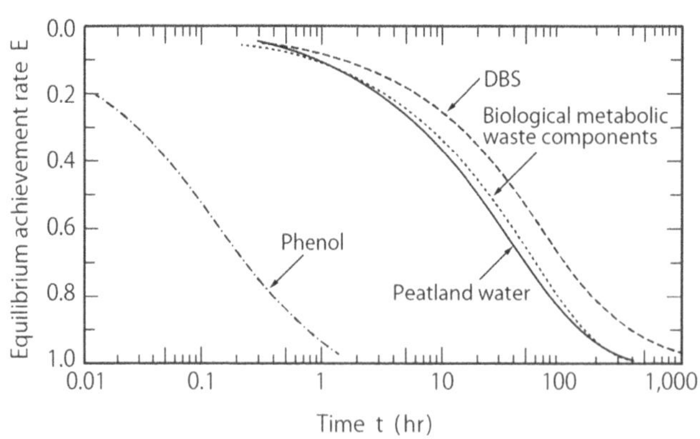

Figure 7.48 The adsorption rates of various components.

substances, which are stable organic coloring components generated by microbial natural purification and one of the major organic elements of municipal wastewater treatment effluent.

When water containing many adsorbed components is treated by addition of powdered activated carbon or fixed bed activated carbon adsorption, the main point is to evaluate the adsorption equilibrium relationship of each adsorbate and how the adsorption rate is affected by adsorption of other components. In the adsorption process of water treatment systems, humic substances, which are the second major organic component, are the dominant competitive components with a concentration difference of two orders of magnitude or more with respect to other trace organic chemical pollutants. Humic substances themselves are organic substances with various functional groups such as carboxyl groups, phenolic hydroxyl groups, and alcoholic hydroxyl groups, and a group of stable organic substances whose structures differ in molecular weight and chemical composition but cannot be clearly identified. Strictly speaking, it is a component that should be handled in a multicomponent system of its own.

Various theoretical kinetic and equilibrium treatments have been proposed as part of multicomponent adsorption theory, but not all of them can be applied to problems affecting the whole environment. In the preceding section, we described the coexistence of polysaccharides as metabolic waste components that are present concurrently with humic substances. In this section, we consider the competitive adsorption behavior of humic substances with trace hazardous organic substances in urban water metabolism systems where humic substances are the main components. This is an extension of the water conversion matrix construction theory.

Ishikari Peaty Ground Water (Hokkaido) was used as a humic-rich sample water, and coagulated settled water with aluminium sulfate was filtered through a 0.2 μm membrane filter and used for the test. This removed all humic substances with a (high) molecular weight of 3000 to 4000 Da or more, leaving only low molecular weight fluvic acids and the like. The test proceeded with UV absorbance $E260_{(1cm\text{-}cell)}$ at approximately 0.1 after diluting the concentration with distilled ion exchanged water to about 1/10 of the concentration of raw peaty groundwater.

As examples of trace organic harmful substances, cymadine (molecular weight: 202) and diazinon (molecular weight: 304), which are heterocyclic compounds widely used as herbicides and insecticides, and tetrachloroethylene (perchloroethylene, PCE; molecular weight: 166) and carbon tetrachloride (molecular weight: 154), which are organic chlorine (volatile) compounds, were selected, distilled and ion exchanged to dissolve them in activated carbon-adsorbed water, and mixed with filtered colored water for testing. Simazine and diazinon concentrations were determined by reversed-phase liquid chromatography using Wacosil-II 5C18-100 packed columns, and tetrachloroethylene and carbon tetrachloride were determined by gas chromatography using ECDs as detectors.

Activated carbon was ground in a ball mill using Cargon Filtrasorb-400 to reduce the time to reach adsorption equilibrium, then washed with ion-exchanged activated carbon adsorption water, wet sieved between 22 to 26 μm sieves, dried, and subjected to testing.

In addition, preloaded activated carbon with a fixed amount of humic material adsorbed on new activated carbon was produced as a base for competitive adsorption. In order to pre-load a multicomponent colorant (peat humics) with as uniform a mixture composition as possible, fixed bed adsorption was undertaken until equilibrium adsorption was reached. The coagulation treatment of the peaty groundwater used in this study removed 90% of the organics by E260 assessment, and the E260 concentration of treated water (make-up water) was about 0.1. In this case, the equilibrium adsorption state can be obtained by passing about 1500 liters of makeup water per 1 g of activated carbon.

As a continuous fixed bed adsorption device for producing preloaded activated carbon, a micro column made of stainless steel having an inner diameter of 2.0 mm × length of 9.8 mm and a large column made of glass having inner diameter of 1.0 cm × length 30 cm, are used. The micro column was filled with fine-particle-size activated carbon of 22–26 μm. The large column was filled with 0.300–0.355 mm Cargon Filtrasorb-400.

This micro-column will be described further below, with details of this device shown in Figures 7.54 and 7.60. The former can be filled with about 50 mg and the latter with about 7 mg of activated carbon. Dry-filled activated carbon was degassed in water, and a water flow test was performed under constant temperature condition at 20°C with a metering pump.

The competitive nature of the adsorption isotherm was next investigated by a batch adsorption test. In the case of cymadine and diazinon, a 200-ml sample was placed in a 500-ml Erlenmeyer flask, and shakne under constant temperature at 20°C for 3 to 6 days until adsorption equilibrium was confirmed. In the case of tetrachloroethylene and carbon tetrachloride, because of their low boiling point and high volatility, a 100 ml vial was used as a reaction vessel in a sealed state, and the shaking duration was set to 24 hours with reference to the results of the preliminary test.

Figures 7.49–7.52 show the relationship between equilibrium concentrations and equilibrium adsorption of the components cymadine, diazinon, tetrachloroethylene and carbon tetrachloride with fresh activated carbon, and the relationship between equilibrium concentrations and equilibrium adsorption of the same four components in the presence of humic substances for preloaded activated carbon. The figures also show the equilibrium adsorption amount obtained by integrating the breakthrough curve of the fixed bed adsorption test described in the next section.

These figures show that the adsorption isotherm can be expressed by the Freundlich adsorption equation, and that the index of the Freundlich adsorption equation (slope of the double logarithmic plot) does not change much when comparing fresh activated carbon with preloaded. In addition, the higher the humic concentration in the water flow when preparing preloaded activated carbon, the lower the equilibrium adsorption amount of harmful trace organic components such as cymadine. The adsorption of humic substances is highly irreversible, as shown in the fixed bed adsorption experiment in the next section, and strongly affects adsorption of other harmful trace components.

Therefore, we next consider how the equilibrium adsorption of trace harmful organic components is reduced by competition with humic substances.

The adsorption of humic substances is thought to result in a single, sequential decrease in the surface area within the pores where trace harmful substances can be adsorbed. In addition, humic substances have

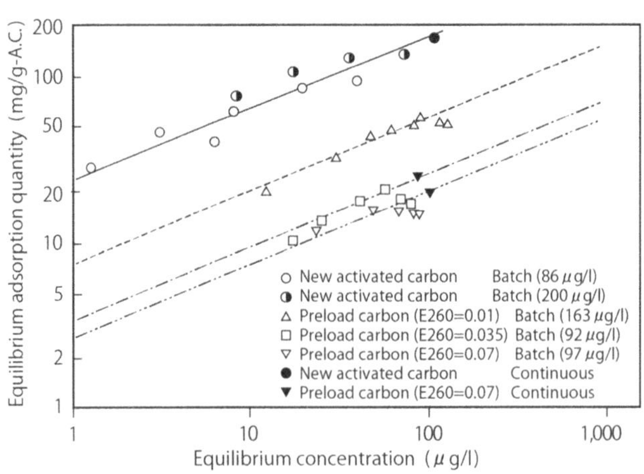

Figure 7.49 The adsorption isotherm for cymadine.

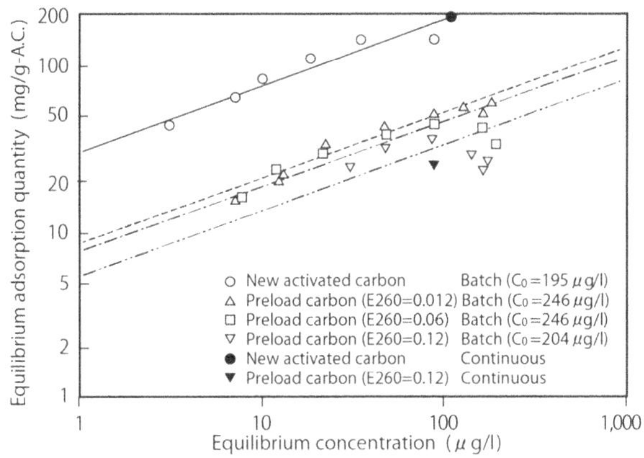

Figure 7.50 The adsorption isotherm for diazinon.

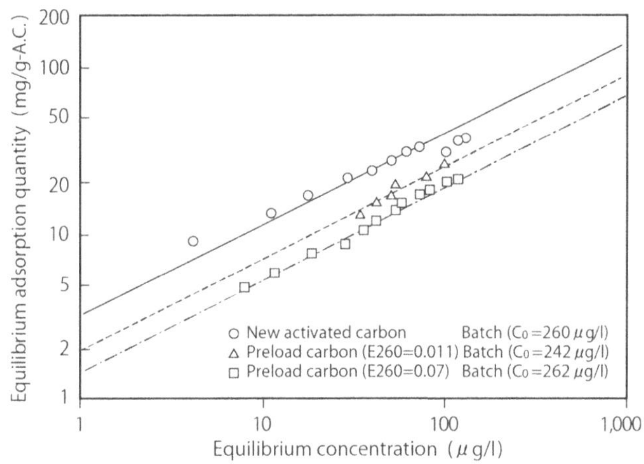

Figure 7.51 The adsorption isotherm for tetrachloroethylene.

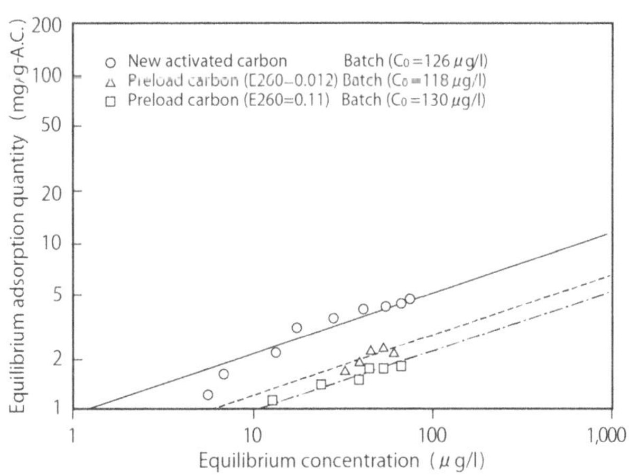

Figure 7.52 The adsorption isotherm for carbon tetrachloride.

molecular masses of 300 Da or above, but trace harmful organic components have molecular masses of a maximum of several hundred Da or so. It can be assumed that the pores have two parts: a pore surface on which adsorption competes, and a pore surface on which only harmful substances can be adsorbed. Therefore, Freunclich's adsorption isotherm can be written as shown in Equation 7.74.

$$q_A = K_{AH} C_A^{1/N} = (1 - \alpha) + \alpha \psi K_A C_A^{1/N} \quad \text{—}(7 \cdot 74) \tag{7.74}$$

q_A: Equilibrium adsorption quantity of harmful components (mg/g)
C_A: Equilibrium concentration of harmful components (μg/l)
K_A: Freundlich adsorption isotherm equation coefficient for harmful components ((mg/g)(μg/l)$^{1/N}$)
K_{AH}: Freundlich adsorption isotherm equation coefficient of harmful components for activated carbon with humic components adsorbed ((mg/g)(μg/l)$^{1/N}$)
N: Exponent of Freundlich adsorption isotherm equation for harmful components (–)
α: Ratio of adsorptive sites in parts that compete with humic components (–)
ψ: Reduction ratio of adsorption sites by humic components adsorption (–)

Equation 7.74 The Freundlich's competitive adsorption isotherm equation.

The experimental results shown in Figures 7.49–7.52 show that the percentage reduction of trace harmful components by preloaded activated carbon produced by previously passing humic substances of the same composition through packed columns can be considered as the function ψ shown in Equation 7.75, which is the function of the amount q_H of humic substances adsorbed on activated carbon determined by the overall concentrations of humic substances.

$$\psi = \text{func.}(q_H) \quad \text{—}(7 \cdot 75) \tag{7.75}$$

q_H: Harmful components adsorption quantity (l/g) expressed by UV absorbance E260

Equation 7.75 The reduction ratio of trace harmful component.

Expressed in terms of the overall humic C_H, the index of the Freundlich adsorption isotherm is considered to have a constant value as described above. From the figures, for coagulated peaty groundwater, the relationship $q_H \propto C_H^{1/3}$ holds, and Equation 7.74 can be written as Equation 7.76.

$$\frac{K_{AH}}{K_A} = \{(1 - \alpha) + \alpha \, \text{func.}(C_H^{1/3})\} \quad \text{—}(7 \cdot 76) \tag{7.76}$$

Equation 7·76 The overall concentration expression of adsorption isotherm equation.

Calculating Equation 7.76 using the results of Figures 7.49–7.52, the degree of adsorption capacity reduction ψ of trace harmful organic substances by humic concentrations in the water environment is shown in Figure 7.53. Pesticides and low-boiling organochlorine compounds appear to constitute two groups that are affected differently by humic substances, and pesticides appear to have a greater impact on the reduction of adsorption capacity by humic substances.

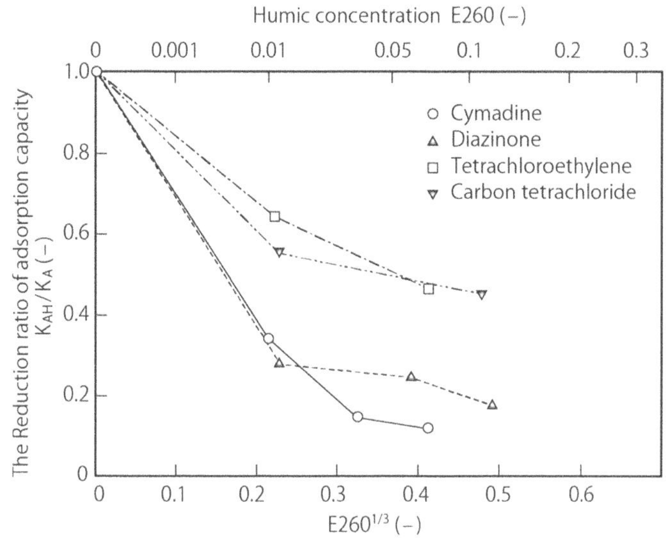

Figure 7.53 The reduction of adsorption capacity of activated carbon of trace organic harmful components by competition with humic substances.

To consider competition between humic substances and trace harmful components in fixed bed adsorption, continuous fixed-layer adsorption experiments were carried out under the conditions shown in Table 7.7, with scale-down mounting values using a microcolumn experiment device as shown in Figure 7.54. For details, such as the percentage similarity of the microcolumn method, see Section 7.1.5.3 below.

The laboratory equipment used utilizes high-pressure quantitative pumps for High-Performance Liquid Chromatography (HPLC). Stainless-steel columns with an inner diameter of 1.4 mm and a length of 10 mm were packed with activated carbon with a fine particle size of 22–26 μm, and operated at a constant temperature of 20°C. The E260 of the adsorption sample water flowing out of the columns was

Table 7.7 Conditions and parameters for the fixed bed adsorption experiment, showing values for an actual fixed bed. Space velocity (SV) = Q/filter volume.

Item	Microcolumn	Actual Device
Activated carbon size d (cm)	$22\sim26 \times 10^{-4}$	0.1
Flow rate (ml/min)	3.37	—
Column diameter (cm)	0.14	
Layer thickness (cm)	1.0	200
Superficial velocity u (cm/sec)	3.65	0.418
S.V. (1/sec)	3.65	0.00209
Layer thickness/Particle size (–)	417	2000
Reynolds number (μd/μ)	0.88	4.17
Time scale	5 (hr)	1 (year)

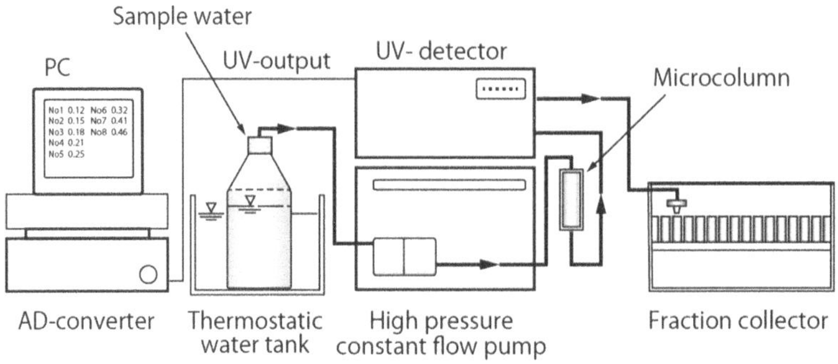

Figure 7.54 A microcolumn and other equipment used for an adsorption experiment.

measured by an ultraviolet absorbance (UV) detector, and the results were continuously recorded on PCs. The sample adsorbed by the UV detector was collected by a fraction collector, and the concentration of a trace component such as cymadine was measured. The humic concentration of adsorbed water was calculated by subtracting the E260 value corresponding to the concentration of cymadine, etc., measured by HPLC, from the measured ultraviolet absorbance E260 value.

Three types of water were used for the test: water containing (1) only harmful organic components or (2) humic substances (coagulated peaty groundwater) and (3) water containing them both at the same time. Two types of activated carbon was used: activated carbon that had reached an equilibrium adsorption state by coagulated peaty ground water, and unused fresh carbon.

Figure 7.55 shows the breakthrough curves for humic substances with ultraviolet absorbance E260 as an index when coagulated peaty groundwater is passed through an activated carbon fixed bed. The dotted line in the figure shows the breakthrough curve for humic substances when cymadine coexists in coagulated peat groundwater. Figure 7.56 shows the breakthrough curves for cymadine alone and for cymadine in coexistence with humic substances in solid and dotted lines, respectively. Figures 7.57 and 7.58 show the breakthrough curves for the fixed bed adsorption test of a single component and coexisting system of diazinon and coagulated peatland treated water, respectively.

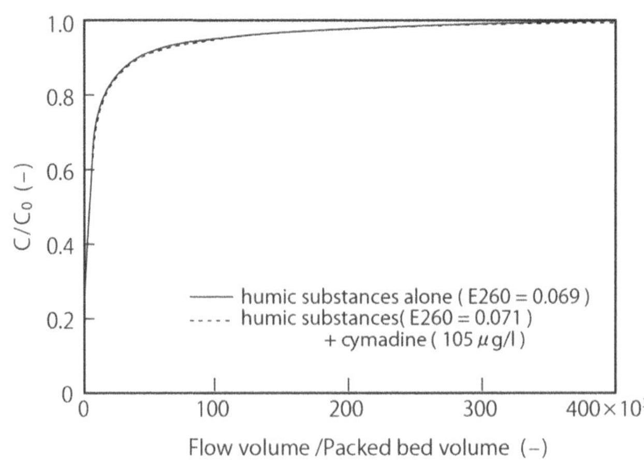

Figure 7.55 Breakthrough curve for humic substances with cymadine.

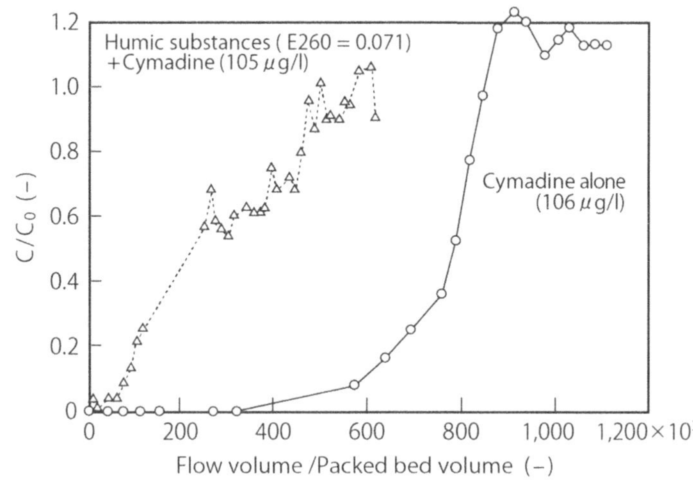

Figure 7.56 Breakthrough curve for cymadine.

The breakthrough of cymadine and diazinon is accelerated by the presence of humic substances, and the total adsorption is reduced to approximately 34% and 26%, respectively, by integrating the entire breakthrough curve. In contrast, the breakthrough curve for humic substances is unaffected by the presence of cymadine and diazinon. Table 7.8 shows the equilibrium adsorption amounts of cymadine and diazinon obtained by integrating the breakthrough curves. The adsorption of humic substances can be extrapolated from the integral of Figure 7.55, which shows the adsorption process of peaty land coagulated water.

The experimental conditions show that cymadine and diazinon have almost the same amount of adsorption on a molar basis as humic substances. Despite the fact that hydrophobic cymadine and

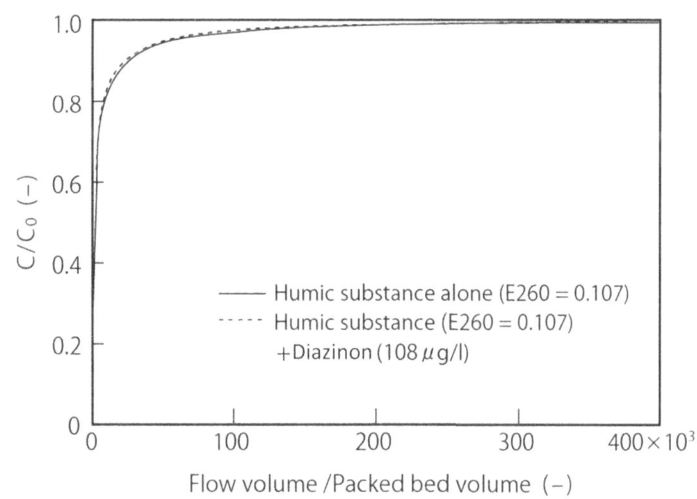

Figure 7.57 Breakthrough curve for humic substance with Diazinon.

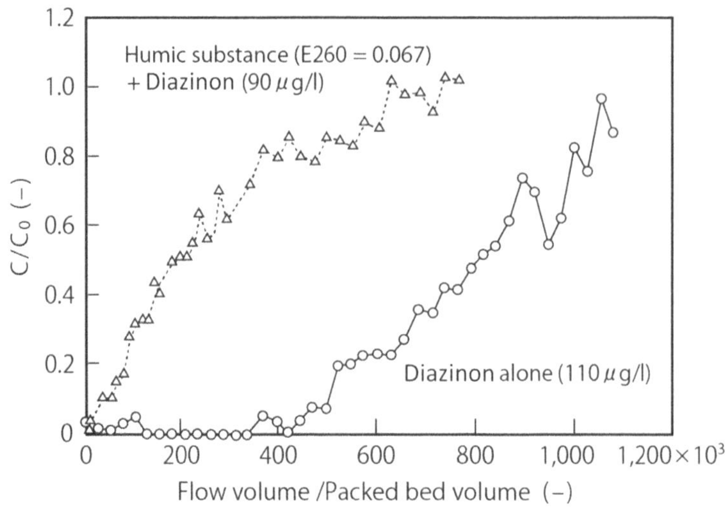

Figure 7.58 Breakthrough curve for Diazinon.

Table 7.8 The equilibrium adsorption quantity of humic and harmful substances.

Item	Equilibrium Concentration	Equilibrium Adsorption Quantity	Equilibrium Concentration μ mol/l	Equilibrium Adsorption Quantity mmol/l
Cymadine	106 μg/l	163 mg/g	0.52	0.81
Diazinone	110 μg/l	192 mg/g	0.36	0.63
Cymadine: Humic substances coexistence E260 = 0.071	105 μg/1	55 mg/g	0.52	0.27
Diajinon: Humic substances coexistence E260 = 0.067	95 μg/l	49 mg/g	0.30	0.16
Coagulated humic substances	0.1 (E260, 1 cm-cell)	7 l/g	8	1

diazinon have low water solubility, high carbon adsorption, and the same level of adsorption equilibrium as humic substances, humic substances have a specific competitive relationship that is hardly affected by adsorption even when cymadine and diazinon coexist. This is an important finding for water treatment processes.

Figure 7.59 shows the breakthrough curve of the fixed bed adsorption of coagulated peaty groundwater and the results of an experiment to confirm the desorption phenomenon by passing distilled water after adsorption saturation has been reached. The figure shows that desorption of humic substances has rarely occurred. Since the adsorption side and desorption side should be the same in the mass transfer process, the adsorption reaction of humic substances is considered to be irreversible, and it can be understood that

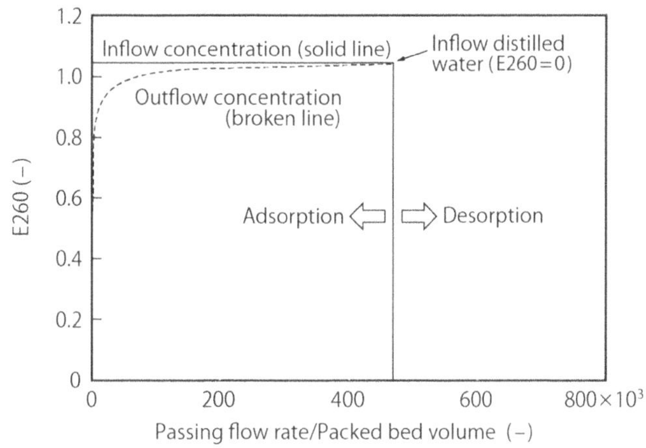

Figure 7.59 Outflow concentrations in fixed bed adsorption and desorption processes for humic substances.

the adsorption of pesticides and the like in a water system in which humic substances are present is unilaterally suppressed on the basis of the main behavior of humic substances.

7.1.5.3 Rapid evaluation of fixed-bed adsorption process by microcolumn method

As described in the previous section, the equilibrium conditions and adsorption rate characteristics of activated carbon adsorption always need to be determined. It is necessary to assume that harmful chemical substances are usually present at the same time in the water with humic substances in a natural water system or urban water metabolism system.

Professors Weber and Crittenden (Crittenden & Weber, 1978a, 1978b; Crittenden *et al.*, 1986, 1987) both proposed rapid estimation of adsorption kinetic properties by short-column methods and microcolumn methods using small-particle-size activated carbon.

Professor Y. Matsui *et al.* (1992) (Hokkaido University) established the experimental microcolumn method which is described here. This microcolumn method enables the collection of adsorption characteristics with a small number of samples in a short period of time by modifying a part of the HPLC equipment, which is now widely used, without undergoing long-term adsorption pilots if the reaction is a physico-chemical one – but it is not useful when including biological reactions. On the other hand, the biological reactions that are the main issues in the advanced treatment of purified water, i.e., the adsorption and decay of odors and tastes, and the nitrification of ammonia, this method can not be used.

The microcolumn laboratory equipment used was equipped with a stainless micro column (inner diameter 2.0 mm, outer diameter 8.0 mm, length 9.8 mm), the cross-section of which is shown in Figure 7.60, filled with fine activated carbon particles ground with a ball mill, on a HPLC device with a configuration such as Figure 7.54. The column effluent water was continuously recorded on a PC with a UV detector (wavelength 260 nm), and finally collected with a fraction collector to measure TOC.

At the same time, a breakthrough experiment was conducted using a real-size activated carbon column (a large column), which is the proto-type of the microcolumn method, and the adsorption and breakthrough characteristics of both columns were compared to establish a relationship between the microcolumn and

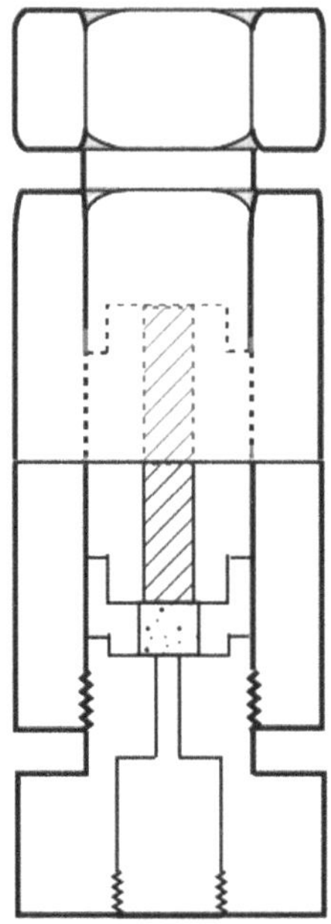

Figure 7.60 The micro column cross-section.

the real-size equipment, and an index was established to return the test results obtained by the microcolumn to the prototype. For this purpose, a large experimental column (glass tube with an inner diameter of 1.0 cm and a length of 30 cm) was used. The ultraviolet 260 nm absorbance and TOC (mg/l) of water adsorption were measured by flowing test water at a predetermined flow rate using a microtube pump and sampling water at regular intervals on an automated water sampler, and comparing the absorbance and TOC with the output water of the microcolumn.

The activated carbon used for the experiments was Cargon Filtrasorb-400, and for the microcolumns, fine particles with a diameter of 22–26 μm were used, which were pulverized in a ball mill and washed with distilled water to remove fine powders. A large column classified into 0.300 to 0.355 mm sieves was used, with the same procedure.

The sample water was prepared by coagulating water from peaty land in the suburbs of Sapporo City containing humic substances, removing colloids of 4000 Da or more almost completely, and filtering the water containing only the low molecular weight portion through a membrane filter of 0.20 μm.

Since the microcolumn is designed to shorten the adsorption process time and predict the progress of adsorption that takes a long time for the prototype, it is necessary to know the similarity rate of each adsorption process (the activated carbon film transfer rate and the mass transfer rate within the activated carbon particles between the microcolumn and the prototype). Since humic substances and surfactants have a low adsorption rate, the axial dispersion of the fixed layer seems to have little effect on the progress of the adsorption reaction, and therefore the phenomenon can be expressed by the surface diffusion rate-limiting model under the extrusion flow of the fixed layer or pore diffusion rate-limiting model (Yuasa & Tambo, 1983a, 1983b, 1983c).

Considering the flow of the packed bed as the extrusion flow, and assuming that intragranular diffusion proceeds with the surface diffusion rate control, the phenomenon can be described by the following equations.

Adsorbed material balance in fixed bed

$$u\frac{\partial C_i}{\partial l}+\rho_b\frac{\partial Q_i}{\partial t}=0 \quad —(7\cdot77)$$

Diffusion inside activated carbon particles (surface diffusion)

$$\frac{\partial q_i}{\partial t}=\frac{1}{r^2}\frac{\partial}{\partial r}\left(r^2 D_{si}\frac{\partial q_i}{\partial r}\right) \quad —(7\cdot78)$$

Boundary condition on activated carbon particle surface

$$\frac{\partial Q_i}{\partial t}=\frac{3D_{si}}{R}\left.\frac{\partial q_i}{\partial r}\right|_{r=R} \quad —(7\cdot79)$$

C_i : Concentration of component i in fixed bed void (m mol/cm^3)
Q_i : Activated carbon adsorbed amount on component i (m mol/g-A.C.)
q_i : Adsorption quantiti of component i inside activated carbon particles (m mol/g-A.C.)
l : Activated carbon packed bed depth (cm)
t : Passing water time (sec)
r : Radial direction distance of activated carbon particles (cm)
R : Activated carbon particle radius (cm)
u : Water flowing velocity (superficial velocity、cm/sec)
ρ_b : Activated carbon packing density (g-A.C./cm^3-bed)
D_{si} : Surface diffusion coefficient of component i (cm^2/sec)

(7.77–7.79)

Equation 7.77 - 7.79 The rate-limiting diffusion model inside activated carbon pore.

To make Equations 7.77–7.79 dimensionless, we introduce the relationships presented in Equations 7.80–7.85.

$$C_i=C_{0i}X_i \quad —(7\cdot80)$$
$$Q_i=Q_{0i}Y_i \quad —(7\cdot81)$$
$$q_i=Q_{0i}y_i \quad —(7\cdot82)$$
$$l=Lz \quad —(7\cdot83)$$
$$t=T\tau \quad —(7\cdot84)$$
$$r=Rs \quad —(7\cdot85)$$

L : Activated carbon layer thickness (cm)
T : Dummy time for dimensioning (sec)
C_{0i} : Inflow concentration of component i (m mol/cm^3)
Q_{0i} : Adsorption quantity in equilibrium with inflow concentration (m mol/g-A.C.)
X_i : Dimensionless concentration (–)
Y_i : Dimensionless adsorption quantity of component i (–)
y_i : Dimensionless adsorption quantity of component i inside activated carbon particles (–)
z : Dimensionless packed bed depth (–)
τ : Dimensionless elapsed time (–)
s : Dimensionless radial direction distance of activated carbon particles (–)

(7.80–7.85)

Equation 7.80 - 7.85 The related equations for dimensionlessning.

As a result, Equations 7.77–7.79 are expressed as Equations 7.86–7.88.

Material balance equation

$$\frac{\partial X_i}{\partial z} + \frac{\rho_b Q_{0i} L}{C_{0i} u T} \frac{\partial Y_i}{\partial \tau} = 0 \quad —(7\cdot86)$$

Intragranular diffusion equation (surface diffusion)

$$\frac{\partial y_i}{\partial \tau} - \frac{T}{R^2} \frac{1}{s^2} \frac{\partial}{\partial s}\left(s^2 D_{si} \frac{\partial y_i}{\partial s}\right) \quad —(7\cdot87)$$

Boundary conditions of activated carbon particle surface

$$\frac{\partial Y_i}{\partial \tau} = \frac{3 D_{si} T}{R} \frac{1}{s} \frac{\partial y_i}{\partial s}\bigg|_{s=1} \quad —(7\cdot88)$$

If the surface diffusion coefficient is constant, the following two dimensionless numbers are induced.

$$N_{Ai} = \frac{uT}{L} \frac{C_{0i}}{\rho_b Q_{0i}} \quad —(7\cdot89)$$

$$N_{Bi} = \frac{D_{si} T}{R^2} \quad —(7\cdot90)$$

N_{Ai} : Dimensionless number on material balance (–)
N_{Bi} : Dimensionless number of surface diffusion (–)

(7.86–7.90)

Equation 7.86 - 7.90 The dimensionless diffusion rate-limiting model of activated carbon intragranular.

Since the time T for dimensionless reduction can be arbitrarily set at present, if T is eliminated by combining Equations 7.89–7.90, Equation 7.91 can be obtained as one similarity rate.

$$N_i = \frac{uR^2}{LD_{si}} \frac{C_{0i}}{\rho_b Q_{0i}} \quad —(7\cdot91)$$

N_i : Similarity rate for a fixed bed assuming surface diffusion limitation under piston flow (–)

(7.91)

Equation 7.91 The similarity rate for the activated carbon fixed bed.

The dimensionless numbers for other mass transfers are determined in a similar manner as follows. Assuming that intragranular diffusion is dominated by pore diffusion, a dimensionless number can be obtained as shown in Equation 7.92. Considering the mass transfer of the liquid boundary film, the dimensionless number can be expressed as shown in Equation 7.93.

$$N_{Ci} = \frac{D_{pi}T}{R^2}\frac{C_{0i}}{\rho_s Q_{0i}} \quad \text{—(7·92)}$$

$$N_{Di} = \frac{k_{fi}T}{R}\frac{C_{0i}}{\rho_s Q_{0i}} \quad \text{—(7·93)}$$

N_{Ci} : Dimensionless number for pore diffusion (–)
N_{Di} : Dimensionless number of liquid boundary layer diffusion (–)
ρ_s : Apparent density of activated carbon particles (g/cm^3)
D_{pi} : Pore inside diffusion coefficient of component i (cm^2/sec)
k_{fi} : Liquid boundary layer mass transfer coefficient of component i (cm/sec)

(7.92–7.93)

Equation7.92 - 7.93 The dimensionless number with respect to pore and liquid boundary layer diffusion

The boundary layer mass transfer coefficients can be summarized as Equations 7.94–7.96 as correlations with the Sherwood, Schmitt, and Reynolds numbers. Substituting Equation 7.94 into Equation 7.93 gives Equation 7.97, or Equation 7.98. Thus, when the dimensionless numbers for intragranular diffusion (N_{Bi}, N_{Ci}) are matched, the dimensionless numbers for boundary layer diffusion are effectively functions of the Reynolds number only.

Sherwood number $Sh_i = \frac{k_{fi}d}{D_{mi}} = F_s(Sc_i, Re, \varepsilon_b)$ —(7·94)

Schmidt number $Sc_i = \frac{\nu}{D_{mi}}$ —(7·95)

Reynolds number $Re = \frac{ud}{\varepsilon_b \nu}$ —(7·96)

$$N_{Di} = \frac{D_{mi}}{2}\frac{T}{R^2}\frac{C_{0i}}{\rho_s Q_{0i}}F_s(Sc_i, Re)$$

$$= \frac{D_{mi}}{D_{si}}\frac{C_{0i}}{\rho_s Q_{0i}}N_{Bi}F_s(Sc_i, Re) \quad \text{—(7·97)}$$

or

$$= \frac{D_{mi}}{D_{pi}}N_{Ci}F_s(Sc_i, Re) \quad \text{—(7·98)}$$

Sh_i : Sherwood number for component i (–)
d : Particle size (cm)
ε_b : Porosity of packed bed (–)
Re : Reynolds number (–)
Sc_i : Schmidt number for component i (–)
ν : Kinematic viscosity (cm^2/sec)
D_{mi} : Molecular diffusion coefficient of component i (cm^2/sec)
D_{pi} : Pore inside diffusion coefficient of component i (cm^2/sec)

(7.94–7.98)

Equation 7.94 - 7.98 The correlation equation for Sherwood·Schmidt·Reynolds number.

The dimensionless Peclet number for the longitudinal dispersion of the packed bed is given by Equation 7.99. The Peclet number on the particle size basis is almost constant for a liquid system, and the Peclet number defined in Equation 7.99 may be considered to be proportional to the number of theoretical plates L/d of the packed bed.

$$Pe = \frac{uL}{\varepsilon_b D_b} = \frac{ud}{\varepsilon_b D_b}\frac{L}{d} \qquad \text{—}(7\cdot 99)$$

D_b : Longitudinal diffusion coefficient of packed bed (cm^2/sec)

(7.99)

Equation 7.99 The Peclet number.

In order to establish the similarity between breakthrough curves of fixed bed activated carbon adsorption processing of solutions with different concentrations, it is necessary to consider not only the adsorption rate but also the similarity of the adsorption equilibrium.

Various experimental and theoretical equations have been proposed as adsorption isotherms. If we apply the idealized solution theory to develop it into a competitive relationship in a multicomponent solution using the Freundlich equation as a single-component adsorption isothermal equation, we obtain Equation 7.100 as a multiple-component adsorption isothermal equation (Crittenden *et al.*, 1985; Matsui *et al.*, 1992; Matsui *et al.*, 2013). Equation 7.100 is converted into dimensionless Equation 7.101 using Equations 7.80–7.82.

$$c_i = \frac{q_i}{\sum_{j=1}^{N} q_j}\left(\frac{\sum_{j=1}^{N} n_j q_j}{n_i K_i}\right)^{n_i} \qquad \text{—}(7\cdot 100)$$

c_i : Equilibrium concentration of component i ($mmol/cm^3$)
q_i : Equilibrium adsorption quantity of component i (mmol/g)
n_i : Power index of Freundlich adsorption isotherm equation of component i (–)
K_i : Coefficient of Freundlich adsorption isotherm equation of component i ($mmol^{1-1/n_i} cm^{3/n_i}/g$)
N : Number of components of adsorbate (–)

(7.100)

Equation 7.100 The multiple-component adsorption isotherm equation.

$$x_i = y_i \frac{\sum_{j=1}^{N} Q_{0j}}{\sum_{j=1}^{N} Q_{0j} y_j}\left(\frac{\sum_{j=1}^{N} n_j Q_{0j} y_i}{\sum_{j=1}^{N} n_j Q_{0j}}\right)^{n_i} \qquad \text{—}(7\cdot 101)$$

(7.101)

Equation 7.101 The dimensionless multiple-component adsorption isotherm equation.

Assuming that all the coefficients n_i of the Freundlich equation have the same values n, Equation 7.102 holds, so Equation 7.101 is transfer to Equation 7.103. When the component ratios of the inflow concentration of the fixed bed are the same, the adsorption isotherms expressed by the dimensionless concentration and the dimensionless adsorption amount are the same.

$$q_i = K_i^n c_i \left(\sum_{j=1}^{N} q_j\right)^{1-n}$$

$$= K_i^n c_i \left(\sum_{j=1}^{N} K_j^n c_j\right)^{(1-n)/n} \quad \text{—} (7 \cdot 102)$$

$$y_i = x_i \left(\frac{\sum_{j=1}^{N} K_j^n \alpha_{0j} x_j}{\sum_{j=1}^{N} K_j^n \alpha_{0j}} \right)^{(n-1)/n} \quad \text{—} (7 \cdot 103)$$

α_{0i} : Ratio to all components of inflow component i (–)

$$\alpha_{0i} = \frac{C_{0i}}{\sum_{j=1}^{N} C_{0j}} \quad \text{—} (7 \cdot 104)$$

(7.102–7.104)

Equation 7.102–7.104 The dimensionless adsorption isotherm with same inflow concentration ratio.

The relationship between total concentration and total adsorption is shown in Equation 7.105 when the Freundlich modulus n for each component is the same value. Therefore, the C_{0i}/Q_{0i} in Equation 7.91 is a function of the total concentration c_T only for the same component ratio, and is shown in Equation 7.106.

$$q_T = \left(\sum_{j=1}^{N} K_j^n C_j\right)^{1/n}$$

$$= \left(\sum_{j=1}^{N} K_j^n \alpha_j\right)^{1/n} c_T^{1/n} \quad \text{—} (7 \cdot 105)$$

q_T : Total adsorption quantity (mmol/g-A.C.)
c_T : Total concentration ($mmol/cm^3$)

(7.105)

Equation 7.105 The total concentration and total adsorption quantity when power factor n is same.

$$\frac{C_{0i}}{Q_{0i}} = K_i^{-n} \left(\sum_{j=1}^{N} K_j^n \alpha_j\right)^{(n-1)/n} c_T^{(n-1)/n} \quad \text{—} (7 \cdot 106)$$

C_{0i} : Inflow concentration of component i ($mmol/cm^3$)
Q_{0i} : Adsorption quantity in equilibrium with inflow concentration (mmol/g-A.C.)

(7.106)

Equation 7.106 Relationship between inflow concentration and equilibrium adsorption quantity.

The total concentration c_T of Equation 7.106 is usually expressed on a molar basis, but it can be expressed in a similar manner to Equation 7.107, using an alternative index such as ultraviolet absorbance.

$$\frac{C_{0i}}{Q_{0i}} = K_i^{-n}\left(\sum_{j=1}^{N} K_j^n \beta_j \gamma_j\right)^{(n-1)/n} E_T^{(n-1)/n} \quad \text{---}(7 \cdot 107) \qquad (7.107)$$

β_i: Conversion factor between molar concentration of component i and UV absorbance (mol/l)
γ_i: Abundance ratio of component i observed by UV absorbance (–)
E_T: Comprehensive concentration observed by UV absorbance (–)

Equation 7.107 The UV absorbance expression for inflow concentration and equilibrium adsorption quantity.

We now work to verify the similarity of the fixed-bed adsorption process for different activated carbon particle sizes. Here, in order to satisfy the condition of Equation 7.91 above, which shows the similarity under the assumption that the surface diffusion rate in the pores is controlled by the extrusion flow, we refer to the operating conditions of a standard advanced treatment plant and conduct a microcolumn process operation similar to that explained above. From Equation 7.91, the similarity rate for the same influent concentration and activated carbon packing density with respect to both large and micro-columns is shown by the simple relationship shown in Equation 7.108.

$$\frac{T_S}{R_S^2} = \frac{T_L}{R_L^2} \quad \text{---}(7 \cdot 108) \qquad (7.108)$$

T_S、T_L: Superficial retention time of microcolumn and large column respectively $= L/u$ (sec)
R_S、R_L: Activated carbon particle size of microcolumn and large column respectively (cm)

Equation 7.108 The similarity rate when inflow concentration and packed density are same.

Under these settings, a set of both large and micro columns were used to perform experiments to evaluate similarity rates under the conditions shown in Table 7.9.

According to Equation 7.108, the duration of the process is reduced (or extended) in proportion to the square of the packed activated carbon particle size. The actual plant values are also included in the table. The breakthrough test, which takes a year in the actual plant, can be evaluated by the microcolumn method in an experiment of several hours. In addition, since the samples necessary can be tested by varying the conditions in a few liters or less, it is likely that this test method would be suitable for setting the physicochemical conditions in a small cycle.

Figures 7.61–7.63 show the results of adsorption experiments of coagulated peaty water using microcolumns and large columns. The horizontal axes of the figures show the amount of water passed per unit weight of activated carbon after correcting for a slight difference in bulk density.

In the large column experiment, the pressure loss of the column increased around 100 hours after the start of water flow, and the experiment was aborted. When TOC is used as the water quality indicator, the breakthrough curves are almost the same within the error range of the TOC meter. A similar breakthrough curve is also drawn when ultraviolet absorbance at 260 nm is used as an index, but the leakage concentration appears to be lower in the large column in the latter half of the breakthrough.

Table 7.9 Experimental conditions (and actual values) required to evaluate a similarity index.

Item	Microcolumn	Large Column	Actual Device
Activated carbon size (cm)	$22\sim26 \times 10^{-4}$	$300–355 \times 10^{-4}$	0.1
Flow rate (ml/min)	6.75	18.5	
Bulk density (g/cm^3)	0.474	0.478	
Column diameter (cm)	0.20	1.00	
Layer thickness (cm)	0.98	20.0	200
Superficial velocity (cm/sec)	3.58	0.393	0.418
S.V. (1/sec)	3.65	0.0196	0.00209
Layer thickness/particle size (–)	410	613	2.000
Reynolds number	0.86	1.28	4.17
Time scale	5 (hr)	39 (day)	1 (year)

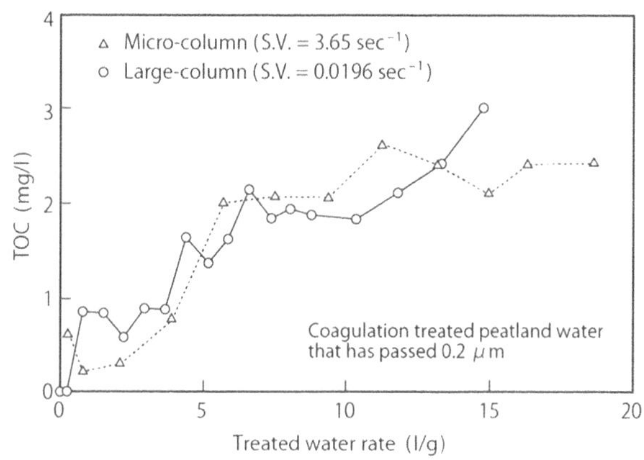

Figure 7.61 A comparison of TOC breakthrough curves.

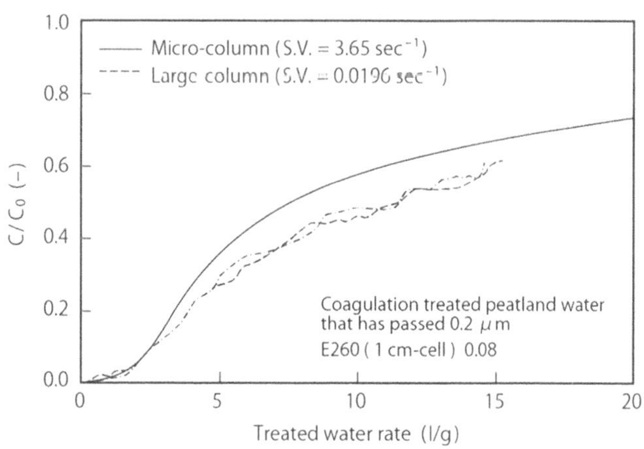

Figure 7.62 Comparison of C/C_0 breakthrough curves, based on E260.

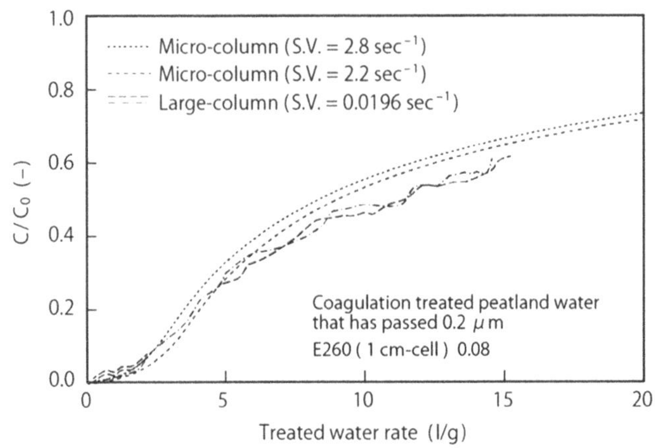

Figure 7.63 Comparison of C/C_0 breakthrough curves, showing the breakthrough curve of the microcolumn without capacity correction.

Three possible causes are:

(1) Biological effects reduce breakthrough concentrations in large columns operated for long periods of time;
(2) Specific adsorption capacity differs during the fine pulverization process of activated carbon; and
(3) There are differences in the intrapore diffusion coefficients.

It is unlikely that the humic color coagulated water will undergo large biodegradation, and a difference may have been introduced in the milling process, affecting the intragranular diffusion coefficient and the surface area (adsorption capacity) in the pores. Therefore, we first measured the area in the pores that governs the adsorption capacity. The specific pore surface area was determined by the nitrogen gas adsorption method based on BET adsorption theory. The microcolumn had 530 m^2/g of small particle activated carbon, whilst the large-diameter activated carbon in the large column had 690 m^2/g. In addition, since the pore distribution by the BET method was the same, it is considered that the difference between the

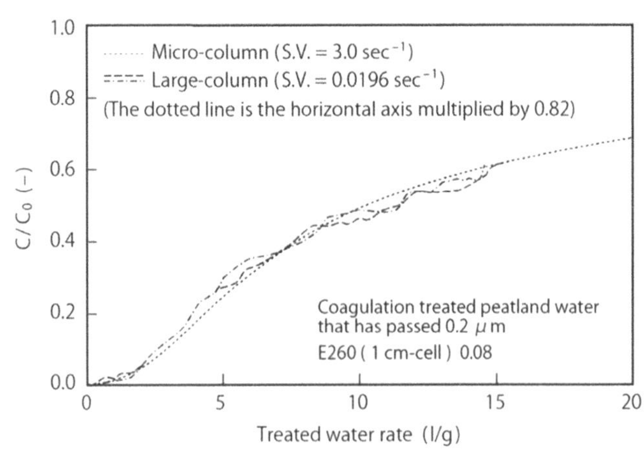

Figure 7.64 The breakthrough curve corrected for adsorption capacity difference.

breakthrough curves of the two columns is due to the difference in surface area within the pores (difference in adsorption capacity). When this correction is added, both breakthrough curves are almost identical, as shown in Figure 7.64, and the law of similarity is established. This confirms that pre-experiments with microcolumns can effectively predict the phenomenon.

To investigate the law of similarity and adsorption isotherms for fixed-bed adsorption processes of multicomponent solutions with different concentrations, we again used the microcolumn method. This time, the component ratio of the treatment raw water was constant and only the concentration differed. Substituting Equation 7.106 into Equation 7.91 yields Equation 7.109, which shows the similarity, and yields the same similarity condition for all components of the adsorbate.

$$\frac{u_1 R_1^2}{L_1 D_{si_1} \rho_{b_1}} C_{T_1}^{(n-1)/n} = \frac{u_2 R_2^2}{L_2 D_{si_2} \rho_{b_2}} C_{T_2}^{(n-1)/n} \quad \text{— (7·109)} \tag{7.109}$$

Subscripts 1 and 2 indicate different nflow concentration.

Equation 7.109 The similarity conditions for adsorbate component.

Assuming that the packed bed thickness, activated carbon particle size, packing density, and surface diffusion coefficient are constant, adsorption tests were performed several times on samples with different concentrations, by changing the water flow rate, and the conditions under which the breakthrough curves drawn by the flow rate match were examined. The sample used was 1/1, 1/2, and 1/5 times diluted Ishikari Peaty Ground Water (the raw water of the sample) and its coagulated water, and the microcolumn experiment was carried out. Figure 7.65 shows the results for the coagulated treated water, and Figure 7.66 shows the results for Ishikari peaty colored water as it is.

In this case, the operating conditions of the experiment in which the breakthrough curve coincided showed that the coefficient n of the Freundlich adsorption isotherm was obtained by Equation 7.109, which yielded $n = 3$ for the coagulated water and $n = 4$ for the non-coagulated water. The results of the measurement of the breakthrough curve of the coagulated water adsorption on the microcolumn are as shown in Figure 7.67, and the results were integrated to determine the equilibrium adsorption amount. The results are given by $E_0 = 0.08$, $Q_0 = 3.8$ (l/g), and the adsorption isotherm plotted using the slope

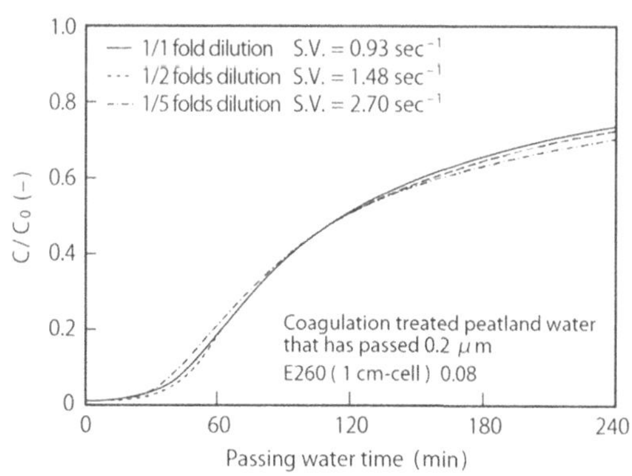

Figure 7.65 Comparison of breakthrough curves of samples with different raw water concentrations (coagulated treated water).

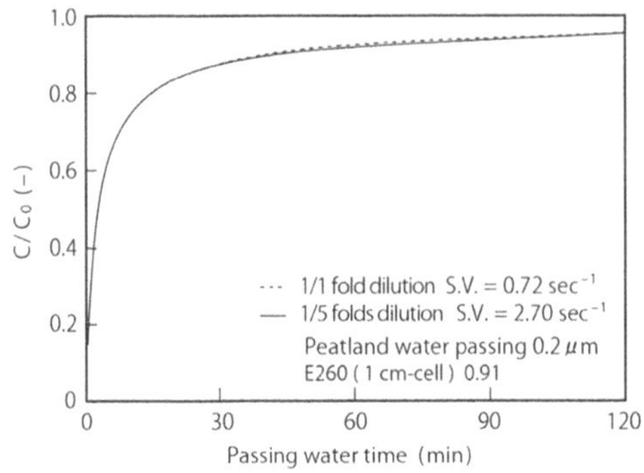

Figure 7.66 Comparison of breakthrough curves of samples with different raw water concentrations (Ishikari peatland water).

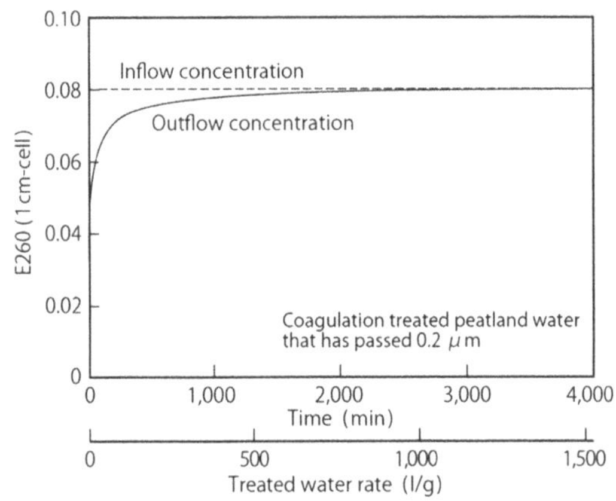

Figure 7.67 The breakthrough curve of the coagulated treated peatland water.

$n = 3$ determined above is shown in Figure 7.68. It agrees well with the white circle data obtained in the batch adsorption test.

Taken together, these results make it possible to explore the physicochemical operating conditions of various fixed bed adsorption processes in a small sample and in a short time, using a microcolumn method using a general-purpose high-performance liquid chromatograph.

Evaluation of fixed-bed adsorption and removal properties when pesticides intermittently flow into water containing natural humic substances was next investigated. Natural water containing humic substances was coagulated to remove high molecular weight color components and then residual low-molecular weight color components (such as fulvic acids) were adsorbed in the activated carbon fixing layer. When there is an intermittent inflow of pesticides and similar into such a system, it is worth noting how the removal of pesticides proceeds, and whether the activated carbon adsorption layer, in which adsorption of low molecular weight humic substances is constantly proceeding, is effective for pesticide removal. Professor Y. Matsui of the Department of Environmental Engineering at Hokkaido University is currently engaged in a variety of activities on this topics, as follows.

The pesticides used in the experiment include cymadine, napropamide, pentazone, asram, and hydroxy isoxazole, which are commonly used as herbicides and insecticides. These pesticides were dissolved in

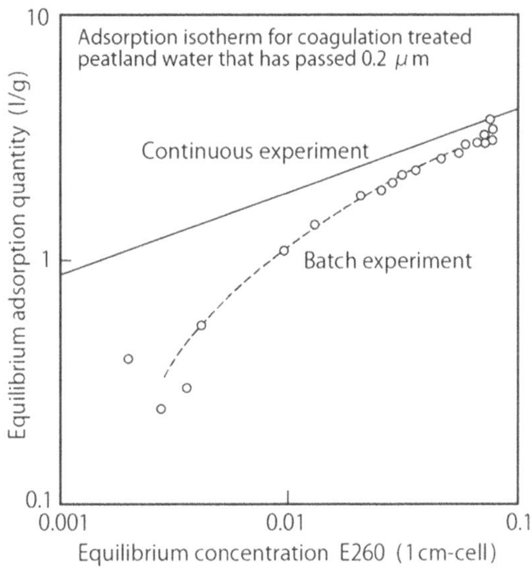

Figure 7.68 The adsorption isotherm for coagulated treated water.

coagulated Ishikari peaty groundwater for the experiment. The adsorption experiment was carried out by the microcolumn method, which can shorten the daily amount of work to about one hour, according to the law of similarity described above. Two types of water were prepared with the same concentration of humic substances: coagulated peaty groundwater containing agricultural chemicals and coagulated peaty groundwater containing no agricultural chemicals. As shown in Figure 7.69, the inflow of water containing pesticides into the activated carbon bed was conducted in a pulsed manner.

The pulse duration of the micro experiment was 3 minutes, and it is assumed that the pulse duration is 3 days in real-time terms. The concentration of humic substances in the effluent was measured continuously at an ultraviolet absorbance of 260 nm, and the concentration of pesticide effluent was measured at 1-minute intervals. Values of 1 to 2 minutes and 2 to 3 minutes were used as the leak concentration from the activated carbon adsorption layer.

Figures 7.70–7.74 show the C/C_0 pesticide spill rates when pesticides such as cymadine, napropamide, pentazone, aslam, and hydroxy isoxazole, which have different water solubilities, are considered to be intermittently flowing in the activated carbon-fixing layer to which humic matter is constantly loaded.

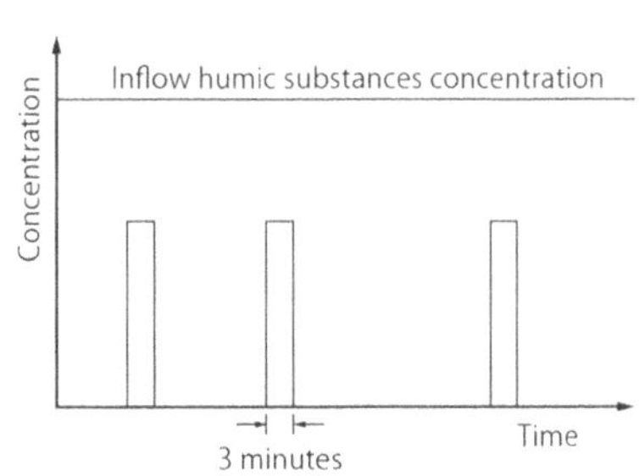

Figure 7.69 Inflow concentration variation pattern.

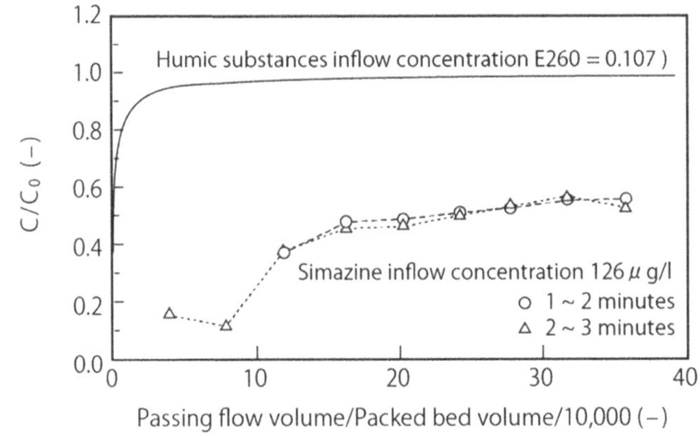

Figure 7.70 Variation of outflow concentration to intermittent inflow of simazine.

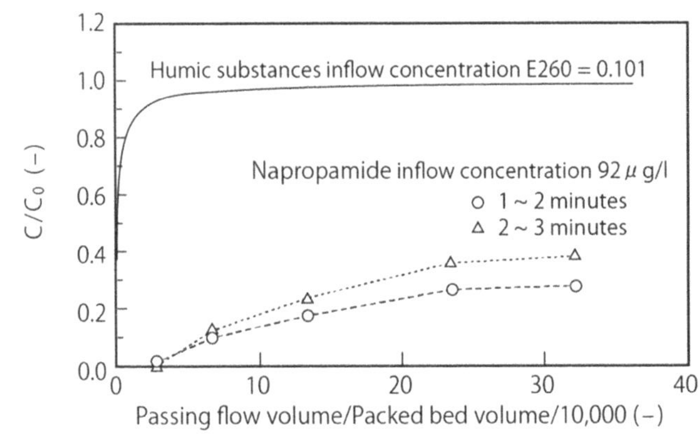

Figure 7.71 Variation of outflow concentration to intermittent inflow of napropamide.

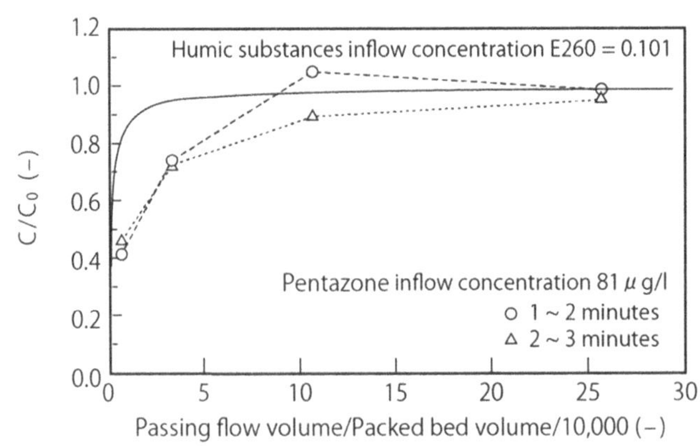

Figure 7.72 Variation of outflow concentration to intermittent inflow of pentazone.

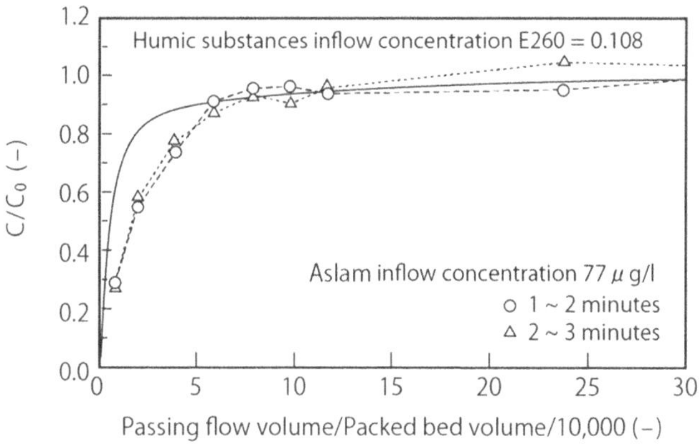

Figure 7.73 Variation of outflow concentration to intermittent inflow of aslam.

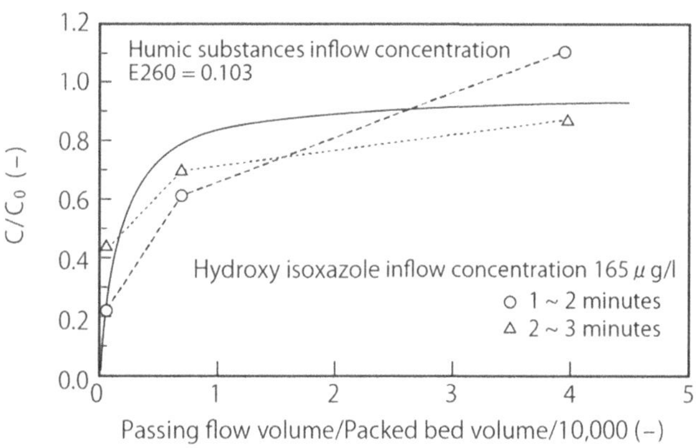

Figure 7.74 Variation of outflow concentration to intermittent inflow of hydroxy isoxazole.

The symbol ○ indicates the rate of outflow based on the concentration 1 to 2 minutes after intermittent inflow, and the symbol Δ indicates the rate of outflow based on the concentration 2 to 3 minutes after intermittent inflow. The symbols indicate the change corresponding to the specific water flow rate of humic-containing water until the time of intermittent inflow of agricultural chemicals. The figures show that the removability of pesticides intermittently flowing into the fixed layer, which permanently passes water containing humic substances, is greatly influenced by the adsorption history of humic substances so far.

7.1.6 Activated carbon treatment for odor and taste

Unpleasant tastes and odors are the most sensitive components of impurities that impair the value of water even in the presence of trace amounts. Sense of smell and sense of taste are possibly the most sensitive indicators by which people notice abnormalities in food and drink (water, in this case), and this ability has likely supported our physiological safety.

The control of odor and taste can be sub-optimally treated using oxidative decomposition by ozone or similar, without removing the causative substance, and in many cases it is widely preferred and widely used because it is cheaper.

7.1.6.1 Measurement and evaluation of odor and taste

Substances that produce odors and tastes are complex organic compounds and are not easily evaluated individually by analytical chemistry, either qualitatively or quantitatively. However, a large percentage of consumers' dissatisfaction with tap water quality, and thus much of their distrust of water supply, is attributable to these trace odors and tastes. The unpleasant taste and odor caused by inorganic components are generally perceived for the first time at concentrations as high as mg/l. However, organic compounds derived from microorganisms of 500 Da or less, which are produced by algal metabolism and decomposition of organic matter by actinomycetes, cause the sense of odors and tastes to be unpleasant even at low concentrations of ng/l. Chemical analysis shows that a large percentage of these components are required to be identified individually by gas chromatography-mass spectrometry (GC-MS). Geosmin and 2-methyl-isoborneol (2-MIB) are treated as the main odorants of soil odor and mold odor, as a target of control and prevailing indicators (Figures 7.75 and 7.76). In addition, various odors such as fish, mold, hay, soil, potato storage tank, fruit, ammonia, disinfectant and hydrogen sulfide odor are also problematic.

In order to test to see if human smell and taste senses are excellent detectors, and with the ultimate objective of ensuring that human discomfort is not felt, a comprehensive evaluation was conducted primarily using human sensory organs as sensors. As a degree of offensive odor, the Threshold Odor Number (TON) has been used as a control indicator in Japan and overseas. The principle is simple, requiring the test water to be diluted multiple times with odorless water (and recording the number of times), until the point at which no odor is sensed. In terms of offensive odor level, $TON = (A + B)/A$ (where A is the amount of test water and B the amount of odorless water (ml).

In general, 200 ml of test water was placed in a 500 ml flask, and the evaporation test was performed by shaking the flask gently at a high temperature of 60°C, which accelerates the evaporation of the offensive odor, to 40°C, which is close to normal temperature. In some cases, it is not easy to obtain diluted water that is odorless. Simple distillation and ion exchange water refined in the field can be used to dilute the taste and

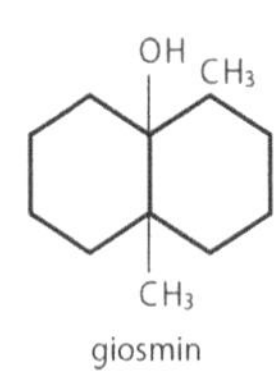

Figure 7.75 Geosmin.

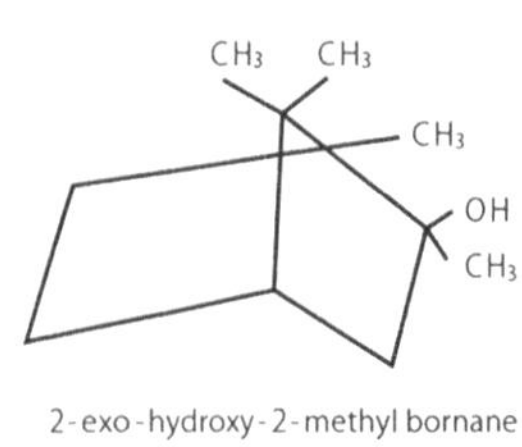

Figure 7.76 2-methyl-isoborneol (2-MIB).

odor, and the acquisition and storage of odorless water requires particular consideration. For ventilation of the laboratory, careful attention should also be paid to air conditioning. In addition, the sensitivity of most people who sense odors differs by a factor of 10, and the sensitivity of certain people can differ by a factor of more than 100. In order to obtain an average value, a panel group of several people in a normal range should be kept on a daily basis to ensure objectivity and to obtain data. In order to detect abnormalities with high sensitivity, it is important to select a person (a master taster) who is particularly sensitive, and uses their sensitivity on a daily basis in their daily work, and ask them to take the lead.

7.1.6.2 Powdered activated carbon (PAC) and granular activated carbon (GAC) treatment

Powdered activated carbon (PAC) is added directly into raw water, rapid mixing ponds, etc., and as long a contact time as possible is allowed before coagulation separation. By contrast, granular activated carbon (GAC) is usually added to fixed bed or fluidized bed adsorption after coagulation sedimentation and filtration. When fluidized bed adsorption is carried out, as in Tokyo Metropolitan Water Works, in some cases the GAC is added and the water then filtered in order to avoid the outflow of microorganisms and crushed charcoal particles generated in the fluidized bed.

In many cases, operating conditions are determined based on the long-term test results from a pilot plant. The preceding microcolumn test is a preliminary test that does not include biological reactions, but is a useful preliminary method. In low-concentration states, such as with 2-MIB or geosmin concentrations of 2–60 ng/l, PAC requires dosing rates of the order of 5–20 mg/l. If the concentration of MIB or geosmin is as high as 100 to 150 ng/l, add more than 20 mg/l of PAC and consider a contact time of about 3 hours.

In the case of fixed bed adsorption (GAC) and PAC adsorption, competition (adsorption inhibition) with humic substances (colored components) occurs in the adsorption of 2-MIB and geosmin, as described in the case of pesticides, etc.

In GAC processes such as in fixed bed adsorption or fluidized bed adsorption, it is observed that the time required to reach adsorption saturation and for breakthrough to occur is much longer than the amount of physical adsorption assumed. This phenomenon originates from the fact that, once adsorbed, taste and odor substances are decomposed by the microbial reaction in the GAC layer, and the activated carbon adsorption capacity is regenerated by the microorganisms. It is considered that biodegradable (slowly degradable) odor and taste substances are removed from the water system by activated carbon adsorption, and then decomposed over time on the inner surface of the pores of the activated carbon to regenerate the inner surface. In this case, GAC may be thought of as a medium providing a reaction site to lend time to microbial degradation of taste and odor substances, and depending on the operating conditions, adsorption saturation may not be reached and regeneration may sometimes be almost unnecessary. However, since theoretical clarification has not been achieved, a pilot test for a long period of time is required, depending on the source water.

7.2 ION EXCHANGE

Ion exchange is widely used for water treatment, when softening of hard water, production of pure water, and complete removal of harmful inorganic substances in wastewater are required. The ion-exchange operation being carried out is the same as in the adsorption process operation itself, in that impurities are removed from water by phase transport from the liquid to the solid phase, and the actual process, system configuration, and speed are almost the same as those of adsorption. However, as a separation process used for water treatment, adsorption mainly deals with physical changes, whilst the ion exchange process basically deals with chemical changes (reactions).

7.2.1 Ion exchange phenomenon

In ion exchange, a reversible ion exchange phenomenon occurs between a solid and a liquid without changing the main structures of the solid. The solid part is referred to as the ion exchanger. Ion exchangers include inorganic and organic exchangers, and organic ion exchangers (mostly ion exchange resins) are used for water treatment. However, we cannot overlook the role of inorganic ion exchangers in the field of environmental engineering, such as the ion exchange phenomenon of clay in soil, ion exchange capacity in the case of coagulation, and the use of natural zeolite in the treatment of radioactive materials.

Ion exchange phenomena were first known in 1850, with the British pair Thompson (1850) and Way (1850) both independently discovering (during the study of ammonium in soil) that ammonia replaces calcium and potassium ions in soils.

Ion exchange reactions have been promoted in research areas related to the soil retention of fertilizing ingredients such as nitrogen, phosphorus, and potassium, which are important for agriculture. One such relationship is the agglomeration structure of soil and the retention of calcium ions. The use of ion exchange for water treatment began in the 20th Century when German chemist R. Gans (1905) used natural and synthetic aluminium silicate compounds for water softening. Since Adams and Holma (1935) developed cation exchange resins from phenols and formaldehydes, a wide range of current ion exchange resin technologies have evolved.

For organic and inorganic ion exchangers to work effectively, it is necessary that (1) the exchanger itself contains ions; (2) the body is insoluble in water; and (3) that it is porous with a three-dimensional network structure.

Ion exchangers have hydrophilic ionic groups such as $-SO_3H$ and $-COOH$, etc. inside the network. When immersed in water, water molecules penetrate into the exchangers, and the ionic groups are ionized and separated into cations and anions. At this time, one of the ions is bound to the network structure and cannot move, and the ion that can move is exchanged with the same kind of ion in the water. The former is called an exchange group ion or a fixed ion, and the latter is called an exchange ion, a movable ion, or a counter ion.

Counter ions move (vibrate) between fixed ions. For example, when ion exchange resins are introduced into an NaCl electrolyte solution, the counter ions stoichiometrically exchange with ions of the same sign in the electrolyte solution. If resin R has a free acid group, the H^+, which is the counter ion of the resin, replaces the Na^+ of the solution, and the exchanged ion changes from H^+ to Na^+. This is shown in a chemical formula as $RH^+ + Na^+ \leftrightarrows RNa^+ + H^+$. And the resin was changed from an H-type to an Na-type by the ion exchange, where R denotes the base of the resin.

Such a reaction is called an ion exchange reaction, and is basically a reversible reaction. Therefore, the resin can be restored to its original ionic form by changing the conditions, and this process is called regeneration.

Figure 7.77 schematically shows how this type of ion exchange occurs in the case of cation exchange. As shown in the figure, an ion exchanger has a fixed neutral base layer (R), which is the matrix of the exchanger, and a fixed anion layer (in the case of a cation exchange resin), which is chemically bonded to the matrix. In order to remain electrically neutral to this fixed ion layer, there is a movable cation layer with the opposite sign, forming an ion double layer. The ions of the mobile ion layer tend to diffuse into the liquid phase by thermal motion, but they are confined to the electrostatic force of the fixed anion layer fixed on the resin

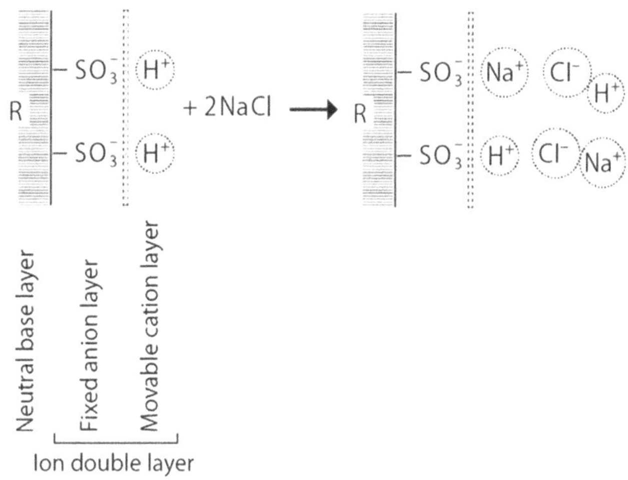

Figure 7.77 A schematic diagram of the exchange action of an ion exchange resin.

surface and exist in a Boltzmann distribution (as described above in Chapter 3; see Figures 3.13 and 3.14) and are attracted from the solution body to the surface of the exchanger to be held in the mesh structure of the exchanger.

7.2.2 Ion exchanger

Many solids, whether natural or synthetic, exhibit ion exchange phenomena. Cell bodies such as soil, humic soil, wool fabric, cotton fabric, and bacteria are not commonly referred to as ion exchangers, but they have ion exchange capabilities. We particularly refer to ion exchangers as substances with ion exchange capacities that can be used for ion exchange-centered treatment. These categories are shown in Table 7.10.

The exchange capacity of inorganic silicates is about 10 meq/100 g, which is remarkably small relative to the 10^0 to 10^1 meq/g of ordinary exchange resins. Natural and synthetic inorganic ion exchangers such as green sands and zeolites have an exchange capacity close to that of ion exchange resins, but their strength is weak, and their usable pH range is as narrow as about 6 to 8. Therefore, the most commonly used exchangers in engineering are based on ion exchange resins. For this reason, we will consider water treatment processes centered on ion exchange resins. However, it should be noted that the presence of a small volume of ion exchange capacity in the natural environment plays an important role in understanding soil pollution, water purification, and pollution processes. In other words, the role of ion exchange in soil against radioactive and heavy metal contamination greatly influences the understanding of the behavior of river and lake sediments and other trace organic and inorganic components.

Table 7.10 Categories of ion exchangers.

Category		Example
Inorganic ion exchanger		Zeolite
Organic ion exchanger	Carbonaceous ion exchanger	Coal sulfide
	Synthetic resin ion exchanger	Cation exchange resin
		Anion exchange resin

7.2.3 Types of ion exchange resins

An ion exchange resin is a spherical solid with a three-dimensional network of hydrocarbons. The solid has ionizable functional groups (exchange groups) on its inner and outer surfaces, which correspond to organic acids and inorganic bases. In this context, the nature of the hydrocarbon framework is related only to the magnitude of the interchangeability, and the chemical nature of what type of exchange is performed depends on the nature of the interchange group (ionizing group) attached to the framework.

Hydrocarbons that form the framework of ion exchange resins, which are widely used in general, are often copolymers of styrene and divinylbenzene (DVB), which affects the resistance of the resins to oxidization, reduction, mechanical abrasion, and fracture. It also affects the difficulty of molding spheres with good hydraulic properties.

There are four major exchange groups that determine the chemical reactivity of ion exchange resins: strong and weakly acidic cation exchange resins, and strong and weakly basic anion exchange resins.

7.2.3.1 Strong-acidic cation exchange resin

Most strong-acidic cation exchange resins are made by suspension polymerization and sulfonation of styrene-DVB. Sulfonic acid resin is the only strong acid resin commercially available, and since it is a strong acid, both acid and base types can be ionized in the entire range of acidity and alkalinity. Therefore, ion exchange is possible in the entire pH range, and neutral salt is decomposed to form an acid as shown in the chemical formula: $(RSO_3)\ H + NaCl \rightarrow (RSO_3)\ Na + HCl$ where R denotes the base of the resin.

This reaction is called the neutral salt decomposition reaction (salt splitting) and is used in the first stage of the total desalination process. The exchange capacity is about 5 meq/g and the reaction rate is high.

7.2.3.2 Strong basic anion exchange resin

Usually, the exchange group used for a strong basic anion exchange resin is a basic quaternary ammonium, such as: $–CH_2^{-}$, $–N^{+}$, $–(CH_3)^{-}$. Together, they are classified into type I and type II. Their structures are shown in Figure 7.78, and type II has one of the methyl groups of type I replaced by an ethyl group.

Quaternary ammonium anion exchange resins are highly ionically dissociated and can be used in the entire pH range. Like strongly acidic cation exchange resins, quaternary ammonium anion exchange resins have a neutral salt decomposition function and make neutral salts strongly basic, as shown in the following reaction formula: $(RNR'_3)\ OH + NaCl \rightarrow (RNR'_3)\ Cl + NaOH$, where R' is an alkyl group, or similar.

By placing such a strongly basic resin in series after a strongly acidic resin, all ions in the water can be removed and used for the production of pure water. It is also used to regenerate uranium, remove hydrogen sulfide and mercaptan (methanethiol) from hydrocarbons, recover chromic acid, neutralize acid, and recover iron from hydrochloric acid solutions. Although this type of resin has a very high exchange rate, it is susceptible to heat and is contaminated by organic matter in water, resulting in rapid deterioration of the exchange capacity.

7.2.3.3 Weakly acid cation exchange resin

Cation exchange resins with carboxylic groups (−COOH) as exchangers, which are highly ionized only as a salt, have an ion exchange capacity which works only on the alkaline side above pH7. It is highly selective to polyvalent metal ions, has great ability to remove basic substances, and is useful for softening of water. Chelating resins resemble this weak acid form and are more selective.

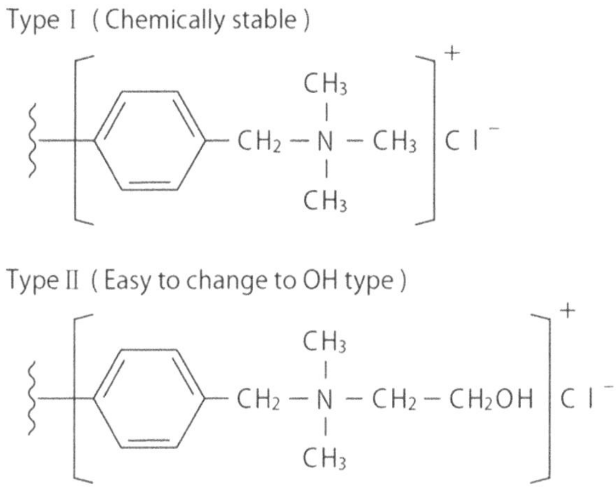

Figure 7.78 Examples of strong basic anion exchange resins.

7.2.3.4 Weakly basic anion exchange resins

Many exchange groups contain polyamines, such as primary, secondary, and tertiary amines (R≡NHOH, R=NH_2OH, R-NHR_3OH), which have representative properties similar to amines or ammonium hydroxide, dissociate only in the form of salts, are less interchangeable on the alkaline side, and are used only in acidic to neutral states.

What is common to the weak acid and weak alkali types is that most of the recycled free resins are in the undissociated state, such as acetic acid and ammonia, and do not act as ionized exchange groups. Therefore, the amount of Na^+ and Cl^- adsorbed varies depending on the pH and concentration of the solution.

Since the dissociation of the exchange group is small, it does not have the neutral salt resolution capability which strong acid and strong alkali types do, and it is used as an adsorption resin to remove free acid from the liquid system. For example, the following reactions can occur: $RNH_2 + HCl \rightarrow RNH_3Cl$ or $RNH_3OH + HCl \rightarrow R\ NH_3Cl + H_2O$.

Although the exchange reaction of weakly basic exchange resins for acids is slower than that of strongly alkaline resins, the reaction of converting one salt to another proceeds rapidly, as in the following example (Equation 7.110)

$$\begin{aligned} &RNH_2 + HCl \xrightarrow{\text{slow}} RNH_3Cl \\ &RNH_3Cl + HNO_3 \underset{}{\overset{\text{fast}}{\rightleftharpoons}} RNH_3HNO_3 + HCl \quad —(7\cdot110) \end{aligned} \tag{7.110}$$

Equation 7.110 The exchange reaction of weakly basic exchange resins for acid.

Table 7.11 shows the operating pH range of each ion exchange resin.

7.2.3.5 Gel and macroreticular (MR) resins

Unlike the classifications based on chemical properties described above, we now consider differences in the characteristics of resin structures.

The basic structure of gel and macroreticular (MR) resins is composed of polymeric electrolyte gels that are sterically linked (cross linked) with DVB, from a few to up to 30% (usually about 10%). The schematic structure of gel resins is shown in Figure 7.79, and the pore diameter of the gel is said to be about 20 Å.

Table 7.11 Type of ion exchange resin and its operating pH range.

Types	Operating pH
Strongly acidic cation exchange resin	pH whole range
Strongly basic anion exchange resin	pH whole range
Weakly acidic cation exchange resin	pH7 or more alkaline region
Weakly basic anion exchange resin	Acidic to neutral region

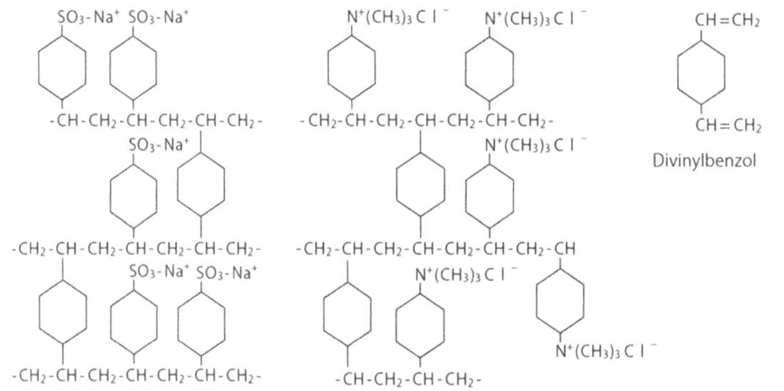

Figure 7.79 Macroreticular (MR) resin structures.

However, it is difficult to accurately measure the diameter of the resin's internal pores in this case because it exists in a swollen state immersed in water and disappears when dried.

In the case of styrene and DVB resins, which are widely used, each monomer is uniformly dissolved and suspension polymerized in water to form resin particles. The ion exchange resin obtained by this method has a homogeneous gel structure, and the degree of cross-linking that governs the three-dimensional structure of the resin matrix changes depending on the DVB content, and the pore size of the resin is thus determined. If the ratio of DVB is large, the cross-linking becomes dense and the resin becomes strong, but when immersed in water, the swelling ratio becomes large and sulfonation becomes difficult. In addition, the internal pores are small and large ions cannot be exchanged, resulting in a decrease in regeneration efficiency. Therefore, in many cases, when constructing a three-dimensional network structure by copolymerization, about 10% of DVB is used.

MR or porous resin is designed to increase the size of the inner pores without decreasing the strength of the resin by a specific polymerization method, and to effectively work against large ions. Taking styrene-DVB MR resin as an example, when copolymerizing styrene and DVB, a third liquid is added which can form a homogeneous solution with the styrene and DVB monomer but is not catalytically active and is chemically inert, and has the property of easily separating from the polymerization when the copolymer is recovered by solid-liquid separation. This liquid forms a homogeneous solution with the monomer, but since it has no solvent action and is inert, the polymerization of the monomer proceeds, and as the solubility decreases, a liquid release phenomenon that separates from the copolymer occurs.

A network structure having a pore diameter of 100 to 200 Å remains inside the particles of the copolymer synthesized in this manner, and this structure does not constitute a part of the gel structure and has high strength, and does not shrink and disappear even when dried. Even organic substances with large molecular weights, such as humic substances, can be adsorbed by anion exchange MR resins, and the diffusion rate in the pores increases. In order to remove humic substances from natural water, this product is used instead of activated carbon adsorption as a pretreatment for a pure water desalination exchanger for the treatment of ultra-high-pressure boiler water at thermal power stations.

7.2.4 Chemical properties of ion exchangers (resins)

7.2.4.1 Ion exchange capacity

Ion exchange capacity is the number of fixed ion sites per dry weight, or the volume of water swelled exchangers involved in ion exchange reactions, expressed as eq/kg-resin, meq/ml resin, or meq/g resin.

For example, when HR-type acid-type resins are titrated with NaOH, the resulting H^+ is neutralized with OH^- to form H_2O, and when the pH reaches the equivalent point, the pH of the system begins to rise. The NaOH equivalent required to reach this equivalence point indicates the exchange capacity of the resins. Measurement of this is called static ion exchange capacity, and when the ion is smaller than the network structure of the resin, it indicates the maximum ion exchange rate – also called total ion exchange capacity.

However, when ion exchange resins are used in actual water treatment processes, ion exchange is usually carried out using a fixed or fluidized bed. Therefore, the breakthrough point at which unreacted ions flow out is obtained by conducting a continuous flow test with a fixed bed in accordance with the operating condition, which reveals the breakthrough capacity or operation exchange capacity for a given operation condition as the exchange capacity (meq/ml). Practical standard values take into account the effects of equipment size, shape, concentration of stock solution, regeneration method, etc.

The ion exchange capacity varies greatly depending on whether the acid or base of the exchange group is a strong acid (strong base) or a weak acid (weak base). This capacity is shown in Figure 7.80 by plotting the pH titration curve for the resin, which limits the range of pH that can be exchanged, but the exchange capacity for weak acid (weak base) exchange resins is markedly greater.

Ion exchange processes for water softening are generally carried out with about 10% NaCl, but seawater is used in coastal areas (in Florida and California, for example) to make economies in regenerating chemicals, and regeneration of the exchange capacity is repeated frequently (in short time cycles).

7.2.4.2 Ion exchange equilibrium and exchange selectivity

Although the ion exchange reaction rate is slow, it eventually reaches equilibrium. This equilibrium is highly reproducible. Generally hysteresis due to operating conditions is negligible, can be taken up stoichiometrically, and can be expressed by the law of mass action.

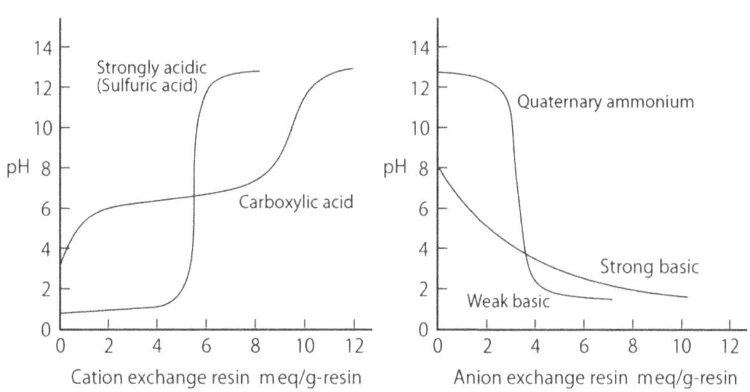

Figure 7.80 The pH titration curves for cation and anion exchange resins.

The amount of HCl in the solution at equilibrium (when equilibrium is reached in a few minutes) is determined by neutralizing titration with a mixture of a strong acid-type R-H cation exchange resin and a NaCl solution. The amounts of NaR, HR, and NaCl are determined by the concentrations of the exchange volumes NaR and HCl (the total concentration of HCl in the solution). If we apply these relationships analogously to the laws of mass action, as in conventional chemical reactions, we arrive at Equation 7.111.

$$R\text{-}H + Na^+Cl^- \rightleftarrows R\text{-}Na + H^+Cl^-$$

$$K_H^{Na} = \frac{[R\text{-}Na]_r}{[R\text{-}H]_r} \times \frac{[H^+]_s}{[Na^+]_s} \quad \text{—(7·111)} \tag{7.111}$$

$[R\text{-}Na]_r$, $[R\text{-}H]_r$: Adsorption quantity inside resins (meq/g)
$[H^+]_s$, $[Na^+]_s$: Ion concentration of solutions (meq/ l)
K_H^{Na} : Selectivity coefficient (-)

Equation 7.111 The selectivity coefficient in strong acid type cation exchange resin.

When the selectivity coefficient obtained in this way is obtained by combining various resins and ions, it is possible to express the magnitude of the ion exchangeability relatively. In general, the relationship shown in Equation 7.112 is approximated by a heterogeneous combination.

$$\frac{K_{M_1}^{M_3}}{K_{M_1}^{M_2}} \fallingdotseq K_{M_2}^{M_3} \quad \text{—(7·112)} \tag{7.112}$$

Equation 7.112 The approximate relationship of selectivity coefficient.

The following rules of thumb relating to the magnitude of the ion exchange properties obtained in this way are available for reference in planning the elementary selection of ion exchange processes:

(1) Exchange ions of dilute aqueous solution at room temperature are more easily exchanged as the charge becomes larger. $Na^+ < Ca^{2+} < Al^{3+} < Th^{4+}$
(2) In the case of a dilute aqueous solution at room temperature, ions having the same charge in the solution are more easily exchanged with ions having a larger atomic number, such that:
$Li^+ < H^+ < Na^+ < K^+ < NH_4^+ < Rb^+ < Cs^+ < Ti^+ < Ag^+$
$Be^{2+} < Mg^{2+} < Ca^{2+} < Sr^{2+} < Ba^{2+}$
$F^- < OH^- < CH_3COO^- < HCO_3^- < Cl^- < NO_2^- < HCO_3^- < HSO_4^- < CN^- < Br^- < NO_3^- < I^- < CNS^-$
However, high-concentration solutions do not work.
(3) In high concentration states, the difference in ease of exchange due to the difference in charge is reduced and sometimes reversed.
(4) The ease with which hydrogen ions and hydroxyl ions are exchanged varies greatly depending on the functional group of the resin, and it is related to the strength of the acid and base produced by the action with the functional group. As the strength of acid or base increases, the ease of exchange decreases.
(5) High-molecular organic ions, metal complex ions, etc. exhibit high exchangeability.
(6) When the degree of cross-linking or the proportion of fixed ions in the ion exchanger is low, the selectivity coefficient approaches 1.

Table 7.12 Rank of ion selectivity.

Cation	Anion
Fe^{3+}	$CrO_4{}^{2-}$
Al^{3+}	$SO_4{}^{2-}$
Pb^{2+}	$SO_3{}^{2-}$
Ba^{2+}	$HPO_4{}^{2-}$
Sr^{2+}	CNS^-
Cd^{2+}	CNO^-
Zn^{2+}	$NO_3{}^-$
Cu^{2+}	$NO_2{}^-$
Fe^{2+}	Br^-
Mn^{2+}	Cl^-
Ca^{2+}	CN^-
Mg^{2+}	$HCO_3{}^-$
K^+	$HSiO_3{}^-$
$NH_4{}^+$	OH^-
Na^+	F^-
H^+	
Li^+	

(7) Ions that do not participate in the exchanging action and counter ions do not significantly affect the selectivity factor. However, this is different when complex ions are formed.

Taken together, these points are summarized in Table 7.12, where the lower part elements and compounds in the table have lower affinity for an exchange resin. Therefore, for cation exchange resins, high concentrations of salt (10%) or hydrochloric acid (1N) are generally used to convert almost all of the resin exchange groups to Na or H and to form R-Na or R-H. The former are called Na type and the latter H type cation exchange resins. For anion exchange resins, treatment is with concentrated saline or sodium hydroxide to form R-Cl or R-OH. These are called Cl-type and OH-type anion exchange resins, respectively. Cyanide, bicarbonate, and silicate ions cannot be removed in Cl type anion exchange resins, and weak acids cannot be treated unless an OH type strong base anion exchange resin is used.

If the removal target is a simple inorganic anion or cation, refer to the selection order and select the appropriate exchange resin from the eight types shown in Table 7.13. Organic iron bound to humic substances in peaty soil is an organic complex, as described in the previous section, and behaves as a negatively charged colloid. Therefore, it is necessary to devise an anion exchange resin (in fact, an MR resin is used because of its large molecular weight).

Whilst referring to Dr Shimizu's original proposals (Shimizu, 1953), we also consider the following points when selecting an appropriate resin:

(1) Weak acid and weak basic ion exchange resin are used for strongly adsorbing ions, and strong acid and strong basic ion exchange resin are used for weakly adsorbing ions, thereby avoiding the difficulty of desorption and regeneration;

Table 7.13 Selection of ion exchange resin type.

Category	Types	Example of Exchange Group
Strongly acidic cation exchange resin	Free acid type	$R\text{-}SO_3H$
	Salt type	$R\text{-}SO_3Na$
Weakly acidic cation exchange resin	Free acid type	R-COOH
	Salt type	R-COONa
Strongly basic anion exchange resin	Free acid type	$R\text{-}NR'_3OH$
	Salt type	$R\text{-}NR'_3Cl$
Weakly basic anion exchange resin	Free acid type	$R\text{-}NH_3OH$
	Salt type	$R\text{-}NH_3Cl$

R′: Methyl group.

(2) For the adsorption of salt constituent ions, an acid type resin and a base type resin are used for the exchange with a base and an acid, respectively. In particular, a resin of a weak acid type and a weak base type can be expected to have a high effect;

(3) For complete desalination, a strong acid type and a strong base type exchange resin are used in combination;

(4) Separation processes must be arranged to make an effective system, in order from weakly exchangeable resins to strongly exchangeable ones.

7.2.5 Ion exchange operation

7.2.5.1 Classification of operation schemes

Once an appropriate ion exchange resin has been selected for the particular water quality factor to be treated, it must be decided in what manner to perform the ion exchange process. The batch and column methods are used for water treatment. In most cases, a column (continuous flow) adsorption method is used.

Usually, the column process is carried out using an ion exchange column, being classified into the following three process methods: fixed-bed; moving-bed; or fluidized-bed. The moving-bed and fluidized-bed methods are countercurrent ion exchange processes.

These methods are classified and illustrated in Figure 7.81; the fixed-bed method is the most commonly used.

In the fixed-bed method, a single-bed type uses only one type of resin, whilst a multi-bed type uses multiple resin beds in series. Single beds do not have a sufficient effect and multiple-bed systems are more often used. The double bed type is used for desalination using resin beds of different anionic and cationic systems in series. A mixed bed is used when the same exchange column is filled with a mixture of cation and anion exchange resins, and different resins are operated in close proximity to each other to produce highly demineralized water.

A moving-bed (fluidized bed) process is used for ion exchange of solutions containing coarse suspensions or for continuous regeneration. The moving bed process is a method used for special purposes.

Of the resins packed into a fixed bed, only a portion of the resin corresponding to the exchange zone which is actually used for ion exchange is fed into the liquid, and the extracted resin is continuously regenerated. This method is useful when it is necessary to use a small amount of resin, and when it is desired not to contact the resin for a long time, such as when radioactive materials are removed.

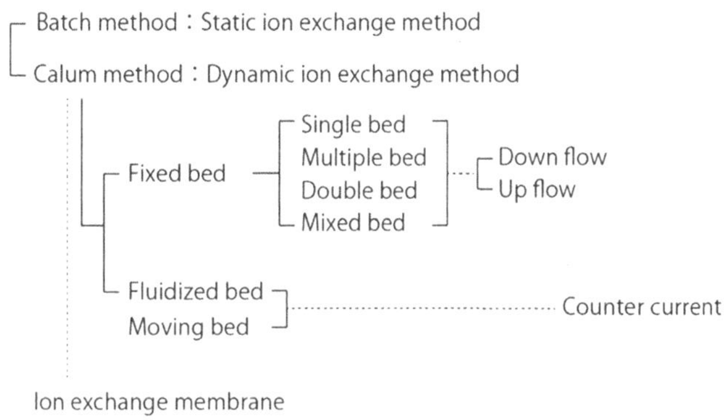

Figure 7.81 Classification of process methods of ion exchange.

7.2.5.2 Fixed-bed ion exchange systems and operation

Fixed-bed ion exchange systems are most widely used, and are typically constructed as shown in Figure 7.82, in which a resin with an effective diameter of about 0.5 mm is filled for 10–30 cm. The ion exchange process is characterized by repeated use of resin. The process repeats the following five steps: (1) water treatment (exhaustion); (2) back washing; (3) regeneration; (4) extrusion (displacement); and (5) wash (rinsing).

For this purpose, in addition to the fixed bed of the ion exchange unit main body, we have a regeneration chemical supply system, a discharge system, and a backwash system. Concentrated solutions containing

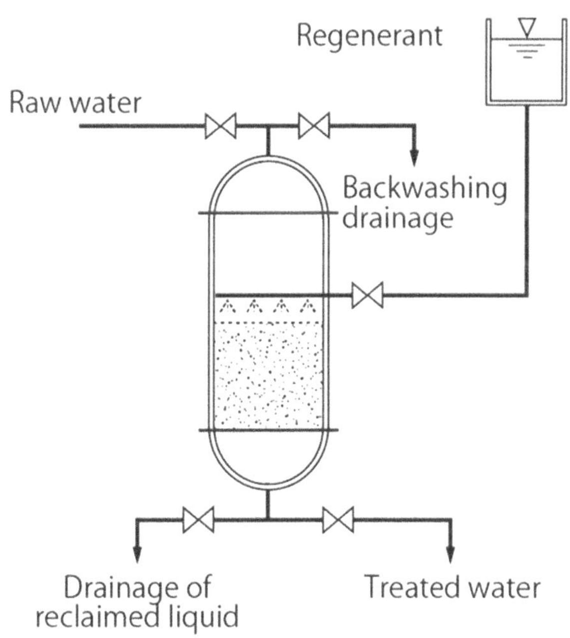

Figure 7.82 Schematic of a fixed bed ion exchange system.

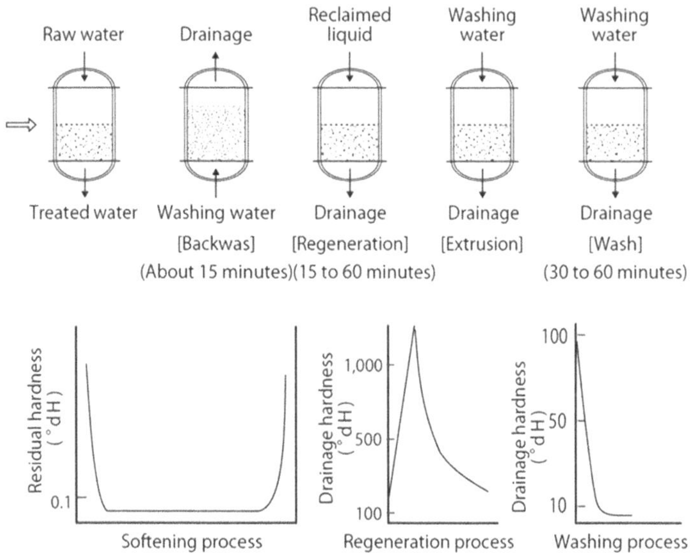

Figure 7.83 The steps involved in fixed bed water softening.

several percent or more of salt, sulfuric acid, hydrochloric acid, and sodium hydroxide are used for regeneration, so corrosion resistance is required for the materials used.

In terms of the flow rate of water through the treatment, space velocity (SV) (1/hr) = flow rate (m^3/hr)/ resin (m^3) of 10–30 (1/hr) is commonly used. Regeneration takes about 1 to 2 hours, including cleaning. To avoid interruption of operation during regeneration, we have devised a method of continuous treatment and regeneration. Figure 7.83 schematically illustrates these steps for an example of water softening.

Water flow for treatment continues until the breakthrough point, at which time the ions to be removed reach the allowable concentrations of the effluent (treated water).

In a fixed bed column, the reaction $BR + A \rightarrow AR + B$ proceeds (where A is the ion to be removed, B is the exchange ion on the exchange resin, and R is the exchange resin). The reaction proceeds only in the relatively narrow exchange zone H_1 to H_2 shown in Figure 7.84; between 0 and H_1 the resin reaches exchange equilibrium, and between H_2 and H it has not yet exhibited exchange action. This is the same as the adsorption zone of a substance with a high adsorption rate, such as phenol, in the activated carbon fixed bed adsorption process.

Leakage starts between H_2 (the end of the exchange zone) and point H (the bottom of the column). When the width of the exchange zone is narrow, all the resin works effectively, and the leakage is clearly observed and rapidly progresses. This is the case when ion A has a large selection factor (selectivity coefficient) for B. However, the thickness of the resin layer is generally desired to be 60 cm or more, although it is influenced by the magnitude of the selection factor. The water flow rate (space velocity; SV) of the treatment is 80 (1/hr) or less in the case of strongly acidic cation resin exchange, and SV ≒ 40 (1/hr) is often adopted in the case of anion exchange resin.

When ion exchange treatment (water flow) is completed at the breakthrough point, ion exchange resins are regenerated. Water is passed upward from the bottom of the resin columns, and the resin bed is fluidized for about 10 minutes (expansion of 50–70%, linear velocity (LV) ≒ 10 m/h) and the suspension retained in the resin bed surfaces and the particle gaps is washed away. In the case of a mixed bed type, by setting the

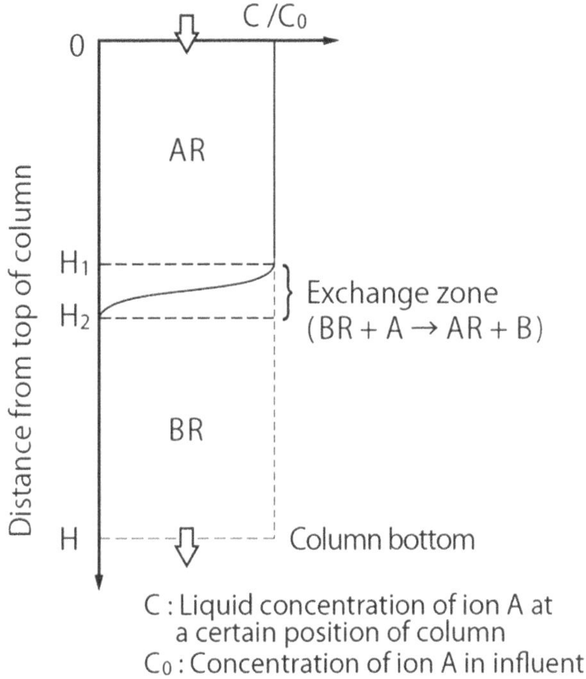

Figure 7.84 How ion exchange proceeds in a fixed bed. (A is the ion to be removed, B is the exchange ion on the exchange resin, and R is the exchange resin).

backwash flow rate appropriately, it is possible to separate a high-density cation exchange resin and a low-density anion exchange resin into two layers for proper regeneration.

After the backwash is completed and the resin has settled and stops, the regeneration liquid is slowly introduced into the exchange bed in a downward flow (SV ≒ 4 to 8 (1/hr)). As regeneration agents, acid is used for H-type exchange resins, chlorine is used for OH-type exchange resins, and salt is used for neutral exchange resins. Strong electrolytes such as hydrochloric acid, caustic soda, and salt are used. In this case, use of regenerants that cause the desorbed ions to precipitate are not used and nitric acid that may damage the ion exchange resin body should be avoided.

Regeneration can be either (1) complete, in which all the resin-like exchange groups are replaced by regeneration agent ions; or (2) partial, in which only a portion is replaced. In the industrial sector, it is common, for increased economy, to use repeated partial regeneration, as appropriate. In partial regeneration, since the exchange capacity varies depending on the regeneration conditions, the balance between the regeneration conditions and the total throughput must be clearly evaluated. Usually, regeneration levels of regenerant per unit of resin (g/l) are shown; levels of 1.5 to 3 times the theoretical amount seem to be used.

When regeneration is completed, the wash solution is poured downward to collect the regenerated solution. The water flow rate of the wash solution should be the same as that for regeneration. This recovered regenerant is reused in the first stage of regeneration in the next cycle. When extrusion is complete, flush the wash water downward to remove the remainder of the recycled water. Normally, this is performed for 30–60 minutes, LV≒10 m/hr.

This completes one cycle and the process then enters the next cycle.

Figure 7.85 shows a series of steps.

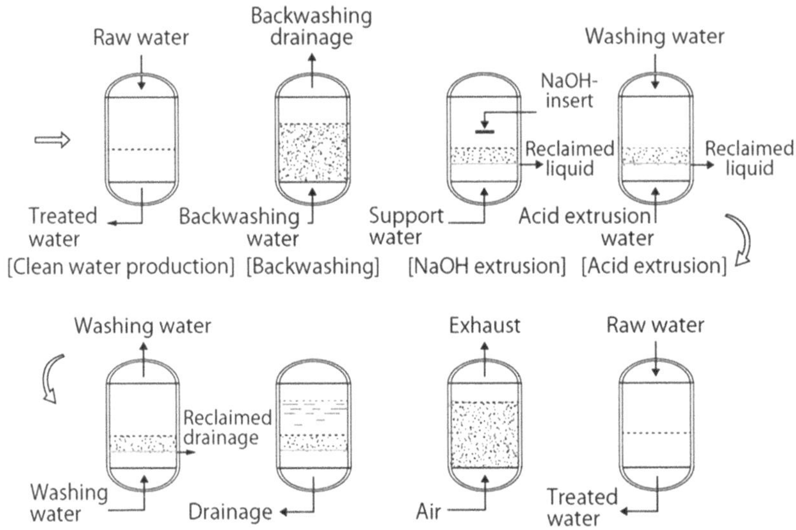

Figure 7.85 A large-scale mixed bed clean water production process, with resin regeneration.

7.2.6 Ion exchange as a water treatment process

The ion exchange process is basically a process to be used for low concentration system. As a water treatment process, it is mostly applied to water purification and industrial water treatment, but it is also a major part of the treatment of radioactive wastewater such as reactor wastewater treatment.

The main targets for which it is used are: water softening, desalination, removal of ammonia, iron, and chlorine, and removal of radioactive materials.

7.2.6.1 Water softening treatment

The most widely used ion exchange process in water treatment removes cations such as Ca, Mg, Fe, and Al by using a single-bed unit of strongly acidic cation exchange resin.

An example of a softening and regeneration reaction is shown for the Na cycle, as follows: $(R\text{-}SO_3Na)_2 + Ca\ (HCO_3)_2 \rightarrow (R\text{-}SO_3)_2Ca + 2NaHCO_3$.

To recycle, the following reaction takes place: $(R\text{-}SO_3)_2Ca + 2NaCl \rightarrow (R\text{-}SO_3Na)_2 + CaCl_2$.

About 10% NaCl solution or 2.5% sulfuric acid is often used for the H cycle.

7.2.6.2 Water softening by dealkalization

The softened water from the Na cycle contains a large amount of sodium bicarbonate and may cause alkali damage. Therefore, the H cycle is used in parallel, as shown in Figure 7.86, to combine the two effluents. The CO_2 from the H cycle neutralizes the alkali from the Na cycle of softening and releases the remaining CO_2 gases.

Use only H-type cation exchange resin such as $2(R\text{-}SO_3H) + Ca\ (HCO_3)_2 \rightarrow (R\text{-}SO_3)_2Ca + 2H_2O + CO_2$.

The CO_2 gases produced by the softening reactions are discharged in a decarboxylation tower and neutralized with sodium hydroxide, or partially softened and neutralized by mixing with a portion of raw water (see Figure 7.86).

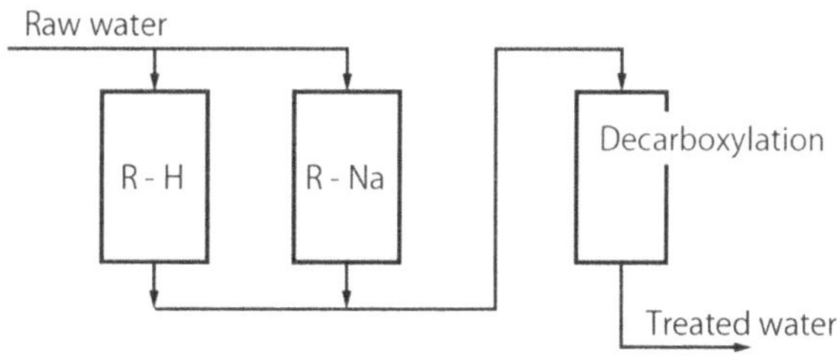

Figure 7.86 The dealkalization water softening method.

7.2.6.3 Desalination for pure water production

In recent years, ultra-high-pressure boilers have been commonly used in thermal power stations to improve thermal efficiency. Therefore, pure water with a conductivity as low as 0.1 μS/cm is required as water for boilers in power plants. Thus, the demand for pure water is increasing in the semiconductor, nuclear, and chemical industries. For these applications, a fully demineralized pure water production process is used in which an H-type cation exchange resin and an OH-type anion exchange resin are used in series.

The following is a typical example of a pure water production process:

(1) Two-bed system, using a strong acid-type cation exchange resin and a strong base-type anion effect resin in series;
(2) A two-bed three-column system in which a decarboxylation tower is placed in the middle of the two-bed system to remove CO_2 contained in the effluent water of the strong acid-type cation exchange resin bed and to suppress the deterioration of the capacity of the strong base-type anion exchange resin bed;
(3) 2-bed, 4-column system in which a strong base anion exchange resin bed is added in addition to a process in which a weak base anion exchange bed to a strong acid cation exchange bed and decarboxylation is performed

7.2.6.4 Zeolite treatment

Natural zeolites that are regenerated by NaCl or HCl have long been used in processes for water softening, iron removal, and manganese removal. Since iron and manganese are ion exchanged prior to calcium and magnesium, they are useful as de-iron and de-manganese methods. Manganese adsorbed by ion exchange on the surface of zeolite is oxidized by manganese dioxide or similar to form a higher oxide of manganese, which is called manganese zeolite, and is used as an effective method for removing iron and manganese.

Zeolite has also been used to remove ammonia that is always present in biological metabolic components. Among natural zeolites, clinoptilite has high selectivity for ammonia as $K > NH_4 > Ca > Mg$ and is useful as one of the processes for recycling sewage after biological treatment.

The ion exchange process is usually used after colloidal components are removed by coagulation, flocculation, sedimentation, sand filtration, etc., and after minute organic matter is removed by activated carbon adsorption. This is because the surface of the ion exchanger is easily contaminated and deteriorated and effective treatment cannot be continued in such a case. It should be noted that the final stage process in a water treatment system described in the Water Quality Conversion Matrix is ion removal.

FURTHER READING

Aris R. and Amundsen N.R. (1960). Some remarks on longitudinal mixing or diffusion in fixed beds. *Journal of American Institute of Chemical Engineers*, **3**(2), 280–282.

Billeo L.J. and Baudet B.A. (1983). Evaluation of activated carbon by dynamic minicolumn adsorption technique in treatment of water by granular activated carbon. In: Advances in Chemistry Series 202, M.J. McGuire and I.H. Suffet (eds), American Chemical Society, Denver USA, pp. 213–246.

Carberry J.J. (1960). A boundary layer model of fluid-particle mass transferring fixed beds. *Journal of American Institute of Chemical Engineers*, **6**(3), 460–463.

Carberry J.J. and Breton R.H. (1957). Axial dispersion of mass in flow through fixed beds. *Journal of American Institute of Chemical Engineers*, **4**(3), 367–375.

Chu J.C., Kalil J. and Wetteroth W.A. (1953). Mass transfer in fluidized bed. *Chemical Engineering Progress*, **49**(3), 141–149.

Chung S.F. and Wen C.Y. (1968). Longitudinal dispersion of liquid flowing through fixed and fluidized bed. *Journal of American Institute of Chemical Engineers*, **14**(6), 857–566.

Crank J. (1948). A diffusion problem in which the amount of diffusing substance is finite II, diffusion with non-linear adsorption. *Philosophical Magazine*, **39**, 140–149.

DiGiano F.A. and Weber W.J., Jr. (1973). Sorption kinetics in infinite batch experiment. *Journal of the Water Pollution Control Federation (W.P.C.F.)*, **45**(4), 713–725.

Hara N. and Takahashi H. (1975). Zeolite. Kodansha Scientific Book, Tokyo, Japan, pp. 1–20, 123–126, 314–320 (In Japanese).

Hashimoto K. (1973). Basic design of adsorption equipment. *Chemical Industry*, **17**(12) (In Japanese).

Hashimoto K., Miura K. and Nagata S. (1975). Intraparticle diffusivities in liquid-phase adsorption with nonlinear isotherm. *Journal of Chemical Engineering of Japan*, **8**(5), 367–373.

Hashimoto K., Miura K. and Tsukano (1977). Experimental verification of design methods for Liquid phase fixed–bed adsorbers. *Journal of Chemical Engineering of Japan*, **10**(1), 27–34.

Hermanowicz S.W. and Roman M. (1980). A comparison of packed-bed & expanded-bed adsorption system, Physicochemical method for water & wastewater treatment. In: Proceedings of the 2nd International Conference, Lublin Poland in 1979, L. Pawlowski (ed.), Pergamon Press, Oxford UK, pp. 141–152.

Idei T. (ed.) (1976). Water Treatment Engineering (Ion Exchange). Gihoudou Publishing Co., Tokyo (In Japanese).

James M. Montgomery Consulting Engineers, Inc. (1985). *Water Treatment Principles & Design, 10: Ion Exchange & Demineralization*. John Wiley & Sons, Inc., New York, USA, pp. 198–236.

Kamei T. and Tambo N. (1977). Gel Chromatography for water quality matrix Assessment. *Journal of Japan Water Works Association*, **519**, 24–41 (In Japanese).

Kataoka T., Yoshida H. and Ueyama K. (1972). Mass transfer in laminar state between liquid & packing material surface in the packed bed. *Journal of Chemical Engineering of Japan*, **5**(2), 132–136.

Kawazoe K. (1965). Mass transfer in adsorption. *Chemical Engineering*, **29**(6), 404–409 (In Japanese).

Keii T. (1966). Adsorption. Kyoritsu books, Tokyo, pp. 26–32 (In Japanese).

Keinath T.M. and Weber W.J., Jr. (1968). A predictive model for the design of fluidized-bed absorbers. *Journal of the Water Pollution Control Federation (W.P.C.F.)*, **40**(5, part 1), 741–765.

Kemmer F.N. (ed.) (1979). The Nalco Water Handbook. MacGraw-Hill, pp. 12–137.

Kennedy S.C. and Bretton R.H. (1966). Axial dispersion of spheres fluidized with liquid. *Journal of American Institute of Chemical Engineers*, **12**(1), 24–30.

Kitagawa, Kunibe, Yanai and Eguchi (1974). Activated Carbon Industry. Heavy Chemical Industry Corporation, Tokyo, pp. 353–539 (In Japanese).

Kunii D. and Levenspiel O. (1969). *Fluidization Engineering*. John Wiley & Sons, Inc. New York, USA, pp. 140–151.

Langmuir I (1918). The adsorption of gases on plane surface of glass, mica & platinum. *Journal of the American Chemical Society*, **40**, 1361–1403.

Miyauchi K. (1972). Particle-to-fluid interface mass transfer coefficient in dilute packed bed. *Chemical Engineering*, **36**, 633–640 (In Japanese).

Miyauchi K. and Nomura T. (1966). Liquid-boundary film mass transfer coefficient of packed layer in low Re number range. *Chemical Engineering*, **35**(11), 9–14 (In Japanese).

Nachod F.C. and Schubert J. (eds) (1956). Ion Exchange Technology. Academic Press. London UK, pp. 7–26, 231–271.

Ohta S. and Mori Y. (1975). Fluidized bed treatment experiment with petroleum-based spherical fine granules. *Journal of Japan Water Works Association* **486**, 39–60 (In Japanese).

Okuyama H., Kamei T. and Tambo N. (1982). Constituent, behavior & properties of metabolic waste components from aerobic biochemical processes (III). *Journal of Japan Sewage Works Association*, **19**(212), 1–4 (In Japanese).

Organo Co. Ltd (1986). Ion Exchange Resin Technology & Applications of Ion Exchange Resin. Organo, Co. Ltd, Tokyo.

Peel R.G. and Benedek A. (1980a). Attainment of equilibrium in activated carbon isotherm studies. *Environmental Science & Technology*, **14**(1), 66–71.

Peel R.G. and Benedek A. (1980b). Dual rate kinetic model of fixed bed adsorber. *Journal. A.S.C.E.*, **106**(EE4), 797–813.

Sharwood T.K., Pigford R.L. and Wilke C.R. (1975). Mass Transfer. McGraw-Hill, New York, London, pp. 241–245.

Snoeyink V.L., Weber W.J., Jr. and Mark H.B., Jr. (1969). Sorption of phenol & nitrophenol by active carbon. *Environmental Science & Technology*, **3**(10), 918–926.

Society of Chemical Engineers (eds) (1964). Physical Property Constants. Society of Chemical Engineers, Japan, pp. 246–265, 321–339, Maruzen Book stores, Tokyo.

Suzuki M. (1965). Fluid mixing in packed bed. *Chemical Engineering*, **29**(4), 253–260 (In Japanese).

Takeuchi (1974). Advanced treatment of wastewater with activated carbon. *Journal of the Society of Petroleum Engineers*, **17**(5), 370–375 (In Japanese).

Tambo N. (1980). Water Supply. New System of Civil Engineering, 88, 7.1 Adsorption (pp. 223–235); 7.4.1 Ion exchange (pp. 254–259). Japan Society of Civil Engineers Edition. Gihoudou Publishing, Tokyo. (In Japanese).

Tambo N. (ed.) (1983). Water Supply & Trihalomethane. Gihoudo Publishing, Tokyo.

Tambo N. and Ito M. (1987). Activated carbon adsorption characteristics of humic substances as trihalomethane & total organic chloride compounds formation precursors. *Journal of Japan Water Works Association*, **632**, 25–38 (In Japanese).

Tambo N. and Kamei T. (1985). New water quality indices of organics for the evaluation of treatment process & self-purification. 5th Asia Pacific Region Water Supply Conference, SS4-4, Seoul, pp. 1–10.

Tambo N. and Kamei T. (1978a). Water quality conversion matrix for treatability evaluation, (I). *Journal of Japan Water Works Association*, **530**, 8–18 (In Japanese).

Tambo N. and Kamei T. (1978b). Water quality conversion matrix for treatability evaluation, (II). *Journal of Japan Water Works Association*, **531**, 15–24 (In Japanese).

Tambo N. and Kamei T. (1978c). Water quality conversion matrix for treatability evaluation, (III). *Journal of Japan Water Works Association*, **532**, 37–44 (In Japanese).

Tambo N. and Tsutsumi Y. (1982). A study on Trihalomethane generation potential (1). *Journal of Japan Water Works Association*, **547**, 21–32 (In Japanese).

Tambo N., Takakuwa T., Fukuhara and Yuasa A. (1976). Dynamics of activated carbon adsorption process [I] evaluation of batch-type adsorption process by pore diffusion model. *Journal of Japan Water Works Association*, **496**, 8–24 (In Japanese).

Tambo N., Kamei T. and Fukushi K. (1979). Water quality conversion matrix III, for treatment evaluation. *Journal of Japan Water Works Association*, **532**, 37–44 (In Japanese).

Tambo N., Kamei T. and Takahashi M. (1981). Constituent, behavior & properties of metabolic waste components from aerobic biochemical processes (I). *Journal of Japan Sewage Works Association*, **18**(210), 1–10 (In Japanese).

The Dow Chemical Co. (1958). *Ion Exchange*. The Dow Chemical Co. Ltd, Midland Michigan, pp. 3–39.

Weber W.J., Jr. and Morris C.J. (1963). Kinetics of adsorption on carbon from solution. *Journal. A.S.C.E.*, **89**(SA2), 31–59.

Weber W.J., Jr. and Morris J.C. (1965). Kinetics of adsorption in columns of fluidized media. *Journal of the Water Pollution Control Federation (W.P.C.F.)*, **37**(4), 425–443.

Weber W. J., Jr. and Rumer R.R., Jr. (1965). Intraparticle transport of sulfonated alkylbenzenes in a porous solid; Diffusion with nonlinear adsorption. *Water Resources Research*, **1**(3), 361–373, 3rd Quarter.

Weber W.J., Jr. and Wang C.K. (1987). A microscale system for estimation of model parameter for fixed-bed adsorbers. *Environmental Science & Technology*, **21**(11), 109–1102.

Weber W. J., Jr., Hopkins C.B. and Bloom R., Jr. (1970). Expanded-bed adsorption for treatment of sewage effluents. *Chemical Engineering Progress Symposium Series*, **67**(107), 541–553.

Wehner J.F. and Wilhelm R.H. (1956). Boundary conditions of flow reactor. *Chemical Engineering Science*, **6**, 89–93.

Wilke R. and Chang P. (1955). Correlation of diffusion coefficients in dilute solution. *Journal of American Institute of Chemical Engineers*, **1**(2), 264–270.

Wilson E.J. and Geankoplis C.J. (1966). Liquid mass transfer at very low Reynolds number in packed bed. *Industrial & Engineering Chemistry Fundamentals*, **5**(1), 9–14.

Yuasa A. and Tambo N. (1978). Dynamics of activated carbon adsorption process [II] evaluation of batch adsorption process by liquid boundary film & intragranular diffusion resistance model in series. *Journal of Japan Water Works Association*, **520**, 14–32 (In Japanese).

Yuasa A., Tambo N. and Hirata S. (1984). Dynamics of fixed-bed adsorption process repeating back-washing. *Journal of Japan Water Works Association*, **596**, 13–26 (in Japanese).

REFERENCES

Adams B.A. and Holmes E.L. (1935). Adsorptive properties of synthetic resins. *Journal of Chemical Society*, **54**, 1–6.

Carson R. (1962). Silent Spring. Houghton Mifflin Comp, Boston, USA.

Crittenden J.C. and Weber W.J., Jr. (1978a). Predictive model for design of fixed–bed adsorbers; Parameter estimation & model development. *Journal A.S.C.E.*, **104**(EE2), 185–197.

Crittenden J.C. and Weber W.J., Jr. (1978b). Predictive model for design of fixed–bed adsorbers; Single compartment model verification. *Journal A.S.C.E.*, **104**(EE23), 433–443.

Crittenden J.C., Berrign J.K. and Hand D.W. (1986). Design of rapid small-scale adsorption test for a constant diffusivity. *Journal J.W.P.C.F*, **58**(4), 312–319.

Crittenden J.C., Berrign J.K., Hand D.W. and Lykins B. (1987). Design of rapid fixed-bed adsorption tests for nonconstant diffusivities. *Journal of Environmental Engineering (Journal ASCE)*, 243–269.

Eguchi Y. (1966). Active Carbon. Kyoritsu books, Tokyo, Japan, pp. 153–182. (In Japanese). [Translated from Hassler J.W. (1951). Active Carbon. Chemical Publishing Co., Inc., Brooklyn].

Gans R. (1905) Zeolites and similar compounds, their constitution and meaning for technology and agrichlture. *Jahr buch der Konigliche Preussischen Geologischen Landesanstlt*. **26**, 179.

Matsui Y., Kamei T., Tambo N. et al. (1992). Rapid evaluation of fixed-bed adsorption process by micro-column method. *Journal of Japan Water Works Association*, **91**, 13–22.

Matsui Y., Nakao S., Taniguchi T. and Matsushita T. (2013). Geosmin 2-methylisoborneol removal using superfine powdered activated carbon: Shell adsorption and branched pore kinetic model analysis and optimal particle size, Water research, Vol 47, pp. 2873–2880.

Shimizu H. (1953). Ion Exchange Resin. Kyoritsu Publishing Co., Tokyo (In Japanese).

Thompson H.S. (1850). On the absorbent power of soils. *Journal of Royal Agricultural Society*, **11**, 68–74.

Way J.T. (1850). On the power of soils to absorb manure. *Journal of Royal Agricultural Society*, **11**, 313–319.

Yuasa A. and Tambo N. (1959). Dynamics of fixed-bed adsorption process with repeated back-washing. *Journal of Japan Water Works Association*, **595**, 13–26 (in Japanese).

Yuasa A. and Tambo N. (1983a). Dynamics of activated carbon adsorption process [I] Analysis methods of fixed-bed adsorption process. *Journal of Japan Water Works Association*, **583**, 43–50 (In Japanese).

Yuasa A. and Tambo N. (1983b). Dynamics of activated carbon adsorption process [II] Parameters of fixed-bed adsorption process. *Journal of Japan Water Works Association*, **584**, 15–29 (In Japanese).

Yuasa A. and Tambo N. (1983c). Dynamics of activated carbon adsorption process [III], Characteristics of break through & adsorption efficiency. *Journal of Japan Water Works Association*, **585**, 14–26 (in Japanese).

Yuasa A., Niino T. and Tambo N. (1983). Dynamics of fluidized bed adsorption process. *Journal of Japan Water Works Association*, **586**, 12–28 (In Japanese).

Chapter 8

Membrane separation

8.1 DEVELOPMENT OF A FUNCTIONAL MEMBRANE SEPARATION PROCESS

Membrane separation is a technology that has rapidly spread worldwide as a core separation method for water treatment in the early 21st Century. In living organisms, regardless of whether they are animals or plants, physiological fluids are selectively transported and distributed between organs through semipermeable and bio-functional membranes. The adoption of this principle to water treatment is expected to remain a major technology for separation operations in engineering systems for some time to come (Tambo, 2012).

From the end of the 19th Century to the present, the water treatment processes of modern urban aquatic systems have matured from slow granular filtration systems, aerobic and anaerobic biological treatment systems that successfully mimic terrestrial ecosystems as separation systems. This is an 'ecology and geoscience' water treatment system that was first proposed in the 18th Century and then became the core system supporting the use and treatment of water in modern society for over 200 years.

We now have a new general separation technology, 'Membrane Separation', for the first time in 200 years. Membrane separation is a central mechanism of biological metabolic pathways that have built up over time to maintain the complexity of an individual organism. It is a general-purpose process used in a range of higher-order tissues, from single cells to humans. This 'biological membrane process' supports the survival of animals and plants by maintaining the high energy efficiency required to drive them, while accompanied by various biological and chemical reactions.

Today, most people know that water use and water treatment technologies based on natural and global ecosystems – which form the core of modern water systems that support human societies – are facing various difficulties due to the abnormal growth of mankind. This difficulty has been caused by the intensive use of energy and materials in today's urbanized world. It is important that a simple urban metabolic system paired with long-distance material transport can find a way to co-exist with the natural ecosystem. However, we

Author(s): Norihito Tambo and Koichi Ogasawara
doi: 10.2166/9781789061307_0605

have all seen that there are many situations that cannot be treated by speeding up or scaling up conventional ecological treatment technologies.

In order to rebuild a high-density society into a sustainable one, it is necessary to make urban areas more closed and independent, to enhance resource conservation and reuse, to replace urban aquatic systems with biologically distributed and self-sustaining metabolic systems, and to coexist with natural ecosystems. When an oversized urban system continues to take over natural ecosystems without resolving overload conditions, our modern high-density civilization will be faced with collapse.

Fortunately, Japan is the world leader in manufacturing technologies and production capacity of separating membranes, and there is a sufficient amount of basic technology to develop this into a self-sustaining and decentralized urban metabolism. The issue is how to remodel modern water systems for the next era. Separation methods, such as coagulation and filtration, used by modern water supply and sewerage systems differ from membrane separation methods in basic performance, just like the navigation systems of propeller and jet aircraft differ: the flight performance of both systems is quite different.

Consider the issue of membrane separation not only as a separation process, but also as a key manipulation of the quality conversion of urban and rural metabolic systems.

During the 1950s and 1960s, US President John F. Kennedy had a sense of crisis that 'the US will experience a significant shortage of freshwater by the end of the 20th Century, and the future of the country will be threatened.' The US Department of Interior established the Office of Saline Water (OSW) and began researching the desalination of seawater as a national project comparable to the National Aeronautics and Space Administration (NASA). 'Reverse osmosis', which is now the mainstream of seawater desalination technologies worldwide, was proposed for the first time by Professor Reid of the University of Florida (Reid & Breton, 1959). At the University of California, S. Loeb and S. Sourirajan (Loeb & Sourirajan, 1963) developed asymmetric semipermeable membranes of cellulose acetate that allowed solute molecules ranging from 100 Å to several Å to be separated from brine and seawater, with fluxes that are practical.

At that time, various proposals for desalination of seawater had already been made, and the distillation method had already been developed as a practical technique. However, in this era, the reverse osmosis (RO) method, which was until then still at a proposal stage, was developed to become the core technology for desalination of seawater for half a century. In the early 1960s, the author (Tambo) was in the Department of Chemistry at the University of Florida, Gainesville, conducting research on water treatment. He put peaty water in a commercially available cellophane bag to concentrate humic substances in the colored water and used it as a semipermeable membrane. He built a sugar pile on the outside, sucked out water from the bag by osmotic pressure of high sugar concentration, and prepared an analytical sample by concentrating humic substances. This is today's (positive) osmotic filtration operation. DuPont Co., Ltd, a large US company, were a pioneer in the development of hollow-fiber membranes by actively promoting the development of new membranes.

Membranes with different pore diameters are made of different materials and have different system configurations for water treatment membrane systems (Japan Water Purification Processes Association (JWPA) 1994a,b, 1997, 2002, 2003, Aptel & Buckley, 1996).

President Kennedy's concerns about the 'American famine of freshwater at the end of the 20th Century' proved unfounded in the United States, but future dangers lie ahead for the Ogallala aquifer which lies under sections of eight states in the USA, and where large amounts of groundwater are being drawn and advanced water treatment technologies (such as the use of recycled sewage) with high energy consumption are taking place. Instead of seawater desalination using membranes, this system is using reuse technologies which are also seen in the upgrading of low-quality freshwater such as for NEWater in Singapore. The greatest application of direct desalination of seawater is in the oil fields of the Arab coastal countries and in remote islands.

Membrane separation for water treatment has attracted attention for its removal of pathogenic microorganisms, some of which have relatively large dimensions (μm) and are present in water at extremely dilute concentrations (1 to 100 cells/l) and are difficult to sterilize. There are also great hopes for the effectiveness of membrane separation for the removal of pollutants in situations which are completely opposite to seawater desalination (i.e. for larger micro particles such as parasite eggs in water in an extremely dilute state).

Beginning in Texas, USA in 1984, there were 13,000 outbreaks of cryptosporidium in Georgia and more than 400,000 outbreaks of cryptosporidium in Milwaukee, USA from the faeces of livestock animals. To control the outbreaks, it is necessary to ensure that eggs (Oocysts, Cryptosporidium parvum: 4–6 μm in diameter) are removed from the water supply, and the introduction of functional membranes with large pore diameters for this purpose marked the introduction of a new technique.

Chlorination (disinfection) is not effective against parasite eggs and spores, and it is difficult to obtain a removal rate of more than two orders of magnitude by coagulation and rapid filtration. Using a microfiltration (MF) membrane with a pore size of 0.1 to 0.3 μm has ensured the removal of parasitic egg contamination (such as Cryptosporidium and Giardia lamblia) around the world (see Figure 8.1), and removal rates of 5 to 6 orders of magnitude (5 log to 6 log) or more can be expected. Large-pore filtration membranes (LP membranes) have also been put into practical use as precision filtration membranes specializing in removing parasite eggs, particularly for special applications in which the pore size is increased to around 1 to 2 μm to reduce drive pressure loss. Tambo *et al.* (1987) experimentally verified that biological filtration by slow sand filtration (Schmutzdecke) removes particles with a leakage ratio of 10^{-3}, confirming the effectiveness of the membrane removal function of slow sand filtration.

A typical MF membrane is a semipermeable membrane with a pore size of 0.1 to 0.3 μm, which corresponds to the size range of the conventional rapid sand filtration method used to remove turbidity and agglomerated color colloid micro-flocs. In addition, membranes with pore sizes of 0.001 μm or less, which can directly cope with colloidal-sized impurities such as turbidity and color components, are successively produced by various manufacturers; such ultrafiltration (UF) membranes are widely used for water treatment.

As described above, the RO membrane for desalination of seawater, which began with the OSW study in the United States, commenced with the proposal of cellulose acetate membranes such as by Loeb and Sourirajan (1963), and has now developed into a wide-spectrum soluble component separating membrane group that includes today's diverse RO and nano-filtration (NF) membranes.

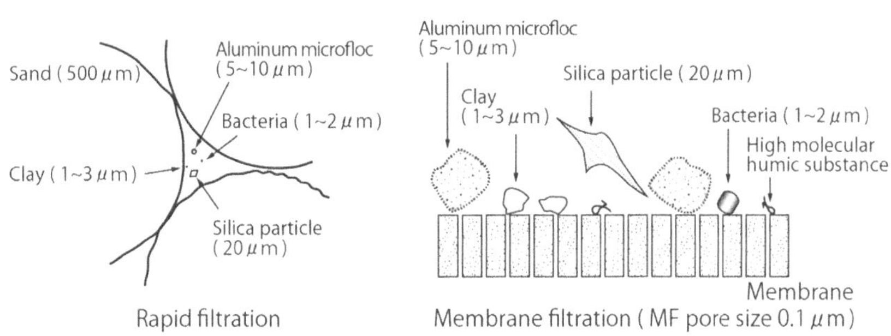

Figure 8.1 Size comparisons of removal media and particles removed (Tambo, 1980).

Depending on the components to be removed, there are seawater RO membranes (SWRO membranes) that can separate high-grade seawater (NaCl removal rate of 99% or more), and brine desalination membranes – brackish water RO (BWRO) membranes – that have slightly lower NaCl removal rates (about 99% or less), which are highly concentrated solutions of monovalent negative and positive ions. These membranes can remove divalent negative and positive ions in the regeneration of wastewater and in the advanced treatment of water, and can also remove low-molecular-weight organic matter. The treatment range of the membrane impurities corresponds to conventional activated carbon adsorption treatment, ion exchange treatment, and hardness treatment. It can also remove most of the organic and inorganic components that cause odors and tastes. NF membranes are essentially relatives of BWRO and are compatible with the treatment of ion-exchange for dissolved components in freshwater systems.

A 'UF membrane + NF membrane system' is now considered almost equivalent to a 'rapid filtration + activated carbon adsorption (ion exchange) system'. Humic substances causing trihalomethane (precursors) can be removed by the NF membrane, and the precursor issue is almost eliminated, and the UF membrane plus NF membrane system also significantly reduces odors and tastes and greatly suppresses odor problems. In addition, the UF membrane plus RO membrane system almost eliminates odors and tastes such as Geosmin and 2-MIB.

Compare the matrices for water quality and treatment in Chapter 3, Figures 3.6, 3.26, 3.28, 3.31, 3.32, 3.38, and 3.39, which are presented for a matrix evaluation of water quality by Tambo and Kamei (1976, 1978), with respect to the interrelationship between conventional water treatment systems and membrane separation water treatment systems.

8.2 MEMBRANE SEPARATIONS

The separation function membrane or separation membrane operation is defined as the operation of selectively separating inflow components present in a mixed state, by means of a kind of thin layer.

The basic items that define the membrane separation operation are:

(1) A driving force for separation;
(2) A membrane structure;
(3) A separation mechanism; and
(4) Whether the separation phase spans only a liquid system or liquid and gas systems.

Table 8.1 summarizes the operating categories for different membranes.

The separating element of the membrane is basically determined by the size/molecular size of the impurities and the separation critical dimension (cut off level) of the membrane used for removal, and is, in principle, straightforward (Mallevialle *et al.*, 1996). A one-dimensional representation is shown in Figure 3.39 in Chapter 3.

8.2.1 Membrane structure

As shown in Figure 3.39 and Table 8.1, the membrane separation operation is largely dominated by the size of the membrane pores and mainly based on the action of sieving, in the case of porous membranes such as MF, UF, NF and dialysis membranes. In these cases, the membrane pores have a distribution of diameters, and the fractional molecular weight curve (which shows the relationship between the molecular weight of a solute and its inhibition rate) is generally as shown in Figure 8.2. The molecular mass (or particle size) that exhibits a 90% inhibition rate is defined as the fractional molecular weight (cut off level), and impurities greater than or equal to that mass (particle size) are virtually 100% separated and removed (Membrane Separation Promotion Association, 2008).

Table 8.1 Characteristics of typical water treatment membrane operating categories.

Membrane Operation	Driving Force	Separation Mechanism	Membrane Structure	Adaptive Phase
Microfiltration (MF)	Pressure	Sieving	Macropore	Liquid-Liquid
Ultrafiltration (UF)	Pressure	Sieving	Mesopore	Liquid-Liquid
Nanofiltration (NF)	Pressure	Sieving + Pore diffusion + Rejection	Micropore	Liquid-Liquid
Reverse osmosis (RO)	Pressure	Sieving + Rejection	High density reticulated polymer	Liquid-Liquid
Stripping	Activity (partial pressure)	Evaporation	Macropore	Liquid-Gas
Dialysis	Activity (concentration)	Diffusion	Mesopore	Liquid-Liquid
Electric dialysis	Potential difference	Ion exchange	Reticulated ion exchanger	Liquid-Liquid

The US classification (IUPAC, 1985) classifies macropores as 50 nm or more, mesopores as 2 to 50 nm, and micropores as 2 nm or less. Non-porous membranes or high-density reticulated polymers such as RO membranes separate components by diffusional migration of the components to be filtered through macro-reticulated membrane structures. The internal structure of NF membranes is intermediate between microporous and non-porous membranes. NF membranes, also called loose RO membranes, are also thought to be a part of the family of RO membranes that target polyvalent ions.

In the pressure-driven membrane group, the amount of permeation decreases in proportion to the membrane's resistance. Therefore, the goal with homogeneous (isotropic) membranes is to make the membrane thinner. In order to stabilize the operation, a thin surface layer – the active skin layer (with a small water flow resistance and a reliable separation function) – is supported by a support layer (with a small water flow resistance with a thick lower part) in case of a non-uniform separation membrane (an anisotropic membrane) A complex membrane (composite membrane) is formed by activating the upper surface of an asymmetric membrane (Figure 8.3) and synthesizing a high-density supporting layer as shown in Figure 8.4.

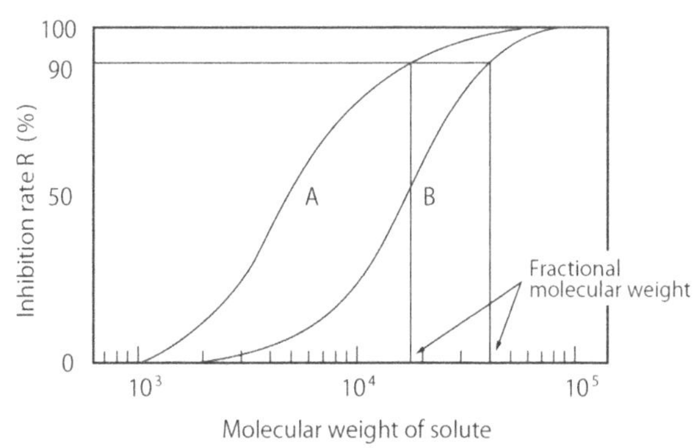

Figure 8.2 The fractional molecular weight curve and cut off level.

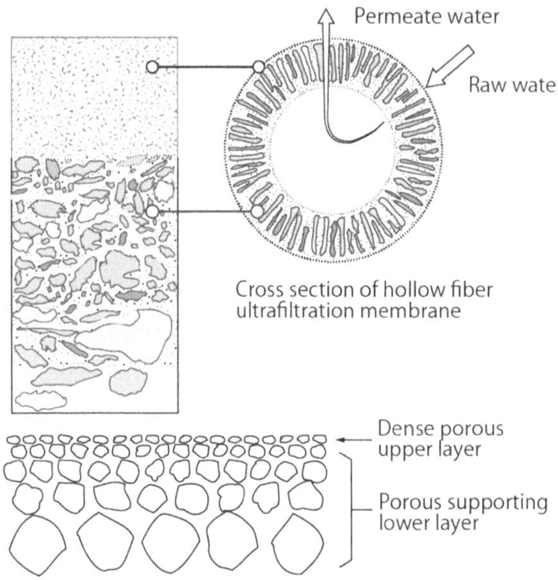

Figure 8.3 Schematic diagram of an asymmetric membrane.

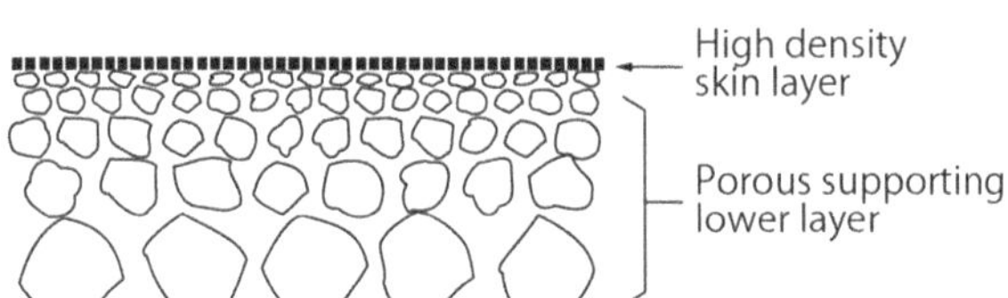

Figure 8.4 Schematic diagram of a composite membrane.

Most separation membranes around the world are made of organic polymers, but inorganic ceramics and other materials are also widely used in Japan. Various materials are selected depending on the application, with emphasis on treatability and durability. The most commonly used cellulose and its derivatives in hydrophobic polymer membranes has also been a traditional material since the research of Loeb and Sourirajan (1963). Chlorine resistance is often required for water treatment because chlorine disinfection is frequently used, and cellulose esters may be used. In polyamides, aromatic polymers are widely used in seawater desalination RO membranes but do not tolerate chlorine well. PAN (polyacrylamide) is used as a UF membrane, and PSf (poly-sulfone) and PES (poly-ether sulfone) membranes are less hydrophobic and tend to be more or less adsorptive, but they are more stable due to temperature resistance and are used as substrates for UF membranes and composite membranes. Due to its stability, polypropylene (PP) is widely used as a material for MF membranes, although it has some minor drawbacks concerning chlorine (Figure 8.5).

Ceramic and other inorganic membranes have excellent chemical resistance, mechanical strength, and temperature stability, and are widely used in MF and UF membranes.

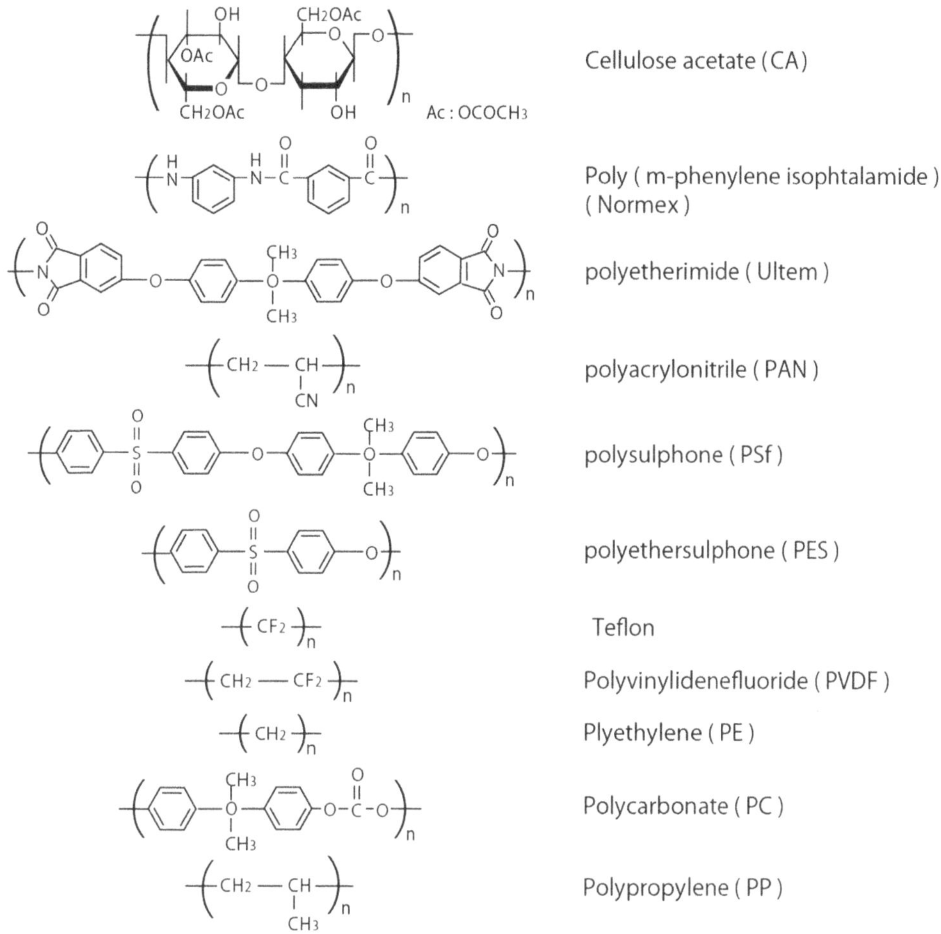

Figure 8.5 The basic molecular structures of polymers that make up organic membranes (Aptel & Buckley, 1996).

To improve membrane function:

(1) Make the transmembrane flux as high as possible per unit of driving force (to minimize power costs);
(2) Membrane fouling (contamination) does not excessively reduce fluxes over time, and is recoverable with simple washing with water and less frequent chemical washing;
(3) Do not lose function due to chemical changes (aging) of the membrane body;
(4) Module is easy to handle on a daily operation and maintenance basis;
(5) Recycling of modules and appropriate disposal processes should be considered.

The membrane apparatus is constructed as a basic unit of operation in the form of a membrane module, and the system is designed by assembling units in various ways. The modules consist of a separation membrane body, pressurized membrane support structure, raw water (pressurized water) supply holes, membrane permeate outflow holes, treated water collection lines, etc.

Design of the membrane separation unit should consider the following points (Peterson, 1993):

(1) Ensure sufficient water circulation (membrane surface flow rate) to prevent a concentration boundary layer or sediment accumulation on the supply surface side of the membrane;
(2) Select a dense modular structure with as large a membrane permeation area per unit volume of the device as possible;
(3) Ensure that the supply side and transmission side are not short-circuited;
(4) Prevent no flow or stagnation of components;
(5) Design a structure to ensure accurate and easy cleaning by water flow and chemical cleaning;
(6) Easiness to assemble and disassemble in due course of maintenance.

Typical modules with such functions include flat membranes (plate and frame), and spiral, tubular, monolithic and hollow-fiber membranes. In addition, there are two main types: cased and non-cased (i.e. with and without a containment vessel).

In the filtration phase (as shown in Figure 8.6), the cross-flow method is a means of measuring membrane permeation by circulating raw water to suppress formation of concentration-limiting membrane on the primary side of the membrane.

The non-cased type filtration system is the one that Japan has pioneered, and the organic hollow-fiber immersion membrane treatment (shown in Figure 8.7) has been developed widely by such organizations

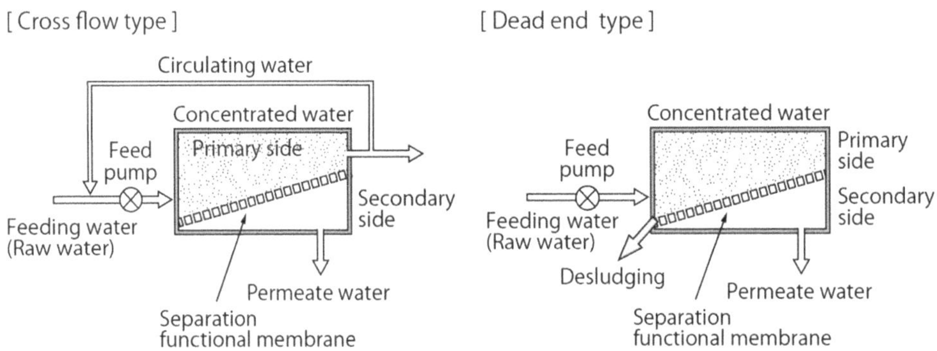

Figure 8.6 The two methods of cased filtration.

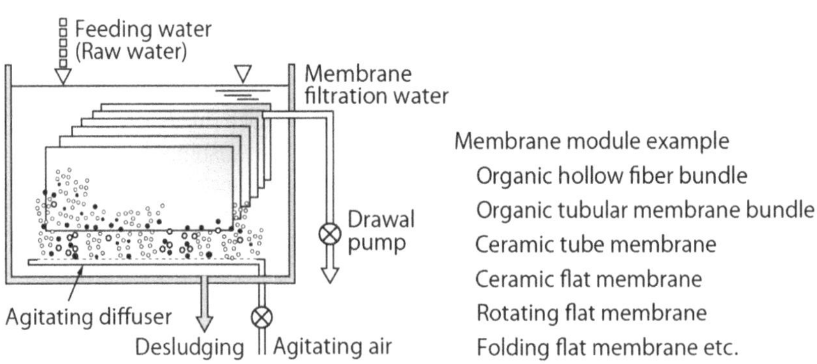

Figure 8.7 The immersion membrane treatment method.

as Kubota Iron Works Co. Ltd. It was developed by Professor Aya of the Faculty of Engineering of the Tokyo University, and Ebara-Infilco Co., Ltd, Japan. The system is deployed in the membrane microbial treatment (membrane bio-reactor) which is described below.

8.2.1.1 Flat membrane (module)

There are a variety of supports and separation membrane combinations, including sheet, plate frame, spiral, rotating flat membrane and pleated. As shown in Figure 8.8, a flat sheet membrane separation system is made up of an arrangement of a large number of unit frame embedded membranes with the main body incorporated in a support frame. It is used for MF and UF membranes, etc.

When flat membranes are used for operations using RO and NF membranes where the driving pressure is high, spiral type modules such as those shown in Figure 8.9 are used to ensure reliable support of the membranes and uniform flux when the flat membranes are molded into the modules. Along with the hollow fiber membrane modules shown in Figure 8.10, they are widely used in seawater desalination and deionization operations.

Spiral-type modules have been widely used since they were developed by Gulf General Atomic with the support of the US Department of Home Affairs, Bureau of Saline Water, in the 1960s. As shown in Figure 8.9(a), two sheets of RO membranes are sandwiched between porous water permeable support membranes and serve as a channel for the collection of permeate water through the membranes. They form a raw water channel between membranes which is around 0.7 mm. Three sides of the group of integrated membranes are welded to form an envelope-like bag, and raw water is flowed into the open end. The envelope is wound around two or three leaves and a core tube (which is bonded to the outlet end of the envelope with a raw water inlet tube) to form a cylindrical module, as shown in Figure 8.9(b).

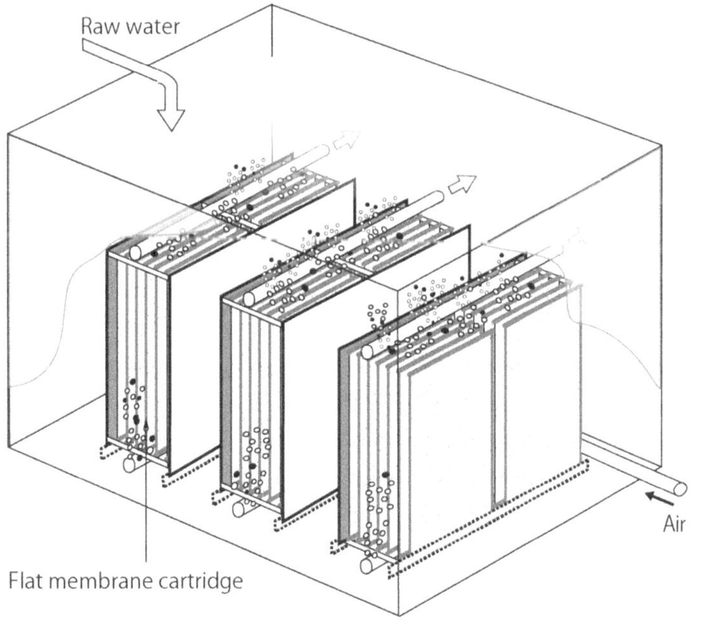

Figure 8.8 An example of a flat membrane cartridge, a type of immersion equipment.

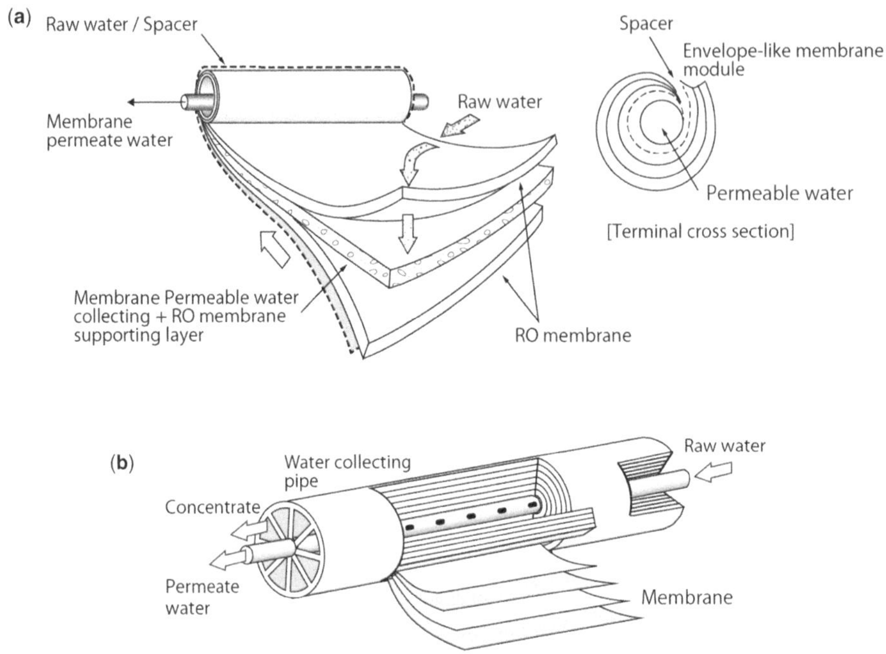

Figure 8.9 A spiral module. (**a**) the membrane and spacer; (**b**) the complete spiral flat membrane module.

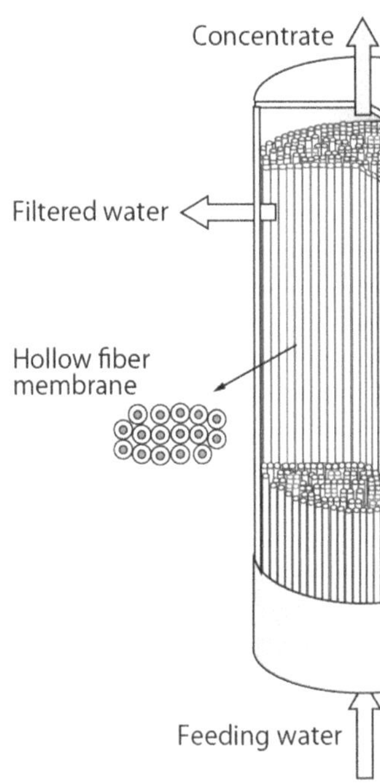

Figure 8.10 Schematic of an internal pressure hollow fiber module.

8.2.1.2 Hollow fiber membranes (cased and non-cased types)

The hollow fibers themselves create the form of a membrane, self-supporting itself, and it is put inside a pressure-resistant casing. Alternatively, a bundle without a casing can be put into an open container and used as a membrane bioreactor (MBR) (in the membrane activated sludge method), or similar. In 1965, Du Pont began developing RO modules, making them one of the mainstream RO and NF membrane systems.

A hollow fiber membrane with an outer diameter of several mm to 250 μm and an inner diameter of about half that can be designed to flow in two directions, either from inside to out, or outside to in. The outer-to-inner type with a high-density layer of membrane on the outside (external pressure water flow system) can increase the area of the active layer by an amount corresponding to the increase in diameter, even with the same hollow fiber. Figure 8.10 shows an example of an internal pressure hollow fiber module.

Figure 8.11 is a conceptual diagram of a non-cased hollow fiber module that is mounted by exposing the hollow fiber bundle to the sedimentation separation portion of a conventional water treatment system. There are a variety of cleaning methods, but the membrane needs to be a strong film that can be self-supporting. The drive pressure difference is small, and a suction type MF membrane system is generally applied for the purpose of solid-liquid separation.

8.2.1.3 Ceramic membranes

In the 1960s, a cylindrical ceramic membrane for uranium enrichment began to be developed in France and multi-channel monolithic membranes were later developed in the middle of the 1980s for water treatment. A thin active membrane is constructed on the inner surface of multiple channels penetrating the ceramic porous support layer to reduce the permeation pressure loss. Aluminium, zirconium, titanium, and other materials can be designed and used over a wide range, including for use in MF, UF, and NF films. It has been clearly evaluated that ceramic membranes have a compatible pH across the entire range of acid-alkali, have a high mechanical strength, and excellent heat resistance and abrasion resistance. As a result, large-sized membrane modules have been manufactured and are widely used in public applications (see Figure 8.12).

8.2.2 Separation mechanism

Membrane treatment is designed with the goal of continuing a necessary removal (maintenance of the required molecular weight fraction and processing flux) with minimal driving force (energy loss of filtration) and stability.

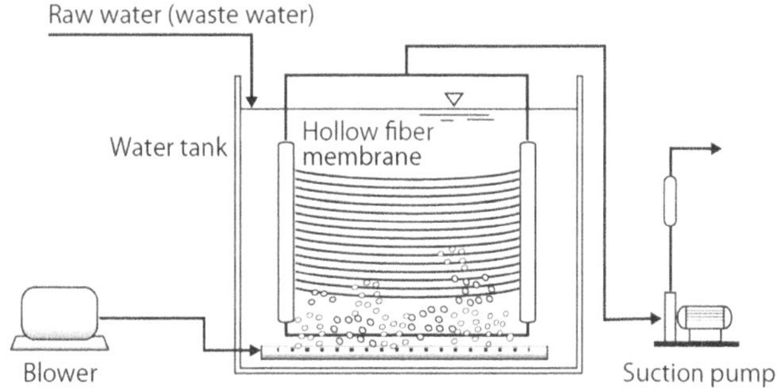

Figure 8.11 An example of a non-cased immersion hollow fiber module.

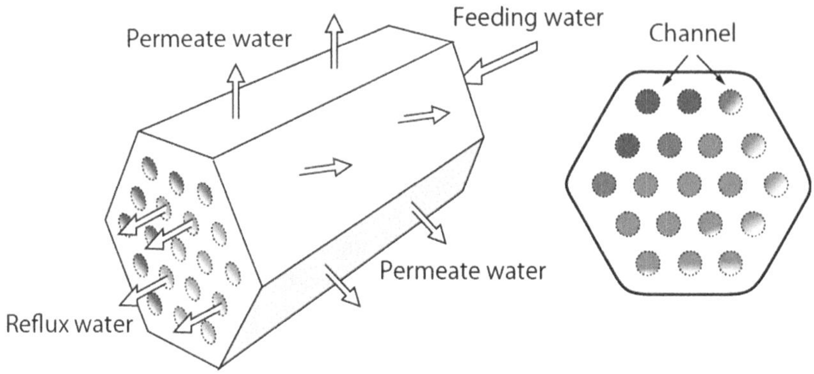

Figure 8.12 Schematic of a ceramic monolithic membrane.

Almost all of the current membrane processes for water treatment using MF, UF, NF and RO membranes are pressure-driven operations. However, the mechanisms of membrane rejection vary greatly depending on the type of membrane. The chemical affinity of an RO membrane with the impurities to be removed is crucial, whilst an MF membrane is almost physically inhibited by sieving (and NF and UF membranes can be considered to have both mechanisms).

The impurity removal rate R of a membrane is defined in Equation 8.1.

$$R = 1 - \frac{C_p}{C_f} \qquad \text{—(8·1)} \tag{8.1}$$

C_p : Permeable water concentration
C_f : Raw water concentration

Equation 8.1 The impurity removal rate of the membrane.

where C_p and C_f are the concentration of raw water and permeate water. As shown in Figure 8.13 (which shows a hollow fiber membrane), polarizations are generated in the raw water which increase Cp concentration.

In actual water treatment, the amount of the components removed is expressed as the amount defined in Equation 8.2, and the R_{mass} in which the water recovery rate r is added is frequently used. Actually, both C_p and C_f change during the process of operation, and are not constant due to fluctuations in the raw water

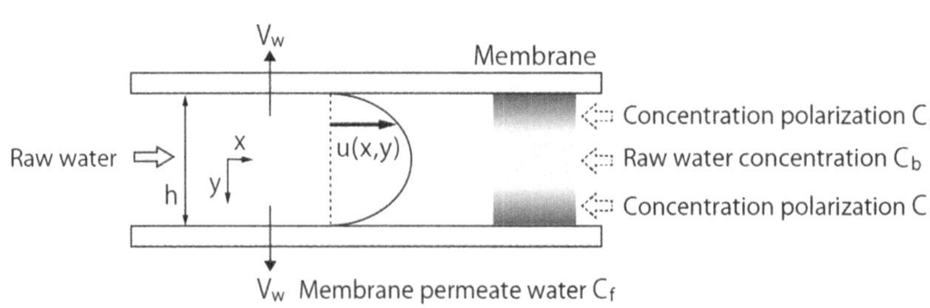

Figure 8.13 The pattern of concentration polarization (boundary layer) in the raw water.

condition and progressive blockage of the membrane. Various conditionals are required to define R and R_{mass}.

$$R_{mass} = 1 - \left(\frac{C_p}{C_f}\right) r \quad \text{—(8·2)} \tag{8.2}$$

r : Ratio of recovered permeate water to raw water introduced (Water recovery rate)

Equation 8.2 The impurity removal rate considering water recovery rate.

The main design and operation factors are how large the membrane permeate flux can be, along with the size of the impurity removal rate R. The permeate flux J is found in Equation 8.3 by Darcy's law, which describes the flow of filtered water in porous material.

$$J = \frac{\Delta p}{\mu R_m} \quad \text{—(8·3)} \tag{8.3}$$

Δp : Pressure difference before and after membrane (driving pressure Pa)
μ : Viscosity of water (Pa·s)
R_m : Fluid resistance of fresh membrane (cm^{-1})

Equation 8.3 The membrane permeate flux (Darcy's law).

If the permeate flux J is larger than the solute flux J_s (e.g., for a dilute system), the permeate flux is expressed by correcting the superficial velocity L_v as shown in Equation 8.4, and the permeate flux is calculated as shown in Equation 8.5 based on Equation 8.4. In addition, the permeable water of the MF and UF membranes flows through a large number of parallel capillaries at a flow rate proportional to the driving pressure with the laminar flow. Therefore, the permeable water flux of the UF and MF membranes can be evaluated by Poiseuille flow, as in Equation 8.6.

$$J \propto L_v(\Delta p - \sigma_k \Delta JI) \quad \text{—(8·4)}$$

$$J = \frac{(\Delta p - \sigma_k \Delta JI)}{\mu R_m} \quad \text{—(8·5)}$$

(8.4)–(8.5)

Δp: Driving pressure for membrane operation
σ : Experiment coefficient
ΔJI : Solution osmotic pressure difference before and after membrane
Proportional to the molar concentration of the solute to be removed
(It can be ignored in MF and UF membranes for colloids, high molecular weight organic substances, etc.)

Equation 8.4, 8.5 The permeate flux of superficial velocity correction.

$$J = \frac{f r_{pore}^2 \Delta p}{8 \mu \theta \delta_m} \quad \text{—(8·6)}$$

f : Abundance ratio of membrane pore section (capillary total water flow cross section)
θ : Pore twisting index
δ_m : Effective thickness of pore
r_{pore} : Pore (capillary) radius
This determines molecular weight cut off (size)

(8.6)

Equation 8.6 The permeable water flux poiseuille flow.

In heterogeneous or complex membranes, flux is dominated by the flux in the effective thickness δ_m of the thin active layer. A support layer with a thick cross section has a large void diameter and a small resistance to water flow, and does not have a large influence on flux.

Although it is possible to increase flux in proportion to the drive pressure Δp, it is also possible to limit the flux to a practically satisfactory range and not to apply excessive drive pressure.

Table 8.2 shows the approximate range of membrane resistance and driving pressure for various types of pressure-driven membrane processing operations.

8.2.3 Membrane degradation and fouling

As operation continues, membrane filtration performance deteriorates, and the treatment efficiency decreases for two reasons. The first of these is the degradation (deterioration) of the material structures of the membrane itself. Long-term pressurization can lead to membrane compaction and creep deformation, damage due to mechanical and physicochemical external forces, changes in materials due to hydrolysis and chemical oxidation, and biological degradation due to microbial utilization. This causes the structure and performance of the membrane itself to change in a direction that is disadvantageous to the application, making it impossible to recover performance. Early cellulose acetate membranes had defects such as increased crystallization under high water temperature conditions and a marked decrease in transmittance.

The second reason for the decrease in treatment efficiency is membrane fouling. The following types of phenomena are conceivable for fouling:

(1) Cake formation; suspension forms a cake or gel layer on the surface of the membrane, and the membrane pores are blocked or the layer is adsorbed within the pores, thereby inhibiting or reducing permeability. In the operation of wastewater treatment by UF and MF membranes, the

Table 8.2 Approximate membrane resistance and membrane driving pressure.

Membrane	Membrane Resistance R_m (cm^{-1})	Membrane Driving Pressure (kPa)
RO	10^{-8}	800 ~ 8000
NA	10^{-6}	350 ~ 1000
UF	10^{-5}	50 ~ 700
MF	10^{-4}	30 ~ 300

cake itself traps fine particles, increases resistance, and lowers flux. It can also cause blockage of membrane pores. In order to prevent this decrease in flux, backflow water cleaning (back flash) is performed frequently (as described in the next section), or the flow velocity of the circulating water on the primary (treated) water side is increased (see Figure 8.13) to perform crossflow operations to suppress the formation of a boundary concentration polarization layer.

(2) Precipitative fouling; as filtration progresses, concentration polarization is formed on the primary side of the membrane. When the concentration exceeds the solubility of salt in the raw water, precipitation occurs, and the phenomena described in Section 6.1 occur near the surface of the membrane. Scale formation is a problem when RO membrane or NF membrane operation is used for solute separation. Hydroxides, carboxides and sulfides of metal ions such as Ca, Mg and Fe are common fouling coagulums. Also, caution should be exercised with respect to water with a high concentration of soluble silica.

(3) Adsorptive fouling; adsorption of impurities that have affinity for the membrane material and the inner surface of the pores reduces water permeability. When humic substances in natural water and colored components in aerobic wastewater treatment are inhibited, a form of flux reduction phenomenon that is quite different from that of inorganic components develops. The authors have experienced MF membrane treatment of clear raw water at the Moiwa purification plant in Sapporo City with E260 sensitive TOCs (as defined in the section on water quality matrices in Chapter 3) that adsorb on the inner surfaces of pores, and show the same behavior as hydrophilic membranes after some operating hours of hydrophobic membranes. This means that a hydrophobic MF membrane could become a hydrophilic-like MF membrane as a result of continued manipulation, rather than membrane degradation, which means that it is difficult for a membrane to remain hydrophobic when handling natural water that contains natural organic matter (NOM).

(4) Pore blocking; in the pores inside the membrane, a layer of deposits may form and the pores may become blocked by fine particles. Especially when a membrane with a fractionation limit of about 0.3 to 0.1 μm is used for MF membrane treatment, the micro-colloid particles enter the pores and cause blocking phenomenon. On the other hand, when using an MF membrane with a smaller fractionation limit size or a UF membrane with a smaller fractionation molecular weight (pore size of 0.01 μm or less), the initial water permeability resistance is large, but, in a long-term experiment or operation of an actual plant, the decrease in flux can often be suppressed after long-term operation.

(5) Membrane biofouling; this is one of the most challenging of these phenomena and it is a phenomenon that exists in all water systems. Bacteria adhere to and grow on the primary side of the membrane to form colonies, and matrices of gels formed by the secretion of extracellular polymers develop on the membrane to form a biofilm. Biofilms continue to grow using assimilable organic carbon (AOCs) in raw water as a source of carbons and energy. Biofilms are identified by microscopic observation or by scraping their surfaces for bacteriological and chemical analysis.

Little quantitative assessment has been made of the reduction in membrane processing efficiency due to biofouling. Biofilm production is also known to shorten the life of membranes. Chemical cleaning is the most reliable method of controlling its production. As an example of cleaning: rinse with membrane treated water for 10–20 minutes; clean by circulating hot water at 30–40°C for 1 to several hours at maximum operating pressure; immerse the membrane in a high-concentration washing solution for 1–24 hour(s); perform a final wash circulation for 20–30 minutes, circulating rinsing with membrane treated water for 10–20 minutes, etc.

The pH of the wash solution is selected according to the type of membrane: pH 6–8 for cellulose acetate membranes, pH 3–12 for polyamides or polysulfones. Table 8.3 shows a typical wash solution for biofouling control.

As the membrane treatment process continues to operate, the membrane permeation flux decreases as the fouling progresses, and the design operating conditions are set so that a predetermined flux can be obtained at a predetermined membrane operating pressure. As shown schematically in Figure 8.14, if the operation is continued at a constant pressure, the flux decreases with the duration of the operation, and if the membrane is not cleaned, the flux becomes lower than 1 m/day in several hours. Therefore, physical cleaning such as backwash with filtered water is performed every 0.5–1 hour to mechanically remove fouling components and discharge them out of the system. However, even if the physical cleaning is repeated, the decrease in flux continues gradually due to the accumulation of the fouling material adhering to the membrane. If it drops to some extent, it is chemically treated with a membrane wash solution to restore the membrane's water permeability. The strongest chemical washes can result in a flux that is nearly close to that of a new membrane.

Physical cleaning is generally performed by setting the operating pressure higher than that of the treatment operation (e.g., by increasing the pressure to 200 kPa during backwash from 100 kPa during water flow operation) in order to allow a large amount of treated water to flow backward. Since a large amount of treatment water (permeate through the membrane) is consumed, the

Table 8.3 Typical wash solutions for biofouling control (Ridgway & Fleming, 1996).

Compound	Name	Adaptive Membrane	Operation Concentration
Detergent	SDS	CA, PA, PS	0.01~2.0%
	SDBS	CA, PA, PS	
	Triton series	CA, PS, PE	
	Quaternary amines	CA, PS, PE	
Destabilizing agent	Urea	CA	6~8 molar
	Guanidium HCl	CA	1~2 molar
Chelating agent	Citrate	All	0.1~1.0%
	EDTA	All	0.1~1.0%
Enzyme	Proteases	All	10~100 mg/l
	Esterases		10~100 mg/l
	Lipases		10~100 mg/l
	Polysaccharidases		10~100 mg/l
	Lysozyme		
Biocide	Sodium bisulfite	All	10~100 mg/l
	Quaternary amines	CA, PS, PE	0.1~1.0%
	Formaldehyde	All	0.1~5.0%
	Glutaraldehyde	All	0.1~5.0%
	Isothiozolinone	All	0.1~1.0%
	(Kathon)	All	0.1~1.0%
	Sodium benzoate	CA, PA, PS	0.5~5.0 mg/l
	Monochloramine	All	0.1~1.0%
	EDTA		

CA: cellulose acetate, PA: polyamide, PS: polysulfone, PE: polyetherurea.

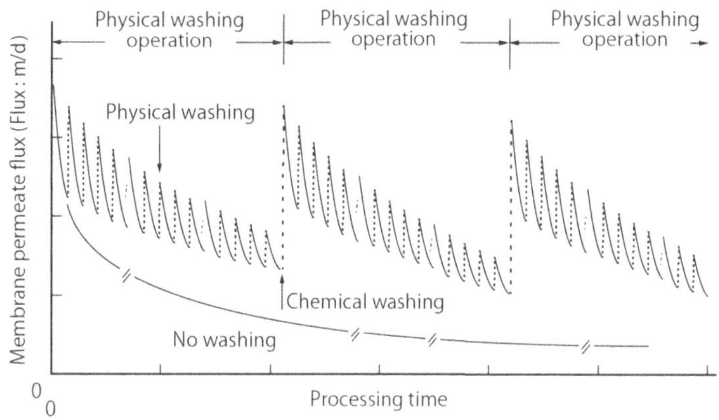

Figure 8.14 Change over time of membrane permeate flux (MF, UF).

interval between backwash times should be as long as practical. By contrast, the backwash time should be as short as possible to keep the yield of treated water high during long-term operation. In addition, chemical cleaning must be carried out at a minimum frequency because it requires a long shutdown for cleaning operations, and it also requires considerable chemical and wastewater treatment costs and energy (Tambo & Repine, 1992).

It is necessary to devise a comprehensive system that enables effective operation of membrane treatment from large-scale systems to distributed small-scale plants, such as the construction of a membrane regeneration system that starts with membrane quality improvement and performs cleaning effectively outside of the processing system.

8.2.4 Membrane resistance factors induced by membrane fouling

When filtering pure water, the resistance of the membrane is written as in Equation 8.4 (above). To assess the various conditions of membrane fouling, we consider a resistance law that can be commonly handled for MF and UF membranes, which represents the conditions that define the flux during actual membrane treatment. The research presented here is taken from the PhD thesis of the late Dr Bian Ru Lin, a graduate student from China at the Hokkaido University, Department of Environmental Energy (in Tambo and Watanabe's laboratory) (Bian, 1998).

When river water is treated with UF membranes, the relationship between changes in permeate flux (decay over time) and filtration resistance were examined considering the increase in filtration resistance due to membrane fouling caused by impurities in water of various sizes and chemical properties.

In addition to the resistance of the membrane itself and the viscous resistance of the suspension, four possible sources of resistance are present:

(1) Boundary-layer resistivity R_B on the primary side of the membrane due to concentration polarization;
(2) Gel-layer resistivity R_G due to gel polarization;
(3) Cake-layer resistivity R_C formed by suspended components adhering to the membrane surface;
(4) Adsorption resistance R_A by components adsorbed to membrane pores.

The permeability flux Equation 8.4 for fouling membranes can therefore be written as Equation 8.7.

$$J = \Delta p / (R_M + R_B + R_G + R_C + R_A)\mu \quad \text{—}(8\cdot7) \tag{8.7}$$

J: Permeate water flux
Δp: Driving pressure for membrane operation

Equation 8.7 The permeate water flux of fouling membrane.

In order to evaluate the occurrence of membrane fouling during one filtration cycle, a batch-type experimental apparatus as shown in Figure 8.15 was used. The system consisted of a nitrogen gas cylinder for pressurization, a stirring cell for membrane filtration, an electronic balance for membrane filtration permeate measurement, and a personal computer for data recording.

A flat membrane with a diameter of 6.1 cm was placed in a stirred membrane permeation cell, 200 ml of the filtrate stock solution was charged, the outflow pressure of the nitrogen gas cylinder was manually set in the range of 10 to 100 kPa to the specified driving pressure, and constant pressure filtration was performed. The permeate volume was measured continuously at 0.6 second intervals and recorded on the computer. Two types of permeable membranes were used: a polysulfone (PS) membrane with a cut-off molecular weight fraction of about 50,000 Da, and a polyacrylnitrile (PAN) membrane with a molecular weight fraction of about 200,000 Da.

In normal continuous membrane filtration operation, physical washing (backwash) is performed every 30 to 60 minutes. If the membrane permeation flux is maintained at about 1.02 m/d for continuous filtration and the physical wash interval is assumed to be 60 minutes, the permeation water volume of the UF membrane in the agitating cell of the experimental unit will be about 150 ml per cycle, with an effective membrane area of 0.0028 m^2. Each test was then terminated when a new membrane was loaded and 150 ml of filtrate was obtained from 200 ml of stock solution. The solution was stirred at about 150 rpm with a magnetic stirrer at a position of 0.3 cm from the membrane surface. The viscosity of the filtrate was measured.

Of the various resistances caused by fouling, it is difficult to experimentally divide the structures of the gel layer and the cake layer to obtain them individually. Therefore, these two resistances ($R_G + R_C$) were evaluated collectively as R_C. R_M, R_B, R_C, and R_A were then calculated in the following steps:

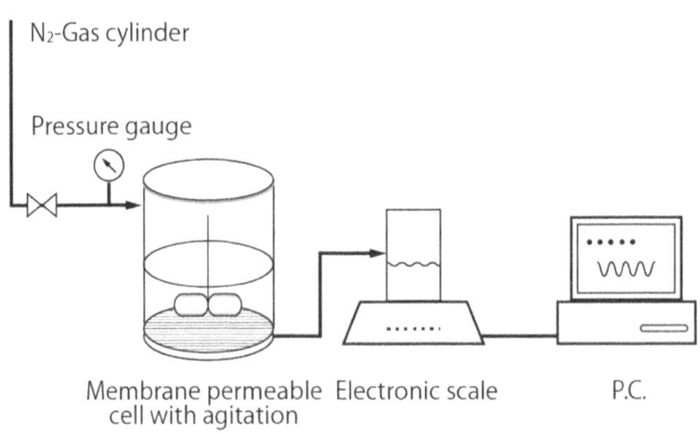

Figure 8.15 A batch-type membrane filtration experimental apparatus.

(1) The distilled water was filtered, and the resistivity R_M of the membrane body calculated using the Darcy equation.
(2) The raw water was filtered by the Darcy method.

$$R_{TO} = R_M + R_B + R_C + R_A$$

(3) Filtration resistance in the absence of concentration polarization of the membrane surface after switching from raw water to distilled water was measured:

$$R_{T1} = R_M + R_C + R_A$$

Resistance due to concentration polarization was measured:

$$R_B = R_{TO} - R_{T1}$$

(4) The deposits on the membrane surface were wiped off by sponge and permeated the distilled water again.
Resistance when the cake layer (+ gel layer) on the membrane surface was removed was:

$$R_{T2} = R_M + R_A$$

As a result, the cake-layer resistance R_C and the adsorption resistance R_A were:

$$R_C = R_{T1} - R_{T2}$$
$$R_A = R_{T2} - R_M$$

Figure 8.16 shows how these resistors are divided according to steps (1) to (2).

The experiment used raw water from the Ebetsu City Water Supply downstream of the Chitose River. This raw water is peaty land river water in Ishikari Low Land near Sapporo City, which constantly has moderate turbidity and high color. A 0.45 μm membrane filter which removes turbidity components was used, so that fouling due to natural humic substances could be evaluated. Table 8.4 presents the water quality of the test waters used. Samples A and B were filtered water from a 0.45 μm membrane filter.

Figure 8.17 shows the results of a test on raw water from Ebetsu City, from the Chitose river, which has both turbidity and humic components, to determine the resistance that produces membrane fouling.

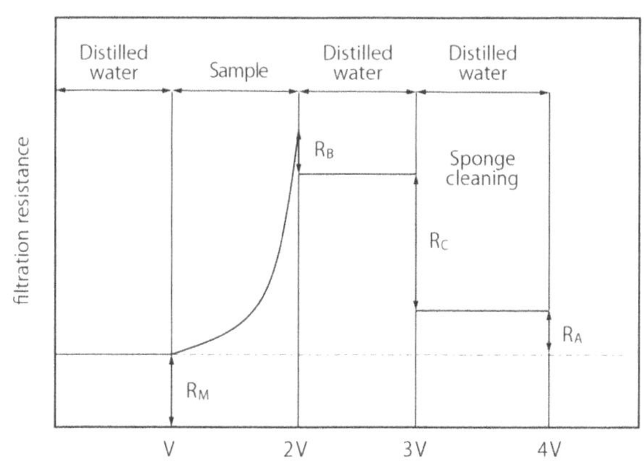

Figure 8.16 Calculation of membrane filtration resistance (membrane fouling).

Table 8.4 Raw water quality of a batch type fouling resistance evaluation test.

Sample	Turbidity TU	E260 1/cm	DOC mg/l	Sample	Turbidity TU	E260 1/cm	DOC mg/l
Raw water A	18.2	0.211	5.7	Raw water G	21.5	0.186	
B	12.7	0.163	3.3	H	58.1	0.234	6.6
C	11.5	0.072		I	18.2	0.210	5.7
D	13.2	0.083		J	15.8	0.150	3.9
E	14.5	0.158		Sample A	0.0	0.604	5.2
F	11.3	0.145		B	0.0	0.302	10.3

Sample A, B: the filtered water of 0.45 μm member filter (Chitose River: Raw water of Ebetsu city water supply).

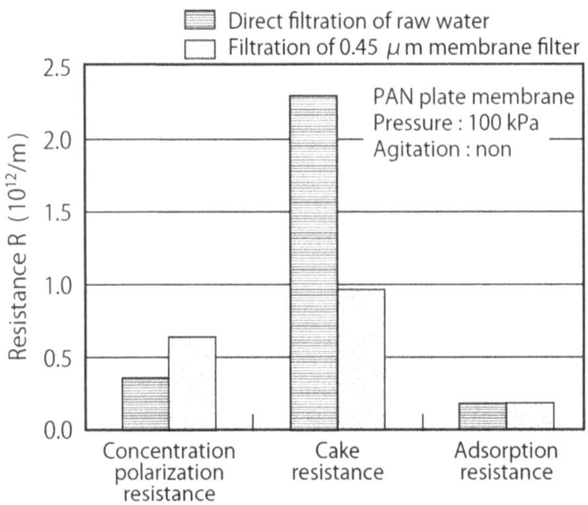

Figure 8.17 Effects of suspended solids and humic components on each resistance.

The cake resistance portion shows a large proportion, indicating that the resistance by suspensions (turbidity) is excellent. However, the color components present in the filtered water of the 0.45 μm membrane filter excluding turbidity also cause resistance, and resistance due to concentration polarization also occurs, suggesting the importance of cross-flow for controlling the solute concentration gradient in the vicinity of the primary interface of the membrane.

It was found that humic substances present in the Chitose River are involved in cake formation and account for a considerable proportion of the fouling. After filtering raw water through a 0.45 μm membrane filter to remove turbidity, supernatant water (sample A) at 12,000 rpm for 10 minutes after centrifugation and water (sample B) obtained by concentrating water permeated through the MF membrane (fractionated pore 0.1 μm) on the primary side with an NF membrane (desalting rate 50%) were prepared, adjusted to pH 7.5, and subjected to a batch permeation test. The results are shown in Figure 8.18, where high molecular weight humic particles (Sample A) were removed by nearly 80–90%, but the removal rate of the low molecular weight side (Sample B) was less than 10%. In addition, the concentration of sample A (which had a higher molecular weight) through the PS membrane (a cutoff molecular weight of 50,000 Da), was adjusted to allow water to flow through the membrane, and the

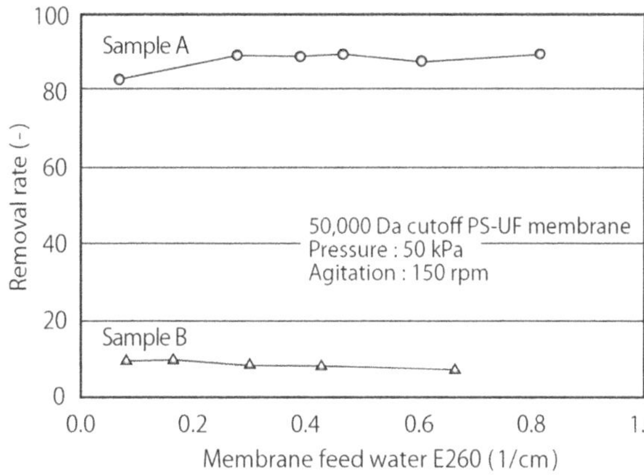

Figure 8.18 Removal rate of E260 by a polysulfone (PS) UF membrane, with a cut-off molecular weight fraction of about 50,000 Da.

situation of fouling caused by higher molecular weight humic substances was observed. The results are shown in Figure 8.19.

The total fouling resistance, cake-layer resistance, and concentration-polarized-layer resistance, which are constituents, increase linearly in proportion to the increase in the deposited amount of the E260 sensitive TOC (humics) of Sample A. The adsorption resistance has hardly been confirmed. This humic cake layer was more difficult to remove by conventional backwash than the clay-based cake layer, and careful manual wiping was required during this batch experiment.

The test results for low molecular weight humic substances (Sample B) are shown in Figure 8.20. The adsorption resistance appears to gradually increase to a constant value on the side of high-concentration adsorption, and the absolute value remained small.

In general situations where turbidity and humic substances coexist, how each of the water quality components of the raw water works on the four resistance components of fouling should be examined.

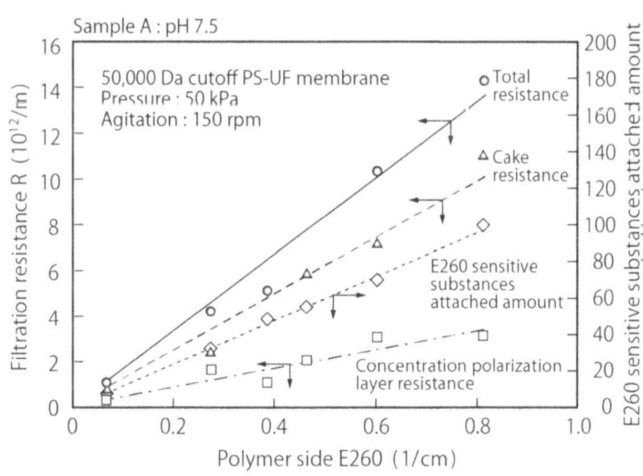

Figure 8.19 The effects of E260 polymer side components on filtration resistance and the deposit quantity of E260 sensitive substances.

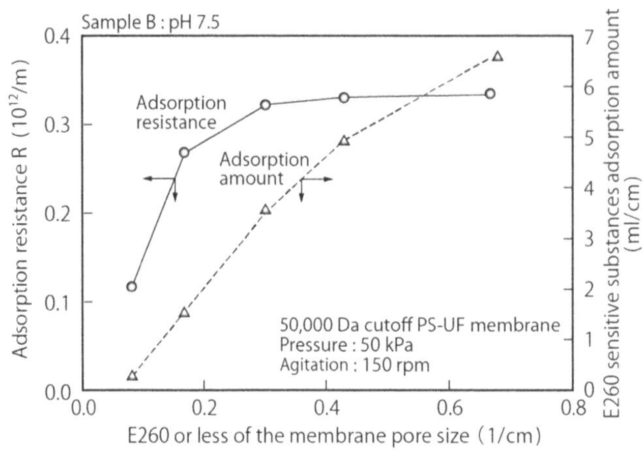

Figure 8.20 The adsorption resistance and adsorption quantity of E260 sensitive substances on low molecular E260 sensitive substances.

Therefore, we proceed below with experiments using PS membranes to investigate the effect on cake resistance and concentration polarization resistance.

8.2.4.1 Cake resistance and concentration polarization resistance

Although the concentration of the experimental raw water varies, there is a good linear relation between the filtration resistance and the concentration of the raw water organic matter sensitive E260 values (humic substance), as shown in Figure 8.21, indicating that the presence of humic substances is the dominant factor in the filtration resistance.

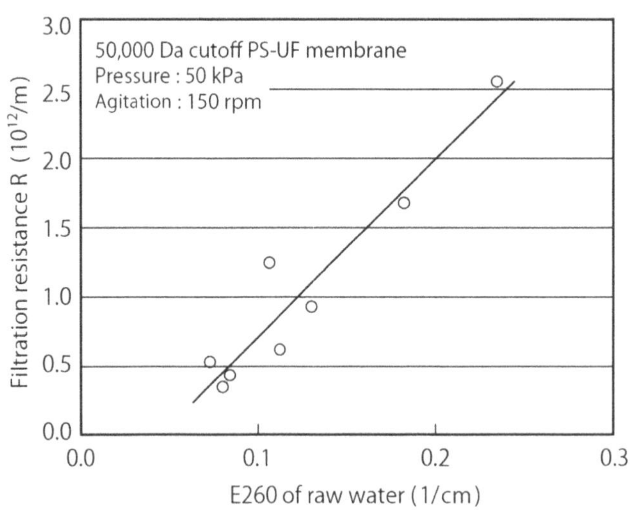

Figure 8.21 The E260 of raw water and total filtration resistance.

In a batch filtration experiment in which the filtration pressure was changed from 10 to 130 kPa, we observed how the membrane resistance varies as a whole and how each element resistance changes as shown in Figure 8.22.

While the resistance and adsorption resistance of the membrane body are hardly affected by the driving pressure, the resistance and concentration polarization resistance of the cake layer, which are the resistive elements of fouling, increase with increasing driving pressure. The increase in concentration polarization resistance will plateau when the driving pressure exceeds 60 kPa, but the cake layer resistance will continue to increase. Colloidal particles in the supply water are transported to the primary surface of the membrane to form a concentration boundary layer consisting of a cake layer and a concentration polarization layer. At the same time, particles in the boundary film are separated from the surface by the shear force of water flow and molecular diffusion, and are transported back to the main body of the primary side. The magnitude of molecular diffusion (inverse) transport due to concentration gradients is large when the impurity particle size is small, but transport due to the shear force of water flow is excellent when the particle size is large. The transport of suspended matter and higher molecular weight humic substances to the membrane surface increases with increasing operating pressure as the flux increases, which naturally increases the cake resistance and polarization resistance. On the high pressure side exceeding 70 kPa, the concentration polarization resistance does not increase, and the cake resistance increases excellently. The distribution of resistance from the polarizing layer to the cake layer increases, suggesting the disadvantage of operating the membrane at higher pressures than necessary.

8.2.4.2 Boundasry layer resistance and boundary layer digging

In order to suppress membrane resistance (cake resistance + concentration polarization resistance) during membrane filtration, the effect of boundary layer scouring by water flow (cross flow, agitation) on the primary side was examined.

If we compare the primary water supply reservoir of the batch membrane experiment system (Figure 8.15) with and without agitation of propeller water flow, we see that (as shown in Figure 8.23)

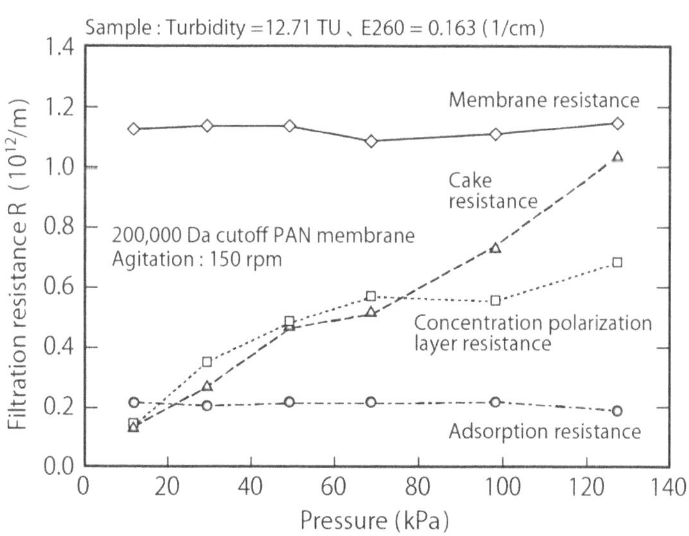

Figure 8.22 The effect of operating pressure on membrane filtration resistance.

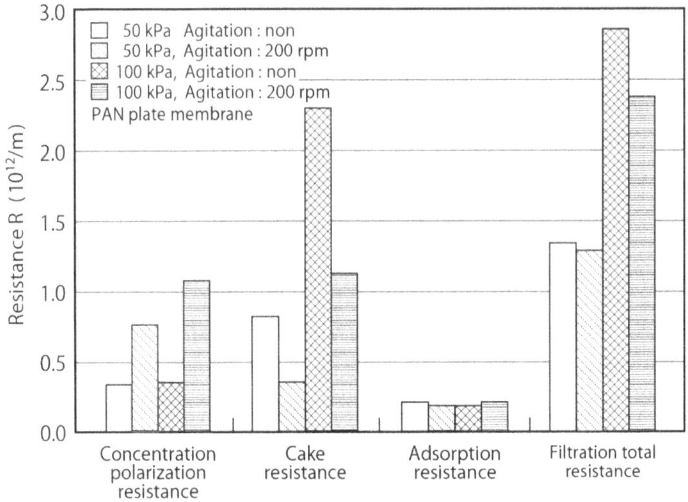

Figure 8.23 The effects of membrane flow conditions on each resistance.

the adsorption resistance is unchanged, but the cake resistance is reduced to about half by agitation, and the effect is remarkable on the high-pressure side. Density polarization resistance is increased by redespersion, but it is negated by a significant decrease in cake resistance, and total resistance is slightly suppressed. It is important to note that resistance suppression during high-pressure operation cannot be adequately achieved with cake layer suppression alone.

8.2.5 Reversible and irreversible resistance

As shown schematically in Figure 8.14 (above), reversible resistance corresponds to concentration polarization resistance and the resistance of the cake layer that is removed by backwashing. Irreversible resistance is the resistance exhibited by the membrane pores themselves and the resistance caused by adsorption and the cake part which is the object of chemical cleaning and which cannot be peeled off by physical cleaning (Tambo *et al.*, 1994).

The sum of the reversible and irreversible resistances is the total resistance of the membrane operation. In opposition to this, driving pressure is applied to ensure a constant permeate flow rate.

Both a constant flow rate operation (variable pressure) and constant pressure operation are adopted, while allowing a certain range of fluctuation. In water treatment, constant flow filtration (with guaranteed permeate volume) is often performed using any increase in operating pressure as control information.

In the following experimental examples, constant flow rate and variable drive pressure filtration are described, unless otherwise stated.

The first experiment was conducted on a four-line constant flow membrane filtration experimental apparatus (Figure 8.24) by the the Department of Environmental Engineering, Hokkaido University, and was the last experimental facility overseen by the author (Tambo) actively working as professor on research. It was installed at the Ebetsu City Water Works Purification Station (near Sapporo), and was studied for the relatively high turbidity and high color river water in the downstream of a peaty area of the Chitose River. In order to compare the water with clean raw water, a comparative test was conducted with Toyohira River raw water for Sapporo City Water Works, with low turbidity and little organic

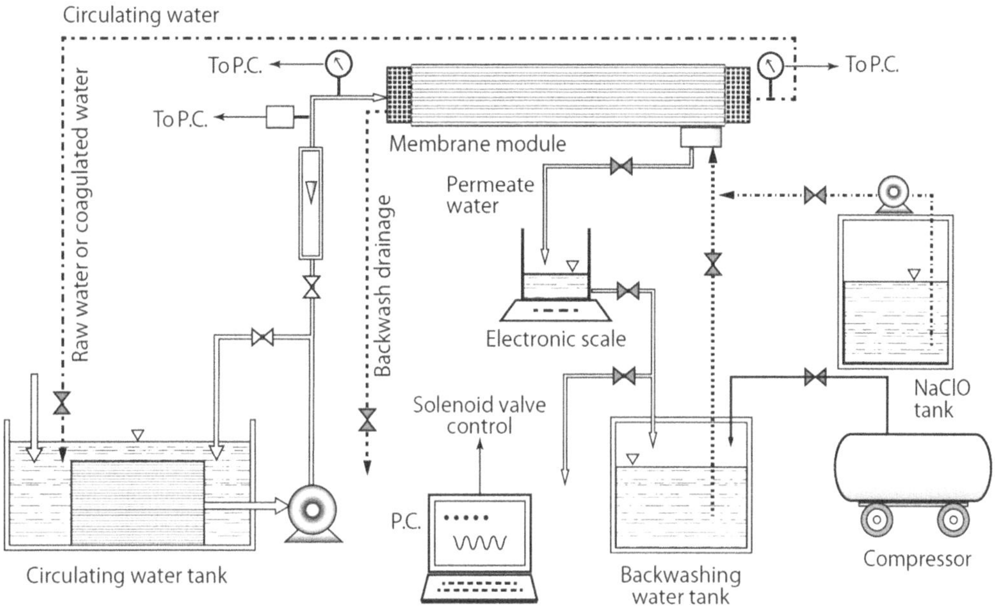

Figure 8.24 A constant flow membrane filtration experimental apparatus.

content, that flowed out from well maintained forests on the Andesite geological region, using a device similar to that installed at the Moiwa purification station in Sapporo City.

In the experiment, the river raw water was used for continuous operation for over 1500 hours. This system was designed to perform two types of test: cross-flow filtration and constant-flow filtration. A hydrophilic CA membrane with a cut off level of 150,000 Da and a hydrophobic PAN membrane with a cut off level of 50,000 Da were mainly used. The operating pressure was set at 30, 50, or 100 kPa for continuous operation.

In order to observe the membrane fouling characteristics of constant flow filtration, the permeation flux was tested at 10 and 2 m/day using a PAN membrane with a nominal pore size of 0.01 μm at an operating pressure of 200 kPa. After cleaning by chemicals, the fluxes were changed to 1.5, 1.0, or 0.5 m/day in response to the high turbidity of the raw water.

For physical cleaning, backwash with membrane permeate was performed at 200 kPa after 60 minutes of filtration, followed by air cleaning at a flow rate of 2.05 m/sec (superficial velocity of air). The backwash sequence was performed with membrane permeate for 60 seconds at 60-minute intervals, air wash was started 20 seconds after the start of backwash, and both water and air wash were used for 40 seconds. 5 mg/l of free chlorine (sodium hypochlorite) was added to the backwash water about 2 to 3 times per week to prevent biological fouling. For the raw water with low load, only backwash water was used.

Figure 8.25 shows the change in filtration resistance of the Chitose River water immediately after the start of filtration using CA and PAN membranes. The increase in the filtration resistance of the PAN membrane (a hydrophobic membrane) immediately after the start of filtration was rapid, and the distinctive difference is that the CA membrane (a hydrophilic membrane) resolved the resistance by physical washing. The irreversible resistance accumulated in the hydrophobic membrane, and the reversible resistance recovery by backwash flushing fluctuated in an unstable manner compared with the CA membrane.

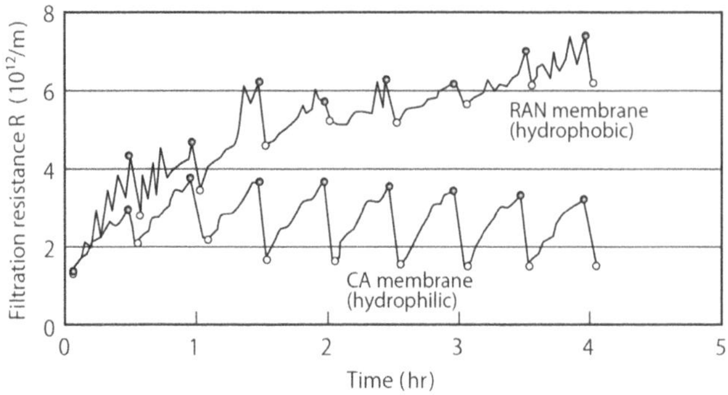

Figure 8.25 Change in filtration resistances in each filtration cycle.

Figure 8.26 shows the effect of physical cleaning of both membranes after 100 hours. After 50 hours, the waveform of the physical cleaning of the PAN membrane increased in 60 minutes and recovered by backwashing in an orderly manner, repeating the same pattern as that of the CA film. However, the absolute value of the reversible resistance increased in the cycle and the magnitude of the recovery span was several times that of the CA membrane. This appears to be due to differences in fractional molecular weights.

The cumulative irreversible resistance of the PAN membrane was 15×10^{12} (1/m) after 64-hour operation, but decreased to 9.5×10^{12} (1/m) at 100 hours. It is considered probable that some of the suspensions and polymer humic substances added to the irreversible resistance components were separated by backwashing, and the hydrophilic components were stably covered by adsorption on the membrane surface. The hydrophilicity of the hydrophobic membrane caused steady-state waveforms after backwashing. An electron microscope (SEM) of the CA membrane showed that the cumulative resistance was caused by an irreversible cake layer of more than 10 μm on the primary surface of the membrane.

Figure 8.27 shows the change in the filtration resistance of the 100 kPa constant pressure cross-flow filtration of the CA and PAN membranes over the entire elapsed time. As noted earlier, the filtration

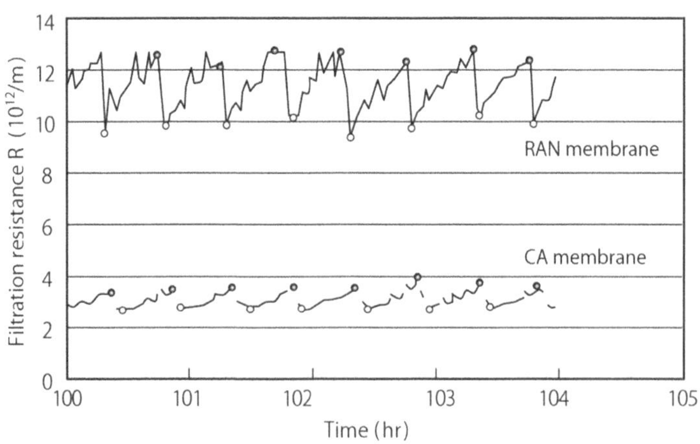

Figure 8.26 Filtration resistances in each filtration cycle.

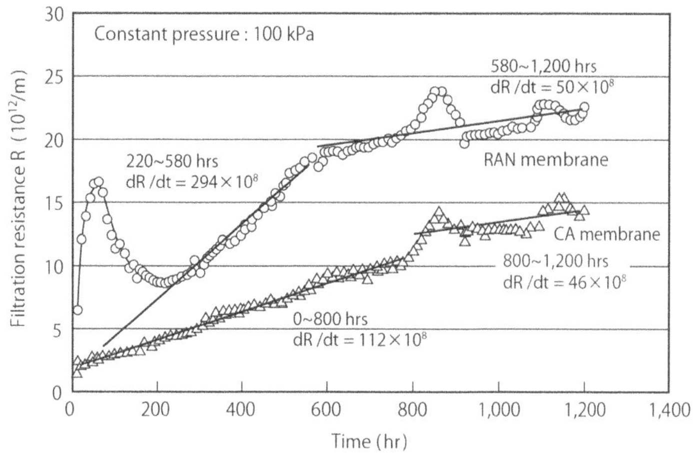

Figure 8.27 Change over time mof membrane filtration resistane for CA and PAN membranes.

resistance of hydrophobic PAN membranes increased sharply over the initial 50 hours, but decreased sharply over 200 hours. Initially, cake formation at the primary interface, such as hydrophilic crude humic substances, increased the resistance. In addition, adsorption to interfaces and pore surfaces of low-molecular humic substances, etc., progressed, and the membrane became hydrophilic by a protective action of adsorbed humic colloids, which facilitated subsequent cake separation of suspensions and coarse humic substance particles by backwashing, so that backwashing worked effectively and, after 200 hours, it could be considered that the membrane behaved as a modified hydrophilic membrane having a smaller molecular weight fraction (greater irreversible resistance) than the CA membrane. The rate of increase in resistance after 600 hours was almost the same for both the CA and PAN films, and only the basic irreversible resistance portion seems to have shown the absolute value. The properties of the PAN membrane, which is highly hydrophobic up to about 500 hours, are still characterized as a hydrophobic membrane that more easily traps humic substances and, after that, it is considered that the membrane behaves as a quasi-hydrophilic membrane covered with humic substances.

We next considered how the difference between clear raw water from the Toyohira River (Sapporo City Water Supply) and raw water from the Chitose River (Ebetsu City Water Supply), which is a peaty land river, affect membrane filtration resistance (permeation flux) and treated water quality.

Figure 8.28 shows the time course of turbidity and color ($E260_{1\ cm\text{-}cell}$) of the raw water from the Toyohira River and the lower stream of the Chitose River during the experiment, and Figure 8.29 shows the change in filtration resistance.

Thirty hours after the start of the experiment, the turbidity and color of the Toyohira River suddenly increased due to a heavy rain fall, and the filtration resistance rapidly increased discontinuously. This showed that, even in the clear river, the surface layer runoff from the fallen organic (leaves) soil layer (A_1 layer) of the forests affected the polymer humic substance concentrations by the rapid increase in runoff. This is a polymeric humic ingredient with a molecular weight of more than (3–4) × 10^3 Da, which exhibits characteristics with a $TOC/E260_{1\ cm\text{-}cell} \fallingdotseq 30$; see Figures 3.36 or 3.41 (in Chapter 3).

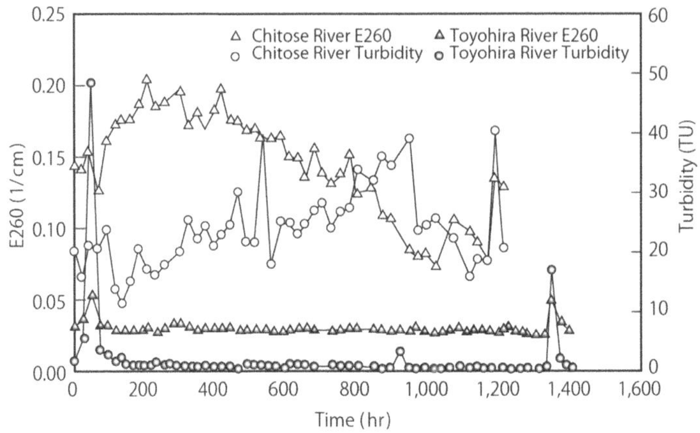

Figure 8.28 The time course of raw water quality for the Toyohira and Chitose rivers.

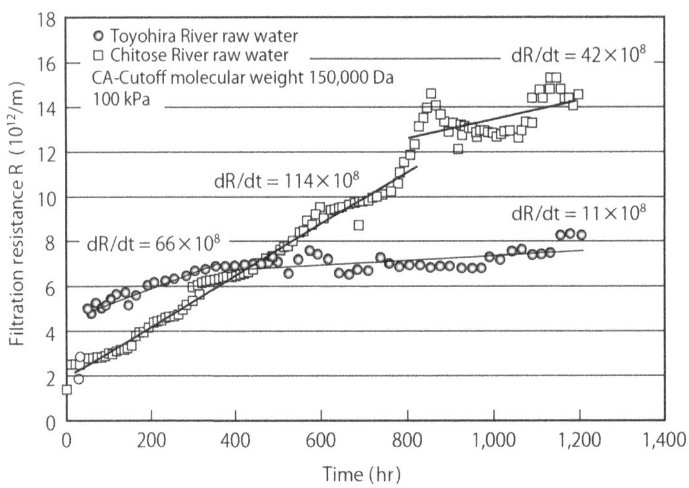

Figure 8.29 The change of filtration resistance of the raw waters from the Toyohira and Chitose rivers.

The membrane permeation resistances of the Chitose and Toyohira rivers at 12 hours immediately after the start of the experiment are shown in Figure 8.30. The filtration resistance of the clear raw water and the filtration resistance recovered by periodic reverse washing (reversible resistance) were about $0.25 \times 10^{12}\,m^{-1}$, whereas the reversible resistance of the Chitose River raw water was about (1.2 to 1.5) $\times\,0^{12}\,m^{-1}$, which is 5 to 6 times greater than that of Toyohira river water. This is the result of a clear difference in the absolute magnitude of reversible resistance, such as cake resistance, due to concentration differences in the raw water, such as turbidity and polymeric humic substances. By contrast, there is no significant difference in the irreversible resistances, and the cumulative values for 10 hours are both around $0.25 \times 10^{12}\,m^{-1}$, and the rate of increase appears to be around $0.25 \times 10^{11}\,m^{-1}\,hr^{-1}$.

Figure 8.31 shows the changes in membrane resistance that occur during a rapid increase in turbidity and color over the entire day (as Figure 8.28 shows for the Toyohira river), which was triggered 30 hours after

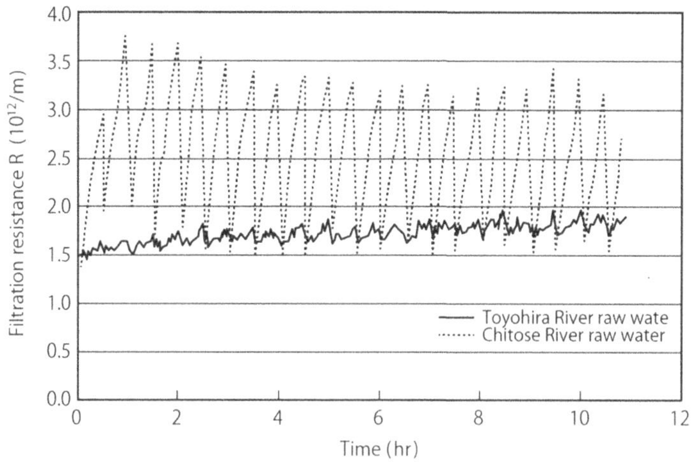

Figure 8.30 The effects of raw water quality on reversible and irreversible resistance of UF membrane filtration.

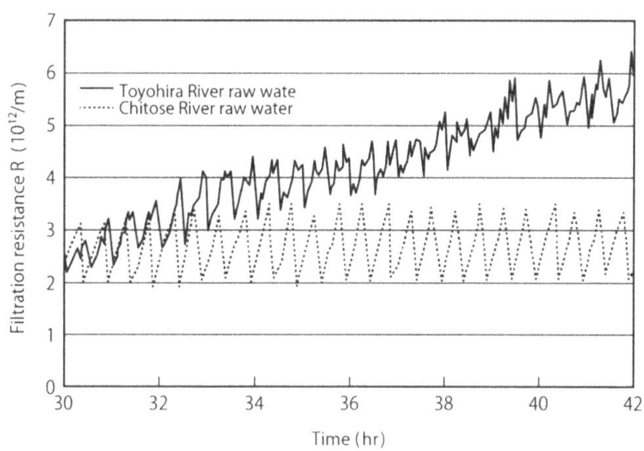

Figure 8.31 The effects of raw water quality on reversible and irreversible resistance of UF membrane filtration.

the start of the experiment, and the change in membrane resistance associated with the progress of filtration of high turbidity and high E260 extinction (humics) of raw water.

The highly turbid water in the Toyohira River has a high concentration of higher molecular weight humic substances. The reversible resistance value is approximately $(1.0\text{–}1.2) \times 10^{12}\ \mathrm{m}^{-1}$, which is similar to the reversible resistance value of the Chitose river raw water at normal times. However, the increase in the irreversible resistance of the filtration membrane of the Chitose river raw water, which has undergone hydrophilization repair, is only small over a period of about 10 hours (as shown in Figure 8.31, but is steadily increasing over a long period of time, as seen after 800 hours in Figure 8.29. On the other hand, in the Toyohira River raw water, the resistance (which rises sharply during high turbidity) increases due to irreversible resistance, indicating that the reversible resistance fluctuates in an oscillatory manner with a constant width. This is probably because humic substances introduced during high turbidity are not recovered by backwashing and accumulate over a short period of time, resulting in irreversible cake resistance or adsorption resistance. When river water returns to normal, the increase falls to a low level.

It should be understood that irreversible increased filtration resistances (membrane clogging) of high turbidity and high E260 (humic) inflows in normally clean rivers and responses to chemical cleaning methods for their restoration occur by a different mechanism than in the case of constant high-color water treatment, such as for peatland river water.

8.2.6 Pre-coagulation treatment for UF membrane treatment

When processing water from a normally clear river such as for Toyohira River raw water, direct treatment with a UF membrane is likely to continue stably for a long period of time. However, in the treatment of river water that flows through the peaty area downstream of the Chitose River, reversible and irreversible resistance increases remarkably at all times, and the involvement of polymeric humic substances, etc. is expected. As described in Chapter 3, coagulation is likely to function effectively as a pretreatment for such water.

Therefore, the contribution of a pre-treatment coagulation to UF membrane filtration was investigated by jar test, batch membrane filtration test, long-term constant pressure cross flow and a constant flow filtration test using the Chitose river water (Ebetsu city Water Works raw water). Polyaluminium chloride (PACl) was used as the coagulant.

In the jar test, PACl was injected into the Chitose river raw water, which was adjusted to a specified pH by 0.1NHCl and NaOH. After rapid agitation at 120 rpm for 3 minutes, the concentration polarization resistance, cake resistance, and adsorption resistance were determined using the batch type membrane filtration experimental device shown in Figure 8.15, and the coagulation effects of the membrane on fouling in one filtration cycle were investigated.

The rates of removal of soluble organic matter by filtration, with and without pre-coagulation, using UF membranes (a) with hydrophobic PAN membranes of cut off level molecular weight 50,000 Da and (b) with hydrophilic CA membranes of 150,000 Da, are shown in Figure 8.32 (PAN) and Figure 8.33 (CA).

In the case of PAN membranes, the coagulation treatment at a PACl injection rate of 2.0 mg/l (as Al) resulted in removal rates of 52, 33, and 30% for E260, DOC and THMFP (trihalomethane formation potential), respectively, which greatly improved the removal rate of 35, 15, and 12% in the case of single

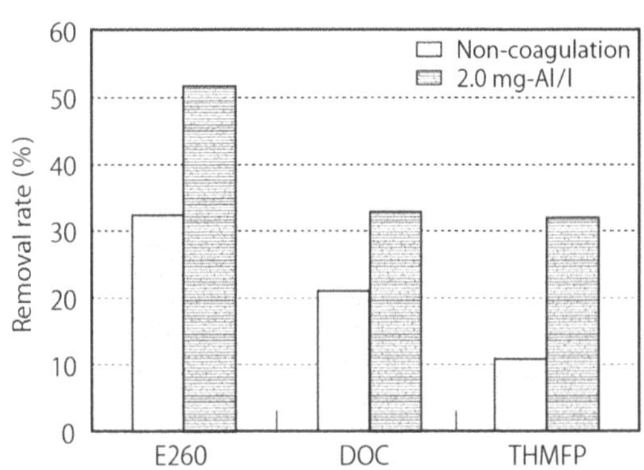

Figure 8.32 The effect of pre-coagulation treatment on removal of soluble organic components in PAN membrane filtration.

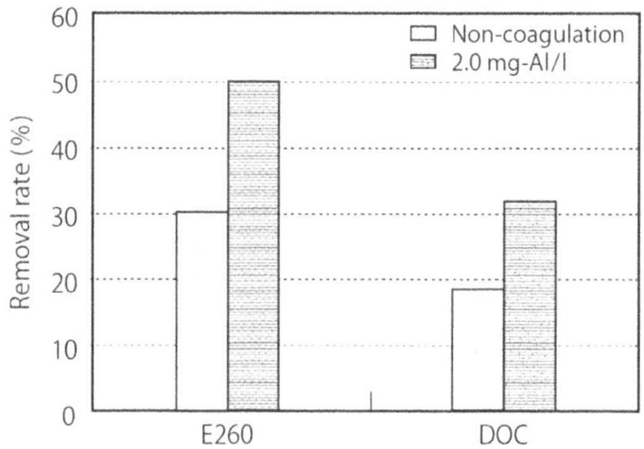

Figure 8.33 The effect of pre-coagulation treatment on removal of soluble organic components in CA membrane filtration.

filtration of PAN membranes. In addition, the removal rate of E260 and DOCs was 50% and 32% when CA membrane treatment alone and pre-coagulation were performed, and the removal rate was 30% and 19% when CA membrane treatment alone was performed. This is an effect of microflocculation, and further coagulation studies for membrane treatment are required. For example, a detailed process study of coagulation and membrane separation that differs from the system of conventional coagulation, flocculation and filtration systems that we have cultivated over many years in rapid sand filtration systems is required, such as of the relationship between the increased removal rate obtained by increasing the addition rate of coagulant from 0.2 to 0.5 mg/l and the inhibitory effect of membrane fouling factors.

Figure 8.34 shows how the elimination rate of various components can be improved by adding PACl coagulant in quantities of 1.0, 5.0, and 7.0 mg/l (as Al) to the treatments with CA membrane alone and with pre-coagulation treatment.

When the aluminium injection rate exceeds 5 mg/l it can be seen that the removal rate of soluble components plateaus and approaches the limit removal rate. The optimal PACl dosing rate for humic

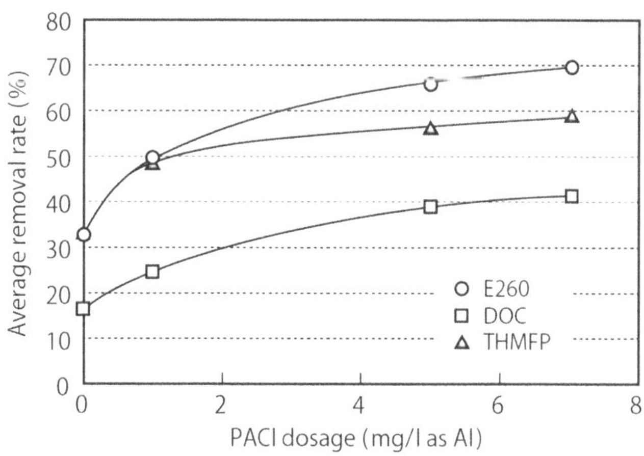

Figure 8.34 The effects of PACl dosage on soluble organic component removal rate.

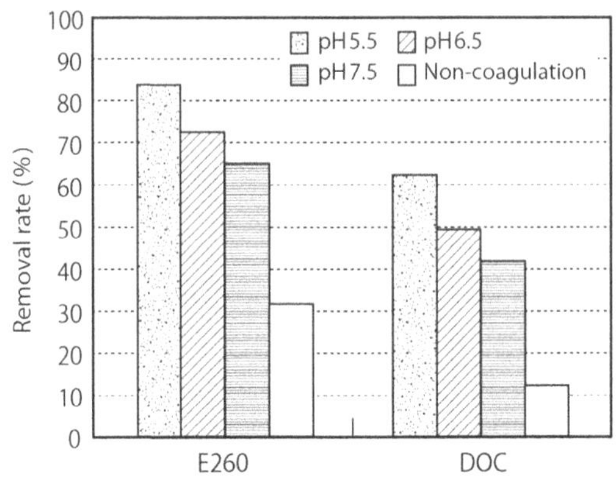

Figure 8.35 The effects of coagulation pH on the removal rate of soluble organic components in coagulation and UF membrane filtration.

removal obtained by jar test for the Chitose river raw water was about 5 mg/l (as Al). The results of the fixed dosage-variable pH jar test with the addition of 5 mg/l (as Al) to determine the optimum pH condition are shown in Figure 8.35. As shown in Tables 5.13 (Chapter 5), the best removal rate is obtained around pH 55, and it can be seen that pre-coagulation for membrane removal can be handled by the same conventional agglomeration mechanism for color removal.

Figure 8.36 shows the results of direct filtration by CA membranes and the apparent molecular mass of components removed by pre-coagulation + UF membranes. The membrane removes 100% of the polymer components of 6000 Da or more. However, the removal rate of the polymer components of 3000 Da or less gradually decreases toward the small Da side for the first time by coagulation, leading to a removal rate of about 50% of 3000 Da. We can read a figure close to the finding that more than 1500 Da (discussed earlier in the matrix section, Section 3.5.4 in Chapter 3), is the limit that metal coagulants can cope with. We believe

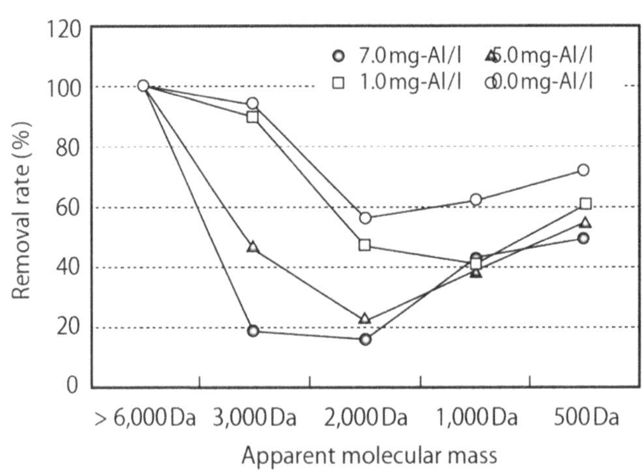

Figure 8.36 The effects of coagulant dosage on the removal rate of humic components, for each molecular weight.

that the creation of a reliable UF membrane with a molecular weight fraction of about 1000 Da and single-stage separation are technical challenges as measures against colored raw water. The decrease in the color component on the low molecular weight side, which cannot be removed by pre-coagulation, seems to be the effect of adsorption, but the relationship with irreversible fouling is not clear. From this result, it is again assumed that a PACl dosing rate of 5 mg/l (as Al) will be the optimal dosage for this water.

In order to investigate the mechanism by which fouling of the membrane proceeds in the operation subjected to a pre-coagulation treatment, three types of experiment were conducted by batch adsorption tests shown in Figures 8.15 and 8.16, including UF membrane filtration, in which higher molecular weight humics were removed by coagulation, UF membrane filtration (in which the low molecular humics was removed by adsorption using powdered activated carbon), and UF membrane filtration, in which the concentration polarization resistance, cake layer resistance, and adsorption resistance were obtained. The results are shown in Figure 8.37.

Although the adsorption resistance was slightly reduced by the powdered activated carbon treatment, the concentration polarization resistance and cake resistance were not changed. On the other hand, when the pre-coagulation treatment was performed, each resistance was remarkably small, and in particular, the concentration polarization resistance and the cake resistance were reduced to almost disappearance.

Figure 8.38 shows the income and expenditure of humic substances, which are the main components of such removals, based on the E260 index of humic substances in water quality matrices by Tambo (1976, 1978). Powder activated carbon adsorption seems to remove about 50% of the humic substances in the raw water (on the low molecular weight side) and about 60% of the remaining polymer (30% in the raw water) is removed by the membrane. In contrast, in the pre-coagulation treatment, about 75% is coagulated (micro-flocculation) and removed by the membrane, and the remainder can be removed by further addition of activated carbon.

We can see that the core of removal in 'micro-flocculation + UF membrane separation' lies in 'coagulation + microflocculation.' The main cause of reversible fouling in this case is cake formation, which seems to be the essence of the operation in the coagulation treatment. In order to minimize irreversible fouling, such as adsorption, in consideration of cake formation with micro flocculation it is likely that consideration will be required to choose (make) a hydrophilic membrane or a membrane that

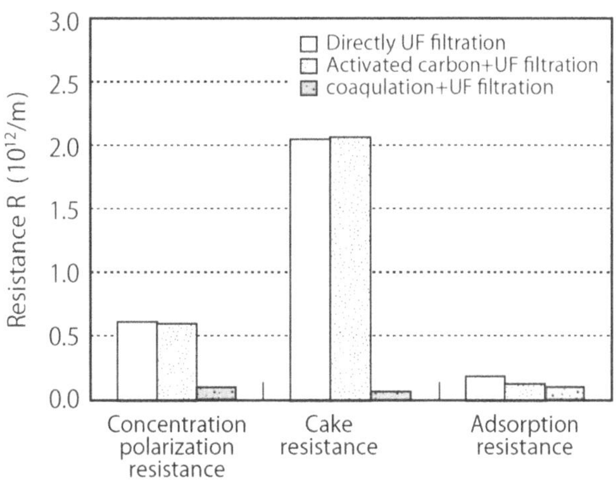

Figure 8.37 The effect of coagulation and activated carbon adsorption on membrane filtration resistances.

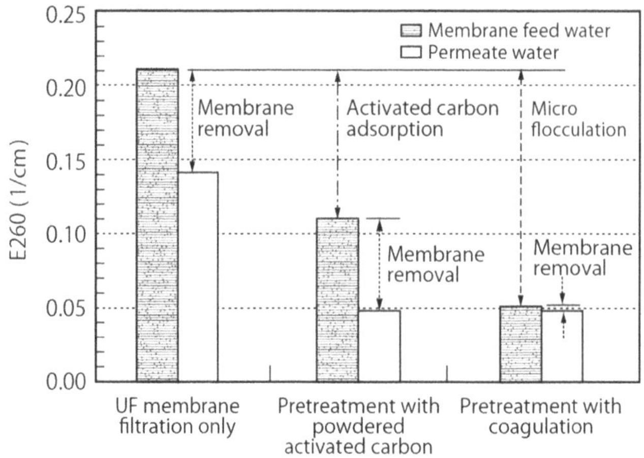

Figure 8.38 A comparison of pretreament and residual E260.

makes it hydrophilic. The key to day-to-day operations is the selection of agglomeration conditions for microfloc formation – which are the intensity of agitation and the conditions for chemical injection.

Figure 8.39 shows an example of a comparison of changes in permeate flux over a long period of operation between single filtration of a CA membrane using 100 kPa constant pressure cross flow and 'coagulation + CA membrane filtration'. After 2000 hours of filtration, both levels decreased to 0.5 m/day, which are equivalent. This is the equilibrium value for irreversible fouling. The system with pre-coagulation had, up to about 1500 hours during this period, a higher flux than the membrane-only system. At 750 hours, the flux was twice as high as that of the membrane-only system, and the cumulative volume of treated water up to chemical cleaning was greatly increased. When the operating pressure of the CA membrane was reduced to 50 kPa, both fluxes decayed, as shown in Figure 8.40. Despite the reduced driving pressure, the final fluxes were larger than 100 kPa, and the effect of

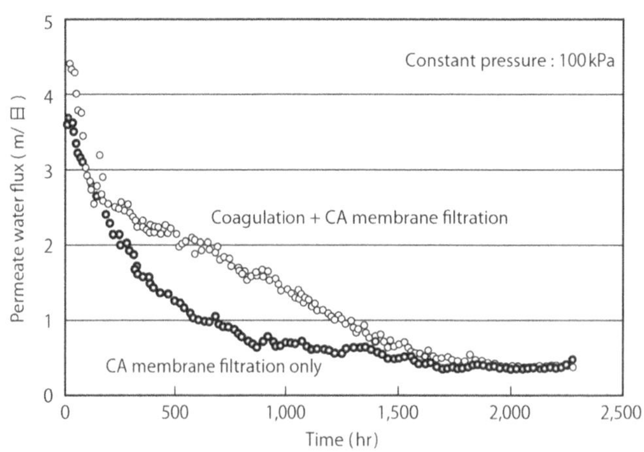

Figure 8.39 The permeate water flux of CA membrane only filtration, and of pre coagulation + CA membrane filtration by constant pressure operation of 100 kPa.

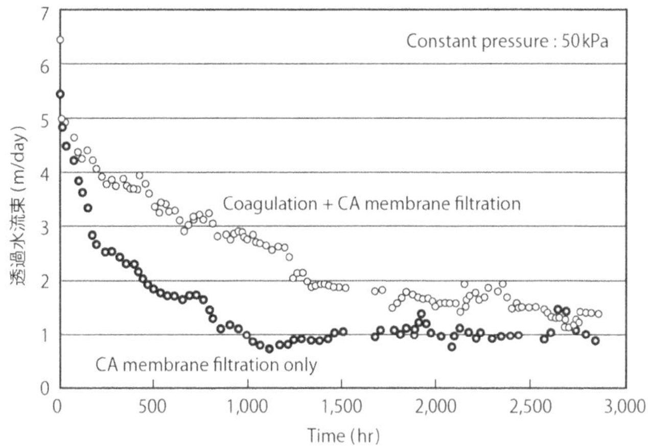

Figure 8.40 The permeate water flux of CA membrane only filtration, and of pre coagulation + CA membrane filtration by reduced constant pressure of 50 kPa.

pre-coagulation was durable and excellent over a long period. The cumulative volume of treated water was doubled in 2500 hours. This is because irreversible fouling (a fixed cake layer) formed by coarse humic substances is suppressed at low operating pressures, and compaction of the cake layer caused by coarse graining by coagulation is also greatly suppressed, thereby reducing the rate negating the positive effect of particle size increase.

Therefore, searching for conditions on the low-pressure side of UF membrane treatment is an important issue in designing the process. In particular, the cake layer of deformable microflocs should be handled with caution in systems where pre-coagulation is added.

Figure 8.41 compares the effect of pre-coagulation of hydrophobic PAN membranes with that of hydrophilic CA membranes shown in Figure 8.39 in long-term tests. In the case of hydrophobic

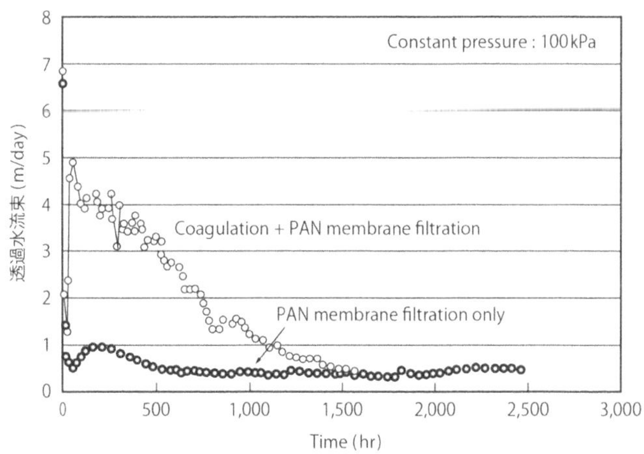

Figure 8.41 The permeate water flux of PAN membrane only filtration and of coagulation-PAN membrane filtration.

membranes alone, the adsorption of humic substances to the membrane surface proceeds rapidly, resulting in irreversible fouling and a decrease in flux in a short time. On the other hand, after micro-flocculation of crude humic substances, the cake layer formed by microflocs becomes a condition of reversible fouling which is easily peeled off by backwashing though, after 1500 hours, the cake layer eventually becomes an irreversible fouling condition. As a result, both fluxes settle down to 0.05 m/day and wait for regeneration by chemical cleaning. It is important to note that the effect of adding pre-coagulation in the case of hydrophobic membrane treatment is very different from that in the case of hydrophilic membranes.

8.2.7 Membrane washing

When river water, groundwater (well water, underflow water), lake water, etc. are filtered through UF and MF membranes, etc., clay particles, humic substances, algae, microorganisms, etc. are retained and adhere to the vicinity of the primary surface, resulting in membrane fouling and an increase in membrane differential pressure (membrane resistance). As a result, the permeation flux is reduced in the case of constant pressure filtration. The raw water for the regeneration of aerobic sewage treatment behaves similarly to peaty colored water. However, when biological treatment is not sufficient, the proportion of nitrogen compounds (proteins) is large, and bacteria grow on the surface of the membrane, which increases the membrane filtration resistance.

In order to prevent such deterioration in membrane filtration performance, the membrane is cleaned after each lapse of a predetermined operation time.

As mentioned earlier, there are two types of membrane cleaning:

(1) Physical clean with a backwash (with membrane permeate to prevent reversible fouling) and a wash with pressurized air, repeatedly at regular intervals of 15 minutes to 1 hour;
(2) Chemical cleaning aimed at almost complete elimination of irreversible fouling by chemicals and return of transmembrane pressure to the initial state for irreversible fouling recovery at the end of filtration (e.g., after 1500 hours to several months) when repeated physical cleaning ceases to be effective.

Physical cleaning requires trial and error investigation to increase the interval and shorten the cleaning time as much as possible, because the recovery rate of membrane-treated water is lowered by shortening the interval and increasing the cleaning time. Washed waste water may contain coagulants or may contain only colloids in raw water. If powdered activated carbon is added to the primary side to precede the batch adsorption reaction, activated carbon is also included in the wastewater. In general, it is temporarily stored in a sump, precipitated with a flocculant, and the supernatant water is returned to the raw water. The sludge produced is concentrated and dewatered and disposed of by landfill.

For chemical cleaning, there are two types:

(1) An on-site method in which the membrane module is not removed from the treatment system, and the chemical solution is circulated in an on-site state to dissolve and remove irreversible fouling substances, followed by rinsing with treated water;
(2) Membrane module is removed from the system and regenerated at the chemical cleaning site.

In large-scale water purification plants, where many membrane modules are operated in parallel, it is possible to connect chemical cleaning systems with valves and pipelines on-site. By constrast, in a small-scale water purification plant, it is possible to ensure the accuracy of regeneration by removing the membrane module and cleaning it in a specialized regeneration plant, and also possible to avoid the difficulty of chemical wastewater treatment. For this purpose, it is necessary to construct a plant in which a large number of small membranes that can be easily removed and operated by one or two people are bundled. It is also necessary to upgrade the regeneration plant to support the membrane filtration system type of the whole region.

Membrane treatment is basically a process that can be developed into a dispersed water treatment system without being strongly influenced by economies of scale. One day, we would like to see a new water supply system in a society with a declining population – which will promote the diffusion of membrane treatment equipment (with automatic operation) as regional membrane treatment plants supported by a regional regeneration center. It would be a disaster-resistant and highly resilient regional system.

Chemicals used for chemical cleaning differ depending on the type of impurity components that cause irreversible contamination of the membrane. The material of the membrane will also vary with regard to what concentration of chemicals it can withstand. Ceramic membranes have advantages in that they can be used in a wide range of pH from acid (pH 2) to alkaline (pH 10 to 12). In the case of an organic membrane, the amount of chemicals that can be used is limited by the material of the film.

Chemical cleaning often involves the use of a mixture of chelating agents and detergents, and some types of cleaning involve oxidizing agents. Phosphoric acid, sulfuric acid, and hydrochloric acid are used as strong acids, and citric acid and oxalic acid are used as organic complex acids to dissolve metal hydroxides such as iron, aluminium, and manganese.

When considering natural rivers, lakes and marshes such as peaty land aquifers, the main components of impurities in water appear to be fine particles of turbidity components and humic substances. In the case of sewage, as described in Chapter 3, the components of the humic system are the main targets of chemical cleaning. Sodium hydroxide (pH 10 to 12) is used for such organic substances, and oxidizing agents such as hydrogen peroxide, sodium hypochlorite, and peracetic acid are used for the oxidative decomposition of organic macromolecular components. Surfactants and detergents are used in combination with chelating agents such as ethylenediamine-tetraacetic acid to reduce interfacial tensions, such as metal complexes. Sodium pyrosulfite is also used for biodegradation protection and reduction of metal oxides.

An example of chemical cleaning of CA membranes is as follows:

(1) First stage wash (1% solution) with anionic surfactant plus water containing chelating agent and 50 mg/l free chlorine;
(2) Second stage wash with a mixed solution of anionic activator and complexing agent such as citric acid, EDTA and phosphoric acid (pH 8, 0.9% solution);
(3) After preliminary rinsing with chlorine-sterilized water for 30 minutes, wash the first stage for 1 hour, and then wash the second stage for 3 hours.

In all three wash steps, chemical circulation by cross flow and contact retention by immersion are performed alternately. Add backwash for 30 to 60 seconds for each wash step and flush 150 to 200 l of rinse water per 1 m^2 membrane as the final step.

The supposed main components of the wash wastewater are COD ≑ 828 g/m^3, BOD ≑ 217 g/m^3, TOC ≑ 218 g/m^3, phosphate ≑ 103 $g(P)/m^3$, anionic surfactant ≑ 56 g/m^3, suspensions SS ≑ 832 g/m^3, and pH ≑ 7.8 to 8.4. In addition, less than 5% of similar components are added to the rinse water. In this case, the amount of chemical cleaning effluent per CA-membrane 1 m^2 seems to be about 4.4 m^3.

Treatment and disposal of chemical cleaning wastewater is the most difficult issue in the treatment of UF and NF membranes. With complex cleaning operations and diverse chemical usage sequences, on-site treatment is nearly impossible. Since the frequency of chemical cleaning is only about once every few months, it is once stored in one-end storage tanks. Then, it is usually treated in the biological treatment process of a sewage treatment plant by diluting it. Little material is available in industries where chemical regeneration is performed commercially.

The standardization of intensive chemical regeneration plants for the popularization of small-scale dispersive membrane treatment plants and their waste liquid treatment will be one of the goals of research and development.

8.3 COMPARISON OF UF AND MF MEMBRANES

The minimum dimensions of UF and NF membranes are close to several nm (or more) and 10 nm (or more), respectively, as shown in Figure 3.39 (in Chapter 3). As shown in Table 8.2, the differential pressure (operating pressure) between the UF membranes required for operation is about 700 kPa or less. If the UF membranes work properly at 50 kPa as described in the previous section, there will be no difference in driving pressure between the two membranes. In addition, there seems to be no reason to choose a large pore MF membrane or an MF membrane with a 0.3 μm cut off level, except for fine-screen applications involving only clay-based suspensions, protozoa, and their eggs. In conclusion, if we consider safety against viruses and bacteria, we think that membranes for water supply and water use should be considered mainly with UF membranes. For example, it has been observed that, even in the case of clean river water in Japan (e.g. Toyohira River raw water), once water contains higher molecular weight humic substances ($DOC/E260_{1\ cm\text{-}cell} \doteqdot 30$) which flow out of the A_1 layer where fallen leaves accumulate on the forest bottom, this causes a sharp increase in irreversible fouling of UF membranes (Figure 8.31).

To compare the properties of MF and UF membranes, a continuous experiment was conducted on Toyohira river raw water using a four-channel system of alumina ceramic MF (tubular: 6 mm inside diameter, 30 mm long) membranes with a 0.1 μm cut off size, and PAN hydrophobic hollow-fiber UF membranes with a cut off molecular weight of 50,000, 13,000, and 6000 Da (inner diameter: 0.8 mm, length: 430 mm).

The system configuration of the experimental equipment was the same as that shown in Figure 8.24. Toyohira river raw water was introduced into a thermostatic tank and pumped by a circulating pump at a constant pressure of 100 kPa to an internal pressure type hollow-fiber module or to a ceramic module. The flow rate in the cylinder was maintained at 0.09 m/sec by a cross-flow method, and membrane filtration for internal and external permeation was carried out. The amount of permeate was measured with an electronic balance and poured into the backwash water tank. For backwash, the backwash pressure water tank was pressurized with a compressor at regular time intervals using membrane permeate water, and then the backwash valve was opened to backwash the treated water from the outside to the inside of the membrane at 200 kPa for 30 seconds to drain the wash water. All operations and records were controlled and recorded by computer. Basically, the characteristics were evaluated by continuous operation for 1200 hours (50 days). The amount of water supplied to the module and the pressure at the inlet of the module were set via the computer. After deployment began, the module supply water volume, module inlet and outlet pressures, membrane permeate water volume, and membrane permeate fluxes were recorded on the computer at 2-minute intervals. The cross flow velocity (circulation flow velocity) and the water recovery rate were calculated and recorded as needed from the results of primary measurement.

The state of water quality conversion was measured over time during the treatment process operation for raw water (coagulated water in the case of treatment with coagulation), as was the water in the circulation loop, the permeated water through the membrane, and backwash wastewater, pH and turbidity. The absorbance ($E260_{1\ cm}$) of the ultraviolet light at 260 nanometers was measured once a day. Other measurements were made at any time when there was a characteristic change. Molecular weight fractionation by high performance liquid chromatography (HPLC) was performed to confirm the behavior of humic substances using E260 as an index.

8.3.1 Pore size and exclusion limit molecular mass for direct filtration

We considered the behavior of MF (ceramic) membranes with a certain pore size and UF (hydrophobic PAN hollow fiber) membranes with a certain exclusion limit molecular weight using clean water from the Toyohira River.

Changes in permeation flux due to differences in membrane pore size were tested using one MF membrane (0.1 μm) and three UF membranes (50,000, 13,000, and 6000 Da).

Figure 8.42 shows the change in membrane permeation flux after direct filtration of Toyohira River raw water. As shown in Figures 8.43 to 8.45, the turbidity removal performance of these low-turbidity raw waters is stable and always lower than 0.1 degrees, and there is no difference depending on the type of membrane. Therefore, the performance of the membrane was evaluated in consideration of the response to the characteristic components of the Toyohira river raw water, such as color, iron, manganese, and arsenic, centering on the conditions for maintaining the flux shown in Figure 8.42.

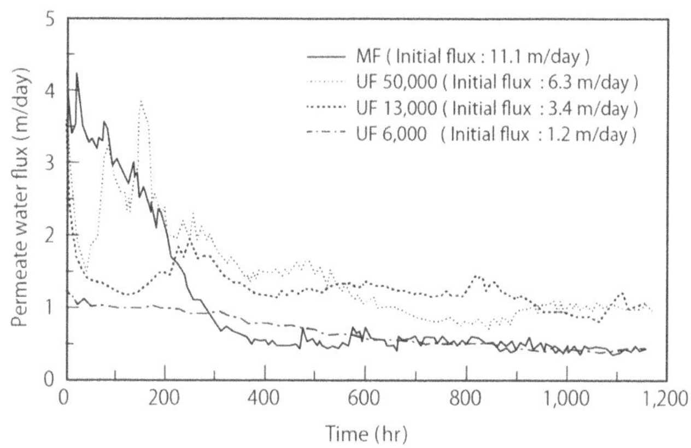

Figure 8.42 Permeate water flux change over time in MF and UF membranes for Toyohira river raw water.

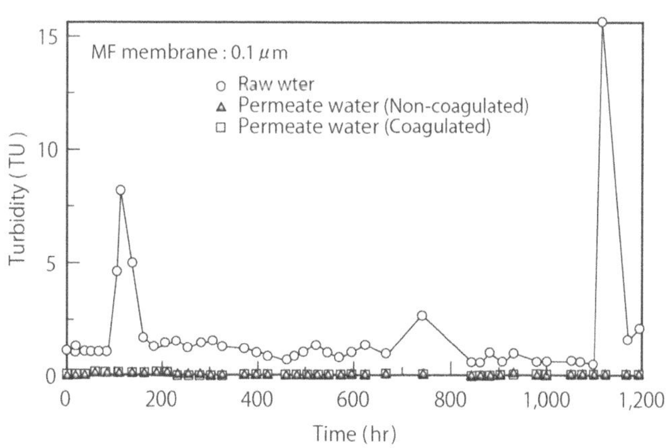

Figure 8.43 Time course of change in turbidity of Toyohira river raw water and membrane permeate water (MF membrane: 0.1 μm).

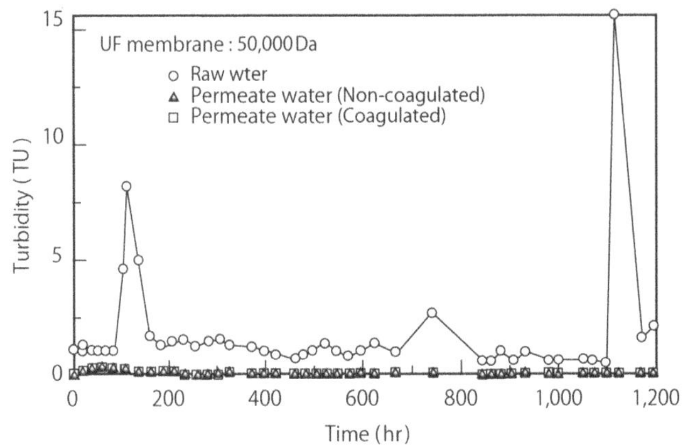

Figure 8.44 Time course of change in turbidity Toyohira river raw water and membrane permeate water (UF membrane: 50,000 Da).

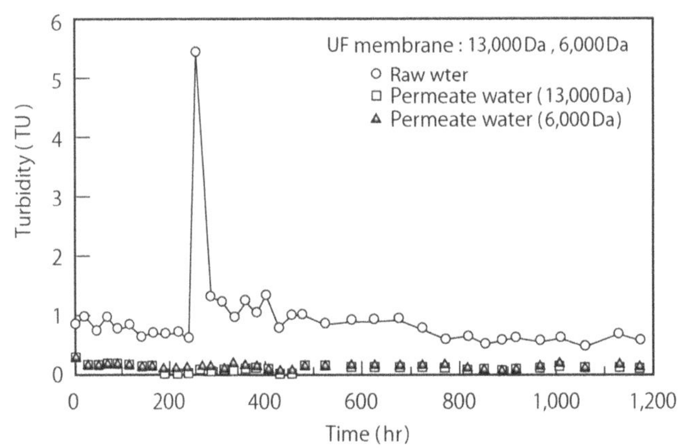

Figure 8.45 Time course of change in turbidity Toyohira river raw water and membrane permeate water (UF membrane: 13,000 Da and 6000 Da).

In the previous section, we explained that when high turbidity (high water) occurs, large quantities of higher molecular weight humic substances flow out of the A_1 layer of the forest bottom, and the cake resistance increases both in coagulated and noncoagulated water, as shown in Figures 8.29 and 8.31. In addition, humic substances on the low molecular weight side can hardly be removed by coagulation even in a normal state, and as a matter of course, the removal rate of humic substances is only about 20% even in the UF membrane with a molecular weight fraction of 6000 Da.

Figure 8.46 shows the time course of E260 removal after filtration of raw water and coagulated water with MF membranes. Figure 8.47 shows the process of removing the E260 of UF membranes with fractionated molecular weights of 13,000 and 6000 Da. The removal of UF membranes on the low-molecular-weight side is about 20% or less. Most of the color components in low-water conditions are low-molecular humins. Figure 8.48 shows the molecular fraction distribution of humic substances (E260) permeating

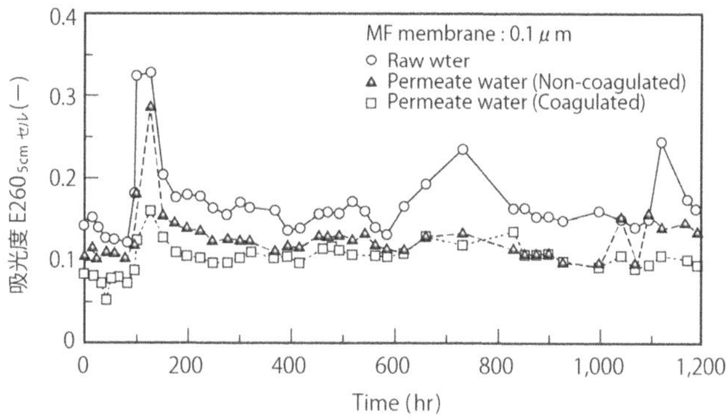

Figure 8.46 Time course of change in E260 of Toyohira river raw water and MF membrane permeate water (MF membrane: 0.1 μm).

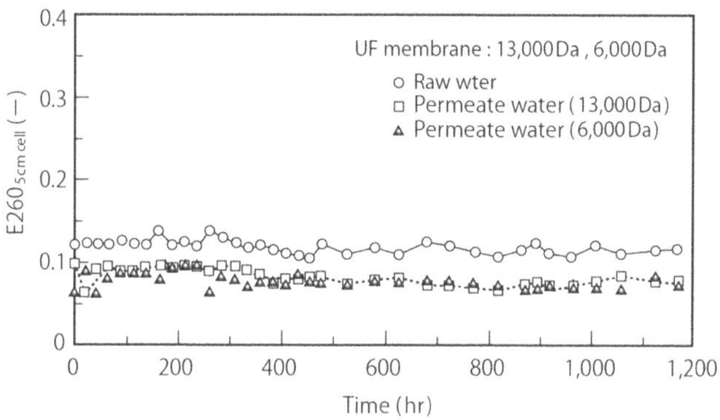

Figure 8.47 Time course of change in E260 of Toyohira river raw water and UF membrane permeate water (UF membrane: 13,000 Da and 6000 Da).

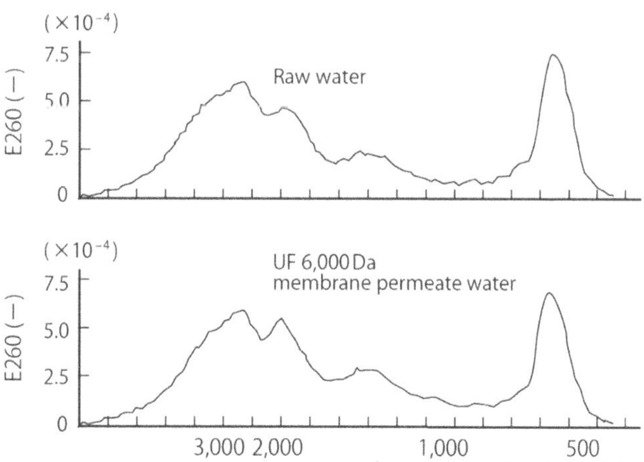

Figure 8.48 Molecular fraction chromatogram of permeate water.

6000 Da of molecular weight in this study. There is little difference between the distribution of raw water and permeating water, indicating that they have passed through the membrane. In order to remove humic substances on the low-molecular-weight side using a membrane, it is necessary to use an NF membrane. Below, we will discuss the removal of inorganic and organic dissolved components in freshwater by using an NF membrane, together with an RO membrane for desalination of seawater.

In the case of results for the removal of iron and manganese, total iron levels in raw water ranged from 0.16 to 0.06 mg/l during the study period, but permeate through the membrane was always at a level of 0.01 mg/l or less, and adequate removal was maintained. Manganese was 0.018 to 0.036 mg/l in raw water and 0.029 mg/l in permeate water until about 15 hours after the start of operation. However, manganese was 0.005 mg/l after the start of operation, indicating stable and good elimination of manganese. The manganese removal rate of the new membrane does not appear to be sufficient. In addition, the manganese removal rate is not sufficient even after chemical cleaning. However, as the iron manganese cake layer is formed over the course of operation, it appears that the removal of iron manganese in all membranes is improved by the same mechanism described in the section on autocatalytic removal of iron manganese in Section 5.3 in Chapter 5.

Arsenic concentrations of up to 0.013 mg/l, usually about 0.006 mg/l, were observed in raw water at the Sapporo City water works. However, arsenic concentrations of up to 0.008 mg/l, often 0.002 to 0.004 mg/l can be maintained in permeate water, and stable results in membrane treatment are expected.

Escherichia coli was not detected in any membrane treatment water.

8.3.2 Selection of appropriate UF membranes

Based on the findings above, the following membrane selection is considered appropriate for river raw waters, such as the clear Toyohira River in Sapporo City.

UF membranes with a slower flux decay seem to be more suitable for surface water treatment than MF membranes with a higher initial flux but faster flux decay over time, regardless of whether they are used for flocculated or non-flocculated source water. By selecting a UF membrane, it is possible that the cumulative amount of water obtained by long-term operation often exceeds that obtained by an MF membrane, and the UF membrane is usually more advantageous (safer) for the removal of other components. This is probably because UF membranes has a smaller pore diameter than MF membranes, and clogging (which may occur when suspended particulates of about 0.45 to 0.1 μm enter the pores of the membrane) is unlikely to occur in a UF membrane. The same phenomenon also occurs with humic substances with a coarse particle size.

In a constant pressure filtration experiment of a ceramic MF membrane conducted by the authors in the past (Kamei *et al.*, 1995a, 1995b), it was confirmed that when the membrane pore diameter was changed to 0.3, 0.2, or 0.1 μm, the initial permeation flux was larger for membranes of a larger pore diameter, but the flux value of the three membranes became almost the same after a certain time elapsed, and the flux of the membrane with the largest pore diameter became smaller when filtration was continued.

Furthermore, the results of experiments in which membrane exclusion limits were changed to 0.1 μm, 50,000, 13.000, and 6000 Da for non-coagulated water are similar to those in the previous experiments, and the membrane permeation flux of a UF membrane with a small exclusion limit is attenuated.

When the same three UF membranes shown in Figure 8.42 are compared,, since the initial flux is small, the flux of the UF 6000-Da membrane and the 50,000 Da membrane approach each other over the filtration duration but do not reverse, and the filtration of the 50,000 Da membrane continues stably around a flux of 1 m/day with the 13,000 Da membrane. The MF membrane, like the 6000 Da UF membrane, eventually drops to a flux of about 0.05 m/day, and only half of the flux of 13,000 Da membrane appears to be available.

Considering these facts, it seems that there is a meaningful UF membrane selection area for water supply using clear river water as raw water of around 10,000 Da. The author Tambo remembers that only one comprehensively used membrane made by Lyonnaise des Eaux in France in 1990 was a UF membrane of a little more than 10,000 Da cut-off.

Comparing coagulated and non-flocculated water, the flocculated raw water provides a slower flux decay and maintains a higher flux. This is probably because the pores of the cakes formed on the membrane surfaces are increased by the aggregation of elementary particles due to coagulation, and the resistivity of the Carman-Kozeny flow process is greatly reduced. In addition, as a result of the aggregation of fine particles, the cakes do not penetrate into the membrane pores, and clogging of the membrane is less likely to occur.

Coagulating metal ions such as iron, manganese and arsenic and E. coli in clean river water flowing under aerobic conditions are reliably removed by appropriate UF membrane treatment. However, in order to reliably remove soluble organic components (low-molecular-weight NOM), pesticides, divalent metal ions, anions, etc., the introduction of NF membrane filtration (described below) must be considered.

8.4 UF AND MF MEMBRANE TREATMENT FACILITIES

UF and MF membranes are widely used to purify natural fresh raw water. They are also used to recycle sewage water and as a pretreatment for RO membrane treatment in the desalination of seawater. In addition, they are widely used as the main player in membrane microbial treatment of sewage in membrane bioreactors (MBRs) (discussed below in Section 8.6). This section explains the basic construction of a treatment facility, focusing on the composition of the water purification plant of the water supply and water system using surface water as raw water (Okazaki *et al.*, 2006).

In many cases, the authors recommend that if a UF membrane with a cut off level of around 10,000 Da or a little larger is used, it is likely that the purpose will be achieved without problem. In addition to ensuring a high flux and achieving an appropriate removal rate, the use of MF membranes should be considered in cases where targets such as the physicochemical strength of the membrane, long-term durability, and the requirement of large pore (LP) membranes with a high flux for the removal of minute parasite eggs in extremely clear raw water (ground water, etc.) are identified.

The Japan Water Purification Processes Association (JWPA) first conducted research on a water purification membrane system in Japan in 1994. As a result, research and development for membrane treatment was advanced under the 'Research on the Development of New Water Purification System Using Membranes (MAC21 Project)' (FY1994–1996) (JWPA, 1994a). Tambo served as the chairman of the Technical Committee, and a membrane filtration facility guideline for small-scale membrane water purification plants was published as the first guideline for water treatment membrane facilities in Japan (JWPA, 1994b). Subsequently, the research continued under ACT21 'Research on Development of High-Efficiency Water Purification Technology' (FY1997–FY2001) (JWPA, 1997), and Tambo also served as the chairman of the Technology Committee. With this research, we were able to lay the foundations for the design of membrane treatment facilities today. In addition, following the e-Water 'R & D on water purification technology for environmental impact reduction' (FY2002–FY2004) (JWPA, 2002), it became possible to develop the technology for large-scale membrane water purification plants.

Based on this series of studies, the membrane filtration system is roughly classified into constant flow filtration and constant pressure filtration. In constant flow filtration, the system is operated so that a constant membrane permeation rate is always obtained during the operation period, and the primary pressure of the drive pump is increased against the membrane resistance which increases as the operation progresses. A constant flux is always obtained, which facilitates water distribution planning, while the transmembrane differential pressure increases as membrane blockage progresses. The increased pressure on the primary side increases the compaction of the cake on the membrane surface and the pushing of fine particles into the pores, further accelerating the increase in membrane resistance. This makes fouling control difficult.

Constant pressure filtration is always operated at a constant pressure during operation. Early fluxes gradually decrease toward end-of-life fluxes over time, but this has the advantage of not forcing the fouling of cake layers and the like. In addition, the membrane system can be operated by constant pressure pump operation or a constant water level difference, which facilitates control. On the other hand, it is indispensable to design a purified water storage resevoir in consideration of the decrease in the flow rate of permeated water through the membrane. In a large-scale water purification plant with multiple membrane systems and destabilization caused by changes in water purification, the two can be mitigated by shifting the timing of membrane cleaning.

A typical system flow for a membrane filtration plant containing cased membranes is shown in Figure 8.49. The immersion tank method, in which the membrane is submerged in a water tank and filtered by a suction pump, is as shown in Figure 8.50.

On the other hand, since the driving pressure is at most 100 kPa (1 atm) or less, the cake layer on the membrane surface is controlled by cross-flow. Dead-end filtration is performed without controlling the concentration-polarization. The driving pressure can be in the zone of 100 to 200 kPa, hence it is possible to use natural potential energy by installing the raw water intake point upstream, with an elevation difference of 10 to 20 m. If cross-flow is required, a small pump for circulating water to the membrane system can be added.

Small-scale water supplies in mountainous areas can easily operate renewable (natural) energy membrane systems, as shown in Figure 8.51.

UF/MF membrane processes that do not have immersed casings are based on Japanese designs, and are basically operated by constant pressure filtration, as shown in Figure 8.52. This type of system can be designed and operated as a very compact and inexpensive system for a small-scale membrane water purification plant.

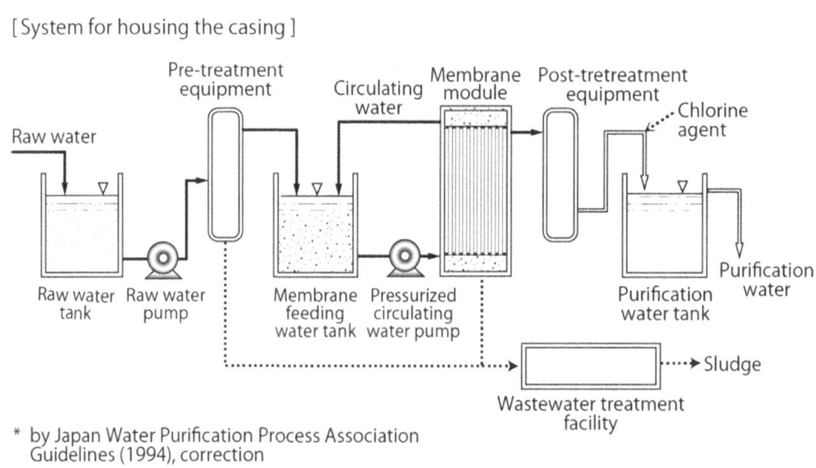

Figure 8.49 An example of a membrane filtration plant.

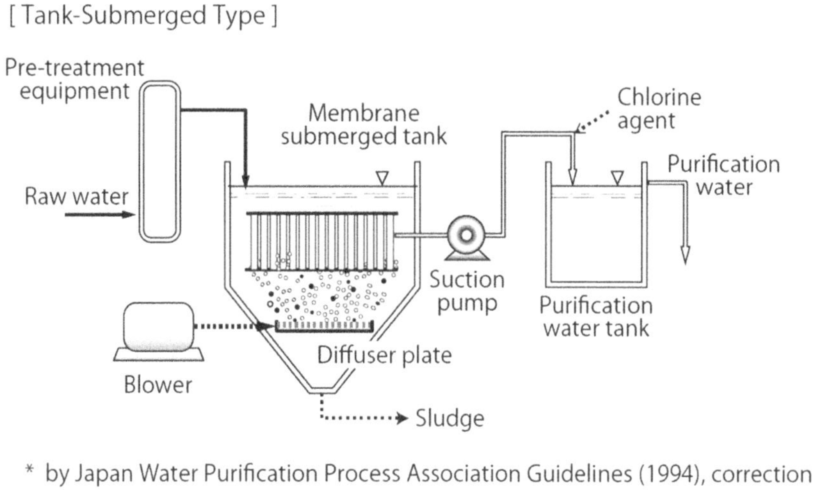

Figure 8.50 An example of a submerged membrane filtration purifcation plant.

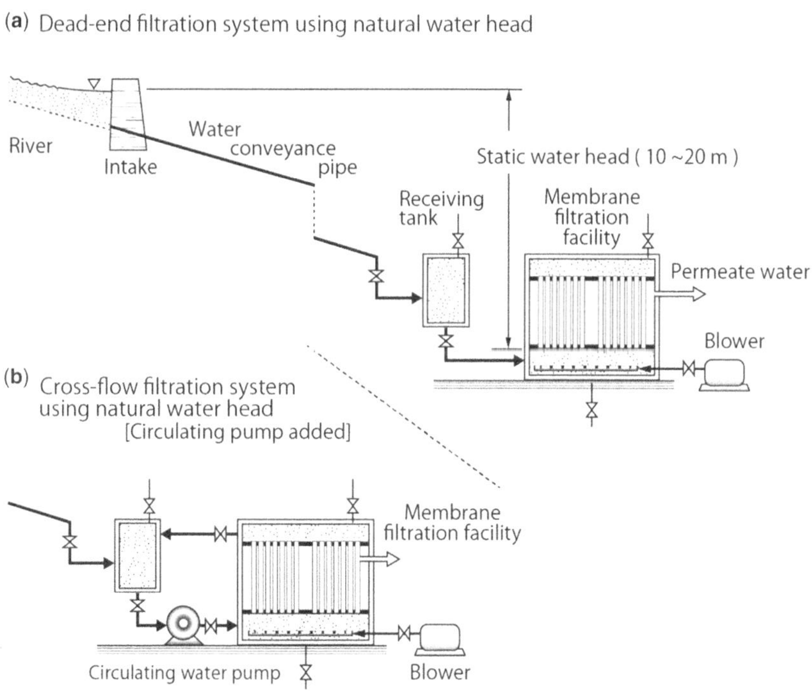

Figure 8.51 Renewable energy membrane systems: (a) a dead-end filtration system using a natural water head; and (b) cross-flow filtration (also with a natural water head but with pump added).

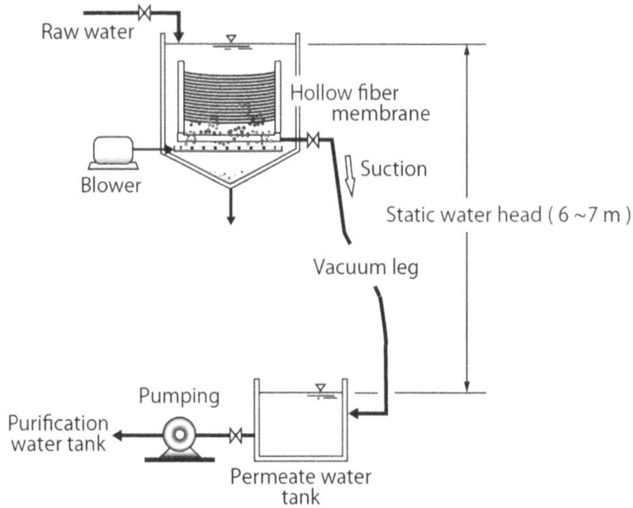

Figure 8.52 A membrane filtration system using vacuum suction.

Tambo believes that, probably in the early 1990s, Assistant Professor Aya of Tokyo University first began talking with engineers at EBARA-Infilco Co. Ltd about the idea of introducing a hollow fiber bundle of MF membranes into a sedimentation tank and suction filtering it together. This process is the dead-end filtration system. As an example, Figure 8.52 shows external and internal hollow fiber suction filtration. Physical cleaning to prevent reversible fouling produced on the surface of the membrane is generally carried out by disturbing the surface of the membrane by generating air bubbles in the lower diffuser plate. The authors think that there is a risk of breakage of the hollow fiber due to the absence of a casing in the cleaning method. In addition, this method is a process that can be used as the core of the Membrane Separation Activated Sludge Method (see below), and further improvements in technology development are desired.

In this case, it is disadvantageous for the equipment to perform suction with a vacuum pump. To avoid this disadvantage, the water level difference between the immersion tank and the permeation tank should be used instead of the vacuum pump as shown in Figure 8.52, and the water from the permeation tank at −60 to −70 kPa (−0.6 to −0.7 atmospheric pressure) is transported by static suction, whilst the water from the permeation tank at the bottom should be fed to the purification tank or the feeding system using a regular pump.

8.5 REVERSE OSMOSIS (RO) AND NANOFILTRATION (NF) MEMBRANES

Reverse osmosis and nanofiltration are typical membrane filtration procedures used in drinking water production. As shown in Table 8.1 (above), the main separation mechanism of RO and NF membranes is the diffusion control in the pores. Separation proceeds by diffusion and chemical exclusion of solutes in the pores. In principle, the separation mechanism of UF and MF membranes is (depending on their sieving action) controlled by the pore size of the membrane.

If the approximate size of the impurities that can be removed is considered, RO membranes can remove impurities with a size of 0.0001 μm (0.1 nm), but NF films are about 0.001 μm (1 nm), which is an order of magnitude different. Being pressure-driven does not change either MF and UF membranes or NF and RO membranes. However, as shown in Table 8.2, the necessary driving pressure is around 100 kPa (1 atm), while operation is on the higher-pressure side – such as around 500 kPa (several atm) for NF membranes and around 6000 kPa (several tens of atm) for RO membranes. Since the expansion of the treatment effect to the low molecular weight side increases energy consumption, the comparison between the two is the key to the operation of treatment at high pressure (Ohya, 1976).

NF and RO membranes form an RO membrane group with similar characteristics in a broad sense. From the viewpoint of the components to be removed, the brine-compatible RO membrane is also used as an NF membrane, and the Brackish water RO membrane (BWRO) is used on the low-pressure side. The Japanese Association of Membrane Separation Technology (AMST) defines a BWRO as a membrane with a sodium elimination rate of 93% or more at operating pressures of 500 to 3000 kPa for brine, with sodium chloride concentrations ranging from 500 to 2000 mg/l.

NF membranes are not much different from BWROs, and remove NOMs from fresh water (such as low-molecular humins) as well as synthetic organic components such as pesticides from divalent metal ions such as hardness components. The operation of NF membranes for these low concentration components in fresh water is carried out at a low pressures of 300–500 kPa in series, downstream of UF membrane operation. The operation of NF membranes provides a multifaceted equivalent of the activated carbon adsorption and hardwater softening processes of conventional systems. In this case, NF membranes are not expected to remove the main components involved with Sea Water RO (SWRO) and BWRO, such as sodium chloride. A freshwater RO membrane compatible with low concentration NaCl water can be considered to be an NF membrane.

In drinking water production, most quality requirements are met by a 'UF membrane + NF membrane' system. However, in the case of ultrapure water, suitable for research and medical use, etc., a series system of 'UF membrane + RO membrane (seawater desalination level)' is required.

SWRO for desalination is required to remove 99% or more of salt from 3.4% or higher concentration seawater. Multi-stage flash distillation systems historically preceded seawater desalination operations. Today, use of SWRO membrane systems has become widespread around the globe, and Japanese SWRO technologies are firmly at the center of the world in terms of both quality and quantity.

The Association of Membrane Separation of Japan (AMST) defined an SWRO membrane as a membrane having a sodium chloride removal rate of 99.0% or more or a removal rate of 98.8% or more for an artificial sodium chloride solution, with a flow rate of 0.01 m^3 or more per 1 m^2 membrane area with an effective pressure 1 MPa under operating pressure of 5.0 to 10.0 0 MPa, and with a sodium chloride concentration or TDS (total dissolved solids) concentration in the range of 3.0×10^4 to 6.0×10^4 mg/l.

Figure 8.53 shows the relationship between raw water concentration and operating pressure to which various types of RO membranes are applied.

8.5.1 Osmotic pressure and RO

Two liquids with different concentrations, one containing a colloidal or molecular solution and the other containing water, were placed in a container with a membrane, as shown in Figure 8.54. In this case, if the pore size of the membrane is small enough to prevent the passage of solutes, water (solvent) on the lower concentration side will move through the membrane to the higher concentration side, and the water level on the higher concentration side will rise and the concentration will decrease. Thus, a membrane that prevents the passage of solutes is called a semipermeable membrane. Various membranes, ranging from MF membranes for water treatment to RO membranes, are semipermeable, with different pour sizes for passage and non-passage (blocking).

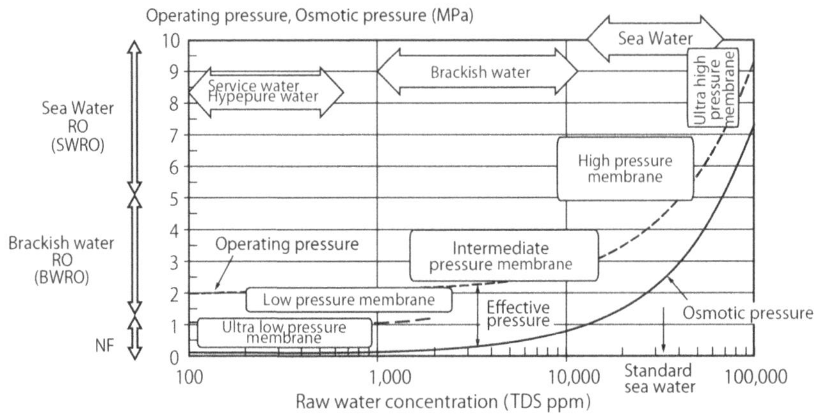

Figure 8.53 Relationship between RO raw water concentration and operating pressure (Japan Water Purification Processes Association (JWPA), 2003).

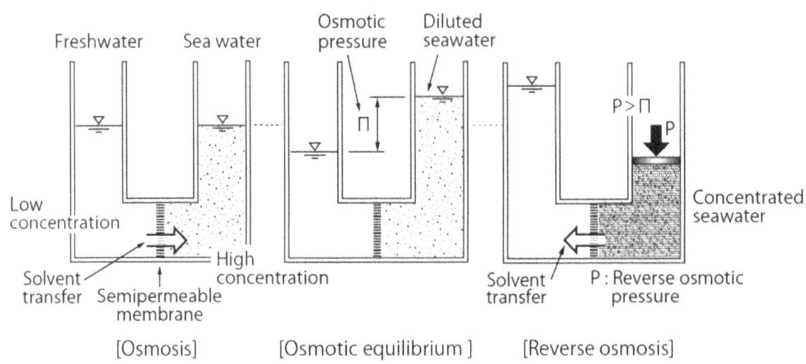

Figure 8.54 Osmotic pressure and RO.

Seawater desalination RO membranes are semipermeable membranes that do not allow the passage of NaCl and other salts in seawater. As shown in Figure 8.54, the water molecule (solvent) concentration in the left compartment is heigher than the right, and the water molecules in the left compartment tend to diffuse through the semipermeable membrane to the seawater side of the right compartment causing the seawater level to rise as it dilutes.

The difference in water level between the left and right fractions, which eventually reaches equilibrium, is the osmotic pressure of the seawater, π. The osmotic pressure of seawater with a salt content of 3.4% is as large as 2500 kPa (2.5 MPa, 250 m water head, 25 atm). If seawater is forced against this osmotic pressure to obtain fresh water (seawater desalination: fresh water formation), the water molecules in the seawater must be forced back through the semipermeable membrane at a high pressure of 25 atm or more.

The osmotic pressure is given by the van't Hoff theory as '$\Pi(\text{atm}) = MRT$' where M is the salt molarity (mol/l), R is the gas constant (8.31451 J/K·mol), and T is the absolute temperature (K).

As this theory shows, the osmotic pressure increases in proportion to the salt concentration, so if half of the seawater is to be desalinated (i.e. a desalination rate of 50%), the necessary RO pressure on the seawater side (primary side) will be over 50 atm. Therefore, if we simply try to obtain a high recovery rate of 70–80%, we need to make the RO pressurization as large as 100 atm (1000 m water head), as shown in Figure 8.53. Seawater desalination is a high-pressure operation and battles with energy consumption rate control.

From a theoretical thermodynamic point of view, the energy required to obtain 1 m^3 of fresh water from seawater seems to be 0.69 kWh/m^3. In reality, RO-membrane desalination systems will need to be constructed by adding seawater intake facilities, flotation to remove NOMs such as fine suspensions and humic substances, and pretreatment such as MF/UF membrane treatment. As a whole, including this, we need to add 0.5–0.8 kWh/m^3 of pretreatment and 2–3 kWh/m^3 of operation for the demineralized SWRO membrane of the main body. The author believes that 3–4 kWh/m^3 is the present total energy required to produce 1 m^3 freshwater by seawater desalination technologies.

As described in the previous section, the energy consumption rate of an SWRO membrane processing unit is generally 7 kWh/m^3, remaining level for up to 20 years ago, depending on the arrangement of membrane modules and how the energy recovery system is assembled (Kinjyou, 1999). At present, the target energy consumption rate is 2–3 kWh/m^3 and the recovery rate is 80% or more.

8.5.2 Formation and structure of RO membranes

As mentioned above, Professor C.E. Reid and E.S. Breton first proposed the principles of RO processes using cellulosic membranes to the US OSW and became the founders of this technology (Reid & Breton, 1959).

The fluxes obtained at that time were as small as 1 gal/ft^2 day (≑ 4.2 cm/day), the technology was in the range of a theoretical proposal, and the way to practical application was still only a possibility. In the early 1950s, in a project designed to operate desalination of salt water by Gibbs's adsorption theory by Professor Samuel Yuster of the University of California, Los Angeles (UCLA), Sidney Loeb and Srinivsa Surirajan developed an RO membrane using cellulose acetate (CA) as a material, with fluxes (5 gal/ft^2/day, 21 cm/day or more) and recovery rates (92–99%) (Loeb & Sourirajan, 1963). The RO process reached a practical level and established the foundation for the seawater desalination system of today.

In 1962–63, when Tambo worked at the University of Florida, US OSW (US Department of Desalination) projects continued, but Reid's research was not continued. However, the desalination of seawater using solar heat was carried out on the south coast of Florida through joint university research. The fact that desalination of coastal groundwater is currently being carried out on a large scale in Florida using BWRO membranes shows the rapid progress of membrane technologies over the past 50 years. At a time when the multi-stage flash evaporation and multi-effect evaporation methods were enthusiastically carried out by people in the chemical engineering field, the development of technology for desalination of seawater was advanced. After my return to Japan, the Department of Sanitary Engineering at Hokkaido University also researched distillation methods, as system research in the chemical engineering field. Although now closed, the Sasakura Iron Works' distillation plant of 0.6 million m^3/day in Hong Kong was the last largest evaporation plant in Asia.

In the US OSW research project conducted during Tambo's time in Florida, a memorable idea for the desalination of solar thermal seawater was a simple transparent plastic bag system that produced fresh water from turbid river water, muddy water, etc. that had been contaminated with radioactivity on a beach where the Marine Corps and others had landed. It is a good idea to know about disaster and marine accident response, so this section introduces a primitive solar desalination system.

Figure 8.55 shows a tube of 40 cm diameter and 5 m length made of transparent flexible polyethylene or vinyl chloride. Folded, it is as large as a washcloth. As shown on the right side of the figure, the upper half of the cylinder was welded with the lower half so that the inner upper side of the lower half protrudes 5 to 10 cm

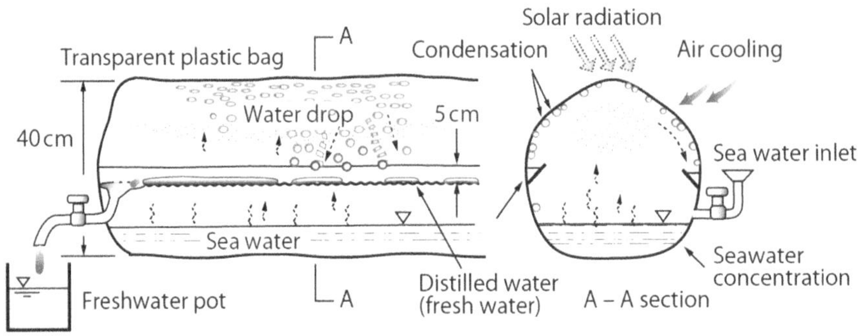

Figure 8.55 Transparent plastic solar thermal desalination equipment (Tambo).

into the cylinder. The part protruding inside the cylinder becomes a rib for fresh water collection, which feeds fresh water in the longitudinal direction, and flows out dropwise to a fresh water reservoir from the outlet at the end. As seen in section A-A, when seawater or radioactively contaminated water is placed in the bottom of the tube and exposed to sunlight, the temperature of the seawater rises due to the greenhouse effect, water vapor rises to the top inner surface of the tube and condenses on the ceiling portion of the plastic cooled by the outside air. Condensation collects on the walls in fresh water collection ribs (channels) of several centimeters, with seams of upper and lower semicircular cross-sections protruding into the lower semicircle. Collected water flows out to the downstream outlet of a desalination tube installed with a gentle gradient, and is collected in an external fresh water reservoir. The author, Tambo, has no memory of data after half a century, but has often recalled the details of the research.

With the aim of obtaining high separation efficiency and high flux with predetermined pressure, membrane structure has been developed following a number of stages of technological evolution:

(1) Uniform membrane (dense membrane);
(2) Asymmetric membrane, the surface dense layer (separation function layer) and its supporting porous layer are made of the same material; and
(3) Complex membrane (composit membrane); the separation functional layer and the support layer on the surface are made of different materials, and each function is more clearly exhibited.

Asymmetric and composite membranes have been used as practical membranes in the NF/RO world. In addition, high-performance composite membranes for SWRO have also been developed, which are difficult to fabricate. Japanese membrane manufacturers have achieved world-leading results.

The first practical application was an asymmetric membrane, the cellulose acetate (CA) membrane of Loeb and Sourirajan. To produce it, dissolve cellulose acetate (a mixture of cellulose and cellulose triacetate) in acetone and dioxane, add chemicals such as CTA and CDA, mix and stir, and allow to stand overnight to remove air bubbles to form a membrane cast stock solution. Spread the casting liquid over one end of a clean glass plate placed horizontally and quickly spread it over the entire glass surface with an applicator (leveling tool) to a uniform thickness. The glass plate and the coarse CA film are peeled off in water after immersing the glass plate horizontally in a flat bath filled with coagulation solution and allowing the solidification to proceed for about 5 to 20 minutes. Next, place the peeled film on an aluminium plate to fix the four corners, wash it at room temperature, and heat-treat it in water at a temperature of 60 to 80°C. This produces an asymmetric RO film, as shown in Figure 8.3 (above).

In the course of solidification, the side of the material in contact with the glass surface becomes the separation active layer with a dense structure, and the rest becomes the support layer. This production requires delicate operation, and the manufacturing process has been automated industrially, but the basic operation and the product structure are the same.

It has become possible to create a composite membrane in which a thin layer (separation function layer: active layer) with a low membrane resistance and a high degree of separation ability is composited onto a porous support layer with a low water resistance. Recently, from a membrane in which an RO active layer was composited onto the surface of a UF membrane, a precise composite membrane was created in which a polysulfone support membrane is placed on the surface of a substrate using a nonwoven fabric, and a polyamide separation function layer (ultra-thin film) expressing RO activation was composited on the surface. Figure 8.56 shows an example of the structure of the composite membrane developed by TORAY.

Figure 8.56(a) shows a monomer-polymerization cross-linked membrane, which is one of the typical methods for producing composite membranes. A polymerizable monomer is applied to or immersed in a support film, and a thin film (active layer) is formed by controlling the thickness of the surface of the support layer to several hundred Å by heating and chemical treatment. A four-layer membrane with a uniform ultrathin film thickness of 300 Å is the core of the separation mechanism. In order to stably maintain the separation performance for a long time, the inner porous surface of the poly-sulfone support layer is also reinforced with a resin, which enhances its pressure and heat resistance. The protective layer is chemically bonded to the thin film on the ultra-thin film layer (active layer) to protect the active ultrathin film. These plastic polymers are formed on a fiber layer and incorporated into the membrane module.

Figure 8.56(b) shows a composite membrane produced by interfacial polymerization, another typical RO membrane process. A water-soluble polymer is coated on a support layer, and an active group (amino group) in the polymer is reacted with a cross-linking agent (acid chloride) to form a thin film. A 100 Å thin film and

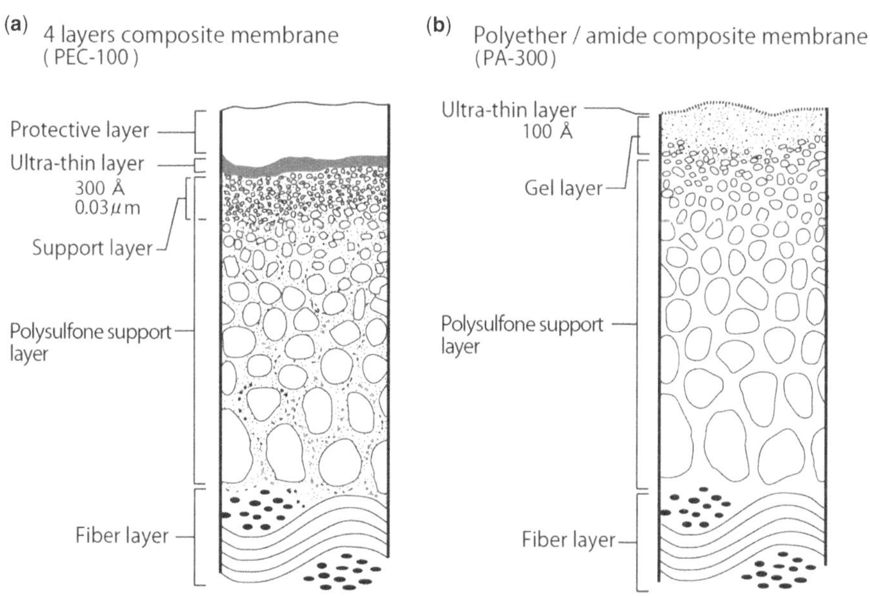

Figure 8.56 An example of the structure of a composite membrane (Dr Suguru Kurihara, TORAY).

a supporting gel layer are formed. An organic amine was added to a water-soluble polymer containing amino groups to form a thin film. As a result, the thickness of the polymer alone was 100 Å, but it could be increased to 300 Å. The desalting rate of 92% of the polymer alone was improved to 99.45–99.60% and a high-performance SWRO membrane was produced.

Another common method for producing composite films is solution-coating (polymer coating), which produces relatively 'thick' films of as much as 1 μm in thickness. It seems that state-of-the-art fabrication is proceeding to produce composite membranes with high-performance thin films as described in the previous section. In addition, new production methods such as plasma polymerization are being tried.

In any case, the number of countries and manufacturers capable of stably producing high performance RO composite membranes in large quantities is limited. Although MF and UF membranes are produced in large quantities in many countries, there seems to be a difference in the capacities of manufacturers for high performance RO membranes.

8.5.3 Mechanism of RO membrane separation

As described above, among the membrane separation methods, MF and UF membranes are considered to separate pollutants (impurities) by the 'sieving' principle of thin pore size control, while NF membranes and RO membranes are considered to separate pollutants under using the 'diffusion rate difference of solute/solvent through the membrane'. In both cases, the operation progresses due to the drive pressure difference. However, the latter is an overwhelmingly high-pressure treatment, as shown in Table 8.2 and Figure 8.53.

When Reid and Breton (1959) first proposed the RO technique, they had already presented models of the mechanism by which salt is removed by RO and only fresh water permeates the membrane. Loeb and Sourirajan (1963) also proposed a model for transmission which explained the mechanism of separation:

(1) If the membrane's chemistry shows negative adsorption to solutes in aqueous solution, the solute concentrations near the membrane pore surfaces rapidly decrease, and a layer of pure water of thickness τ is formed at the interface between the solution and the membrane pores, adsorbed by the membrane (bound water);
(2) When pressure is applied to the membrane pores to extrude the solution in the pores, both the adsorbed pure water layer and the solution in the center of the pores are extruded. However, if the pore diameter D is twice the thickness τ of the adsorbed pure water layer, only pure water exists in the pores, and only pure water permeates the membrane, resulting in all solutes being impermeable to the RO membrane under consideration;
(3) If the pore diameter D is twice (2τ) or less than the pure water layer thickness τ adsorbed on the inner surface, the solute (salt) removal rate of the RO membrane is 100%;
(4) If the pore size is distributed and there are pores with pore size $D > 2\delta$, the solute-containing solution (primary raw water) permeates the membrane from the cylindrical portion with diameter $(D - 2\tau)$ at the center of the pore, and the solute removal rate decreases accordingly;
(5) According to this model, making a membrane with a large number of uniform pore diameters of $D = 2\delta$ from a material with high water molecule adsorption (large τ) is likely to be the key to making a good RO membrane.

RO membrane separations can be expressed commonly for RO membrane (SWRO and BWRO), NF membrane (BWRO and soluble organic matter treatment), and hardness treatment applications.

In semipermeable membranes such as in RO, the rate of solute removal can be approximated to 100%, and the basic equation of the solution-diffusion type, which expresses the water flux J_w and the salt flux J_s

(both of g/cm^2s) that permeate through the membrane, are shown in Equations 8.8 and 8.9.

Water flux $J_w = k_w[\Delta P - \Delta \Pi]$ —(8·8)

Salt flux $J_s = k_s \Delta C$ —(8·9)

J_w : Membrane permeate water flux
J_s : Membrane permeate salt flux
k_w : Permeability coefficient of pure water ($g/cm^2 \cdot kPa$)
k_s : Permeability coefficient of solute (salt) (cm/s)
ΔP : Operating pressure (pressurizing force、kPa)
$\Delta \Pi$: Osmotic pressure (kPa)
ΔC : Solute concentration difference between raw water and permeate water (g/cm^3)

(8.8)–(8.9)

Equation 8.8, 8.9 The water flux and salt flux.

Since salt is removed from the RO membrane and only water permeates through the membrane, the salt removed during operation concentrates near the membrane surface on the primary side, resulting in a concentration polarization phenomenon that is partially higher than the salt concentration of the raw water. Figure 8.13 describes the suppression of concentration polarization by introducing cross-flow to reduce concentration polarization and cake formation in the UF membrane or to prevent coagulation. For RO membranes and NF membranes, the suspensions are removed by pretreatment for the purpose of desalination, resulting in concentration polarization of almost only the solutes. In diffusion type separation, the concentration polarization directly influences the concentration difference ΔC between before and after the membrane that controls the salt flux, shown in Equation 8.9, unlike the fouling inhibition of UF and MF membranes.

A schematic depiction of concentration polarization is shown in Figure 8.57. The actual primary concentration of membrane permeation is not the raw water (bulk) concentration C_1, but the concentration C_2 at the interface with the membrane surface. Therefore, if the secondary bulk

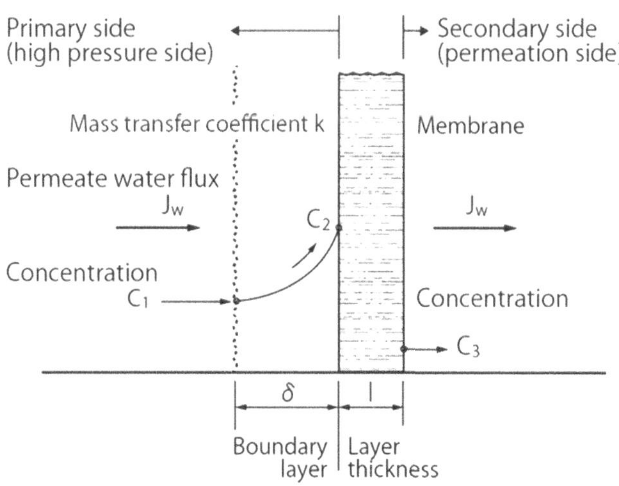

Figure 8.57 Salt concenration distribution near the membrane.

concentration is C_3, he true desalination rate of the membrane in practice is $R = (C2 - C_3)/C_2$, and the apparent (bulk) desalination rate $R_{obs} = (C_1 - C_3)/C_1$.

In contrast, $R \geqq R_{obs}$, which requires more stringent operations.

Considering the material balance of the boundary layer with concentration polarization as shown in Figure 8.57, the permeate flux J_w of water is shown in Equation 8.10 by compensating for the amount of $-D(dC/dx)$ that returns to the bulk water due to back-diffusion by the concentration gradient.

$$J_w C_3 = J_w C_1 - D\frac{dC}{dx} \quad \text{---(8·10)}$$

J_w : Membrane permeate water flux
C_1 : Primary bulk concentration
C_3 : Secondary bulk concentration
D : Diffusion coefficient (m^2/s)
x : Distance from membrane surface

Integrate Eq.8·11

$$\frac{C_2 - C_3}{C_1 - C_3} = \exp\left(\frac{J_w}{k}\right) \quad \text{---(8·11)} \qquad (8.10)\text{–}(8.11)$$

Boundary conditions of integration :
$C = C_2$ at membrane surface x=0
$C = C_1$ at the surface of the boundary layer thickness $x = \delta$
k : Mass transfer coefficient (D/δ, cm/s)

Eq.8·10, Eq.8·11 The Permeate water flux under the concentration polarization

Equation 8.11 shows the relation between the apparent desalination rate R_{obs} and the true demineralization rate R as shown in Equation 8.12. It is necessary to experimentally calculate the mass transfer coefficient k in consideration of polarization. In order to set the value of $k = D/\delta$ to be large, it is necessary to reduce the thickness of the polarization boundary layer, and the problem is dealt with by adding cross flow to the membrane surface on the primary side (Wiesner & Aptel, 1996).

$$\frac{1 - R_{obs}}{R_{obs}} = \frac{1 - R}{R}\exp\left(\frac{J_w}{k}\right) \quad \text{---(8·12)} \qquad (8.12)$$

R_{obs} : Apparent desalting rate
R : True desalting rate

Equation 8.12 The apparent desalting rate and true desalting rate.

If the membrane is operated for a long time, the flux will decrease. The problem of membrane clogging is as described above, but depending on the membrane material, hydrolysis of the crystal structure progresses and densification is reported, which is considered to be a kind of negative aging effect. Also, RO membranes that have been exposed to high pressure for long periods of time are problematic in the development of physically irreversible consolidation (creep). Physical and chemical destruction of the active layer is also difficult to detect in detail, which may cause malfunction of normal operation. If the pre-treatment is insufficient, fouling phenomena such as adsorption and cake formation occur on the primary side of the RO membrane as described for UF membrane treatment, and flux for the original desalination cannot be obtained.

It is necessary to grasp various phenomena qualitatively and to evaluate them quantitatively. It is important for both manufacturers and users to develop detection methods as well as to study emergency measures.

8.5.4 Configuration of RO membrane devices (process operation)

With the seawater desalination spiral module which is most commonly used in mind, we now consider how the membrane treatment process is configured.

We have sought high desalination and water recovery rates through the design of the membrane itself, and we have been making continuous efforts to achieve desalination of seawater and brine with as little energy (driving pressure) as possible. In addition, there is another important step beyond chemical engineering: how to devise an array of different membrane modules, drive energy systems, and energy recovery systems in conjunction with the evolution of a single membrane.

As shown in Equations 8.9 and 8.10, the flux for fresh water production is proportional to the driving pressure that drives the osmotic membrane operation (transmembrane differential pressure), but the passage of salt (removal rate) is proportional to the transmembrane concentration difference and is not governed by the driving pressure.

The water recovery rate in membrane operation, $R = (Q_p/Q_f) \times 100 = [Q_p/(Q_p + Q_c)] \times 100$. This is expressed as the percentage of the amount of fresh water produced Q_p to the amount of water supplied (seawater) Q_f. However, since the amount of raw water supplied will eventually become the amount of fresh water produced Q_p and the amount of wastewater discharged Q_c, the key to how to make the amount of brine highly concentrated and small is how to make the amount of fresh water.

In this case, the degree of concentration, $C_{Fn} = 1/(1 - r)$. When the product is operated at a recovery rate of 0.5 (50%), the concentration becomes 2. The pressure required for RO desalination of seawater requires a high pressure of more than 50 atm ($2 \times 25 \times 10^2$ kPa), which is more than twice the 25 atm of seawater osmotic pressure.

Consider a simple seawater desalination system that operates with a 50% recovery rate of pre-treated seawater. When membrane permeation fresh water 0.5 is produced relative to raw water 1, twice as much concentrated seawater is produced as waste brine. In this situation, the driving pressure of the high-pressure pump will need approximately twice of the seawater osmotic pressure, i.e. 6 MPa. However, in order to operate the high-pressure pump with a small margin, a 7 MPa high-pressure pump (or similar) will need to be prepared. When the unit is operated as it is, concentrated high-pressure water on the primary side of the membrane is discharged to the atmosphere at a very high pressure of 6 to 7 MPa (600–700 m of water head), and all of its energy (pressure) is lost. For this reason, in seawater desalination operations, the energy possessed by the primary side water has been recovered and most of it has been reused as driving energy.

Using high-pressure brine as a driving source, high-pressure multi-stage turbines and Pelton turbines can achieve energy conversion efficiencies as high as 88% and 82%, respectively, and can drive the motor with recovered power to return it to high-pressure pump power. As a result, about 2.7 to 2.8 kWh/m^3 of power consumption of high-pressure pumps, which required power consumption of about 4.5 kWh/m^3, is recovered. Seawater desalination plants along the Mediterranean coast, in the Middle East, and in Australia are expected to consume 0.6 to 0.7 kWh/m^3 of electricity, resulting in energy consumption of about 2 to 3 kWh/m^3. If energy recovery is not performed, energy consumption of 6 to 7 kWh/m^3 is required, which is the level of old seawater desalination systems such as at Okinawa and Fukuoka in Japan.

Single-stage systems are common in facilities with small operating conditions, such as 40% recovery and 5–6 MPa driving pressure, due to scale generation, membrane pressure resistance problems, and temporal degradation.

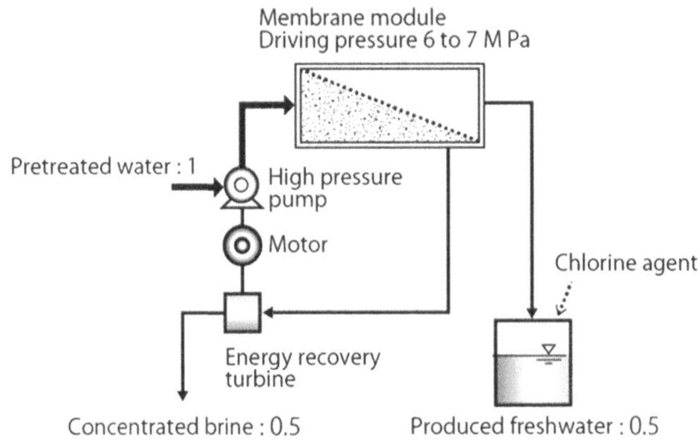

Figure 8.58 The energty recovery path of the RO membrane system.

Figure 8.58 schematically illustrates the energy recovery path of the RO membrane system.

In recent years, new proposals have been made for energy-recovery equipment, from a conventional pressure-conversion system using turbine pumps to a method of direct pressure recovery from high-pressure concentrated water. The concentrated seawater is contacted with the seawater (pretreatment raw water) of the direct supply side, and the high pressure (6 to 7 MPa) of the brine is efficiently delivered to the supply side seawater.

Another pressure exchanger was presented by S. Hayashi (2004) at Toyobo Co., Ltd. The center of the pressure exchanger structure is the ceramic rotor shown in Figure 8.59. Like the rotary magazine of a handgun, the rotor has several dozen holes through it, formed in parallel to the axis of the rotor. The rotor floats through a thin layer of concentrated seawater in an outer cylinder made of fiber reinforced plastic, with pressure dissipated while the axis of rotation is vertical. This thin layer of concentrated seawater becomes a lubricating layer and enables smooth rotation of the rotor.

In the low-pressure area (gray area in Figure 8.60), seawater is pumped into the lower part of the PX rotor at a pressure of 0.2 to 0.3 MPa, and concentrated seawater accumulated through the several dozen holes is pushed out from the other side of the holes by the flowing low-pressure seawater. Seawater in the

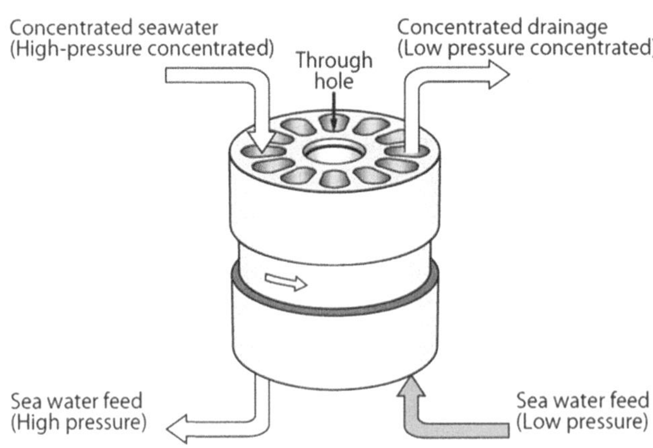

Figure 8.59 Hayashi pressure exchanger: PX rotor (Hayashi, 2004).

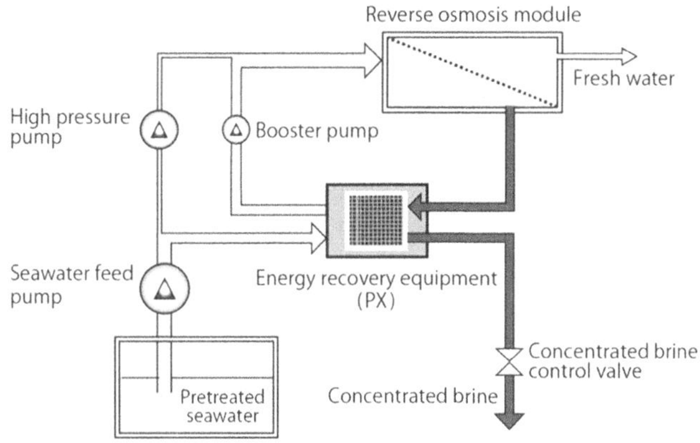

Figure 8.60 Hayashi pressure exchanger: basic system flow during PX operation (Hayashi, 2004).

low-pressure region is connected through these holes to the brine pipes by the rotation of the rotor and move to the high-pressure region. In the treatment-pressure region, high-pressure brine concentrated by the RO membrane flows in through the holes, and the pretreated seawater (raw water) accumulated in them is pushed out to the opposite side by the high pressure. This allows a large amount of low-pressure seawater (raw water) to compensate (boost) the slightly reduced brine system pressure to increase the RO pressure specified by the high-pressure pump on the primary side, thereby driving the RO membrane. The through-holes filled with concentrated brine come into contact with the raw seawater in the low-pressure region due to the rotation of the rotor, and the through-holes replace the low-pressure water.

By rotating the rotor at an appropriate speed, steps 1 to 4 can be performed continuously, and the high-pressure energy of the concentrated seawater (brine) can be continuously supplied to the raw seawater (Figure 8.40).

The results of the above-mentioned system experiments from Hayashi's report for each stage of the process are shown in Table 8.5 (Hayashi, 2004).

The direct operating pressure of the RO membrane was 5.34 MPa, as shown at [7] in Table 8.5. For the recovery rate: 'Membrane permeate [7]/pretreated raw water [1]' = 24.60 l/min/74.32 l/min = 0.33

Table 8.5 Hayashi PX pressure exchanger test results (by Hayashi Shigetoshi)

December 10,1999 Iwakuni Functional Membrane Factory	Pressure (MPa)	Flow Rate (l/min)
① Filtered seawater	ATM	74.32
② High pressure pump side seawater	0.31	27.94
③ PX side seawater	0.31	46.38
④ Concentrated drainage	0.22	49.72
⑤ PX high pressure discharge seawater	4.98	59.36
⑥ Concentrated water	5.10	62.70
⑦ RO feed seawater	5.34	87.30
⑧ Permeate water	ATM	24.60

Recovery rate 33% RO: HOLLOSEP (TOYOBO).

(33%). When energy is not recovered, the energy consumption is 6.27 Wh/m^3, whereas when energy is recovered using PX rotors, the energy can be reduced by 50% at a of 3.24 kWh/m^3.

The breakdown of energy recovery by the pressure exchanger is as follows: High-pressure energy at the inlet of concentrated seawater [6]: E_i (high-pressure energy × flow rate) = 5.10 MPa × 62 7 l/min/60 = 5.33 kW.

Then, the amount of energy on the high-pressure side of the seawater outlet [5]: E_0 (low-pressure-side energy × flow rate) = 4.98 MPa × 59.36 l/min/60 = 4.93 kW. In terms of energy recovery: $(E_0/E_i) \times 100\% = 4.93/5.33 \times 100 = 92.4\%$. Higher values have also been achieved.

Advances in energy recovery systems have resulted in a kWh/m^3 energy consumption rate of 3 kWh/m^3 for ROs. However, if the recovery rate (desalination rate) is between 30–40%, much of the structure intake volume and the throughput of pretreatment facilities will be wasted, and the economic efficiency of the entire system will be low. Therefore, increasing the recovery rate of membrane permeate water (fresh water) is attempted by using the RO membrane in multiple stages. Needless to say, in order to further squeeze out the concentrated seawater, the RO membrane at the subsequent stage will be treated at a higher pressure, and new challenges will be the development of membrane materials capable of withstanding the driving energy load and high pressure, the evolution of modules, and the suppression of scale precipitation on the membrane surface.

Figure 8.61 shows an example of a high-efficiency RO membrane system with 60% recovery through a two-stage process. A major goal is to reduce the cost of water production. Such a system as shown in the figure does not seem to cause a scale problem that increases polarization. Cellulose acetate (CA) and polyamide (PA) membranes, which withstand high pressures, have been developed to enable continuous operation at high pressures, such as second-stage 9 MPa.

The outline of the process operation shows raw seawater that has undergone pretreatment:

(1) Raw seawater is pumped to the RO membrane of the first stage with a high-pressure pump of 6 MPa to produce 40% fresh water;
(2) Concentrated seawater that is 60% raw water with 5.8% salinity and 6.5 MPa pressure is pressurized again by a booster pump to add 5 MPa, and forced into the primary side of the the second stage RO membrane with 9.0 MPa;
(3) The RO membrane in the second stage produces 20% fresh water from raw seawater (33% of concentrated seawater), and then (with the 40% fresh water from the first stage) a total of 60% fresh water from raw seawater (60% recovery rate) is obtained; as a result:
(4) Brine concentrated to a salt concentration of 8.8%, with 40% of the raw seawater, is discharged

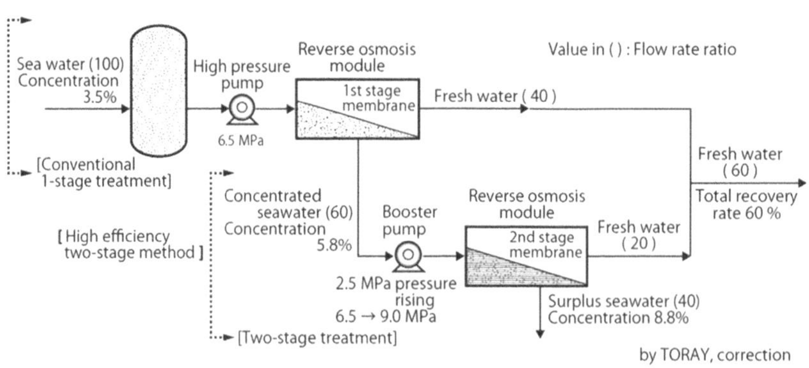

Figure 8.61 A two-stage RO membrane desalination system.

The first stage uses a conventional high-pressure pump as a drive pump. The second stage boost pump uses a pressure recovery unit such as a turbocharger that uses the residual pressure of concentrated water (8.8% brine, 9.0 MPa) in the second stage.

In addition, we can arrange the membranes in a multistage transient (Christmas tree) configuration, as shown in Figure 8.62, to increase the recovery rate. If the raw water salt concentration is high, such as in seawater, two or three stages may be used at best, but in some cases additional stages may be used.

The structure of the separation system described above centers on the RO membrane process. Actual seawater/brine desalination systems include water intake facilities, pretreatment equipment, concentrated brine disposal facilities, chemical treatment facilities, machinery, electricity, electronics, information and measurement facilities.

If clean underground seawater and brine water that has passed through sand layers can be taken as raw water by means of a water collection infiltration gallery or brine well if possible, the construction and operation of the following pre-treatment system will be very easy. If surface-flow seawater is directly taken from a water intake pipe or tower, a full-specification solid-liquid separation operation is required to remove suspended and collidal matters.

Pretreatment systems require full-specification rapid filtration systems, such as screens, coagulation–flocculation, disolved air flotation (sedimentation is disadvantageous because of high seawater density), and sand filtration, for surface-flow seawater intake. If underground seawater intake can be selected, only the UF membrane treatment may be required.

When brine is returned to the sea, it may cause a density current and adversely affect the ecosystem. Therefore, it is necessary to take measures to disperse brine into the sea surface in a plume-like manner. If a large amount of brine is released into a small area such as in the Red Sea, the coastal ecosystem may be adversely affected in the long term.

The biggest problem with desalination of seawater is the need for a large amount of driving energy, which should be considered in combination with nuclear power plants, wind energy in wind-powered areas such as the coast of continents and islands with permanent winds, and solar energy in desert areas.

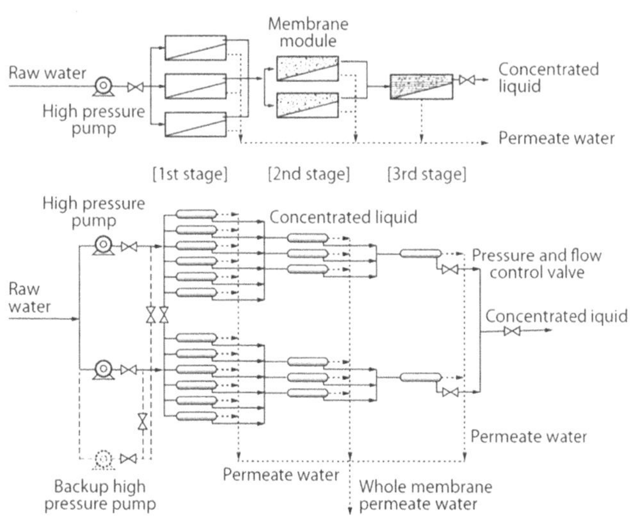

Figure 8.62 Multistage transient (Christmas tree arrangement) membrane filtration.

8.5.5 RO membrane for NF membrane manipulation

Previously, NF membranes were similar to RO membranes, and RO membranes with different desalination rates were produced together with RO membranes for seawater and brackish water desalination. When a similar RO membrane is used for advanced treatment of water purification (to remove trace contaminants i.e. micropollutants) and elimination of unpleasant components such as bad odors and tastes in fresh water, the driving force for osmotic pressure (RO) need hardly be considered. We have also explained that NF membranes generally have a slightly larger micropore diameter (cut off level) than seawater RO membranes. In many cases, this is intended to reduce the membrane resistivity as much as possible in order to reduce the driving pressure, because it is often intended for use against impurities such as minerals, organics, and viruses, which are slightly larger than those of Na^+ and Cl^- (hydrated ions).

On the other hand, the use of RO membranes for brackish water and seawater is not problematic for the purpose of removing trace components, and the increase of the driving pressure for maintaining the water flux is the only aspect to be considered. If low-concentration NOMs, trace organic synthetic chemicals, trace toxic heavy metals, and harmful inorganic ions in fresh water are to be removed, the hegh RO pressure (which is higher than 6 MPa for seawater desalination) is unnecessary, and the driving pressure can be 0.4 to 0.5 MPa (400 to 500 kPa) or less. The removal rate obtained differs slightly depending on the desalination rate, but most of the impurities that are problematic are removed, resulting in a lower concentration in chemical analysis than in natural water, and the lack of mineral components is sometimes more problematic for drinking than for safety.

The largest examples of a treatment operation that reaches drinkable levels with such sewage treatment water is the Singapore Public Utility Board (PUB) NEWater system. Sewage is pretreated with an MF membrane or a UF membrane treatment before treatment with the activated sludge method, then RO membranes remove most of the impurities, and the sewage is finaly treated by ultraviolet (UV) disinfection. In this example of RO membrane treatment, it seems that a full-scale RO demineralization membrane with a demineralization rate of 99.65% or more with low fouling specifications operates at a driving pressure of 15 MPa.

Each sewage collection area in Singapore has a water regeneration plant, and a water reservoir is constructed by closing a bay port in each basin to serve as a water source for the regeneration plant. 30% of Singapore's water demand is supplied by this regeneration sewage RO membrane system. At present, the main use is for industrial water supplies, but bottled NEWater is occasionally supplied to elementary school pupils. In addition, it seems that about 1% of the inflow is added to four reservoirs, such as the Selecta reservoir, located in the natural reserves around the Bukitema highlands, which are the main reservoirs of Singapore water supply. The author supposes that they will be carefully prepared for drinking in the future.

For all water quality items, NEWater clears WHO/USEPA standards. However, if the intra-regional circulation of 30% of water continues for a long period of time, issues of stable metabolic waste components such as low molecular weight humic substances, as well as the accumulation of boron, all of which cannot be removed by RO membranes, will need to be kept in mind for microbial treatment (Nakagaki & Shimizu, 1992).

The author would like to underline his respect for the work of the Singapore PUB (Ministry of Public Works), which has a much higher ethical and economic profile than just that of a seawater desalination RO membrane system – in terms of the improvement of the level of environmental management in the area by using the water basin as a water source, reducing the energy needed for water regeneration, the improvement and stabilization of supply water quality, and the achievement of sewage regeneration through the use of NF or RO membranes.

As a method of alleviating the accumulation of residual components, which is a problem when a regeneration cycle is repeated, fresh seawater is mixed with treated sewage water. By using a mixing ratio that lowers the RO pressure to the brine level, it is also possible to think about the driving pressure of the brine desalination level as a 'treated sewage water + clean seawater' desalination plant. This reduces fouling caused by NOMs in the RO membranes, doubles the amount of raw water to a level equivalent to the amount of sewage treatment, and reduces the RO operating pressure to nearly half. Considering the relative gains of straight seawater, this desalination and mixing process is also likely to be meaningful as a project.

8.6 MEMBRANE BIOREACTORS (MBRs)

8.6.1 The membrane separation activated sludge method

The membrane separation activated sludge method is basically the same as the conventional activated sludge process, except that a final sedimentation tank is replaced by membrane filtration. Figure 8.63 shows the composition of the conventional (standard) activated sludge method. As shown in Figure 8.63, a membrane bioreactor (MBR) separation activated sludge method has emerged as a new activated sludge method, using MF or UF membranes to separate solids/liquids after microbial reaction. Cross flow MBR was proposed in the 1970s as an earlier approach to membrane separation, and immersion (submerged; S-type) MBR originated in Japan in the 1990s, as mentioned earlier (see Figures 8.11, 8.50 and 8.51). Today, immersion MBR has spread to a wide range of aspects of sewage, organic wastewater, and sewage recycling. In water treatment plants and industrial water treatment, more precise management of membrane modules are now preferred, and cross-flow type external membrane separation is used instead of sand filtration (Kimura, 2000).

The cross-flow and immersion MBR configurations and operating characteristics are summarized in Figure 8.63.

As mentioned above, the main target MBR for organic wastewater seems to be immersion MBR, which started in Japan. Since the MF or UF membrane is the main component of solid-liquid separation and is placed in an aeration tank, the MF or UF membrane is constantly exposed to a high-concentration

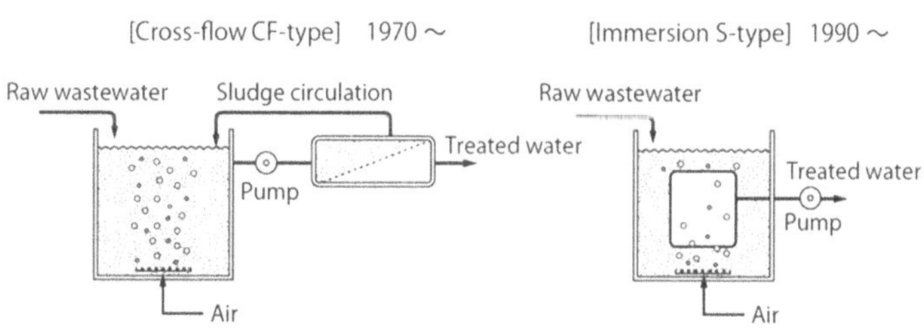

	CF - MBR	S - MBR
Energy consumption rate (kWh/m³)	2 ~ 10	0.2 ~ 0.4
Operating pressure (kPa)	100 ~ 200	< 50
Permeation flux (m³/m²/day)	1 ~ 2	0.1 ~ 0.5
Membrane type	UF	MF

Figure 8.63 Cross-flow and immersion (S-type) membrane bioreactor (MBR) process configurations (Kimura & Watanabe, 2008).

suspension of 10^3 mg/l of activated sludge. The nature of the solution system that coexists with the activated sludge produced varies depending on the type of raw water, and a detailed and exhaustive study of the membrane permeability and fouling made need to be undertaken.

It is also possible to install an aeration tank mainly for aerobic microbial reaction and an immersion membrane separation process in two stages. In this case, the operating conditions for the aerobic activated sludge reaction (air flow rate, air bubble miniaturization, structure and arrangement of the air diffuser) and the stirring and washing conditions of the membrane separation tank can be optimized independently, which is effective from the viewpoint of improving the operability (energy efficiency).

Even if some membrane damage (breakage) is taken into account, an MBR is expected to maintain the separation efficiency of its activated sludge in the final sedimentation tank by two orders of magnitude or more at all times, and the water quality of the aerobic microbial degradation limit can be maintained almost constantly. As previously reported by Tambo *et al.* (1981), and as explained in Section 3.5 on the water quality matrix treatment, most of the carbon-based BOD is removed by aerobic microbial treatment, and 3–5% of the carbon-based TOC remains in the liquid system as a color solubility component of humic origin. As a result, the COD removal rate is at best 97−95%. When expressed in terms of TOC, TOCs that are biologically persistent are hardly decomposed by the aerobic activated sludge method, so the removal rate is slightly lower. As explained in Section 8.1 above, humic substances (E260 sensitive TOCs or COD-BOD), which are components of microbial metabolic waste, cannot be separated at all by MF, and only a small percentage of them can be separated by UF. Therefore, the separation of humic components (NOMs) with a relatively large molecular mass of up to 1500 Da also requires precise coagulation treatment. Therefore, when organic matter suspended in an MBR is removed, the required separation operation can be performed with an MF of about 0.1 μm in consideration of the membrane resistance etc., and there is no point in carefully using UF.

Figure 8.64 shows actual data from the Japanese Sewerage Agency on the removal of organic matter using BOD, COD, and TOC as indexes. The data seem to demonstrate the above-mentioned mechanism for the formation of metabolic waste components (humic substances) expressed by Tambo and colleagues, that a small amount (a few percent) of humic components (E260 sensitive TOC) is generated in the water quality matrix after the aerobic microbial process.

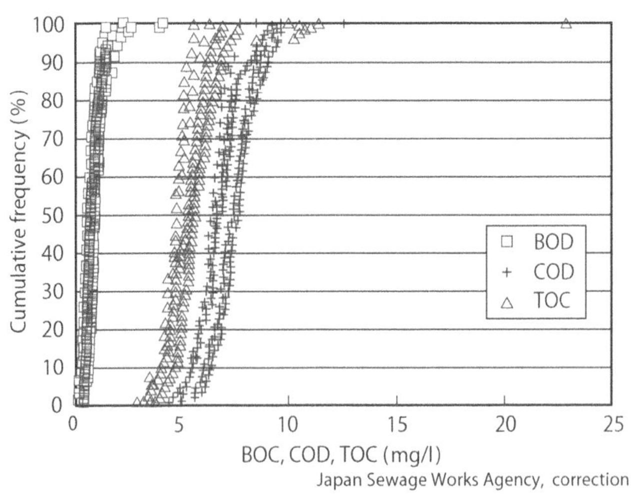

Figure 8.64 Cumulative frequency of organic matter removal rates (Japan Sewage Works Agency).

An MBR, which is treated with the prior coagulation of iron and aluminium salts, is effective as a system for the purpose of water production of recycled sewage water, by reducing the fouling of membranes and enhancing the removal of phosphorus.

8.6.2 Denitrification by MBR (aerobic and anaerobic treatment)

Using an immersion S-type MBR, it is possible to perform nitrification reactions and denitrification in a compact and accurate manner in a double tank (an anoxic tank + a nitrification tank) or a single tank loaded with a separation partition. The results of research on nitrification and denitrification MBR processes at the Hokkaido University by Professors Y. Watanabe and K. Kimura are presented below.

Nitrogen removal was performed using the same composition as for the conventional activated sludge method, using an immersion S-type MBR, a two-tank system consisting of an anaerobic (agitation-free) denitrification tank (as shown in Figure 8.65) and an immersion S-type MBR (in which subsequent aerobic nitrification reactions took place). A three-tank system consisting of a nitrification tank and an S-type membrane separator was also used; however, since the nitrification reaction proceeds more slowly than in the case of an aerobic BOD removal reaction, the amount of air diffused is small. Therefore, the main system used was a two-layer type in which the final stage can be treated in a single tank after examining conditions. Up to 80% of ammoniacal nitrogen in sewage was nitrified by aerobic processes, then released into the atmosphere as N_2 gas through anaerobic processes, and removed from the liquid system.

In order to further reduce the volume of the system, a denitrification S-type MBR with a single tank of intermittent aeration type (as shown in Figure 8.66) can be considered. However, although the system is compact, intermittent aeration interrupts the membrane filtration operation of the S-type MBR during

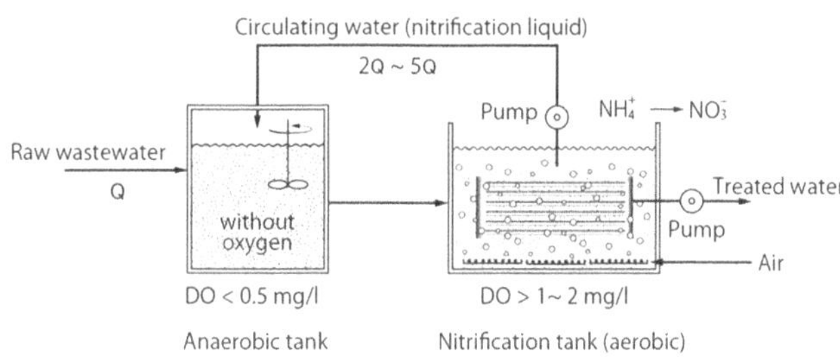

Figure 8.65 A two-tank denitrification S-type MBR system.

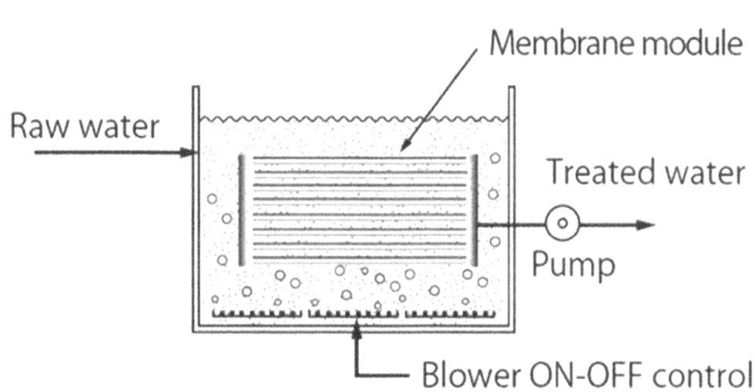

Figure 8.66 An intermittent aeration S-type denitrification MBR.

non-aeration (anaerobic conditions), resulting in intermittent membrane filtration only under aerobic conditions where aeration occurs, resulting in the drawback that the total amount of treated water cannot be increased.

Kimura and Watanabe (2008) proposed a single-tank MBR nitrification and denitrification system (a baffled membrane bioreactor: BMBR) with a baffle partition wall, as shown in Figure 8.67. The unit is a micro-organism reactor using a single-vessel MF membrane, and it is equipped with a draft tube baffle partition (flow-conducting) that opens at the top and bottom of the MF membrane, or with a tube around the MF membrane.

The operation of the system is as follows:

(1) The S-type MF membrane contained in the tube is subjected to membrane filtration while constantly receiving aeration, and the treated water is continuously taken out at a predetermined flow rate Q.
(2) On the other hand, by intermittently varying the inflow amount of raw wastewater, when raw water is received at a flow rate exceeding the flow rate Q that can be treated by the membrane, the water level in the tank rises to a full level, as shown in Figure 8.67a, and aerobic mixed liquor suspended solids (MLSS), which has been sufficiently aerated, circulates in all the tank over the upper end of the baffle partition wall. At this time, the nitrification reaction vigorously progresses, and aerobically treated water (water permeated through the MF membrane) with a flow rate of Q is obtained.
(3) When the tank reaches the maximum water level, the inflow of raw water is stopped and only the membrane permeation is allowed to proceed. The water level drops to a position slightly lower than the upper end of the baffle partition, as shown in Figure 8.67b.
(4) In this situation, the zone where the membrane separates inside the baffle septum continues to be aerobic, so that the permeate becomes water drawn from the aerobic MLSS.
(5) By contrast, the zone outside the baffle wall is deprived of air supply, and the microbial community is anaerobic and the denitrification reaction proceeds.
(6) When the water level inside the baffle partition reaches the set minimum water level, the inflow of raw water is resumed.
(7) Added organic carbon in the raw water effectively promotes the denitrification reaction because anaerobic microorganisms outside the flow barrier supply the substrate required.

In this way, the single-tank MBR nitrification/denitrification system with a baffle partition (BMBR) enables aerobic and anaerobic conditions to be created in the outer zones of the baffle partition (the draft tube) at

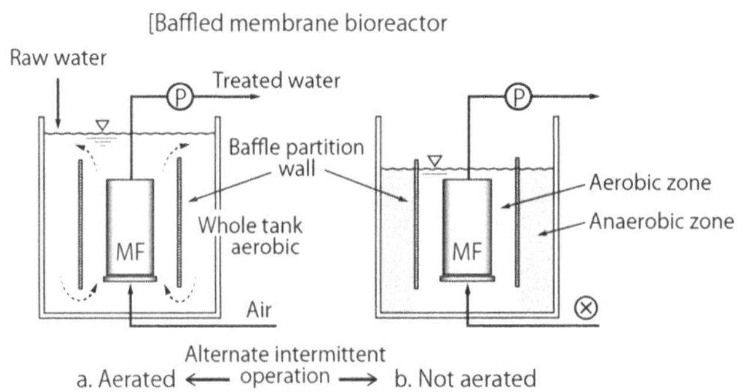

Figure 8.67 A single-tank MBR nitrification/denitrification system with baffle partition walls.

regular intervals, enabling the design of a compact denitrification system that can continuously produce a constant flow of membrane permeate (treated water). The time interval between the anaerobic and aerobic conditions of the system is the most important.

The pilot experiments conducted by Kimura and Watanabe at the Sewage Treatment Plant in Sapporo City gave the following results. The MF membrane used for the separation was a polyvinyl-denfloride (PVDF) membrane with a fractional size of 0.1 μm and a membrane area of 6.8 m^2, and was immersed in a rectangular reactor with a size of 80 cm × 50 cm × 178 cm. The outside of the membrane module was surrounded by a 55 × 28 × 110 cm draft tube (with a baffle) to form an anaerobic separation space. The maximum water volume is 500 l. This system is a single-vessel MBR nitrification/denitrification system with a baffle partition (BMBR), which is a multi-stage variable-volume reactor.

As a result of a long-term 5-day/week operation test in which the cycle time was changed from 12 to 50 minutes in the preliminary test, it was inferred that the hydraulic retention time (HRT) was 4.7 hours and the sludge retention time (SRT) was 36 days with a membrane flux of 0.04 m/m^2/day, and as a result, the MLSS was about 13 to 20 g/l. After 3 months of operation to acclimatise the microorganisms, membrane fouling was removed by chemical cleaning (NaClO) and a continuous long-term test was conducted with 16 g/l of activated sludge suspended matter (MLSS) with cycle time of 12 minutes.

The operation was carried out at a low water temperature of 10 to 15°C. On day 37 of operation, the entire BMBR tank was in an aerobic environment for about the first 1.4 to 4 minutes. However, after 4 minutes when the water level in the tank dropped to the upper end of the baffle partition, the aerobic state inside the baffle partition was strengthened, whereas the outside was rapidly deprived of oxygen, and after 6 minutes, the water was completely anaerobic.

Figure 8.68 shows the time-dependent changes in dissolved oxygen inside and outside the baffle bulkhead (draft tube) over a 12-minute cycling period. The results are shown for water levels of 0.1, 0.6, and 1.0 m, from the top of the baffle bulkhead.

Figure 8.69 shows the results of T-N and T-P removal and the effect of TOC removal when a long-term test was performed under the above operating conditions (cycle time) considered appropriate in the preliminary test.

Both TOC and T-P in treated water were removed to small values of 3.6 and 0.009 mg/l, respectively. Overall, nearly 80% of the T-N of the raw wastewater can be removed with a residence time of 4.7 hours

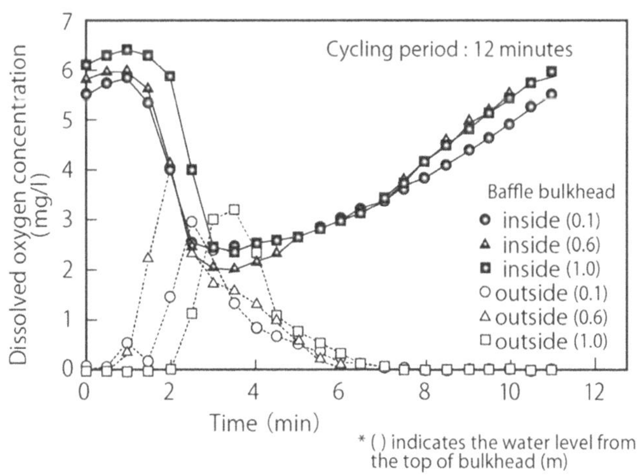

Figure 8.68 Changes in dissolved oxygen inside and outside the draft tube, over time.

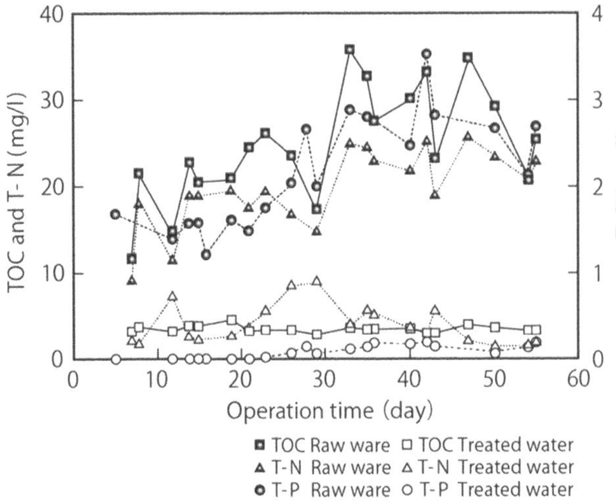

Figure 8.69 TOC, T-N and T-P removal under long-term conditions in the BMBR.

without the forced circulation of sludge using a large amount of energy as in conventional denitrification methods. Moreover, the addition of substrate carbon as a nutrient source is not required. The total amount of nitrogen removal is 46.9% of the total contained in the biomass, which is obtained by solid-liquid separation using a membrane, and which is obtained by nitrification and anaerobic denitrification.

Figure 8.70 shows how the form of nitrogen changes over time. The figure shows that the nitrification reaction was sufficiently carried out, but NO_3-N concentration increased around 30 days after the reaction. From this point on, the anaerobic denitrification reaction became the rate-limiting reaction, and the process to N_2 gasification did not progress.

The reduction in the nitrogen-removal rate in this BMBR over a period of time during the course of operation may be attributed to the fact that the aerobic condition of the MLSS of the inner zone becomes stronger, and the MLSS of the outer anaerobic zone formed by lowering the water level to below the

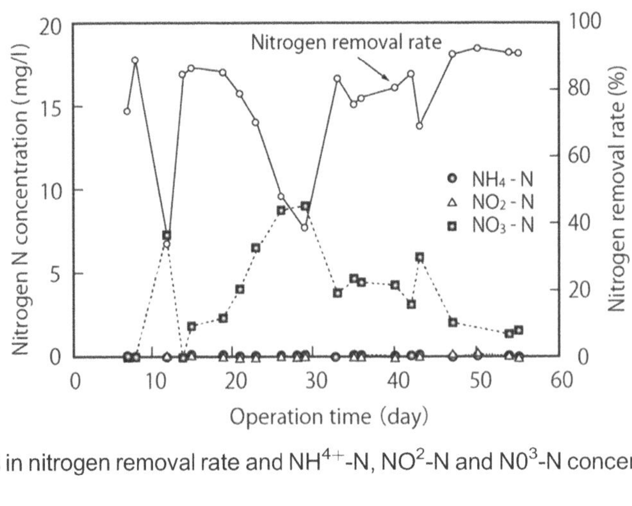

Figure 8.70 Changes in nitrogen removal rate and NH^{4+}-N, NO^2-N and $N0^3$-N concentration of treated water by operation time.

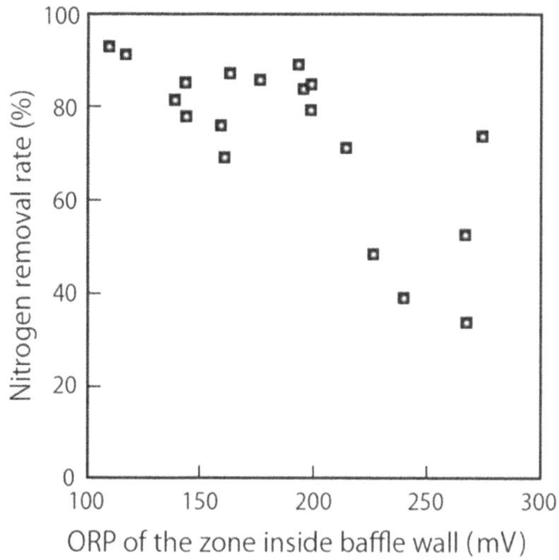

Figure 8.71 Daily average value of redox potential (ORP) and nitrogen removal rate in the zone inside the baffle wall.

upper edge of the draft tube cannot be sufficiently converted to an anaerobic state. Figure 8.71 shows the relationship between the redox potential (oxidation reduction potential; ORP) and the nitrogen removal rate in the zone inside the baffle wall, showing that the higher the ORP (the stronger the aerobic state), the lower the nitrogen removal rate. The ORPs in the outer zone should be measured if possible, but the values in the inner zone are inferred because measurements in MLSS without agitation are difficult to continue due to sticky substances adhering to the sensors.

As shown in Figure 8.72, it can be inferred that this excess-aerobic condition is caused by an increase in the ORP of the inner zone when the raw wastewater load is low for the same aeration volume, and that the

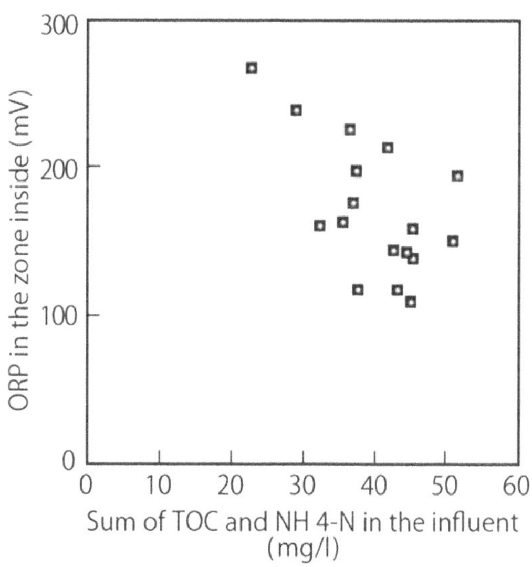

Figure 8.72 Raw wastewater load and redox potential (ORP) in the zone inside the baffle wall.

outer zone is not sufficiently anaerobic. For this reason, programmatic control in conjunction with the ORP of the internal zone is now likely to be easier in order to ensure appropriate anaerobic conditions of the external zone in response to varying raw wastewater concentrations and compositions.

In these operations, total phosphorus (P) is removed at a higher rate, the volume of biomass taken up by biomass is about 187 mg P/g-SS, and the amount of inorganic phosphorus remaining in the liquid system is only about 11 mg P/g-SS. The removal rate is high because the sludge system (solid system) is largely transferred by membrane separation. Thus, an BMBR is an effective treatment for relatively low organic wastewater, such as sewage, with the purpose of controlling both N and P nutrients.

Preceding coagulation and sedimentation separation using polymerized iron salts, aluminium salts, etc., as described in the section on coagulation and flocculation above, can reduce the BMBR load of high-concentration organic wastewater systems, enabling stable operation of nitrification and denitrification reactions, and suppressing fouling in membrane separation. In order to promote this operation stably, the author thinks it is necessary to further examine the maintenance of the component balance in the anaerobic denitrification so that the removal of the substrate does not proceed excessively.

FURTHER READING

Bian R. L., Watanabe Y., Ozawa G. and Tambo N. (1998). Fouling associated with UF membrane treatment by river water: bach experiment evaluation. *Journal of Japan Water Works Association*, (760), 24–33 (In Japanese).

Kimura K., Yamato N., Yamamura H. and Watanabe Y. (2005). Membrane fouling in pilot-scale membrane bioreactors (MBRs) treating municipal wastewater. *Environmental Science & Technology*, **39**, 6293–6299.

Kurihara Y. (1983). Basics of polymer membrane materials. In: the Japanese Society of Membrane Science, Kitami Books, Tokyo, pp. 94–115 (In Japanese).

Kurihara Y. (1984). Manufacturing and Function of Artificial Membranes: Asymmetric Membranes and Composite Membranes. *Chemical Review* No.45 (Functional Organic Thin Films I), pp. 30–38, edited by the Japan Chemical Society, Academic Society Publishing Center (In Japanese).

Taniguchi M. (2009). Future and Prospects of Seawater Desalination Technology Using RO Membranes. *Journal of Seawater Society of Japan*, **63**(4), 214–220 (In Japanese).

REFERENCES

Aptel P. and Buckley C. A. (1996). Categories of membrane operations (Chapter 2). In: Water Treatment Membrane Processes, J. Mallevialle, P. E. Odendaal and M. R. Wiesner (eds), American Water Works Association Research Foundation. Lyonnaise des Eaux, Water Research Commission of South Africa, McGraw-Hill, USA.

Bian R. L. (1998). Study on UF-Membrane Filtration Characteristics of River Water. PhD thesis, Hokkaido University.

Hayashi S. (2004). Pressure-converting energy-recovery equipment-pressure exchanger. *Journal of Japan Society of Sea Water*, **58**(3), 273–278 (In Japanese).

Japan Water Purification Processes Association (JWPA). (1994a). 'Research on the Development of New Water Purification System Using Membranes (MAC21 Project)' (FY1994–1996). Gihodo, Tokyo (In Japanese).

Japan Water Purification Processes Association. (1994b). Guidelines for introduction of Membrane Filtration Facilities in Small-scale Water Supply. Gihodo, Tokyo (In Japanese).

Japan Water Purification Processes Association (JWPA). (1997). ACT21 'Research on Development of High-Efficiency Water Purification Technology' (FY1997–FY2001). Gihodo, Tokyo (In Japanese).

Japan Water Purification Processes Association (JWPA). (2002). 'R & D on water purification technology for environmental impact reduction' (FY2002–FY2004). Gihodo, Tokyo (In Japanese).

Japan Water Purification Processkes Association. (2003). Water purification membrane (in Japanese). Gihodo, Tokyo.

IUPAC (International Union of Pure and Applied Chemistry). (1985). Reporting physisorption data. *Pure Applied Chemistry*, **57**, 603.

Kamei T., Matsui Y., Nakanowatari K. and Tambo N. (1995a). Removal of High Color Components by External Pressure Type Ceramic Membranes. *Journal of Japan Water Works Association*, (730), 34–39 (In Japanese).

Kamei T., Matsui Y., Nakanowatari K., Tambo N. and Yuasa Y. (1995b). External Pressure Type Ceramic Membrane Treatment Characteristics Combined with coagulation Treatment. *Journal of Japan Water Works Association*, (731), 26–33 (In Japanese).

Kimura K. (2000). A Novel biofilm-membrane reactor for advanced drinking water treatment. PhD thesis, Hokkaido University.

Kimura K. and Watanabe Y. (2008). Baffled membrane bioreactor (BMBR) for efficient nutrient removal from municipal wastewater. *Water Research*, **42**, 625–632.

Kinjyou Y. (1999). Current Status and Future Technological Trends of Seawater Desalination Facilities as Water Supply Sources. *Journal of Seawater Society of Japan*, **53**(6), 432–438 (In Japanese).

Loeb S. and Sourirajan S. (1963). Sea Water Desalination by means of an Osmotic Membrane. *American Chemical Society, Advances in Chemistry Series*, **ACS38**, 117–132.

Mallevialle J., Odendaal P. E. and Wiesner M. R. (eds) (1996). Water Treatment Membrane Processes. AWWA Research Foundation, Lyonnaise des Eaux, Water Research Commission of South Africa. McGraw-Hill, New York.

Membrane Separation Promotion Association. (2008). Water Purification Membrane, 20th edn. Gihoudou Publishing, Tokyo (In Japanese).

Nakagaki M. and Shimizu H. (eds) (1992). Membrane Treatment Technology System, Vol. 1. In: Ohi, T., Yamazaki, T., Kosaka, T. Composition of Boron Isotopes in Seawater near the Coast. *Journal of Seawater Society of Japan*, **46**(6), 385–390 (In Japanese).

Ohya H. (1976). Theory of Reverse Osmosis Method and Ultrafiltration Method I. Saiwai Shobo, Tokyo (In Japanese).

Okazaki H., Taniguchi R. and Suzuki H. (2006). Illustration and Water Treatment Membrane. Nikkan Kogyo Shimbun Co., Ltd., Tokyo (In Japanese).

Peterson R. J. (1993). Composite reverse osmosis and nanofiltration membranes. *Journal of Membrane Science*, **83**, 81–150.

Reid C. E. and Breton E. J., Jr. (1959). Water and ion flow across Cellulosic Membrane. *Journal of Applied Polymer Science*, **1**, 133–143.

Ridgway H. F. and Fleming H.-C. (1996). Membrane biofouling. In: Water Treatment Membrane Processes, J. Mallevialle, P. E. Odendaal and M. R. Wiesner (eds), AWWA Research Foundation, Lyonnaise des Eaux, Water Research Commission of South Africa. McGraw-Hill, New York, pp. 6.1–6.62.

Tambo N. and Kamei T. (1976). Evaluation of Urban Water Metabolism by Matrix. *Journal of Japan Water Works Association*, (502), 2–24 (In Japanese).

Tambo N. and Kamei T. (1978). Treatability evaluation of organic matter, Matrix conception and its application for a regional water and waste water system. *Water Research*, **12**, 931–950.

Tambo N., Kamei T., Ohnishi T. and Sekiguchi A. (1987). A study on the hazard of tap water from Echinococcus eggs. *Journal of Japan Water Works Association*, (638), 28 (In Japanese).

Tambo N., Kamei T. and Takahashi (1981). Journal of Japan Sewage Works Association, 210 (In Japanese).

Tambo N. (1980). Water Supply (New System Civil Engineering 88) (7.4.1 Membrane Treatment). Japan Society of Civil Engineers, Gihoudou Publishing, Tokyo. pp. 259–266, (In Japanese).

Tambo N. (2012). History and Technologies of Urban and Regional Water Metabolism Systems. Kashima Press, Tokyo, pp. 120–123, 233, 284–306, 319, 369, 355–370.

Tambo N. and Repine A. (1992). Treatment of River Water with Hollow Fiber UF Membranes-Evaluation in the Toyohira River, Sapporo City. *Journal of Japan Water Works Association*, (698), 19–26 (In Japanese).

Tambo N., Ozawa G., Iketani Y., Ebara Y. Y. and Abe N. (1994). Evaluation of treatment characteristics of MF membranes and UF membrane systems with different separation dimensions: Toyohira river in Sapporo City. *Journal of Japan Water Works Association*, (718), 16–2 (In Japanese).

Wiesner M. R. and Aptel P. (1996). Mass transport and permeate flux and fouling in pressure-driven processes (Chapter 4). In: Water Treatment Membrane Processes, J. Mallevialle, P. E. Odendaal and M. R. Wiesner (eds), American Water Works Association Research Foundation. Lyonnaise des Eaux, Water Research Commission of South Africa. McGraw-Hill, New York.

Index

IWA Publishing's authorised EU representative for General Product Safety Regulations is Diane D'Arras, 15 rue Duret, 75116 Paris, France, e-mail: safety@iwap.co.uk.

Printed and bound by CPI Group (UK) Ltd, Croydon, CR0 4YY
17/09/2026
02218800-0001